TABLE B.2

SI Unit Prefixes

Prefix	Symbol	Factor
tera	T	10^{12} = 1,000,000,000,000
giga	G	10^{9} = 1,000,000,000
mega	M	10^{6} = 1,000,000
kilo	K	10^{3} = 1,000
hecto	H	10^{2} = 100
deka	da	10^{1} = 10
deci	d	10^{-1} = 0.1
centi	c	10^{-2} = 0.01
milli	m	10^{-3} = 0.001
micro	μ	10^{-6} = 0.000 001
nano	n	10^{-9} = 0.000 000 001
pico	p	10^{-12} = 0.000 000 000 001

Note: The use of the prefixes hecto, deka, and centi is not recommended. However, they are sometimes encountered in practice.

PLATES AND SHELLS
Theory and Analysis

CRC Series in

APPLIED and COMPUTATIONAL MECHANICS

Series Editor: J.N. Reddy, *Texas A&M University*

PUBLISHED TITLES

ADVANCED MECHANICS OF CONTINUA
Karan S. Surana

ADVANCED THERMODYNAMICS ENGINEERING, SECOND EDITION
Kalyan Annamalai, Ishwar K. Puri, and Miland Jog

APPLIED FUNCTIONAL ANALYSIS
J. Tinsley Oden and Leszek F. Demkowicz

COMBUSTION SCIENCE AND ENGINEERING
Kalyan Annamalai and Ishwar K. Puri

COMPUTATIONAL MODELING OF POLYMER COMPOSITES: A STUDY OF CREEP AND ENVIRONMENTAL EFFECTS
Samit Roy and J. N. Reddy

CONTINUUM MECHANICS FOR ENGINEERS, THIRD EDITION
Thomas Mase, Ronald Smelser and George F. Mase

DYNAMICS IN ENGINEERING PRACTICE, ELEVENTH EDITION
Dara W. Childs and Andrew Conkey

EXACT SOLUTIONS FOR BUCKLING OF STRUCTURAL MEMBERS
C. M. Wang, C. Y. Wang, and J. N. Reddy

THE FINITE ELEMENT METHOD FOR BOUNDARY VALUE PROBLEMS: MATHEMATICS AND COMPUTATIONS
Karan S. Surana and J. N. Reddy

THE FINITE ELEMENT METHOD FOR INITIAL VALUE PROBLEMS: MATHEMATICS AND COMPUTATIONS
Karan S. Surana and J. N. Reddy

THE FINITE ELEMENT METHOD IN HEAT TRANSFER AND FLUID DYNAMICS, THIRD EDITION
J. N. Reddy and D. K. Gartling

MECHANICS OF MATERIALS
Clarence W. de Silva

MECHANICS OF SOLIDS AND STRUCTURES
Roger T. Fenner and J. N. Reddy

MECHANICS OF LAMINATED COMPOSITE PLATES AND SHELLS, SECOND EDITION
J. N. REDDY

MICROMECHANICAL ANALYSIS AND MULTISCALE MODELING USING THE VORONOI CELL FINITE ELEMENT METHOD
Somnath Ghosh

NUMERICAL AND ANALYTICAL METHODS WITH MATLAB®
William Bober, Chi-Tay Tsai, and Oren Masory

NUMERICAL AND ANALYTICAL METHODS WITH MATLAB® FOR ELECTRICAL ENGINEERS
William Bober and Andrew Stevens

PLATES AND SHELLS: THEORY AND ANALYSIS, FOURTH EDITION
Ansel C. Ugural

PHYSICAL COMPONENTS OF TENSORS
Wolf Altman and Antonio Miravete

PRACTICAL ANALYSIS OF COMPOSITE LAMINATES
J. N. Reddy and Antonio Miravete

SOLVING ORDINARY AND PARTIAL BOUNDARY VALUE PROBLEMS IN SCIENCE AND ENGINEERING
Karel Rektorys

PLATES AND SHELLS
Theory and Analysis

Fourth Edition

ANSEL C. UGURAL

CRC Press is an imprint of the
Taylor & Francis Group, an **informa** business

CRC Press
Taylor & Francis Group
6000 Broken Sound Parkway NW, Suite 300
Boca Raton, FL 33487-2742

Printed on acid-free paper

International Standard Book Number-13: 978-1-138-03245-3 (Hardback)

Library of Congress Cataloging-in-Publication Data

Names: Ugural, A. C., author.
Title: Plates and shells : theory and analysis / by Ansel C. Ugural.
Other titles: Stresses in plates and shells
Description: Fourth edition. | Boca Raton : CRC Press, [2018] | Series:
Applied and computational mechanics | Original edition published
under the title: Stresses in plates and shells / Ansel C. Ugural. |
Includes bibliographical references and index.
Identifiers: LCCN 2017014434| ISBN 9781138032453 (hardback : alk. paper) |
ISBN 9781315104621 (ebook)
Subjects: LCSH: Plates (Engineering) | Shells (Engineering) | Girders. |
Strains and stresses.
Classification: LCC TA660.P6 U39 2018 | DDC 624.1/776--dc23
LC record available at https://lccn.loc.gov/2017014434

Visit the Taylor & Francis Web site at
http://www.taylorandfrancis.com

and the CRC Press Web site at
http://www.crcpress.com

Printed and bound in the United States of America by Sheridan

Contents

Section III Shells

Preface

Introduction

The subject matter of the text is usually covered in one-semester senior and one-semester graduate-level courses dealing with the *analysis of beams, plates and shells, stress analysis, pressure vessels, analysis of thin-walled structures, advanced statics,* or *special topics in solid and structural mechanics.* As sufficient material is provided for a full year of study, this book may stimulate the development of courses in *advanced engineering design.* The coverage presumes knowledge of the elementary mechanics of materials.

The text is intended to serve a twofold purpose: to complement *classroom lectures* and to accommodate the needs of *practicing engineers* in the analysis of beam, plate, and shell structures. The material presented is applicable to aeronautical, astronautical, chemical, civil, mechanical, and ocean engineering; engineering mechanics; and science curricula. This volume attempts to provide synthesis and analysis that cut through the clutter and save time for readers. It is hoped that clarity of presentation is maintained, as well as simplicity as permitted by the nature of the subject, unpretentious depth, an effort to encourage intuitive understanding, and a shunning of the irrelevant.

Approach

Emphasis is given to computer-oriented *numerical finite difference and finite element techniques* in the solution of problems resisting *analytical approaches.* The reader is helped to realize that a firm grasp of *fundamentals* is necessary to perform critical interpretations, so important when computer-based solutions are employed. However, the stress placed on numerical methods is not intended to deny the merit of *classical analysis,* which is given a rather full treatment. To facilitate the study of the subject, a *dual identification system* for equations is used in this text. Important formulas are marked in sequence in each chapter. On the other hand, equations that are mostly used in derivations are marked with *letters* in each section *only.* The volume attempts to fill what the writer believes is a void in the world of texts on the subject.

This book offers a simple, comprehensive, and methodical presentation of the principles of beam, plate, and shell theories and their applications to numerous structural elements, including domes, pressure vessels, tanks, and pipes. Theories of failures are employed in predicting the behavior of beams, plates, and shells under combined loading. Above all, an effort is made to provide a visual interpretation of the basic equations and of the means by which loads are resisted in shells and plates. A balance is presented between the theory necessary to gain insight into the mechanics and the numerical solutions, both so useful in performing stress analysis in a more realistic setting. Throughout the text, the author has attempted to provide the fundamentals of theory and application necessary to prepare students for more advanced study and for professional practice. Development of the physical and mathematical aspects of the subject is deliberately pursued.

The physical significance of the solutions and practical applications are given emphasis. With regard to application, often classical engineering examples are used to maintain simplicity and lucidity. The author has made a special effort to illustrate important principles and applications with numerical examples. A variety of problems is provided for solution by the student. *Answers to selected problems* are given at the end of this book. A *sign convention,* consistent with vector mechanics, is employed throughout for loads, internal force resultants, and stresses. This convention conforms to that used in most classical mechanics of materials and elasticity texts as well as to that most often employed in numerical analysis of complex structures. The International System of Units (SI) is used.

The expression defining the small lateral deflection of the midplane of a thin plate is formulated in two ways. The first utilizes the fundamental assumptions made in the customary *theory of beams and plates.* The second is based on the differential equations of equilibrium for the *three-dimensional stress.* The former approach, which requires less mathematical rigor but more physical interpretation, is regarded as more appealing to the engineer and is equally used in the case of thin shells. Emphasized also are the energy aspects of plate and shell bending and buckling because of the importance of *energy methods* in the solution of many real-life problems and in modern computational techniques. Because of the applied nature of this book, the classical approaches requiring extensive mathematical background are not treated.

Recent publications dealing with shell-bending theory include analytical presentations generally valid for any shell under any kind of loading. These formulations usually necessitate the employment of tensor notation, vector analysis, and a system of curvilinear coordinates. The theory introduced in this text is a special case of the above. The equations governing shells are developed only to the extent necessary for solving the more usual engineering problems. The finite element method is applied to treat bars, trusses, beams, plates of nonuniform thickness, as well as to represent shells of arbitrary form subjected to variable loading.

Text Arrangement

Most chapters are substantially self-contained. Hence, the order of presentation can be smoothly altered to meet an instructor's preference. It is suggested, however, that Chapters 1 through 3 be studied first. The remaining chapters may be taken in any sequence except that Chapters 12 and 13 should be read before Chapters 14 and 15. Numerous alternatives are possible in making selections from this book for two single-semester courses. The chapters have been arranged in a sequence compatible with an orderly study of the analysis of beams, plates, and shells. *Section I* and a number of sections marked with an asterisk (*) are *optional* and can be omitted without loss of continuity in the text. Some chapters are carefully integrated by means of cross-referencing.

New to This Edition

The fourth edition of *Plates and Shells: Theory and Analysis* seeks to preserve the objectives and emphasis of the previous editions. It is extended to include classification of structures and materials, stresses in bars, and thin-walled beams, analysis of composite plates and shells, selective *case studies*, and some other additional topics described below. A major effort has been made to provide a more comprehensive and modern text. The concept of "design to meet strength requirements" as those requirements relate to individual members, elastic design criteria, basic design process, and design factor of safety are briefly discussed. The examples and problems that appear through the book illustrate the close link between analysis and design.

Practical stress analysis is usually done using widely available and highly sophisticated software. This demands that engineering courses emphasize more the *fundamentals* and the basis of the *computer-oriented methods*. To this end, great care is taken to explain the *three aspects of solid mechanics problems*, basic assumptions made in developing engineering theory of typical structural members, and the implications and limitations of the method.

Included are the new discussions and topics dealing with fundamental principles of stress analysis; engineering materials; thermal stresses in plates; rectangular orthotropic plates; laminated composite and sandwich plates; shells of general shape; single and multisphere shells; laminated composite cylindrical shells; approximate and energy methods for beams, plates and shells; and tables of area characteristics, material properties, units, and beam and plate stresses and deflections.

The entire text has been re-examined, and many major improvements are made throughout by elimination, rearrangement, and the addition of numerous sections in Chapters 1 and 2, Appendices B, C, and D. Some sections are expanded to improve on previous expositions. The temptation to greatly increase the material covered is resisted. However, it is considered desirable to add numerous new real-life examples and problems. Most problems can be readily modified for in-class tests. References (identified in *brackets*), listed at the end of each chapter, provide readily available sources where additional information can be obtained.

This text offers a wide range of over 110 fully worked-out illustrative examples, various case studies, and 370 problem sets, many of which are drawn from engineering practice; a multitude of formulas and tabulations of beam, plate, and shell theory solutions from which direct and practical design calculations can be made; analysis of plates and shells made of isotropic as well as composite materials under ordinary and high-temperature loadings; numerical methods amenable to computer solution; and applications of the formulas developed and of the theories of failures to increasingly important structural members.

Supplements

Solutions Manual *for Instructors* is available to adopters through the publisher. Written and class-tested by the author, it features complete solutions to problems in the text.

Optional Material is available from the CRC Website: http://www.crcpress.com/product/isbn/9781439887806. This includes solutions using MATLAB® for a variety of examples of and case studies of practical importance. This book is **independent** of any **software** package.

Acknowledgments

To acknowledge everyone who contributed to this book in some manner is clearly impossible, but a major debt is owed to the readers and reviewers who offered constructive suggestions and made detailed comments on the previous editions. These particularly include the following: R. H. Gallagher, University of Arizona; C. R. Steele, Stanford University; R. Lipp, University of New Orleans; S. K. Fenster, New Jersey Institute of Technology; B. Lefkowitz, Fairleigh Dickinson University; E. M. Dombourian, California State University, Northridge; D. A. Danielson, University of California, San Diego; G. E. O. Widera, University of Illinois; S. Dharmarajan, San Diego State University; J. D. Masson, Texas Instruments Co.; B. Koo, University of Toledo; T. C. Kennedy, Oregon State University; H. Gesund, University of Kentucky; H. Saunders, General Electric Co.; K. Chandrashekhara, University of Missouri; C. Feng, University of Colorado; LeRoy A. Lutz, Marquette University; T. D. Hinnerichs, Air Force Institute of Technology; D. M. Blackketter, University of Idaho.

This edition of *Plates and Shells: Theory and Analysis* have been significantly influenced by the publisher's reviewers. Their general comments and suggestions were sound and worthwhile. These reviewers were G. H. Paulino, University of California, Davis; R. E. Dippery, Jr., GMI Engineering and Management Institute; A. Kalnins, Lehigh University; L. Godoy, University of Puerto Rico; J. D. Vasikalis, Renesselaer Polytechnic Institute; S. Yim, Oregon State University; K. Rother, Munich University of Applied Sciences; J. N. Reddy, Texas A&M University. The author owes a sincere gratitude to these people for their invaluable attention and advice.

I am indebted to my colleagues and to Jonathan Plant, Taylor & Francis senior executive engineering editor, who have encouraged the development of this edition. Accuracy-checking of the examples and solutions manual, proofreading, and solutions of MATLAB® problems on the website were done by my former graduate student

Youngjin Chung

In addition, contributing considerably to this volume with computer work, typing new inserts, proofreading, assisting with a variety of figures, and cover design was

Errol A. Ugural

Their work is very much appreciated.

Lastly, thanks are due for the understanding and encouragement of my wife, Nora, daughter, Aileen, and son, Errol, during the preparation of the manuscript.

Ansel C. Ugural
Holmdel, New Jersey

Author

Ansel C. Ugural has been a research and visiting professor of mechanical and civil engineering at the New Jersey Institute of Technology. He has taught in the engineering mechanics department at the University of Wisconsin. Dr. Ugural has held various faculty and administrative positions in mechanical engineering at Fairleigh Dickinson University. He has considerable and diverse industrial experience in both full-time and consulting capacities as a design, development, and research engineer.

Professor Ugural earned his MS in mechanical engineering and PhD in engineering mechanics from the University of Wisconsin–Madison. Dr. Ugural was a National Science Foundation (NSF) fellow. He has been a member of several professional societies, including the American Society of Mechanical Engineers and the American Society of Engineering Education. He is also listed in the *Who's Who in Engineering.*

Dr. Ugural is the author of several books, including *Mechanical Design: An Integrated Approach* (McGraw-Hill, 2004); *Mechanical Design of Machine Components* (2nd ed., CRC Press, 2016); *Stresses in Plates and Shells* (McGraw-Hill, 1999); *Stresses in Beams, Plates, and Shells* (CRC Press, 3rd ed., 2010); *Mechanics of Materials* (McGraw-Hill, 1990); *Mechanics of Materials* (Wiley, 2nd ed., 2008). Most of these texts have been translated into Korean, Chinese, and Portuguese. Professor Ugural is also the coauthor (with S.K. Fenster) of *Advanced Mechanics of Materials and Applied Elasticity* (5th ed., Prentice Hall, 2012). In addition, he has published numerous articles in trade and professional journals.

Symbols

Roman Letters

A	area, constant
a, b	dimensions, outer and inner radii of annular plate
c	distance from neutral axis to outer fiber
D	flexural rigidity
d	diameter, distance
$[D]$	elasticity matrix
E	modulus of elasticity
F	resultant external loading on shell element, concentrated forces
f	frequency
G	modulus of elasticity in shear
g	acceleration of gravity
h	mesh width, numerical factor, depth
I	moment of inertia
J	polar moment of inertia
k	modulus of elastic foundation, numerical factor, axial load factor for slender members in compression
$[k]$	stiffness matrix of finite element
$[K]$	stiffness matrix of whole structure
L	length, span
m	integer, numerical factor
M	moment per unit distance, moment, moment-sum
$\bar{m}$	mass per unit area
M^*	thermal moment resultant per unit distance
M_x, M_y	bending moments per unit distance on x and y planes, moments
M_{xy}	twisting moments per unit distance on x plane
M_r, M_θ	radial and tangential moments per unit distance
$M_{r\theta}$	twisting moment per unit distance on radial plane
M_ϕ	meridional bending moment per unit distance on parallel plane
M_s	meridional bending moment per unit distance on parallel plane of conical shell
$M_{x\theta}$	twisting moment per unit distance on axial plane of cylindrical shell
n	factor of safety, integer, numerical factor
N	normal force per unit distance
N_{cr}	critical compressive load per unit distance
N^*	thermal force resultant per unit distance
N_x, N_y	normal forces per unit distance on x and y planes
N_{xy}	shear force per unit distance on x plane and parallel to y axis
N_r, N_θ	radial and tangential forces per unit distance

N_ϕ	meridional force per unit distance on parallel plane
$N_{\phi\theta}$	shear force per unit distance on parallel plane and perpendicular to meridional plane
$N_{x\theta}$	shear force per unit distance on axial plane and parallel to y axis of cylindrical shell
N_s	normal force per unit distance on parallel plane of conical shell
p	intensity of distributed transverse load per unit area or length, pressure, surface force per unit area
p^*	equivalent transverse load per unit area
P	concentrated force
Q	first moment of area
$\{Q\}$	nodal force matrix of finite element
Q_x, Q_y	shear force per unit distance on x and y planes
Q_r, Q_θ	radial and tangential shear forces per unit distance
Q_θ	shear force per unit distance on plane perpendicular to the axial plane of cylindrical shell
Q_ϕ	meridional shear force per unit distance on parallel plane
R	reactive forces
r	radius, radius of gyration
r, θ	polar coordinates
r_x, r_y	radii of curvature of midsurface in xz and yz planes
r_1, r_2	radii of curvature of midsurface in meridional and parallel planes, principal radii of curvature
s	distance measured along generator in conical shell
S	elastic section modulus
T	kinetic energy, temperature, torque
t	thickness
u, v, w	displacements in x, y, and z directions; axial, tangential, and radial displacements in shell midsurface
U	strain energy
U_o	strain energy density
V	shear force, volume
V_x, V_y	effective shear force per unit distance on x and y planes
V_r, V_θ	radial and tangential effective shear forces per unit distance
W	work, weight
x, y, z	distances, rectangular coordinates

Greek Letters

α	angle, coefficient of thermal expansion, numerical factor
β	angle, cylinder geometry parameter, numerical factor
γ	shear strain, weight per unit volume or specific weight
γ_{xy}, γ_{yz}, γ_{zx}	shear strains in the xy, yz, and zx planes
$\gamma_{r\theta}$	shear strain in the $r\theta$ plane
δ	deflection, finite difference operator, numerical factor, variational symbol

$\{\delta\}$	nodal displacement matrix of finite element
ε	normal strain
ϵ	transmissibility
$\varepsilon_x, \varepsilon_y, \varepsilon_z$	normal strains in x, y, and z directions
$\varepsilon_r, \varepsilon_\theta$	radial and tangential normal strains
$\varepsilon_\theta, \varepsilon_\phi$	normal strain of the parallel circle and of the meridian
θ	angle, angular nodal displacement
κ	curvature
λ	sphere or cylinder parameter, numerical factor
ν	Poisson's ratio
Π	potential energy
ρ	density (mass per unit volume)
σ	normal stress
$\sigma_x, \sigma_y, \sigma_z$	normal stresses on the x, y, and z planes
σ_r, σ_θ	radial and tangential normal stresses
σ_ϕ	meridional normal stress on parallel plane
$\sigma_1, \sigma_2, \sigma_3$	principal stresses
σ_{cr}	compressive stress at critical load
σ_u	ultimate tension stress
σ_{uc}	ultimate compression stress
σ_{yp}	yield stress
τ	shear stress
$\tau_{xy}, \tau_{yz}, \tau_{zx}$	shear stresses on the x, y, and z planes and parallel to the y, z, and x directions
$\tau_{r\theta}$	shear stress on radial plane and parallel to the tangential plane
τ_u	ultimate stress in shear
τ_{yp}	yield stress in shear
ϕ	angle, numerical factor, angle of twist
Φ	stress function
χ	change of curvature in shell
ω	natural circular frequency, rad/s

Section I

Fundamentals

George Washington Bridge, spanning the Hudson River between New Jersey and New York. Chapters 1 and 2 are concerned with forces, stresses, and deflections occurring in many of the members contained in such a structure. These include bars, beams, frames, columns, and connections. A beam resists loads with bending and shear stresses. In designing a sandwich and honeycomb structure, we will apply the theory of thin-walled beams.

1

Basic Concepts

1.1 Introduction

A *beam* is a *bar*, possessing length significantly greater than the depth and width. *Plates* and *shells* are initially flat and curved surface structures, respectively, whose thicknesses are slight compared to their other dimensions. Beams are usually loaded in a direction normal to the longitudinal axis, while bars are axially loaded or twisted. Because of the one-, two-, and three-dimensional (3D) load-carrying capacity, the foregoing members are extensively used in various applications in all fields of engineering.

Load-supporting action of plates resembles, to a certain extent, that of beams. However, the load-carrying mechanism of a shell differs from that of other structural forms. This book will primarily deal with the theory and analysis of plate and shell-like structures. Emphasis will be placed on the clear presentation of the basic principles and methods rather than on achieving exhaustive coverage of an inherently large body of subject matter.

The text consists of three sections. *Section I* is about *fundamentals* that form the cornerstone of the theories of beams, plates, and shells. It has to be reviewed carefully by readers who have never had a modern treatment of this material or who have forgotten it and need a refresher on these topics. This section begins by introducing basic concepts, scope of treatment, some equations of *mechanics of materials* or *technical theory* and the *theory of elasticity* in a condensed form, properties of common engineering materials, rational procedure in design, and factor of safety. It then proceeds with stress analysis of bars, beams, and simple pressure vessels. There is also a discussion on material failure criteria. A thorough grasp of the fundamentals will prove of great value in attacking new and unfamiliar problems.

Sections II and III are on stresses and deformations in *plates and shells* due to bending, shear, tension, or compression loads. In analyzing such cases, unless otherwise specified, we shall assume that the members are made of homogeneous and isotropic materials. But composite and anisotropic plates and shells will also be considered. It should be noted that for the understanding of the technical theory of the bending of beams, plates, and shells, knowledge of mechanics of materials is considered sufficient. A number of *selected references* are identified in *brackets* and listed at the end of each chapter for those seeking a more extensive treatment.

1.2 Methods of Analysis

Mechanics of materials and *theory of elasticity* deal with the internal behavior of variously loaded solid bodies. The former uses assumptions based on experimental evidence along with engineering experience to make a reasonable solution of the practical problem possible. The latter concerns itself largely with more mathematical analysis of the "exact" stress distribution on a loaded body. In general, however, theory of elasticity can provide solutions with increased considerable difficulty. The differences between these approaches lie primarily in the extent to which strains are described and in the nature of simplifications used [1,2].

Formulas of the mechanics of materials give *average* stresses at a section. Since irregular stresses arise at concentrated forces and abrupt changes in the cross section, only *at distances about equal to the depth of the member* from such disturbances are the stresses in agreement with the mechanics of materials theory. This is in accordance with *St. Venant's Principle*: The stress of a member at points sufficiently away from points of load application may be determined on the basis of a statically equivalent loading system. Thus, the mechanics of materials approach is best suited for relatively *slender members* under consideration in this text.

The classical theory of plates or shells, which formulates and solves problems from the point of view of rigorous mathematical analysis, is an important application of the theory of elasticity. The study of mechanics of materials and the theory of elasticity is based on an understanding of equilibrium of bodies under action of forces. While *statics* treats the external behavior of bodies that are assumed to be ideally rigid and at rest, mechanics of materials and the theory of elasticity are concerned with the relationships between external loads and internal forces and deformations induced in the body.

Mechanics of materials and theory of elasticity methods are used to determine strength, stiffness, and stability of variously loaded members, or possible *modes of failure*. In most general terms, *failure* refers to any action resulting in an inability on the part of the structure to function in the manner intended. The ability of a member to resist yielding or fracture is called *strength*. The *stiffness* refers to the ability of a member to resist deformation. It is often necessary to limit the magnitude of displacement in the member for it to function in normal service. The ability of the structure to retain its equilibrium configuration under loading is termed *stability*. Instability occurs if the loading produces an abrupt shape change of a member.

The complete analysis of a load-carrying member by the so-called *method of equilibrium* involves three *basic principles*. These so-called *three aspects of solid mechanics problems* can be outlined as follows:

1. *Statics*. The equilibrium conditions of forces must be satisfied.
2. *Deformations*. Stress–strain or force–deformation relations (e.g., Hooke's law) must apply to the behavior of the material.
3. *Geometry*. The conditions of geometric fit, or compatibility of deformations, must be satisfied.

The solutions based on this procedure must satisfy the *boundary conditions*. The preceding requirements will be expressed mathematically and used in problems presented as the subject unfolds.

Alternatively, the analysis of stress and deformation can be accomplished through the use of *energy methods*, which are based on the concept of strain energy. The roles of the equilibrium and energy approaches are twofold. They can provide solutions of acceptable accuracy, where configurations of loading and member are regular, and they can be employed as a basis of *numerical methods* for more complex problems.

1.2.1 Case Studies in Analysis

Case studies are taken from real-life situations and may come in the varieties, such as the history of an engineering activity, illustration of some form of engineering process, an exercise (such as stress and deformation analysis), a proposal of problems to be solved, or a preliminary design project. Through case studies, we can create a link between systems theory and actual design plans. A general case study in *structural analysis* may include the step by step process through the problem formulation and solution stages, outlined in Section 1.17.

The basic geometry and loading on a member must be prescribed before any analysis can be done. For instance, the stress that would result in a bar subjected to a load would depend on whether the loading gives rise to tension, transverse shear, direct shear, torsion, bending, or contact stresses. In this case, *uniform stress* patterns may be more efficient at carrying the load than others. Thus, by making a careful study of the types of loads and stress patterns that can arise in structures or machines, considerable insight can be gained into improved shapes and orientations of components. A few case studies presented in this text involve situations found in the analysis of simple structures.

1.3 Loading Classes and Equilibrium

External forces acting on a member may be classified as *surface forces* and *body forces*. A surface force is of the *concentrated* type when it acts at a point; a surface force may also be *distributed* uniformly or nonuniformly over a finite area. Body forces (gravitational, magnetic, and inertia forces) are associated with the mass of a member rather than surfaces and are distributed throughout the volume of a member. They are specified in terms of force per unit volume. All forces acting on a member, including the reactions, are considered external forces. *Internal forces* are the forces that hold together the particles forming the member. Unless otherwise stated, we assume in this text that the body forces can be neglected and that the force is applied steadily and slowly in *static loading*.

A *structure* is a unit composed of interconnected members supported in a manner that is capable of resisting applied forces in static equilibrium (Section 2.2). Adoption of thin-walled behavior allows certain assumptions to be made in the structural analysis (Sections 3.3 and 13.1). *Self-weight* of a structure, sometimes referred to as the *dead load*, is fixed in position and constant in magnitude over the life of the structure. On the other hand, a *live load* is a movable or moving load that may vary in magnitude. Examples include occupancy load, snow load, wind load, earthquake load, highway line load to represent a vehicle, and impact and dynamic loads [3]. The American Society of Civil Engineers (ASCE) lists design loads for buildings and other common structures.

1.3.1 Conditions of Equilibrium

When a system of forces acting on a body has zero resultant, the body is said to be in *equilibrium*. That is, equilibrium of forces is the state in which the forces applied on a body are in balance. In a *3D problem, conditions of equilibrium* require the fulfillment of the following *equations of statics*:

$$\sum F_x = 0 \qquad \sum F_y = 0 \qquad \sum F_z = 0$$
$$\sum M_x = 0 \qquad \sum M_y = 0 \qquad \sum M_z = 0 \tag{1.1}$$

These state that the sum of all forces acting on a body in any direction is zero; the sum of all moments about any axis is zero.

In a *two-dimensional (2D)* or *planar problem*, where all forces act in a single (xy) plane, there are only three independent equations of statics:

$$\sum F_x = 0 \qquad \sum F_y = 0 \qquad \sum M_A = 0 \tag{1.2a}$$

By replacing a force summation by an equivalent moment summation in Equations 1.2a, the following *alternative* sets of conditions are obtained:

$$\sum F_x = 0 \qquad \sum M_A = 0 \qquad \sum M_B = 0 \tag{1.2b}$$

provided that the line connecting the points A and B is *not* perpendicular to the x axis, or

$$\sum M_A = 0 \qquad \sum M_B = 0 \qquad \sum M_C = 0 \tag{1.2c}$$

where points A, B, and C are *not* collinear.

The preceding equations are directly applicable to deformable solid bodies. The deformations tolerated in engineering structures are usually disregarded when compared to the overall dimensions of structures. Thus, for the purposes of force analysis in members, the *initial* undeformed *dimensions* of members *are used* in computations.

A structure is said to be *statically determinate* when all forces on its members can be obtained by using only the equilibrium conditions. Otherwise, the structure is called *statically indeterminate*. The degree of *static indeterminacy* is equal to the difference between the number of unknown forces and the number of pertinent equilibrium equations. Any reaction that is in excess of those that can be found by statics alone is referred to as *redundant*. Thus, the number of redundants is the same as the degree of indeterminacy.

1.3.2 Free-Body Diagrams

Use of equilibrium conditions requires a complete specification of all loads and reactions that act on a structure. It is usually necessary to make simplifying idealizations of the system or of the nature of the forces acting on the system. These allow the construction of a *free-body diagram* (FBD) to which the equations of statics can be applied. The FBD is a sketch of the isolated body and all external forces acting on it. Complete, carefully drawn, free-body diagrams (FBDs) ease visualization.

When internal forces are of concern, an imaginary cut is made through the body at the section of interest. Obviously, the prudent selection of the free body to be employed is of prime significance. The reader is strongly urged to adopt the habit of using clear and complete FBDs in the solution of problems concerning equilibrium. Examples 1.1 and 2.1 will illustrate the construction of FBDs and application of the equations of statics.

1.4 Units and Conversion

The units of the physical quantities used in engineering calculations are of major importance. The most recent universal system is the *International System of Units* (SI). The *U.S. Customary System (USCS) of units* has long been utilized by engineers in the United States. While both systems of units are reviewed briefly, this text uses SI units. A few fundamental quantities in SI units are listed in Table 1.1.

We observe from Table 1.1 that, in SI, force F is a derived quantity (obtained by multiplying the mass m by the acceleration a, hence Newton's second law, $F = ma$). However, in the USCS, the situation is reversed, with mass being the derived quantity. It is found from Newton's second law, as $\text{lb} \cdot \text{s}^2/\text{ft}$, sometimes called the *slug*. The USCS units are foot (ft) for length, pound (lb) for force, second (s) for time, and degree Fahrenheit (°F) for temperature.

Temperature is expressed in SI by a unit called kelvin (K), but customarily, the degree Celsius (°C) is used (as shown in Table 1.1). The relationship between the two units: temperature in Celsius = temperature in kelvins −273.15. Conversion formulas between the temperature scales are given by

$$t_c = \frac{5}{9}(t_f - 32) \tag{a}$$

and

$$t_k = (t_f - 32) + 273.15 \tag{b}$$

where t denotes the temperature. Subscripts c, f, and k represent the Celsius, Fahrenheit, and kelvin, respectively.

TABLE 1.1

SI Units and Symbols

Class	Quantity	Name of Unit	SI Symbol	Unit Formula
Base	Length	meter	m	–
	Mass	kilogram	kg	–
	Time	second	s	–
	Temperature	kelvin	K	–
		Celsius	C	–
Derived	Area	square meter	m^2	–
	Volume	cubic meter	m^3	–
	Force	newton	N	$kg \cdot m/s^2$
	Stress, pressure	pascal	Pa	N/m^2
	Work, energy	joule	J	$N \cdot m$
	Power	watt	W	$N \cdot m/s$
Supplementary	Plane angle	radian	rad	–

The *acceleration due to gravity*, designated by the letter g, changes with elevation and latitude on the Earth. For ordinary purposes, the following *approximate* value can be assumed:

$$g = 9.81\ \mathrm{m/s^2} \quad (\text{or } 32.2\ \mathrm{ft/s^2})$$

The nominal "average" value of $g = 9.8066$ (or 32.174) at the Earth's surface is referred to as *standard gravity*. Therefore, from Newton's second law, the weight W of a body of mass 1 kg is equal to $W = mg = (1\ \mathrm{kg})\ (9.81\ \mathrm{m/s^2}) = 9.81\ \mathrm{N}$.

The unit of force is of particular importance in engineering analysis and design, because it is involved in calculations of the force, moment, torque, stress (or pressure), work (or energy), power, and elastic modulus. Interestingly, in SI units, a newton equals approximately the weight of (or Earth's gravitational force on) an average apple. Because the newton is a small quantity, the kilonewton (kN) is frequently used in practice.

Tables B.1 and B.2 list usually employed conversion factors and SI prefixes. The use of prefixes avoids unusually large or small numbers. Note that, a dot is to be used to separate units that are multiplied together. So, for instance, a newton meter is written N·m and must not be confused with mN, which stands for millinewton. The reader is urged always to check the units in any equation written for a problem solution. When properly written, an equation should cancel all units across the equal sign.

1.5 Stress Defined

Consider a body in equilibrium, subject to the system of external forces (Figure 1.1a). To investigate the internal forces at some point Q in the interior of the body, we cut the body through Q by an imaginary plane, dividing the body into two portions. The equilibrium of the forces acting on one portion alone requires the presence of internal forces on the sectioning plane. These internal forces, applied to both portions, are distributed continuously over the cut surface. The preceding process, which is referred to as the *method of sections*, will be relied on as a first step in solving all problems where internal forces are being investigated.

The isolated left portion of the body is depicted in Figure 1.1b. An element of area ΔA, located at the point Q on the cut surface, is acted on by force ΔF. In general, ΔF does not

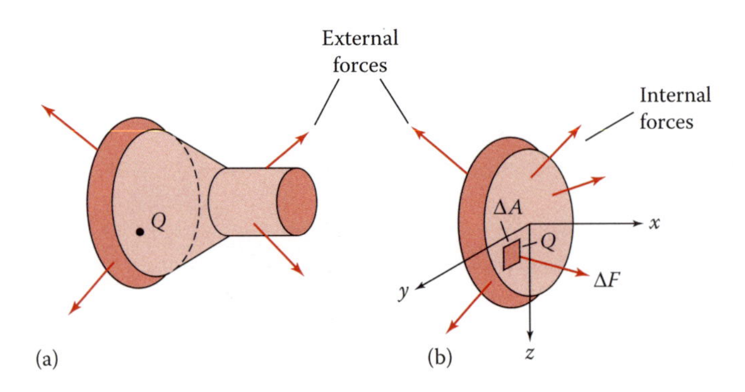

FIGURE 1.1
(a) Sectioning of a body; (b) FBD of the part of the body.

lie along x, y, or z. Let the components of this force be ΔF_x normal, ΔF_y, ΔF_z tangent to ΔA. Then the *normal stress* σ_x *and the shear stresses* τ_{xy} and τ_{xz} may be defined as follows:

$$\sigma_x = \lim_{\Delta A \to 0} \frac{\Delta F_x}{\Delta A} = \frac{dF_x}{dA}$$
$$\tau_{xy} = \lim_{\Delta A \to 0} \frac{\Delta F_y}{\Delta A} = \frac{dF_y}{dA} \tag{1.3}$$
$$\tau_{xz} = \lim_{\Delta A \to 0} \frac{\Delta F_z}{\Delta A} = \frac{dF_z}{dA}$$

These expressions provide the stress components at a point Q to which the area ΔA is reduced in the limit. Note that $\Delta A \to 0$ is an idealization because the surface itself is not continuous on an atomic scale. But our consideration is with the stress on areas where sizes are large compared with the distance between atoms in the solid body. Therefore, stress is an adequate definition for engineering purposes.

Since stress (σ or τ) is obtained by dividing the force by the area, it has units of force per unit area. In SI units, the stress is measured in newtons per square meter (N/m^2), or pascals. The pascal is a very small quantity, and the megapascal (MPa) is commonly used. Typical prefixes of the SI units are given in Table B.2.

1.5.1 Components of Stress

A cube of infinitesimal dimensions isolated from a solid would expose the general case of *3D state of stress* (Figure 1.2). The stresses shown are considered to be the same on the mutually parallel faces and uniformly distributed on each face. However, in general, the stresses would vary from one face to a parallel face and would also vary over a particular face. A *face* or *plane* is usually *identified by the axis normal to it*; for example, the x faces are perpendicular to the x axis. A total of nine scalar stress components, defining the state of stress at a point, can be assembled in the form:

$$[\tau_{ij}] = \begin{bmatrix} \tau_{xx} & \tau_{xy} & \tau_{xz} \\ \tau_{yx} & \tau_{yy} & \tau_{yz} \\ \tau_{zx} & \tau_{zy} & \tau_{zz} \end{bmatrix} = \begin{bmatrix} \sigma_x & \tau_{xy} & \tau_{xz} \\ \tau_{yx} & \sigma_y & \tau_{yz} \\ \tau_{zx} & \tau_{zy} & \sigma_z \end{bmatrix} \tag{1.4}$$

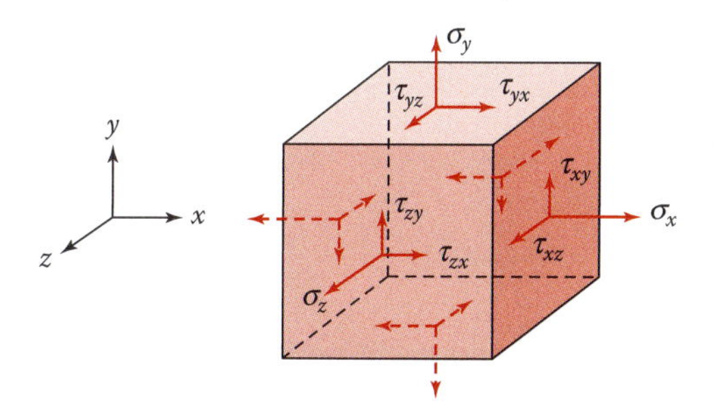

FIGURE 1.2
Element in 3D stress. All stresses have positive sense.

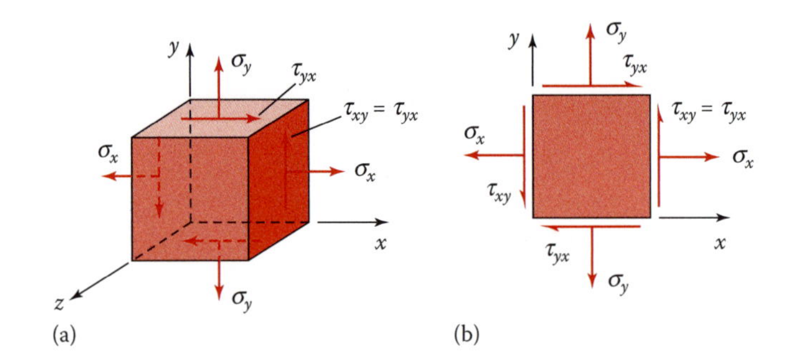

FIGURE 1.3
(a) Element in plane stress; (b) 2D representation of plane stress element.

This is a matrix representation of the *stress tensor.* It is a second-rank tensor requiring two indices to identify its elements. For instance, a vector is a tensor of first rank, and a scalar is of zero rank.

The *double subscript notation* is interpreted as follows: The first subscript denotes the direction of a normal to the face on which the stress component acts; the second denotes the direction of the stress. Repetitive subscripts will be avoided in this text so that the normal stresses will be designated σ_x, σ_y, and σ_z, as shown in Equation 1.4. In Section 1.6, it is shown rigorously that $\tau_{xy} = \tau_{yx}$, $\tau_{yz} = \tau_{zy}$, and $\tau_{xz} = \tau_{zx}$, or $\tau_{ij} = \tau_{ji}$.

In the case of *2D* or *plane stress*, only the x and y faces of the element are subjected to stresses, and all the stresses act parallel to the x and y axes (Figure 1.3a). For convenience, we often draw a *2D view* of the plane stress element, as shown in Figure 1.3b. When only two normal stresses are present, the state of stress in termed *biaxial.*

1.5.2 Sign Convention

The face of an element whose outward normal is along the positive (negative) direction of a coordinate axis is defined to be a positive (negative) face. When a stress component acts on a positive face in a positive coordinate direction, the stress component is *positive.* In addition, a stress component is considered positive when it acts in a negative face in the negative coordinate direction. But a stress component is considered *negative* when it acts in a positive face in a negative coordinate direction (or vice versa).

Accordingly, tensile stresses are always positive, and compressive stresses are always negative. The sign convention can also be stated as follows: A stress component is positive if *both* the outward normal of the plane on which it acts and its direction are in coordinate directions of the *same* sign; otherwise, it is negative. Figures 1.2 and 1.3 show positive normal and shear stresses.

1.6 Internal-Force Resultants

Forces distributed within a member can be represented by a statically equivalent force and a moment vector acting at any arbitrary point of a section. These internal-force resultants, also called stress resultants, consist of axial force, shear force, and moments. They are exposed by an imaginary cutting plane usually including the centroid C through

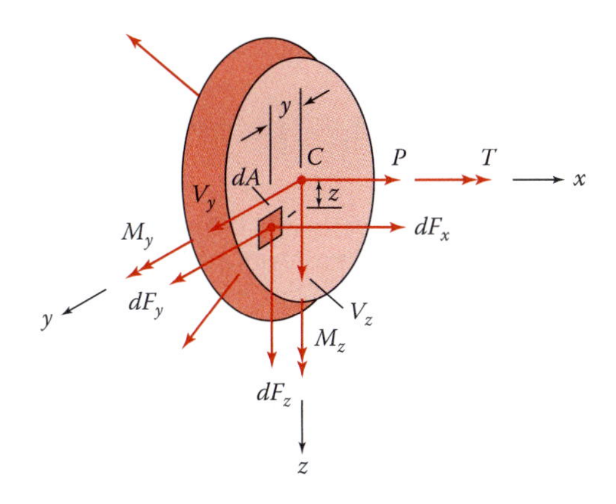

FIGURE 1.4
Positive internal-force resultants and components of dF at a point on cut section of isolated part of body shown in Figure 1.1a.

the member and resolved into components normal and tangential to the cut section (Figure 1.4). Note that the sense of moments follows the right-hand screw rule and, for convenience, is frequently represented by double-headed vectors, as indicated in the figure.

Each internal force and moment component reflects a different effect of the applied loading on a member. These effects are as follows, with reference to Figure 1.4:

Axial force P tends to elongate or contract the member.

Shear forces V_y and V_z tend to shear one part of the member relative to the adjacent part.

Torque T is responsible for twisting the member.

Bending moments M_y and M_z cause the member to bend.

Generally, it is convenient to work with the preceding *modes of force transmission* separately and, if relevant, to combine the results for the final solution. A member may be under any or all of the modes simultaneously.

The *sign convention* adopted for the stress in Section 1.5.2 also applies to the force and moment components. That is, a positive force (or moment) component acts on the positive face in the positive coordinate direction or on a negative face in the negative coordinate direction and so on. The foregoing notation for internal-force resultants is consistent with that commonly used in the analysis of most structural members.

To establish the relationships between the *stress components and internal-force resultants*, consider an infinitesimal area dA of the cut section shown in Figure 1.4. This typical area is acted on by the components of dF, expressed by Equations 1.3 as $dF_x = \sigma_x dA$, $dF_y = \tau_{xy} dA$, and $dF_z = \tau_{xz} dA$. Obviously, the stress components on the cut section cause the internal-force resultants on that section. Therefore, the sum of the incremental forces in the x, y, and z directions are

$$P = \int \sigma_x dA \qquad \int V_y = \int \tau_{xy} dA \qquad V_z = \int \tau_{xz} dA \tag{1.5a}$$

Similarly, the sums of the moments of the same forces about the x, y, and z axes give

$$T = \int (\tau_{xz}y - \tau_{xy}z)dA \qquad M_y = \int \sigma_x z dA \qquad M_z = -\int \sigma_x y dA \tag{1.5b}$$

Here integrations are over the area A of the cut section. We shall observe first in Section 2.2 how Equations 1.5 connect internal-force resultants and state of stress in a specific case.

EXAMPLE 1.1: Force Analysis of a Pipe

An L-shaped pipe of mass 1.1 kg/m, securely fastened to a rigid wall, is subjected to a vertical force 60 kN and a torque 100 N·m as shown in Figure 1.5a. Determine the axial force, shear forces, and moments acting on the cross section at point C.

Solution

The weight of segments AD and DC of the pipe are

$$W_{ad} = 1.1(1.0)(9.81) = 10.79 \text{ N}$$
$$W_{cd} = 1.1(0.8)(9.81) = 8.63 \text{ N}$$

We cut the pipe at C and obtain two parts. Considering the FBD of ADC (Figure 1.5b), we write

$$\sum F_x = 0: \qquad P = 0$$
$$\sum F_y = 0: \qquad V_y = 0$$
$$\sum F_z = 0: \quad V_z + 10.79 + 8.63 + 60 = 0 \quad V_z = -79.42 \text{ N}$$

and

$$\sum M_x = 0: \quad T + 10.79(0.5) + 60(1.0) = 0 \qquad T = -65.4 \text{ N}\cdot\text{m}$$
$$\sum M_y = 0: \quad M_y + 100 - 60(0.8) - 10.79(0.8) - 8.63(0.4) = 0 \quad M_y = -39.92 \text{ N}\cdot\text{m}$$
$$\sum M_z = 0: \qquad M_z = 0$$

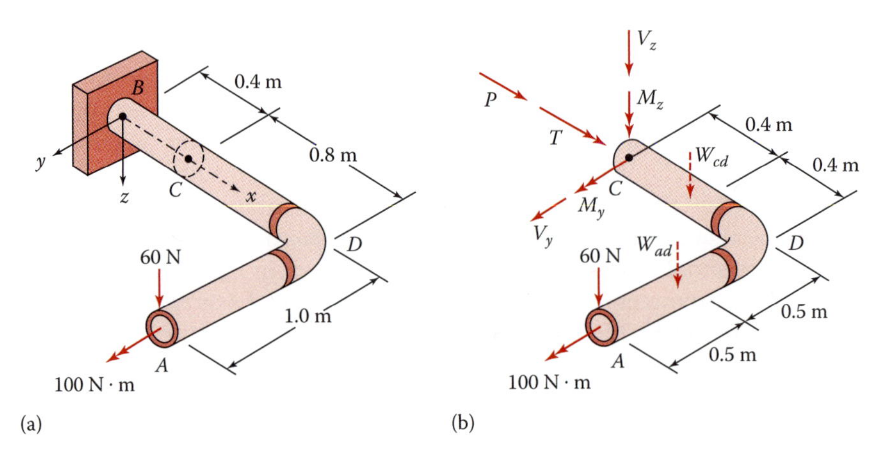

FIGURE 1.5
(a) An L-shaped pipe ADB; (b) free-body ADC.

The negative values found for V_z, T, and M_y mean that the shear force, torque, and moment vectors are directed opposite to those shown in the figure.

Comment: MATLAB® solution of this sample problem and many others are on the website (see Appendix D).

1.7 Differential Equations of Equilibrium

As previously noted, the components of stress generally vary from point to point in a loaded body. Such variations of stress, considered by the theory of elasticity, are governed by the equations of statics. Fulfillment of these conditions leads to the *differential equations of equilibrium*.

For a 2D case, the stresses acting on an element of sides dx, dy, and of unit thickness are shown in Figure 1.6. The x and y components of the body forces per unit volume, F_x and F_y, are independent of z, and the z component of the body force $F_z = 0$. Consider, for example, the variation of one stress component, say σ_x, from point to point or between the left and right faces of the element. This may be expressed by a truncated Taylor's expansion

$$\sigma_x + \left(\frac{\partial \sigma_x}{\partial x}\right) dx \tag{1.6}$$

The partial derivatives is required because σ_x varies with x and y. The components σ_y and τ_{xy} change in an analogous manner.

We now require that the stress element (Figure 1.6) satisfy the equilibrium condition $\sum M_z = 0$. Thus,

$$\left(\frac{\partial \sigma_y}{\partial_y} dxdy\right)\frac{dx}{2} - \left(\frac{\partial \sigma_x}{\partial x} dxdy\right)\frac{dy}{2} + \left(\tau_{xy} + \frac{\partial \tau_{xy}}{\partial x} dx\right) dxdy$$
$$-\left(\tau_{yx} + \frac{\partial \tau_{yx}}{\partial y} dy\right) dxdy + F_y dxdy\frac{dx}{2} - F_x dxdy\frac{dy}{2} = 0$$

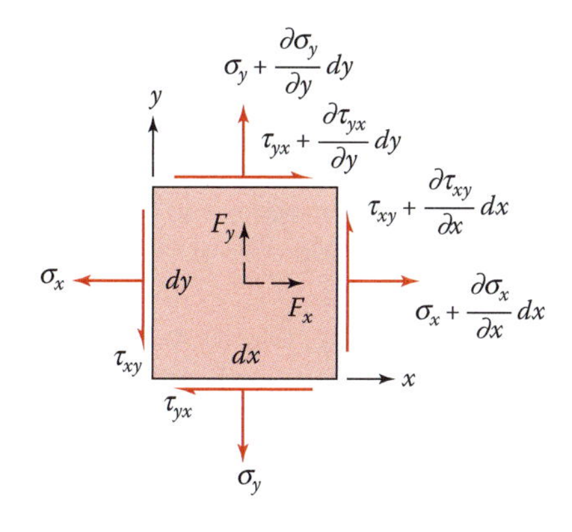

FIGURE 1.6
Element with stresses and body forces.

Neglecting the triple products involving dx and dy, we have

$$\tau_{xy} = \tau_{yx} \tag{1.7}$$

Likewise, for a general state of stress, it can be verified that $\tau_{yz} = \tau_{zy}$ and $\tau_{xz} = \tau_{zx}$. Hence, the subscripts for the shear stresses are commutative, and the stress tensor is symmetric. This means that *shear stresses in mutually perpendicular planes* of the element are *equal.*

The equilibrium condition that x-directed forces be zero, $\sum F_x = 0$, leads to the expression

$$\left(\sigma_x + \frac{\partial \sigma_x}{\partial x} dx\right) dy - \sigma_x dy + \left(\tau_{xy} + \frac{\partial \tau_{xy}}{\partial y} dy\right) dx - \tau_{xy} dx + F_x dx dy = 0$$

A similar equation is written for $\sum F_y = 0$. Simplifying these expressions, we obtain the differential equations of equilibrium for a *2D* stress in the form [2]

$$\begin{aligned} \frac{\partial \sigma_x}{\partial x} + \frac{\partial \tau_{xy}}{\partial y} + F_x = 0 \\ \frac{\partial \sigma_y}{\partial y} + \frac{\partial \tau_{xy}}{\partial x} + F_y = 0 \end{aligned} \tag{1.8}$$

The differential equations of equilibrium may be generalized by considering the *3D* counterpart of Figure 1.6 with the following result:

$$\begin{aligned} \frac{\partial \sigma_x}{\partial x} + \frac{\partial \tau_{xy}}{\partial y} + \frac{\partial \tau_{xz}}{\partial z} + F_x = 0 \\ \frac{\partial \sigma_y}{\partial y} + \frac{\partial \tau_{xy}}{\partial x} + \frac{\partial \tau_{yz}}{\partial z} + F_y = 0 \\ \frac{\partial \sigma_z}{\partial z} + \frac{\partial \tau_{xz}}{\partial x} + \frac{\partial \tau_{yz}}{\partial y} + F_z = 0 \end{aligned} \tag{1.9}$$

Here each body force component may be a function of x, y, and z. We note that, in many practical applications, the weight of the member is the only body force. If we take the y axis upward and denote by ρ the mass density per unit volume of the member and by g, the gravitational acceleration, then $F_x = F_z = 0$ and $F_y = -\rho g$ in Equations 1.8 and 1.9.

Interestingly, since the *two* equilibrium conditions of Equations 1.8 involve the *three* unknowns (σ_x, σ_y, τ_{xy}) and the *three* relations of Equations 1.9 contain the *six* unknown stress components, the problems in stress analysis are *internally statically indeterminate.* In the mechanics of materials approach, this indeterminacy is eliminated by introducing simplifying assumptions regarding the distribution of strain and hence stress and by considering the equilibrium of the *finite segments* of a load-carrying structural member.

1.8 Transformation of Stress

So far, the stresses on planes perpendicular to the coordinates describing a member have been considered. We shall here deal with the states of stress at points located on *inclined planes*. The components of stress generally also depend on the position of the point in a loaded member, as shown in Section 1.7. Our discussions are limited to *2D*, or *plane, stress*.

Let us consider stress components σ_x, σ_y, τ_{xy} at a point in a body represented by a 2D stress element (Figure 1.7a). An infinitesimal wedge, cut from this element, is shown in Figure 1.7b. The angle θ from x to x' axis is assumed positive when measured in the counterclockwise direction. The side AB is normal to the x' axis, and in accordance with the sign convention (see Section 1.5.2), the stresses are indicated as positive values. It can be verified that equilibrium of the forces caused by all the stress components gives the transformation equations for stress [1]:

$$\sigma_{x'} = \sigma_x \cos^2\theta + \sigma_y \sin^2\theta + 2\tau_{xy} \sin\theta\cos\theta \tag{1.10a}$$

$$\tau_{x'y'} = \tau_{xy}(\cos^2\theta - \sin^2\theta) + (\sigma_y - \sigma_x)\sin\theta\cos\theta \tag{1.10b}$$

The stress $\sigma_{y'}$ is found by substituting $\theta + \pi/2$ for θ in the equation for $\sigma_{x'}$ (Figure 1.7c):

$$\sigma_{y'} = \sigma_x \sin^2\theta + \sigma_y \cos^2\theta - 2\tau_{xy} \sin\theta\cos\theta \tag{1.10c}$$

The preceding expressions can be written in the following convenient form:

$$\sigma_{x'} = \frac{1}{2}(\sigma_x + \sigma_y) + \frac{1}{2}(\sigma_x - \sigma_y)\cos 2\theta + \tau_{xy} \sin 2\theta \tag{1.11a}$$

$$\tau_{x'y'} = -\frac{1}{2}(\sigma_x - \sigma_y)\sin 2\theta + \tau_{xy} \cos 2\theta \tag{1.11b}$$

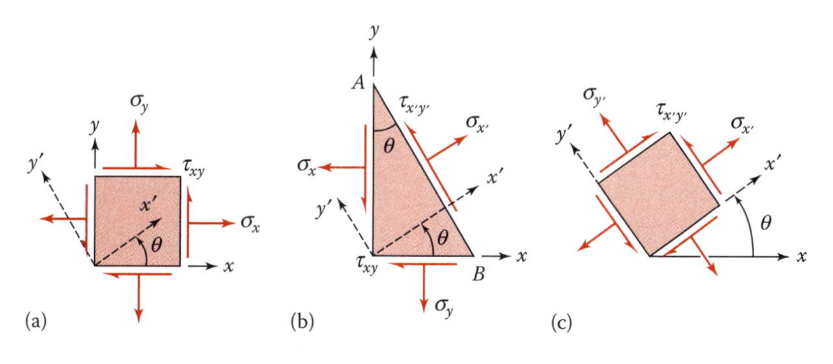

FIGURE 1.7
Elements in plane stress.

$$\sigma_{y'} = \frac{1}{2}(\sigma_x + \sigma_y) = -\frac{1}{2}(\sigma_x - \sigma_y)\cos 2\theta - \tau_{xy}\sin 2\theta \tag{1.11c}$$

The two perpendicular directions (θ_p' and θ_p'') of planes on which the shearing stress vanishes and the normal stress has extreme values can be obtained from

$$\tan 2\theta_p = \frac{2\tau_{xy}}{\sigma_x - \sigma_y} \tag{1.12}$$

The angle θ_p defines the orientation of the *principal planes*. Introducing the values of the sine and cosine corresponding to the double angle given by Equation 1.12 into Equations 1.11a,c results in

$$\sigma_{\max,\min} = \sigma_{1,2} = \frac{\sigma_x + \sigma_y}{2} \pm \sqrt{\left(\frac{\sigma_x - \sigma_y}{2}\right)^2 + \tau_{xy}^2} \tag{1.13}$$

The algebraically larger stress given in the preceding is the *maximum principal stress*, designated σ_1. The *minimum principal stress* is denoted by σ_2. It is necessary to substitute one of the values θ_p into Equation 1.11a to find which of the two corresponds to σ_1.

The general variation of the normal and shear stresses is illustrated in Figure 1.8. This sketch of $\sigma_{x'}$ and $\tau_{x'y'}$ versus the angle θ (from Equations 1.10 or 1.11), is plotted for the particular case of $\sigma_y = 0.2\sigma_x$ and $\tau_{xy} = 0.8\sigma_x$. Observe that the stresses differ continuously as the orientation of the element is changed. At certain angles, the normal stress is a maximum or minimum value; while at other angles, it becomes zero. Likewise, the shear stress has maximum or zero values at certain angles.

1.8.1 Mohr's Circle for Stress

Equations 1.11 can be graphically represented in a Mohr's circle (Figure 1.9a) with σ and τ as coordinate axes. The center C is at (σ', 0), and the circle radius r equals length CA.

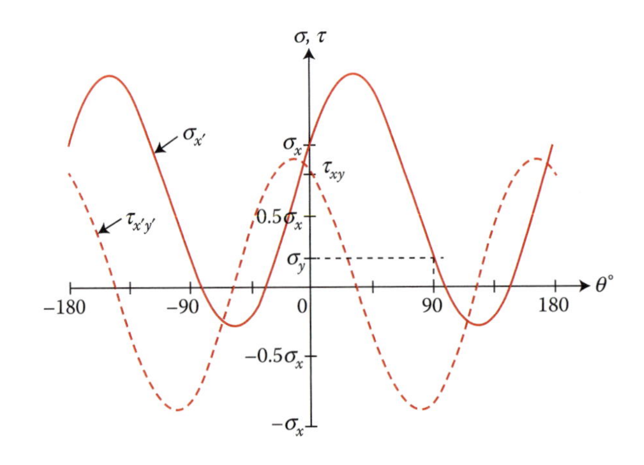

FIGURE 1.8
Graphical representation of normal stress $\sigma_{x'}$ and shear stress $\tau_{x'y'}$ with angle θ for $\sigma_y = 0.2\sigma_x$ and $\tau_{xy} = 0.8\sigma_x$.

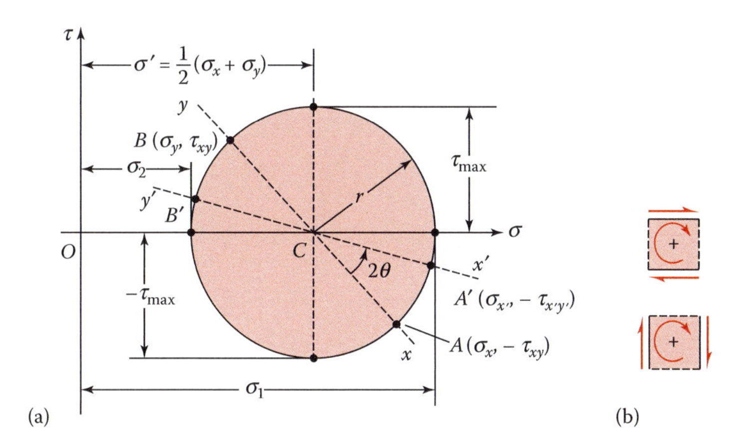

FIGURE 1.9
(a) Mohr's circle of stress; (b) interpretation of positive shear stress.

An angle of 2θ on the circle corresponds to an angle of θ on the element. The coordinates for a point A on the circle correspond to the stresses in Figure 1.7a on the x face of the element. Likewise, a point A' on Mohr's circle defines the stress components $\sigma_{x'}$ and $\tau_{x'y'}$, which act on the x' plane. Clearly, points B and B' on the circle located the stresses on the y and y' planes, respectively.

In Mohr's circle representation, the *normal stresses* obey the sign convention of Section 1.5.2. However, the *shear stresses* on the y faces of the element are taken as positive (as before), but those on the x faces are now *negative*, as depicted in Figure 1.9b. This special sign convention for shear stress is required for the purposes *only* of constructing and reading values of stress from a Mohr's circle.

The magnitude of the *maximum shear stress* is equal to the radius r of the circle. From the geometry of Figure 1.9a, we have

$$\tau_{\max} = \pm\sqrt{\left(\frac{\sigma_x - \sigma_y}{2}\right)^2 + \tau_{xy}^2} = \pm\frac{1}{2}(\sigma_1 - \sigma_2) \tag{1.14}$$

We observe from the Mohr's circle that the planes of maximum shear are always oriented at 45° from planes of principal stress (Figure 1.10). Note that a diagonal of a stress element along which the algebraically larger principal stress acts is called the shear diagonal. The maximum shear stress acts toward the shear diagonal. The normal stress occurring on planes of maximum shearing stress equals

$$\sigma' = \frac{1}{2}(\sigma_x + \sigma_y) \tag{1.15}$$

It can readily be shown by using Mohr's circle that, on any mutually perpendicular planes,

$$I_1 = \sigma_x + \sigma_y = \sigma_{x'} + \sigma_{y'} \qquad I_2 = \sigma_x\sigma_y - \tau_{xy}^2 = \sigma_{x'}\sigma_{y'} - \tau_{x'y'}^2 \tag{1.16}$$

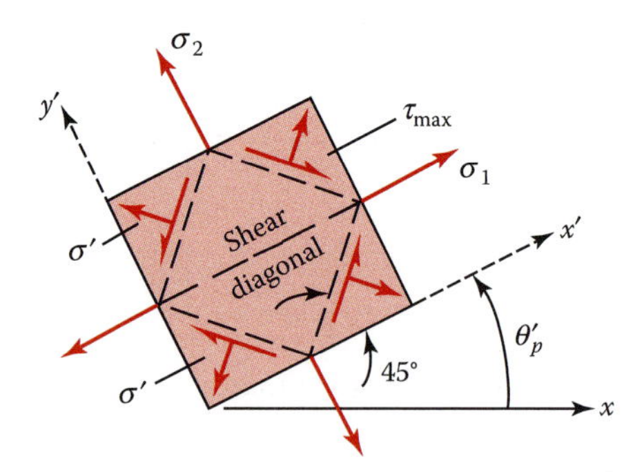

FIGURE 1.10
Planes of maximum stresses.

The foregoing quantities I_1 and I_2 are called 2D stress invariants because they remain the same regardless of the angle θ. This is particularly helpful in checking the results of a stress transformation.

As already pointed out in Section 1.5, quantities like stress (strain, moment of inertia, curvature, and moment) that are subject to transformations such as defined by Equations 1.11 are tensors of second-rank. *Mohr's circle* is therefore a *graphical* representation of a *tensor transformation*.

EXAMPLE 1.2: Principal Stresses in a Plate

The state of stress at a point in a loaded plate is shown in Figure 1.11a. Using Mohr's circle, determine: (a) the principal planes and the principal stresses; (b) the maximum shear stress and the associated normal stress. In each case, sketch the results on a properly oriented element.

Solution

Mohr's circle constructed from the given numerical values is shown in Figure 1.11b. The radius is calculated as

$$r = \sqrt{10^2 + 24^2} = 26 \text{ MPa}$$

a. The principal stresses are represented by the abscissa of point A_1 and B_1. Thus,

$$\sigma_1 = OA_1 = 40 + 26 = 66 \text{ MPa}$$

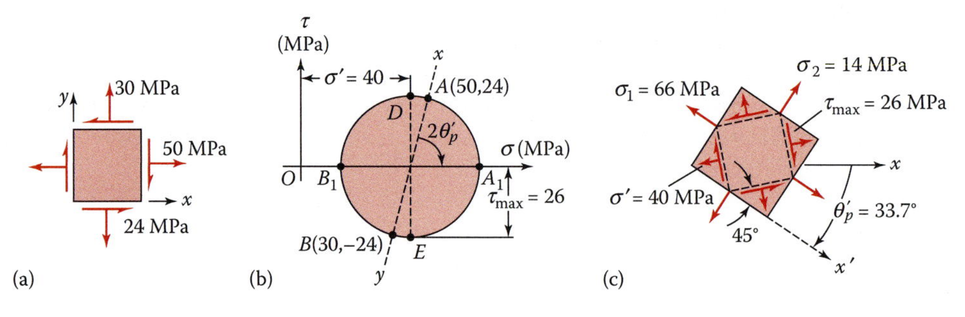

FIGURE 1.11
Stresses by Mohr's circle.

$$\sigma_2 = OB_1 = 40 - 26 = 14 \text{ MPa}$$

The plane of the maximum principal stress is given by

$$\theta_p' = \frac{1}{2}\tan^{-1}\frac{24}{50-40} = 33.7°$$

The required sketch is shown in Figure 1.11c.

b. The coordinate of point D (or E) on Mohr's circle represents the maximum shear stress and the corresponding normal stress, respectively:

$$\tau_{\max} = r = 26 \text{MPa} \quad \sigma' = 40 \text{ MPa}$$

These stresses are exerted on an element that is oriented as shown by the dashed lines in Figure 1.11c.

1.9 Strain Defined

As a result of *deformation*, the *extension, contraction, or change in shape* of a body may occur. To determine the actual stress distribution within a body, it is necessary to understand the type of deformation occurring in that body. This requires the description of the *concept of strain*. The components of displacement at a point within a body in the x, y, and z directions are described by u, v, and w, respectively. With the exception of Chapter 10, only *small displacements* are considered in this text. The *strains* resulting from small deformations are *small compared with unity*, and their products, *higher-order terms*, are *neglected*.

The preceding assumption leads to one of the fundamentals of solid mechanics called the *principle of superposition*. This rule is valid whenever the quantity (deformation or stress) to be obtained is directly proportional to the applied loads. In such situations, the total quantity due to all loads acting simultaneously on a member may be found by *separately* determining the quantity due to each load and then *combining* the results obtained. The principle of superposition will be repeatedly used in this book. It allows a complex loading to be replaced by two or more simpler loads and therefore presents a problem more amenable to solution.

The concept of normal strain may be illustrated by considering the deformation of the prismatic bar shown in Figure 1.12a. A prismatic bar is a straight member having constant cross-sectional areas throughout its length. The initial length of the member is L. Following the application of the load, total deformation is δ. The normal strain is defined as the unit change in length:

$$\varepsilon = \frac{\delta}{L} \tag{1.17}$$

A *positive* sign means *elongation*; a *negative* sign, *contraction*. The preceding state of strain is termed *uniaxial strain*.

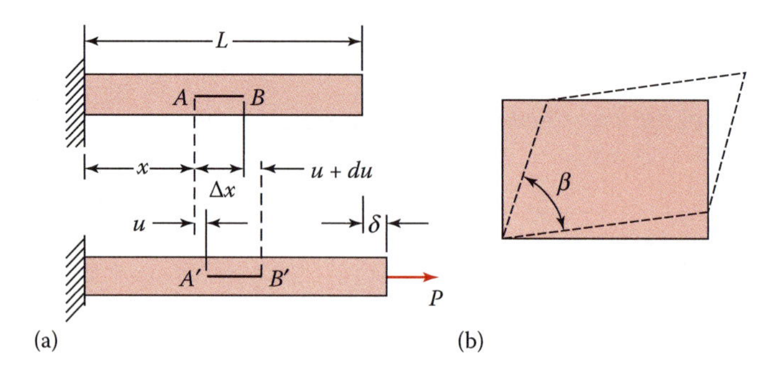

FIGURE 1.12
(a) Deformation of a prismatic bar; (b) distortion of a rectangular plate.

When an unconstrained body is subjected to a temperature change ΔT, a normal strain also develops. The *thermal strain* for a homogeneous and isotropic material is given by

$$\varepsilon_t = \alpha(\Delta T) \tag{1.18}$$

The coefficient of expansion α is approximately constant over a moderate temperature change. It represents a quantity per degree Celsius (1/°C) when ΔT is measured in °C.

Shear strain is the tangent of the total change in angle taking place between two perpendicular lines in a body during deformation. For small displacements, we can set the tangent of the angle of distortion equal to the angle. Therefore, for a rectangular plate (Figure 1.12b), the *shear strain* measured in radians is defined as

$$\gamma = \frac{\pi}{2} - \beta \tag{1.19}$$

where β represents the angle between the two rotated edges. The shear strain is positive if the right angle between the reference lines decreases, as shown in the figure; otherwise, the shear strain is negative. We shall indicate both normal and shear strains as dimensionless quantities. They seldom exceed values of 0.002, or 2000 μm, in the elastic range of most engineering materials. We read this as "2000 micros."

1.10 Components of Strain

When uniform deformation does not occur, the strains vary from point to point in a body. Then the expression for strain must relate to a line AB of length Δx that stretches by an amount u under the axial load, as shown in Figure 1.12a. On this basis, the normal strain is defined by

$$\varepsilon_x = \lim_{\Delta x \to 0} \frac{\Delta u}{\Delta x} = \frac{du}{dx} \tag{1.20}$$

The foregoing represents the strain at a point to which Δx shrinks.

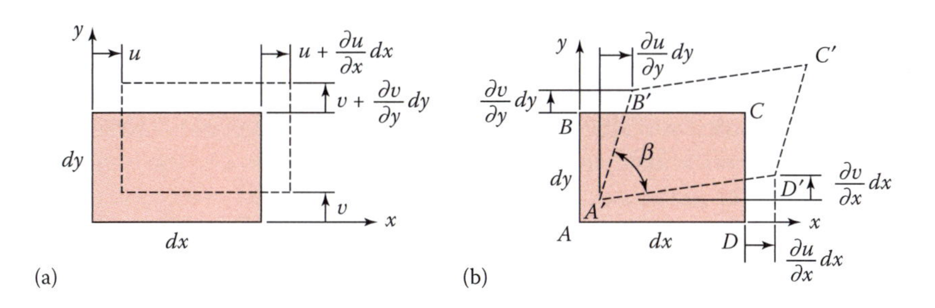

FIGURE 1.13
Deformations of 2D element: (a) linear strain; (b) shear strain.

When 2D or *plane strain* occurs, all points in the body, before and after application of load, remain in the same plane. Thus, the deformation of an element of dimensions dx, dy, and of unit thickness can contain linear strains (Figure 1.13a) and a shear strain (Figure 1.13b). Here the partial derivative notation is used since the displacements u and v are functions of x and y. Referring to Figure 1.13, Equations 1.19 and 1.20, we write

$$\varepsilon_x = \frac{\partial u}{\partial x} \qquad \varepsilon_y = \frac{\partial v}{\partial y} \qquad \gamma_{xy} = \frac{\partial u}{\partial y} + \frac{\partial v}{\partial x} \tag{1.21a}$$

Clearly, $\gamma_{xy} = \pi/2 - \beta$ represents the shear strain between the x and y axes (or y and x axes), and hence $\gamma_{xy} = \gamma_{yx}$.

In the case of a thin rectangular element of sides Δx and Δy, an approximate version of Equations 1.21a may be expressed in the form

$$\varepsilon_x = \frac{\Delta u}{\Delta x} \qquad \varepsilon_y = \frac{\Delta v}{\Delta y} \qquad \gamma_{xy} = \frac{\Delta u}{\Delta y} + \frac{\Delta v}{\Delta x} \tag{1.22}$$

These represent Equations 1.17 and 1.19. Equations 1.22 may be applied to obtain the plane strain components readily from the given displacements.

In an analogous manner, strains at a point in a rectangular prismatic element of sides dx, dy, and dz are obtained in terms of the displacements u, v, and w. The *3D strain* components are ε_x, ε_y, γ_{xy}, and

$$\varepsilon_z = \frac{\partial w}{\partial z} \qquad \gamma_{yz} = \frac{\partial v}{\partial z} + \frac{\partial w}{\partial y} \qquad \gamma_{xz} = \frac{\partial u}{\partial z} + \frac{\partial w}{\partial x} \tag{1.21b}$$

where $\gamma_{yz} = \gamma_{zy}$ and $\gamma_{xz} = \gamma_{zx}$. Equations 1.21 represent the components of a *strain tensor*, which is analogous to the stress tensor already discussed.

1.10.1 Conditions of Compatibility

The preceding definitions show that the six strain components depend linearly on the derivatives of the three displacement components. Thus, the strains cannot be independent of one another. *Six* equations, referred to as the *conditions of compatibility*, can be developed, showing the interrelationships between ε_x, ε_y, ε_z, γ_{xy}, γ_{yz}, and γ_{xz} [2]. The number of such equations becomes *one* for a 2D problem. The conditions of compatibility predicate that the displacements are continuous. Physically, this means that the *body must be pieced together.*

It is to be noted that the method of the theory of elasticity is based on the requirements strain compatibility and stress equilibrium (Equations 1.9) generalized Hooke's law (Equations 1.34), and boundary conditions for a given problem. On the other hand, in the method of mechanics of materials approach, basic assumptions are made concerning the distribution of strains in the body as a whole or in the finite portion of the body. Hence, solutions of Equations 1.21, the conditions of compatibility, and Equations 1.9 are not required.

1.10.2 Large Strains

As pointed out in Section 1.9, the small displacements are considered in most applications of this text. This is consistent with the magnitude of deformation commonly found in engineering structures. The following more general *large strain–displacement relationships* are introduced here so that the reader may better understand the approximations leading to the relations of small-deformation theory.

When displacements are relatively large, the strain components are prescribed in terms of the square of the element length instead of the length itself. Thus, with reference to Figure 1.13b,

$$\varepsilon_x = \frac{(A'D')^2 - (AD)^2}{2(AD)^2} \tag{1.23}$$

In the foregoing, we have

$$\begin{aligned}(A'D')^2 &= \left(dx + \frac{\partial u}{\partial x}dx\right)^2 + \left(\frac{\partial v}{\partial x}dx\right)^2 \\ &= \left[1 + 2\frac{\partial u}{\partial x} + \left(\frac{\partial u}{\partial x}\right)^2 + \left(\frac{\partial v}{\partial x}\right)^2\right](dx)^2\end{aligned}$$

and $AD = dx$.

Substituting the preceding terms in Equation 1.23 results in 2D *finite* or *large normal strain*:

$$\varepsilon_x = \frac{\partial u}{\partial x} + \frac{1}{2}\left[\left(\frac{\partial u}{\partial x}\right)^2 + \left(\frac{\partial v}{\partial x}\right)^2\right] \tag{1.24a}$$

Similarly,

$$\varepsilon_y = \frac{\partial v}{\partial y} + \frac{1}{2}\left[\left(\frac{\partial u}{\partial y}\right)^2 + \left(\frac{\partial v}{\partial y}\right)^2\right] \tag{1.24b}$$

It can also be shown that [4]

$$\varepsilon_{xy} = \frac{1}{2}\left[\frac{\partial v}{\partial x} + \frac{\partial u}{\partial y} + \frac{\partial u}{\partial x}\frac{\partial u}{\partial y} + \frac{\partial v}{\partial x}\frac{\partial v}{\partial y}\right] \tag{1.24c}$$

where $\varepsilon_{xy} = \gamma_{xy}/2$.

Comments: For small strains, the higher-order terms are neglected. Doing so reduces Equations 1.24 to Equations 1.21a, as expected. The expressions for the 3D state of strain may readily be generalized from the foregoing equations.

1.11 Transformation of Strain

A *2D*, or *plane, strain* occurs when all points in the body, before and after application of load, remain in the same plane. Thus, in the xy plane, three strain components, ε_x, ε_y, and γ_{xy}, may exist. The strains taking place in the cross sections of a slender member subjected to lateral loading may exemplify an essentially plane-strain distribution. In this section, we shall briefly discuss transformation of strain from one set of rotated axes to another.

Mathematically, in every respect, strain transformation is identical to stress transformation. It can verified [2] that transformations of stress are converted into strain relationships by replacing

$$\sigma \text{ with } \varepsilon \quad \text{and} \quad \tau \text{ with } \gamma/2$$

These substitutions are made in all analogous relations.

Therefore, *equations for transformation of strain* proceed from Equations 1.11:

$$\varepsilon_{x'} = \frac{1}{2}(\varepsilon_x + \varepsilon_y) + \frac{1}{2}(\varepsilon_x - \varepsilon_y)\cos 2\theta + \frac{\gamma_{xy}}{2}\sin 2\theta \tag{1.25a}$$

$$\gamma_{x'y'} = -(\varepsilon_x - \varepsilon_y)\sin 2\theta + \gamma_{xy}\cos 2\theta \tag{1.25b}$$

$$\varepsilon_{y'} = \frac{1}{2}(\varepsilon_x + \varepsilon_y) - \frac{1}{2}(\varepsilon_x - \varepsilon_y)\cos 2\theta - \frac{\gamma_{xy}}{2}\sin 2\theta \tag{1.25c}$$

Similarly, the *principal strain directions* are found from Equation 1.12:

$$\tan 2\theta_p = \frac{\gamma_{xy}}{\varepsilon_x - \varepsilon_y} \tag{1.26}$$

From Equation 1.13, the magnitudes of the *principal strains* are

$$\varepsilon_{1,2} = \frac{\varepsilon_x + \varepsilon_y}{2} \pm \sqrt{\left(\frac{\varepsilon_x - \varepsilon_y}{2}\right)^2 + \left(\frac{\gamma_{xy}}{2}\right)^2} \tag{1.27}$$

Likewise, the magnitude of the *maximum shear strain*, by Equation 1.14, is

$$\gamma_{\max} = \pm 2\sqrt{\left(\frac{\varepsilon_x - \varepsilon_y}{2}\right)^2 + \left(\frac{\gamma_{xy}}{2}\right)^2} = \pm(\varepsilon_1 - \varepsilon_2) \tag{1.28}$$

which occurs on planes 45° relative to the principal planes. The transformation of 3D strain also proceeds in an analogous manner.

In *Mohr's circle for strain*, the normal strains are measured on the horizontal axis, positive to the right. The center of the circle is at $(\varepsilon_x + \varepsilon_y)/2$. The vertical axis is measured in terms of $\gamma/2$. When the shear strain is *positive*, the point representing the x axis strain is plotted a distance $\gamma/2$ *below* the ε axis and vice versa when shear strain is negative. Obviously, this convention for shearing strain, used only in constructing and reading values from Mohr's circle, concurs with the convention used for stress in Section 1.8.1.

1.12 Engineering Materials

In the case of the one-dimensional problem of an axially loaded member, stress–load and strain–displacement relations represent *two* equations involving *three* unknowns—stress σ, strain ε, and displacement u or v. These relationships are obtained by satisfying the first and third principles of analysis: equilibrium conditions and geometric deformation (see Section 1.2). The insufficient number of available equations is compensated for by a material-dependent relationship connecting stress and strain: the third principle of analysis. Accordingly, the loads acting on a member, the resulting displacements, and the *material properties* can be associated.

Ordinarily used *engineering materials* include various metals, plastics, wood, ceramics, glass, and concrete, as shown in Table 1.2. Average properties of some materials are listed in Table B.3 [5,6]. In this book, we shall refer to these ordinary characteristics of materials. It is to be noted that the analysis and design of plate and shell-like elements, for example, as components of a missile or space vehicle, embody an unusual integration of materials, having properties dependent on environmental conditions [7].

An *elastic material* is one that returns to its original dimensions on removal of applied loads. Generally, the elastic range includes a region throughout which stress and strain have

TABLE 1.2
Typical Engineering Materials

Ferrous Metals	Nonferrous Metals
Metallic Materials	
Cast iron	Aluminum
Cast steel	Copper
Plain carbon steel	Lead
Steel alloys	Magnesium
Stainless steel	Nickel
Special steels	Platinum
Structural steel	Silver
Nonmetallic Materials	
Graphite	Plastics
Ceramics	Brick
Glass	Stone
Concrete	Wood

a linear relationship. This part ends at a point called the *proportional limit*. Our considerations will be limited to such *linearly elastic materials*. A *plastically* deformed member does not return to its initial size and shape when the load is removed. In a *viscoelastic solid*, the state of stress is a function not only of the strain but also of the time rates of change of stress and strain.

A *ductile material* (e.g., many alloys, nylon) is capable of substantial elongation prior to failure. The converse applies to a *brittle material* (e.g., cast iron, concrete); a brittle material experiences little deformation before rupture. A member that ruptures is said to be *fractured*. Ductile materials usually fail in shear, while brittle materials fail in *tension*. The distinction between brittle and ductile behavior is taken arbitrarily at 5% total elongation in the tensile test.

A *homogeneous* solid displays identical properties throughout. If the properties are the same in all directions, the material is *isotropic*. Nonisotropic or *anisotropic* solids display direction-dependent properties. The properties of an *orthotropic* material differ in two mutually perpendicular directions. A *composite* material is made of two or more distinct constituents. With the exception of Chapter 8 and Sections 15.11 and 15.12, we assume in this text that the material is isotropic.

1.12.1 Stress–Strain Diagrams

The stress–strain diagram is used to explain a number of material properties useful in the study of mechanics of materials and the theory of elasticity. Data for this diagram is usually obtained from a *tension test*. In such a test, a specimen of the material, usually in the form of a round bar, is mounted in the grips of a testing machine and subjected to tensile loading, applied slowly and steadily, or *statically*.

The tension-test procedure consists of applying successive increments of load while taking corresponding extensometer readings of the elongation between the two gage marks—gage length—on the bar. The stress in the bar is found by dividing the force by the cross-sectional area, and the strain is found by dividing the elongation by the gage length. In this way, a *complete stress–strain diagram, a plot* of strain as abscissa and stress as the ordinate, can be obtained for the material.

A typical shape of a stress–strain plot for a *ductile material* such as structural steel is shown in Figure 1.14a. Curve *OABCDE* represents an engineering of a *conventional stress–strain diagram*. The other curve, *OABCE′*, is the true stress–strain diagram. The *true stress* equals the load divided by the actual instantaneous cross-sectional area of the bar; the *true strain* is the sum of the elongation increments divided by the corresponding momentary length. Engineering stress equals the load divided by the initial cross-sectional area; the engineering strain is defined by Equation 1.17. In practice, the conventional stress–strain diagram provides satisfactory information for use in design.

The *elastic range* is represented by the part *OA* of the diagram. The linear variation of stress–strain ends at the *proportional limit* (point *A*). For most cases, this point and the *yield point* (point *B*) are assumed as one: $\sigma_{pl} \approx \sigma_{yp}$. The portion of the stress–strain curve extending from point *A* to the point of *fracture* (*E*) is the *plastic range*. In the range *CD*, an increase in stress is required for a continued increase in strain; this effect is called *strain hardening*.

The engineering stress diagram for the material when strained beyond *C* displays a typical maximum referred to as the *ultimate stress* (*D*), σ_u, and a lower value, the *fracture stress* (*E*), σ_f. Failure at *E* occurs by separation of the bar into two parts, along a cone-shaped surface forming an angle 45° with its axis. In the vicinity of the ultimate stress, the reduction of the cross-sectional area or the lateral contraction becomes clearly visible, and a pronounced *necking* of the bar occurs in the range *DE*.

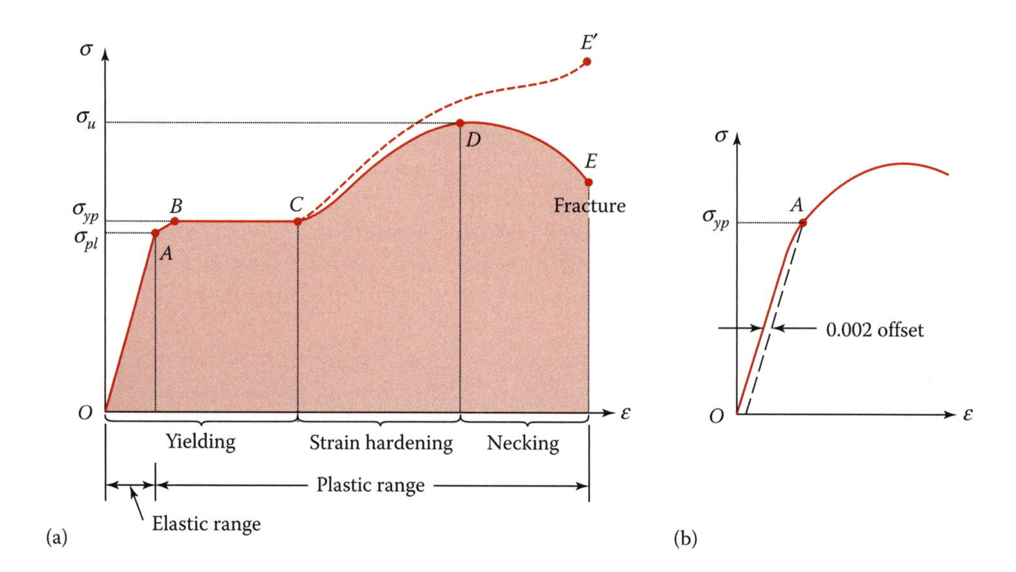

FIGURE 1.14
(a) Typical stress–strain diagram for structural steel; (b) determination of arbitrary yield strength by the offset method.

Certain materials, such as magnesium, aluminum, and copper, do not show a distinctive yield point, and it is usual to use an *arbitrary* yield point. According to the so-called *0.2% offset method*, a line is drawn through a strain of 0.002, parallel to the initial slope at point *O* of the curve, as shown in Figure 1.14b. The intersection of this line with the stress–strain curve defines the arbitrary *yield stress* or *yield strength* (point *A*).

Diagrams analogous to those in tension may also be obtained for a variety of materials in *compression* as well as in *torsion*. For some ductile materials, it is found that the yield point stress is about the same in tension and in compression. However, many materials, such as cast iron and concrete, have characteristic stresses in compression that are much greater than in tension.

1.13 Hooke's Law, Poisson's Ratio

For the straight-line portion of the diagram of Figure 1.14a, the stress is directly proportional to the strain. Therefore, we have

$$\sigma = E\varepsilon \tag{1.29}$$

This relationship is known as *Hooke's law*. The constant E is called the *modulus of elasticity* or Young's modulus. Since ε is a dimensionless quantity, E has units of σ, that is, newtons per square meter, or pascals. It is to be emphasized that Hooke's law is valid only up to the proportional limit of the material. The modulus of elasticity represents the *slope* of the stress–strain curve in the linearly elastic range and is different for various materials.

Similarly, linear elasticity can be measured in a member subjected to shear loading. Referring to Equation 1.29, we now write

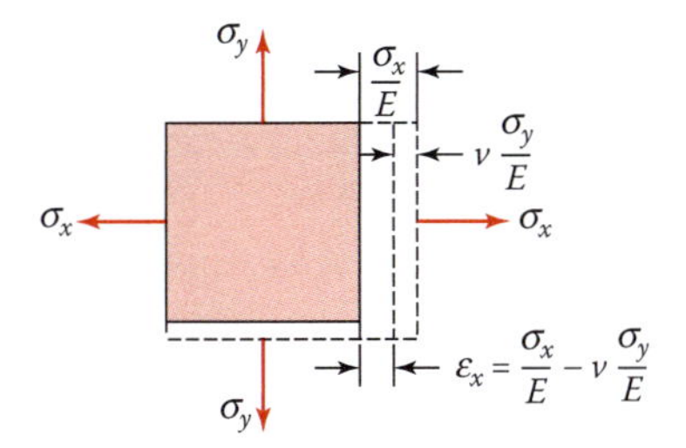

FIGURE 1.15
Element deformations caused by biaxial stresses.

$$\tau = G\gamma \tag{1.30}$$

This relation is Hooke's law for shear stress τ and shear strain γ. The constant G is called the *shearing modulus of elasticity* or *modulus of rigidity* of the material, expressed in the same units as E: pascals. Values of E and G for common materials are listed in Table B.3.

In the elastic range, the ratio of the lateral strain to the axial strain is constant and is known as *Poisson's ratio*:

$$\nu = \left|\frac{\textit{lateral strain}}{\textit{axial strain}}\right| = -\frac{\textit{lateral strain}}{\textit{axial strain}} \tag{1.31}$$

Here the minus sign means that the lateral strain is of a sense opposite to that of the axial strain. Experiments show that in most common materials, the values of ν are in the range 0.25–0.35. For steels, Poisson's ratio is often assumed as 0.3. Extreme cases include $\nu = 0.1$ for some concretes and $\nu = 0.5$ for rubber. Note, however, that there are some solids with negative Poisson's ratio. Such materials become fatter in the cross section when stretched [8].

Consider now an element of unit thickness subjected to a biaxial state of stress (Figure 1.15). Because of the stress σ_x, not only the direct strain σ_x/E occurs but a y contraction occurs as well, $-\nu\sigma_x/E$, shown by the dashed lines in Figure 1.15. Likewise, if σ_y only acts, an x contraction $-\nu\sigma_y/E$ and a y strain σ_y/E result. Hence, simultaneous action of both stresses σ_x and σ_y gives their following x- and y-directed strains:

$$\varepsilon_x = \frac{\sigma_x}{E} - \nu\frac{\sigma_y}{E} \qquad \varepsilon_y = \frac{\sigma_y}{E} - \nu\frac{\sigma_x}{E} \tag{1.32a, b}$$

The elastic stress–strain relation for the state of pure shear, by Equation 1.30, is

$$\gamma_{xy} = \frac{\tau_{xy}}{G} \tag{1.32c}$$

The inverse relationships, giving the stresses in terms of the strains, are

$$\begin{aligned} \sigma_x &= \frac{E}{1-\nu^2}(\varepsilon_x + \nu\varepsilon_y) \\ \sigma_y &= \frac{E}{1-\nu^2}(\varepsilon_y + \nu\varepsilon_x) \\ \tau_{xy} &= G\gamma_{xy} \end{aligned} \tag{1.33}$$

Equations 1.32 and 1.33 represent *Hooke's law for 2D stress*.

The foregoing procedure is readily extended to the case of 3D stress state (Figure 1.2). Strain–stress relations, often referred to as the *generalized Hooke's law*, are then

$$\varepsilon_x = \frac{1}{E}[\sigma_x - \nu(\sigma_y + \sigma_z)] \tag{1.34a}$$

$$\varepsilon_y = \frac{1}{E}[\sigma_y - \nu(\sigma_x + \sigma_z)] \tag{1.34b}$$

$$\varepsilon_z = \frac{1}{E}[\sigma_z - \nu(\sigma_x + \sigma_y)] \tag{1.34c}$$

$$\gamma_{xy} = \frac{\tau_{xy}}{G} \qquad \gamma_{yz} = \frac{\tau_{yz}}{G} \qquad \gamma_{xz} = \frac{\tau_{xy}}{G} \tag{1.34d, e, f}$$

The shearing modulus of elasticity G is related to the modulus of elasticity E and Poisson's ratio ν as

$$G = \frac{E}{2(1+\nu)} \tag{1.35}$$

Thus, for an isotropic material, there are only two independent elastic constants.

EXAMPLE 1.3: Strains and Deformations of a Plate

An aluminum alloy plate ($E = 70$ GPa, $\nu = 0.3$), of thickness $t = 8$ mm, width $b = 150$ mm, and length $a = 300$ mm, is subjected to a state of stress having $\sigma_y = 120$ MPa (Figure 1.16). Determine: (a) the value of the stress σ_x for which length a remains unchanged; (b) the final thickness t' and width b'; and (c) the normal strain for the diagonal AC.

Solution

Inasmuch as the length does not change, we have $\varepsilon_x = 0$. In addition, for plane stress $\sigma_z = 0$. Equations 1.34a–c then reduce to

$$\sigma_x = \nu\sigma_y \tag{a}$$

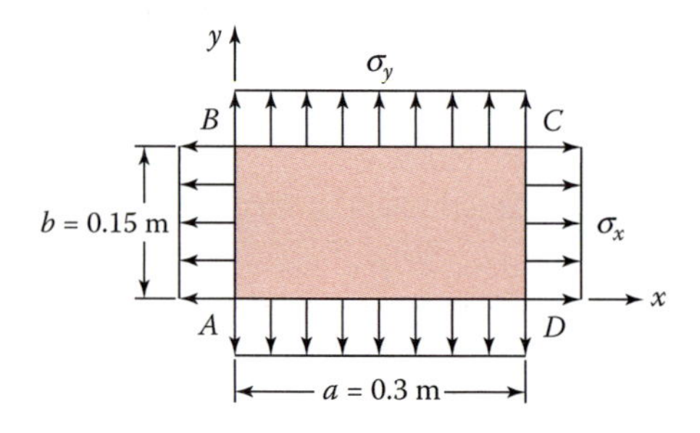

FIGURE 1.16
Plate in plane stress.

$$\varepsilon_y = \frac{1}{E}(\sigma_y - \nu\sigma_x)$$
$$\varepsilon_z = -\frac{\nu}{E}(\sigma_x + \sigma_y) \tag{b}$$

a. The given numerical values are carried into Equation a to yield

$$\sigma_x = 0.3(120 \times 10^6) = 36 \text{ MPa}$$

b. From Equations b,

$$\varepsilon_y = \frac{1}{70(10^3)}[120 - 0.3(36)] = 1560\ \mu\text{m}$$

$$\varepsilon_z = -\frac{0.3}{70(10^3)}(36 + 120) = -669\ \mu\text{m}$$

Here a minus sign indicates a decrease in the thickness. Therefore,

$$t' = t(1 + \varepsilon_z) = 8(0.99933) = 7.995 \text{ mm}$$

$$b' = b(1 + \varepsilon_y) = 150(1.00156) = 150.234 \text{ mm}$$

c. The initial and final lengths of the diagonal are, respectively,

$$AC = (150^2 + 300^2)^{1/2} = 335.4102 \text{ mm}$$
$$A'C' = (150.234^2 + 300^2)^{1/2} = 335.5149 \text{ mm}$$

The normal strain for the diagonal is thus,

$$\varepsilon_{AC} = \frac{335.5149 - 335.4102}{335.4102} = 312\ \mu\text{m}$$

Alternatively, let x' coincide with the line AC and hence $\theta = \tan^{-1}(1.5/3) = 26.56°$ between the x and x' axes. Using Equation 1.25a, we then have

$$\varepsilon_{AC} = \frac{1560}{2} - \frac{1560}{2}\cos 2(26.56°) = 312\ \mu\text{m}$$

as before.

1.14 General Properties of Materials

The brief discussion that follows attempts to provide some general information for readers to help identify the types of a few frequently used materials. The four ordinary groups of materials are of engineering interest: metals, plastics, ceramics, and composites. Each class often has similar properties such as chemical make up and atomic structure, processing routes, and applications. There are a variety of engineering materials (see Table 1.2). Proper material selection is very important in structural and machine designs. It should be noted that the common properties alone are not sufficient for selecting a material for a particular application. Rather one or several combinations of properties are needed.

1.14.1 Metals

Often, metals are preferred as structural materials. Metals can be made stronger as well as corrosion resistant by *alloying* and by various *mechanical* and heat *treatments.* Most metals are ductile and good conductors of electricity and heat. However, certain cast metals can have very low ductility. Interestingly, cast iron and steel are iron alloys containing over 2% carbon and less than 2% carbon, respectively. Cast irons comprise a whole family of materials including carbon. Gray cast iron is a widely used form of cast iron.

Steels can be classed as plain carbon steels, alloy steels, high-strength steels, cast steels, and special purpose steels. Low carbon steels or mild steels are also referred to as the *structural steels.* There are many effects of adding any alloy to a basic carbon steel. *Stainless steels* (in addition to carbon) contain at least 12% chromium as the basic alloying element. Note that, most aluminum and magnesium alloys have a high strength-to-weight ratio.

1.14.2 Plastics

Plastics are synthetic materials, also called *polymers.* They are used increasingly in structures and various different types are available. Table B.4 lists several common plastics. The mechanical characteristics of these materials vary considerably, with some plastics being brittle and others ductile. Resistance to environmental degradation, such as the photomechanical effects of sunlight, of plastics is moderate. Polymers are of two groups: *thermoplastics* and *thermosets.*

Thermoplastics include acetyl, acrylic, nylon, teflon, polypropylene, polystyrene, polyvinyl chloride (PVC), and saran. Thermosets embody epoxy, polyester, polyurethane, and bakelite. Thermoplastic materials often soften when heated and harden when cooled; they may be formed into a variety of shapes by the simple application of heat and pressure. There are also highly elastic flexible materials called thermoplastic *elastomers.* A typical elastomer is a rubber band.

Thermosets preserve structural change during processing; they can be formed only by cutting or machining. Large elastic deflections permit the design of polymer components that snap together, which makes assembly fast and cheap. Polymers are corrosion resistant, have low coefficient of friction, and demonstrate remarkable resistance to wear. Technical information related to engineering plastics may be found, for example, at www.dupont.com/enggpolymers and www.ge.com/plastics.

1.14.3 Ceramics

Ceramics are usually compounds of nonmetallic as well as metallic elements, mostly oxides, nitrides, and carbides. Frequently, silica and graphite ceramics dominate the industry. Glasses are too made up metallic and nonmetallic elements just as are ceramics. But, glasses and ceramics have different structural forms. Ceramic materials have great thermal stability and are resistant to corrosion.

Glass ceramics are in widespread usage as electrical, electronic, and laboratory ware. Ceramics have high hardness and brittleness, high compressive but low tensile strengths. High temperature and chemical resistance, high dielectric strength, and low weight characterize many of these materials. So, they are considered as an important class of engineering materials for use in machine and structural parts.

1.14.4 Composites

A *composite material* consists of two or more distinct constituents. A material of this type ordinarily exhibits a relatively large strength-to-weight ratio in comparison to a homogeneous material. Also referred to as *advanced materials*, composite materials mainly have other desirable characteristics and are widely used in a variety of structures, pressure vessels, and machine components.

A composite is designed to display a combination of the best characteristics of each component material. For instance, graphite reinforced epoxy acquires strength from graphite fibers while the epoxy protects the graphite from oxidation. In addition, epoxy helps to support shear stresses and provides toughness to the material. Our discussions will include isotropic composites, like reinforced concrete and multilayered members, sandwich plate, filament-wound anisotropic cylinders, as well as laminated shells.

It is recalled from Section 1.12 that a material whose properties depend upon the direction is called anisotropic. An important class of anisotropic materials is termed *fiber-reinforced composites*. There are many situations which require the use of such composite materials. Examples include thick-walled vessels under high pressure, marine and aircraft wind shields, portions of space vehicles, and components of many other machines and structures. Engineering publications associated with the theory and applications of composites are extensive [9].

A fiber-reinforced composite is obtained by embedding (noncontinuous or continuous) *fibers* of a strong, stiff material into a weaker reinforcing material or *matrix*. Common materials employed for fibers include carbon, glass, polymers, graphite, and some metals, while a variety of resins are used as a matrix (such as in glass filament/epoxy rocket motor cases). Figure 1.17a depicts the cross section of a fiber-reinforced composite material. Fiber length is an important parameter for strengthening and stiffening of fiber-reinforced composites. Note that, for a number of glass-and-carbon-fiber-reinforced composites the fiber length is about 1 mm, or 20–150 times its diameter.

A layer or *lamina* of a composite material consists of a large number of parallel fibers embedded in a matrix and a *laminate* consist of arbitrarily oriented various bonded layers

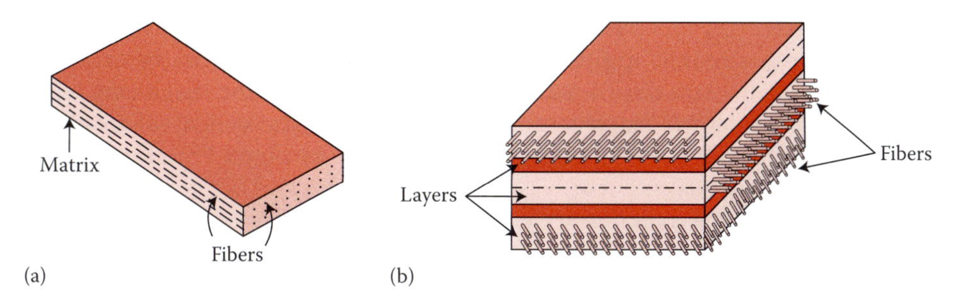

FIGURE 1.17
Fiber-reinforced materials: (a) Single-layer materials; (b) three-layer materials made up with different fiber orientations.

or laminas. Each layer may possess a different thickness, orientation of fiber directions, and anisotropic properties. Positioning some layers so that their fibers are oriented usually at 30°, 45°, or 60° with respect to one another increases the resistance of the laminate to the loading. When the fibers in all layers are given the same orientation the laminate is *orthotropic.* A commonly used composite often consists of bonded three-layer orthotropic materials. Figure 1.17b illustrates a three-ply laminate where the fibers of the midlayer and top layer are arbitrarily oriented along the axial and lateral directions, while the midlayer is arbitrarily oriented, respectively.

1.15 Engineering Design

In engineering, the word *design* conveys the meaning of planning and drafting the details of a *component*, or the creation of a *system*. The latter consists of a combination of several different elements arranged to work together as a whole. *Design function* is the process in which mathematics, computers, and graphics are used to produce a plan. *Mechanical design* means the design of components and systems of a mechanical nature—machines, structures, devices, and instruments. For the most part, it utilizes stress analysis methods and materials engineering.

Generally, *structural design* interacts with many engineering disciplines that require a structural member or system. Design is the essence, art, and intent of engineering. A good design meets performance, cost, and safety requirements. An *optimum design* is the best solution to a design problem within given constraint(s). Design analysis has its objective satisfactory performance as well as durability with minimum weight and competitive cost.

The complete *design process* may be outlined by *design flow diagrams* with feedback [10]. The process begins with a *recognition of a need* and a decision to do something about it. Definition of the problem and all *specifications* must be carefully spelled out. The next phase of the design flow is to make a *feasibility study*: to verify the possible success or failure of a proposal. The *synthesis* of the solution represents perhaps the most challenging and interesting part of design. Here, the designer combines the separate parts of elements to form a complex whole of various ideas and concepts.

The synthesis cannot occur without both *analysis and optimization.* If the design fails, the synthesis procedure must begin again. *Evaluation* of the product is the final proof of a successful design and usually involves the testing of a prototype in the laboratory that provides the analysis database. The design process ends by the *presentation* of the plans for satisfying the need.

The *design analysis* attempts to predict the stress (or deformation) in the component in order that it may safely carry the loads that will be imposed on it and that it may last for the expected life of the system. If the structure is simple enough, theoretical solutions for basic configurations may be adequate for obtaining the stresses involved. For more complicated structures, *finite-element models* not only can estimate the stresses but also can utilize them to evaluate the failure criteria for each element in a member.

1.15.1 Design Procedure

A rational method of design attempts to take the results of fundamental tests such as tension, compression, and fatigue and apply them to the situations encountered in structures and machines. The *rational procedure in design* of a load-carrying member may be outlined as follows:

1. Evaluate the mode of possible failure of the member.
2. Obtain a relationship between the applied load and stress or deformation.
3. Obtain the maximum usable value of stress or deformation that could cause failure. Use this value in connection with the formula found in step 2, if required, in any equation of failure criteria.
4. Select a factor of safety.

The foregoing process provides an elementary treatment of the concept of *design to meet strength requirements* of structural members. That is, the shape and material of a member are preselected and the applied loads or deformations specified. Then the basic formulas are applied to determine adequate size in each case.

Suffice it to say that design solutions are *not* unique, involve a consideration of many factors, and often require a *trial-and-error process*. Structural analysis and structural design are interlocked subjects, as will be observed from the examples and problems that appear in the following chapters. To conclude, the rational design of beams, plates, and shells relies greatly on their stress and deformation analysis, to which this text is directed.

Comments: It should be pointed out that, the design of numerous structures such as pressure vessels, space missiles, aircrafts, dome roofs, ships, and bridge decks is based upon the theories of plates and shells. Often, simplifying realistic assumptions are made to adopt the theories to the design of a variety of structures. For instance, a water storage tank can be satisfactorily designed using the shell-membrane criterion. However, the design of a missile casing demands a more precise bending theory in order to minimize weight and materials. Likewise, the design of a nozzle-to-cylinder junction in a nuclear reactor may necessitate an elaborate finite element analysis.

1.16 Factor of Safety

Occasionally, it is difficult to determine accurately the various factors that are involved in the phases of design structures. A significant area of *uncertainty* is related to the assumptions made in stress and deformation analysis. An equally important item is the *nature of failure*. If failure is caused by ductile *yielding*, the consequences are likely to be less than if caused by brittle *fracture*. A design must also take into account such matters as the following: types of service loads, variations in the properties of the material, effects of time and size on the material strength, consequences of failure, human safety, and economics.

Engineers employ a so-called *design factor of safety* or simply *factor of safety* to ensure against the foregoing uncertainties involving strength and loading. The factor of safety, n, is defined as the ratio of the maximum load that produces failure of the member to the load allowed under service conditions:

$$n = \frac{\textit{failure load}}{\textit{allowable load}} \tag{1.36}$$

This ratio must always be greater than unity. Inasmuch as the allowable service load is a known quantity, the usual design procedure is to multiply this by the safety factor to obtain the failure load.

A *common method of design* is to use a safety factor with respect to strength of the member. In most problems, a linear relationship exists between the load and the stress produced by the load. Then the factor of safety may also be defined as

$$n = \frac{\textit{material strength}}{\textit{allowable stress}} \tag{1.37}$$

In this equation, material strength represents either *yield stress* or the *ultimate stress*. The allowable stress is the *working* or *design stress*. When the factor of safety is too low and the allowable stress is too high, the structure may prove weak in service. The preceding definitions of factor of safety are used for any type of member and loading condition (e.g., axial, bending, shear).

1.16.1 Selection of a Factor of Safety

Modern engineering design provides a rational accounting for all factors possible, leaving relatively few items of uncertainty to be covered by a factor of safety. Usually, the safety factors are based on the yield strength σ_y of a *ductile material*. When they are used with a *brittle material* and the ultimate strength σ_u, the factors must be approximately doubled. The following few numerical values of the factor of safety are introduced as a guide for solving problems in this text [10]:

1. $n = 1.25$–1.5 is for exceptionally reliable materials used under controllable conditions and subjected to loads and stresses that can be found with certainty. It is used almost invariably where low weight is an important consideration.
2. $n = 1.5$–2 is for well-known materials under moderately constant environmental conditions, subjected to loads and stresses that can be found readily.
3. $n = 2$–2.5 is for average materials operated in ordinary environments and subjected to loads and stresses that can be obtained.
4. $n = 2.5$–4 is for less-tried (or 3–4 for untried) materials subjected to average conditions of environment, load, and stress.
5. $n = 3$–4 is also for better-known materials used in uncertain environments or under uncertain stresses.

Usually, where factors of safety higher than 4 might appear to be desirable, a more thorough analysis of the problem should be undertaken before deciding on their use.

In the *aircraft industry*, where it is necessary to reduce the weight of the structures as much as possible, the margin of safety is commonly used rather than the factor of safety. The *margin of safety* is defined as the factor of safety minus one: $n - 1$. In the *nuclear reactor industries*, the safety factor is of prime importance in the face of many unknown effects, and hence the factor of safety may be as high as 5.

For most applications, relevant factors of safety are found in various construction and manufacturing codes. The use of factor of safety in design is a reliable, time-proven approach. When properly employed, sound and safe designs are obtained by using it.

A concept closely related to safety factor is called *reliability*: the statistical measure of the probability that a member will not fail in service.

1.17 Problem Formulation and Solutions

A basic method of attack for analysis of beam, plate, and shell problems is to define the problem. Formulation of the problem requires consideration of the physical situations and an idealized description by the FBDs of the actual member involved. The following outline may be helpful in *formulation and solution* of a problem:

1. Define the problem and state briefly what is known.
2. State what is to be determined; if appropriate.
3. List simplifying idealizations to be made.
4. Apply the relevant equations to determine the unknowns; and, when appropriate, discuss the results briefly.

Clearly, *assumptions* expand on the given information to further constrain the problem. For example, we might take the effects of friction to be negligible, or the weight of the member can be ignored in a particular case. The reader needs to understand what idealizations are made in solving a problem. Solutions must be based on the basic concepts, formulas, charts, and diagrams. Problem statements should indicate precisely what information is required. *FBDs* must be complete, showing all essential quantities involved.

1.17.1 Numerical Accuracy and Significant Digits

In practical engineering problems, the data are seldom known with an *accuracy* greater than 0.2%. Thus, it is seldom justified to write the answers to such problems with an accuracy greater than 0.2%. Since calculations are often performed by electronic calculators and computers (usually carrying eight or nine digits), the possibility exists that numerical results will be reported to an accuracy that has no physical meaning. For consistency throughout this text, we generally follow a common *engineering rule* to report the *final results* of calculations:

Numbers beginning with "1" are recorded to *four* significant digits.

All other numbers (that begin with "2" through "9") are recorded to *three* significant digits.

Therefore, a force of 15 N, for example, should read 15.00 N, and a force of 65 N should read 65.0 N. *Intermediate results*, if kept for further calculations, are recorded to several extra digits to save the numerical accuracy. We note that the value of π and *trigonometric functions* are calculated to many significant digits (10 or more) within the calculator or computer.

1.17.2 Computational Tools

Various computational tools can be used to perform analysis calculations with success. A *scientific calculator* may be the best tool for solving most of the problems in this text. General purpose analysis tools such as spreadsheets and equation solvers have particular merit for certain computational tasks. Some mathematical software packages of these types are MATLAB (see: Appendix D), TK Solver, and MathCAD. These allow the user to document and save completed work.

Computer-aided drafting or design (CAD) software packages can produce realistic 3D representations of a member. Most CAD packages provide an interface to one or more finite element analysis, or FEA, programs (Chapter 7). They permit direct transfer of the member's geometry to an FEA package for static, dynamic, or thermal analysis of stress, as well as fluid analysis.

The foregoing computer-based software may be used as tools to assist readers with lengthy problems. But computer output providing analysis results must not be accepted on faith alone; the analyst must always make checks on computer solutions. It is important that fundamentals of analysis of structural members be thoroughly understood.

Problems

Sections 1.1 through 1.8

1.1 The frame of Figure P1.1 consists of three pin-connected bars. Determine: (a) the support reactions; (b) the forces acting on the horizontal member at points A and C; (c) the axial force, shear force, and moment acting on the cross section at point E.

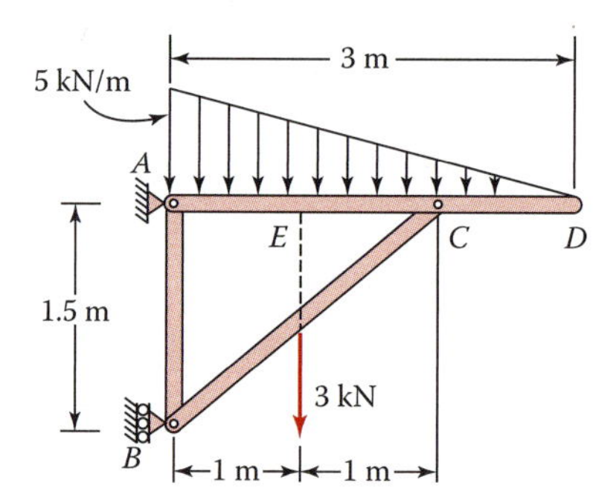

FIGURE P1.1

1.2 A bent beam ABC is supported and loaded as shown in Figure P1.2. Determine: (a) an expression for the bending moment $M(x)$ for part BC and its maximum value; (b) the shear force and bending moment acting on the cross section at point D.

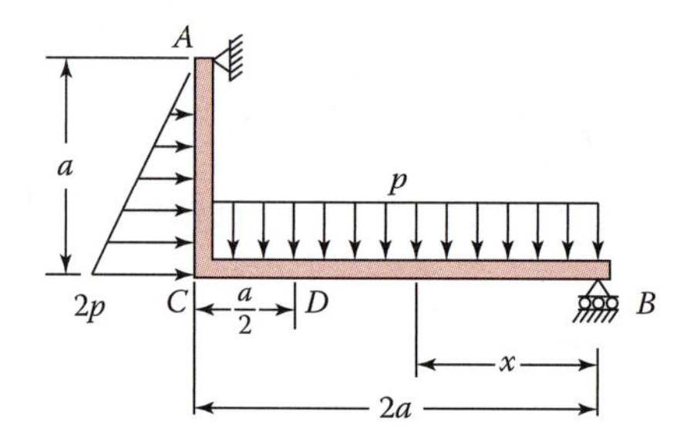

FIGURE P1.2

1.3 Figure P1.3 depicts an engine system consisting of a piston attached to a connecting rod *AB*, which in turn connected to a crank arm *BC*. The piston slides without friction in a cylinder and is under a force *P*. *Given*: The crank arm is exerting a torque $T = 3\ \text{kN}\cdot\text{m}$. *Find*, for the position shown: (a) the force *P* required to hold the system in equilibrium; (b) the axial force in the rod *AB*.

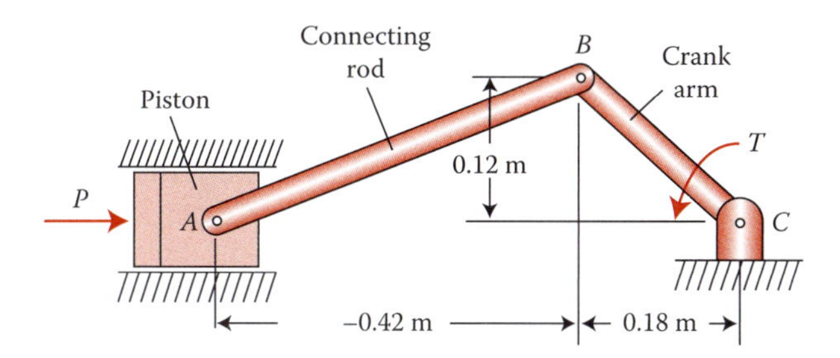

FIGURE P1.3

1.4 For the structure shown in Figure P1.4, calculate: (a) the reactions at *A*, *B*, and *C*; (b) the axial force, shear force, and moment acting on the cross section at point *D*.

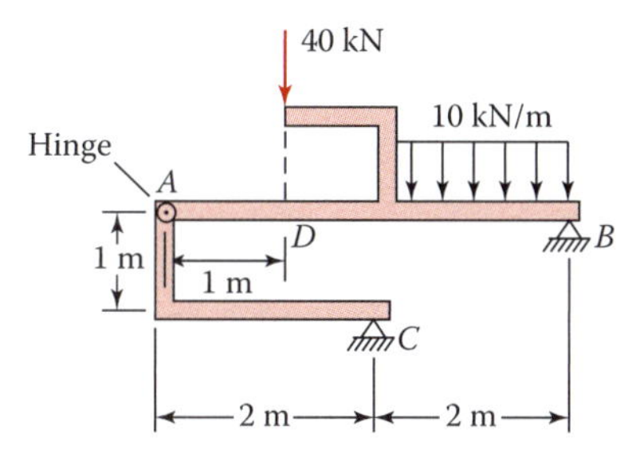

FIGURE P1.4

1.5 A structure constructed by joining beams *AE* and *AC* with bar *BD* by hinges carries a vertical load *P* at point *E* as shown in Figure P1.5. Determine: (a) the reactions at supports *A* and *C*; (b) the axial force, shear force, and moment acting on the cross section *O*.

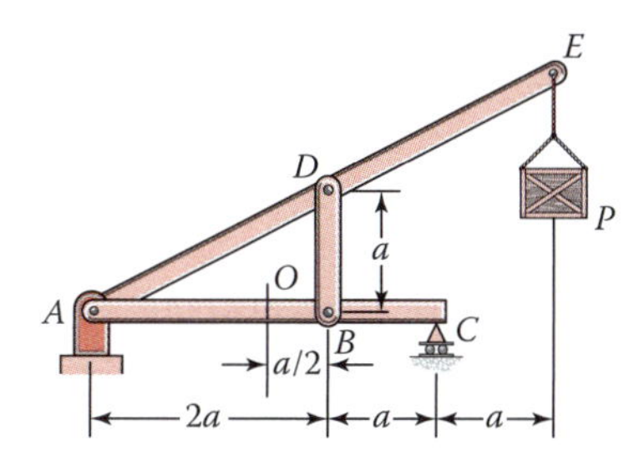

FIGURE P1.5

1.6 The state of stress at a point in a loaded structural component is represented in Figure P1.6. Calculate the normal and shear stresses acting on the indicated inclined plane.

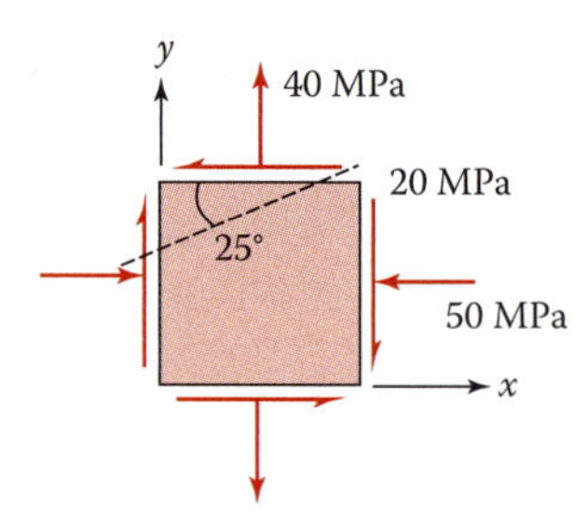

FIGURE P1.6

1.7 Stresses acting uniformly at a triangular plate is illustrated in Figure P1.7. Calculate $\sigma_{x'}$, $\sigma_{y'}$, and $\tau_{xy'}$. *Sketch* the results on a properly oriented element.

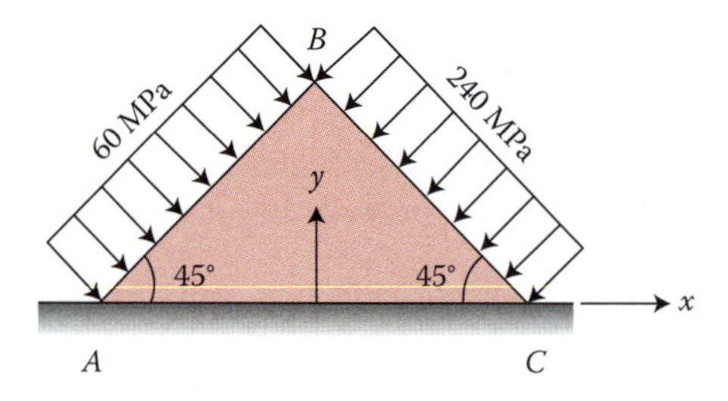

FIGURE P1.7

1.8 and 1.9 The stresses at two points in a loaded beam are represented in Figures P1.8 and P1.9. For each point, using Mohr's circle, determine and sketch the results on a properly oriented element for: (a) the principal stresses; (b) the maximum shear stress with the associated normal stresses.

FIGURE P1.8

FIGURE P1.9

1.10 At a particular point in a structure state of stress is given in Figure P1.10. Find, for that point: (a) the magnitude of the shear stress τ if the maximum principal stress is not to exceed 150 MPa; (b) the corresponding maximum shearing stresses and the planes at which they act.

1.11 Given the stresses acting uniformly over sides of a skewed plate as depicted in Figure P1.11. Find: (a) the stress components on a plane parallel to *a–a*; (b) the magnitude and orientation of maximum shear stress and associated normal stresses. Sketch the results on properly oriented elements.

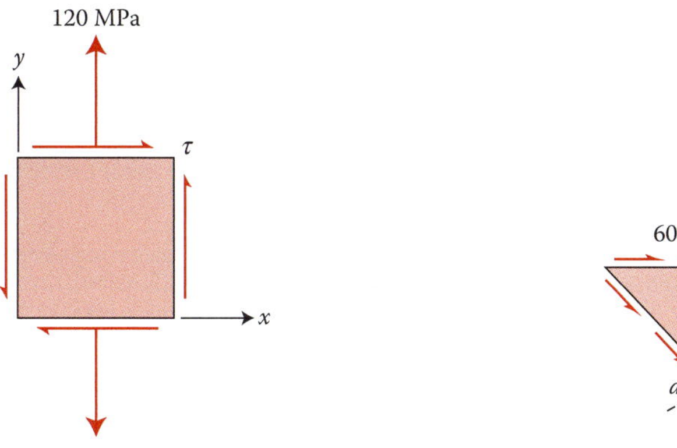

FIGURE P1.10

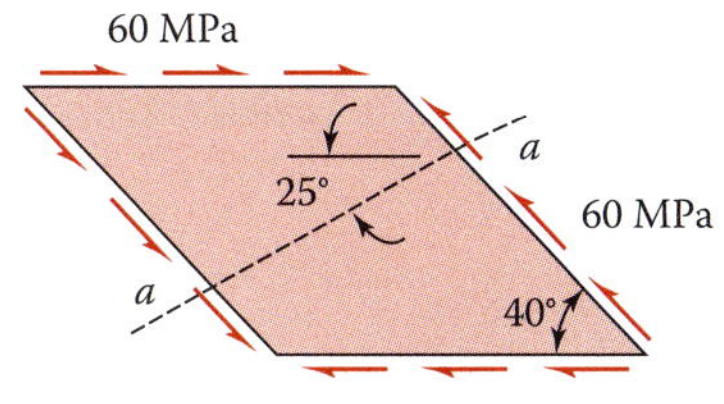

FIGURE P1.11

1.12 The stresses acting uniformly at the edges of a wall panel of a flight structure are shown in Figure P1.12. Calculate the stress components on planes parallel and perpendicular to *a*–*a*. Sketch the results on a properly oriented element.

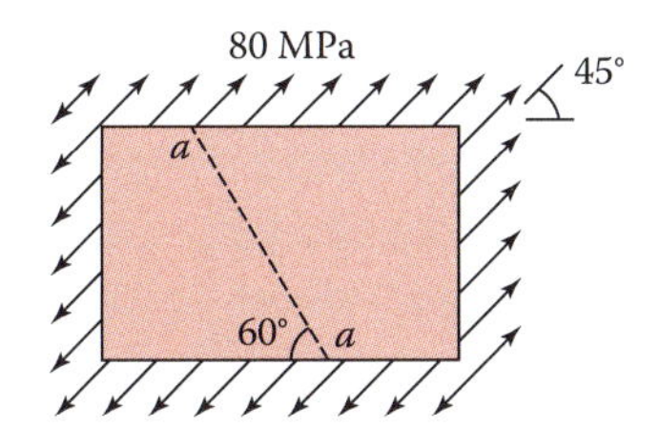

FIGURE P1.12

1.13 A rectangular plate is subjected to uniformly distributed stresses acting along its edges (Figure P1.13). Determine the normal and shear stresses on planes parallel and perpendicular to *a*–*a*. Sketch the results on a properly oriented element.

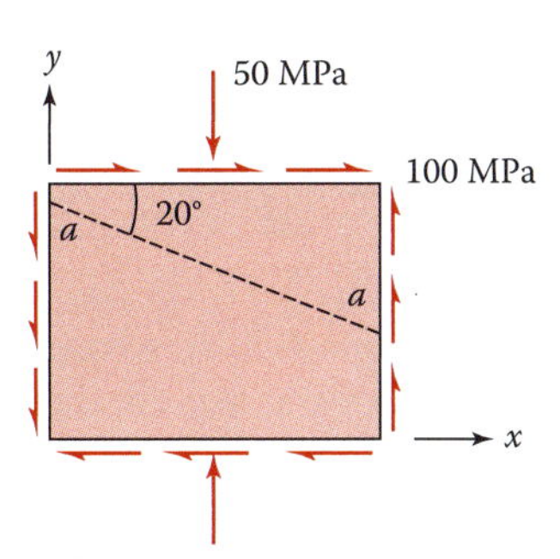

FIGURE P1.13

1.14 Stresses (in MPa) acting uniformly over sides of a skewed plate are shown in Figure P1.14. Determine: (a) the stress components on a vertical plane; (b) the magnitude and orientations of the principal stresses. Sketch the results on properly oriented elements.

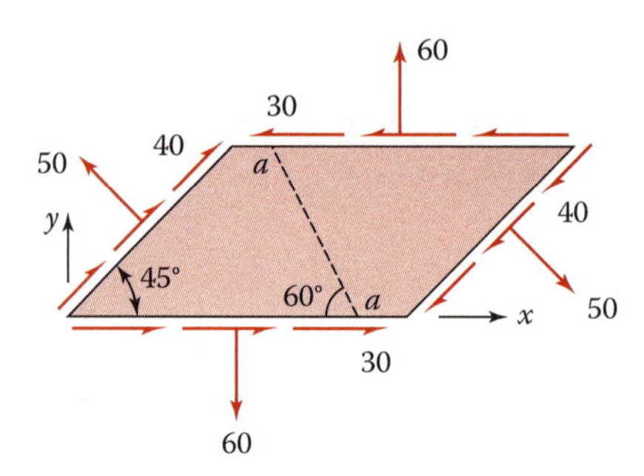

FIGURE P1.14

1.15 A welded plate is subjected to biaxial stresses with $\sigma_x = -10$ MPa and $\sigma_y = 30$ MPa, as shown in Figure P1.15. Calculate: (a) the normal stress acting perpendicular to the weld; (b) the shear stress acting parallel to the weld.

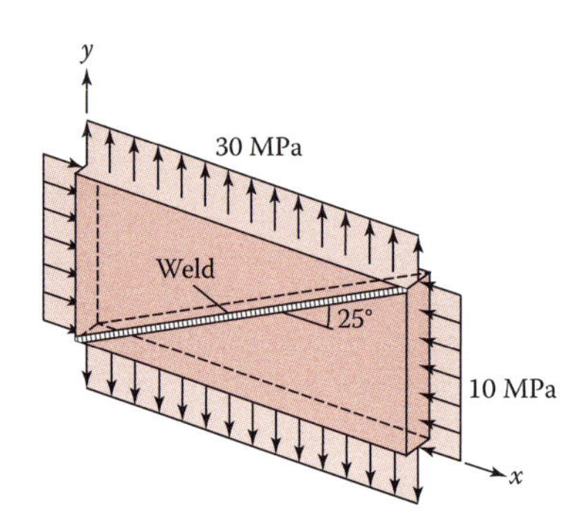

FIGURE P1.15

Sections 1.9 through 1.17

1.16 A slender horizontal bar ($\alpha = 11.7 \times 10^{-6}/°C$) of length $L = 250$ mm is fixed at its left and free of axial load, $P = 0$ (Figure 1.12a). Calculate the elongation δ of the bar: (a) for a uniform temperature change $\Delta T = 80°C$; (b) for a linearly varying temperature change $\Delta T = (80x/L)°C$.

1.17 For the skewed plate ($E = 70$ GPa, $\nu = 0.3$) shown in Figure P1.14. Determine: (a) the normal strain in the direction of the line a–a; (b) the maximum shear strain.

1.18 The strain at a point on a loaded steel plate ($E = 210$ GPa, $\nu = 0.03$) has components $\varepsilon_x = 50$ μm, $\varepsilon_y = 250$ μm, and $\gamma_{xy} = -150$ μm. Calculate: (a) the principal strains; (b) the maximum principal stresses.

1.19 As a result of loading, the thin rectangular plate of Figure P1.19 deforms into a parallelogram in which sides AB and CD shorten 0.004 mm and rotate 1000 μrad counterclockwise, while sides AD and BC elongate 0.006 mm and rotate 500 μrad clockwise. Determine at corner point A: (a) the normal strains ε_x and ε_y and the shear strain γ_{xy}; (b) the maximum shear strain. Use $a = 50$ mm and $b = 25$ mm.

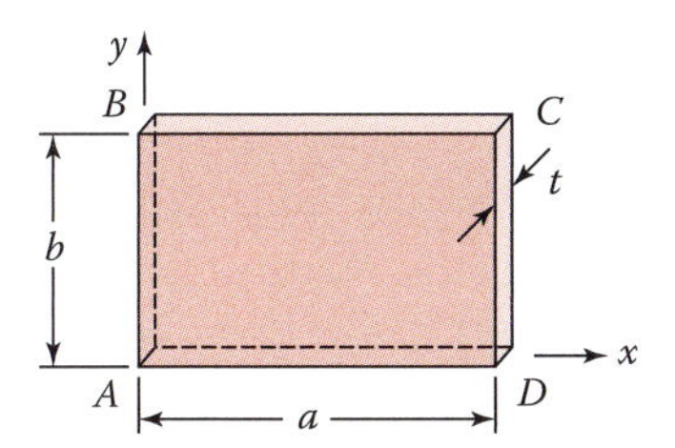

FIGURE P1.19

1.20 The rectangular plate shown in Figure P1.19 is acted on by a stress distribution that results in the uniform strains: $\varepsilon_x = 400\ \mu$, $\varepsilon_y = 800\ \mu$, and $\gamma_{xy} = 200\ \mu$. For $a = 30$ mm and $b = 40$ mm, determine: (a) the normal strain parallel to the direction of diagonal *BD*; (b) the maximum principal strain and its direction.

1.21 A bronze rectangular plate ($E = 100$ GPa, $\nu = 1/3$) is subjected to uniform stresses $\sigma_x = 150$ MPa and $\sigma_y = -90$ MPa along its edges (Figure P1.19). If it has a length $a = 100$ mm, width $b = 50$ mm, and thickness $t = 10$ mm prior to loading, calculate subsequent to loading:

(a) the dimensions *a*, *b*, and *t*; (b) the length of diagonal *AC*.

1.22 Resolve Problem 1.21, assuming that the plate is subjected to a uniform pressure of $p = 120$ MPa on all its faces.

1.23 Uniformly distributed loading at the edges of a rectangular aluminum plate ($E = 69$ GPa, $\nu = 0.3$) is illustrated Figure P1.23. What are the values of p_x and p_y (in kilonewtons per meter) that produce a change in length in the *x* direction of 2 mm and in the *y* direction of 2.5 mm? *Given*: $a = 2.5$ m, $b = 4$ m, and $t = 5$ mm.

1.24 Redo Problem 1.23, for the case in which the plate is made of brass ($E = 105$ GPa, $\nu = 0.3$). *Given*: $a = 4$ m, $b = 5$ m, $t = 7$ mm.

1.25 A 120 × 90-mm rectangular plate with $E = 210$ GPa and $\nu = 1/3$ carries the uniform stresses depicted in Figure P1.25. Calculate the deformation of the diagonal *BD*.

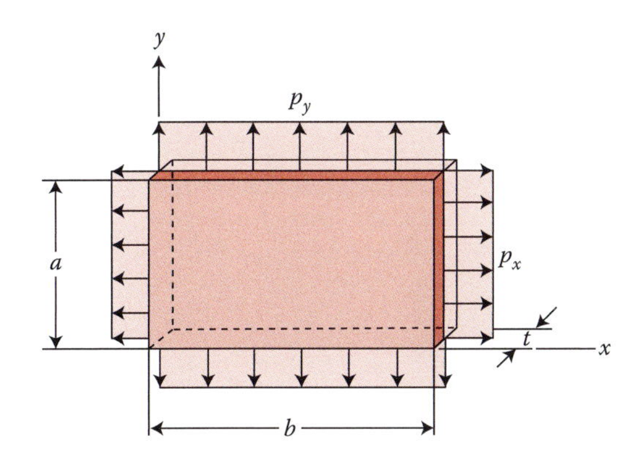

FIGURE P1.23

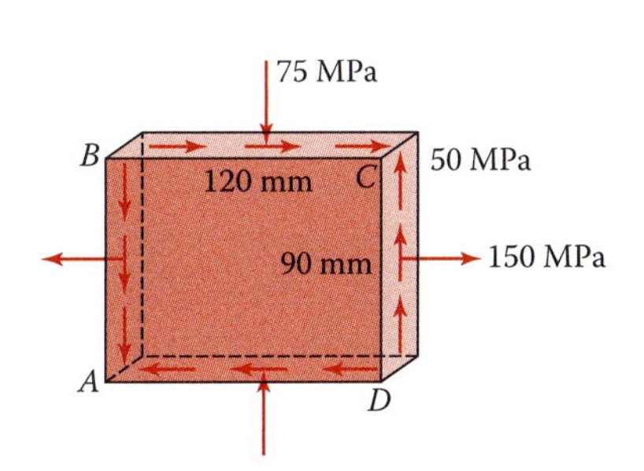

FIGURE P1.25

References

1. A. C. Ugural, *Mechanics of Materials*, 2nd ed., Wiley, Hoboken, NJ, 2008.
2. A. C. Ugural and S. K. Fenster, *Advanced Mechanics of Materials and Applied Elasticity*, 5th ed., Chapers 2 and 3, Prentice-Hall, Upper Saddle River, NJ, 2012.
3. H. H. West and L. F. Geschwindner, *Fundamentals of Structural Analysis*, 2nd ed., Wiley, Hoboken, NJ, 2002.
4. J. N. Reddy, *Theory and Analysis of Elastic Plates and Shells*, 2nd ed., CRC Press, Boca Raton, FL, 2007.
5. E. A. Avallone, T. Baumeister III, and A. Sadegh (eds.), *Mark's Standard Handbook for Mechanical Engineers*, 11th ed., McGraw-Hill, New York, 2006.
6. G. S. Brady and H. R. Clauser, *Materials Handbook*, 15th ed., McGraw-Hill, New York, 2002.
7. S. Zhang and D. Zhao, *Aerospace Materials Handbook*, CRC Press, Boca Raton, FL, 2012.
8. R. S. Lakes, Advances in negative Poisson's ratio materials, *Advanced Materials*, 5, 1993, 293–295.
9. J. N. Reddy, *Mechanics of Laminated Composite Plates and Shells, Theory and Analysis*, 2nd ed., CRC Press, Boca Raton, 2004.
10. A. C. Ugural, *Mechanical Design of Machine Components*, 2nd ed., CRC Press, Boca Raton, FL, 2015.

2

Simple Structural Members

2.1 Introduction

This chapter's main objective is to provide a review of and insight into the stress analysis of variously loaded *simple structural members* that appear in diverse forms in structures and machines. Discussions will include both statically determinate and indeterminate *bars* and *beams* and *pressure vessels*. We shall employ the mechanics of materials approach, where simplified assumptions related to the deformation pattern are made so that strain distributions for a cross section of a member can be determined.

A fundamental assumption of mechanics of materials is that *plane sections remain plane*. This hypothesis can be shown to be exact for axially loaded prismatic bars, circular torsion members, and beams, plates, and shells subjected to pure bending [1–3]. It is approximate for other stress analysis problems. Note, however, that there are extraordinarily many cases where applications of the *basic formulas of mechanics of materials* or *technical theory* lead to useful results for slender members under any type of loading. According to *St. Venant's Principle*, the actual stress distribution closely approximates that given by these formulas, except near the restraint or geometric discontinuities on the member.

Our coverage of the subject presumes the knowledge of mechanics of materials procedures for determining stress and displacement in simple structural members. Basic formulas are introduced in Sections 2.3 through 2.10, with the emphasis being on the underlying assumptions used in their derivations. Failure criteria are considered in Section 2.11. The chapter concludes with a discussion of the analysis of strain energy and displacement by a widely used energy method. Although here we discuss only typical structural elements, we will see later that these same topics are also important for plates and shells. Analysis of any member requiring determination of displacement will involve all three principles outlined in Section 1.2.

Note that, with the exception of Chapter 10, each formula presented in this text gives elastic stress (or small displacement) as directly proportional to the magnitude of a single force, torque, moment, or pressure at a section of a *linearly elastic* structural member. When a member is acted on by two or more types of load simultaneously, causing various internal forces on a section, it is assumed that each load produces the stress that it would if it were the only load acting on the member. The final or *combined stress* is then found by *superposition* of several states of stress that do not exceed the proportional limit of the material.

2.2 Types of Structures

As pointed out in Section 1.3, a *structure* consists of interconnected members supporting loads in static equilibrium. A variety of structures appears in nature or has been built by human beings. The constituents of such a units or systems are bars, beams, rings, plates, and shells or their combinations. Numerous structures are employed in various fields of engineering. The form that a structure takes depends on many considerations.

Often the functional requirements of a structure will narrow the possible shape that can be used. In addition, the aesthetic requirements, material availability, economic limitations, and foundation conditions may be significant factors in establishing the structural form. We shall here conduct a brief survey of how structural members or units are classified geometrically, beginning from the most elementary shapes and loading to what is actually called a plate or shell. In this context, the three types of structures are illustrated in Table 2.1.

TABLE 2.1

Some Common Structural Forms

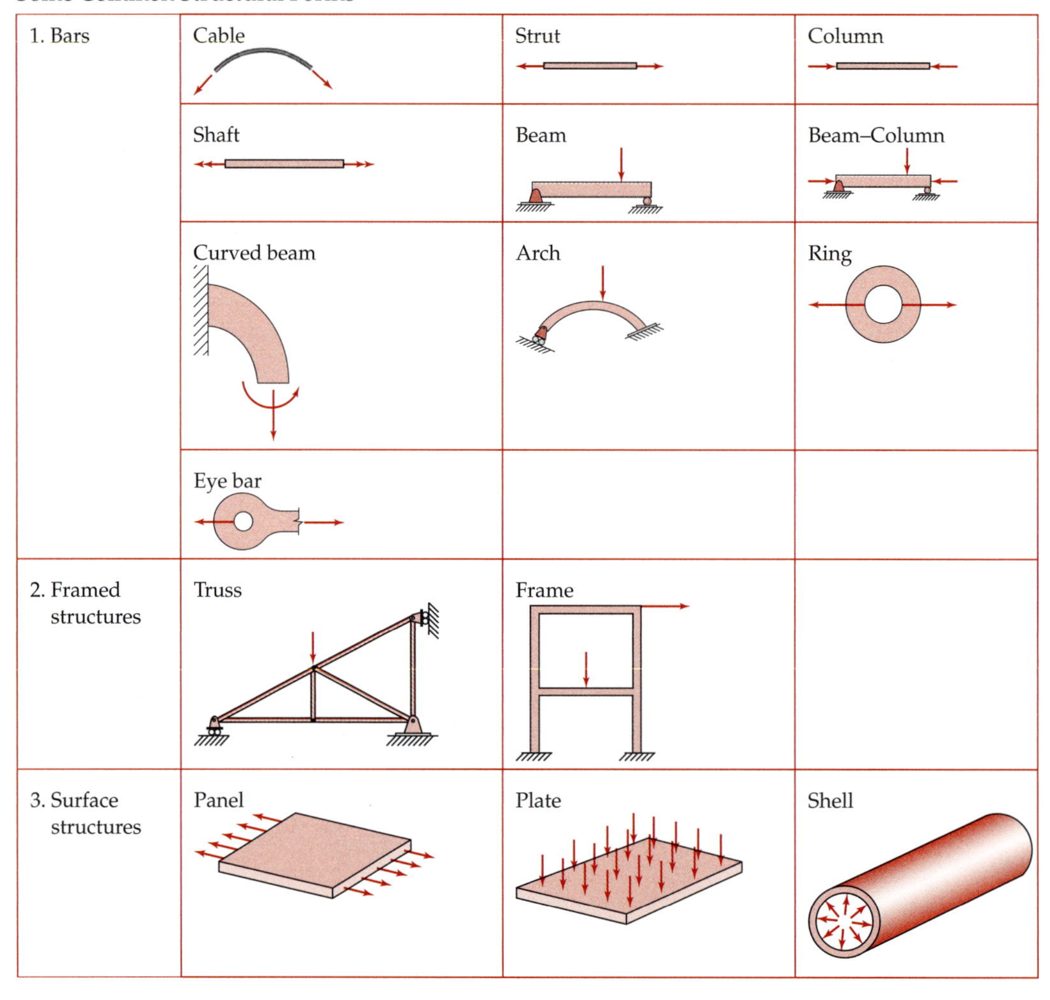

1. Bars	Cable	Strut	Column
	Shaft	Beam	Beam–Column
	Curved beam	Arch	Ring
	Eye bar		
2. Framed structures	Truss	Frame	
3. Surface structures	Panel	Plate	Shell

Bars. The simplest structure is a bar. Recall from Section 1.1 that bars and beams are members having one dimension significantly larger than the other two. A host of other familiar structural members can be grouped under this broad category. Usually, a *cable* consists of a large number of wires wound in a special manner. A typical cable or so-called wire rope is formed by six or more strands wound helically, with each strand consisting of many wires, also wound helically. *Rings* are closed or curved members. Examples include link plates or bars of chains and eye bar structures.

Framed structures. Frames are formed through the assemblage of two or more members. When bars are attached by "frictionless" hinges with members subjected to only axial forces, the configuration is called a *truss*. However, if the members are rigidly connected, supplementary loads can also be resisted in addition to those that act axially; such structures are referred to as *frames*. Machines are similar to frames in which any of the elements may be multiforce members. A *machine*, however, is designed to transmit and modify forces (or energy) and always contains moving parts.

Surface structures. Members that have two dimensions that are large relative to the third are known as planar or curved structures. *Panels* are subjected to in-plane loads. *Plates* are similar to panels with the exception that the loads are applied transversely or out of plane. *Shells* may be considered as curved plates acting mainly to resist tensile and compressive forces. We shall see in Section 3.1 that plates are further distinguished as being thin or thick, depending on thickness t relative to the width a. Likewise, shells (see Section 12.1) are often referred to as being thin or thick, depending on thickness t relative to the radius of curvature r.

2.3 Axially Loaded Members

Consider a homogeneous prismatic bar *loaded by axial forces* P at the ends (Figure 2.1a). To obtain an expression for the normal stress, an imaginary cut (section a–a) through the member is made at right angles to its axis. An FBD of the isolated portion is shown in Figure 2.1b. Here the stress is substituted on the cut section as a replacement for the effect of the removed portion.

The equilibrium of the axial forces, the first of Equations 1.5a, requires that $P = \int \sigma_x dA$ or $P = A\sigma_x$. The *normal stress* is thus

$$\sigma_x = \frac{P}{A} \tag{2.1}$$

wherein A is the cross-sectional area of the bar. Because V_y, V_z, and T are all equal to zero, the second and third of Equations 1.5a and the first of Equations 1.5b are satisfied with

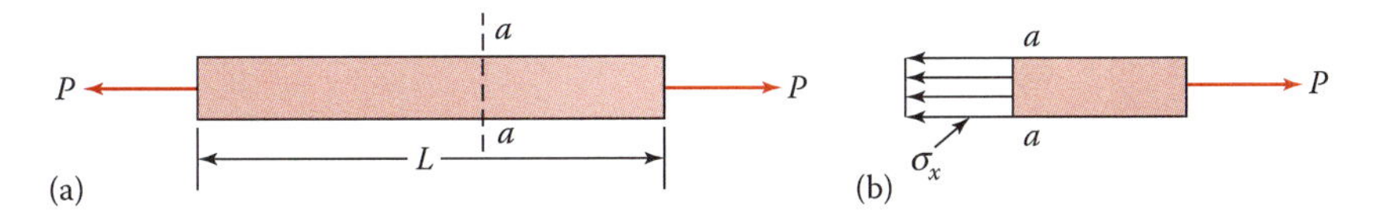

FIGURE 2.1
Prismatic bar in tension.

$\tau_{xy} = \tau_{xz} = 0$. In addition, $M_y = M_z = 0$ in the last two of Equations 1.5b requires only that $P = \int \sigma_x dA$ be symmetrically distributed about the y and z axes, as depicted in Figure 2.1b.

Equation 2.1 represents the value of the *uniform stress* over a cross section of the bar. If the member is being stretched as shown in the figure, the resulting stress is a *uniaxial tensile stress*; if the direction of the forces is reversed, the bar is in compression and *uniaxial compressive stress* takes place. In the latter case, Equation 2.1 is applicable only to chunky and short members.

When the resulting axial stress $\sigma_x = P/A$ does not exceed the proportional limit of the material, we may apply Hooke's law and write $\varepsilon_x = \sigma_x/E$. The axial strain is also defined by $\varepsilon_x = \delta/L$, in which L represents the length of the member (see Section 1.9). These expressions can be combined to yield the *axial deformation*, a contraction or expansion, of the bar:

$$\delta = \frac{PL}{AE} \tag{2.2}$$

The product AE is called the *axial rigidity* of the bar.

A change in temperature of ΔT degrees causes a strain of $\varepsilon_t = \alpha\Delta T$, given by Equation 1.18. The formula for *thermal axial deformation* produced by a uniform temperature is then

$$\delta_t = \alpha(\Delta T)L \tag{2.3}$$

Thermal stresses are produced when thermal expansion or contraction of a body is restricted. Applying Hooke's law gives

$$\sigma_t = \alpha(\Delta T)E \tag{2.4}$$

It is seen from this relationship that a high modulus of elasticity E and high coefficient of expansion α for the material increase the thermal stress σ_t.

EXAMPLE 2.1: Design of Hoist

A hoist made of pin-connected round structural steel eyebar AC and square aluminum alloy 6061-T6 post BC supports a vertical load $P = 50$ kN at C as depicted in Figure 2.2a. *Find:* The cross-sectional areas of each member on the basis of a factor of safety of $n = 2.5$. *Given*: The maximum strengths of the steel and aluminum (from Table B.3) are 400 and 300 MPa, respectively. *Assumptions:* Friction in the joints and possibility of member BC buckling are disregarded. The load acts in the plane of the hoist. Weights of the bars are insignificant compared to the applied load and ignored.

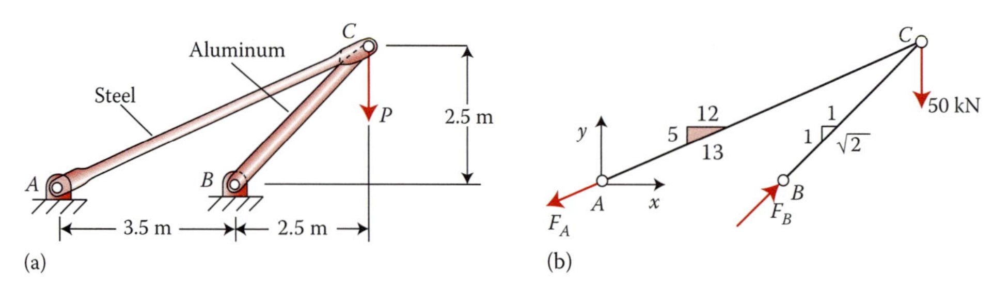

FIGURE 2.2
(a) A pin-connected hoist; (b) FBD of entire structure.

Solution

Members *AC* and *BC* are under axial loading with relative dimensions shown by small triangles in Figure 2.2b. Observe that the slopes of the forces F_A and F_B are 5/12 and 1/1, respectively. Hence, $F_{Ay}/5 = F_A/13$ and $F_{By} = F_B/\sqrt{2}$. Applying equations of equilibrium to the FBD of Figure 2.2b, we have

$$\sum M_B = -50(2.5) + \frac{5}{13}F_A(3.5) = 0 \qquad F_A = 92.86 \text{ kN}$$

$$\sum M_A = -50(6) + \frac{1}{\sqrt{2}}F_B(3.5) = 0 \qquad F_B = 121.2 \text{ kN}$$

Note, as a check, that $\sum F_x = 0$.

The allowable stresses, from design procedure (steps 3 and 4 of Section 1.15.1), equal to

$$(\sigma_{all})_{AC} = \frac{400}{2.5} = 160 \text{ MPa}, \qquad (\sigma_{all})_{BC} = \frac{300}{2.5} = 120 \text{ MPa}$$

By Equation 2.1, the required cross-sectional areas of the bars are

$$A_{BC} = \frac{121.2(10^3)}{120} = 1010 \text{ mm}^2, \qquad A_{AC} = \frac{92.86(10^3)}{160} = 580.4 \text{ mm}^2$$

Comment: A 28-mm diameter steel eyebar and a 32-mm by 32-mm aluminum post should be used.

EXAMPLE 2.2: Analysis of Steel Tube and Brass Rod Assembly

A brass rod of length L_b, cross-sectional area A_b, and modulus of elasticity E_b is inserted into a thick-walled steel tube having a cross-sectional area A_s, modulus of elasticity E_s, and length L_s (Figure 2.3). Develop a formula for the axial deformation of each member resulting from axial load P that is large enough to close the small gap Δ.

Solution

We denote the deformations, subsequent to loading, of the brass rod and steel tube δ_b and δ_s, respectively.

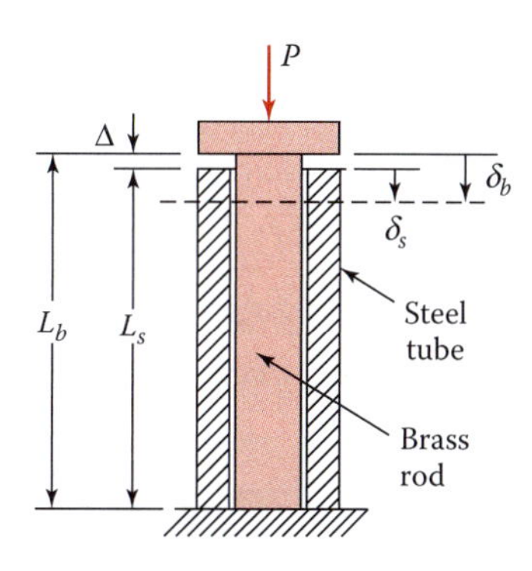

FIGURE 2.3
Axially loaded statically indeterminate system.

Statics. An axial force P_b will be produced in the brass and an axial load force P_s in the steel. Equilibrium conditions, Equations 1.2a, give

$$P_b + P_s = P \tag{a}$$

It is observed that the system is *statically indeterminate* to the first degree.

Deformations. From Equation 2.2, we obtain the contractions of the members as

$$\delta_b = \frac{P_b L_b}{A_b E_b} \qquad \delta_s = \frac{P_s L_s}{A_s E_s} \tag{b}$$

Geometry. The axial deformation of the brass rod equals that of the steel tube plus the gap, $\delta_b = \delta_s + \Delta$. That is,

$$\frac{P_b L_b}{A_b E_b} = \frac{P_s L_s}{A_s E_s} + \Delta \tag{c}$$

Solution of Equations a and c simultaneously results in the compressive forces:

$$\begin{aligned} P_b &= \frac{PL_s + A_s E_s \Delta}{L_s + (A_s E_s L_b / A_b E_b)} \\ P_s &= \frac{PL_b + A_b E_b \Delta}{L_b + (A_b E_b L_s / A_s E_s)} \end{aligned} \tag{2.5}$$

Carrying these expressions into Equation b gives the axial deformations of the brass rod and the steel tube.

2.3.1 Columns

A *prismatic bar* loaded in compression is called a *column*. *Buckling* can be defined as the sudden, large, lateral deflection of a column due to a small increase in the existing compressive load of the member. The ratio of the length L to the *radius of gyration* r of the cross-sectional area A of the column is called *slenderness ratio* L/r.

Columns with low slenderness ratios showing no buckling phenomenon are called *short columns* or *struts* (e.g., steel rods with $L/r < 30$). For short columns, failure is a result of material failure. Thus, $\sigma_{max} = P/A$, where σ_{max} is the strength limit of the column, that is, the yield stress or ultimate stress. At small L/r ratios, ductile materials "squash out" and support large loads.

A column that buckles elastically is usually referred to as a *long column* (Figure 2.4a). Refer to Figure 2.4b, where load P has been increased sufficiently to cause a *very small* lateral deflection. This condition is called the *neutral equilibrium* and the corresponding value of the load is the critical load. It can be shown that, the *critical load* for a long column with *pinned ends* is expressed in the form

$$P_{cr} = \frac{\pi^2 EI}{L^2} \tag{2.6}$$

The foregoing is known as the *Euler buckling load*. Here $I = Ar^2$ represents the centroidal moment of inertia of the cross-sectional area. The *critical stress* corresponding to the foregoing load is given by *Euler's formula*:

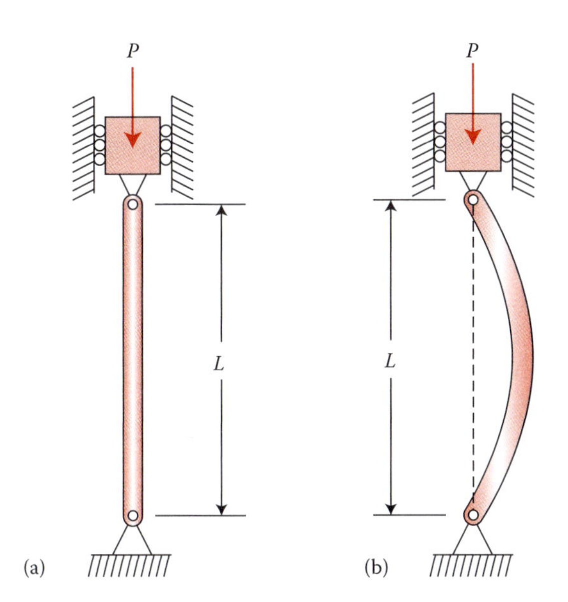

FIGURE 2.4
Column with pinned ends: (a) Initially straight column; (b) buckled shape.

$$\sigma_{cr} = \frac{P_{cr}}{A} = \frac{\pi^2 E}{(L/r)^2} \tag{2.7}$$

For columns with *various* combinations of fixed, free, and pinned *supports,* one need only replace L by L_e in Equations 2.6 and 2.7, where L_e is termed the *effective length*. The column effective length is the distance between the *inflection points* of its small lateral elastic curve of zero moments. We have

$$\begin{aligned} L_e &= 2L && \text{(fixed – free)} \\ L_e &= L && \text{(pinned – pinned)} \\ L_e &= 0.7L && \text{(fixed – pinned)} \\ L_e &= 0.5L && \text{(fixed – fixed)} \end{aligned} \tag{d}$$

Observe that only column with pinned–pinned ends has an effective length same as the actual length L (Figure 2.4b). The designer tries to configure the ends of columns using pins, bolts, or weld to achieve required ideal end conditions.

To ascertain the lowest possible critical load or stress, r should be taken as the *least radius of gyration* of the section. We note that the smallest value of the slenderness ratio the Euler's formula applies is found by equating to the proportional limit of a given material. For example, in the case of a structural steel with $E = 210$ GPa and $\sigma_{pl} = 250$ MPa, Equation 2.7 gives $(L/r)_c = 91$.

In *intermediate-length columns,* which lie between short and long columns, failure occurs by inelastic buckling. Substituting *tangent modulus* E_t, the slope of the stress–strain curve beyond the yield point, for the elastic modulus E, we may apply Equation 2.7 in the inelastic range. Numerous other empirical formulas have also been developed for columns of intermediate length.

2.4 Stress Concentration Factors

High localized stresses created by an abrupt change in geometry are known as *stress concentrations*. The abrupt change of form or *discontinuity* occurs in such frequently encountered configurations as holes, notches, and fillets. The formulas of mechanics of materials apply as long as the material remains linearly elastic and shape variations are *gradual.* In some cases, the stress and accompanying deformation near a discontinuity can be analyzed by applying the theory of elasticity. However, it is more usual to rely on experimental techniques and/or finite element analysis.

A stress concentration effect is conveniently expressed by the ratio of the maximum stress $\sigma_{\max}$ or $\tau_{\max}$ at the discontinuity to the nominal stress σ_{nom} or τ_{nom}. This is called the theoretical or geometric *stress concentration factor* K. Therefore,

$$K = \frac{\sigma_{\max}}{\sigma_{nom}} \quad \text{or} \quad K = \frac{\tau_{\max}}{\tau_{nom}} \tag{2.8}$$

The *nominal stresses* are stresses that would occur if the abrupt change in the cross section did not exist or had no influence on stress distribution. A stress concentration factor is applied to the stress computed for the *net* or *reduced* cross section. Many stress concentration factors are available in the technical literature [4] as functions of the geometrical parameters of members. All results indicate the advisability of streamlining junctures and transition portions that make up a structural member.

It should be noted that, for the case of *static loading*, stress concentration is important only for brittle materials. However, for some brittle materials having internal irregularities such as cast iron, stress raisers usually have little effect, regardless of the nature of loading. Hence, use of stress concentration factors appears to be unnecessary for cast iron. It is customary to *ignore* stress concentration in static loading of *ductile materials.*

In the case of *repeated* or *fatigue loading*, most materials may fail as a result of propagation of cracks originating at the point of high stress. The presence of stress concentration in the case of fatigue loading must thus be taken into account, regardless of whether the material is brittle or ductile. This consideration also applies to shock or impact loading.

The matter of *ductile–brittle transition* of metals has important applications where the operating environment includes a wide variation in temperature or when the rate of dynamic loading changes. The transition temperature is roughly the temperature at which a material's behavior changes from ductile to brittle. Stress raisers also have a significant effect on the transition from brittle to ductile failure [2]. In summary, stress concentration is the primary cause of fatigue failures and of static failures in brittle materials.

2.5 Torsion of Circular Bars

Torsion of circular prismatic *bars* or *shafts* caused by a torque T results in a shearing stress τ and an angle of twist or angular deformation ϕ. The *basic assumptions* of the formulations of the torsional loading of a circular bar are that plane sections perpendicular to the longitudinal axis of the bar remain plane and undisturbed after the application of the torques, shear strains γ vary linearly from zero at the center axis to a maximum at the periphery, and the material is homogeneous and obeys Hooke's law.

2.5.1 Shear Stress

For equilibrium, internal resisting torque produced by stress distribution must be equal to the applied torque T (Figure 2.5). Thus,

$$T = \int \rho(\tau\, dA) = \int \rho\left(\frac{\rho}{r}\tau_{\max}\right)dA \tag{a}$$

where r is the radius of the bar. The preceding expression may be written as

$$T = \frac{\tau_{\max}}{r}\int \rho^2 dA \tag{b}$$

The polar moment of inertia J of the cross-sectional area is defined by

$$J = \int \rho^2 dA \tag{2.9}$$

For solid shafts, we have $J = \pi r^4/2$. In the case of a circular shaft of inner radius r_i and outer radius r_o, $J = \pi\left(r_o^4 - r_i^4\right)/2$.

The maximum shear stress on members with circular solid and hollow cross sections is thus given by the *torsion formula*:

$$\tau_{\max} = \frac{Tr}{J} \tag{2.10}$$

The shear stress at a distance ρ from the center is given by

$$\tau = \frac{T\rho}{J} \tag{2.11}$$

The *transverse* shear stress obtained by either of these equations is accompanied by a *longitudinal* shear stress of equal value (Figure 2.5).

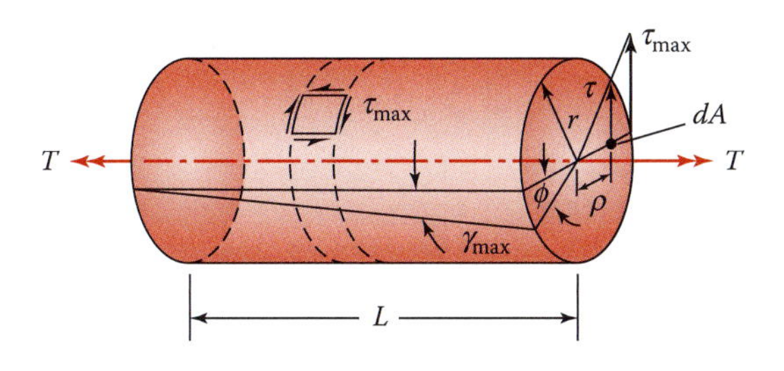

FIGURE 2.5
Circular member in pure torsion.

2.5.2 Angle of Twist

From Hooke's law, $\gamma_{max} = \tau_{max}/G$ or, according to the torsion formula, $\gamma_{max} = Tr/GJ$. Here G is the modulus of elasticity in shear. For small deformations, by taking $\tan \gamma_{max} = \gamma_{max}$, we also have $\gamma_{max} = r\phi/L$ (Figure 2.5). These relationships give the *total angle of twist* of a circular shaft, the angle through which one end cross section of the shaft rotates with respect to another:

$$\phi = \frac{TL}{GJ} \tag{2.12}$$

Angle ϕ is measured in radians. Here GJ is called the *torsional rigidity* of the shaft.

EXAMPLE 2.3: Deformation of Fixed-Ended Torsion Bar

A torque of T is applied to a circular steel bar ($G = 79$ GPa), as shown in Figure 2.6a. The ends A and B of the bar are fixed and hence prevented from rotating. Calculate the angle of twist at the center C of the bar for $a = 1.6$ m, $b = 0.8$ m, $d = 25$ mm, and $T = 200$ N·m.

Solution

FBD of Figure 2.5b indicates that the problem is statically indeterminate to the first degree; the one equation of equilibrium available is not sufficient to obtain the two unknown reactions T_A and T_B.

Statics. From the condition of equilibrium of torques, we obtain (Figure 2.6b)

$$T_A + T_B = T \tag{c}$$

Deformations. For the angle of twist at section D for the left and right segments of the bar,

$$\phi_{AD} = \frac{T_A a}{GJ} \qquad \phi_{BD} = \frac{T_B b}{GJ} \tag{d}$$

Geometry. Continuity of the bar at section D requires that

$$\phi_{AD} = \phi_{BD} \quad \text{or} \quad T_A a = T_B b \tag{e}$$

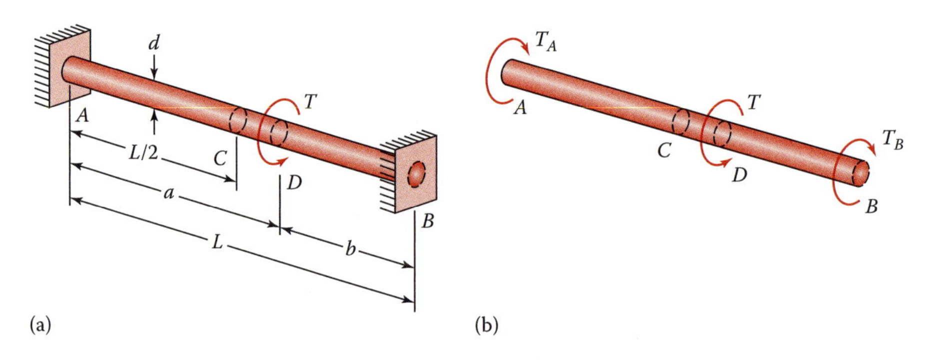

FIGURE 2.6
Statically indeterminate bar in torsion.

Equations c and e can be solved simultaneously to give

$$T_A = \frac{Tb}{L} \qquad T_B = \frac{Ta}{L} \tag{f}$$

The angle of rotation at midsection C is therefore

$$\phi_C = \frac{T_A(L/2)}{GJ} = \frac{Tb}{2GJ} \tag{2.13}$$

Introducing the given numerical values into this equation, we have

$$\phi_C = \frac{200(0.8)}{2(79\times10^9)(\pi/2)(0.0125)^4} = 0.0264 \text{ rad} = 1.51°$$

2.6 Rectangular Torsion Bars

In dealing with torsion of *noncircular prismatic bars*, the cross sections, initially plane, experience out-of-plane deformation. Then, only in some cases, the problem can be analyzed by using the methods of the theory of elasticity. Generally, it can be handled experimentally or by the finite element method. Torsional shear stress and displacement for a number of noncircular sections are listed in various references [5,6].

Consider a bar having a *rectangular cross section* subjected to a torque T. The following approximate formula for maximum shear stress is of interest:

$$\tau_{\max} = \frac{T}{ab^2}\left(3 + 1.8\frac{b}{a}\right) \tag{2.14}$$

In the foregoing, the quantities a and b denote the lengths of the long and short sides of a rectangular cross section, respectively. The shear stress distribution along three radial lines initiating from the center are depicted in Figure 2.7, where the largest stresses are along the center line of each face. The difference in this stress distribution in comparison

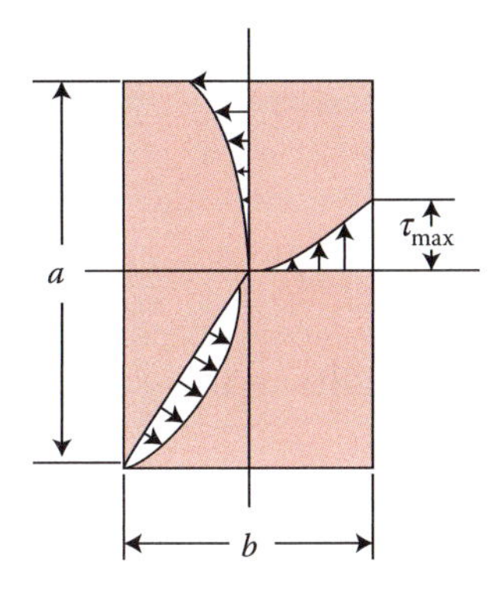

FIGURE 2.7
Shear stress distribution in the rectangular cross section of a torsion bar.

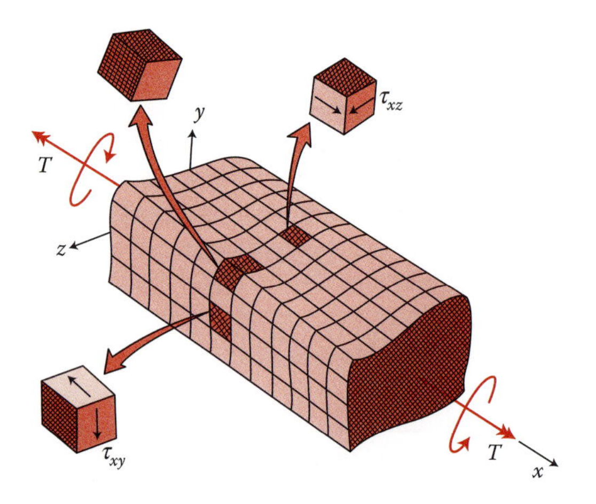

FIGURE 2.8
Deformation and stresses in a rectangular bar segment in torsion. Observe that the original plane cross sections warped out of their own plane.

with that of a circular section is very clear. For the latter, the stress is a maximum at most remote point, while for the former; the stress is zero at the most remote point.

We note that, a *corner element* of the cross section of a rectangular shaft under torsion does not distort at all, and consequently the shear stresses are zero at the corners, as portrayed in Figure 2.8. This is possible because the outside surfaces are free of all stresses. The identical considerations can be applied to the other points on the boundary. The shear stresses acting on three *outermost* cubic elements isolated from the bar are depicted in the figure, where the stress-free surfaces are indicated as shaded. It is obvious that, all shear stresses τ_{xy} and τ_{xz} in the plane of a cut near the boundaries act on them.

2.7 Theory of Beams

A *beam* is a bar supporting loads applied laterally or transversely to its longitudinal or axial axis. The flexure member is commonly used in structures. Examples are the main members supporting floors of buildings, automobile axles, and leaf springs. We shall observe in Chapter 3 that the following formulas for stresses and deflections of beams can readily be reduced from those of rectangular plates.

The fundamental *assumptions* of the *Bernoulli–Euler* or *ordinary theory of beams* is based on geometry of deformations. These may be outlined as follows:

1. The *deflection* of the beam axis is *small* compared with its span; the slope of the deflection curve is thus very small, and its square is negligible in comparison with unity.
2. *Plane sections* through a beam taken normal to its axis *remain plane* after the beam is subjected to bending.

A generalization of the foregoing hypotheses forms the basis for the classical theories of plates and shells.

Interestingly, the *Timoshenko theory of beams* constitutes an improvement over the ordinary theory. In *static case*, the difference between the two hypotheses is that the former includes the effect of shear stresses on the deformation by assuming a constant shear over the beam height. The latter ignores the influence of transverse shear on beam deformation. Thus, the Timoshenko beam theory is an extension of the Bernoulli–Euler theory to allow for the effect of the transverse shear deformation while relaxing the assumption 2 stated above. The Timoshenko theory is highly suited for describing the behavior of short and sandwich beams. *In dynamic case, the theory incorporates shear deformation as well as rotational inertia effects, and it will be more accurate for very* slender beams.

2.8 Stresses in Beams

We now briefly discuss stresses associated with the bending moments and shear forces that act laterally to the longitudinal axis of the beams. Considerations are limited to only beams having initially straight longitudinal axis. Note that it may often be necessary to distinguish between pure bending and nonuniform bending of beams. *Pure* bending refers to flexure of a beam under a constant bending moment. In contrast, *nonuniform* bending refers to flexure in the presence of shear forces.

2.8.1 Normal Stress

Figure 2.9a shows a linearly elastic beam having the z axis as a vertical axis of symmetry. On a cross section such as A–B (Figure 2.9b), the normal stress σ_x acting on a longitudinal fiber is provided by the *flexure formula*:

$$\sigma_x = \frac{Mz}{I} \tag{2.15}$$

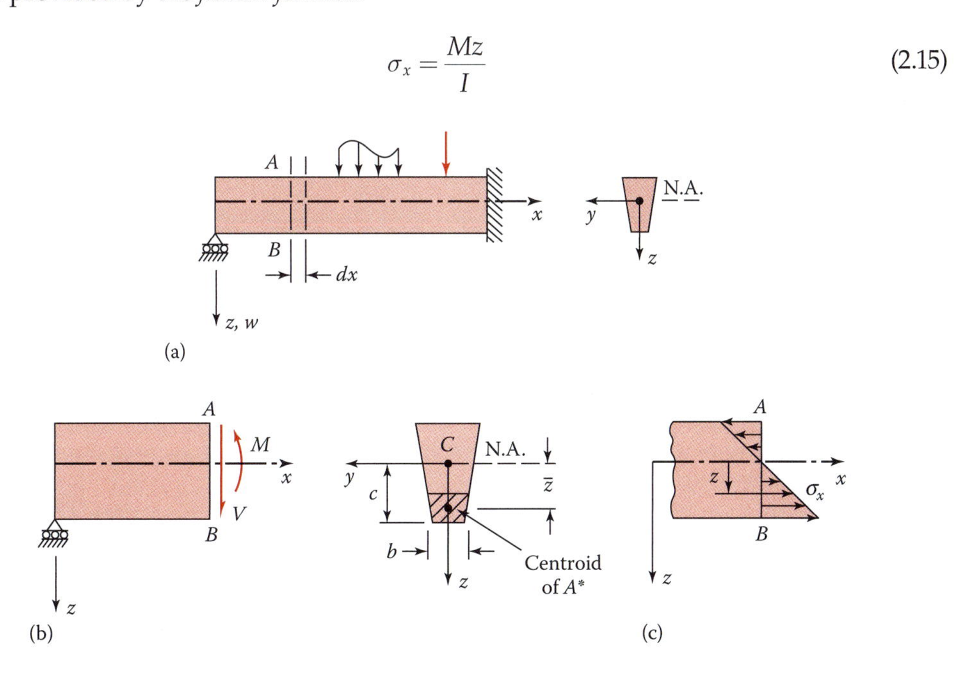

FIGURE 2.9
(a) Beam subjected to transverse loading; (b) FBD of beam segment; (c) normal stress distribution for section A–B.

Bending moment at the section A–B is M and I represents the moment of inertia of the cross section, both with respect to the neutral axis (N.A.). The normal stresses, like the deformations (strains), vary linearly with the distance z from the neutral axis as depicted in Figure 2.9c. They are positive (tensile) on one side of the neutral axis and negative (compressive) on the other side.

The *maximum stress*, occurring at the outmost fibers of the beam, is expressed as

$$\sigma_{\max} = \frac{Mc}{I} = \frac{M}{S} \tag{2.16}$$

Here c is the distance from the neutral axis to the outmost fiber. The quantity $S = I/c$ is known as the *section modulus* of the cross-sectional area. Note that the flexure formula also applies to a beam of *unsymmetrical* cross-sectional area, provided that I is a principal moment of inertia and M is a moment around a *principal axis.*

2.8.2 Shear Stress

The vertical stresses acting on a cross section of a beam are called *shear stresses.* The vertical shear stress τ_x at any point on the cross section is numerically equal to the horizontal shearing stress at the same point (see Section 1.5). The shear stresses as well as the normal stresses are taken to be uniform across the width of the beam. The shear stress $\tau_{xz} = \tau_{zx}$ at any point of a cross section (Figure 2.9b) is given by the shear formula

$$\tau_{xz} = \frac{VQ}{Ib} \tag{2.17}$$

where

V = the shearing force at the section,

b = the width of the section measured at the point in question, and

Q = the first moment with respect to the neutral axis of the (shaded) area beyond the point at which the shearing stress is required.

We have

$$Q = \int_{A^*} z dA = \bar{z} A^* \tag{2.18}$$

By definition, the area A^* represents the area of the part of the section *below* the point in question, and $\bar{z}$ is the distance from the neutral axis to the centroid of A^*. Clearly, if $\bar{z}$ is measured *above* the neutral axis, Q represents the *first moment of the area* above the level where the shear stress is to be found.

The shear stress varies in accordance with the shape of the cross section. The distribution of the shear stress on a cross section of a *rectangular beam* is parabolic. The stress is zero at the top and bottom of the section and has its *maximum* value at the neutral axis as shown in Figure 2.10. Thus,

$$\tau_{\max} = \frac{V}{Ib} A^* \bar{z} = \frac{V(bh/2)(h/4)}{Ib} = \frac{3}{2}\frac{V}{A} \tag{2.19}$$

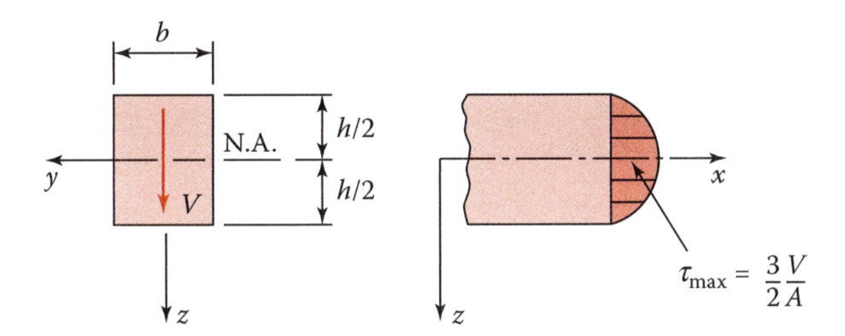

FIGURE 2.10
Shear stresses in a beam of rectangular cross section.

in which $A = bh$ is the cross-sectional area. For *narrow* rectangular beams, the preceding equation gives "exact" results.

It should be pointed out that the *shear formula*, also known as *shear stress formula*, for beams is derived on the basis of the flexure formula. Hence, the limitations of the bending formula apply. A variety of cross sections are treated by the following procedures identical to that for the rectangular section. Table B.6 lists some typical cases. Observe from this table that shear stress can always be expressed as a constant times the average stress (P/A), where the constant is a function of the cross-sectional shape.

2.8.3 Shear Flow

The shear force acting across the width of the beam *per unit length* long the beam axis may be found by multiplying τ_{xz} in Equation 2.17 by b (Figure 2.9b). This quantity is denoted by q, referred to as the shear flow:

$$q = \frac{VQ}{I} \tag{2.20}$$

The preceding equation is very useful in obtaining the necessary interconnections between the two or more pieces of element making up a *built-up beam*.

EXAMPLE 2.4: Stresses in T-Shaped Cross-Sectional Beam

A beam *ABC* of T-shaped cross section carries a concentrated load of 4 kN at the end of the overhand (Figure 2.11a). Calculate: (a) the maximum shear stress; (b) the shear flow q_j and the shear stress τ_j in the joint between the flange and the web; and (c) the maximum bending stress.

Solution

Referring to Figure 2.11b, the distance $\bar{z}$ to the centroid is

$$\bar{z} = \frac{A_1 z_1 + A_2 z_2}{A_1 + A_2} = \frac{20(100)90 + 80(20)40}{20(100) + 80(20)} = 67.8 \text{ mm}$$

The moment of inertia I with respect to the neutral axis, according to the parallel-axis theorem, is

$$\begin{aligned} I &= \frac{1}{12}(100)(20)^3 + 20(100)(22.2)^2 + \frac{1}{12}(20)(80)^3 + 20(80)(27.8)^2 \\ &= 3.142 \times 10^6 \text{ mm}^4 \end{aligned}$$

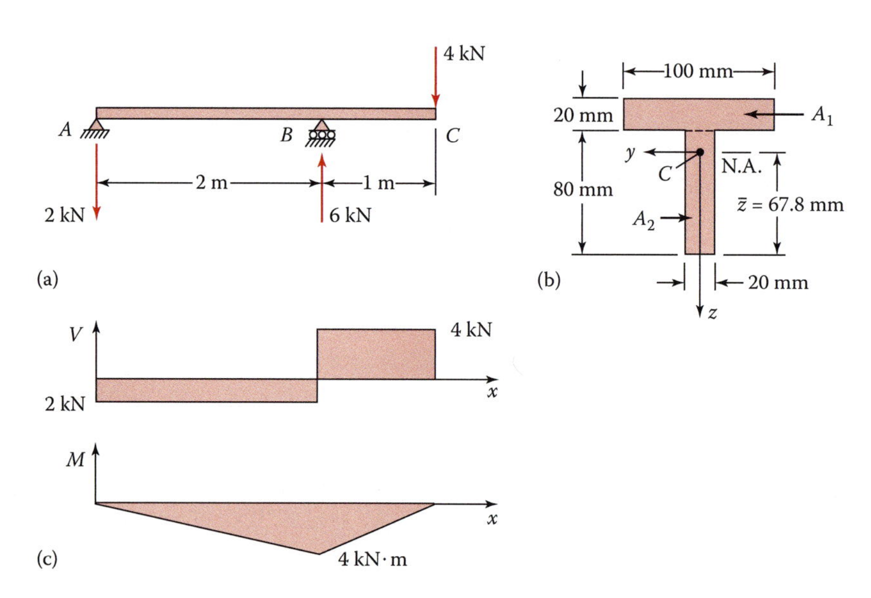

FIGURE 2.11
(a) Transversely loaded beam; (b) beam cross section; (c) shear and moment diagrams.

The *shear and moment diagrams*, as drawn by the *method of sections*, are shown in Figure 2.11c.

a. The maximum shear stress in the beam will take place at the neutral axis on the cross section supporting the largest shear force V. Thus,

$$Q_{N.A.} = 67.8(20)33.9 = 45{,}968 \text{ mm}^3$$

Inasmuch as the maximum shear force equals 4 kN (Figure 2.11c),

$$\tau_{\max} = \frac{V_{\max} Q_{N.A.}}{Ib} = \frac{4(10^3)(45{,}968)10^{-9}}{3.142(10^{-6})(0.02)} = 2.93 \text{ MPa}$$

b. The first moment of the area of the flange about the neutral axis is found as

$$Q_f = 20(80)22.2 = 35{,}520 \text{ mm}^3$$

By Equations 2.20 and 2.17

$$q_j = \frac{VQ_f}{I} = \frac{4(35.52 \times 10^{-6})}{3.142(10^{-6})} = 45.22 \text{ kN/m}$$

and

$$\tau_j = \frac{q_j}{b} = \frac{45{,}220}{0.02} = 2.261 \text{ MPa}$$

c. The largest moment occurs at B (Figure 2.11c). Therefore, through the use of Equation 2.16,

$$\sigma_{\max} = \frac{Mc}{I} = \frac{4(10^3)(0.0678)}{3.142(10^{-6})} = 86.31 \text{ MPa}$$

CASE STUDY 2.1 Shear Stress Distribution in Flanged Beams

The distribution of shear stresses in beams with *thin-walled* flanges (*W* beams and *I* beams) may be obtained by using Equation 2.17. An identical procedure previously applied to rectangular beams is employed here. As before, it is assumed that the shear stresses act parallel to the vertical *axis of symmetry* (y) and are *uniform* across the width of the beam. But, because of the shape, the distribution of the shear stresses is much more involved than in the case of a rectangular beam.

Interestingly, given any cross-sectional form, one point may be found in the plane of the cross section through which the resultant of the transverse shear stresses passes. A transverse load applied on the beam must act through the point, termed the *shear center* or flexure center, if no twisting is to occur. On the contrary, the *location* of shear center for a typical thin-walled open cross section (such a channel section) needs to be determined [2]. The theory of thin-walled beams fits well with thin-walled plates, sandwich, and honeycomb plates (Section 8.2) and thin-walled shells as well.

Shear Stresses in Web

Shown in Figure 2.12a is a flanged beam subjected to a vertical shear force V. We wish to find the shear stress at locations of ef in the web of the beam. The area used in calculating the first moment Q becomes the area between ef and the upper edge of the cross section (indicated shaded in the figure). This contains of the flange and the web areas, respectively,

$$A_f^* = b(c - c_1) \qquad A_w^* = t_w(c_1 - y_1) \tag{a}$$

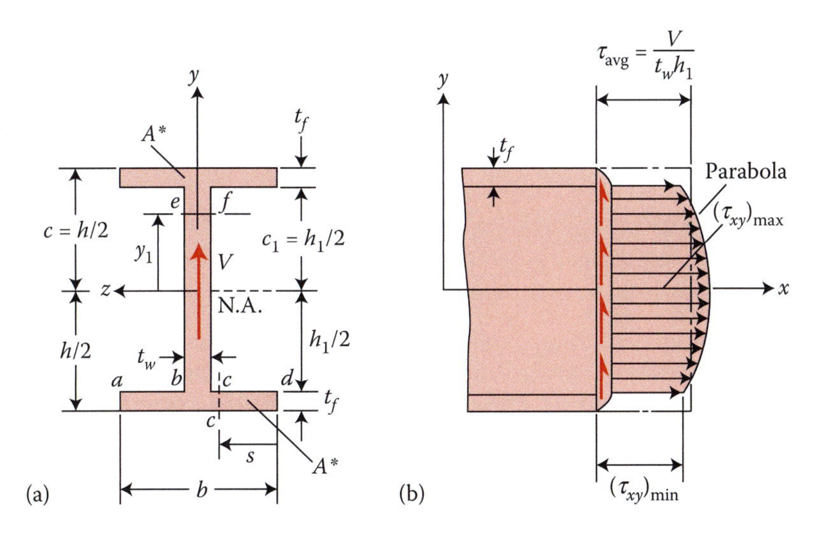

FIGURE 2.12
(a) Shear force V in a wide-flange beam; (b) shear stress distribution at a cross section.

The distances from the z axis to the centroids of the areas equal

$$\bar{y}_f = c_1 + \frac{c - c_1}{2} \qquad \bar{y}_w = y_1 + \frac{c_1 - y_1}{2} \tag{b}$$

The shear stress τ_{xy} is taken as a constant over the web thickness. So, for $0 < y_1 < c_1$, we have

$$\sigma_x = \frac{V}{Ib} = \frac{V}{It_w}\left(A_f^* \, \bar{y}_f + A_w^* \, \bar{y}_w\right) \tag{c}$$

After inserting Equations a and b and simplifying, Equation c results in the following expression for the shear stress in the web:

$$\tau_{xy} = \frac{V}{2It_w}\left[b\left(c^2 - c_1^2\right) + t_w\left(c_1^2 - y_1^2\right)\right] \tag{2.21}$$

Here the moment of inertia of the entire cross-sectional area about the neutral axis is

$$I = \frac{(2c)^3}{12} - \frac{(b - t_w)(2c)^3}{12} = \frac{2}{3}\left(bc^3 - bc_1^3 + t_w c_1^3\right) \tag{2.22}$$

Equation 2.21 indicates that shear stress τ_{xy} varies *parabolically* throughout the height of the web (Figure 2.12b). The *maximum shear stress* occurs at the neutral axis ($y_1 = 0$) and *minimum shear stress* in the web takes place at the juncture with the flange ($y_1 = \pm c_1$). Thus, using Equation 2.21, we have

$$\begin{aligned}
(\tau_{xy})_{\max} &= \frac{V}{2It_w}\left(bc^2 - bc_1^2 + t_w c_1^2\right) \\
(\tau_{xy})_{\min} &= \frac{Vb}{2It_w}\left(c^2 - c_1^2\right)
\end{aligned} \tag{2.23}$$

Clearly, when $t_w \ll b$, the values of these stresses do not differ noticeably.

Shear Stresses in Flanges

An expression for the shearing stress in the beam flanges can be written in a like manner to that described in the preceding equation. Thus, for $c_1 < y_1 < c$:

$$\tau_{xy} = \frac{V}{2It_w}\left[b(c - y_1)\left(y_1 + \frac{c - y_1}{2}\right)\right]$$

Simplifying, we have

$$\tau_{xy} = \frac{V}{2I}\left(c^2 - y_1^2\right) \tag{2.24}$$

The foregoing equation represents a *parabolic* variation of stress, as portrayed in Figure 2.12b.

It is seen from Figure 2.12b that, for a thin flange, the stress is *very small* as compared with the shear in the web. Hence, the approximate *maximum* or *average shear stress in the beam* may be found by dividing shear force V by the area (ht_w) of the web:

$$\tau_{avg} = \frac{V}{A_{web}} \tag{2.25}$$

This, shown by the dotted lines in the figure, is frequently used in design work.

The magnitude of the horizontal shearing stress τ_{xz} across any section c–c of the flange (Figure 2.12a) may be obtained using Equation 2.17, as was indicated in Section 2.7.2. The cut c–c is located a distance s from the free edge. The first moment of area $A^* = st_f$ about the z axis, $Q = st_f\,(c - t_f/2)$. The shearing stress at c–c is therefore

$$\tau_{xz} = \frac{VQ}{It_f} = \frac{V}{I}\left(c - \frac{t_f{}^2}{2}\right)s \tag{2.26}$$

The distribution of shear stress on the flange is *linear* with s, as Equation 2.26 shows. *Maximum shearing stress* takes place at $s = b/2$. We thus have

$$(\tau_{xz})_{\max} = \frac{V}{2I}\left(c - \frac{t_f}{2}\right)b \tag{2.27}$$

Limitations of the Shear Stress Formula

The shearing stress equations developed in this section are quite *accurate* when used for determining shear stresses in the *web*. But, the assumption that shear stress is constant across the width b of the *flanges* cannot be made. This is because at the flange-web junction there is a sudden cross-sectional change, where a stress concentration occurs (Figure 2.12a). Moreover, for $y_1 = h_1/2$ the stress distribution given by Equation 2.24 is fictitious. This is true due to observation that the inner surfaces (*ab* and *cd*) of the flanges must be *free of shearing stress* ($\tau_{xy} = 0$), since they are *load-free boundaries* of the beam, whereas across the junction *cd* the shear stress is $(\tau_{xy})_{\min}$.

Comments: The preceding contradiction cannot be resolved by the mechanics of materials or the technical theory. A correct solution must be determined by applying the theory of elasticity or FEA. As already noted, the web carries almost all of the shear force, the vertical shear stresses in the flanges are small. The distribution of shearing stresses τ_{xy} at the junction of the web and the flange is quite involved. In practice, beams are provided with fillets at these points to reduce concentrations of stress.

Shear Flow

The variation of the vertical and horizontal shear stresses, or *shear flow* over the entire cross-sectional area in a flanged beam is portrayed in Figure 2.13. Clearly, the *sense* of the horizontal shear stress τ_{xz} may readily be found from the sense of the vertical shearing stress τ_{xy} in the web. That is, the latter is the *same* as the sense of the shear force V. We note that, the vertical shear stresses τ_{xy} in the flanges are small compared with the horizontal shear stresses τ_{xz} and are not indicated in the figure.

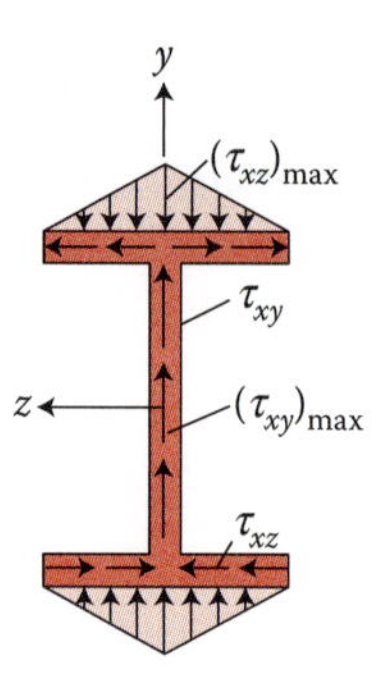

FIGURE 2.13
Shear flow in a flanged beam.

2.9 Deflection of Beams

Our analysis of *beam deflection* will be based on the fundamental assumptions of beam theory outlined in Section 2.7. Mainly deflections caused by bending will be considered. For slender members, contribution of shear to deflection is disregarded since for static bending problems, the shear deflection represents only a few percent of the total deflection. Two common procedures for determining elastic beam deflection are discussed in the following sections.

2.9.1 Method of Integration

The basic *differential equation* relating the deflection w to the internal bending moment M in a linearly elastic beam whose cross section is symmetrical about the plane (xz) of loading is [7]

$$\frac{d^2w}{dx^2} = -\frac{M}{EI} \tag{2.28}$$

where EI is termed the *flexural rigidity*. The sign convention for applied loading and the internal forces (Section 1.6) is shown in Figure 2.14. The angle of rotation or *slope* θ (in radians) of the deflection curve is given by

$$\theta = -\frac{dw}{dx} = -w' \tag{2.29}$$

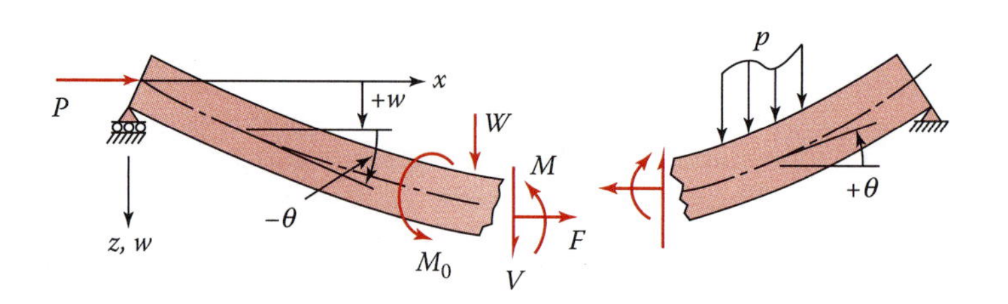

FIGURE 2.14
Deflection of a beam. All loads and internal-force resultants (F, V, M) have positive sense.

Positive and negative θ, like moments, follow the right-hand rule, as seen in Figure 2.14. Accordingly, the slope is *positive* when it is measured *counterclockwise* from the x axis.

Consider equilibrium of the vertical forces and moments acting on a beam element of length dx (Figure 2.9a) subjected to a distributed loading. It can be shown that internal forces V and M and the load intensity p can be related by

$$V = \frac{dM}{dx} = -\frac{d}{dx}\left(EI\frac{d^2w}{dx^2}\right) \tag{2.30a}$$

$$p = -\frac{dV}{dx} = \frac{d^2}{dx^2}\left(EI\frac{d^2w}{dx^2}\right) \tag{2.30b}$$

For beams with *constant* flexural rigidity EI, Equations 2.30 combine with Equation 2.28 to yield the following useful sequence of expressions:

$$M = -EI\frac{d^2w}{dx^2} = -EIw'' \tag{2.31}$$

$$V = -EI\frac{d^3w}{dx^3} = -EIw''' \tag{2.32}$$

$$p = EI\frac{d^4w}{dx^4} = EIw^{\mathrm{IV}} \tag{2.33}$$

The deflection w of a beam is obtained from any one of these equations by successive integrations. The *choice of equation* depends on mathematical and individual preference. Note that, in the case of *wide beams* ($b \gg h$), the preceding equations must be modified by replacing $EI/(1 - \nu^2)$ as shown in Section 3.5. Thus, a wide beam has approximately 10% *greater stiffness* than a narrow beam.

The constants of the integration are determined from known conditions on the ends of the beam, that is, the *boundary conditions*. The three common boundary conditions that may apply at the end ($x = a$) of a beam are shown in Figure 2.15. Other boundary conditions may be represented in an analogous manner. We observe from the figure that the *force* (static) variables M and V and the *geometric* (kinematic) variables w, θ are zero for frequently occurring cases. A more detailed discussion on boundary conditions is given in Section 3.9.

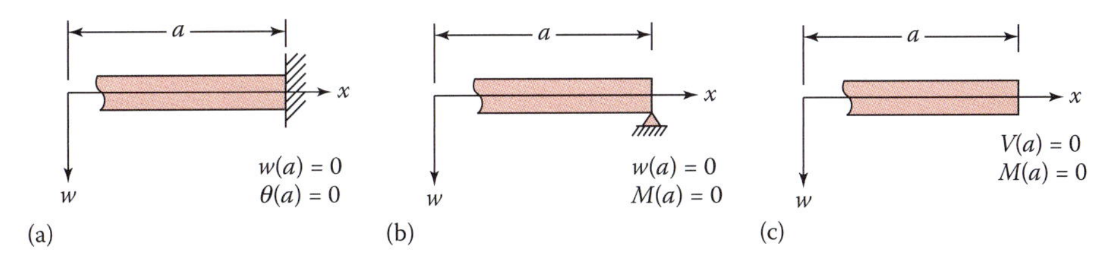

FIGURE 2.15
Boundary conditions: (a) Fixed end; (b) simply supported end; (c) free end.

EXAMPLE 2.5: Displacements of a Simply Supported Beam

A simple beam supports a concentrated load P acting at the midpoint as shown in Figure 2.16a. Determine: (a) the equations of the elastic curve and slope; (b) the deflection at C and end slopes at A and B.

Solution

The reactions are noted in Figure 2.16a.

a. The expression for the moment in segment AC is $M = Px/2$. Carrying this into Equation 2.31,

$$EIw'' = -\frac{Px}{2} \tag{a}$$

Integrating the foregoing,

$$EIw' = -\frac{Px^2}{4} + c_1 \qquad EIw = -\frac{Px^3}{12} + c_1x + c_2 \tag{b}$$

There is no displacement at support A. In addition, because of *symmetry*, the maximum deflection occurs at midspan, where the slope is zero (Figure 2.16b). These conditions are applied to Equations b,

$$w(0) = (0) = c_2 \quad \text{and} \quad \theta\left(\frac{L}{2}\right) = 0 = -\frac{PL^2}{16} + c_1$$

or

$$c_1 = \frac{PL^2}{16}$$

Substitution of the preceding values for c_1 and c_2 into Equations b provides the expressions for the deflection and slope:

$$\begin{aligned} w &= \frac{P}{48EI}(3L^2x - 4x^3) \\ \theta &= \frac{P}{16EI}(L^2 - 4x^2) \end{aligned} \tag{2.34}$$

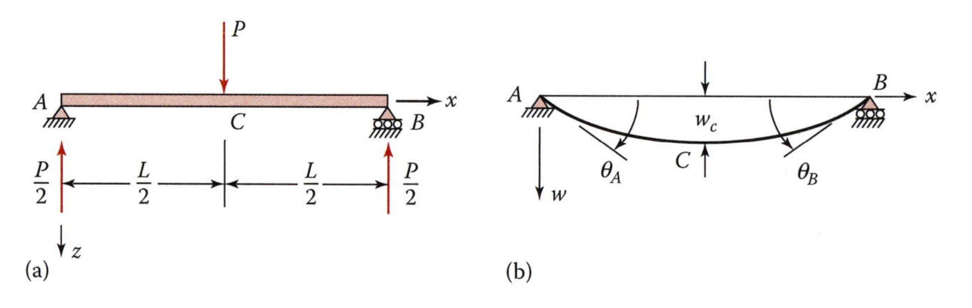

FIGURE 2.16
Simple beam with a concentrated load.

b. Making $x = L/2$ and $x = 0$ in each of Equations 2.34, we obtain the maximum deflection and end slopes, respectively:

$$w_C = w_{\max} = \frac{PL^3}{48EI}$$
$$\theta_A = -\theta_B = \frac{PL^2}{16EI} \tag{2.35}$$

EXAMPLE 2.6: Reactions of a Statically Indeterminate Beam

A propped cantilever beam AB with a triangular load of maximum intensity p_0 is shown in Figure 2.17a, with reactions also noted. Determine: (a) the equation of the elastic curve; (b) the reactions.

Solution

Reactions R_A, R_B, and M_B are indicated in Figure 2.17a. We note that these are *statically indeterminate* because there are only two equilibrium conditions ($\sum F_z = 0, \sum M_y = 0$).

a. The differential equation of the beam, with $p = xp_0/L$ (Figure 2.17b), is

$$EIw^{\text{IV}} = \frac{x}{L} p_0$$

Successive integrations give

$$\begin{aligned} EIw''' &= \frac{x^2}{2L} p_0 + c_1 \\ EIw'' &= \frac{x^3}{6L} p_0 + c_1 x + c_2 \\ EIw' &= \frac{x^4}{24L} p_0 + \frac{1}{2} c_1 x^2 + c_2 x + c_3 \\ EIw &= \frac{x^5}{120L} p_0 + \frac{1}{6} c_1 x^3 + \frac{1}{2} c_2 x^2 + c_3 x + c_4 \end{aligned} \tag{c}$$

where c_1, c_2, c_3, and c_4 are the constants of integration.

Boundary conditions are

$$\begin{aligned} w &= 0 \qquad w'' = 0 \qquad (x = 0) \\ w &= 0 \qquad w' = 0 \qquad (x = L) \end{aligned} \tag{d}$$

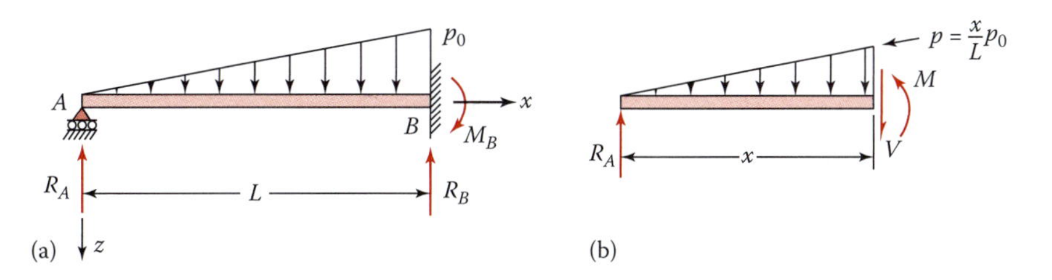

FIGURE 2.17
Propped-up cantilever beam under triangular load.

Applying these requirements to be last three of Equations c gives

$$c_4 = 0 \qquad c_2 = 0 \tag{e}$$

and

$$\frac{p_0 L^4}{120} + \frac{L^3 c_1}{6} + c_3 L = 0$$
$$\frac{p_0 L^3}{24} + \frac{L^2 c_1}{2} + c_3 = 0$$

Solving gives

$$c_1 = -\frac{1}{10} p_0 L \qquad c_3 = \frac{1}{120} p_0 L^3 \tag{f}$$

Introducing Equations e and f into the fourth of Equations c, we have

$$w = \frac{p_0}{120EIL}(x^5 - 2L^2x^3 + L^4x) \tag{2.36}$$

b. Equations c show that c_1 and c_2 are static boundary conditions, while c_3 and c_4 represent kinematic boundary conditions. Thus,

$$R_A = -c_1 = \frac{1}{10} p_0 L \tag{2.37a}$$

The remaining reactions are

$$R_B = \frac{2}{5} p_0 L \qquad M_B = \frac{1}{15} p_0 L^2 \tag{2.37b}$$

as obtained from the equations of equilibrium.

2.9.2 Method of Superposition

The deflections of a simply loaded beam are obtained by using the integration procedures and are readily available (Tables B.7 and B.8). Therefore, in practice, displacements of combined loading conditions are usually synthesized from the simpler loadings, employing the principle or *method of superposition*. The principle is valid when the displacements are linearly proportional to the applied loads. This is true only if Hooke's law holds for the material and when the deflections are small.

In the cases of *statically determinate* beams, the total deflection and the angle of rotation at any point of the beam can be found by the algebraic sum of the given deflection at that point due to each loading. For the situations where the beam is *statically indeterminate*, the *redundant* reactions are considered as unknown loads, and the corresponding supports are removed or modified accordingly. Superposition is used then, the diagrams

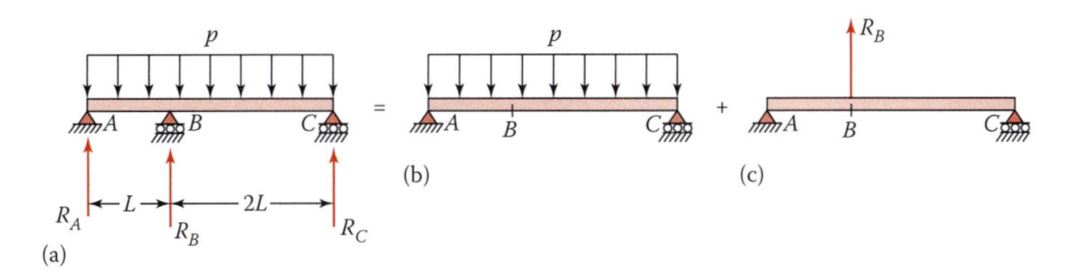

FIGURE 2.18
Two-span continuous beam.

are drawn, and equations are written for the displacements caused by the individual loads (both known and unknown). Finally, the redundant reactions are obtained by satisfying the geometric boundary conditions.

Consider, as an illustration of the foregoing procedure, the case of a beam indeterminate to the first degree (Figure 2.18a). Reaction R_B is designated as redundant and is treated as an unknown load by eliminating the support at B. Decomposition of the loads is depicted in Figure 2.18b and c. The deflections due to the uniform load p and the redundant R_B are (see cases 5 and 4 of Table B.7), respectively,

$$(w)_p = \frac{11pL^4}{12EI} \qquad (w)_R = -\frac{4R_BL^3}{9EI} \tag{a}$$

The equation of compatibility pertaining to the vertical deflection at point B of the original beam is

$$w_B = (w)_p - (w)_R = 0 \tag{b}$$

Inserting Equations a gives

$$\frac{11pL^4}{12EI} - \frac{4R_BL^3}{9EI} = 0$$

Solving,

$$R_B = \frac{33}{16}pL \tag{c}$$

The remaining reactions, as determined from the static equilibrium, are given by

$$R_A = \frac{1}{8}pL \qquad R_C = \frac{13}{16}pL \tag{d}$$

With the reactions known, deflection can be obtained by applying the method discussed in Section 2.9.1.

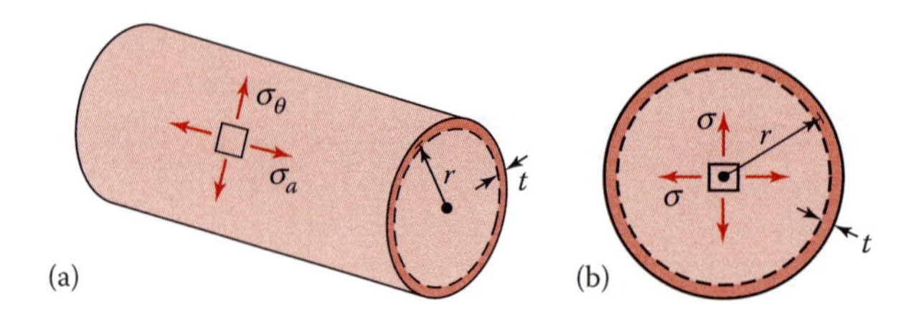

FIGURE 2.19
Thin-walled pressure vessels: (a) Cylinder; (b) sphere.

2.10 Thin-Walled Pressure Vessels

Pressure vessels are closed structures that contain liquids or gases under pressure. Detailed treatment of these members is given in Chapters 12 and 14. We shall here briefly discuss only two common types of thin-walled pressure vessels: cylindrical and spherical. A *thin-walled vessel* has thickness less than about one-tenth its inner radius. Hence, we can take $r_i \approx r_o \approx r$, where r_i, r_o, and r refer to inner, outer, and average radii of the vessel, respectively.

The stresses acting in the plane of the thin-walled vessels are called *membrane stresses.* Application of the equations of equilibrium to an appropriate portion of a thin-walled vessel suffices to determine these stresses. For a *thin-walled cylindrical vessel* with internal pressure p (Figure 2.19a) the *axial* or longitudinal and *tangential* or circumferential *stresses*, from Equations 12.10 are respectively,

$$\sigma_a = \frac{pr}{2t} \qquad \sigma_\theta = \frac{pr}{t} \tag{2.38a, b}$$

Note that $\sigma_\theta = 2\sigma_a$.

The circumferential stresses σ_θ act in the plane of the wall of a *thin-walled spherical vessel* and are identical in all directions under internal pressure p (Figure 2.19b). These are one-half the magnitude of the circumferential stresses of the cylinder, as given by Equation 12.4. The *stresses in spherical vessels* are thus

$$\sigma = \frac{pr}{2t} \tag{2.39}$$

Comments: It should be mentioned that the stress acting in the *radial* direction on the wall of a cylinder or sphere varies from $-p$ at the inner surface of the vessel to zero at the outer surface. For thin-walled vessels, *radial stress* is much smaller than the membrane stresses and is generally omitted. Therefore, the *state of stress in the wall* of a vessel is taken to be biaxial.

EXAMPLE 2.7: Pressure Capacity of Vessel

A cylindrical pressure vessel having a radius of 250 mm and a wall thickness of 5 mm is welded along a helical seam making an angle of 50° with respect to the axial axis (Figure 2.20a). Knowing that the allowable tensile stress acting perpendicular to the weld is 70 MPa, calculate the maximum value of internal pressure $p_{\max}$ and the corresponding shear stress in the weld.

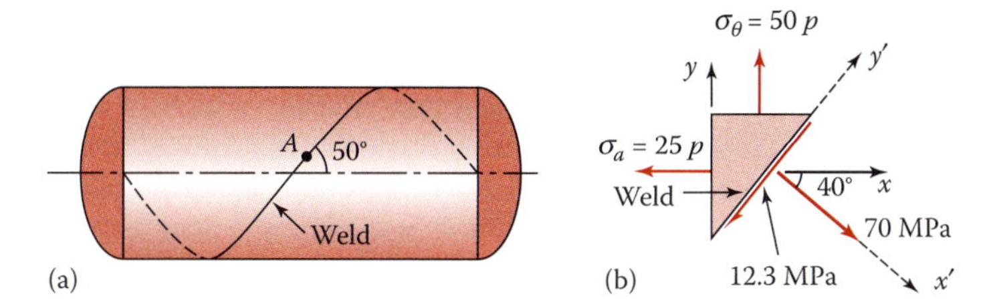

FIGURE 2.20
Cylindrical vessel with a helical weld.

Solution

The stresses acting on the wall element at a point A (Figure 2.20b) are

$$\sigma_a = \frac{pr}{2t} = \frac{p(250)}{2(5)} = 25p$$
$$\sigma_\theta = 2\sigma_a = 50p$$

The angle $\theta = -40°$ locates the x' axis, which is perpendicular to the plane of weld. The tensile stress acting normal to the weld, using Equation 1.11a, is

$$\sigma_{x'} = \frac{1}{2}(25p + 50p) + \frac{1}{2}(25p - 50p)\cos 2(-40°) = 38.9p \le 70\,\text{MPa}$$

from which

$$p_{\max} = 1.8\,\text{MPa}$$

Therefore, through the use of Equation 1.11b, we have

$$\tau_{x'y'} = -\frac{1}{2}(25p - 50p)\sin 2(-40°) = -12.3\,\text{MPa}$$

Comment: The complete answer is shown in Figure 2.20b.

2.11 Yield and Fracture Criteria

Unless we are content to overdesign structural members, it is necessary to predict the most probable modes of failure. We observe from the formulas derived that the failure is likely to originate at certain locations of the components. However, the formulas alone cannot predict at what loading failure will occur. The material strength and an appropriate *theory of failure* consistent with the type of material strength, whether brittle or ductile, must be considered. *Yielding* of members and initiation of *fracture* are the basis of *elastic design criteria*.

Here attention is directed toward two fracture theories (Sections 2.11.1 and 2.11.2) and two yielding theories (Sections 2.11.3 and 2.11.4), converting uniaxial to combined stress data. Discussions are limited to the case of *plane stress*. As before, the algebraically largest and smallest principal stresses are denoted by σ_1 and σ_2, respectively. The yield tension, yield shear, ultimate tension, and ultimate compression strengths of material are designated by σ_{yp}, τ_{yp}, σ_u, and σ_{uc}, respectively. Note that the actual failure mechanism in a component may be quite complicated; each failure theory is only an attempt to model that mechanism for a given class of materials. In each case, a factor of safety is employed.

2.11.1 Maximum Principal Stress Theory

This criterion predicts that a brittle material fails by fracturing when the maximum (or minimum) principal stress exceeds the ultimate strength in tension. That is, at the beginning of fracture

$$|\sigma_1| = \frac{\sigma_u}{n} \quad \text{or} \quad |\sigma_2| = \frac{\sigma_u}{n} \tag{2.40}$$

for safety factor n. Figure 2.21 is a plot of Equations 2.40 for $n = 1$. The boundary of the square indicates the onset of failure by fracture. The area within the boundary is thus a region of no failure.

2.11.2 Coulomb–Mohr Theory

This criterion may be used to predict the effect of a given state of plane stress on a brittle material having different properties in tension and compression. The Mohr's circles for the uniaxial tension and compression tests are used to define failure by the Coulomb–Mohr theory (Figure 2.22a).

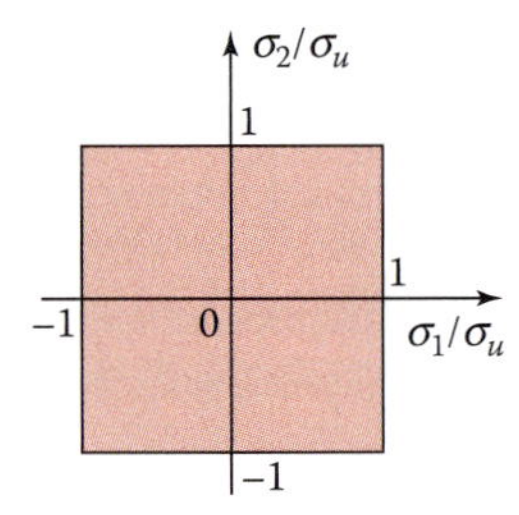

FIGURE 2.21
Fracture criterion based on maximum principal stress.

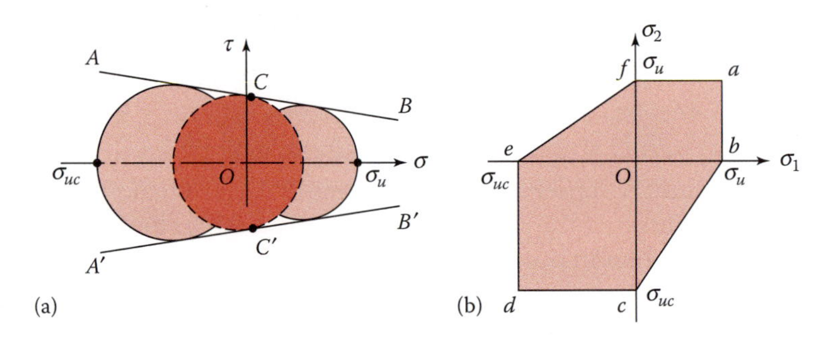

FIGURE 2.22
(a) Straight-line Mohr's envelope; (b) Coulomb–Mohr fracture criterion.

The points of contact of the *straight-line envelopes* (AB and $A'B'$) with the stress circles define the state of stress at a fracture. For example, if such a point is C (or C'), the stresses and planes on which they act can be obtained by using the established procedure for Mohr's circle of stress. When σ_1 and σ_2 have opposite signs, that is, one tensile, the other compressive, it can be verified that [2,8]

$$\frac{\sigma_1}{\sigma_u} - \frac{\sigma_2}{|\sigma_{uc}|} = \frac{1}{n} \tag{2.41a}$$

In the case of *biaxial tension*, we have

$$\sigma_1 = \frac{\sigma_u}{n} \quad \text{or} \quad \sigma_2 = \frac{\sigma_{uc}}{n} \tag{2.41b}$$

Similarly, for *biaxial compression*,

$$\sigma_2 = -\frac{\sigma_{uc}}{n} \quad \text{or} \quad \sigma_1 = -\frac{\sigma_{uc}}{n} \tag{2.41c}$$

The preceding expressions are illustrated in Figure 2.22b, for $n = 1$. Lines *ab* and *af* represent Equation 2.41b, and lines *dc* and *de* represent Equation 2.41c. The boundary *bc* is found by using Equation 2.41a. Line *ef* completes the hexagon. Points within the shaded area represent states of nonfailure, in accordance with the Coulomb–Mohr theory. The boundary of the figure shows the onset of failure due to fracture.

2.11.3 Maximum Shear Stress Theory

This criterion states that yielding will begin when the maximum shear stress in the material is equal to the maximum shear stress at yielding in a simple tension test:

$$\frac{1}{2}|\sigma_1 - \sigma_2| = \tau_{yp} = \frac{1}{2}\sigma_{yp}$$

Therefore, the maximum shear stress theory,

$$|\sigma_1 - \sigma_2| = \frac{\sigma_{yp}}{n} \tag{2.42a}$$

for safety factor n. We observe that if σ_1 and σ_2 are of the *opposite sign*, the maximum shear stress is $(\sigma_1 - \sigma_2)/2$. Hence, the condition of the inelastic action is given by Equation 2.42a. When σ_1 and σ_2 carry the *same sign*, we then obtain the yielding conditions

$$|\sigma_1| = \frac{\sigma_{yp}}{n} \quad \text{and} \quad |\sigma_2| = \frac{\sigma_{yp}}{n} \tag{2.42b}$$

for $|\sigma_1| > |\sigma_2|$ and $|\sigma_1| > |\sigma_3|$, respectively.

For $n = 1$, Equation 2.42 is represented graphically in Figure 2.23. It is noted that Equation 2.42a applies to the second and fourth quadrants, while Equation 2.42b applies to the first and third quadrants. The boundary of the hexagon thus marks the onset of yielding, with points outside the shaded region showing a yielded state.

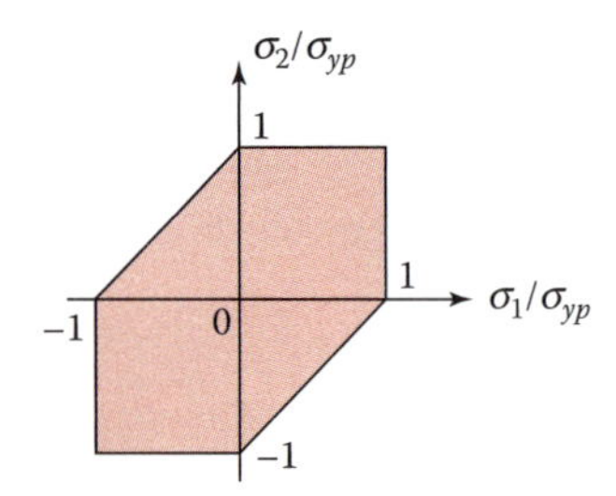

FIGURE 2.23
Yield criterion based on maximum shear stress.

2.11.4 Maximum Distortion Energy Theory

This criterion states that failure by yielding occurs when, at any location in the member, the distortion energy per unit volume in a state of combined stress will be equal to that associated with yielding in a simple tension test. Thus,

$$\sigma_1^2 - \sigma_1\sigma_2 + \sigma_2^2 = \left(\frac{\sigma_{yp}}{n}\right)^2 \tag{2.43}$$

for safety factor n.

The foregoing expression defines the ellipse shown in Figure 2.24. As in the case of the maximum shear stress criterion, points within the shaded area represent states of nonyielding. The boundary of the ellipse indicates the onset of yielding, with the points outside the shaded areas representing a yielded state.

The maximum energy of distortion theory agrees best with experimental data for ductile materials, and its use in design practice is increasing. But the maximum shear stress theory has been employed in some design codes because it is simple to apply and offers a conservative result. Good agreement between Coulomb–Mohr theory and the experiment has been observed for brittle materials in all ranges of stresses. This conclusion is also valid for the maximum principal stress theory, provided that a tensile principal stress exists.

2.11.5 A Typical Case of Combined Loadings

Consider a structural member subject to combined bending and torsion or bending and transverse shear, so that $\sigma_y = \sigma_z = \tau_{yz} = \tau_{xz} = 0$. The principal stresses are now given by

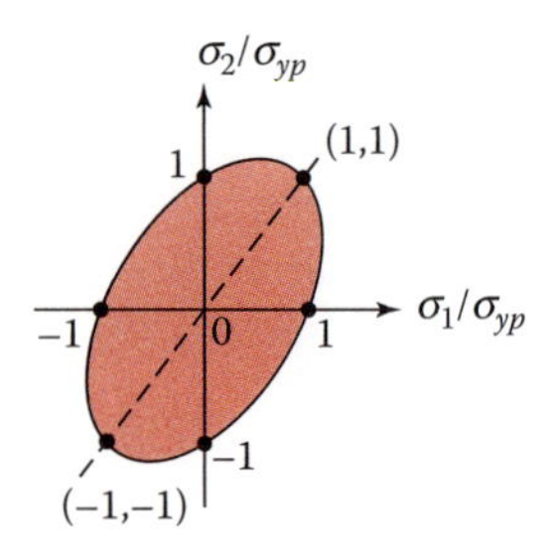

FIGURE 2.24
Yield criterion based on maximum energy of distortion.

$$\sigma_{1,2} = \frac{\sigma_x}{2} \pm \tau_{max} \tag{2.44}$$

where

$$\tau_{max} = \left[\left(\frac{\sigma_x}{2} \right)^2 + \tau_{xy}^2 \right]^{1/2} \tag{2.45}$$

Substituting this into Equation 2.43, we have

$$\frac{\sigma_{yp}}{n} = \left(\sigma_x^2 + 3\tau_{xy}^2 \right)^{1/2} \tag{2.46}$$

on the basis of the *maximum energy of distortion theory*. Likewise, introducing Equation 2.44 into Equation 2.42a gives

$$\frac{\sigma_{yp}}{n} = \left(\sigma_x^2 + 4\tau_{xy}^2 \right)^{1/2} \tag{2.47}$$

based on the maximum shear stress theory.

EXAMPLE 2.8: Failure by Rupture of a Structural Component

The stresses shown in Figure 2.25 are expected in a cast-iron structural component for which $\sigma_u = 170$ MPa and $\sigma_{uc} = 340$ MPa. Determine whether rupture of casting will occur using the following criteria for $n = 1$: (a) the maximum principal stress; (b) Coulomb–Mohr.

Solution

Applying Equation 1.13 with $\sigma_x = 0$, $\sigma_y = -200$ MPa, and $\tau_{xy} = -120$ MPa, we have

$$\sigma_{1,2} = -\frac{200}{2} \pm \sqrt{\left(\frac{200}{2} \right)^2 + (-120)^2} = -100 \pm 156$$

from which

$$\sigma_1 = 56 \text{ MPa} \qquad \sigma_2 = -256 \text{ MPa} \tag{a}$$

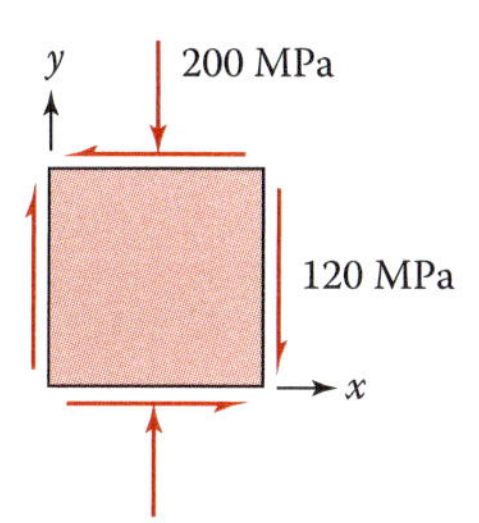

FIGURE 2.25
State of stress at a critical point.

a. *Maximum principal stress theory*: Inasmuch as 56 < 170, the first expression of Equations 2.40 predicts that rupture will *not occur.* Because $\sigma_u \neq \sigma_{uc}$, the second expression of Equations 2.40 does not apply.
b. *Coulomb–Mohr theory*: Through the use of Equations a and 2.41a, we have

$$\frac{56}{170} - \frac{-256}{340} = 10.8$$

As the preceding number exceeds unity, this theory predicts that rupture *will occur.* We note that the Coulomb–Mohr criterion is most reliable when compressive stresses are dominant and when $\sigma_{uc} \gg \sigma_u$, as is the case in this example.

EXAMPLE 2.9: Load-Carrying Capacity of an Axially Laded Cylinder

A circular cylinder, fabricated of a ductile material having tensile yield strength σ_{yp}, carries to an axial tensile force P. Calculate the torque T that can be applied simultaneously to the cylinder on the basis of the maximum energy of distortion criteria of failure. Take a safety factor 1.4, $P = 300$ kN, diameter $d = 60$ mm, and $\sigma_{yp} = 280$ MPa.

Solution

The critical stresses occur on the elements at the surface of the shaft. According to the maximum energy of distortion theory, by Equation 2.46,

$$\tau_{xy} = \frac{1}{\sqrt{3}}\left[\left(\frac{\sigma_{yp}}{n}\right)^2 - \sigma_x^2\right]^{1/2} \tag{b}$$

Here

$$\sigma_x = \frac{P}{A} = \frac{4P}{\pi d^2} \qquad \tau_{xy} = \frac{Tr}{J} = \frac{16T}{\pi d^3} \tag{c}$$

With substitution of the given data, Equation b results in

$$\tau_{xy} = \frac{1}{\sqrt{3}}\left\{\left(\frac{280\times 10^6}{1.4}\right)^2 - \left[\frac{4\times 300\times 10^3}{\pi(0.06)^2}\right]^2\right\}^{1/2} = 97.88 \text{ MPa}$$

We thus have, from the second expression of Equations c,

$$T = \frac{\pi(0.06)^3(97.88\times 10^6)}{16} = 4.151 \text{ kN}\cdot\text{m}$$

Comment: The foregoing represents the maximum torque that can be applied without producing permanent deformation.

EXAMPLE 2.10: Failure by Yielding of Steel Plate

The state of stress shown in Figure 2.26 takes place at a critical point of a plate having $\sigma_{yp} = 240$ MPa. Calculate the factor of safety n with respect to yielding, employing the

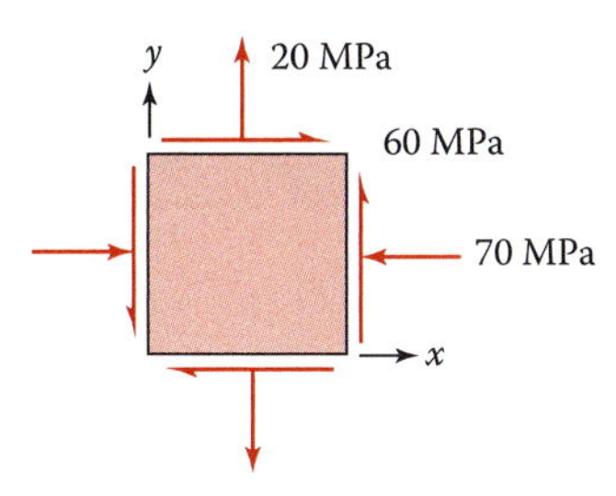

FIGURE 2.26
Element in plane stress.

following failure criteria: (a) the maximum shear stress; (b) the maximum energy of distortion.

Solution

Through the use of Equation 1.13 with $\sigma_x = -70$ MPa, $\sigma_y = 20$ MPa, and $\tau_{xy} = 60$ MPa

$$\sigma_{1,2} = \frac{-70+20}{2} \pm \sqrt{\left(\frac{-70-20}{2}\right)^2 + 60^2} = -25 \pm 75$$

or

$$\sigma_1 = 50 \text{ MPa} \qquad \sigma_2 = -100 \text{ MPa}$$

a. *Maximum shear stress theory*: From Equation 2.42a,

$$|50+100| = \frac{240}{n}$$

The foregoing gives $n = 1.6$.

b. *Maximum energy of distortion*: By Equation 2.43,

$$(50)^2 - (50)(-100) + (-100)^2 = \left(\frac{240}{n}\right)^2$$

Solving, we have $n = 1.81$.

2.12 Strain Energy

As an alternative to the equilibrium method outlined in Section 1.2, analysis of displacement and force can be accomplished by using the energy methods. The latter are based on the concept of *strain energy*. Energy techniques apply effectively to the cases involving members of variable cross sections and to the problems dealing with elastic stability. They also can offer concise and relatively simple approaches for computation displacement of slender members under combined loading.

The area under the stress–strain diagram is the energy absorbed per unit volume, or *strain energy density*, designated U_0, of a tensile specimen. For a linearly elastic

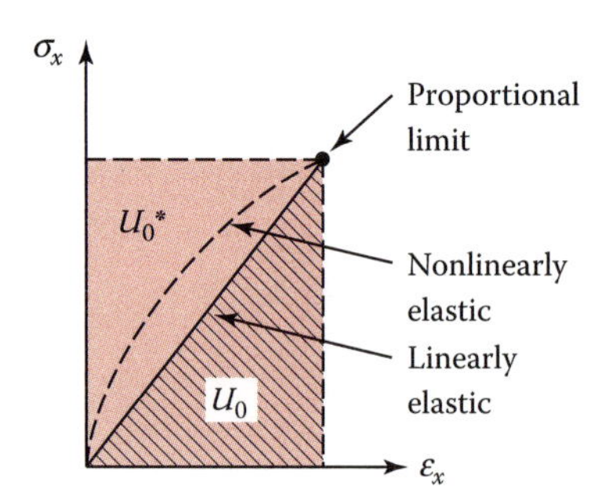

FIGURE 2.27
Stress–strain curves.

material (Figure 2.27) from origin up to the proportional limit, by replacing ε_x by σ_x/E, we have

$$U_0 = \frac{1}{2}\sigma_x\varepsilon_x = \frac{1}{2E}\sigma_x^2 \tag{2.48}$$

The quantity so found is called the *modulus of resilience*. It represents the ability of the material to absorb energy within the elastic range. The area *above* the stress–strain curve is called the *complementary energy density*, denoted by U_0^*, as shown in the figure.

By reasoning analogous to that preceding, the quantity termed *toughness* refers to the ability of the material to absorb energy without fracturing. The *modulus of toughness* is equal to the area below the entire stress–strain curve. Most of the energy is *dissipated* in permanently deforming the material and is lost in heat.

In SI units, the strain energy density is measured in joules per cubic meter [(J/m³)]. This is equal to $\mathrm{N\cdot m/m^3}$, or pascals. For example, steel with a proportional limit of 210 MPa and modulus of elasticity of $E = 210$ GPa has a modulus of resilience of $\sigma_x^2/2E = 105\,\mathrm{kPa}$.

In a like manner, strain energy density for *shear stress* is given as follows:

$$U_0 = \frac{1}{2}\tau_{xy}\gamma_{xy} = \frac{1}{2G}\tau_{xy}^2 \tag{2.49}$$

For a body under a *general state of stress*, the total strain energy density is the sum of the expressions identical to the preceding equations. That is,

$$U_0 = \frac{1}{2}(\sigma_x\varepsilon_x + \sigma_y\varepsilon_y + \sigma_z\varepsilon_z + \tau_{xy}\gamma_{xy} + \tau_{yz}\gamma_{yz} + \tau_{xz}\gamma_{xz}) \tag{2.50}$$

Introducing the generalized Hooke's law into this equation, we obtain

$$\begin{aligned} U_0 &= \frac{1}{2E}\left[\sigma_x^2 + \sigma_y^2 + \sigma_z^2 - 2\nu(\sigma_x\sigma_y + \sigma_y\sigma_z + \sigma_x\sigma_z)\right] \\ &\quad + \frac{1}{2G}\left(\tau_{xy}^2 + \tau_{yz}^2 + \tau_{xz}^2\right) \end{aligned} \tag{2.51}$$

If the *principal axes* are used as coordinate axes, the shearing stresses vanish, and this expression reduces to

$$U_0 = \frac{1}{2E}\left[\sigma_1^2 + \sigma_2^2 + \sigma_3^2 - 2\nu(\sigma_1\sigma_2 + \sigma_2\sigma_3 + \sigma_1\sigma_3)\right] \tag{2.52}$$

Here, σ_1, σ_2, and σ_3 are the principal stresses.

The elastic *strain energy* stored within an entire body can be found by integrating the strain energy density over the volume. Thus,

$$U = \iiint U_0\, dx\, dy\, dz = \int U_0\, dV \tag{2.53}$$

The form of equation of the strain energy is convenient in applications. We note that the strain energy is a *nonlinear* (quadratic) function of load or deformation. Hence, the principle of superposition does *not* apply to the strain energy.

Expression for the *strain energy in a straight or curved member* subjected to an axial force P, bending moment M, and torque T can be obtained on substituting Equations 2.1, 2.15, and 2.11 into Equation 2.53. In so doing, we have [7]

$$U = \int \frac{P^2 dx}{AE} + \int \frac{M^2 dx}{EI} + \int \frac{T^2 dx}{GJ} \tag{2.54}$$

where the integrations proceed over the length L of the bar. Recall that the term given by the last integral is valid only for a circular cross-sectional area.

2.13 Castigliano's Theorem

There are two theorems, proposed in 1879 by A. Castigliano (1847–1884). The first relies on a virtual (imaginary) variation in deformation. The second, which we will refer to as *Castigliano's theorem* and which we deal with in this text, is concerned with the finite deformation experienced by a load-carrying member. Castigliano's theorem is widely used in the analysis of structural displacements and forces. It applies with ease to a variety of statically determinate as well as indeterminate structures composed of linearly elastic materials.

We observe from Equation 2.51 that $\partial U_0/\partial \tau_{ij} = \varepsilon_{ij}$, where τ_{ij} and ε_{ij} designate stress and strain components, respectively. Derivatives of this type are useful in connection with energy methods. An alternative form of the foregoing is Castigliano's theorem [9,10]:

$$\delta_i = \frac{\partial U}{\partial P_i} \tag{2.55}$$

This states that for a linear structure, *the partial derivative of the strain energy with respect to an applied force is equal to the component of displacement at the point of application and in the direction of that force.*

Castigliano's theorem can be similarly demonstrated to be valid for applied moments M_i (or torques T_i) and the resulting slope θ_i (or angle of twist) of the structure. Therefore,

$$\theta_i = \frac{\partial U}{\partial M_i} \tag{2.56}$$

In employing Castigliano's theorem, the strain energy must be expressed in terms of the external forces or moments.

For a *beam*, we have the strain energy expressed by $U = \int M^2\,dx/2EI$. To ascertain the deflection w corresponding to load P_i, it is often much simpler to differentiate under the integral sign:

$$w_i = \frac{\partial U}{\partial P_i} = \int_0^L \frac{M}{EI}\frac{\partial M}{\partial P_i}dx \tag{2.57}$$

In an analogous manner, the slope may be written as

$$\theta_i = \frac{\partial U}{\partial M_i} = \int \frac{M}{EI}\frac{\partial M}{\partial M_i}dx \tag{2.58}$$

The total strain energy in a *straight or curved member* subjected to an axial force P, bending moment M, and torque T is given by Equation 2.54. Thus, by applying Equation 2.55, the displacement at any point in the member is obtained in the following convenient form:

$$\delta_i = \frac{1}{AE}\int P\frac{\partial P}{\partial P_i}dx + \frac{1}{EI}\int M\frac{\partial M}{\partial P_i}dx + \frac{1}{GJ}\int T\frac{\partial T}{\partial P_i}dx \tag{2.59}$$

Similarly, an expression may be written for the angle of rotation.

We note that, to determine the displacement at a point where *no* corresponding *load acts*, the problem is dealt with as follows. Place a *fictitious* force Q (or couple C) at the point in question in the direction of the desired displacement δ (or θ). Then apply Castigliano's theorem; the fictitious force Q (or C) is set to zero in order to obtain the desired results.

EXAMPLE 2.11: Displacements by the Energy Method

The cantilever beam AB supports a uniformly distributed load of intensity p (Figure 2.28a). Determine at the free end A: (a) the deflection; (b) the angle of rotation.

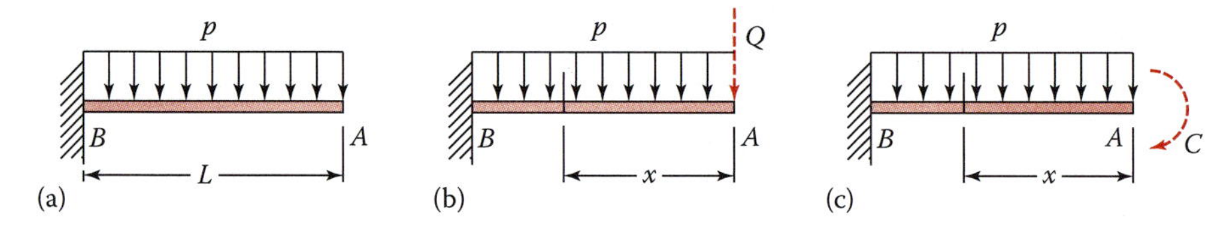

FIGURE 2.28
Cantilever beam under uniform load.

Solution

a. As the deflection is sought, a fictitious force Q is applied at A (Figure 2.28b). The bending moment M at a distance x from A is

$$M = -Qx - \frac{1}{2}px^2$$

and hence $\partial M/\partial Q = -x$. Substituting the preceding expressions into Equation 2.57 and setting $Q = 0$ gives

$$w_A = \frac{1}{EI}\int_0^L \left(-\frac{1}{2}px^2\right)(-x)dx = \frac{pL^4}{8EI} \tag{2.60}$$

The positive sign of w_A indicates that the deflection has the same sense, that is, *downward*, as Q.

b. We now apply a fictitious couple C at A (Figure 2.28c) and express the moment as

$$M = -C - \frac{1}{2}px^2$$

from which $\partial M/\partial C = -1$. Carrying these into Equation 2.58 and making $C = 0$, we have

$$\theta_A = \frac{1}{EI}\int_0^L \left(-\frac{1}{2}px^2\right)(-1)dx = \frac{pL^3}{6EI} \tag{2.61}$$

Since the fictitious couple was clockwise, the positive sign means that the angle θ is also clockwise.

2.13.1 Statically Indeterminate Structures

Castigliano's theorem may be applied as a supplement to the equations of statics in solving support reactions of a statically indeterminate structure. In the case of a structure indeterminate to the nth degree, we select n reactions as the redundants (or unknown loads) by removing the corresponding supports. All external forces, including both loads and redundant reactions, must produce displacements compatible with the original supports. We first express the strain energy U due to the combined action of R_n and the given loads. Equation 2.55 may then be applied at the removed supports and equated to the given displacements. If these displacements are zero, we have

$$\frac{\partial U}{\partial R_1} = 0, \ldots, \quad \frac{\partial U}{\partial R_n} = 0 \tag{2.62}$$

By solving these equations simultaneously, the magnitudes of the redundant reactions are determined. The remaining reactions are obtained from the equations of statics. We note that Equation 2.52, Castigliano's theorem, is also known as the *principle of least work* in some literature.

The analytical technique is illustrated in the solution of the following sample problem.

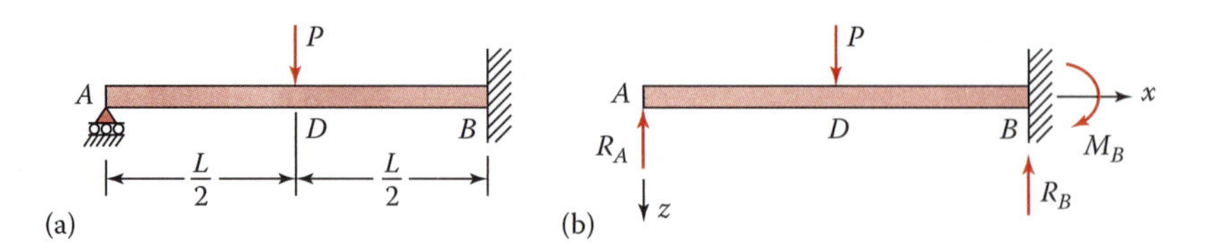

FIGURE 2.29
A propped-up cantilever carries concentrated load.

EXAMPLE 2.12: Reactions by the Energy Method

Determine the support reactions for the beam of Figure 2.29a.

Solution

The FBD of Figure 2.29b shows that the problem is statically indeterminate to the first degree. We select R_A as redundant, and hence the expressions for the moment are

$$M_{AD} = R_A x \quad M_{DB} = R_A x - P\left(x - \frac{L}{2}\right)$$

Note that the *remaining unknowns*, R_B and M_B, *do not appear* in the preceding equations.

The deflection w_A at A must be zero. Thus, Equation 2.62, together with Equation 2.57, gives

$$w_A = \frac{1}{EI}\left\{\int_0^{L/2} (R_A x)x dx + \int_{L/2}^{L}\left[R_A x - P\left(x - \frac{1}{2}\right)\right]x dx\right\} = 0$$

The preceding, after integration, results in

$$R_A = \frac{5}{16}P \tag{2.63a}$$

The other two reactions are

$$R_B = \frac{11}{16}P \quad M_B = \frac{3}{16}PL \tag{2.63b}$$

as determined from the equations of equilibrium.

Problems

Sections 2.1 through 2.8

2.1 A vertical bar *CBD* is supported by a hinge at *C* and by a member *AB* of cross-sectional area 500 mm, as depicted in Figure P2.1. Calculate the maximum value of the load *P* if maximum normal stress is limited to 50 MPa in the member *AB*.

2.2 Figure P2.2 portrays the piston, connecting rod, and crank of an engine system. A force $P = 10$ kN acts as indicated in the figure. *Find*: (a) the torque *T* required to hold the system in equilibrium; (b) the normal stress in the rod *AB* if its cross-sectional area is 500 mm^2.

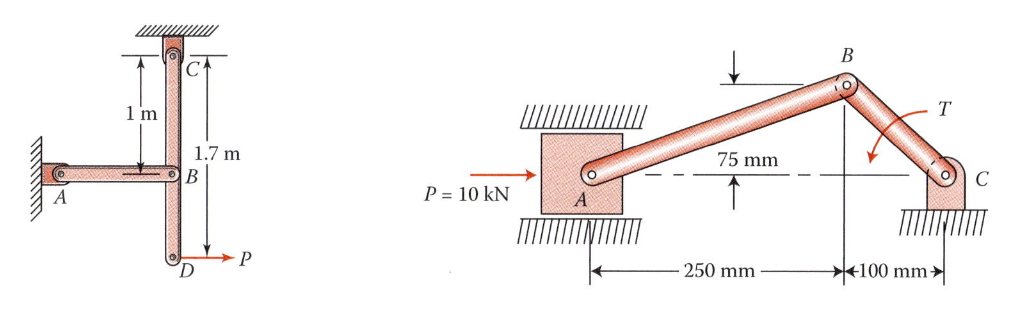

FIGURE P2.1

FIGURE P2.2

2.3 Calculate the required cross-sectional area of rod *AC* of the monoplane wing of Figure P2.3 for an allowable stress of 80 MPa. *Assumption*: The air load is linearly changing along the span of the wing as shown in the figure.

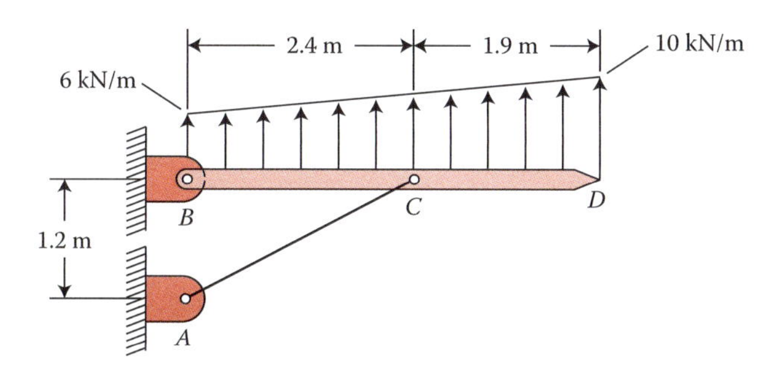

FIGURE P2.3

2.4 A tapered plate of uniform thickness τ is subjected to an axial load *P*, as shown in Figure P2.4. Determine the elongation δ in terms of *P*, *L*, *a*, *t*, and *E*, as required.

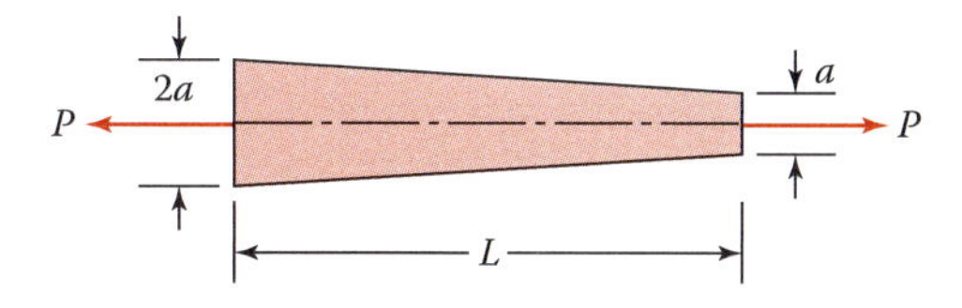

FIGURE P2.4

2.5 A pipe of 150-mm outside diameter and 10-mm thickness is fabricated with a helical weld making an angle of 50° with the longitudinal axis (Figure P2.5). If the allowable tensile stress acting perpendicular to the weld is 200 MPa, calculate the maximum torque *T* that may be applied to the pipe.

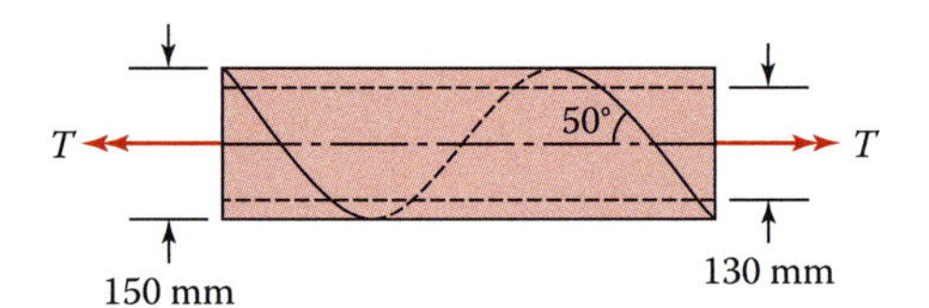

FIGURE P2.5

2.6 A brass tube ($G_b = 42$ GPa) has an outer diameter of 60 mm and an inner diameter of 40 mm. The end B of the tube is attached to a fixed support, and a torque of $T = 1$ kN·m is applied at end A, as shown in Figure P2.6. Determine the depth b to which it has to be filled with a steel plug ($G_S = 80$ GPa) so that the rotation of end A is not to exceed 0.01 rad. Assume that the tube and the plug are firmly bonded together.

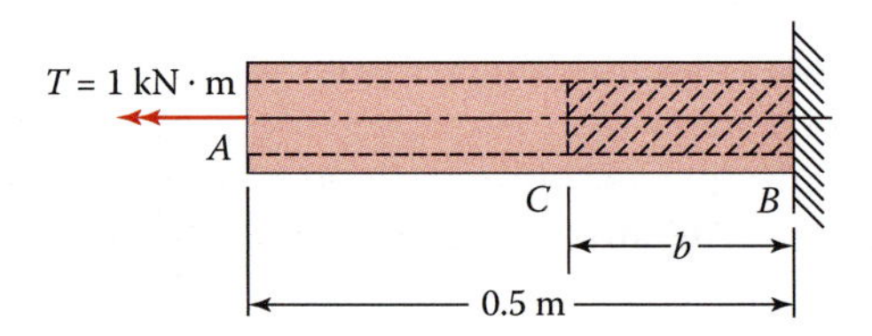

FIGURE P2.6

2.7 The strain in the 45° direction at a point A on the surface of a hollow circular copper tube ($E = 120$ GPa, $\nu = 0.33$) is 1900 μm when the torque $T = 150$ N·m (Figure P2.7). Calculate the inside diameter d.

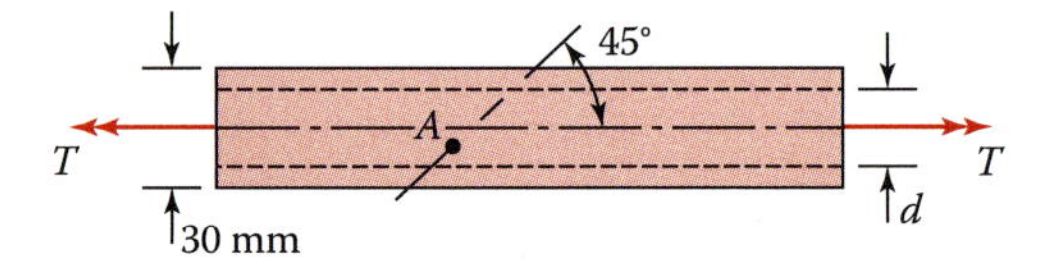

FIGURE P2.7

2.8 A simply supported wooden beam, with $\sigma_{all} = 9$ MPa and $\tau_{all} = 1.5$ MPa, carries a uniform load of intensity p (Figure P2.8). *Given*: $b = 50$ mm and $h = 150$ mm. *Determine*: (a) the maximum permissible length L; (*b*) the maximum permissible distributed load p.

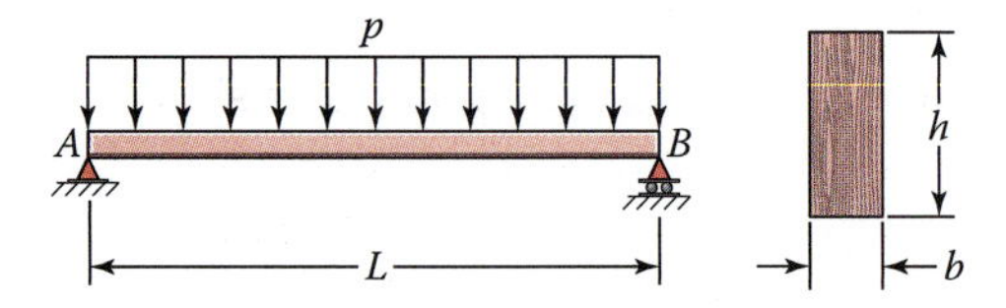

FIGURE P2.8

2.9 What is the ratio of the maximum shearing stress to the largest bending stress for the simple rectangular beam in Figure P2.8 under a uniformly distributed load of intensity p?

2.10 A simply supported beam of circular cross section carrying a concentrated load P at its midspan as shown in Figure P2.10. *Find*: The ratio of the largest shear stress to the maximum bending stress.

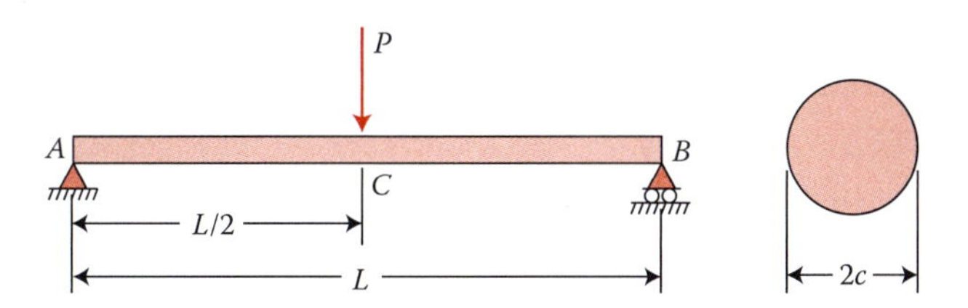

FIGURE P2.10

2.11 A hollow aluminum box beam with rectangular cross section depicted in Figure P2.11 carries a shear force V. *Calculate*: (a) the maximum shear stress τ_{max} in the webs of the beam; (b) the minimum shear stress τ_{min} in the webs of the beam. *Given*: $b = 120$ mm, $h = 160$ mm, $t = 12$ mm, $t_1 = 18$ mm, $V = 250$ kN.

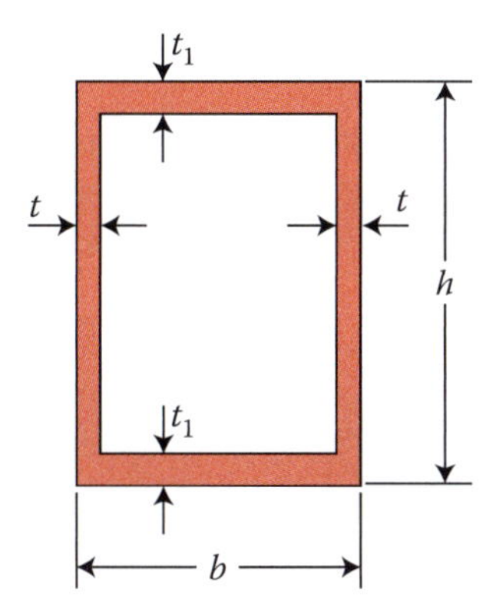

FIGURE P2.11

Section 2.9

2.12 Redo Problem 2.11 for the case in which a by a hollow square box beam of wall thickness t is acted upon by a shear force V. *Given*: $a = 200$ mm, $t = 15$ mm, $V = 120$ kN.

2.13 When the allowable normal stress at a point D in simply supported beam is 15 MPa (Figure P2.13), what is the maximum allowable value of load P?

2.14 A cantilever beam AB supports a parabolically varying load of intensity $p = p_0(L^2 - x^2)/L^2$, as depicted in Figure P2.14. Employing the fourth-order differential equation for the deflection, determine the expression for the elastic curve w. What are the deflection w_B and slope θ_B at the free end?

2.15 An overhanging beam supports a uniformly distributed load of intensity p (Figure P2.15). Using the method of superposition, determine the deflection at the free end.

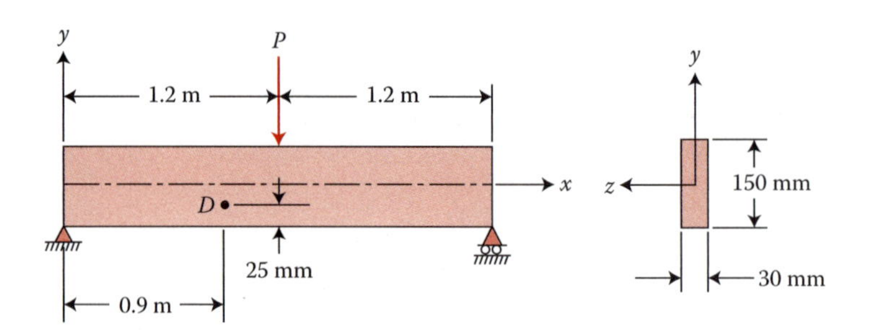

FIGURE P2.13

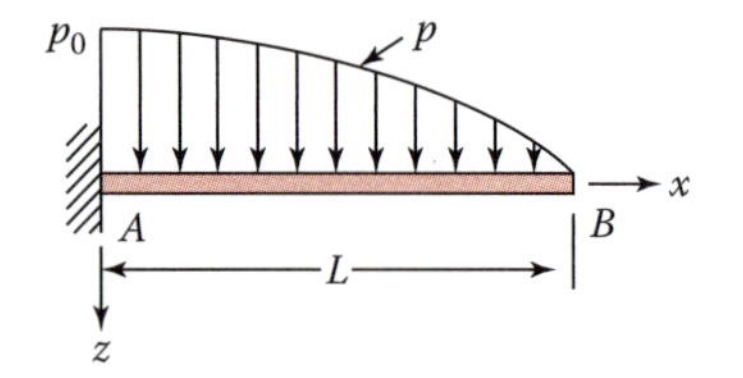

FIGURE P2.14

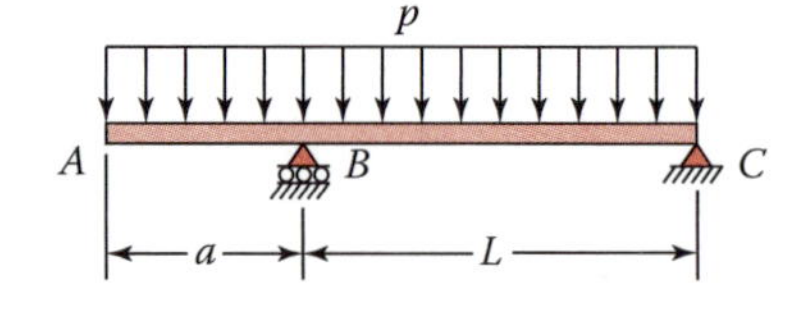

FIGURE P2.15

2.16 A simply supported beam is subjected to a distributed load of intensity $p = p_0 \sin \pi x/L$, as shown in Figure P2.16. Apply the fourth-order differential equation for the deflection to derive the expression for the elastic curve w. Determine the slopes at the ends.

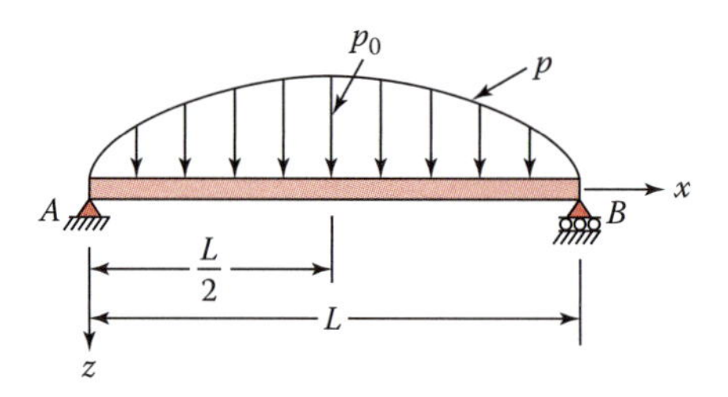

FIGURE P2.16

2.17 For the beam and loading shown in Figure P2.17, apply the integration method to determine: (a) the reactions; (b) the equation of the elastic curve w; (c) the maximum deflection.

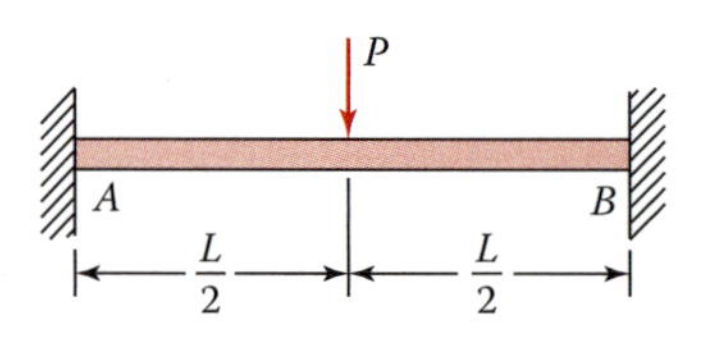

FIGURE P2.17

Sections 2.10 and 2.11

2.18 A cylindrical vessel of thickness $t = 5$ mm and diameter $d = 400$ mm is made of steel with $\sigma_{yp} = 250$ MPa. Determine the internal pressure p that can be carried by the vessel based on a factor of safety of $n = 2$ on yielding.

2.19 A cylindrical tank of a diameter $d = 1$ m and wall thickness $t = 5$ mm is simply supported as shown in Figure P2.19. The tank and its contents weigh 1.2 kN/m of length, and the contents exert a uniform internal pressure of $p = 40$ kPa on the tank. Calculate, at points A and B on the surface of the tank, the maximum shear stress and its orientation.

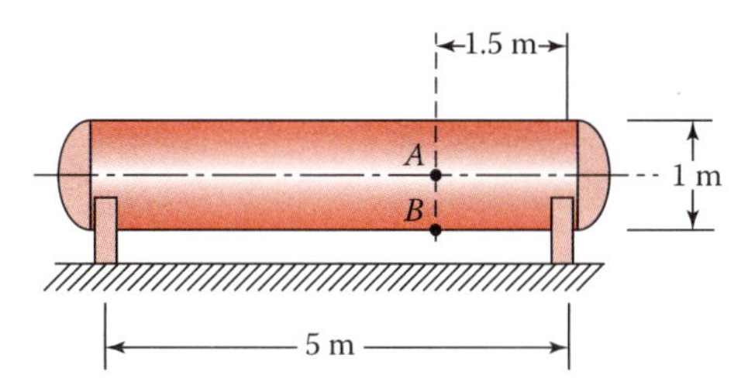

FIGURE P2.19

2.20 A closed-ended cylinder 250 mm in internal radius and 10 mm thick is under an internal pressure of 4 MPa and axial tension of 500 kN. Calculate for the shell wall: (a) the maximum tensile stress; (b) the maximum shearing stresses and orientation of their planes.

2.21 A thin-walled cylindrical pressure vessel ($\sigma_u = 240$ MPa, $\sigma_{uc} = 600$ MPa) with 200-mm diameter and 8-mm thickness is subjected to an internal pressure $p = 4$ MPa, a torque of 30 kN·m and an axial tension $P = 50$ kN. Apply the following failure criteria to evaluate the ability of the vessel to resist failure by fracture: (a) maximum principal stress; (b) Coulomb–Mohr.

2.22 The state of stress shown in Figure P2.22 occurs at a critical point in a cast-iron structure ($\sigma_u = 150$ MPa, $\sigma_{uc} = 600$ MPa). Determine the factor of safety n with respect to fracture, applying the following criteria: (a) maximum principal stress; (b) Coulomb–Mohr.

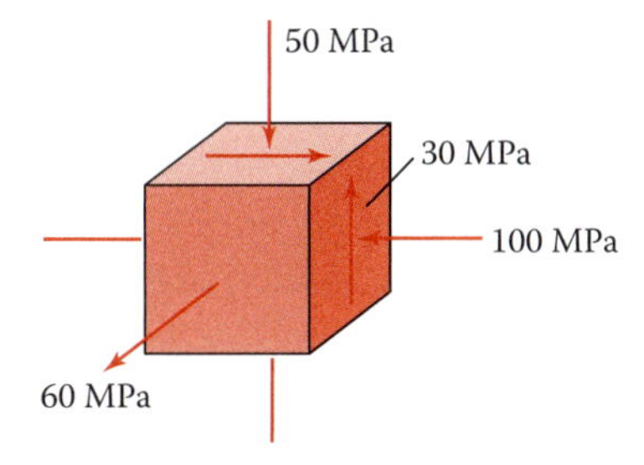

FIGURE P2.22

2.23 A closed end tube of thickness $t = 3.6$ mm and diameter $d = 1.8$ m is fabricated of steel having yield strength in tension $\sigma_{yp} = 250$ MPa. What is the internal pressure p that can he carried by the shell with a safety factor of $n = 2.2$ on yielding? *Assumption*: The vessel is supported at ends by cradles that act as simple supports.

2.24 A cylindrical pressure vessel (Figure 2.19a) of radius $r = 0.3$ m and wall thickness $t = 12$ mm is made of steel with a tensile strength of $\sigma_{yp} = 260$ MPa. *Find*: The allowable pressure p the shell can carry on the basis of a safety factor of $n = 1.8$. *Apply*: (a) the maximum shear stress theory; (b) von Mises theory.

2.25 A solid steel shaft ($\sigma_{yp} = 250$ MPa) of 75-mm diameter supports end tensile loads $P = 40$ kN and torque $T = 6$ kN·m. Calculate the factor of safety n on the basis of the following failure criteria: (a) maximum energy of distortion; (b) maximum shear stress.

2.26 Figure P2.26 depicts a $d = 60$-mm-diameter steel shaft of yield strength in tension σ_{yp} and yield strength in shear τ_{yp} subjected to a load R and $P = Q = 0$. Based upon a factor of safety $n = 1.9$, determine the largest permissible value of R according to: (a) the maximum shear stress theory; (b) von Mises theory. *Given*: $\sigma_{yp} = 240$ MPa, $\tau_{yp} = 130$ MPa.

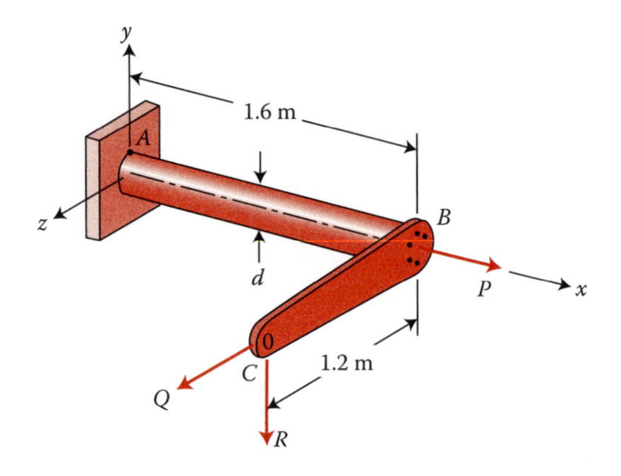

FIGURE P2.26

2.27 Resolve Problem 2.26 for the case in which $P = 50\,R$ and $Q = 0$.

Sections 2.12 and 2.13

2.28 Determine the *deflection* w_D at point D of a simply supported beam carrying a uniformly distributed load of intensity p (Figure P2.28). Use Castigliano's theorem.

2.29 A beam supported and loaded is shown in Figure P2.29. Apply Castigliano's theorem to determine the deflection at point C.

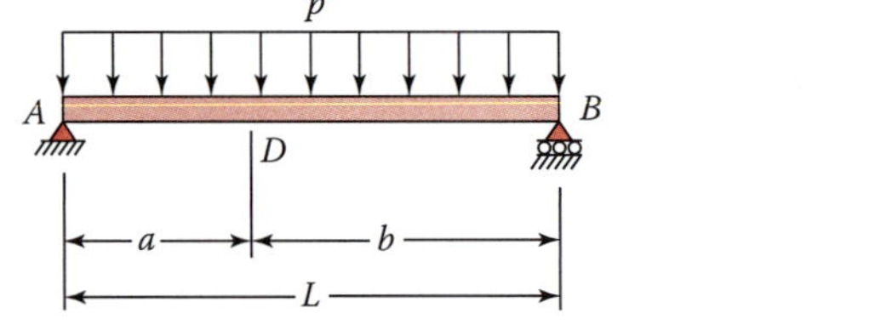

FIGURE P2.28

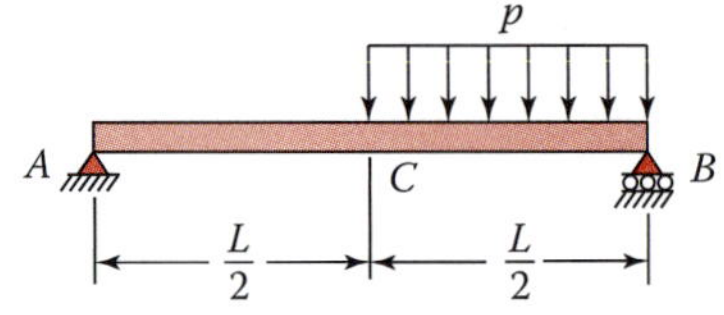

FIGURE P2.29

2.30 A bracket of constant flexural rigidity EI is shown in Figure P2.30. Using Castigliano's theorem, determine at the free end: (a) the vertical deflection; (b) the horizontal deflection; (c) the angular rotation.

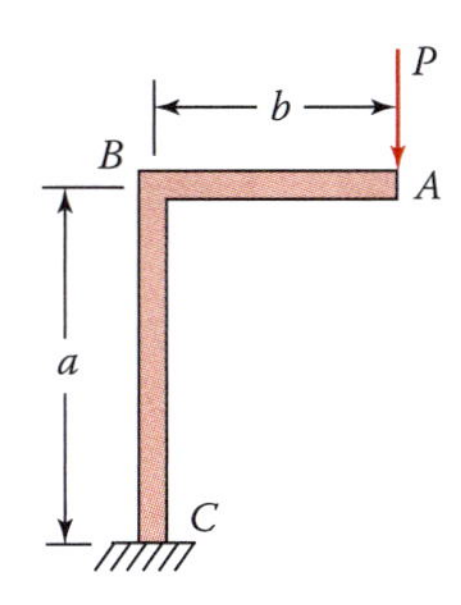

FIGURE P2.30

2.31 Redo Problem 2.17a and c, applying Castigliano's theorem.

2.32 For the beam and loading shown in Figure P2.32, employing Castigliano's theorem, determine the reactions at each support.

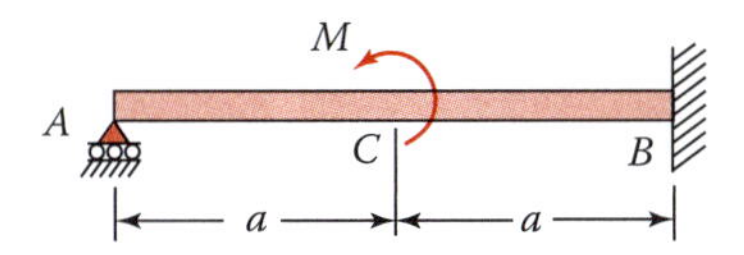

FIGURE P2.32

References

1. H. H. West and L. F. Geshwindner, *Fundamentals of Structural Analysis*, 2nd ed., Wiley, Hoboken, NJ, 2002.
2. A. C. Ugural and S. K. Fenster, *Advanced Mechanics of Materials and Applied Elasticity*, 5th ed., Prentice-Hall, Upper Saddle River, NJ, 2012.
3. S. Timoshenko and J. M. Goodier, *Theory of Elasticity*, 3rd ed., McGraw-Hill, New York, 1970.
4. W. D. Pilkey, *Peterson's Stress Concentration Factors*, 3rd ed., Wiley, Hoboken, NJ, 2008.
5. W. C. Young, R. C. Budynas, and A. H. Sadegh, *Roark's Formulas for Stress and Strain*, 8th ed., McGraw-Hill, New York, 2011.
6. A. P. Boresi and R. J. Schmidth, *Advanced Mechanics of Materials*, 6th ed., Wiley, Hoboken, NJ, 2003.
7. A. C. Ugural, *Mechanics of Materials*, Wiley, Hoboken, NJ, 2008.
8. J. Marin, *Mechanical Behavior of Materials*, Prentice-Hall, Upper Saddle River, NJ, 1962.
9. J. N. Reddy, *Energy and Variational Methods in Applied Mechanics*, Wiley, Hoboken, NJ, 2002.
10. A. C. Ugural, *Mechanical Design of Machine Components*, 2nd ed., CRC Press, Boca Raton, FL, 2015.

Section II

Plates

A large cruise ship, Norwegian Escape, under construction in Papenburg, Germany (2015). Many components of this modern structure (such as hull, bulkheads, decks, watertight compartments, engine parts, windows, balconies, lifeboats) represent *plates* with a variety of loadings and edge conditions. A number of plates can be observed in the shipyard as well. A plate acts in a manner analogous to a beam, through bending and shear stresses. The practical applications of plate theory and analysis can be found in most engineering fields. We will demonstrate that a plate has about ten percent greater stiffness than that of a beam.

3

Plate-Bending Theory

3.1 Introduction

Plates are initially flat structural elements, having thicknesses much smaller than their other dimensions. Included among the more familiar examples of plates are tabletops, street manhole covers, side panels and roofs of buildings, turbine disks, bulkheads, and tank bottoms. Plates are extensively used in architectural structures, airplanes, bridges, missiles, submarines, ships, and machine components. Many practical engineering problems fall into the categories "plates in bending" or "shells in bending." In Chapters 3 through 11 we treat plates. It is common to divide the thickness t of a plate into equal halves by a plane parallel to the faces. This plane is called the *midplane* of the plate. The plate thickness is measured in a direction normal to the midplane. The flexural properties of a plate depend greatly on its thickness in comparison with its other dimensions.

Plates may be classified into three groups: *thin* plates with small deflections, *thin* plates with large deflections, and *thick* plates. According to the criterion often applied to define a *thin plate* (for purposes of technical calculations), the ratio of the thickness to the smaller span length should be less than 1/20. With the exception of Chapter 10, we shall consider only small deflections of thin plates, a simplification consistent with the magnitude of deformation commonly found in plate structures. For clarity, however, the deflections and thicknesses of plates will be shown greatly exaggerated on some diagrams. As noted previously, unless otherwise specified, we shall assume that plate and shell materials are homogeneous and isotropic.

Ascertaining the distribution of stress and displacement for a plate subject to a given set of forces requires consideration of the *basic principles* outlined in Section 1.2, pertaining to certain physical laws, material properties, geometry, and surface forces. These conditions, stated mathematically in this chapter, are used to solve the bending problems of plates in the chapters to follow. It is often advantageous, where the shape of the plate or loading configuration precludes a theoretical solution or where verification is sought, to employ experimental methods. The approximate numerical and energy approaches (Chapter 7 and Section 3.12, respectively) are also efficient for this purpose. For the cases of nonuniformly heated and orthotropic plates (Chapters 11 and 8, respectively), rederivation of some basic relationships and the governing equation are required.

3.2 Historical Development of Plate and Shell Theory

The development of plate and shell theory is a blend of experiment and theory. It has a rich history with contribution being made by well-known mathematicians, scientists, and engineers. Investigation into plate behavior began by exploring free vibrations. Leonard Euler (1707–1783) was the first to start such inquiries in 1776; these inquiries were further extended by his student and, as it turns out, grandson-in-law, Jacob (II) Bernoulli (1759–1789). In 1811, the renowned mathematician Joseph-Louis Lagrange (1736–1813) received credit in developing a fourth-order partial differential equation to describe plate vibrations.

Many other prominent scientists, such as Simeon Denis Poisson (1781–1840) and Claude-Louis Navier (1785–1836), were involved in the development of plate theory as well. In the field of mathematics, the former is noted for his work in partial differential equations and in the fields of physics and engineering is credited with the Poisson's ratio; the latter engineer and physicist is famous for his development of relations describing the motion of fluids and gases, the so-called Navier–Stokes equations.

The German physicist well known for his work on electric circuits, Gustav Robert Kirchhoff (1824–1887), published in 1890 an important thesis on the theory of thin plates wherein the first three basic assumptions described in Section 3.3 were made. The British physicist William Thomas Kelvin (1824–1907), better known under his knighted name Sir Lord Kelvin, first demonstrated that torsional moments acting at the edges of plates could be decomposed into shearing forces. The prominent English mathematician Augustus Edward Hough Love (1863–1940) introduced simple analysis of shells, known as Love's approximate shell theory.

> I cannot doubt that these things, which now seem to be mysterious, will be no mysteries at all; that the scales will fall from our eyes; that we shall learn to look on things in a different way—when that which is now a difficulty will be the only common sense and intelligible way of looking at the subject.
>
> **W. T. Kelvin**

The first significant treatment of plates occurred in the 1800s. Since then, a great many cases of plate-bending problems have been worked out: the fundamental theory (principally by Navier, Kirchhoff, and Lévy) and numerical approaches (by Galerkin and Wahl and others). The literature related to plate and shell analysis is extensive [1–16].

Love applied Kirchhoff's classical large deflection analysis to thick plates [17]. Prescott initiated a more accurate theory for the bending of plates by considering the strains in the midplane [18]. A rigorous plate theory that takes into account the deformations caused by the transverse shear forces was introduced by Reissner [19]. Historical reviews of the subject are given in References 20 through 22.

In 1956, Turner, Clough, Martin, and Topp introduced the *finite element method,* which permits the numerical solution of complex plate and shell problems in an economical way [23]. Numerous contributions in this area are due to Argyris [24] and Zienkiewitcz. The recent trend in the development of the plate theories is characterized by heavy reliance on high-speed computers and by the introduction of more rigorous theories. The numerical methods introduced in Chapter 7 and applied in the following chapters have clear application to computation by means of electronic digital computers.

3.3 General Behavior and Theory of Plates

Consider a load-free plate, shown in Figure 3.1a, in which the xy plane coincides with the *midplane*, and hence the z deflection is zero. The components of displacement at a point, occurring in the x, y, and z directions, are denoted by u, v, and w, respectively. When, because of lateral loading, deformation takes place, the *midsurface* at any point $A(x_a, y_a)$ has deflection w (Figure 3.1b).

The *fundamental assumptions* of the *classical* or the Kirchhoff *theory* for elastic, thin plates are based on the geometry of deformations. They may be stated as follows:

1. The deflection of the midsurface is small compared with the thickness of the plate. The slope of the deflected surface is therefore very small, and the square of the slope is a negligible quantity in comparison with unity.
2. The midplane remains unstrained subsequent to bending.
3. Plane sections (mn) initially normal to the midsurface *remain plane* and normal to that surface after the bending. This means that the vertical shear strains γ_{xz} and γ_{yz} are negligible. The deflection of the plate is thus associated principally with bending strains. It is deduced, therefore, that the normal stain o_z. resulting from transverse loading may also be omitted.
4. The stress normal to the midplane, σ_z, is small compared with the other stress components and may be neglected. This supposition becomes unreliable in the vicinity of highly concentrated transverse loads.

These assumptions, known as the *Kirchhoff hypotheses*, are the extension of those associated with the Euler–Bernoulli theory of beams. Small- and large-scale tests have shown their validity. In the vast majority of engineering applications, adequate justification may be found for the simplifications stated with respect to the state of deformation and stress. Because of the resulting decrease in complexity, a 3D plate problem reduces to one involving only two dimensions. Consequently, the governing plate equation can be derived in a concise and straightforward manner.

When the deflections are *not* small, the bending of plates is accompanied by strain in the midplane, and assumptions 1 and 2 are inapplicable. An exception, however, applies when a plate bends into a *developable surface* (e.g., surfaces of cones and cylinders; Section 10.2).

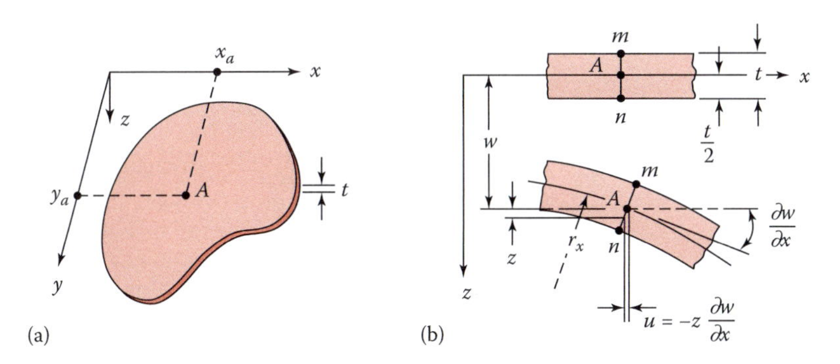

FIGURE 3.1
(a) A plate of constant thickness; (b) part of the plate before and after deflection.

This type of surface can be bent back to a plane without variation in the distances between any two points on the surface. If the midsurface of a freely or simply supported and loaded plate has a developable form, it remains unstrained even for deflections that are *equal* to or *larger* than its thickness but are still small as compared with the other dimensions of the plate. Only under this limitation on the deflections will the squares of slopes be small compared with unity; hence, the approximate expression used for the curvatures (Section 3.4) is sufficiently accurate.

In *thick* plates, the shear stresses are important, as in short, deep beams. Such plates are treated by means of a more general theory because assumptions 3 and 4 are no longer appropriate. The modified plate theories and the effect of shear on-plate deflection are discussed in Sections 3.10 and 4.6.

3.4 Strain–Curvature Relations

To gain insight into the plate-bending problem, consideration is now given to the geometry of deformation. As a consequence of assumption 3 of the foregoing section, the strain–displacement relations, Equations 1.21, reduce to

$$\varepsilon_x = \frac{\partial u}{\partial x} \qquad \varepsilon_z = \frac{\partial w}{\partial z} = 0$$

$$\varepsilon_y = \frac{\partial v}{\partial y} \qquad \gamma_{xz} = \frac{\partial w}{\partial x} + \frac{\partial u}{\partial z} = 0 \tag{3.1a–f}$$

$$\gamma_{xy} = \frac{\partial u}{\partial y} + \frac{\partial v}{\partial x} \qquad \gamma_{yz} = \frac{\partial w}{\partial y} + \frac{\partial v}{\partial z} = 0$$

where $\gamma_{ij} = \gamma_{ji}$ $(i, j = x, y, z)$. Note that these expressions are also referred to as the *kinematic* relations, treating the *geometry* of strain rather than the matter of cause and effect. Integrating Equation 3.1d, we obtain

$$w = w(x, y) \tag{a}$$

indicating that the lateral deflection does not vary over the plate thickness. In a like manner, integration of the expressions for γ_{xz} and γ_{yz} gives

$$u = -z\frac{\partial w}{\partial x} + u_0(x, y) \qquad v = -z\frac{\partial w}{\partial y} + v_0(x, y) \tag{b}$$

It is clear that $u_0(x, y)$ and $v_0(x, y)$ represent, respectively, the values of u and v on the midplane. Based on assumption 2 of Section 3.3, we conclude that $u_0 = v_0 = 0$. Thus,

$$u = -z\frac{\partial w}{\partial x} \qquad v = -z\frac{\partial w}{\partial y} \tag{3.2}$$

The above expression for u is represented in Figure 3.1b at section mn passing through arbitrary point $A(x_a, y_a)$. A similar illustration applies for v in the zy plane. We see that Equations 3.2 are consistent with assumption 3. Substitution of Equations 3.2 into the first three expressions of Equations 3.1 yields

$$\varepsilon_x = -z\frac{\partial^2 w}{\partial x^2} \qquad \varepsilon_y = -z\frac{\partial^2 w}{\partial y^2} \qquad \gamma_{xy} = -2z\frac{\partial^2 w}{\partial x\,\partial y} \tag{3.3a}$$

These formulas provide the strains at any point in the plate.

The *curvature* (equal to the reciprocal of the *radius of curvature*) of a plane curve is defined as the *rate* of change of the slope angle of the curve with respect to distance along the curve. Because of assumption 1 of Section 3.3, the square of a slope may be regarded as negligible, and the partial derivatives of Equations 3.3a represent the curvatures of the plate. Therefore, the curvatures κ (kappa) at the midsurface in planes *parallel* to the xz, yz, and xy planes are, respectively,

$$\begin{aligned} \frac{1}{r_x} &= \frac{\partial}{\partial x}\left(\frac{\partial w}{\partial x}\right) = \kappa_x \\ \frac{1}{r_y} &= \frac{\partial}{\partial y}\left(\frac{\partial w}{\partial y}\right) = \kappa_y \\ \frac{1}{r_{xy}} &= \frac{\partial}{\partial x}\left(\frac{\partial w}{\partial y}\right) = \kappa_{xy} \end{aligned} \tag{3.4}$$

where $\kappa_{xy} = \kappa_{yx}$.

The curvature κ_x and radius of curvature r_x at the midsurface in the xz plane is shown in Figure 3.1b. Similarly, the κ_y and r_y may be depicted in the yz plane. The designation r_{xy} is chosen to bring it in line with radii of curvature; r_{xy} has the dimension of a *length* like r_x and r_y. Clearly, Equations 3.4 are the *rates* at which the slopes vary over the plate. The last of these expressions is also referred to as the *twist* of the midplane with respect to the x and y axes. The local twist of a plate element is shown in Figure 3.2.

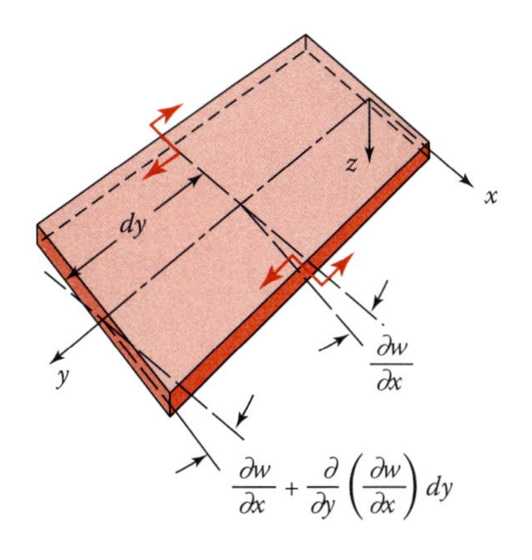

FIGURE 3.2
Twist of a plate element.

We note that, if a positive slope $\partial w/\partial x$ of the elastic surface *becomes more positive* as x increases, the curvature κ_x is positive (Figure 3.1b). An identical conclusion applies to the κ_y and κ_{xy}. Thus, the curvature of the midsurface is *positive* when the plate is bent *concave downward* and negative when it is bent concave upward.

The strain–curvature relations, by means of Equations 3.3a and 3.4, may be expressed in the form

$$\varepsilon_x = -z\kappa_x \qquad \varepsilon_y = -z\kappa_y \qquad \gamma_{xy} = -2z\kappa_{xy} \tag{3.3b}$$

Since no mechanical properties are involved in deriving the preceding equations, these relationships can be used for *inelastic* as well as for *elastic* problems. Equations 3.3 state that the *strains* in the plate *vary linearly* with the distance z from the midplane.

3.4.1 Mohr's Circle of Curvature

An examination of Equations 3.3 shows that a circle of curvature can be constructed similarly to *Mohr's circle* of strain. The curvatures therefore transform in the same manner as the strains. Figure 3.3 shows a plate element and a circle of curvature in which n and t represent perpendicular directions at a point on the midsurface. The *principal* or maximum and minimum curvatures are indicated by κ_1 and κ_2. The planes associated with these curvatures are called the *principal planes* of curvature. The curvature and the twist of a surface vary with the angle θ, measured in the clockwise direction from the set of axes xy to the $x'y'$ set. The product of the principal curvatures defines a so-called *Gaussian curvature.*

Mohr's circle shows that the maximum twist, represented by the radius of the circle, occurs at 45° from the planes of principal curvature. It is seen that when the two principal curvatures are the same, Mohr's circle shrinks to a point. This means that the curvature is the same in all directions; there is no twist in any direction. The surface is purely spherical at that point. From the circle, we have (see Problem 3.3)

$$\kappa_x + \kappa_y = \kappa_{x'} + \kappa_{y'} \tag{3.5}$$

The sum of the two curvatures in perpendicular directions at a point, called the *average curvature,* is thus *invariant* with respect to rotation of the coordinate axes. This assertion is valid at any location on the midsurface. Interestingly, a developable surface has zero Gaussian curvature, $\kappa_1 \kappa_2 = 0$ (see Equation P3.4).

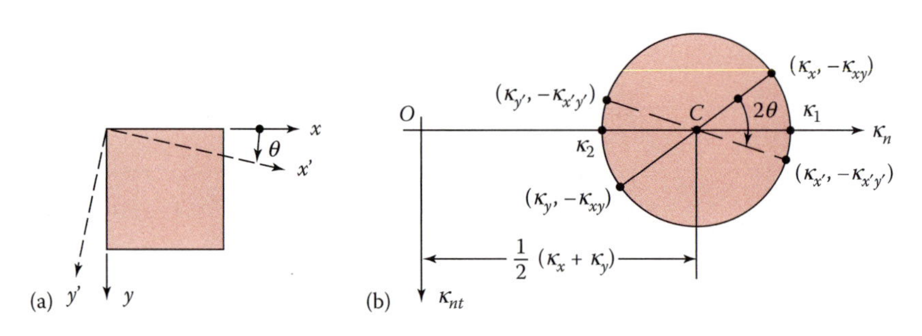

FIGURE 3.3
Mohr's circle for curvatures.

3.5 Stresses and Stress Resultants

In the case of a 3D state of stress, stress and strain are related by the generalized Hooke's law, valid for an isotropic homogeneous material

$$\begin{aligned} \varepsilon_x &= \frac{1}{E}\left[\sigma_x - \nu(\sigma_y + \sigma_z)\right] & \gamma_{xy} &= \frac{\tau_{xy}}{G} \\ \varepsilon_y &= \frac{1}{E}\left[\sigma_y - \nu(\sigma_x + \sigma_z)\right] & \gamma_{xz} &= \frac{\tau_{xz}}{G} \\ \varepsilon_z &= \frac{1}{E}\left[\sigma_z - \nu(\sigma_x + \sigma_y)\right] & \gamma_{yz} &= \frac{\tau_{yz}}{G} \end{aligned} \tag{1.34}$$

where $\tau_{ij} = \tau_{ji}(i, j = x, y, z)$. The constants E, ν, and G represent the modulus of elasticity, Poisson's ratio, and the shear modulus of elasticity, respectively. The connecting expression is

$$G = \frac{E}{2(1+\nu)} \tag{1.35}$$

The double *subscript notation* for stress is interpreted as follows (see Section 1.5): The first subscript indicates the direction of a normal to the plane or face on which the stress component acts, and the second subscript relates to the direction of the stress itself. Repeated subscripts will be omitted in this text. That is, the normal stresses σ_{xx}, σ_{yy}, and σ_{zz} are designated by σ_x, σ_y, and σ_z (Figure 3.4a). *A face, plane, or surface is usually identified by the axis normal to it*; for example, the x faces are perpendicular to the x axis.

Recall from Section 1.5 that the *sign convention* for stresses relies on the relationship between the direction of an *outward normal* drawn to a particular surface and the directions of the stress components on the same surface. Accordingly, if *both* the outer normal and the stress component are in a positive (or negative) direction relative to the coordinate axes, the stress is positive. If the outer normal points in a positive direction while the stress points in a negative direction (or vice versa), the stress is negative. On this basis, all stress components shown in Figure 3.4a are positive.

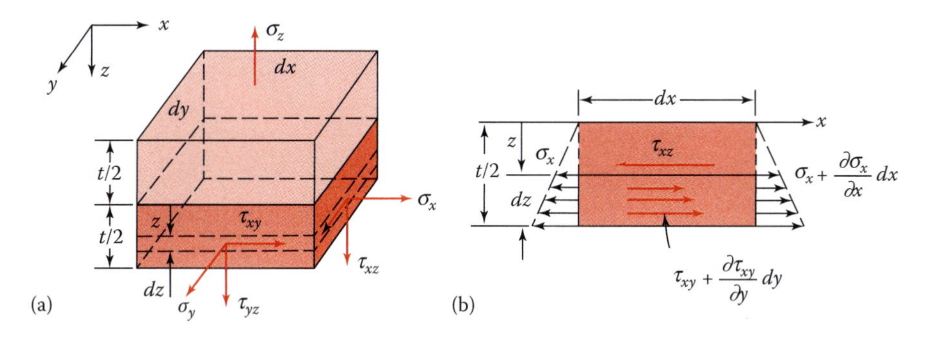

FIGURE 3.4
Stresses in the bottom half and stress normal to midplane of a plate element.

Substitution of $\varepsilon_z = \gamma_{yz} = \gamma_{xz} = 0$ into Equations 1.34 yields the following stress–strain relations for a thin plate:

$$
\begin{aligned}
\sigma_x &= \frac{E}{1-\nu^2}(\varepsilon_x + \nu\varepsilon_y) \\
\sigma_y &= \frac{E}{1-\nu^2}(\varepsilon_y + \nu\varepsilon_x) \\
\tau_{xy} &= G\gamma_{xy}
\end{aligned}
\tag{3.6}
$$

Introducing the plate curvatures, Equations 3.3b and 3.4, we find that the above appear in the form

$$
\begin{aligned}
\sigma_x &= -\frac{Ez}{1-\nu^2}(\kappa_x + \nu\kappa_y) = -\frac{Ez}{1-\nu^2}\left(\frac{\partial^2 w}{\partial x^2} + \nu\frac{\partial^2 w}{\partial y^2}\right) \\
\sigma_y &= -\frac{Ez}{1-\nu^2}(\kappa_y + \nu\kappa_x) = -\frac{Ez}{1-\nu^2}\left(\frac{\partial^2 w}{\partial y^2} + \nu\frac{\partial^2 w}{\partial y^2}\right) \\
\tau_{xy} &= -\frac{Ez}{1-\nu}\kappa_{xy} = -\frac{Ez}{1+\nu}\frac{\partial^2 w}{\partial x \partial y}
\end{aligned}
\tag{3.7}
$$

We observe from these formulas that the stresses vanish at the midsurface and vary *linearly* over the thickness of the plate.

The stresses distributed over the thickness of the plate produce bending moments, twisting moments, and vertical shear forces. These moments and forces *per unit length* are also called *stress resultants*. Referring to Figure 3.4a, we have

$$
\int_{-t/2}^{t/2} z\sigma_x \, dy \, dz = dy \int_{-t/2}^{t/2} z\sigma_x \, dz = M_x \, dy
$$

Similarly, expressions for the other resultants are derived as follows:

$$
\begin{Bmatrix} M_x \\ M_y \\ M_{xy} \end{Bmatrix} = \int_{-t/2}^{t/2} \begin{Bmatrix} \sigma_x \\ \sigma_y \\ \tau_{xy} \end{Bmatrix} z \, dz
\tag{3.8a}
$$

where $M_{xy} = M_{yx}$ and

$$
\begin{Bmatrix} Q_x \\ Q_y \end{Bmatrix} = \int_{-t/2}^{t/2} \begin{Bmatrix} \tau_{xz} \\ \tau_{yz} \end{Bmatrix} dz
\tag{3.8b}
$$

The sign convention for shear force is the same as that for shear stress. A positive moment is one that results in positive stresses in the bottom half of the plate. Accordingly, all moments and shear forces acting on the element in Figure 3.6 are positive.

Thus, the curvature induced by a positive moment is *opposite* to that associated with the positive curvature of the elastic surface.

It is important to mention that while the theory of thin plates omits the effect of the strain components $\gamma_{xz} = \tau_{xz}/G$ and $\gamma_{yz} = \tau_{yz}/G$ on bending, vertical forces Q_x and Q_y are not negligible. In fact, they are of the same order of magnitude as the surface loading and moments and are included in the derivation of the equilibrium equations (Section 3.7).

Substituting Equations 3.7 into Equation 3.8a, we derive the following formulas for the bending and twisting moments in terms of the curvatures and the deflection:

$$\begin{aligned} M_x &= -D(\kappa_x + \nu\kappa_y) = -D\left(\frac{\partial^2 w}{\partial x^2} + \nu\frac{\partial^2 w}{\partial y^2}\right) \\ M_y &= -D(\kappa_y + \nu\kappa_x) = -D\left(\frac{\partial^2 w}{\partial y^2} + \nu\frac{\partial^2 w}{\partial x^2}\right) \\ M_{xy} &= -D(1-\nu)\kappa_{xy} = -D(1-\nu)\frac{\partial^2 w}{\partial x\,\partial y} \end{aligned} \tag{3.9}$$

where

$$D = \frac{Et^3}{12(1-\nu^2)} \tag{3.10}$$

is the *flexural rigidity* of the plate. The negative sign agrees with the convention for moment and curvature. The vertical shear forces Q_x and Q_y are related to w on derivation of the equilibrium equations.

Interestingly, it is noted that if a plate element of *unit width* and parallel to the x axis were free to move sideways under transverse loading, the top and bottom surfaces would be deformed into saddle-shaped or anticlastic surfaces of curvature κ_y. The flexural rigidity would then be $Et^3/12$, as in the case of a beam. The remainder of the plate prevents the anticlastic curvature, however. Because of this action, a *plate manifests greater stiffness* than a beam by a factor of $1/(1 - \nu^2)$, or approximately 10%.

The 2D stress components are found from Equations 3.7 by substituting Equations 3.9 and by employing Equation 3.10. In this way, we obtain

$$\sigma_x = \frac{12M_x z}{t^3} \qquad \sigma_y = \frac{12M_y z}{t^3} \qquad \tau_{xy} = \frac{12M_{xy} z}{t^3} \tag{3.11}$$

The *maximum* stresses occur on the *bottom* and *top* surfaces (at $z = \pm t/2$) of the plate. In SI, the stress is measured in *newtons* per square meter (N/m^2) or pascals (Pa), as described in Section 1.5. The moments and forces per unit length have units $N \cdot m/m$, or simply N, and N/m, respectively.

3.6 Equations for Transformation of Moment

It is observed from Equations 3.8a and 3.11 that there is a direct correspondence between the moments and stresses. Hence, transformation equations for stress and moment per

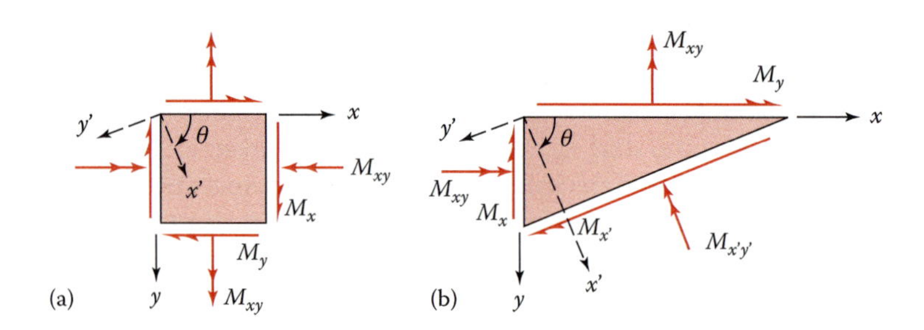

FIGURE 3.5
Moment components on plate elements.

unit length are analogous. The Mohr's circle analysis and all conclusions drawn for stress therefore apply to the moments.

The transformation expressions for stress are converted into moment relationships by replacing (see Section 1.8)

$$\sigma \text{ with } M_x \text{ or } M_y \quad \text{and} \quad \tau \text{ with } M_{xy}$$

Then, referring to Figure 3.5 and similarity with Equations 1.11, the moments shown by double-headed vectors acting on an inclined x' plane are expressed as

$$M_{x'} = \frac{1}{2}(M_x + M_y) + \frac{1}{2}(M_x - M_y)\cos 2\theta + M_{xy}\sin 2\theta \tag{3.12a}$$

$$M_{x'y'} = -\frac{1}{2}(M_x - M_y)\sin 2\theta + M_{xy}\cos 2\theta \tag{3.12b}$$

Similarly,

$$M_{y'} = \frac{1}{2}(M_x + M_y) - \frac{1}{2}(M_x - M_y)\cos 2\theta - M_{xy}\sin 2\theta \tag{3.12c}$$

From Equation 1.12, the orientation of the principal axes is given by

$$\tan 2\theta_p = \frac{2M_{xy}}{M_x - M_y} \tag{3.13}$$

The *principal moments* and the *maximum twisting moment*, by Equations 1.13 and 1.14, are

$$M_{1,2} = \frac{M_x + M_y}{2} \pm \sqrt{\left(\frac{M_x - M_y}{2}\right)^2 + M_{xy}^2} \tag{3.14a}$$

and

$$(M_{xy})_{\max} = \sqrt{\left(\frac{M_x - M_y}{2}\right)^2 + M_{xy}^2} \tag{3.14b}$$

Subscripts 1 and 2 refer to the maximum and minimum values, respectively.

It should be noted that with *moments per unit length, we do not use the vector convention for signs.* However, the sense of moments follows the right-hand screw rule, as before. Figure 3.5 shows that now the new set of axes $x'y'$ is inclined at an angle of θ *clockwise* to the xy set. An angle of 2θ on Mohr's circle for moments corresponds to this angle.

3.7 Variation of Stress within a Plate

The components of stress (and thus the stress resultants) generally vary from point to point in a loaded plate. These variations are governed by the *conditions of equilibrium* of *statics*. Fulfillment of these conditions establishes certain relationships known as the equations of equilibrium. We shall eventually reduce the latter system of equations to a single relationship expressed in terms of moments.

Consider an element $dx\,dy$ of the plate subject to a uniformly distributed load per unit area p (Figure 3.6). We assume that inclusion of the plate weight, a small quantity, in the load p cannot affect the accuracy of the result. Note also that, as the element is very small, for the sake of simplicity the force and moment components may be considered to be distributed uniformly over each face. In the figure they are shown by a single vector, representing the *mean values* applied at the center of each face.

With change of location, as, for example, from the upper left corner to the lower right corner, one of the moment components, say M_x, acting on the negative x face, varies in value relative to the positive x face. This variation with position may be expressed by a truncated Taylor's expansion:

$$M_x + \frac{\partial M_x}{\partial x}dx$$

The partial derivative is used because M_x is a function of x and y. Treating all the components similarly, the state of stress resultants shown in the figure is obtained.

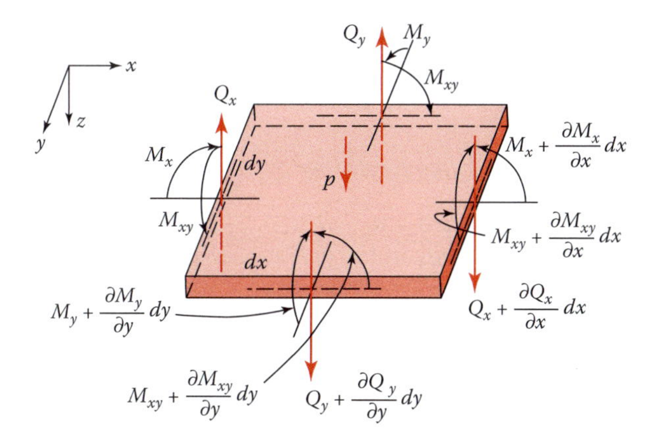

FIGURE 3.6
Positive stress resultants and load on a plate element.

The condition that the sum of the z-directed forces equals zero leads to

$$\frac{\partial Q_x}{\partial x} dx\, dy + \frac{\partial Q_y}{\partial y} dx\, dy + p\, dx\, dy = 0$$

from which

$$\frac{\partial Q_x}{\partial x} + \frac{\partial Q_y}{\partial y} + p = 0 \tag{a}$$

The equilibrium of moments about the x axis is governed by

$$\frac{\partial M_{xy}}{\partial x} dx\, dy + \frac{\partial M_y}{\partial y} dx\, dy - Q_y\, dx\, dy = 0$$

or

$$\frac{\partial M_{xy}}{\partial x} + \frac{\partial M_y}{\partial y} - Q_y = 0 \tag{b}$$

Products of infinitesimal terms, such as the moment of p and the moment due to the change in Q_y, have been omitted.

Similarly, from the equilibrium of moments about the y axis, we have

$$\frac{\partial M_{xy}}{\partial y} + \frac{\partial M_x}{\partial x} - Q_x = 0 \tag{c}$$

Finally, introduction of the expression for Q_x and Q_y from Equations b and c into Equation a yields

$$\frac{\partial^2 M_x}{\partial x^2} + 2\frac{\partial^2 M_{xy}}{\partial x\, \partial y} + \frac{\partial^2 M_y}{\partial y^2} = -p \tag{3.15}$$

This is the *differential equation of equilibrium* for bending of thin plates.

Expressions for vertical shear forces Q_x and Q_y may now be written in terms of deflection w, from Equations b and c together with Equations 3.9:

$$\begin{aligned} Q_x &= -D\frac{\partial}{\partial x}\left(\frac{\partial^2 w}{\partial x^2} + \frac{\partial^2 w}{\partial y^2}\right) = -D\frac{\partial}{\partial x}(\nabla^2 w) \\ Q_y &= -D\frac{\partial}{\partial y}\left(\frac{\partial^2 w}{\partial x^2} + \frac{\partial^2 w}{\partial y^2}\right) = -D\frac{\partial}{\partial y}(\nabla^2 w) \end{aligned} \tag{3.16a, b}$$

where

$$\nabla^2 = \frac{\partial^2}{\partial x^2} + \frac{\partial^2}{\partial y^2} \tag{d}$$

is the *Laplace operator.*

Comments: Since one equation (Equation 3.15) for three unknown moments M_x, M_y, and M_{xy} is not sufficient to obtain a solution, plate problems are *internally statically indeterminate*. Reduction of unknowns to one, which follows, is made on utilization of the moment–displacement relations.

3.8 Governing Equation for Deflection of Plates

The basic differential equation for the deflection of plates may readily be derived on the basis of the results obtained in the preceding sections. Introducing into Equation 3.15 the first expressions for M_x, M_y, and M_{xy} given by Equations 3.9, we have

$$\frac{\partial^2 \kappa_x}{\partial x^2} + 2\frac{\partial^2 \kappa_{xy}}{\partial x \partial y} + \frac{\partial^2 \kappa_y}{\partial y^2} = \frac{p}{D} \tag{a}$$

The above expresses the plate equilibrium in terms of the curvatures. An alternate form of Equation a is determined by inserting the definition of curvatures from Equations 3.4:

$$\frac{\partial^4 w}{\partial x^4} + 2\frac{\partial^4 w}{\partial x^2 \partial y^2} + \frac{\partial^4 w}{\partial y^4} = \frac{p}{D} \tag{3.17a}$$

This equation, first derived by Lagrange in 1811, can also be written in concise form:

$$\nabla^4 w = \frac{p}{D} \tag{3.17b}$$

in which $\nabla^4 = \nabla^2\nabla^2 = (\nabla^2)^2$. When there is no lateral load acting on the plate,

$$\frac{\partial^4 w}{\partial x^4} + 2\frac{\partial^4 w}{\partial x^2 \partial y^2} + \frac{\partial^4 w}{\partial y^4} = 0 \tag{3.18}$$

Expressions (3.17) is the *governing differential equation for deflection* of thin plates. To determine w, it is required to integrate this equation with the constants of integration dependent on the appropriate boundary conditions (discussed in Section 3.9).

3.8.1 Reduction of Plate-Bending Problem to That of Deflection of a Membrane

It is significant to note that the sum of the bending-moment components defined by Equations 3.9 is invariant. That is,

$$M_x + M_y = -D(1+\nu)\left(\frac{\partial^2 w}{\partial x^2} + \frac{\partial^2 w}{\partial y^2}\right) = -D(1+\nu)\nabla^2 w$$

By letting M denote the *moment function* or so-called *moment sum*,

$$M = \frac{M_x + M_y}{1+\nu} = -D\nabla^2 w \tag{3.19}$$

the expressions for shear forces can be written as

$$Q_x = \frac{\partial M}{\partial x} \qquad Q_y = \frac{\partial M}{\partial y} \tag{b}$$

and we may represent Equation 3.17a as follows:

$$\frac{\partial^2 M}{\partial x^2} + \frac{\partial^2 M}{\partial y^2} = -p \tag{3.20}$$

$$\frac{\partial^2 w}{\partial x^2} + \frac{\partial^2 w}{\partial y^2} = -\frac{M}{D} \tag{3.21}$$

The plate equation, $\nabla^4 w = p/D$ is thus reduced to two second-order partial differential equations that are sometimes preferred, depending on the method of solution to be employed. This reduction was first introduced by Marcus [12]. Given the loading and the boundary conditions, one can solve M from Equation 3.20, then Equation 3.21 leads to w. This deformation is caused by bending moments (M_x, M_y) and the shear forces (Q_x, Q_y). The torsion results from the twisting moments (M_{xy}, M_{yx}).

Very thin plates with *no flexural rigidity* are called *membranes*. This type of plate is extensively used in machine design for pumps, compressors, and pressure regulators. It can be demonstrated that Equations 3.20 and 3.21 are of the same form as the equation describing the deflection of a uniformly stretched and laterally loaded membrane [23]. Hence, an *analogy* exists between the bending of a plate and membrane problems, serving as the basis of a number of experimental and approximate numerical techniques. The latter are discussed in Chapter 7.

In conclusion, we note that the governing equations for the deflection developed in the foregoing are appropriate for plates of constant thickness and material properties. However, plates with *variable flexural rigidity D* are sometimes encountered in the design of structural and machine parts. If the *flexural rigidity* of the plate is a *function of x and y*, substituting Equations 3.9 into Equation 3.15, we have

$$\frac{\partial^2}{\partial x^2}\left[D\left(\frac{\partial^2 w}{\partial x^2} + \nu\frac{\partial^2 w}{\partial y^2}\right)\right] + 2(1-\nu)\frac{\partial^2}{\partial x \partial y}\left(D\frac{\partial^2 w}{\partial x \partial y}\right) + \frac{\partial^2}{\partial y^2}\left[D\left(\frac{\partial^2 w}{\partial y^2} + \nu\frac{\partial^2 w}{\partial x^2}\right)\right] = p$$

This may be rewritten as follows:

$$\nabla^2(D\nabla^2 w) - (1-\nu)\left(\frac{\partial^2 D}{\partial x^2}\frac{\partial^2 w}{\partial y^2} - 2\frac{\partial^2 D}{\partial x \partial y}\frac{\partial^2 w}{\partial x \partial y} + \frac{\partial^2 D}{\partial y^2}\frac{\partial^2 w}{\partial x^2}\right) = p \tag{3.22}$$

When D is constant, the preceding reduces to Equations 3.17. Equation 3.20 may also be replaced in a like manner.

3.9 Boundary Conditions

The differential equation of equilibrium that must be satisfied within the plate is derived in Section 3.8. The distribution of stress in a plate must also be such as to accommodate the conditions of equilibrium with respect to prescribed forces or displacements at the boundary.

For a plate, solution of Equations 3.17 requires that two boundary conditions be satisfied at each edge. These may be a given deflection and slope, or force and moment, or some combination. The basic difference between the boundary conditions applied to plates and those of beams is the existence of twisting moments along the plate edges. It is demonstrated below that these moments may be replaced by equivalent forces. Such a substitution causes an alteration of the distribution of stress and strain only in the immediate region of the boundary, in accordance with St. Venant's Principle (see Section 1.2).

We now treat the boundary conditions or a rectangular plate with edges a and b parallel to the x and y axes, as shown in Figure 3.7. Consider two successive elemental lengths dy on edge $x = a$ (Figure 3.7). It is seen that, on the right-hand element, a twisting moment M_{xy} dy acts, while the left-hand element is subjected to $[M_{xy} + (\partial M_{xy}/\partial y)dy]dy$.

In the figure, the moments are indicated as replaced by statically equivalent force couples. Thus, in an *infinitesimal region* of the edge shown within the dashed line, we see that an upward-directed force M_{xy} and a downward-directed force $M_{xy} + (\partial M_{xy}/\partial y)dy$ act. The algebraic sum of these forces may be added to the shearing force Q_x to produce an *effective transverse force per unit length* for an edge parallel to the y axis, V_x. That is,

$$V_x = Q_x + \frac{\partial M_{xy}}{\partial y} = -D\left[\frac{\partial^3 w}{\partial x^3} + (2-\nu)\frac{\partial^3 w}{\partial x\,\partial y^2}\right] \tag{3.23a}$$

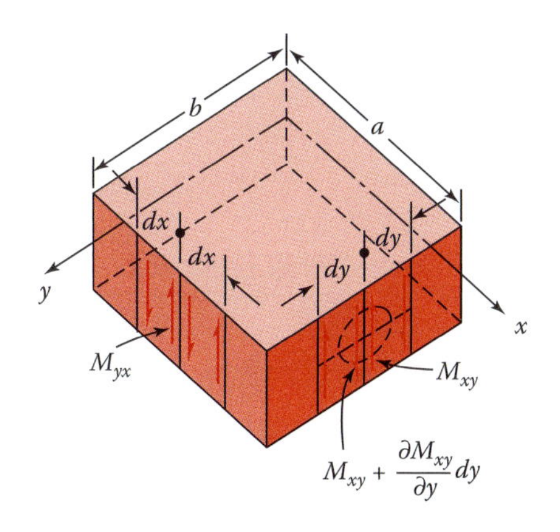

FIGURE 3.7
Edge effect of twisting moment.

Similarly, it can be shown that, for an edge parallel to the x axis,

$$V_y = Q_y + \frac{\partial M_{xy}}{\partial x} = -D\left[\frac{\partial^3 w}{\partial y^3} + (2-\nu)\frac{\partial^3 w}{\partial x^2 \partial y}\right] \tag{3.23b}$$

Expressions (3.23) is due to *Kirchhoff*: A distribution of M_{xy} along an edge is statically equivalent to a distribution of vertical shear forces.

In addition to the edge forces described above, there may be *concentrated forces* F_c produced *at the corners*. Consider, as an example, the case of a *uniformly loaded* rectangular plate with *simply supported* edges (Figure 3.7). At the corner (a, b), the above-discussed action of twisting moments (because $M_{xy} = M_{yx}$) results in

$$F_c = 2M_{xy} = -2D(1-\nu)\frac{\partial^2 w}{\partial x\, \partial y} \qquad (x = a, \quad y = b) \tag{3.24}$$

The negative sign indicates an upward direction. Because of the symmetry of the uniform loading, this force must have the same magnitude and direction at all corners of the plate. Thus, if no anchorage is provided, the corners of the plate described tend to rise (Example 5.2).

The additional corner force for plates having various edge conditions may be determined similarly; for instance, when two adjacent plate edges are *fixed* or *free*, we have $F_c = 0$ since along these edges no twisting moments exist.

We can now formulate a variety of commonly encountered situations. The *boundary conditions* that apply along the edge $x = a$ of the rectangular plate with edges parallel to the x and y axes (Figure 3.8) are as follows:

Clamped, fixed, or built-in edge (Figure 3.8a). In this case, both the deflection and the slope must vanish. That is,

$$w = 0 \qquad \frac{\partial w}{\partial x} = 0 \qquad (x = a) \tag{3.25}$$

Simply supported edge (Figure 3.8b). At the *simple support* considered, both the deflection and the bending moment are zero. Hence,

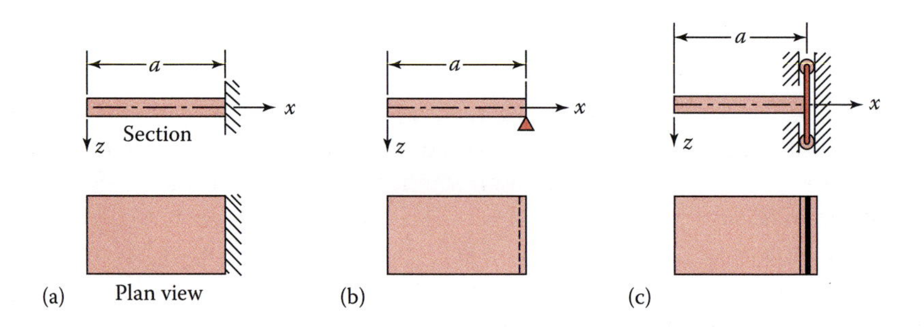

FIGURE 3.8
Various boundary conditions: (a) clamped edge; (b) simply supported edge; (c) sliding edge.

$$w = 0 \qquad M_x = -D\left(\frac{\partial^2 w}{\partial x^2} + \nu \frac{\partial^2 w}{\partial y^2}\right) = 0 \qquad (x = a), \tag{3.26a}$$

The first of these equations implies that along edge $x = a$, $\partial w/\partial y = 0$, $\partial w^2/\partial y^2 = 0$. Therefore, conditions expressed by Equations 3.26a may be written in the following equivalent form:

$$w = 0 \qquad \frac{\partial^2 w}{\partial x^2} = 0 \qquad (x = a) \tag{3.26b}$$

Free edge. Such an edge at $x = a$ is free of moment and vertical shear force. That is,

$$\frac{\partial^2 w}{\partial x^2} + \nu \frac{\partial^2 w}{\partial y^2} = 0 \qquad \frac{\partial^3 w}{\partial x^3} + (2 - \nu)\frac{\partial^3 w}{\partial x\,\partial y^2} = 0 \qquad (x = a) \tag{3.27}$$

Sliding edge (Figure 3.8c). In this case, the edge is free to move vertically, but the rotation is prevented. The support is not capable of resisting any shear force. Thus,

$$\frac{\partial w}{\partial x} = 0 \qquad \frac{\partial^3 w}{\partial x^3} + (2 - \nu)\frac{\partial^3 w}{\partial x\,\partial y^2} = 0 \qquad (x = a) \tag{3.28}$$

Some other types of boundary conditions may be treated similarly. It is observed that the boundary conditions are of two basic kinds: A *geometric* or *kinematic* boundary condition describes the end constraint pertaining to deflection or slope, and a *static* boundary condition equates the internal forces (and moments) at the edges of the plate to the given external forces (and moments). Accordingly, in Equations 3.25, both conditions are kinematic; in Equation 3.27, both are static; and in Equations 3.26 and 3.28, the conditions are *mixed*.

In addition to the *homogeneous* boundary conditions described above, it is, of course, possible to have prescribed shear, moment, rotation, or displacement at the boundary. The latter cases, *nonhomogeneous* boundary conditions, are expressed by replacing the zeros in Equations 3.25 through 3.28 with the specified quantity as shown in Table 11.1.

3.10 Exact Theory of Plates

Recall from the beam-bending theory that the effect of the transverse shear force on deformation and stress is small and can be neglected if the beam depth is small in comparison with beam span. The same conclusion also applies in the case of plates (Section 4.6). We should mention here, however, that in vibration at higher modes and in wave propagation, the effect of shear is of great importance in slender as well as in other beams and plates.

The *exact theory of plates* is exclusively governed by the theory of elasticity [17]. It can be verified [1] that, by considering Equations 1.9 together with the six conditions of compatibility (see Section 1.10), in the case of bending of plates resulting in a *plane stress distribution*, the deflections w rigorously satisfy Equations 3.17 and 3.9. Examples of such bending include (a) annular plates and plates of polygonal shape simply supported and bent by moments uniformly distributed along the boundary and (b) rectangular plates clamped along one edge and uniformly loaded on the surface or along the opposite edge. In these cases, general solutions obtained by the exact theory of plates *coincide* with those given by the customary plate theory.

Rigorous solutions of various other problems indicate that classical theory of thin plates is *accurate* enough for practical purposes, with the exception of the following cases:

1. *In the vicinity of a highly concentrated transverse load*
2. *In narrow edge zones*, particularly near the *corners* of plates and *around holes* with diameter of the order of magnitude of the plate thickness itself

The exact theory generally comprises a classical theory solution, corrected in those areas where the foregoing discontinuity effects are pronounced.

In the first case, the normal stress to the midplane σ_z and the transverse shear stresses τ_{yz} and τ_{xz} must be regarded as equally significant in their effect on the deformation of the plate. In such circumstances, the *thick-plate theory* proves most convenient for the solution of the problem. The maximum plane stresses at the point of application of a concentrated load are discussed in Section 4.7.

In the second case, the effect of σ_z on the deformation becomes secondary as compared with the effect of τ_{yz} and τ_{xz}. The *modified* or *refined thin-plate theories* [1,19], which take into account primarily the latter effect, are better suited for the solution of the problem than the more rigorous thick-plate theory. Stress concentration around holes in a plate is taken up in Section 6.6. Because of space limitations, the extended methods are not included in this book. Following is a discussion on the distribution of the transverse normal and shear stresses that are omitted in the customary plate theory.

Determination of the stress components σ_z, τ_{xz}, and τ_{yz} through the use of Hooke's law is not possible since according to Equations 3.1 they are not related to strains. The differential equations of equilibrium for a plate element under a general state of stress serves well for this purpose, however. These equations are, by neglecting the body forces in Equations 1.9,

$$\begin{aligned}
&\frac{\partial \sigma_x}{\partial x}+\frac{\partial \tau_{xy}}{\partial y}+\frac{\partial \tau_{xz}}{\partial z}=0\\
&\frac{\partial \sigma_y}{\partial y}+\frac{\partial \tau_{xy}}{\partial x}+\frac{\partial \tau_{yz}}{\partial z}=0\\
&\frac{\partial \sigma_z}{\partial z}+\frac{\partial \tau_{xz}}{\partial x}+\frac{\partial \tau_{yz}}{\partial y}=0
\end{aligned} \tag{a}$$

From the first two of the above expressions and Equations 3.7, the shear stresses τ_{xz} (Figure 3.4b) and τ_{yz} are, after integration,

$$\begin{aligned}
\tau_{xz} &= \int_z^{t/2}\left(\frac{\partial \sigma_x}{\partial x}+\frac{\partial \tau_{xy}}{\partial y}\right)dz = -\frac{E}{2(1-\nu^2)}\left(\frac{t^2}{4}-z^2\right)\left[\frac{\partial}{\partial x}\left(\frac{\partial^2 w}{\partial x^2}+\frac{\partial^2 w}{\partial y^2}\right)\right]\\
\tau_{yz} &= \int_z^{t/2}\left(\frac{\partial \sigma_y}{\partial y}+\frac{\partial \tau_{xy}}{\partial x}\right)dz = -\frac{E}{2(1-\nu^2)}\left(\frac{t^2}{4}-z^2\right)\left[\frac{\partial}{\partial y}\left(\frac{\partial^2 w}{\partial x^2}+\frac{\partial^2 w}{\partial y^2}\right)\right]
\end{aligned} \tag{3.29}$$

It is observed that the distribution of components τ_{xz} and τ_{yz} over the plate thickness varies according to a *parabolic* law. The component σ_z is readily determined by using the third of Equations a on substitution of τ_{xz} and τ_{yz} from Equations 3.29 and integration:

$$\sigma_z = -\frac{E}{2(1-\nu^2)}\left(\frac{t^3}{12}-\frac{t^2 z}{4}+\frac{z^3}{3}\right)\left[\left(\frac{\partial^2}{\partial x^2}+\frac{\partial^2}{\partial y^2}\right)\left(\frac{\partial^2 w}{\partial x^2}+\frac{\partial^2 w}{\partial y^2}\right)\right] \quad (3.30)$$

The normal stress σ_z thus varies as a cubic parabola over the thickness of the plate. This stress, according to assumption 4 of Section 3.3, is negligible. The z-directed shear stress components are also regarded as very small when compared with the remaining plane stresses.

Substituting Equations 3.16 and 3.17 into Equations 3.29 and 3.30, the stress components τ_{xz}, τ_{yz}, and σ_z are as follows:

$$\tau_{xz} = \frac{3Q_x}{2t}\left[1-\left(\frac{2z}{t}\right)^2\right] \qquad \tau_{yz} = \frac{3Q_y}{2t}\left[1-\left(\frac{2z}{t}\right)^2\right]$$
$$\sigma_z = -\frac{3p}{4}\left[\frac{2}{3}-\frac{2z}{t}+\frac{1}{3}\left(\frac{2z}{t}\right)^3\right] \quad (3.31)$$

The maximum shear stress, as in the case of a beam of rectangular section, occurs at $z = 0$ and is represented by the formulas

$$\tau_{xz,\max} = \frac{3}{2}\frac{Q_x}{t} \qquad \tau_{yz,\max} = \frac{3}{2}\frac{Q_y}{t} \quad (3.32)$$

Thus, the *key to determining the stress components*, using the formulas derived, is the solution of Equations 3.17 for w.

Comments: We mention that an alternative derivation of Equations 3.17 results from equating the stress normal to the plate to the surface loading per unit area at the upper face of the plate. In this way, from Equations 3.30, letting $z = -t/2$ and $\sigma_z = -p$, we obtain

$$\frac{Et^3}{12(1-\nu^2)}\nabla^4 w = p$$

which, with Equation 3.10, yields Equations 3.17.

3.11 Methods for Solution of Plate Deflections

Except for simple types of loading and shapes, such as axisymmetrically loaded circular plates, the governing plate equation $\nabla^4 w = p/D$ yields plate deflections only with considerable difficulty. It is common to attempt a solution by the inverse method. The inverse method relies on assumed solutions for w that satisfy the governing equation and the boundary conditions. Some cases may be treated by using polynomial expressions for w in x and y and undetermined coefficients. Usually, choosing the acceptable series form is laborious and requires a systematic approach. The most powerful such method is the Fourier series, where, once a solution has been found for sinusoidal loading, any other loading can be handled by infinite series (Sections 5.2 and 5.4). This approach offers as an important advantage the fact that a single expression may apply to the entire surface of the plate.

Energy methods (Section 3.12) should be included in a treatment of a general approach. These may be employed to develop solutions, often in the form of infinite series.

The role of the foregoing methods is twofold. They can provide "exact" answers where configurations of loading and shape are simple, and they can be used as the basis of approximate techniques through numerical analysis for more practical problems.

Another approach to overcoming the difficulty involved in the solution of the governing equation is to use finite differences (Chapter 7). In this case, Equations 3.17 or Equations 3.20 and 3.21 are replaced by finite difference expressions that relate the w (and M) at nodes that are removed from one another by finite distances. The resulting equations, however, serve only for numerical treatment.

3.11.1 Cylindrical Bending of Plate Strips

Two cases of plates that can be considered as one-dimensional problems: beams; cylindrical bending of *plate strips.* If the width of a plate is very small compared to the length, it is treated as a beam. In cylindrical bending the plate is taken to be a plate strip that is very long along the x axis and narrow along the y axis, as depicted in Figure 3.9.

The transverse loading p is assumed to be constant at any section parallel to the y axis, that is, $p = p(y)$, and all derivatives with respect to x are zero. Determination of deflections and moments in such plates will be illustrated in Examples 3.1, 5.7, 5.9, and 6.1. As will become evident in the foregoing situations, a deformation pattern can be established from symmetry considerations alone.

EXAMPLE 3.1: Plate Strip under Nonuniform Load

Determine the deflection and stress in a very long and narrow rectangular plate if it is simply supported at edges $y = 0$ and $y = b$ (Figure 3.10). (a) The plate carries a nonuniform loading expressed by

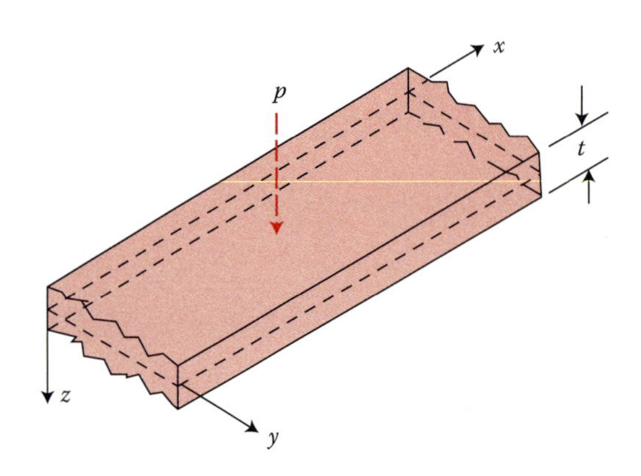

FIGURE 3.9
A plate strip.

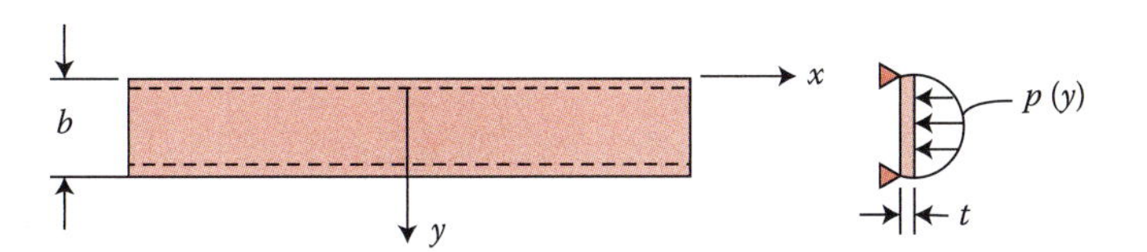

FIGURE 3.10
Simply supported plate strip.

$$p(y) = p_0 \sin \frac{\pi y}{b} \quad \text{(a)}$$

where the constant p_0 represents the load intensity along the line passing through $y = b/2$, parallel to the x axis. (b) The plate is under a uniform load p_0. Let $p_0 = 10$ kPa, $b = 0.4$ m, $t = 10$ mm, $\nu = 1/3$, and $E = 200$ GPa.

Solution

Clearly, the loading described deforms the plate into a *cylindrical surface* possessing its generating line parallel to the x axis: The slope of the deflected plate along the x axis is zero. We thus have $\partial w/\partial x = 0$. It follows that $\partial^2 w/\partial x \partial y = 0$, and Equations 3.9 yield

$$M_x = -\nu D \frac{d^2 w}{dy^2} \qquad M_y = -D \frac{d^2 w}{dy^2} \quad (3.33)$$

Equations 3.17 reduces to

$$\frac{d^4 w}{dy^4} = \frac{p}{D} \quad (3.34)$$

This expression is of the same form as the beam equation. Hence, the solution proceeds as in the case of a beam.

Comment: Note that since the bent plate surface is of *developable* type and the edges are free to move horizontally, the formulas derived in this example also hold for large deflections ($w \geq t$ but $w < b$).

a. Substituting Equation a into Equation 3.34, integrating successively four times, and satisfying the boundary conditions ($w = 0$ and $d^2 w/dy^2 = 0$ at $y = 0$ and $y = b$), we have

$$w = \left(\frac{b}{\pi}\right)^4 \frac{p_0}{D} \sin \frac{\pi y}{b} \quad (3.35)$$

The maximum stresses in the plate are obtained by substituting the above with $\nu = 1/3$ into Equations 3.11, 3.31, and 3.32:

$$\sigma_{x,\max} = 0.2\, p_0 \left(\frac{b}{t}\right)^2 \qquad \sigma_{y,\max} = 0.6\, p_0 \left(\frac{b}{t}\right)^2 \qquad \left(z = \frac{t}{2}, y = \frac{b}{2}\right)$$

$$\sigma_{z,\max} = -p_0 \qquad \left(z = -\frac{t}{2}\right)$$

$$\tau_{xy} = 0 \qquad \tau_{xz} = 0$$

$$\tau_{yz,\max} = 0.5 p_0 \left(\frac{b}{t} \right) \qquad (z = 0, y = 0)$$

To gauge the magnitude of the deviation between the stress components, consider the ratios

$$\frac{\sigma_{z,\max}}{\sigma_{x,\max}} = 5 \left(\frac{t}{b} \right)^2 \qquad \frac{\tau_{yz,\max}}{\sigma_{x,\max}} = 2.5 \left(\frac{t}{b} \right)$$

If, for example, $b = 20t$, the above quotients are only 1/80 and 1/8, respectively. For a thin plate, $t/b < (1/20)$, and it is clear that stresses σ_z and τ_{yz} are very small compared with the normal stress components in the xy plane.

b. Now Equation 3.34 with $p = p_0$, on successively integrating four times and satisfying $w = 0$ and $d^2w/dy^2 = 0$ at $y = 0$ and $y = b$, yields

$$w = \frac{p_0 b^4}{24D} \left(\frac{y^4}{b^4} - 2\frac{y^3}{b^3} + \frac{y}{b} \right) \tag{3.36}$$

This represents the deflection of a uniformly loaded and simply supported plate strip parallel to the y axis. The maximum deflection of the plate is found by substituting $y = b/2$ in Equation 3.36, yielding $w_{\max} = 5p_0b^4/384D$. The largest moment and stress also occur at $y = b/2$, in the direction of the shorter span b. These are readily calculated (by means of Equations 3.36, 3.9, and 3.11) as $p_0b_2/8$ and $3p_0b_2/4t^2$, respectively. It is observed that for very long and narrow plates, the supports along the short sides have little effect on the action in the plate, and hence the *plate behaves as would a simple beam* of span b.

Introducing the given numerical values into Equation 3.10,

$$D = \frac{200(10^9)(0.01)^3}{12(1 - (1/9))} = 18{,}750$$

Similarly,

$$w_{\max} = \frac{5p_0b^4}{384D} = \frac{5(10 \times 10^3)(0.4)^4}{384(18{,}750)} = 0.18 \text{ mm}$$

$$\sigma_{y,\max} = \frac{3p_0b^2}{4t^2} = \frac{3(10 \times 10^3)}{4}\left(\frac{400}{10} \right)^2 = 12 \text{ MPa}$$

$$\sigma_{x,\max} = \nu\sigma_{y,\max} = 4 \text{ MPa}$$

Interestingly, Hooke's law, with $\sigma_z = 0$, gives

$$\varepsilon_{y,\max} = \frac{1}{200(10^9)}\left(12 - \frac{4}{3} \right)10^6 = 53\,\mu\text{m}$$

and, from Equation 3.3b, the radius of curvature is

$$r_y = -\frac{0.01}{2(53\times 10^{-6})} = -94.3 \text{ m}$$

Thus, $w_{\max}/t = 0.018$ and $r_y/b = 236$. These results show that the deflection curve is *extremely flat*, as is usually the case for small deformations.

3.11.2 Variously Loaded Plates

The procedure given in Section 3.11.1 can be used for very long plates under a variety of loads and with any set of boundary condition. Derivation of the expressions for the deflection surface of a *circular plate* under uniform loading is demonstrated in Example 3.2. A plate that transmits a constant bending moment is said to be in *pure bending*. We consider this and another simple practical case in Examples 3.3 and 3.4 to follow.

EXAMPLE 3.2: Deflection Plate of Clamped Circular Plate

A circular plate, clamped at the edge, is under a uniform pressure of intensity p_0 (Figure 3.11). Derive an expression for the surface deflection.

Solution

The equation of the boundary of the plate is

$$1 - \frac{x^2}{a^2} - \frac{y^2}{a^2} = 0 \tag{b}$$

The conditions for the edge

$$w = 0 \qquad \frac{\partial w}{\partial x} = \frac{\partial w}{\partial y} = 0 \qquad (r = a) \tag{c}$$

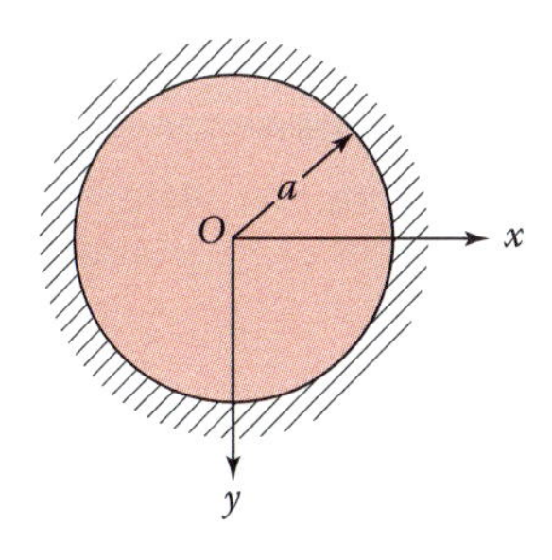

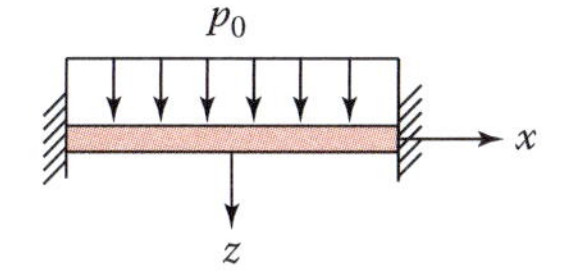

FIGURE 3.11
Circular plate carries uniform load.

are satisfied by taking the deflection w in the form

$$w = k\left(1 - \frac{x^2}{a^2} - \frac{y^2}{a^2}\right)^2 \quad \text{(d)}$$

Here k represents an unknown constant. Note that Equation d and its first derivatives with respect to x and y become zero at the boundary by virtue of Equation b. The fourth derivatives of w are

$$\frac{\partial^4 w}{\partial x^4} = 24\frac{k}{a^4} \qquad \frac{\partial^4 w}{\partial x^2 \partial y^2} = 8\frac{k}{a^4} \qquad \frac{\partial^4 w}{\partial y^4} = 24\frac{k}{a^4}$$

Substitution of these into Equations 3.17 leads to

$$\frac{k}{a^4}(24 + 16 + 24) = \frac{p_0}{D}$$

or

$$k = \frac{p_0 a^4}{64D} \quad \text{(e)}$$

The deflection is thus

$$w = \frac{p_0 a^4}{64D}\left(1 - \frac{x^2}{a^2} - \frac{y^2}{a^2}\right)^2 \quad (3.37)$$

The moments and the stresses corresponding to Equation 3.37 are derived from Equations 3.9 and 3.7, respectively.

Similarly, the deflection of a uniformly loaded elliptical plate with clamped edge is readily derived (see Section 6.4). However, the circular plate problems are usually treated by employing polar coordinates, as will be illustrated in Chapter 4.

EXAMPLE 3.3: Corner Displacements of Rectangular Plate

Determine the displacement of a rectangular plate with free edges of lengths a and b and subjected to transverse corner forces of magnitude P, as shown in Figure 3.12.

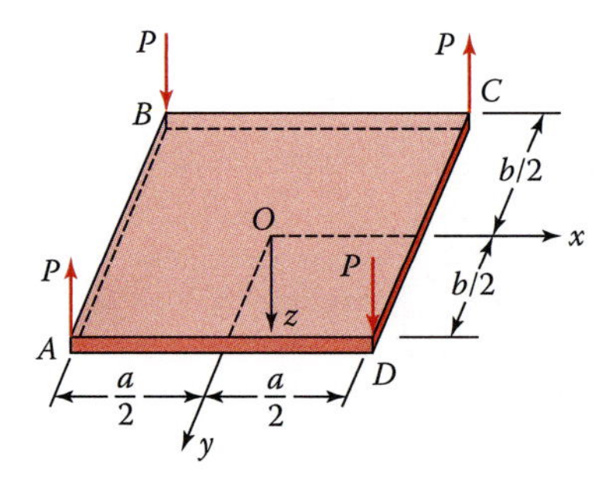

FIGURE 3.12
Plate is under corner loads.

Solution

The plate is free from surface loading, and hence $p = 0$. The boundary conditions are represented by

$$\begin{aligned} M_x &= 0 \qquad V_x = 0 \qquad \left(x = \pm\frac{a}{2}, y\right) \\ M_y &= 0 \qquad V_y = 0 \qquad \left(x, y = \pm\frac{b}{2}\right) \end{aligned} \tag{f}$$

Inasmuch as the center of the plate is free of displacement, we may assume the deflection w in the form

$$w = c_1 xy \tag{g}$$

where c_1 is a constant. It is readily shown that Equations 3.18 and f are fulfilled by this expression. On the basis of Equations 3.24, 3.9, and g, the corner conditions, $F_c = -2M_{xy}$, lead to $P = 2D(1-\nu)c_1$ or $c_1 = P/2D(1-\nu)$.

Substituting this value of c_1 into Equation g, the surface deflection is found to be

$$w = \frac{P}{2b(1-\nu)} xy \tag{3.38}$$

The corner displacements are then

$$w_B = w_D = -w_A = -w_C = \frac{Pab}{8D(1-\nu)} \tag{3.39}$$

Here the negative sign indicates an upward direction.

EXAMPLE 3.4: Pure Bending of Large Rectangular Plate

A rectangular bulkhead of an elevator shaft is subjected to uniformly distributed bending moments $M_x = M_b$ and $M_y = M_a$, applied along its edges (Figure 3.13a). Derive the equation governing the surface deflection for two cases: (a) $M_a \neq M_b$, (b) $M_a = -M_b$.

Solution

a. We have positive edge moments applied to the plate (Figure 3.13a). Substituting $M_b = M_x$ and $M_a = M_y$ into Equations 3.9 gives

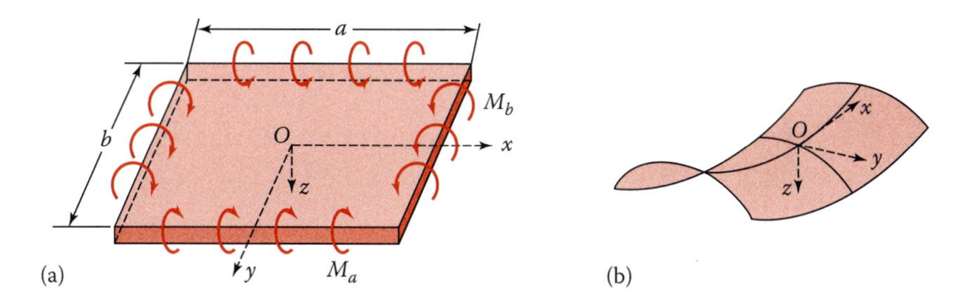

FIGURE 3.13
(a) Plate with edge moments; (b) saddle-shaped deflection surface.

$$\frac{\partial^2 w}{\partial x^2} = -\frac{M_b - \nu M_a}{D(1-\nu^2)} \quad \frac{\partial^2 w}{\partial y^2} = -\frac{M_a - \nu M_b}{D(1-\nu^2)} \quad \frac{\partial^2 w}{\partial x \partial y} = 0 \tag{h}$$

Integrating the above leads to

$$w = -\frac{M_b - \nu M_a}{2D(1-\nu^2)}x^2 - \frac{M_a - \nu M_b}{2D(1-\nu^2)}y^2 + c_1 x + c_2 y + c_3$$

If the origin of xyz is located at the center and midsurface of the *deformed plate*, the constants of integration vanish, and we have

$$w = -\frac{M_b - \nu M_a}{2D(1-\nu^2)}x^2 = \frac{M_a - \nu M_b}{2D(1-\nu^2)}y^2 \tag{3.40}$$

b. Now y edges are subjected to negative moments $-M_a$. Thus, by letting $M_a = -M_b$ in Equations h, the result is

$$\kappa_x = -\kappa_y = \frac{\partial^2 w}{\partial x^2} = -\frac{M_b}{D(1-\nu)} \tag{i}$$

This reveals that there is a saddle point 0 at the center of the plate (Figure 3.13b). Integrating and locating the origin xyz as before, Equation i leads to

$$w = -\frac{M_b}{2D(1-\nu)}(x^2 - y^2) \tag{3.41}$$

Comments: It is clear that the above expression represents an *anticlastic* or saddle-shaped surface with a negative Gaussian curvature. We note that in the particular case where $M_a = M_b$, Equation 3.40 yields a *paraboloid of revolution* (see Problem 3.18).

3.12 Strain Energy of Plates

The strain energy stored in an elastic body, for a general state of stress, is expressed by substituting Equation 2.50 into Equation 2.53. In so doing, we obtain

$$U = \frac{1}{2}\iiint_V (\sigma_x \varepsilon_x + \sigma_y \varepsilon_y + \sigma_z \varepsilon_z + \tau_{xy}\gamma_{xy} + \tau_{xz}\gamma_{xz} + \tau_{yz}\gamma_{yz})dx\,dy\,dz \tag{3.42}$$

Integration extends over the entire body volume. According to the assumptions of Section 3.2, for thin plates, σ_z, γ_{xz}, and γ_{yz} can be omitted. Thus, by Hooke's law and Equations 3.6, the above expression reduces to the following form involving only stresses and elastic constants:

$$U = \iiint_V \left[\frac{1}{2E}\left(\sigma_x^2 + \sigma_y^2 - 2\nu\sigma_x\sigma_y\right) + \frac{1}{2G}\tau_{xy}^2\right] dx\,dy\,dz \tag{3.43}$$

For a plate of *uniform thickness*, Equation 3.43 may be written in terms of deflection w by use of Equations 3.7 and 3.10 as follows:

$$U = \frac{1}{2}\iint_A D\left[\left(\frac{\partial^2 w}{\partial x^2}\right)^2 + \left(\frac{\partial^2 w}{\partial y^2}\right)^2 + 2\nu\frac{\partial^2 w}{\partial x^2}\frac{\partial^2 w}{\partial y^2} + 2(1-\nu)\left(\frac{\partial^2 w}{\partial x\,\partial y}\right)^2\right] dx\,dy$$

or, alternately,

$$U = \frac{1}{2}\iint_A D\left\{\left(\frac{\partial^2 w}{\partial x^2} + \frac{\partial^2 w}{\partial y^2}\right)^2 - 2(1-\nu)\left[\frac{\partial^2 w}{\partial x^2}\frac{\partial^2 w}{\partial y^2} - \left(\frac{\partial^2 w}{\partial x\,\partial y}\right)^2\right]\right\} dx\,dy \tag{3.44}$$

where A is the area of the plate surface.

The second term in Equation 3.44 is known as the *Gaussian curvature*. We observe that the strain energy is a *nonlinear* (quadratic) function of deformation or stress. The principle of superposition is therefore *not* valid for the strain energy.

Comment: In the case of a plate experiencing a temperature change, the strain energy that results from heating or cooling must also be included in the above expressions.

3.13 Energy Methods in Theory of Plates: Variational Principles

It is recalled from Section 1.2 that, as an alternative to the *equilibrium methods*, the analysis of deformation and stress in an elastic body can be accomplished by employing *energy methods*. These two techniques are, respectively, the *Newtonian* and *Lagrangian* approaches to mechanics. The last is predicated on the fact that the governing equation of a deformed elastic body is derivable by minimizing the energy associated with deformation and loading. Applications of energy methods are effective in situations involving irregular shapes, nonuniform loads, variable cross sections, and anisotropic materials.

Equations 3.42 through 3.44 are useful in the formulation of various energy techniques and numerical finite element approaches. We shall review some commonly employed strain-energy methods based on the potential and a variation of deformation on an elastic body [25–29]. Next, we introduce three commonly used variational principles of solid mechanics.

3.13.1 The Principle of Virtual Work

Suppose that an elastic body undergoes an *arbitrary incremental displacement* or so-called *virtual displacement*. This displacement need not actually occur and need not be infinitesimal. When the displacement is taken to be infinitesimal, as is often done, it is reasonable to consider the system of forces acting on the body as *constant*.

The virtual work done by surface forces p per unit area on the body in the process of bringing the body from the initial state to the equilibrium state is expressed as

$$\delta W = \int_A (p_x\,\delta u + p_y\,\delta v + p_z\,\delta w)\,dA \tag{3.45}$$

In the preceding, A is the boundary surface, and δu, δv, and δw are the x-, y-, and z-directed virtual displacements. The notation δ denotes the *variation* of a quantity. The strain energy δU acquired by a body of volume V as a result of virtual straining is

$$\delta U = \int_V (\sigma_x\,\delta\varepsilon_x + \sigma_y\,\delta\varepsilon_y + \sigma_z\,\delta\varepsilon_z + \tau_{xy}\,\delta\gamma_{xy} + \tau_{xz}\,\delta\gamma_{xz} + \tau_{yz}\,\delta\gamma_{yz})\,dV \tag{3.46}$$

The total work done during the virtual displacement is zero: $\delta W - \delta U = 0$. The *principle of virtual work* for an elastic body is thus represented as

$$\delta W = \delta U \tag{3.47}$$

It should be noted that the virtual work done by virtual displacements is the work done by actual forces in displacing the body through virtual displacements that are consistent with the geometric constraints.

3.13.2 The Principle of Minimum Potential Energy

A special case of the principle of virtual displacements dealing with linear as well as non-linear elastic bodies referred to as the *principle of minimum potential energy*. Inasmuch as the virtual displacements do not alter the shape of the body and the surface forces are regarded as constants, Equation 3.47 can be written as follows:

$$\delta\Pi = \delta(U - W) = 0 \tag{3.48}$$

In this expression,

$$\Pi = U - W \tag{3.49}$$

denotes the *potential energy* of the body, also known as the *total potential energy*. Equation 3.48 represents the condition of stationary potential energy of the system. It can be shown that, for *stable equilibrium*, the potential energy is a minimum. For all displacements satisfying given boundary conditions and the equilibrium conditions, the potential energy will assume a minimum value. This is referred to as the *principle of minimum potential energy*.

The potential energy stored in a plate under a *distributed lateral load* $p(x, y)$ is

$$\Pi = \frac{1}{2}\iiint_V (\sigma_x\varepsilon_x + \sigma_y\varepsilon_y + \tau_{xy}\gamma_{xy})\,dx\,dy\,dz - \iint_A (pw)\,dx\,dy \tag{3.50}$$

For the case of *constant* plate *thickness*, the above may be written as

$$\Pi = -\frac{1}{2}\iint_A (M_x\kappa_x + M_y\kappa_y + 2M_{xy}\kappa_{xy})\,dx\,dy - \iint_A (pw)\,dx\,dy \tag{3.51}$$

A physical explanation of the terms of U in this expression is as follows. As $\partial^2 w/\partial x^2 = \kappa_x$ represents the curvature of the plate in the xz plane, the angle corresponding to the moment $M_x\,dy$ equals $-(\partial^2 w/\partial x^2)dx$. The strain energy or work done by the moments $M_x\,dy$ is thus $-(1/2)M_x\kappa_x\,dx\,dy$. The strain energy due to $M_y\,dx$ and $M_{xy}\,dy$ are interpreted similarly. The principle of potential energy, referring to Equation 3.51, is expressed in the form

$$\delta\Pi = -\iint_A \left(M_x\,\delta\kappa_x + M_y\,\delta\kappa_y + 2M_{xy}\,\delta\kappa_{xy}\right)dx\,dy - \iint_A \left(p\,\delta w\right)dx\,dy = 0 \tag{3.52}$$

3.13.3 The Ritz Method

A number of approximate methods that can be used to solve differential equations exist. These include finite difference and finite element methods. The so-called *Ritz method* is a direct convenient procedure for determining solutions by the principle of minimum potential energy. The essence of this approach is described for the case of elastic bending of plates as follows:

First, choose a solution for the deflection w in the form of a series containing undetermined parameters $a_{mn}(m,n=1,2,\ldots)$. The deflection so selected must satisfy the geometric boundary conditions. The static boundary conditions need not be fulfilled. A variety of methods for choosing suitable deflections for bending analysis of plates by the Ritz approach are available [2].

Clearly, a proper choice of the deflection expression is important to ensure good accuracy for the final solution. Thus, it is desirable to assume an expression for w that is nearly identical with the true bent surface of the plate. Next, employing the selected solution, determine the potential energy Π in terms of a_{mn}. (This demonstrates that the a_{mn}s govern the variation of the potential energy.)

In order that the *potential energy be a minimum at equilibrium,*

$$\frac{\partial\Pi}{\partial a_{11}} = 0,\ldots, \qquad \frac{\partial\Pi}{\partial a_{mn}} = 0 \tag{3.53}$$

The foregoing represents a system of algebraic equations that are solved to yield the parameters a_{mn}. Introducing these values into the assumed expression for deflection, one obtains the solution for a given problem. In general, a_{mn} includes only a finite number of parameters, and the final results are therefore only approximate. Of course, if the assumed w should happen to be the "exact" one, the solution will then be "exact."

Advantages of the Ritz approach lie in the relative ease with which mixed edge conditions can be handled. This method is among the simplest for solving plate and shell deflections by means of a hand calculation.

The applications of the strain-energy techniques in the treatment of bending, stretching, as well as buckling problems of plates and shells will be discussed throughout the text.

3.14 *Natural Frequencies of Plates by the Energy Method

The energy methods are also applicable for determining natural frequencies since the classical theory of *plate vibrations* is based on the assumptions of Section 3.3. In this dynamic treatment, the only inertia forces considered are those due to the lateral translation of the plate elements; the effect of rotary inertia is neglected [30]. Clearly, now the total energy given by Equation 3.49 must be augmented by an additional term, representing the kinetic energy.

Let us designate by $\bar{m}$ the *mass per unit area* and the velocity by $\dot{w}$, which is the first derivative of the displacement w with respect to time. For a vibrating plate of uniform thickness, the *kinetic energy* is thus

$$T = \frac{\bar{m}}{2} \iint \dot{w}^2 \, dx \, dy \tag{3.54}$$

Integrations proceed over the surface area A of the plate.

The reader will recall from a study of dynamics that the frequency of a free vibration of a system is called its natural frequency (f_n) and is measured in cycles per second (Hz). The lowest natural frequency is the fundamental mode; this has the greatest amplitudes and usually the greatest stresses. The displacements continue to decrease for higher modes. Note that the time variation of the configuration is referred to as a *mode of vibration*. The natural frequency is often used for appraisal of the dynamic characteristics of structural members and complex systems.

The ratio of the maximum dynamic to static displacements, termed dynamic load factor or *transmissibility* ϵ, for lightly damped systems may be approximated by

$$\epsilon = \frac{1}{1-(f/f_n)^2} \tag{3.55}$$

Here f represents the frequency of the simple harmonic disturbance. If the f/f_n ratio approaches unity, the amplitude of the forced motion becomes infinitely large. This condition, which is most feared by structural engineers, is called *resonance*. A static load per area equivalent of *dynamic load per unit area* p_d can be obtained from

$$p_d = \bar{m} g_{in} \epsilon \tag{3.56}$$

where g_{in} is the input acceleration.

Vibration problems in plates are of practical interest in connection with the application to the telephone receiver, printed-circuit boards used in electronic systems [31], and other devices. By using the Ritz method together with the expressions for T and U, the natural frequencies of vibration can readily be obtained. Then, replacing p_0 by p_d in the formulas derived by static analysis, dynamic stresses and deflections are calculated, as will be shown for some particular cases in Examples 4.6 and 5.17.

Problems

Sections 3.1 through 3.7

3.1 A rectangular sheet metal of 5 mm thickness is bent into a circular cylinder having radius r. Calculate the diameter of the cylinder and the maximum moment developed in the metal if the allowable stress is not to exceed 96 MPa, $E = 70$ GPa, and $\nu = 0.3$.

3.2 A 20-mm-wide and 0.5-mm-thick steel band saw ($E = 210$ GPa, $\nu = 0.3$) runs over two pulleys of 0.5-m diameter (Figure P3.2). Calculate the maximum bending strain and maximum bending stress in the saw as it goes over and confirms the radius r of a pulley.

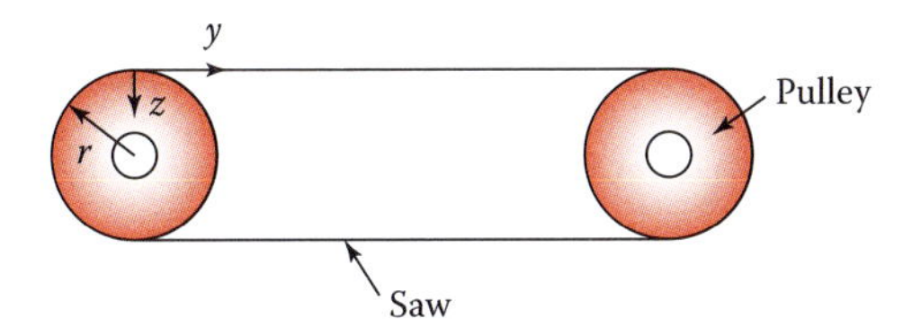

FIGURE P3.2

3.3 Verify the result given by Equation 3.5 by employing Mohr's circle.

3.4 By means of Mohr's circle, show that the Gaussian curvature is equal to

$$\kappa_x \kappa_y - \kappa_{xy}^2 = \kappa_1 \kappa_2 \tag{P3.4}$$

3.5 Demonstrate that at a corner of a polygonal, simply supported plate, $M_{xy} = 0$ unless the corner is 90°.

Sections 3.8 through 3.14

3.6 A steel plate ($E = 200$ GPa, $\nu = 0.3$) of thickness $t = 12$ mm covers a circular opening having a diameter of 280 mm. The plate is clamped at the edge and is under a uniform load of intensity $p_0 = 5$ MPa. Calculate: (a) the maximum deflection; (b) the maximum strain at the center; (c) the maximum stress $\sigma_{x,\max}$ at the edge.

3.7 For the plate described in Example 3.1b, assuming that $p_0 = 10$ kPa, $b = 0.4$ m, $t = 10$ mm, $E = 200$ GPa, and $\nu = 1/3$, calculate the maximum deflection, maximum bending stress, maximum strain (by taking $\sigma_z = 0$), and radius of curvature at center.

3.8 A long and narrow rectangular plate carries a nonuniform loading $p = p_0 \sin(\pi y/b)$ and is clamped at edges $y = 0$ and $y = b$ (Figure P3.8). (a) Derive an expression for the deflection surface w. (b) Determine the maximum deflection and maximum

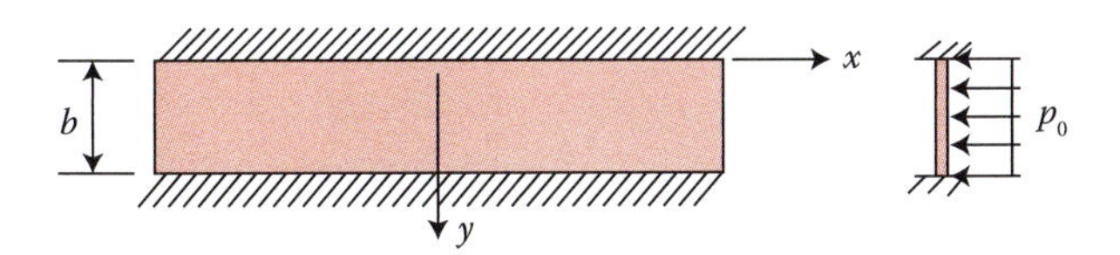

FIGURE P3.8

strain. (c) Find the radius of curvature at $y = b/3$ by taking $b = 0.6$ m, $t = 15$ mm, $E = 200$ GPa, $\nu = 1/3$, and $p_0 = 50$ kPa.

3.9 The uniform load p_0 acts on a long and narrow rectangular plate with edge $y = 0$ simply supported and edge $y = b$ clamped (Figure P3.9). Determine: (a) the equation of the surface deflection w; (b) the maximum bending stress; (c) the maximum bending strain at the center for $\nu = 1/4$.

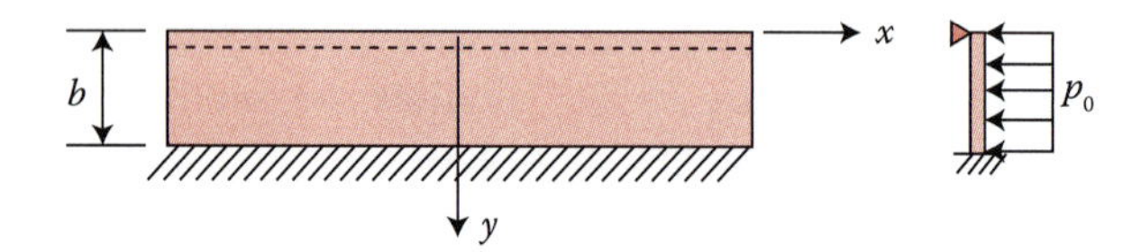

FIGURE P3.9

3.10 Redo Problem 3.8 for the case in which the plate is subjected to a uniform load p_0.

3.11 A long and narrow rectangular aluminum alloy plate is subjected to a hydrostatic loading expressed by $p = p_0 y/b$ and support ed as shown in Figure P3.11. (a) Derive the equation of the deflection surface w. (b) Find the slope at the simply supported edge. (c) Calculate the maximum stress $\sigma_{y,\max}$ for the following data: $b = 0.5$ m, $t = 10$ mm, $p_0 = 100$ kPa.

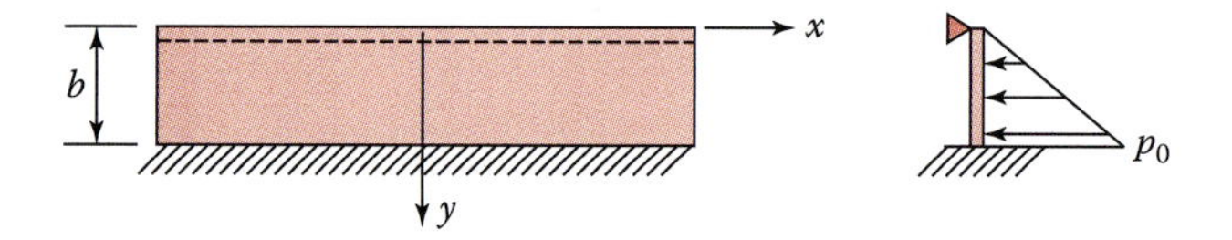

FIGURE P3.11

3.12 The lateral deflection of a rectangular plate (Figure P3.12), with built-in edges of lengths a and b and subjected to a uniform load p_0, is given by

$$w = c_0\left(\frac{x^4}{a^4} - 2\frac{x^3}{a^3} + \frac{x^2}{a^2}\right)\left(\frac{y^4}{b^4} - 2\frac{y^3}{b^3} + \frac{y^2}{b^2}\right)$$

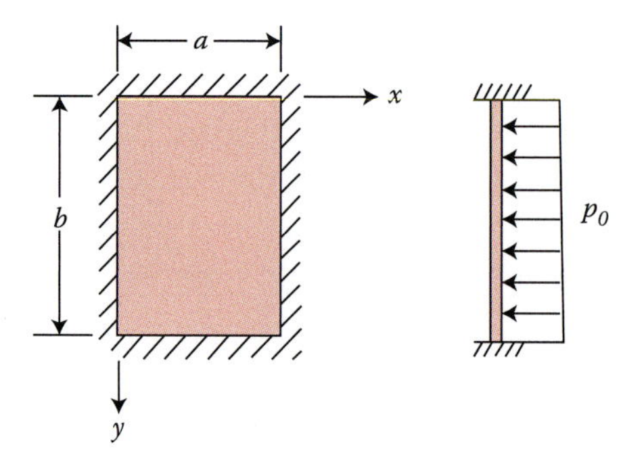

FIGURE P3.12

where c_0 is a constant. Determine: (a) whether this deflection satisfies the boundary conditions of the plate; (b) the maximum plane stress components σ_x and τ_{xy} at the center, for $a = b$.

3.13 A rectangular plate has two opposite edges $y = 0$ and $y = b$ simply supported, the third edge $x = 0$ clamped, and the fourth edge $x = a$ free (Figure P3.13). An approximate expression for the de-flection surface is

$$w = c\left(\frac{x}{a}\right)^2 \sin\frac{\pi y}{b} \tag{P3.13}$$

where c is a constant. Determine: (a) whether this deflection fulfills all boundary conditions of the plate; (b) the approximate maximum plate strain components at the center, for $a = b$ and $\nu = 1/3$.

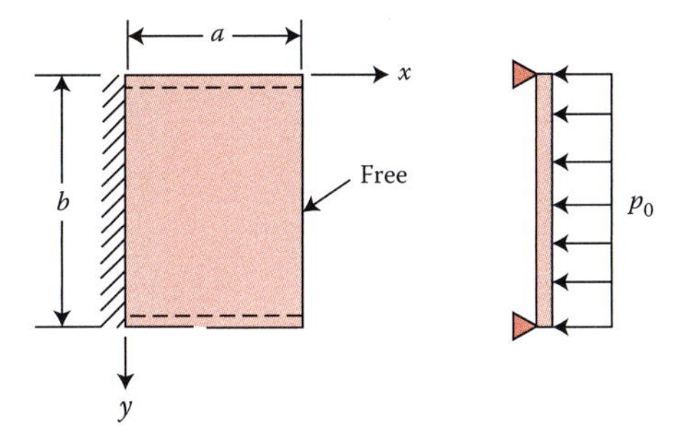

FIGURE P3.13

3.14 Redo Problem 3.13a for the case in which the plate is square ($a = b$).

3.15 An approximate expression for the deflection surface of a clamped triangular plate (Figure P3.15) is given in the form

$$w = cx^2y^2\left(1 - \frac{x}{a} - \frac{y}{b}\right)^2 \tag{P3.15}$$

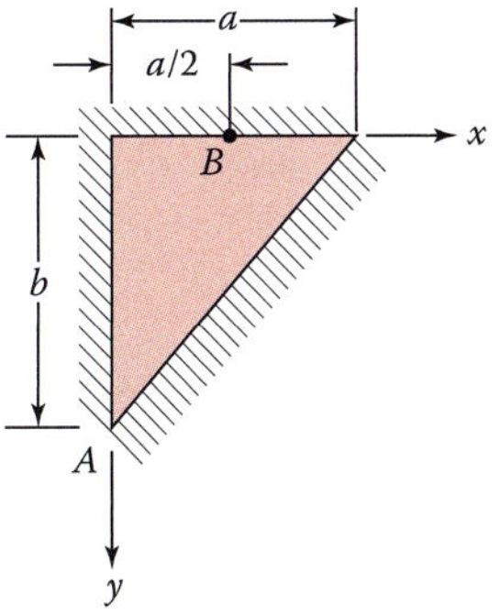

FIGURE P3.15

where c is a constant. (a) Verify that Equation P3.15 satisfies the boundary conditions. (b) Determine the approximate maximum plane stress components at points A and B.

3.16 A square spacecraft panel (Figure P3.16) is subjected to uniformly distributed twisting moment $M_{xy} = M_0$ at all four edges. Determine an expression for the deflection surface w.

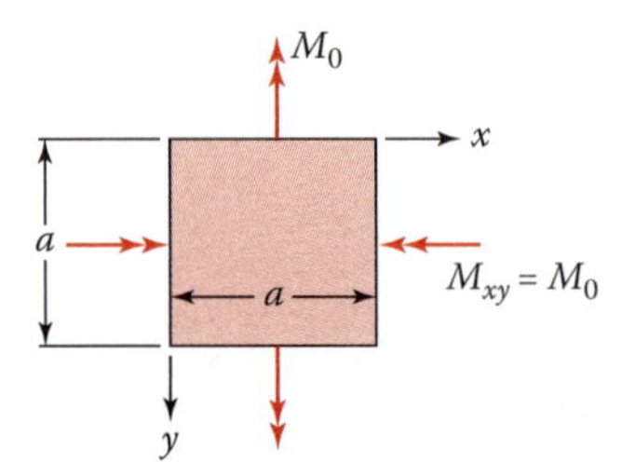

FIGURE P3.16

3.17 Determine whether the following expression satisfies the boundary conditions of a simply supported very long and narrow plate (Figure 3.10) carrying a concentrated load P at a point (0, $b/2$):

$$w = \frac{Pb^2}{2\pi^3 D} \sum_{n=1}^{\infty} \frac{1}{n^3} (1 + \alpha x) e^{-\alpha x} \sin \frac{n\pi}{2} \sin \alpha y \tag{P3.17}$$

Here $\alpha = n\pi/b$ and $x \geq 0$. Using $\nu = 1/3$, obtain the corresponding expressions for: (a) the moment M_x in terms of M defined by Equation 3.19; (b) the maximum stress σ_x for $n = 1$.

3.18 Consider the bending of the rectangular plate (Figure 3.13) for the particular case in which $M_x = M_y = M_0$. (a) Verify that, in this case, even for a plate of *any shape*, the bending moments M_0 are uniform along the boundary and the twisting moments vanish. (b) Show that Equations 3.40 and 3.9 yield, respectively,

$$w = -\frac{M_0(x^2 + y^2)}{2D(1+\nu)}, \qquad \kappa_x = \kappa_y = -\frac{M_0}{D(1+\nu)} \tag{P3.18}$$

It is seen that the first of the above expressions represents a paraboloid of revolution, while the second implies that the surface is spherical. Explain why the results are inconsistent.

3.19 Determine an expression of the strain energy for the plate described in Problem 3.12.

3.20 Resolve Problem 3.19 for the case in which the plate is a square ($a = b$).

References

1. S. Timoshenko and S. Woinowsky-Krieger, *Theory of Plates and Shells*, McGraw-Hill, New York, 1959.
2. R. Szilard, *Theory and Analysis of Plates*, Prentice-Hall, Upper Saddle River, NJ, 2004.

3. J. R. Vinson, *Structural Mechanics: The Behavior of Plates and Shells*, Wiley, Hoboken, NJ, 1974.
4. D. MacFarland, B. L. Smith, and W. D. Bernhart, *Analysis of Plates*, Spartan Books, Washington, DC, 1972.
5. B. Aalami and D. G. Williams, *Thin Plate Design for Transverse Loading*, Wiley, Hoboken, NJ, 1975.
6. J. N. Reddy, *Theory and Analysis of Elastic Plates and Shells*, 2nd ed., CRC/Taylor & Francis, Boca Raton, FL, 2007.
7. C. P. Heins, *Applied Plate Theory for Engineers*, Lexington Books, Lexington, MA, 1976.
8. H. Altenbach and G. I. Mikhasev (eds.), *Shell and Membrane Theories in Mechanics and Biology*, Springer, Switzerland, 2015.
9. E. Ventsel and T. Krauthammer, *Thin Plates and Shells: Theory, Analysis and Applications*, CRC Press, Boca Raton, FL, 2001.
10. W. Flügge (ed.), *Handbook of Engineering Mechanics*, Chapter 39, McGraw-Hill, New York, 1984.
11. A. Nadai, *Die elastichen Platten*, Springer, Berlin, 1924.
12. H. Marcus, *Die Theorie elastichen Gewebe und ihre Anwedung auf die Berechnung biegsamer Platten*, Springer, Berlin, 1924.
13. L. H. Donnell, *Beams, Plates and Shells*, McGraw-Hill, New York, 1976.
14. J. R. Vinson, *The Behavior of Thin Walled Structures: Beams, Plates, and Shells*, Kluwer, Dordrecht, The Netherlands, 1989.
15. D. E. Beskos, *Boundary Element Analysis of Plates and Shells*, Springer, Berlin, 1991.
16. G. Kirchhoff, *Vorlesungen über mathematische Physik*, Vol. 1, B. G. Teubner, Leipzig, 1876.
17. A. E. H. Love, *A Treatise on the Mathematical Theory of Elasticity*, Dover, New York, 1944.
18. J. Prescott, *Applied Elasticity*, Dover, New York, 1946.
19. E. Reissner, The effect of transverse shear deformation on the bending of elastic plates, *Journal of Applied Mechanics*, 12, 1954, A69–A77.
20. S. Timoshenko, *History of Strength of Materials*, Dover, New York, 1983.
21. I. Todhunter and K. Pearson, *A History of the Theory of Elasticity and the Strength of Materials*, Vols. I and II, Dover, New York, 1960.
22. D. M. Boyajian, D. C. Weggel, and S. Chen, Instituting seminal teaching-Datums: Examples from plate and shell theory, *World Transactions on Engineering and Technology Education*, 5(1), 2006.
23. M. J. Turner, R. W. Clough, G. C. Martin, and L. J. Topp, Stiffness and deflection analysis of complex structures, *Journal of Aeronautical Science*, 23, 1956, 805–823.
24. J. H. Argyis, Energy theorems of structural analysis, *Aircraft Engineering*, 26, 1954, 347–356 and 383–387.
25. A. C. Ugural and S. K. Fenster, *Advanced Mechanics of Materials and Applied Elasticity*, 5th ed., Prentice-Hall, Upper Saddle River, NJ, 2012.
26. J. N. Reddy, *Energy and Variational Methods in Applied Mechanics*, Wiley, Hoboken, NJ, 2002.
27. E. Kreyszic, *Advanced Engineering Mathematics*, 10th ed., Wiley, Hoboken, NJ, 2011.
28. A. C. Ugural, *Mechanical Design Machine Components*, CRC Press, Boca Raton, FL, 2015.
29. J. T. Oden and E. A. Rippinger, *Mechanics of Elastic Structures*, 2nd ed., McGraw-Hill, New York, 1981.
30. S. Timoshenko, D. H. Young, and W. Weaver Jr., *Vibration Problems in Engineering*, 4th ed., Wiley, Hoboken, NJ, 1974.
31. D. S. Steinberg, *Vibration Analysis for Electronic Equipment*, 2nd ed., Wiley, Hoboken, NJ, 1988.

4

Circular Plates

4.1 Introduction

In practice, members that carry transverse loads, such as end plates and closures of pressure vessels, pump diaphragms, telephone and loudspeaker diaphragms, thrust-bearing plates, piston heads, diffusers, clutches, springs made of assemblages of plates, turbine disks, manhole covers, and so on, are usually circular in shape. Thus, many of the significant applications fall within the scope of the formulas derived for circular plates. A number of circular-plate problems have stress distributions that are symmetrical about the center.

We shall treat the bending of circular plates of constant thickness subject to symmetrical loading and boundary conditions in Sections 4.5 through 4.11. There is also a brief discussion on plates of variable thickness. Sections 4.12 and 4.13 deal with the situations involving asymmetrical loading. In all cases, the basic relationships in polar coordinates, developed in Sections 4.2 through 4.4, are employed.

4.2 Basic Relations in Polar Coordinates

In general, polar coordinates are preferred over Cartesian coordinates (used exclusively thus far), where a degree of axial symmetry exists either in geometry or in loading. Examples include a circular plate and a large thin plate containing holes.

The polar coordinate set (r, θ) and the Cartesian set (x, y) are related by the equations (Figure 4.1a):

$$x = r\cos\theta \qquad r^2 = x^2 + y^2$$

$$y = r\sin\theta \qquad \theta = \tan^{-1}\frac{y}{x}$$

Referring to the above,

$$\frac{\partial r}{\partial x} = \frac{x}{r} = \cos\theta \qquad \frac{\partial r}{\partial y} = \frac{y}{r} = \sin\theta$$

$$\frac{\partial \theta}{\partial x} = -\frac{y}{r^2} = -\frac{\sin\theta}{r} \qquad \frac{\partial \theta}{\partial y} = \frac{x}{r^2} = \frac{\cos\theta}{r}$$

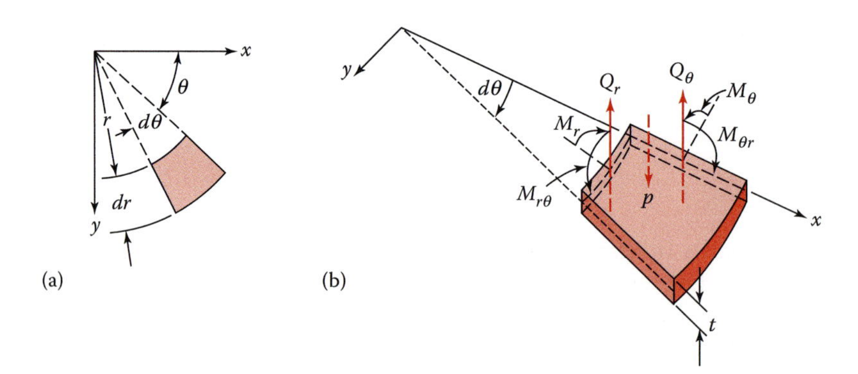

FIGURE 4.1
(a) Polar coordinates; (b) circular plate element.

Inasmuch as the deflection is a function of r and θ, the *chain rule*, together with the above relationship, leads to

$$\frac{\partial w}{\partial x} = \frac{\partial w}{\partial r}\cos\theta - \frac{1}{r}\frac{\partial w}{\partial \theta}\sin\theta \tag{a}$$

To evaluate to expression $\partial^2 w/\partial x^2$ we use Equation a, this time applied to $\partial w/\partial x$ rather than to w:

$$\begin{aligned}\frac{\partial^2 w}{\partial x^2} &= \cos\theta\frac{\partial}{\partial r}\left(\frac{\partial w}{\partial x}\right) - \frac{1}{r}\sin\theta\frac{\partial}{\partial \theta}\left(\frac{\partial w}{\partial x}\right)\\ &= \frac{\partial^2 w}{\partial r^2}\cos^2\theta - 2\frac{\partial^2 w}{\partial\theta\partial r}\frac{\sin\theta\cos\theta}{r} + \frac{\partial w}{\partial r}\frac{\sin^2\theta}{r}\\ &\quad + 2\frac{\partial w}{\partial \theta}\frac{\sin\theta\cos\theta}{r^2} + \frac{\partial^2 w}{\partial\theta^2}\frac{\sin^2\theta}{r^2}\end{aligned} \tag{b}$$

Similarly,

$$\begin{aligned}\frac{\partial^2 w}{\partial y^2} = \frac{\partial^2 w}{\partial r^2}\sin^2\theta + 2\frac{\partial^2 w}{\partial\theta\partial r}\frac{\sin\theta\cos\theta}{r} + \frac{\partial w}{\partial r}\frac{\cos^2\theta}{r}\\ -2\frac{\partial w}{\partial \theta}\frac{\sin\theta\cos\theta}{r^2} + \frac{\partial^2 w}{\partial\theta^2}\frac{\cos^2\theta}{r^2}\end{aligned} \tag{c}$$

$$\begin{aligned}\frac{\partial^2 w}{\partial x\partial y} = \frac{\partial^2 w}{\partial r^2}\sin\theta\cos\theta + \frac{\partial^2 w}{\partial r\partial\theta}\frac{\cos 2\theta}{r} - \frac{\partial w}{\partial \theta}\frac{\cos 2\theta}{r^2}\\ -\frac{\partial w}{\partial r}\frac{\sin\theta\cos\theta}{r} - \frac{\partial^2 w}{\partial\theta^2}\frac{\sin\theta\cos\theta}{r^2}\end{aligned} \tag{d}$$

On substitution of Equations b and c into Equation d of Section 3.6, the Laplacian operator becomes

$$\nabla^2 w = \frac{\partial^2 w}{\partial r^2} + \frac{1}{r}\frac{\partial w}{\partial r} + \frac{1}{r^2}\frac{\partial^2 w}{\partial \theta^2} \tag{4.1}$$

Determination of the fundamental equations of a laterally loaded plate in polar coordinates requires only that the appropriate relationships of Chapter 3 be transformed from Cartesian to polar coordinates [1–3]. Consider now the state of moment and shear force on an infinitesimal element of thickness t, described in polar coordinates (Figure 4.1b). Note that, to simplify the derivations, the x axis is taken in the direction of radius r; that is, $\theta = 0$.

The *radial, tangential, twisting moments*, M_r, M_θ, $M_{r\theta} = M_{\theta r}$, and the *vertical shear forces*, Q_r, Q_θ, then have the same values as the moments M_x, M_y, M_{xy} and the shears Q_x, Q_y at the same point in the plate. Thus, letting $\theta = 0$ in Equations b, c, and d and substituting the resulting expressions into Equations 3.9 and 3.16, we have

$$M_r = -D\left[\frac{\partial^2 w}{\partial r^2} + \nu\left(\frac{1}{r}\frac{\partial w}{\partial r} + \frac{1}{r^2}\frac{\partial^2 w}{\partial \theta^2}\right)\right]$$

$$M_\theta = -D\left[\frac{1}{r}\frac{\partial w}{\partial r} + \frac{1}{r^2}\frac{\partial^2 w}{\partial \theta^2} + \nu\frac{\partial^2 w}{\partial r^2}\right]$$

$$M_{r\theta} = -(1-\nu)D\left(\frac{1}{r}\frac{\partial^2 w}{\partial r\,\partial \theta} - \frac{1}{r^2}\frac{\partial w}{\partial \theta}\right) \tag{4.2}$$

$$Q_r = -D\frac{\partial}{\partial r}(\nabla^2 w)$$

$$Q_\theta = -D\frac{1}{r}\frac{\partial}{\partial \theta}(\nabla^2 w)$$

Similarly, formulas for the *plane stress components*, from Equations 3.11, are written in the following form:

$$\sigma_r = \frac{12M_r z}{t^3} \qquad \sigma_\theta = \frac{12M_\theta z}{t^3} \qquad \tau_{r\theta} = \frac{12M_{r\theta} z}{t^3} \tag{4.3}$$

where M_r, M_θ, and $M_{r\theta}$ are defined by Equations 4.2. Clearly, the *maximum stresses* take place on the surfaces (at $z = \pm t/2$) of the plate.

The *effective transverse force* per unit length (Equation 3.23a) for an edge at $r = a$ and any θ becomes

$$V_r = Q_r + \frac{1}{r}\frac{\partial M_{r\theta}}{\partial \theta} = -D\left[\frac{\partial}{\partial r}(\nabla^2 w) + \frac{1-\nu}{r}\frac{\partial}{\partial \theta}\left(\frac{1}{r}\frac{\partial^2 w}{\partial r\,\partial \theta} - \frac{1}{r^2}\frac{\partial w}{\partial \theta}\right)\right] \tag{4.4a}$$

Expression (3.23b) appears as

$$V_\theta = Q_\theta + \frac{\partial M_{r\theta}}{\partial r} = -D\left[\frac{1}{r}\frac{\partial}{\partial \theta}(\nabla^2 w) + (1-\nu)\frac{\partial}{\partial r}\left(\frac{1}{r}\frac{\partial^2 w}{\partial r \partial \theta} - \frac{1}{r^2}\frac{\partial w}{\partial \theta}\right)\right] \tag{4.4b}$$

describing the effective transverse force on the edge at $\theta = \theta_0$ (constant) and any r.

On introduction of Equations b through d into Equation 3.17, the *governing differential equation* for plate deflection in polar coordinates is derived:

$$\nabla^4 w = \left(\frac{\partial^2}{\partial r^2} + \frac{1}{r}\frac{\partial}{\partial r} + \frac{1}{r^2}\frac{\partial^2}{\partial \theta^2}\right)\left(\frac{\partial^2 w}{\partial r^2} + \frac{1}{r}\frac{\partial w}{\partial r} + \frac{1}{r^2}\frac{\partial^2 w}{\partial \theta^2}\right) = \frac{p}{D} \tag{4.5}$$

Let w_h denote the *solution of the homogeneous equation*

$$\left(\frac{\partial^2}{\partial r^2} + \frac{1}{r}\frac{\partial}{\partial r} + \frac{1}{r^2}\frac{\partial^2}{\partial \theta^2}\right)\left(\frac{\partial^2 w_h}{\partial r^2} + \frac{1}{r}\frac{\partial w_h}{\partial r} + \frac{1}{r^2}\frac{\partial^2 w_h}{\partial \theta^2}\right) = 0 \tag{4.6}$$

and w_p the *particular solution* of Equation 4.5. The complete solution is expressed by

$$w = w_h + w_p \tag{e}$$

Determination of the particular solution, to be illustrated for a prescribed loading in Section 4.12, is relatively easy.

We assume the homogeneous or complementary solution to be expressed by the following series [1–3]:

$$w_h = \sum_{n=0}^{\infty} f_n \cos n\theta + \sum_{n=1}^{\infty} f_n^* \sin n\theta \tag{4.7}$$

where f_n and f_n^* are functions of r only. Substituting Equation 4.7 into Equation 4.6 and noting the validity of the resulting expression for all r and θ leads to two ordinary differential equations with the following solutions (Problem 4.18):

$$\begin{aligned}
f_0 &= A_0 + B_0 r^3 + C_0 \ln r + D_0 r^2 \ln r \\
f_1 &= A_1 r + B_1 r^3 + C_1 r^{-1} + D_1 r \ln r \\
f_n &= A_n r^n + B_n r^{-n} + C_n r^{n+2} + D_n r^{-n+2s} \\
f_1^* &= A_1^* r + B_1^* r^3 + C_1^* r^{-1} + D_1^* r \ln r \\
f_n^* &= A_n^* r^n + B_n^* r^{-n} + C_n^* r^{n+2} + D_n^* r^{-n+2}
\end{aligned} \tag{4.8}$$

In the preceding $A_n, \ldots, D_n^*$ are C constants, determined by satisfying the boundary conditions of the plate. On introducing these expressions for f_n and f_n^* into Equation 4.7, we obtain the solution of Equation 4.6 in a general form.

TABLE 4.1

Various Boundary Conditions for Circular Plates

Edge	Clamped	Simply Supported	Free	Sliding
At $r = r_0$ and any θ	$w = 0$	$w = 0$	$M_r = 0$	$\frac{\partial w}{\partial r} = 0$
	$\frac{\partial w}{\partial r} = 0$	$\frac{\partial^2 w}{\partial r^2} + \frac{\nu}{r}\frac{\partial w}{\partial r} = 0$	$V_r = 0$	$V_r = 0$

The *boundary conditions* at the edges of an annular circular plate of outer radius a and inner radius b may readily be written by referring to Equations 3.25 through 3.28, 4.2, and 4.4. They are listed in Table 4.1. It is noted that the inner (or outer) radius is represented by r_0. Clearly, nomenclature employed in the table parallels that defined in connection with rectangular plates in Section 3.9.

4.3 The Axisymmetrical Bending

The deflection w of a plate will depend on radial position r only when the applied load and end restraints are independent of the angle θ. The situation described is the axisymmetrical bending of the plate. That is, axisymmetry infers to the following:

1. The plate is continuous in the θ direction (in the region $0 \le \theta \le 2\pi$).
2. The loading is not a function of θ.
3. The boundary conditions do not vary around the circumference.

As a result of symmetry, all $(\partial/\partial\theta)[f(w)] = 0$, where $w = w(r)$ only on account of symmetry. We conclude therefore that

$$\frac{\partial(\ )}{\partial\theta} = \frac{\partial^2(\ )}{\partial\theta^2} = 0$$

and hence $M_{r\theta} = Q_\theta = 0$.

In the case of symmetry, only M_r, M_θ, and Q_r act on the circular plate element shown in Figure 4.1b. The moments and shear force, in an axisymmetrically loaded circular plate, are found from Equations 4.2 as follows:

$$\begin{aligned} M_r &= -D\left(\frac{d^2w}{dr^2} + \frac{\nu}{r}\frac{dw}{dr}\right) \\ M_\theta &= -D\left(\frac{1}{r}\frac{dw}{dr} + \nu\frac{d^2w}{dr^2}\right) \\ Q_r &= -D\frac{d}{dr}\left(\frac{d^2w}{dr^2} + \frac{1}{r}\frac{dw}{dr}\right) = -D\frac{d}{dr}\left[\frac{1}{r}\frac{d}{dr}\left(r\frac{dw}{dr}\right)\right] \end{aligned} \tag{4.9a–c}$$

The differential equation of the surface deflection (Equation 4.5) now reduces to

$$\nabla^4 w = \left(\frac{d^2}{dr^2} + \frac{1}{r}\frac{d}{dr}\right)\left(\frac{d^2 w}{dr^2} + \frac{1}{r}\frac{dw}{dr}\right) = \frac{p}{D} \tag{4.10a}$$

where p is the intensity of the lateral load.

The expressions for stress are readily obtained by substituting Equations b, c, and d, of Section 4.2 into Equations 3.7 and setting θ equal to zero:

$$\begin{aligned}\sigma_r &= -\frac{Ez}{1-\nu^2}\left(\frac{d^2w}{dr^2} + \frac{\nu}{r}\frac{dw}{dr}\right)\\ \sigma_\theta &= -\frac{Ez}{1-\nu^2}\left(\frac{1}{r}\frac{dw}{dr} + \nu\frac{d^2w}{dr^2}\right)\end{aligned} \tag{4.11}$$

To write Hooke's law in polar coordinates, it is necessary to replace subscripts x by r and y by θ in Equations 1.43 through 1.45 with the result that

$$\begin{aligned}\varepsilon_r &= \frac{1}{E}(\sigma_r - \nu\sigma_\theta)\\ \varepsilon_\theta &= \frac{1}{E}(\sigma_\theta - \nu\sigma_r)\\ \gamma_{r\theta} &= \frac{\tau_{r\theta}}{G}\end{aligned} \tag{a}$$

Introducing the identity

$$\nabla^2 w = \frac{d^2w}{dr^2} + \frac{1}{r}\frac{dw}{dr} = \frac{1}{r}\frac{d}{dr}\left(r\frac{dw}{dr}\right)$$

change Equation 4.10a to the form

$$\frac{1}{r}\frac{d}{dr}\left\{r\frac{d}{dr}\left[\frac{1}{r}\frac{d}{dr}\left(r\frac{dw}{dr}\right)\right]\right\} = \frac{p}{D} \tag{4.10b}$$

The deflection w is obtained by successive integrations when $p(r)$ is given

$$w = \int\frac{1}{r}\int r\int\frac{1}{r}\int\frac{rp}{D}\,dr\,dr\,dr\,dr \tag{4.12}$$

If the plate is under a *uniform loading* $p = p_0$, the general solution of Equations 4.10 is (Problem 4.19)

$$w = w_h + w_p = c_1 \ln r + c_2 r^2 \ln r + c_3 r^2 + c_4 + \frac{p_0 r^4}{64D} \tag{4.13}$$

where the cs are constants of integration. It is seen from a comparison of Equation 4.13 and the first of Equations 4.8 that the homogeneous solution f_0 represents the case of axisymmetrical bending of circular plates.

4.4 Equations of Equilibrium for Axisymmetrically Loaded Circular Plates

In discussing the condition of equilibrium of the plate element, we must take into consideration the variations in the stress resultants with the radial position r, as shown in Figure 4.2a. Referring to the figure, the sum of the moments of the forces about the y axis is expressed by

$$\left(M_r + \frac{dM_r}{dr} dr\right)(r + dr)\,d\theta - M_r r\,dr - Q_r r\,d\theta\,dr - M_\theta\,dr\,d\theta = 0 \tag{a}$$

The moment of p and the moment due to the change in Q_r have been omitted, as in Section 3.6. Note that the two moment *vectors* $M_\theta\,dr$ acting on the straight edges of the element are not parallel. They form a small angle $d\theta$ with one another and therefore have a resultant expressed by the last term in the preceding equation. Omission of the quantities of higher order in Equation a gives

$$M_r - M_\theta + r\frac{dM_r}{dr} - Q_r r = 0$$

or, in a concise form,

$$M_\theta - \frac{d(rM_r)}{dr} = -Q_r r \tag{4.14}$$

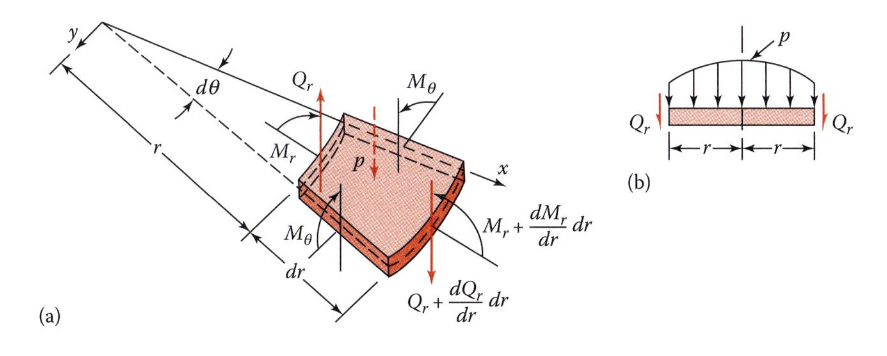

FIGURE 4.2
(a) Axisymmetrically loaded circular plate element; (b) FBD of the plate segment.

This is the *differential equation of equilibrium* for bending of axisymmetrically loaded circular plates. By introducing Equations 4.9a and 4.9b into Equation 4.14, we obtain Equation 4.9c.

The transverse shear force Q_r, per unit length of a circular section of radius r, is related to the loading as follows:

$$Q_r = -\frac{1}{2\pi r}\int_0^r p2\pi r\,dr = -\frac{1}{r}\int_0^r pr\,dr \tag{4.15}$$

The foregoing is readily verified by the free-body equilibrium diagram (Figure 4.2b).

The basic relationships derived thus far are appropriate for circular plates of uniform thickness and material properties. In the cases where the *flexural rigidity* D of the plate *varies with the radius* r, it is convenient to represent the formulas (Equations 4.9a and 4.9b) as

$$\begin{aligned} M_r &= D\left(\frac{d\theta}{dr} + \frac{\nu}{r}\theta\right) \\ M_\theta &= D\left(\frac{1}{r}\theta + \nu\frac{d\theta}{dr}\right) \end{aligned} \tag{4.16a, b}$$

Here θ denotes the *slope* of the deflection surface:

$$\theta = -\frac{dw}{dr} \tag{4.17}$$

Substitution of Equations 4.16 into Equation 4.14 results in

$$D\frac{d}{dr}\left(\frac{d\theta}{dr} + \frac{\theta}{r}\right) + \frac{dD}{dr}\left(\frac{d\theta}{dr} - \nu\frac{\theta}{r}\right) = Q_r \tag{4.18}$$

Clearly, if the flexural rigidity is constant, the preceding reduces to Equation 4.9c.

Determination of the deflection surface and stresses in circular plates subjected to various loads is illustrated in the following sections.

4.5 Uniformly Loaded Circular Plates

Consider the case of a circular plate of radius a under a uniformly distributed load p_0. The lateral displacement w is expressed by Equation 4.13. The constants of integration (the cs) in this equation and then stresses are determined for a number of particular cases described here and in Sections 4.8 and 4.9.

4.5.1 Plate with Clamped Edge

For a circular plate with built-in edge (Figure 4.3), the deflection and slope are zero at the edge. The boundary conditions are therefore

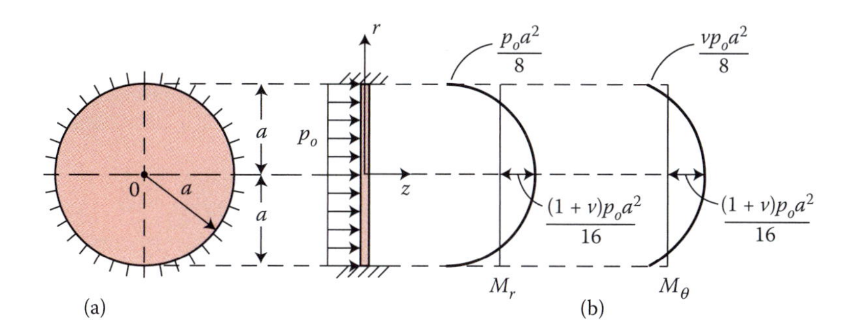

FIGURE 4.3
(a) Uniformly loaded plate with clamped edge; (b) variation of moments M_r and M_θ with radius r of the plate.

$$w = 0 \qquad \frac{dw}{dr} = 0 \qquad (r = a) \tag{a}$$

The terms involving logarithms in Equation 4.13 yield an infinite displacement at $r = 0$ for all values of c_1 and c_2 except zero; therefore, $c_1 = c_2 = 0$. Satisfying Equation a, we obtain

$$c_3 = -\frac{p_0 a^2}{32D} \qquad c_4 = \frac{p_0 a^4}{64D}$$

The deflection is then

$$w = \frac{p_0}{64D}(a^2 - r^2)^2 \tag{4.19}$$

The *maximum displacement* occurs at the center of the plate:

$$w_{\max} = \frac{p_0 a^4}{64D} \tag{b}$$

Expressions for the bending moments are calculated by means of Equations 4.19 and 4.9 as follows:

$$\begin{aligned} M_r &= \frac{p_0}{16}[(1+\nu)a^2 - (3+\nu)r^2] \\ M_\theta &= \frac{p_0}{16}[(1+\nu)a^2 - (1+3\nu)r^2] \end{aligned} \tag{4.20}$$

The stresses from Equations 4.19 and 4.11 are

$$\begin{aligned} \sigma_r &= \frac{3p_0 z}{4t^3}\left[(1+\nu)a^2 - (3+\nu)r^2\right] \\ \sigma_\theta &= \frac{3p_0 z}{4t^3}\left[(1+\nu)a^2 - (1+3\nu)r^2\right] \end{aligned} \tag{4.21}$$

Algebraically extreme values of the moments are found at the center and at the edge. At the edge ($r = a$), Equations 4.20 lead to

$$M_r = -\frac{p_0 a^2}{8} \qquad M_\theta = -\frac{\nu p_0 a^2}{8}$$

while at $r = 0$, $M_r = M_\theta = (1 + \nu)p_0 a^2/16$. It is observed that the maximum moment course at the edge. Thus, we have

$$\sigma_{r,\max} = \frac{6M_r}{t^2} = -\frac{3p_0}{4}\left(\frac{a}{t}\right)^2 \tag{c}$$

4.5.2 Plate with Simply Supported Edge

As in the case of a clamped edge plate, displacement must be finite at $r = 0$. The values of c_1 and c_2 in Equation 4.13 are therefore zero. Referring to Figure 4.4, the boundary conditions at the simply supported edge are

$$w = 0 \qquad M_r = 0 \qquad (r = a)$$

These yield the following respective expressions:

$$c_3 a^2 + c_4 + \frac{p_0 a^4}{64D} = 0 \qquad c_3 = -\frac{p_0 a^2}{32D}\frac{3+\nu}{1+\nu}$$

Solving, we have $c_4 = p_0 a^4(5 + \nu)/64(1 + \nu)D$. The plate deflection then equals

$$w = \frac{p_0 a^4}{64D}\left(\frac{r^4}{a^4} - 2\frac{3+\nu}{1+\nu}\frac{r^2}{a^2} + \frac{5+\nu}{1+\nu}\right) \tag{4.22}$$

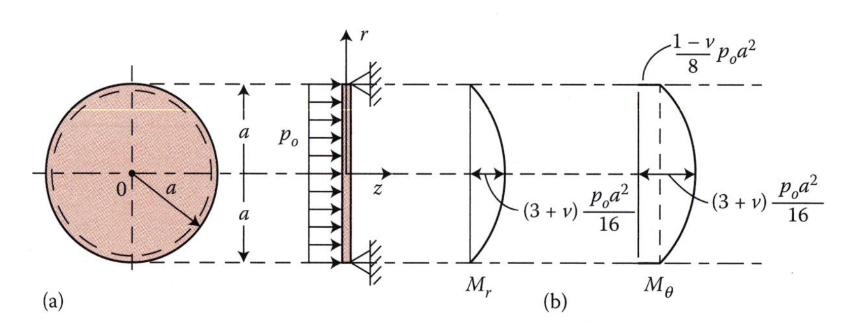

FIGURE 4.4
(a) Uniformly loaded plate with simply supported edge; (b) variation of the moments M_r and M_θ with radius r of the plate.

The maximum deflection, which occurs at $r = 0$, is thus

$$w_{\max} = \frac{p_0 a^4}{64D}\frac{5+\nu}{1+\nu} \tag{d}$$

Given the deflection curve w, the distribution of moments can readily be obtained in the form

$$\begin{aligned} M_r &= \frac{p_0}{16}(3+\nu)(a^2 - r^2) \\ M_\theta &= \frac{p_0}{16}[(3+\nu)a^2 - (1+3\nu)r^2] \end{aligned} \tag{4.23}$$

The stresses are therefore

$$\begin{aligned} \sigma_r &= \frac{3p_0 z}{4t^3}(3+\nu)(a^2 - r^2) \\ \sigma_\theta &= \frac{3p_0 z}{4t^3}[(3+\nu)a^2 - (1+3\nu)r^2] \end{aligned} \tag{4.24}$$

The maximum stress takes place at the center of the plate ($r = 0$) and is given by

$$\sigma_{r,\max} = \sigma_{\theta,\max} = \frac{3(3+\nu)p_0}{8}\left(\frac{a}{t}\right)^2 \tag{e}$$

4.5.3 Comparison of Deflections and Stresses in Built-in and Simply Supported Plates

Deflections: If $\nu = 1/3$, comparing Equations b and d, we see that the maximum deflection for a simply supported plate is about four times as great as that for the plate with a clamped edge. So, when the load resistance of a plate is limited by the large deflections, the *results related to the simply supported case will be conservative*. In addition, the analysis of a simply supported plate is, in general, much *simpler* than that of a plate with restrained edges, recommending its use in practice. Actually, many support members tolerate some degree of flexibility, and a condition of *true edge fixity* is especially difficult to obtain. As a result, a partially restrained plate exhibits deformations nearly identical with those of a hinged plate.

Based on these considerations, a *designer* sometimes simplifies the model of the original clamped plate, using instead a hinged plate. To attain more accurate results, however, the effect of a definite amount of edge yielding or relaxation of the fixing moment can often be accommodated by the formulas for edge slope and the method of superposition [4].

Stresses: Shown in Figure 4.5 the variation of stress with the ratio r/a in uniformly loaded clamped circular plates (upper base line). The curves are parabolas expressed by Equations 4.21 with $\nu = 1/3$. The stresses (Equations 4.24) in a simply supported circular plate are displayed (lower base line) in Figure 4.4 as functions of radial location (for $\nu = 1/3$). Observe from Figure 4.5 that the *ratio of the maximum stresses* for the simply supported and the clamped plates is equal to 5/3. Interestingly, the corresponding ratio for a simply supported beam and a fixed beam is equal to 1.5.

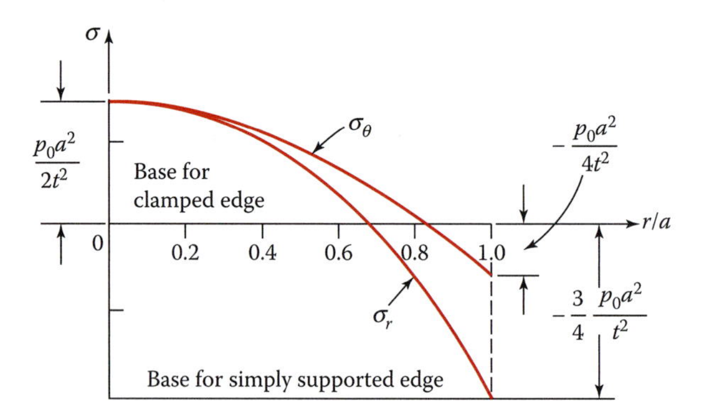

FIGURE 4.5
Variation of stresses σ_r and σ_θ with the radius r in the bottom half of the plate.

It is noted, however, that because of a small degree of yielding or loosening at a nominally clamped edge, the stresses will be considerably lessened there, while the deflection and stress will increase at the center. Thus, for uniformly loaded ordinary plate structures of clamped edge, the maximum stress will be somewhat *higher* than obtained above. This is also valid for the plates with clamped edges of *any* other shape.

EXAMPLE 4.1: Load Capacity of Clamped Plate

A circular clamped edge panel of an instrument is subjected to uniform pressure differential p_0 between the inside and the outside. The plate is made of an aluminum alloy of tensile yield strength σ_{yp}, thickness t, and radius a. Determine: (a) the load-carrying capacity of the plate, using the maximum energy of distortion theory based on a factor of safety of n; (b) the value of p_0 for the data *given*: $a = 200$ mm, $t = 5$ mm, $\sigma_{yp} = 260$ MPa, $n = 2.4$.

Solution

The principal stress components at the built-in edge, referring to Figure 4.5, are

$$\sigma_2 = \sigma_{r,\max} = -\frac{3}{4}\frac{p_0a^2}{t^2} \qquad \sigma_1 = \sigma_{\theta,\max} = -\frac{p_0a^2}{4t^2}$$

a. According to Equation 2.43

$$\left(-\frac{p_0a^2}{4t^2}\right)^2 - \left(-\frac{p_0a^2}{4t^2}\right)\left(-\frac{3}{4}\frac{p_0a^2}{t^2}\right) + \left(-\frac{3}{4}\frac{p_0a^2}{t^2}\right)^2 = \left(\frac{\sigma_{yp}}{2.4}\right)^2$$

The foregoing results in

$$p_0 = 1.512\left(\frac{t}{a}\right)^2\left(\frac{\sigma_{yp}}{n}\right) \tag{f}$$

as the value of pressure differential governing the onset of the yielding action.

b. Substituting the given numerical into Equation f, we obtain

$$p_0 = 1.512\left(\frac{5}{200}\right)^2\left(\frac{260\times 10^6}{2.4}\right) = 102.4 \text{ kPa}$$

4.6 *Effect of Shear on the Plate Deflection

In Section 3.3 and Example 3.1, we have observed that in the bending of plates, the influences of the shear stress τ_{rz} and the normal stress σ_z are neglected. We now demonstrate that the solutions for deflection of thin plates based on the bending strains only, yield results of acceptable accuracy.

Denoting the deflection of the midsurface due to the shear alone by w_s, the shear strain $\gamma_{rz} = dw_s/dr$, referring to Equations 3.32, is expressed as

$$\frac{dw_s}{dr} = \frac{3}{2}\frac{Q_r}{Gt}$$

Consider, for example, a *simple supported circular plate* under uniform load p_0 (Figure 4.4). The *deflection due to the shear* is

$$w_s = -\frac{3}{2Gt}\int_a^r Q_r\, dr \tag{a}$$

where the vertical shear force per unit circumferential length, at a distance r from the center, from Equation 4.15 is $Q_r = -rp_0/2$. Introducing this value of Q_r into Equation a, we have, on integration,

$$w_s = \frac{3p_0}{8Gt}(a^2 - r^2) = \frac{p_0 t^2}{16(1-\nu)D}(a^2 - r^2) \tag{b}$$

The *total deflection* w_t is obtained by addition of the deflections associated with bending and shear (Equations 4.22 and b):

$$w_t = \frac{p_0 a^4}{64D}\left[\frac{r^4}{a^4} - 2\frac{3+\nu}{1+\nu}\frac{r^2}{a^2} + \frac{5+\nu}{1+\nu} + \frac{4t^2}{a^2(1-\nu)}\left(1 - \frac{r^2}{a^2}\right)\right] \tag{4.25}$$

The maximum deflection occurs at the center of the plate:

$$w_{\max} = \frac{p_0 a^4}{64D}\left[\frac{5+\nu}{1+\nu} + \frac{4t^2}{(1-\nu)a^2}\right] \tag{c}$$

The ratio of the shear deflection to the bending deflection at $r = 0$ provides a measure of plate slenderness. This ratio is, for $\nu = 1/3$,

$$\frac{p_0 t^2 a^2/16(1-\nu)D}{p_0 a^4(5+\nu)/64(1+\nu)D} = \frac{3}{2}\left(\frac{t}{a}\right)^2$$

In the case of a plate of radius 10 times its thickness, the above is small, 1/67. For thin plates ($t/r < 0.1$), we thus conclude that the bending theory yields a result of sufficient accuracy. For *thick plates*, the thickness of the plate is not very small compared with its radius, and the effect of shear on total deflection becomes of practical importance.

4.7 Local Stresses at the Point of Application of a Concentrated Load

It will be observed that all equations for the moments and the stresses due to a concentrated load P give infinitely large values for the bending moments and the stresses in a plate. This result is due to the assumption that the load is concentrated at a point. If the distribution is taken over a small circle, the stresses become finite, as is illustrated in Sections 4.8.1 and 4.8.2.

Analysis by an elaborate method [5] indicates that the actual stress caused by P on a very small area of radius r_c can be obtained by replacing the *actual* r_c by a so-called *equivalent radius* r_e. The latter is given by the approximate formula [6]

$$r_e = \sqrt{1.6r_c^2 + t^2} - 0.675t \qquad (r_c < 0.5t) \tag{4.26a}$$

or

$$r_e = r_c \qquad (r_c \geq 0.5t) \tag{4.26b}$$

where t is the thickness of plate.

The preceding equation applies to a *plate of any shape*. Use of r_e makes it possible to calculate the *finite maximum stresses* caused by a concentrated load, whereas the ordinary formulas would show that these stresses are infinite. It should be mentioned that the "exact" distribution of stress near the point of application of the concentrated load is obtained by employing the method of the theory of elasticity [7,8].

We also note that, although the compressive stresses at the top of the plate may be larger than the tensile stresses at the bottom (e.g., Figure 4.6), they have localized character. The local inelastic deformation occurring in the case of a ductile material will not affect the general yielding of the plate if the tensile stresses at the bottom of the plate remain within allowable values. Inasmuch as the compressive strength of a brittle material is often quite larger than its tensile strength, a plate of brittle material will also be safe if the tensile stress at the bottom is within its permissible value. Thus, in *design for strength* of a circular plate subject to a concentrated load, only the maximum tensile bending stresses at the bottom of the plate need be considered.

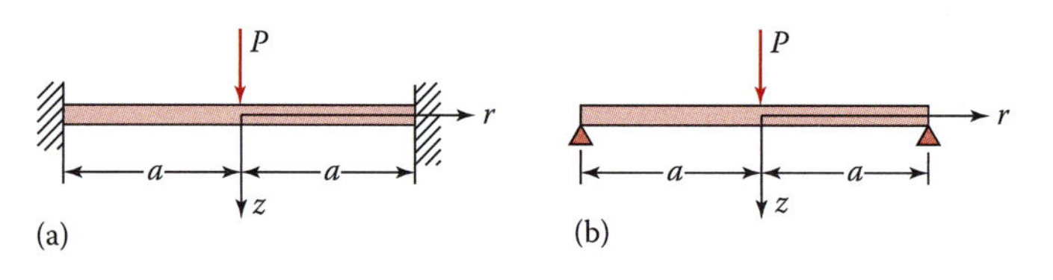

FIGURE 4.6
Concentrated center force on a circular plate: (a) With clamped edge; (b) with simply supported edge.

4.8 Circular Plates under a Concentrated Load at the Center

When a concentrated or a (nominal) point load P acts at the plate, one can set $p_0 = 0$ in Equation 4.13. The value of c_1 must be taken as zero, as in Section 4.5, in order that the deflection be finite at $r = 0$. The term involving c_2 must now be retained because of the very high shear forces present in the vicinity of the center. The deflection surface of the plate is then represented:

$$w = c_2 r^2 \ln r + c_3 r^2 + c_4 \tag{4.27}$$

The constants c_2, c_3, and c_4 are calculated for two particular cases, described below.

4.8.1 Plate with Clamped Edge

Referring to Figure 4.6a, the boundary conditions are: $w = 0$ and $dw/dr = 0$ at $r = a$. Introducing these into Equation 4.27, lead to

$$\begin{aligned} &c_2 a^2 \ln a + c_3 a^2 + c_4 = 0 \\ &c_2 a(2\ln a + 1) + 2c_3 = 0 \end{aligned} \tag{a}$$

The additional condition is that the vertical shear force Q_r must equal $-P/2\pi r$. Thus, from Equations 4.9c, and 4.27, we obtain

$$\frac{4D}{r} c_2 = \frac{P}{2\pi r} \tag{b}$$

Solving Equations a and b gives the constants of integration as

$$c_2 = \frac{P}{8\pi D} \qquad c_3 = -\frac{P}{16\pi D}(2\ln a + 1) \qquad c_4 = \frac{Pa^2}{16\pi D}$$

The foregoing, on substitution into Equation 4.27, provides the following expression for the deflection:

$$w = \frac{P}{16\pi D}\left(2r^2 \ln\frac{r}{a} + a^2 - r^2\right) \tag{4.28}$$

The maximum deflection occurs at $r = 0$ and is given by

$$w_{\max} = \frac{Pa^2}{16\pi D} \tag{c}$$

The stresses corresponding to Equation 4.28, derived from Equation 4.11, are

$$\begin{aligned} \sigma_r &= \frac{3Pz}{\pi t^3}\left[(1+\nu)\ln\frac{a}{r} - 1\right] \\ \sigma_\theta &= \frac{3Pz}{\pi t^3}\left[(1+\nu)\ln\frac{a}{r} - \nu\right] \end{aligned} \tag{4.29}$$

On introduction of the equivalent radius from Equations 4.26, $r_e = r$ in Equations 4.29, the maximum finite stresses produced by the concentrated load can be calculated.

4.8.2 Plate with Simply Supported Edge

In this case (Figure 4.6b), the deflection and the radial moment of the plate vanish at the edge and the applied load $P = -2\pi r\, Q_r$. That is,

$$(w)_{r=a} = 0 \qquad (M_r)_{r=a} = 0 \qquad Q_r = -\frac{P}{2\pi r} \tag{d}$$

Substitution of Equations 4.27 and 4.9 into the above gives three linear equations for c_2, c_3, and c_4. Solving for the constants leads to the following relationships for deflection and stress:

$$w = \frac{P}{16\pi D}\left[2r^2 \ln\frac{r}{a} + \frac{3+\nu}{1+\nu}(a^2 - r^2)\right] \tag{4.30}$$

$$\begin{aligned} \sigma_r &= \frac{3Pz}{\pi t^3}(1+\nu)\ln\frac{a}{t} \\ \sigma_\theta &= \frac{3Pz}{\pi t^3}\left[(1+\nu)\ln\frac{a}{t} + 1 - \nu\right] \end{aligned} \tag{4.31}$$

At $r = 0$, we have the maximum deflection as

$$w_{\max} = \frac{Pa^2}{16\pi D}\frac{3+\nu}{1+\nu} \tag{e}$$

To find the largest finite stress, r_e is substituted for r in Equations 4.31.

For certain plate problems, the solution can be obtained by the application of fictitious forces to create the desired deflection patterns, as illustrated in the following example.

EXAMPLE 4.2: Deflection of Fixed-Ended Plate Under Central Moment

A plate with clamped edge is subjected to a moment at its center (Figure 4.7). Determine an expression for the deflection.

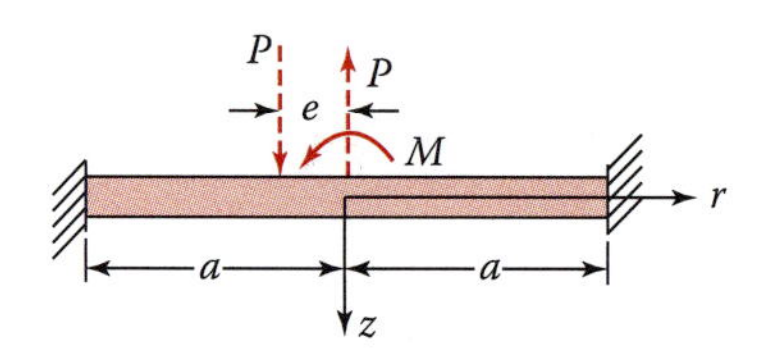

FIGURE 4.7
Circular plate with built-in edge.

Solution

The action of the moment is equivalent to that of the two forces P (shown by the dashed lines in the figure) if Pe approaches M while e approaches zero. Using Equation 4.28, we thus have

$$w = \frac{M}{16\pi D}\lim_{e\to 0}\left\{\frac{1}{e}\left[2(r+e)^2\ln\left(\frac{r+e}{a}\right)-(r+e)^2-2r^2\ln\frac{r}{a}+r^2\right]\right\}$$
$$= \frac{M}{16\pi D}\frac{d}{dr}\left(2r^2\ln\frac{r}{a}-r^2\right)$$

Performing the differentiation, we obtain

$$w = \frac{M}{4\pi D}r\ln\frac{r}{a} \tag{4.32}$$

Similarly, the case of a simply supported plate can be treated by using the solution given by Equation 4.30.

4.8.3 A Short Catalog of Solutions

Several other cases of practical importance can also be treated on the basis of the mathematical analyses described in the foregoing sections. For reference purposes, Table 4.2 lists the equations of deflection surfaces for variously loaded circular plates (Problem 4.15).

It is of interest to note that a more accurate solution for a concentrated load P acting at the center of the circular plate can be found from the stress formulas of Cases 5 or 8 of Table 4.2, setting $P = \pi c^2 p_0$. By diminishing the radius c of the circle over which the load is distributed, we approach the condition of a concentrated load: $c = r_e$, where r_e is defined by Equations 4.26. The stresses at the center increase as c decreases but remain finite as long as c is finite.

4.9 Annular Plates with Simply Supported Outer Edges

In this section, we discuss the bending of circular plates with a concentric circular hole, so-called annular plates. The inner and outer radii of a plate are denoted by b and a, respectively. Two typical cases of loading of considerable interest in the design of machine parts are illustrated below.

4.9.1 Plate Loaded by Edge Moments

To arrive at a solution for the deflection we utilize Equation 4.9c. Inasmuch as $Q_r = 0$ (Figure 4.8b), we have

$$\frac{d}{dr}\left[\frac{1}{r}\frac{d}{dr}\left(r\frac{dw}{dr}\right)\right]=0 \tag{a}$$

TABLE 4.2

Formulas for Circular Plates

Case No.	Support and Loading	Maximum Bending Stress σ, Deflection w at Center, and Slope θ at Edge
1.	Edge simply supported (or no support); load uniform along edge 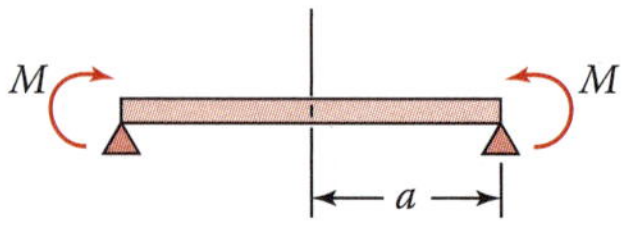	$\sigma = 6\frac{M}{t^2}$ (uniform) $w = 6(1-\nu)\frac{Ma^2}{Et^3}$ $\theta = 12(1-\nu)\frac{Ma}{Et^3}$
2.	Edge simply supported; load uniform along a circle of radius c 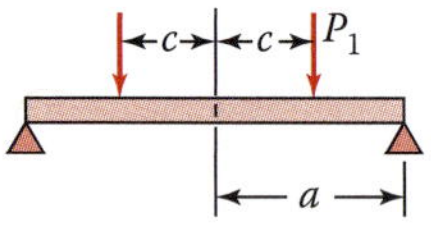	$\sigma = \frac{3}{2}\frac{P_1 c}{t^2}\left[(1-\nu)\left(1-\frac{c^2}{a^2}\right)+2(1+\nu)\ln\frac{a}{c}\right]$ (at center) $w = \frac{3(1-\nu)}{2}\frac{P_1 c}{Et^3} \times \left[(3+\nu)(a^2-c^2)-2(1+\nu)c^2\ln\frac{a}{c}\right]$ $\theta = 6(1-\nu)\frac{P_1 ac}{Et^3}\left(1-\frac{c^2}{a^2}\right)$
3.	Concentrated load at a distance c from the center 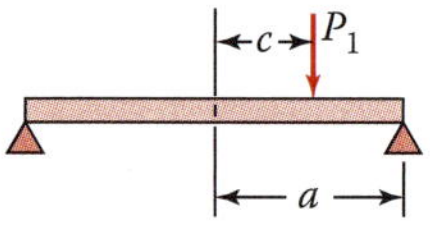	Deflection w at center approximately same as case 2 for edge simply supported and same as case 6 for edge fixed
4.	Edge simply supported; load uniform 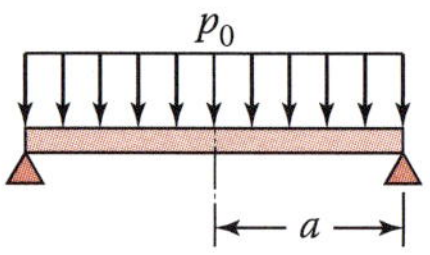	$\sigma = \frac{3(3+\nu)}{8}\frac{p_0 a^2}{t^2}$ (at center) $w = \frac{3(1-\nu)(5+\nu)}{16}\frac{p_0 a^4}{Et^3}$ $\theta = \frac{3(1-\nu)}{2}\frac{p_0 a^3}{Et^3}$
5.	Edge simply supported; uniform load on circular area of radius c 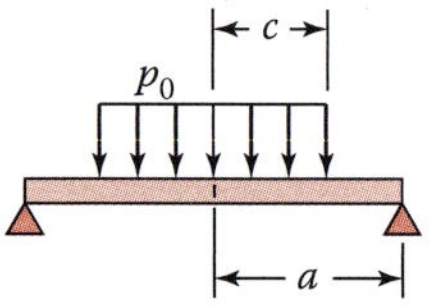	$\sigma = \frac{3}{8}\frac{p_0 c^2}{t^2}\left[4-(1-\nu)\frac{c^2}{a^2}+4(1+\nu)\ln\frac{a}{c}\right]$ (at center) $w = \frac{3(1-\nu)}{16}\frac{p_0 c^2}{Et^3} \times \left[4(3+\nu)a^2-(7+3\nu)c^2-4(1+\nu)c^2\ln\frac{a}{c}\right]$ $\theta = \frac{3(1-\nu)}{2}\frac{p_0 ac^2}{Et^3}\left(2-\frac{c^2}{a^2}\right)$

(Continued)

TABLE 4.2 (*Continued*)

Formulas for Circular Plates

Case No.	Support and Loading	Maximum Bending Stress σ, Deflection w at Center, and Slope θ at Edge
6.	Edge fixed; load uniform along a circle of radius c	$\sigma = \frac{3(1+\nu)}{2}\frac{P_1 c}{t^2}\left(\frac{c^2}{a^2} + 2\ln\frac{a}{c} - 1\right)$ (at center) $\sigma = 3\frac{P_1 c}{t^2}\left(1 - \frac{c^2}{a^2}\right)$ (at edge) $w = \frac{3(1-\nu^2)}{2}\frac{P_1 c}{Et^3}\left(b^2 - c^2 - 2c^2\ln\frac{a}{c}\right)$
7.	Edge fixed; load uniform	$\sigma = \frac{3}{4}\frac{p_0 a^2}{t^2}$ (at edge) $w = \frac{3(1-\nu^2)}{16}\frac{p_0 a^4}{Et^3}$[a]
8.	Edge fixed; load uniform over a circular area of radius c	$\sigma = \frac{3(1+\nu)}{8}\frac{p_0 c^2}{t^2}\left(\frac{c^2}{a^2} + 4\ln\frac{a}{c}\right)$ (at center) $\sigma = \frac{3}{4}\frac{p_0 c^2}{t^2}\left(2 - \frac{c^2}{a^2}\right)$ (at edge) $w = \frac{3(1-\nu^2)}{4}\frac{p_0 c^2}{Et^3}\left(a^2 - \frac{3}{4}c^2 - c^2\ln\frac{a}{c}\right)$

Notation: a = radius of plate; p_0 = uniform load per unit area; P_1 = load per unit length, uniformly distributed along a centric circle of radius c; P = concentrated load; M = moments per unit length uniformly distributed along edge; ν = Poisson's ratio; E = modulus of elasticity.

[a] For thicker plates ($a < 10t$), a modified bending theory gives $w = (3/16)\alpha(1 - \nu^2)(p_0a^4/Et^3)$, where the constant $\alpha = 1 + 5.72(t/a)^2$.

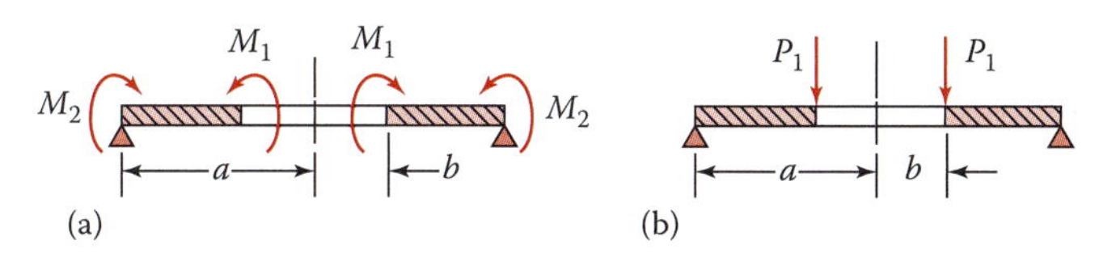

FIGURE 4.8
Simply supported annular plate: (a) With moments; (b) under shear force at inner edge.

Three successive integrations of the above give

$$w = \frac{1}{4}c_1 r^2 + c_2 \ln r + c_3 \tag{b}$$

Referring to Figure 4.8a, the boundary conditions are represented by

$$\begin{aligned} w &= 0 \qquad M_r = M_2 \qquad (r = a) \\ M_r &= M_1 \qquad\qquad\quad\; (r = b) \end{aligned} \tag{c}$$

where the radial bending moment M_r is given by Equation 4.9a. The constants are evaluated by substitution of Equation b into Equations c. Expressions for deflection and moments are then

$$
\begin{aligned}
w &= \frac{1}{2}\frac{r^2-a^2}{a^2-b^2}\frac{M_1b^2-M_2a^2}{(1+\nu)D}+\frac{a^2b^2}{a^2-b^2}\frac{M_1-M_2}{(1+\nu)D}\ln\frac{r}{a} \\
M_r &= \frac{M_1b^2-M_2a^2}{a^2-b^2}+\frac{a^2b^2(M_1-M_2)}{r^2(a^2-b^2)} \\
M_\theta &= -\frac{M_1b^2-M_2a^2}{a^2-b^2}-\frac{a^2b^2(M_1-M_2)}{r^2(a^2-b^2)}
\end{aligned} \tag{4.33}
$$

The stress components are calculated by means of Equations 4.33 and 4.3.

4.9.2 Plate Loaded by Shear Force at Inner Edge

We now consider a plate under shear force P_1 per unit circumferential length, uniformly distributed over the inner edge, resulting in a total load $2\pi bP_1$ (Figure 4.8b). This must be equal to $2\pi rQ_r$, the total shear force at a distance r from the center. We thus have $Q_r = -P_1b/r$ and Equation 4.9c becomes

$$
\frac{d}{dr}\left[\frac{1}{r}\frac{d}{dr}\left(r\frac{dw}{dr}\right)\right]=\frac{P_1b}{Dr} \tag{4.34}
$$

The displacement w is obtained by successive integrations of Equation 4.34:

$$
w=\frac{P_1br^2}{4D}(\ln r-1)+\frac{c_1r^2}{4}+c_2\ln r+c_3 \tag{d}
$$

The constant c_1, c_2, and c_3 are determined from the following conditions at the outer and inner edges:

$$
\begin{aligned}
&w=0 \qquad M_r=0 \qquad (r=a) \\
&M_r=0 \qquad\qquad\quad\ (r=b)
\end{aligned} \tag{e}
$$

On substitution of the constants into Equations d, the following expression for the plate deflection is obtained:

$$
w=\frac{P_1a^2b}{4D}\left\{\left(1-\frac{r^2}{t^2}\right)\left[\frac{3+\nu}{2(1+\nu)}-\frac{b^2}{a^2-b^2}\ln\frac{b}{a}\right]+\frac{r^2}{a^2}\ln\frac{r}{a}+\frac{2b^2}{a^2-b^2}\frac{1+\nu}{1-\nu}\ln\frac{b}{a}\ln\frac{r}{a}\right\} \tag{4.35}
$$

Comment: It is observed that if the radius of the hole becomes infinitesimally small, b^2 ln (b/a) vanishes, and Equation 4.35, for $P_1 = P/2\pi b$, reduces to Equation 4.30, as expected.

EXAMPLE 4.3: Deflection of Annular Plate

An annular pump diaphragm with outer edge ($r = a$) simply supported and inner edge ($r = b$) free is under a uniform pressure p (Figure 4.9). Outline the derivation for the deflection surface w.

Solution

The boundary conditions (Table 4.1) are

$$
w=0 \qquad M_r=0 \qquad (r=0) \tag{f}
$$

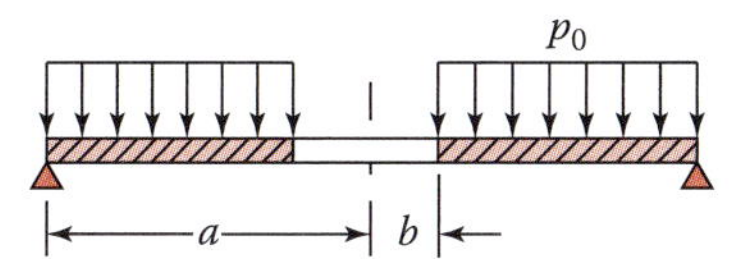

FIGURE 4.9
Simply supported annular plate with uniform load.

$$Q = 0 \qquad M_r = 0 \qquad (r = b) \tag{g}$$

Formulas for deflection w, moment M_r, moment M_θ, and shear Q_r are given by Equations 4.13 and 4.9a–c, respectively. Substitution *of* Equations f and g *into the preceding formulas gives the values of* c_1, c_2, c_3, and c_4 (see Problem 4.16). Expression for the deflection is then found from Equation 4.13.

EXAMPLE 4.4: Moment Resultants in Annular Plate

Consider a simply supported annular plate with linearly varying thickness and rod assembly (Figure 4.10a). The rod is subjected to an axial load P as shown in the figure. Determine expressions for the bending moments in the plate, taking $\nu = 1/3$.

Solution

The plate carries a line load $P_1 = P/2\pi b$, uniformly distributed along the edge of the hole (Figure 4.10b). We denote the flexural rigidity at $r = b$ by

$$D_1 = \frac{Et_1^3}{12(1-\nu^2)} = \frac{3}{32}Et_1^3 \tag{4.36}$$

where t_1 represents the inner edge thickness at any distance r from the center, $t = t_1 r/b$, and hence

$$D = D_1\frac{r^3}{b^3} \tag{h}$$

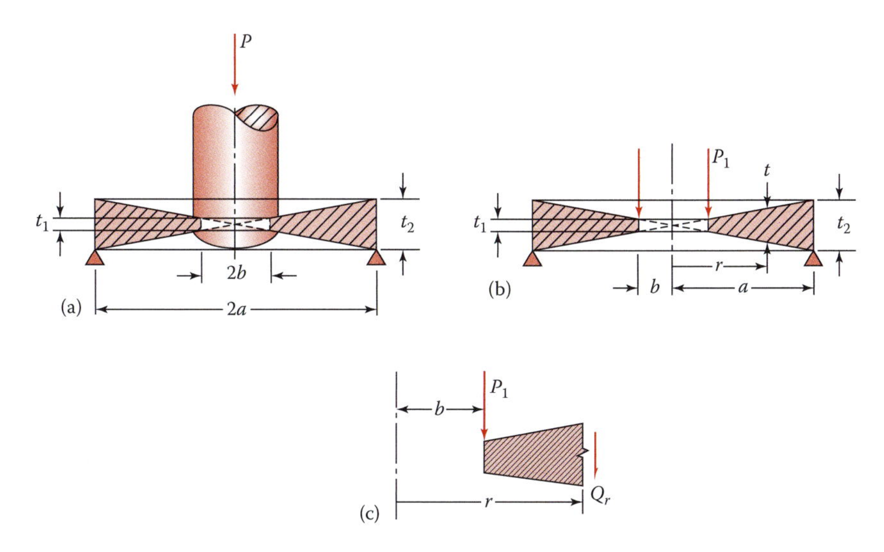

FIGURE 4.10
Annular plate with linearly varying thickness under an axial load.

Referring to the FBD (Figure 4.10c), equilibrium of the vertical forces gives

$$\sum F_z = 0: \quad Q_r\, 2\pi r + P_1\, 2\pi b = 0$$

in which $P = P_1(2\pi b)$. The transverse shear is thus

$$Q_r = -\frac{P}{2\pi r} \tag{i}$$

Inserting Equations h and i into Equation 4.18, we have

$$r^2 \frac{d^2\theta}{dr^2} + 4r\frac{d\theta}{dr} + (3-1)\theta = -\frac{Pb^3}{2\pi D_1 r^3} \tag{j}$$

The general solution of this equidimensional equation is (Problem 4.25)

$$\theta = c_1 r^{m_1} + c_2 r^{m_2} + \frac{Pb^3}{6\pi(1-\nu)D_1 r^3} \tag{k}$$

where

$$m_{1,2} = -1.5 \pm \sqrt{3.25 - 3\nu}$$

Here c_1 and c_2 are arbitrary constants. With $\nu = 1/3$, Equation k is simplified to

$$\theta = c_1 + \frac{c_2}{r^3} + \frac{Pb^3}{4\pi D_1 r^2} \tag{l}$$

The boundary conditions are represented by

$$M_r = 0 \quad (r = a, r = b)$$

Substitution of Equation 4.16a together with Equation l into the preceding gives

$$c_1 = -\frac{5P}{4\pi D_1}\frac{b^3(a-b)}{a^3-b^3} \qquad c_2 = \frac{5P}{32\pi D_1}\frac{ab^4(a^2-b^2)}{a^3-b^3}$$

The slope of the deflection surface is therefore

$$\theta = -\frac{Pb^3}{4\pi D_1}\left[\frac{5(a-b)}{a^3-b^3} - \frac{5ab(a^2-b^2)}{(a^3-b^3)r^3} - \frac{1}{r^2}\right] \tag{4.37}$$

The bending moments, as calculated by means of Equations 4.16, are expressed in the forms

$$\begin{aligned} M_r &= \frac{5P}{12\pi}\left[\frac{ab(a^2-b^2)}{(a^3-b^3)r} + \frac{(a-b)r^2}{a^3-b^3} - 1\right] \\ M_\theta &= \frac{P}{4\pi}\left[\frac{5(a-b)r^2}{a^3-b^3} + \frac{1}{3}\right] \end{aligned} \tag{4.38}$$

The equations for the stresses are determined in the usual way, by introducing Equation 4.17 into Equation 4.11 or from Equations 4.38 and 4.3.

4.10 Deflection and Stress by Superposition

The integration procedures discussed in the foregoing sections for determining the elastic deflection and stress of loaded plates are generally applicable to other cases of plates. It is noted, however, that the solutions to numerous problems with simple loadings are already available. For complex configurations of loads, therefore, the method of superposition may be used to good advantage to simplify the analysis.

The method is illustrated for the case of the annular plate shown in Figure 4.11a. The plate is simply supported along its outer edge and is subjected to a uniformly distributed load p_0 at its surface. Shown in Figure 4.11b is a circular plate under a uniform load p_0. Figure 4.11c is an annular plate carrying along its inner edge a shear force per unit circumferential length $p_0\, b/2$ and a radial bending moment, defined by Equation 4.23, $p_0(3+\nu)(a^2-b^2)/16$. The solutions for each of the latter two cases are known from Sections 4.5 and 4.9. Hence, the deflection and stress *at any point* of the plate in Figure 4.11a can be found by the superposition of the results at that point for the cases indicated in Figure 4.11b and c (Problem 4.23).

4.10.1 Design Tables for Annular Plates

By employing similar procedures, annular plates with various load and edge conditions may be treated. Several situations of practical importance shown in Table 4.3 have been investigated by Wahl and Lobo [9]. In all these cases, the maximum stresses and deflections can be expressed in the form

$$\sigma_{\max} = k\frac{pa^2}{t^2} \quad \text{or} \quad \sigma_{\max} = k\frac{P}{t^2} \tag{4.39a}$$

and

$$w_{\max} = h\frac{pa^4}{Et^3} \quad \text{or} \quad w_{\max} = h\frac{Pa^2}{Et^3} \tag{4.39b}$$

The preceding values of the coefficients k and h are furnished in Table 4.4 for six values of the ratio a/b and for a Poisson's ratio of $\nu = 0.3$. Such problems are of practical importance in mechanical design. Interestingly, Case 8 in Table 4.3 furnishes solution of the plate shown in Figure 4.12.

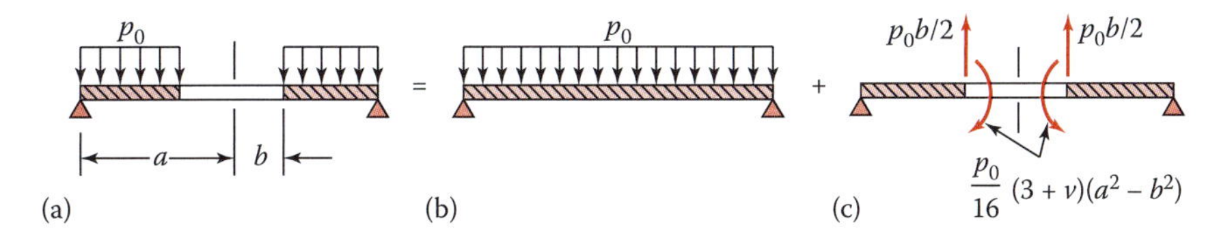

FIGURE 4.11
Method of superposition.

TABLE 4.3

Variously Supported and Uniformly Loaded Annular Plates

Case No.	Support	Load Diagram	Maximum Stress σ_{max}	Deflection between Edges w_{max}
1.	Inner edge fixed	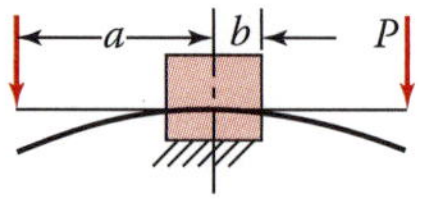	$k_1 \frac{P}{t^2}$	$h_1 \frac{Pa^2}{Et^3}$
2.	Inner edge fixed	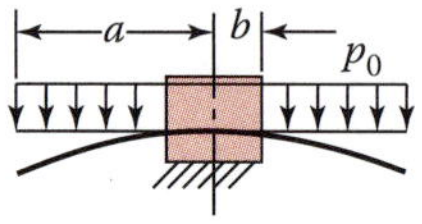	$k_2 \frac{p_0 a^2}{t^2}$	$h_2 \frac{p_0 a^4}{Et^3}$
3.	Inner edge simply supported	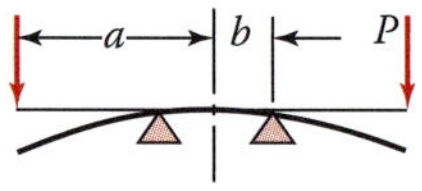	$k_3 \frac{P}{t^2}$	$h_3 \frac{Pa^2}{Et^3}$
4.	Inner edge simply supported	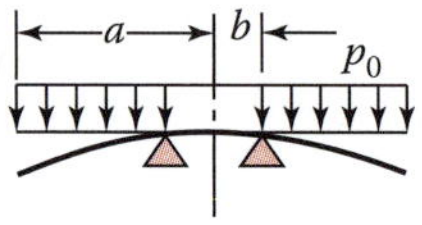	$k_4 \frac{p_0 a^2}{t^2}$	$h_4 \frac{p_0 a^4}{Et^3}$
5.	Outer edge simply supported	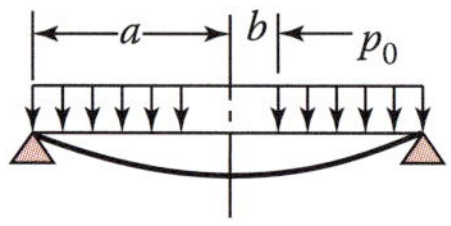	$k_5 \frac{p_0 a^2}{t^2}$	$h_5 \frac{p_0 a^4}{Et^3}$
6.	Outer edge simply supported, inner-edge rotation prevented	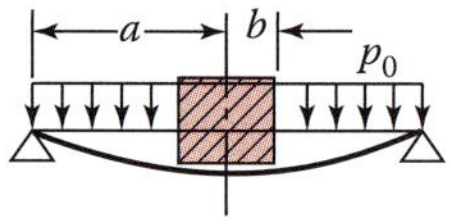	$k_6 \frac{p_0 a^2}{t^2}$	$h_6 \frac{p_0 a^4}{Et^3}$
7.	Inner edge fixed, outer-edge rotation prevented	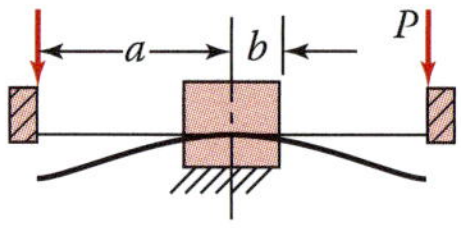	$k_7 \frac{P}{t^2}$	$h_7 \frac{Pa^2}{Et^3}$
8.	Inner edge fixed, outer-edge rotation prevented	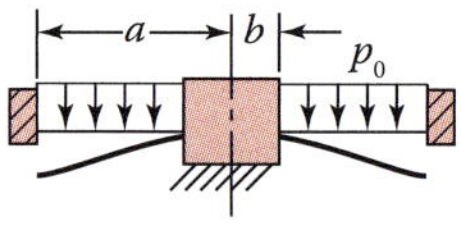	$k_8 \frac{p_0 a^2}{t^2}$	$h_8 \frac{p_0 a^4}{Et^3}$
9.	Outer edge fixed	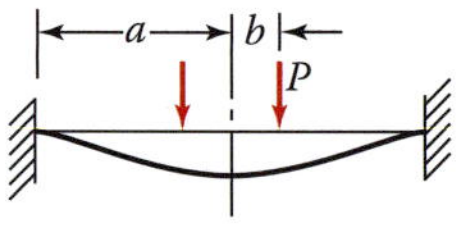	$k_9 \frac{P}{t^2}$	$h_9 \frac{Pa^2}{Et^3}$
10.	Outer edge fixed	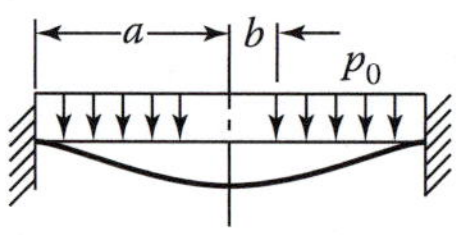	$k_{10} \frac{p_0 a^2}{t^2}$	$h_{10} \frac{p_0 a^4}{Et^3}$

Notation: a = outer radius of plate; b = inner radius of plate; t = thickness of plate; E = modulus of elasticity; P = total load uniformly distributed along outer or inner edge; p_0 = uniform load per unit surface area of plate.

TABLE 4.4

Coefficients k and h for Cases 1 through 10 in Table 4.3

a/b	1.25	1.50	2.00	3.00	4.00	5.00
k_1	0.227	0.428	0.753	1.205	1.514	1.745
h_1	0.0051	0.0249	0.0877	0.209	0.293	0.350
k_2	0.135	0.410	1.04	2.15	2.99	3.69
h_2	0.0023	0.0183	0.0938	0.2925	0.448	0.564
k_3	1.10	1.26	1.48	1.88	2.17	2.34
h_3	0.341	0.519	0.672	0.734	0.724	0.704
k_4	0.66	1.19	2.04	3.34	4.30	5.10
h_4	0.202	0.491	0.902	1.220	1.300	1.310
k_5	0.592	0.976	1.440	1.880	2.080	2.19
h_5	0.1841	0.4139	0.6640	0.8237	0.8296	0.813
k_6	0.122	0.336	0.74	1.21	1.45	1.59
h_6	0.0034	0.0313	0.1250	0.291	0.417	0.492
k_7	0.115	0.220	0.405	0.703	0.933	1.13
h_7	0.0013	0.0064	0.0237	0.0619	0.0923	0.114
k_8	0.090	0.273	0.71	1.54	2.23	2.80
h_8	0.0008	0.0062	0.0329	0.1096	0.1792	0.2338
k_9	0.194	0.320	0.454	0.673	1.021	1.305
h_9	0.00504	0.0242	0.0810	0.172	0.217	0.288
k_{10}	0.105	0.259	0.480	0.657	0.710	0.730
h_{10}	0.00199	0.0139	0.0575	0.130	0.162	0.175

Table 4.5 contains some cases corresponding to Table 4.3 for the plates with linearly varying thickness, involving various edge conditions and loads. Table 4.6 lists the values of coefficients h and k for $\nu = 0.3$. The maximum stresses and deflection are now given by the formulas (Figure 4.10)

$$\sigma_{\max} = k\frac{pa^2}{t_2^2} \quad \text{or} \quad \sigma_{\max} = k\frac{P}{t_2^2} \tag{4.40a}$$

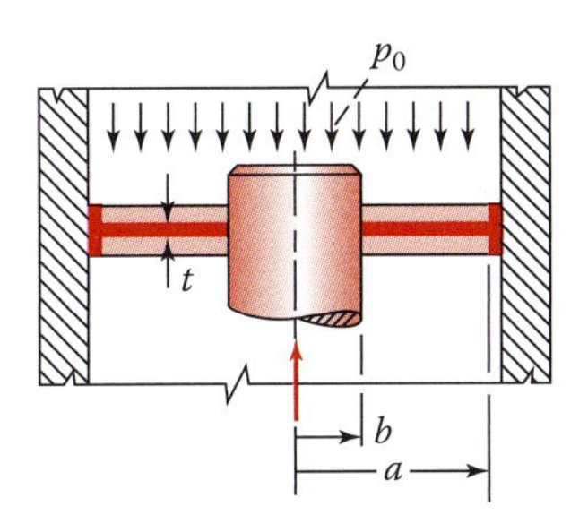

FIGURE 4.12
Use of annular plate in machine design.

TABLE 4.5

Variously Supported and Uniformly Loaded Annular Plate with Linearly Varying Thickness

Case No. (Corresp. to Table 4.3)	Support	Load Diagram	Maximum Stress σ_{max}	Deflection between Edges w_{max}
1.	Inner edge fixed	a, b, P, t_2	$k_1 \frac{P}{a^2}$	$h_1 \frac{Pa^2}{Et_2^3}$
2.	Inner edge fixed	a, b, p_0, t_2	$k_2 \frac{p_0 a^2}{t_2^2}$	$h_2 \frac{p_0 a^4}{Et_2^3}$
6.	Outer edge simply supported, inner-edge rotation prevented	a, b, p_0, t_2	$k_6 \frac{p_0 a^2}{t_2^2}$	$h_6 \frac{p_0 a^4}{Et_2^3}$
7.	Inner edge fixed, outer-edge rotation prevented	a, b, P, t_2	$k_7 \frac{P}{t_2^2}$	$h_7 \frac{Pa^2}{Et_2^3}$
8.	Inner edge fixed, outer-edge rotation prevented	a, b, p_0, t_2	$k_8 \frac{p_0 a^2}{t_2^2}$	$h_8 \frac{p_0 a^4}{Et_2^3}$

Notation: a = outer radius of plate; b = inner radius of plate; t = outer edge thickness of plate; E = modulus of elasticity; P = total load uniformly distributed along outer edge; p_0 = uniform load per unit surface area of plate.

TABLE 4.6

Coefficients k and h for Cases 1, 2, 6, 7, and 8 in Table 4.5

a/b	1.25	1.50	2.00	3.00	4.00	5.00
k_1	0.353	0.933	2.63	6.88	11.47	16.51
h_1	0.0082	0.0583	0.345	1.358	2.39	3.27
k_2	0.249	0.638	3.96	13.64	26.0	40.6
h_2	0.0037	0.0453	0.401	2.12	4.25	6.28
k_6	0.149	0.991	2.23	5.57	7.78	9.16
h_6	0.0055	0.0564	0.412	1.673	2.79	3.57
k_7	0.159	0.396	0.091	3.31	6.55	10.78
h_7	0.0017	0.0112	0.0606	0.261	0.546	0.876
k_8	0.1275	0.515	2.05	7.97	17.35	30.0
h_8	0.0011	0.0115	0.0934	0.537	1.261	2.16

Note: Poisson's ratio $\nu = 1/3$.

and

$$w_{\max} = h\frac{pa^4}{Et_2^3} \quad \text{or} \quad w_{\max} = h\frac{Pa^2}{Et_2^3} \tag{4.40b}$$

Numerous other solutions for symmetrical bending of circular plates are given in References [4,7]. Design calculations are often facilitated by this type of compilation.

4.11 The Ritz Method Applied to Bending of Circular Plates

In the problem discussed thus far, support was provided at the plate edges, and the plate was assumed to undergo no deflection at these supports. We now consider the cases of a plate on an elastic foundation and a plate with vibrating end supports. Examples 4.5 and 4.6 illustrate the application of the Ritz method to the foregoing bending problems of circular plates.

Note that the foundation reaction forces will be taken to be *linearly proportional* to the plate deflection at any point, that is, wk. Here w is the plate deflection, and k is a constant, termed the *modulus of the foundation* or *bedding constant* of the foundation material, having the dimensions or force per unit surface area of plate per unit of deflection (e.g., Pa/m). The above assumption with respect to the nature of the support not only leads to equations amenable to solution but also closely approximates many real situations [10]. Examples of this type of plate include concrete slabs, bridge decks, floor structures, and airport runways.

EXAMPLE 4.5: Deflection of Plate by Energy Method

Figure 4.13 depicts a circular plate of radius a rests freely on an elastic foundation and is subjected to a center load P. Determine the maximum deflection.

Solution

In this case of axisymmetrical bending, the expression for strain energy given by Equation P4.20 reduces to

$$U_1 = \pi D\int_0^a \left[\left(\frac{d^2w}{dr^2} + \frac{1}{r}\frac{dw}{dr}\right)^2 - \frac{2(1-\nu)}{r}\frac{dw}{dr}\frac{d^2w}{dr^2}\right] r\,dr \tag{4.41}$$

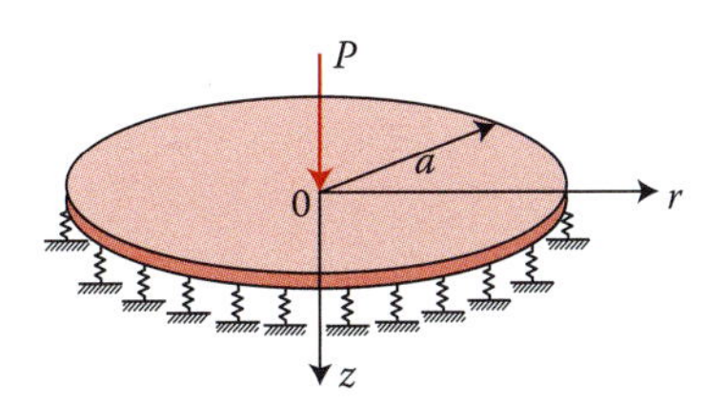

FIGURE 4.13
Schematic representation of plate on elastic foundation with a concentrated load.

A solution can be assumed in the form of a series,

$$w = c_0 + c_2 r^2 + \cdots + c_n r^n \tag{4.42}$$

in which c_n are to be determined from the condition that the potential energy Π of the system in stable equilibrium is minimum.

If we retain, for example, only the first two terms of Equation 4.42,

$$w = c_0 + c_2 r^2 \tag{4.43}$$

the strain energy, from Equation 4.41, is then

$$U_1 = 4c_2^2 D\pi a^2 (1+\nu) \tag{a}$$

The strain energy due to the deformation of the elastic foundation is determined as follows:

$$U_2 = \int_0^{2\pi}\int_0^a \frac{1}{2} k w^2 r\, dr\, d\theta = \frac{1}{2}\pi k \left(c_0 a^2 + c_0 c_2 a^4 + \frac{1}{3} c_2^2 a^6 \right) \tag{b}$$

The work done by the load is given by

$$W = P(w)_{r=0} = P c_0 \tag{c}$$

The potential energy, $\Pi = U_1 + U_2 - W$, is this

$$\Pi = 4c_2^2 D\pi a^2 (1+\nu) + \frac{\pi k}{2}\left(c_0^2 a^2 + c_0 c_2 a^4 + \frac{1}{3} c_2^2 a^6 \right) - P c_0$$

Applying the minimizing condition, $\partial \Pi / \partial c_n = 0$, we find that

$$\begin{aligned} c_0 &= \frac{P}{\pi k a^2}\left[1 + \frac{1}{(1/3) + 32D(1+\nu)/ka^4} \right] \\ c_2 &= -\frac{P}{\pi k a^4 \left[(1/6) + 16D(1+\nu)/ka^4 \right]} \end{aligned} \tag{d}$$

Then, by substituting Equations d into Equation 4.43, we obtain the maximum deflection at the center ($r = 0$):

$$w_{\max} = \frac{P}{\pi k a^2}\left[1 + \frac{1}{(1/3) + 32D(1+\nu)/ka^4} \right] \tag{4.44}$$

Comment: An improved approximation results from retention of more terms of the series given by Equation 4.42.

EXAMPLE 4.6: Natural Frequency of Plate

(a) Determine the lowest natural frequency of a circular clamped plate (Figure 4.14a). (b) If the plate is mounted in a rigid chassis that is subjected to a harmonic vibration with peak input $g_{in} = 5\,g$ at a frequency of $f = 450$ Hz, calculate the maximum dynamic deflection and stress in the plate. The plate is made of steel ($E = 200$ GPa, $\nu = 0.3$, density $\rho = 7.86$ Mg/m^3) having radius $a = 100$ mm and thickness $t = 2$ mm.

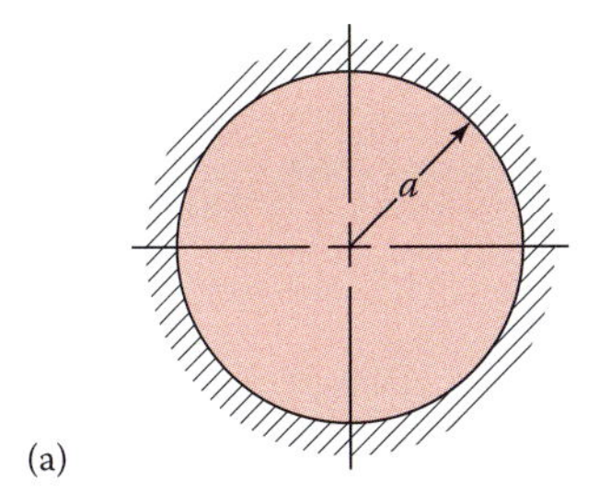

(a)

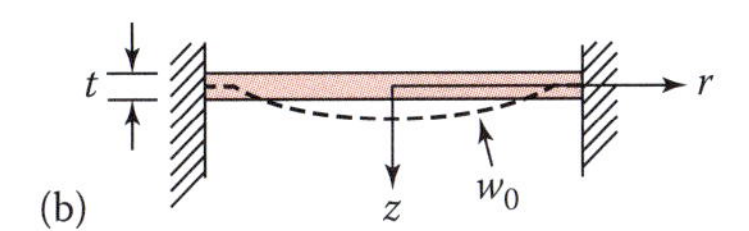

(b)

FIGURE 4.14
(a) Plate with built-in edge; (b) first mode of vibration.

Solution

a. In the case of the lowest mode of vibration, the shape of the plate is symmetrical about the center of the plate (Figure 4.14b). We assume that the deflections of the plate, while it is vibrating, are given by

$$w = w_0 \cos \omega_n t \tag{e}$$

Here w_0 is a function of r only, t is time in seconds, and ω_n represents the *natural circular frequency* in radians per second (rad/s). The boundary conditions,

$$w = 0 \qquad \frac{dw}{dr} = 0 \qquad (\text{at}\, r = a) \tag{f}$$

and condition of symmetry, $dw/dr = 0$ at $r = 0$, are satisfied by taking w_0 in the form of a series [11]:

$$w_0 = c_1\left(1 - \frac{r^2}{a^2}\right)^2 + c_2\left(1 - \frac{r^2}{a^2}\right)^3 + \cdots \tag{4.45}$$

According to Equation f, the last integral in Equation 4.41 vanishes, and we have

$$U = \pi D \int_0^a \left(\frac{d^2w}{dr^2} + \frac{1}{r}\frac{dw}{dr}\right)^2 r\, dr \tag{4.46}$$

Equation 3.54 in polar coordinates becomes

$$T = \frac{\bar{m}}{2} \int_0^{2\pi} \int_0^r \dot{w}^2 r\, d\theta\, dr \tag{4.47}$$

The foregoing equation for the kinetic energy, in *symmetrical cases*, reduces to

$$T = \pi \bar{m} \int_0^a \dot{w}^2 r\, dr \tag{4.48}$$

Substituting Equation e into Equations 4.46 and 4.48, we obtain the following expressions for the maximum strain or potential energy and kinetic energy of vibration:

$$U_{\max} = \pi D \int_0^a \left(\frac{d^2 w_0}{dr^2} + \frac{1}{r}\frac{dw_0}{dr} \right)^2 r\, dr \tag{4.49}$$

and

$$T_{\max} = \pi \bar{m} \omega_n^2 \int_0^a w_0^2 r\, dr \tag{4.50}$$

By retaining only one term of the series in Equation 4.45 and inserting it into Equations 4.50, from minimizing condition $\partial(U_{\max} - T_{\max})/\partial c_1 = 0$, we have

$$\frac{96}{9a^2} - \frac{\omega_n^2 \bar{m}}{D}\frac{a^2}{10} = 0$$

or

$$\omega_n = \frac{10.33}{a^2}\sqrt{\frac{D}{\bar{m}}} \tag{g}$$

Similar calculations conducted for the first two terms of the series of Equation 4.45 result in closer approximation to the circular frequency of the *lowest mode of vibration* of the plate:

$$\omega_n = \frac{10.21}{a^2}\sqrt{\frac{D}{\bar{m}}} \tag{4.51}$$

The natural frequency is thus,

$$f_n = \frac{\omega_n}{2\pi} = \frac{1.625}{a^2}\sqrt{\frac{D}{\bar{m}}} \tag{4.52}$$

Comment: We observe from this result that the frequency is directly proportional to the square root of the modulus of rigidity and inversely proportional to the square of the length of radius of the plate and to the square root of the mass per unit area.

b. For the given data, we have

$$D = \frac{Et^3}{12(1-\nu^2)} = \frac{200\times 10^9(0.002)^3}{12(1-0.09)} = 146.52\ \text{N}\cdot\text{m}$$

$$\bar{m} = \rho t = 7.86\times 10^6(0.002) = 15.72\ \text{kg}/\text{m}^2$$

Introducing the values into Equation 4.52 gives

$$f_n = \frac{1.625}{(0.1)^2}\sqrt{\frac{146.52}{15.72}} = 496.1\ \text{Hz}$$

Equation 3.55 is used to ascertain the transmissibility at 450 Hz:

$$\epsilon = \frac{1}{1-(f/f_n)^2} = \frac{1}{1-(0.907)^2} = 5.64$$

The dynamic load from Equation 3.56 is

$$p_d = \bar{m} g_{in} \epsilon = 15.72(5\times 9.81)5.64 = 4.349\ \text{kPa}$$

The maximum dynamic deflection can thus be found from Equation b of Section 4.5 as

$$w_{\max} = \frac{p_d a^4}{64D} = \frac{4.349\times 10^3(0.1)^4}{64(146.52)} = 0.05\ \text{mm}$$

Similarly, using Equation c of Section 4.5, we obtain

$$\sigma_{r,\max} = \frac{3}{4}p_d\left(\frac{a}{t}\right)^2 = \frac{3}{4}(4.349\times 10^3)(50)^2 = 8.15\ \text{MPa}$$

for the maximum dynamic stress in the plate.

4.12 Asymmetrical Bending of Circular Plates

In the foregoing sections, our concern was with circular plates loaded axisymmetrically. We now turn to asymmetrical bending. For analysis of deflection and stress, we must obtain appropriate solutions of the governing differential Equation 4.5.

Consider the case of a *clamped* circular plate of radius a and subjected to a linearly varying or hydrostatic loading represented by

$$p = p_0 + p_1\frac{r}{a}\cos\theta \tag{4.53}$$

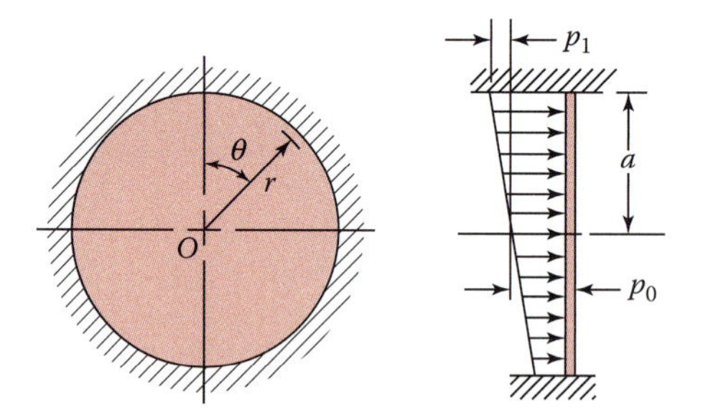

FIGURE 4.15
Hydrostatic load on a clamped plate.

as shown in Figure 4.15. The boundary conditions are

$$w = 0 \qquad \frac{\partial w}{\partial r} = 0 \qquad (r = a) \tag{a}$$

where $w = w_p + w_h$.

The particular solution corresponding to p_0, referring to Section 4.5, is $w_p' = p_0 r^4/64D$. For the linear portion of the loading,

$$w_p'' = A\frac{p_1 r^5 \cos\theta}{a}$$

Introduction of the above into Equation 4.5 yields $A = 1/192\,D$. We thus have

$$w_p = \frac{p_0 r^4}{64D} + \frac{p_1 r^5 \cos\theta}{192aD} \tag{b}$$

Comment: It is noted that the general method of obtaining the particular solution of Equation 4.5, given in Problem 4.33, follows a procedure identical with that described in Section 5.4 (Lévy's solution for rectangular plates).

The homogeneous solution w_h will be symmetrical in θ; thus, f_n^* in Equation 4.7 vanishes. Because of the nature of p and w_p, we take only the terms of series (Equation 4.8) containing the functions f_0 and f_1. The deflection w_h and its derivatives (or slope, moment, and shear) must be finite at the center ($r = 0$). It follows that $C_0 = D_0 = C_1 = D_1 = 0$ in expressions for f_0 and f_1. Hence,

$$w_h = A_0 + B_0 r^2 + (A_1 r + B_1 r^3)\cos\theta \tag{c}$$

Equations a combined with Equations b and c yield two equations in the four unknown constants:

$$\frac{p_0 a^4}{64} + A_0 + B_0 a^2 + \left(\frac{p_1 a^4}{192D} + A_1 a + B_1 a^3\right)\cos\theta = 0$$

$$\frac{4p_0 a^3}{64D} + 2B_0 a + \left(\frac{5p_1 a^3}{192D} + A_1 + 3B_1 a^2\right)\cos\theta = 0$$

Since the term in each pair of parentheses is independent of cos θ, a solution exists for all θ provided that

$$\frac{p_0 a^4}{64D} + A_0 + B_0 a^2 = 0 \qquad \frac{p_1 a^4}{192D} + A_1 a + B_1 a^3 = 0$$
$$\frac{4p_0 a^3}{64D} + 2B_0 a = a \qquad \frac{5p_1 a^3}{192D} + A_1 + 3B_1 a^2 = 0$$

which, on solution, leads to

$$A_0 = \frac{p_0 a^4}{64D} \qquad B_0 = -\frac{p_0 a^2}{32D} \qquad B_1 = -\frac{2p_1 a}{192D} \qquad A_1 = \frac{p_1 a^3}{192D} \tag{d}$$

The deflection is therefore

$$w = \frac{p_0}{64D}(a^2 - r^2)^2 + \frac{p_1}{192D}\frac{r}{a}(a^2 - r^2)^2 \cos\theta \tag{4.54}$$

The center displacement is

$$w_c = \frac{p_0 a^4}{64D} \tag{e}$$

We observe that, when the loading is uniform, $p_1 = 0$, and Equation 4.54 reduces to Equation 4.19 as expected. The expressions for the bending and twisting moments are, from Equations 4.54 and 4.2,

$$M_r = \frac{p_0}{16}[a^2(1+\nu) - r^2(3+\nu)] - \frac{p_1}{48}\left[\frac{r^3}{a}(5+\nu) - ar(3+\nu)\right]\cos\theta$$
$$M_\theta = \frac{p_0}{16}[a^2(1+\nu) - r^2(1+3\nu)] - \frac{p_1}{48}\left[\frac{r^3}{a}(5\nu+1) - ar(3\nu+1)\right]\cos\theta \tag{4.55}$$
$$M_{r\theta} = \frac{1-\nu}{48} p_1 ra\left(1 - \frac{r^2}{a^2}\right)\sin\theta$$

Comment: The case of the simply supported plate under hydrostatic loading can be treated in a similar way.

4.13 *Deflection by the Reciprocity Theorem

Presented in this section is a practical approach for computation of the center deflection of a circular plate with symmetrical edge conditions under asymmetrical or nonuniform

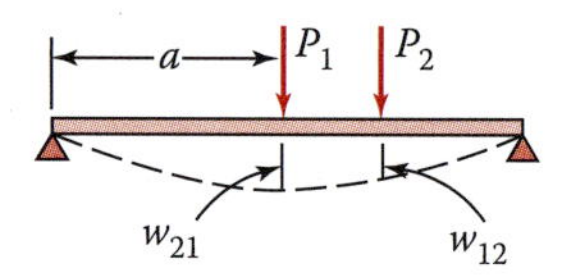

FIGURE 4.16
Simply supported circular pate under centric and eccentric concentrated loads.

loading. The method utilizes the reciprocity theorem together with expressions for deflection of axisymmetrically bent plates.

Consider, for example, the forces P_1 and P_2 acting at the center and at r (any θ) of a circular plate with simply supported edge (Figure 4.16). According to the reciprocity theorem, introduced by E. Betti and Lord Rayleigh, we may write

$$P_1 w_{21} = P_2 w_{12} \tag{a}$$

That is, the work done by P_1 due to displacement w_{21} due to P_2 is equal to the work done by P_2 due to displacement w_{12} due to P_1.

For the sake of simplicity let $P_1 = 1$, $w_{21} = w_c$, and $w_{12} = \bar{w}(r)$. The deflection at the *center* w_c of a circular plate with a nonuniform loading $p(r, \theta)$ but symmetric boundary conditions may therefore by determined through application of Equation a as follows:

$$w_c = \int_0^{2\pi}\int_0^{a} p(r,\theta)\bar{w}(r) r\,dr\,d\theta \tag{4.56}$$

Clearly, $\bar{w}(r)$ is the deflection at r due to a *unit force* at the center. In the cases of fixed and simply supported plates, $\bar{w}(r)$ is given by the expressions obtained by setting $P = 1$ in Equations 4.28 and 4.30, respectively.

To illustrate the application of the approach, reconsider the bending of the plate described in Section 4.12. On substituting $p(r, \theta) = p_0 + p_1(r/a)\cos\theta$ and Equation 4.28 into Equation 4.56, setting $P = 1$ and integrating, we have

$$w_c = \frac{1}{16\pi D}\int_0^{2\pi}\int_0^{a}\left(p_0 + p_1\frac{r}{a}\cos\theta\right)\left(2r^2\ln\frac{r}{a} + a^2 - r^2\right) r\,dr\,d\theta = \frac{p_0 a^4}{64D}$$

This is identical to the value given by Equation e of the preceding section.

Problems

Sections 4.1 through 4.10

4.1 A pressure control system includes a thin steel disk that is to close an electrical circuit by deflecting 1 mm at the center when the pressure attains a value of 3 MPa.

Calculate the required disk thickness if it has a radius of 0.030 m and is built-in at the edge. Let $\nu = 0.3$ and $E = 200$ GPa.

4.2 Redo Problem 4.1 for the case in which the plate is simply supported.

4.3 A cylindrical thick-walled vessel of 0.25-m radius and flat, thin plate head is subjected to an internal pressure of 1 MPa. Determine: (a) the thickness of the cylinder head if the allowable stress is limited to 90 MPa; (b) the maximum deflection of the cylinder head. Use $E = 200$ GPa and $\nu = 0.3$.

4.4 Rework Example 4.1, for the case in which the plate is simply supported. Use $\nu = 1/3$.

4.5 A structural, ASTM-A36 steel (see Table B.3) disk valve of 0.5-m diameter and 20-mm thickness is under a uniform pressure of 600 kPa. What is the factor of safety, based on maximum shear stress theory of failure?

4.6 An aluminum alloy (6061-T6) flat, simply supported disk valve of 0.2-m diameter and 10-mm thickness is subject to a water pressure of 0.5 MPa. What is the factor of safety, assuming failure to take place according to the maximum principal stress theory? The yield strength of the material is 241 MPa, and $\nu = 0.3$.

4.7 A high-strength, ASTM-A242 steel plate covers a circular opening of diameter $d = 2a$. *Assumptions:* The plate is fixed at its edge and carries a uniform pressure p. *Data:* $E = 200$ GPa, $\sigma_{yp} = 345$ MPa (Table B.3), $\nu = 0.3$, $t = 12$ mm, $a = 150$ mm. Find: (a) the pressure p_0 and maximum deflection at the onset of yield in the plate; (b) allowable pressure, based on a safety factor of $n = 1.5$ with respect to yielding of the plate.

4.8 The flat head of a piston is considered to be a clamped circular plate of radius a. The head is under a pressure

$$p = p_0\left(\frac{r}{a}\right)^2$$

where p_0 is constant. Derive the equation

$$w = \frac{p_0 a^4}{576D}\left[\left(\frac{r}{a}\right)^6 - 3\left(\frac{r}{a}\right)^2 + 2\right] \tag{P4.8}$$

for the resulting displacement.

4.9 A circular plate of thickness t is to be used under a concentrated center load P acting on an area of radius $0.4t$ and has its edge simply supported (Figure 4.6b). Calculate the maximum deflection and stress for $t = a/30$ and $\nu = 0.3$.

4.10 A clamped circular plate of thickness t carries a concentrated center load P acting on an area of radius $0.6t$ (Figure 4.6a). Determine the maximum value of P, knowing that the allowable stress is not to exceed 60 MPa, $t = 15$ mm, $a = 180$ mm, and $\nu = 0.3$.

4.11 A cast-iron valve ($E = 100$ GPa, $\nu = 0.2$) that can be approximated as a simply supported circular plate of 280 mm in diameter is under a uniform pressure supplied by a head of 50 m of water. If $\sigma_{all} = 12$ MPa, calculate: (a) the required valve thickness; (b) the maximum deflection of the valve. Specific weight of water $\gamma = 9.81$ kN/m^3.

4.12 Redo Problem 4.11 for the case in which the plate is approximated as clamped at its edge.

4.13 A circular steel plate ($E = 210$ GPa, $\nu = 0.3$) of thickness t is subjected to a moment M_0 per unit circumferential length, uniformly distributed around the entire edge. What thickness is required for the plate: (a) if the allowable defection is 1.5 mm, $a = 0.3$ m, and $M_0 = 500$ N; (b) if the allowable stress is limited to 40 MPa and $M_0 = 800$ N?

4.14 An annular aluminum alloy plate ($E = 70$ GPa, $\nu = 0.3$) of outer diameter 0.8 m and inner diameter 0.6 m is to support only a uniform edge moment $M_1 = 1.5$ kN (Figure 4.8a). Determine: (a) The minimum required thickness of the plate if the bending and shear stresses are not to exceed $\sigma_{\text{all}} = 180$ MPa and $\sigma_{\text{all}} = 100$ MPa; (b) The maximum allowable deflection for $E = 70$ GPa and $\nu = 0.3$.

4.15 For the circular plate loaded as in Case 2 of Table 4.2, verify the result provided for the maximum deflection at the center.

4.16 An annular plate loaded and supported as shown in Figure 4.9. Determine the constants c_1, c_2, c_3, and c_4 for the deflection (w) and stress resultants (M_r, M_θ, and Q_r).

4.17 A simply supported circular plate is under a rotationally symmetric lateral load that increases from the center to the edge (Figure P4.17):

$$p = p_0 \frac{r}{a}$$

Show that the expression

$$w = \frac{p_0 r^5}{225D} + c_4 + c_3 r^2 \tag{P4.17}$$

where

$$c_4 = \frac{p_0 a^4}{45D}\left[\frac{4+\nu}{2(1+\nu)} - \frac{1}{5}\right]$$

$$c_3 = -\frac{p_0 a^2}{90D}\left(\frac{4+\nu}{1+\nu}\right)$$

represents the resulting displacement.

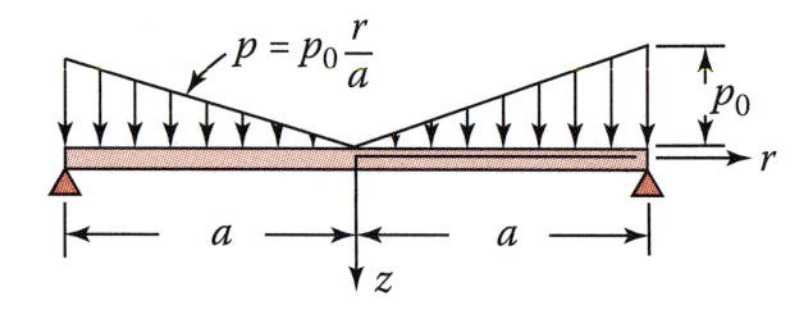

FIGURE P4.17

4.18 Verify the result given by Equation 4.8. *Hint*: Introduction of Equation 4.7 into Equation 4.6 leads to

$$\frac{d^4 f_n}{dr^4}+\frac{2}{r}\frac{d^3 f_n}{dr^3}-\frac{1+2n^2}{r^2}\frac{d^2 f_n}{dr^2}+\frac{1+2n^2}{r^3}\frac{df_n}{dr}+\frac{n^2(n^2-4)}{r^4}f_n=0 \quad (n=0,1,2,\ldots)$$

$$\frac{d^4 f_n^*}{dr^4}+\frac{2}{r}\frac{d^3 f_n^*}{dr^3}-\frac{1+2n^2}{r^2}\frac{d^2 f_n^*}{dr^2}+\frac{1+2n^2}{r^3}\frac{df_n^*}{dr}+\frac{n^2(n^2-4)}{r^4}f_n^*=0 \quad (n=1,2,\ldots) \tag{P4.18}$$

Solution of these equidimensional equations can be taken as [12]: $f_n^*(r)=b_n r^\lambda$, and $f_n(r)=a_n r^\lambda$. Here a_n and b_n are constants, and the λ are the roots of auxiliary equation of Equations P4.18.

4.19 Verify the result given by Equation 4.13: (a) by integrating Equation 4.12; (b) by expanding Equation 4.10a, setting $t = \ln r$, and thereby transforming the resulting expression into an ordinary differential equation with constant coefficients.

4.20 Show that Equation 3.44 for the strain energy results in the following form in terms of the polar coordinates:

$$U=\frac{D}{2}\iint_A\left[\left(\frac{\partial^2 w}{\partial r^2}+\frac{1}{r}\frac{\partial w}{\partial r}+\frac{1}{r^2}\frac{\partial^2 w}{\partial \theta^2}\right)^2-2(1-\nu)\frac{\partial^2 w}{\partial r^2}\left(\frac{1}{r}\frac{\partial w}{\partial r}+\frac{1}{r^2}\frac{\partial^2 w}{\partial \theta^2}\right)+2(1-\nu)\left(\frac{1}{r}\frac{\partial^2 w}{\partial r\,\partial\theta}-\frac{1}{r^2}\frac{\partial w}{\partial \theta}\right)^2\right]r\,dr\,d\theta \tag{P4.20}$$

4.21 Calculate the maximum deflection w in the annular plate loaded as shown in Figure 4.8a by setting $a = 2b$, $M_1 = 2M_2$, and $\nu = 0.3$.

4.22 Determine the maximum displacement in the annular plate loaded as shown in Figure 4.8b by setting $a = 2b$ and $\nu = 0.3$.

4.23 A pump diaphragm can be approximated as an annular plate under a uniformly distributed surface load p_0 and with outer edge simply supported (Case 5 in Table 4.3). Compute, using the method of superposition, the maximum plate deflection for $b = a/4$ and $\nu = 0.3$. Compare the result with that given in the table.

4.24 Resolve Problem 4.23 for the case in which $b = 0.5a$.

4.25 Derive the solution given by Equation k of Example 4.4 by setting $r = e^\alpha$ and thereby transforming the resulting expression into an ordinary differential equation with constant coefficients.

4.26 Consider a simply supported solid plate with profile as shown in Figure P4.26. Take the variation of the thickness in the form

$$t=t_0\left(1-\frac{r}{d}\right) \tag{a}$$

in which d is the altitude of the triangle. Using $\nu = 0.3$, verify that the homogeneous solution of the slope of the deflection surface is

$$\theta_h = c_1\left(\frac{d+2r}{r}\right) + c_2\left[\frac{(3d-2r)r}{d^2(1-r/d)^2}\right] \quad \text{(P4.26)}$$

where c_1 and c_2 are arbitrary constants, to be ascertained from the boundary conditions.

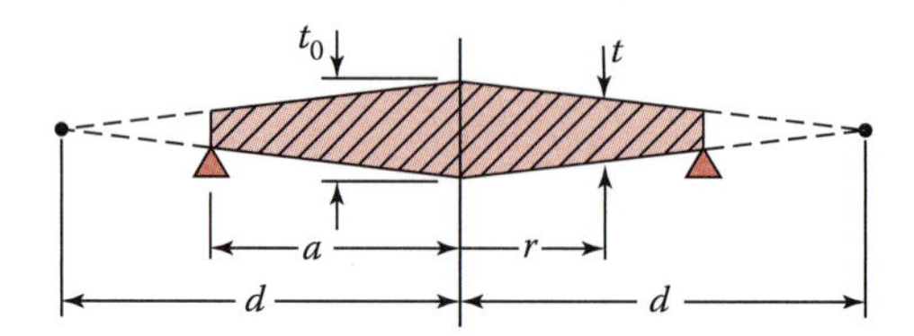

FIGURE P4.26

Sections 4.11 through 4.13

4.27 An aircraft window is approximated as a simply supported circular plate of radius a. The window is subject to a uniform cabin pressure p_0. Determine its maximum deflection, assuming that a diametrical section of the bent plate is parabolic. Employ the Ritz method. Take $\nu = 0.3$.

4.28 Redo Problem 4.27 for a simply supported circular plate that is loaded only by a concentrated center force P.

4.29 Determine the maximum deflection of a structural steel circular plate with free end resting on a gravel–sand mixture foundation and submitted to a load P at its center. Use the Ritz method, taking the first three terms in Equation a of Section 4.11. Let $k = 200$ MPa/m, $a = 0.5$ m, $t = 40$ mm, $E = 200$ GPa, and $\nu = 0.3$.

4.30 In the case of thin plates, where the mass of the liquid (or the air) in which the plate is vibrating may affect the natural frequency, Equation 4.52 is replaced by [11]

$$f_n = \frac{1.625}{a^2\sqrt{1+\beta^2}}\sqrt{\frac{D}{\bar{m}}} \quad \text{(P4.30)}$$

Here

$$\beta = 0.6689\left(\frac{\rho_1}{\rho}\right)\frac{a}{t}$$

and ρ_1/ρ represents the ratio of the density of the fluid to the density of the material of the plate. Rework Example 4.6b on the basis of Equation P4.30, assuming that the steel plate ($\rho = 7.86$ Mg/m^3) is vibrating in water ($\rho_1 = 1$ Mg/m^3) and the rigid chassis is subjected to peak input $g_{\text{in}} = 3\,g$ at a frequency $f = 18$ Hz.

4.31 A simply supported circular plate is loaded by asymmetrically distributed edge couples described by

$$M_r = \sum_{n=0,1,\ldots}^{\infty} M_n \cos n\theta \quad (r = a)$$

In this case, it is observed that $w_p = 0$, and thus $w = w_h$ reduces to

$$w = \sum_{n=0,1,\ldots}^{\infty} \left(A_n r^n + c_n r^{n+2}\right)\cos n\theta$$

Verify that the resulting deflection is

$$w = -\frac{1}{2D}\sum_{n=0,1\ldots}^{\infty} \frac{M_n(r^{n+2} - a^2 r^n)}{a^n(2n+1+\nu)}\cos n\theta \tag{P4.31}$$

4.32 Determine the expression for the radial stress in the plate described in Problem 4.31 by taking $n = 0, 1$.

4.33 The particular solution of Equation 4.5 for an arbitrary loading $p(r, \theta)$ expanded in a Fourier series

$$p(r,\theta) = p_0(r) + \sum_{n=1}^{\infty}\left[P_n(r)\cos n\theta + R_n(r)\sin n\theta\right] \tag{a}$$

where

$$P_n(r) = \frac{1}{\pi}\int_{-\pi}^{\pi} p(r,\theta)\cos n\theta \, d\theta \qquad (n = 0, 1, \ldots)$$

$$R_n(r) = \frac{1}{\pi}\int_{-\pi}^{\pi} p(r,\theta)\sin n\theta \, d\theta \qquad (n = 1, 2, \ldots)$$

is expressed in the general form [2]

$$w_p = F_0(r) + \sum_{n=1}^{\infty}\left[F_n(r)\cos n\theta + G_n(r)\sin n\theta\right] \tag{P4.33}$$

In the preceding, the quantities $F_0(r)$, $F_n(r)$, and $G_n(r)$ are functions of r. Demonstrate that substitution of Equations a and P4.33 into Equation 4.5 leads to

$$\begin{aligned}
\frac{d^4F_0}{dr^4} + \frac{2}{r}\frac{d^3F_0}{dr^3} - \frac{1}{r^2}\frac{d^2F_0}{dr^2} + \frac{1}{r^3}\frac{dF_0}{dr} &= \frac{p_0}{D} \\
\frac{d^4F_0}{dr^4} + \frac{2}{r}\frac{d^3F_n}{dr^3} - \frac{1+2n^2}{r^2}\frac{d^2F_n}{dr^2} + \frac{1+2n^2}{r^3}\frac{dF_n}{dr} + \frac{n^2(n^2-4)}{r^4}F_n &= \frac{P_n}{D} \\
\frac{d^4G_n}{dr^4} + \frac{2}{r}\frac{d^3G_n}{dr^3} - \frac{1+2n^2}{r^2}\frac{d^2G_n}{dr^2} + \frac{1+2n^2}{r^3}\frac{dG_n}{dr} + \frac{n^2(n^2-4)}{r^4}G_n &= \frac{R_n}{D}
\end{aligned} \tag{b}$$

Thus, for a prescribed loading $p(r, \theta)$, Equations b are solved for F_0, F_n, and G_n. The particular solution is then obtained from Equation P4.33.

4.34 A simply supported circular plate is subjected to hydrostatic loading. Employ the reciprocity theorem to find the maximum deflection.

4.35 A clamped circular plate carries a concentrated downward load P at a point located at a distance b from its center. Apply the reciprocity theorem to obtain the center deflection.

4.36 Redo Problem 4.35 for the case of the plate with simply supported edge.

References

1. R. Szilard, *Theory and Analysis of Plates—Classical and Numerical Methods*, Section 1.14, Prentice-Hall, Upper Saddle River, NJ, 2004.
2. D. McFarland, B. L. Smith, and W. D. Bernhart, *Analysis of Plates*, Spartan Books, Washington, DC, 1972.
3. W. Flügge (ed.), *Handbook of Engineering Mechanics*, Chapter 39, McGraw-Hill, New York, 1984.
4. W. C. Young, R. C. Budynas, and A. Sadegh, *Roark's Formulas for Stress and Strain*, 8th ed., Chapter 10, McGraw-Hill, New York, 2011.
5. A. Nadai, *Die Elastischen Platten*, Springer, Berlin, 1925 and 1968.
6. H. M. Westergaard, Stresses in concrete pavements computed by theoretical analysis, Public Roads, U.S. Department of Agriculture, Bureau of Public Roads, Vol. 7, no. 2, 1926.
7. S. Timoshenko and S. Woinowsky-Krieger, *Theory of Plates and Shells*, 2nd ed., Chapter 3, McGraw-Hill, New York, 1959.
8. A. C. Ugural and S. K. Fenster, *Advanced Mechanics of Materials and Applied Elasticity*, 5th ed., Chapter 3, Prentice-Hall, Upper Saddle River, NJ, 2012.
9. A. M. Wahl and G. Lobo, Stresses and deflections in flat circular plates with circular holes, *Transactions of ASME*, 52, 1930, 29–43.
10. G. S. Brady and H. R. Clauser, *Materials Handbook*, 15th ed., McGraw-Hill, New York, 2002.
11. S. Timoshenko, D. H. Young, and W. Weaver Jr., *Vibration Problems in Engineering*, 4th ed., Wiley, New York, 1974.
12. E. Kreyszig, *Advanced Engineering Mathematics*, 10th ed., Wiley, Hoboken, NJ, 2011.

5

Rectangular Plates

5.1 Introduction

In this chapter, consideration is given to stresses and deflections in thin rectangular plates. As observed in Chapter 3, the rectangular-plate element is an excellent model for the development of the basic relationships in Cartesian coordinates. On the other hand, for most rectangular-plate problems, there is no exact solution to the governing equations. Instead, recall from Section 3.11 that solutions for deflections and moments may be obtained by employing various infinite series formulations. The numerical values are much easier to determine with the use of the computer.

Rectangular plates are generally classified in accordance with the type of support used. We are here concerned with the bending of simply supported plates, clamped or built-in plates, plates having *mixed* support conditions, plates on an elastic foundation, and continuous plates [1–9]. The term *continuous plates* often refer to structures consisting of a single plate supported by intermediate beams or columns. All cases are treated by relationships derived in Chapter 3. The strip method of Section 5.8 is a discussion of the bending of rectangular plates based on elementary beam theory.

5.2 Navier's Solution for Simply Supported Rectangular Plates

Consider a simply supported rectangular plate of sides a and b, subjected to a distributed load $p(x, y)$. The origin of coordinates is placed at the upper left corner of the plate as shown in Figure 5.1a. In general, solution of the bending problem employs the following *Fourier series* (Appendix A) for load and deflection:

$$p(x,y)=\sum_{m=1}^{\infty}\sum_{n=1}^{\infty} p_{mn}\sin\frac{m\pi x}{a}\sin\frac{n\pi y}{b} \tag{5.1a}$$

$$w(x,y)=\sum_{m=1}^{\infty}\sum_{n=1}^{\infty} a_{mn}\sin\frac{m\pi x}{a}\sin\frac{n\pi y}{b} \tag{5.1b}$$

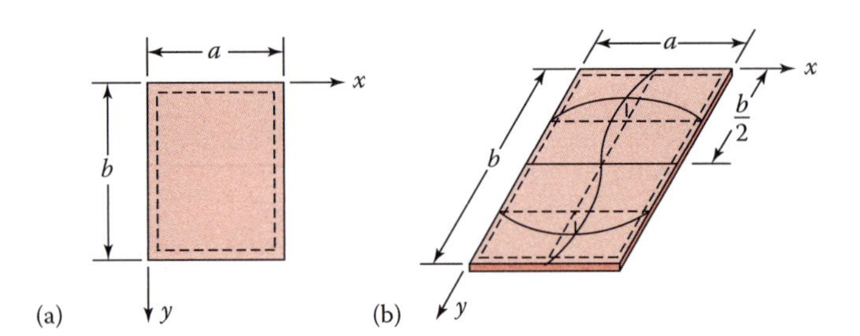

FIGURE 5.1
(a) Location of coordinate system for Navier's method; (b) deflection of the simply supported plate into half-sine curves of $m = 1$ and $n = 2$.

where p_{mn} and a_{mn} represent coefficients to be determined. This approach was introduced by *Navier* in 1820. The deflection must satisfy the differential equations 3.17 with the following boundary conditions (Equations 3.26b):

$$\begin{aligned} &w = 0 \qquad \frac{\partial^2 w}{\partial x^2} = 0 \qquad (x = 0, x = a) \\ &w = 0 \qquad \frac{\partial^2 w}{\partial y^2} = 0 \qquad (y = 0, y = b) \end{aligned} \tag{a}$$

Clearly, these edge restraints are fulfilled by Equation 5.1b, and the coefficients a_{mn} must be such as to satisfy Equation 3.17. The solution corresponding to loading $p(x, y)$ thus requires determination of p_{mn} and a_{mn}.

As a *physical interpretation* of Equation 5.1b, consider the true deflection surface of the plate to be the superposition of sinusoidal curves of m and n different configurations in the x and y directions, respectively. The coefficients a_{mn} of the series are the maximum central coordinates of the sine curves, and the ms and the ns indicate the number of half-sine curves in the x and y directions, respectively. For example, the term $a_{12} \sin(\pi x/a) \sin(2\pi y/b)$ of the series is illustrated in Figure 5.1b. By increasing the number of terms in the series, the accuracy can, of course, be improved.

We proceed by dealing first with a general load configuration, subsequently treating specific loadings. To determine the *Fourier coefficients* p_{mn}, each side of Equation 5.1a is multiplied by

$$\sin\frac{m'\pi x}{a}\sin\frac{n'\pi y}{b}\,dx\,dy$$

and integrated between limits 0, a and 0, b:

$$\begin{aligned} &\int_0^b\int_0^a p(x,y)\sin\frac{m'\pi x}{a}\sin\frac{n'\pi y}{b}\,dxdy \\ &= \sum_{m=1}^{\infty}\sum_{n=1}^{\infty} p_{mn}\int_0^b\int_0^a \sin\frac{m\pi x}{a}\sin\frac{n\pi y}{b}\sin\frac{m'\pi x}{a}\sin\frac{n'\pi y}{b}\,dxdy \end{aligned} \tag{b}$$

It can be shown by direct integration that

$$\int_0^a \sin\frac{m\pi x}{a}\sin\frac{m'\pi x}{a}dx = \begin{cases} 0 & (m \neq m') \\ a/2 & (m = m') \end{cases}$$
$$\int_0^b \sin\frac{n\pi y}{b}\sin\frac{n'\pi y}{b}dy = \begin{cases} 0 & (n \neq n') \\ b/2 & (n = n') \end{cases} \tag{5.2}$$

The coefficients of the double Fourier expansion are therefore (see Section A.3)

$$p_{mn} = \frac{4}{ab}\int_0^b\int_0^a p(x,y)\sin\frac{m\pi x}{a}\sin\frac{n\pi x}{b}dxdy \tag{5.3}$$

Evaluation of a_{mn} in Equation 5.1b requires substitution of Equations 5.1 into Equations 3.17, with the result

$$\sum_{m=1}^{\infty}\sum_{n=1}^{\infty}\left\{a_{mn}\left[\left(\frac{m\pi}{a}\right)^4 + 2\left(\frac{m\pi}{a}\right)^2\left(\frac{n\pi}{b}\right)^2 + \left(\frac{n\pi}{b}\right)^4\right] - \frac{p_{mn}}{D}\right\}\sin\frac{m\pi x}{a}\sin\frac{n\pi y}{b} = 0$$

This equation must apply for all x and y. We conclude therefore that

$$a_{mn}\pi^4\left(\frac{m^2}{a^2} + \frac{n^2}{b^2}\right)^2 - \frac{p_{mn}}{D} = 0$$

from which

$$a_{mn} = \frac{1}{\pi^4 D}\frac{p_{mn}}{[(m/a)^2 + (n/b)^2]^2} \tag{5.4}$$

Substituting Equation 5.4 into Equation 5.1b, we find the equation of the deflection surface of the plate becomes

$$w = \frac{1}{\pi^4 D}\sum_{m=1}^{\infty}\sum_{n=1}^{\infty}\frac{p_{mn}}{\left[(m/a)^2 + (n/b)^2\right]^2}\sin\frac{m\pi x}{a}\sin\frac{n\pi y}{b} \tag{5.5}$$

in which P_{mn} is given by Equation 5.3. It can be shown, by noting that $|\sin(m\pi x/a)| \leq 1$ and $|\cos(n\pi y/b)| \leq 1$ for every x and y and for m and n, that the series (Equation 5.5) is convergent. Thus, Equation 5.5 is a valid solution for bending for simply supported rectangular plates under various kinds of loadings. Application of Navier's method to several particular cases is presented in the next section.

5.3 Simply Supported Rectangular Plates under Various Loadings

When a rectangular plate is subjected to a *uniformly distributed load* $p(x, y) = p_0$, the results of the previous section are simplified considerably. Now Equation 5.3, after integration, yields

$$p_{mn} = \frac{4p_0}{\pi^2 mn}(1 - \cos m\pi)(1 - \cos n\pi) = \frac{4p_0}{\pi^2 mn}[1-(-1)^m][1-(-1)^n]$$

or

$$p_{mn} = \frac{16p_0}{\pi^2 mn} \qquad (m, n = 1, 3, \ldots) \tag{a}$$

It is observed that because $p_{mn} = 0$ for even values of m and n, these integers assume only odd values. Introducing p_{mn} into Equation 5.5, we have

$$w = \frac{16p_0}{\pi^6 D}\sum_m^{\infty}\sum_n^{\infty}\frac{\sin(m\pi x/a)\sin(n\pi y/b)}{mn\left[(m/a)^2 + (n/b)^2\right]^2} \qquad (m, n = 1, 3, \ldots) \tag{5.6}$$

Clearly, on the basis of physical considerations, the uniformly loaded plate must deflect into a symmetrical shape. Such a configuration results when m and n are *odd*. The maximum deflection occurs at the center of the plate ($x = a/2$, $y = b/2$) and its value, from Equation 5.6, is

$$w_{\max} = \frac{16p_0}{\pi^6 D}\sum_m^{\infty}\sum_n^{\infty}\frac{(-1)^{[(m+n)/2]-1}}{mn[(m/a)^2 + (n/b)^2]^2} \tag{b}$$

Note that in Equation 5.6, $\sin m\pi/2$ and $\sin n\pi/2$ are replaced by $(-1)^{(m-1)/2}$ and $(-1)^{(n-1)/2}$, respectively.

By introducing Equation 5.6 into Equations 3.9, the components of the moment are derived:

$$\begin{aligned}
M_x &= \frac{16p_0}{\pi^4}\sum_m^{\infty}\sum_n^{\infty}\frac{(m/a)^2 + \nu(n/b)^2}{mn[(m/a)^2 + (n/b)^2]^2}\sin\frac{m\pi x}{a}\sin\frac{n\pi y}{b} \\
M_y &= \frac{16p_0}{\pi^4}\sum_m^{\infty}\sum_n^{\infty}\frac{\nu(m/a)^2 + (n/b)^2}{mn[(m/a)^2 + (n/b)^2]^2}\sin\frac{m\pi x}{a}\sin\frac{n\pi y}{b} \\
M_{xy} &= -\frac{16p_0(1-\nu)}{\pi^4 ab}\sum_m^{\infty}\sum_n^{\infty}\frac{1}{[(m/a)^2 + (n/b)^2]^2}\cos\frac{m\pi x}{a}\cos\frac{n\pi y}{b}
\end{aligned} \tag{5.7}$$

We observe that the bending moments M_x and M_y are both zero at ($x = 0, x = a$) and ($y = 0$, $y = b$), respectively. However, the twisting moment M_{xy} does not vanish at the edges and at the corners of the plate. The presence of M_{xy} causes a modification of the distribution of the reactions on the supports (Section 3.9). Recall, however, that St. Venant's principle permits us to regard the stress distribution unaltered for sections away from the edges and the corners.

Situations involving simply supported rectangular plates carrying sinusoidal, partial, and concentrated loadings are discussed in Examples 5.2 through 5.4. The reader will find further applications of Navier's method in several sections of this text and in a number of References 1 through 4.

EXAMPLE 5.1: Analysis of Plate under Uniform Pressure

A square wall panel is taken to be simply supported on all edges and subjected to a uniform load differential p_0. Determine the maximum deflection, moment, and stress.

Solution

The first term ($m = 1, n = 1$) of Equation b yields, for $a = b$,

$$w_{\max} = 0.00416\, p_0 \frac{a^4}{D}$$

Very rapid convergence of Equation b is demonstrated by noting that retaining the first four terms ($m = 1, n = 1, 3; m = 3, n = 1, 3$) results in what is essentially the "exact" solution, $w_{\max} = 0.00406 p_0 a^4/D$.

The maximum bending moments, found at the center of the plate, are determined by applying Equations 5.7. The first term of the series yields, for $\nu = 0.3$:

$$M_{x,\max} = M_{y,\max} = 0.0534\, p_0 a^2$$

while the first four terms result in

$$M_{x,\max} = M_{y,\max} = 0.0469\, p_0 a^2 \tag{c}$$

Comments: It is thus observed from a comparison of the above that the series for the bending moments given by Equations 5.7 does not converge as rapidly as that of Equation b. The maximum bending stress produced by the moment of Equation c, by application of Equations 3.11, is determined to be $\sigma_{\max} = 0.281\, p_0 a^2/t^2$.

EXAMPLE 5.2: Support Reactions on Plate

A rectangular warehouse floor slab of sides a and b is simply supported on all edges. Determine the reactions at the supports if the material is stored on the entire floor in such a way that the loading is expressed in the following approximate form:

$$p(x, y) = p_0 \sin\frac{\pi x}{a} \sin\frac{\pi y}{b} \tag{d}$$

Here p_0 represents the intensity of the load at the center of the plate, as shown in Figure 5.2.

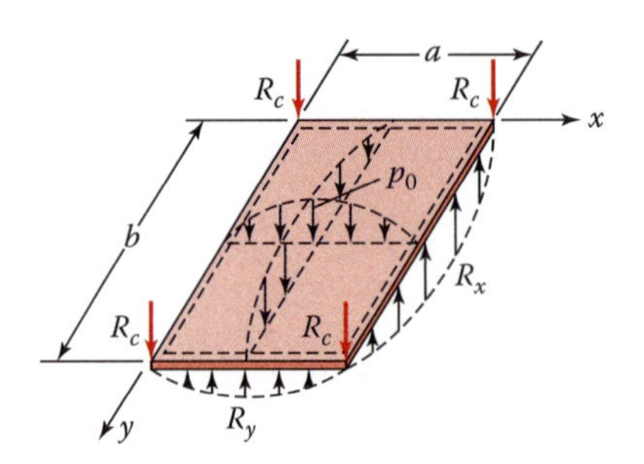

FIGURE 5.2
Distributed and concentrated edge reactions on a pate under sinusoidal load.

Solution

Substituting Equation d into Equation 5.3 and letting $m = n = 1$, we have $p_{mn} = p_0$. Equation 5.5 in this case leads to

$$w = \frac{p_0}{\pi^4 D(1/a^2 + 1/b^2)^2} \sin\frac{\pi x}{a} \sin\frac{\pi y}{b}$$

Introducing w from the above into Equations 3.9, we have

$$\begin{aligned} M_x &= \frac{p_0}{\pi^2(1/a^2 + 1/b^2)^2}\left(\frac{1}{a^2} + \frac{\nu}{b^2}\right)\sin\frac{\pi x}{a}\sin\frac{\pi y}{b} \\ M_y &= \frac{p_0}{\pi^2(1/a^2 + 1/b^2)^2}\left(\frac{\nu}{a^2} + \frac{1}{b^2}\right)\sin\frac{\pi x}{a}\sin\frac{\pi y}{b} \\ M_{xy} &= -\frac{p_0(1-\nu)}{\pi^2(1/a^2 + 1/b^2)^2 ab}\cos\frac{\pi x}{a}\cos\frac{\pi y}{b} \end{aligned} \tag{e}$$

The shear forces are determined from Equations 3.16:

$$\begin{aligned} Q_x &= \frac{p_0}{\pi a(1/a^2 + 1/b^2)}\cos\frac{\pi x}{a}\sin\frac{\pi y}{b} \\ Q_y &= \frac{p_0}{\pi b(1/a^2 + 1/b^2)}\sin\frac{\pi x}{a}\cos\frac{\pi y}{b} \end{aligned} \tag{f}$$

The total load carried by the plate, neglecting its weight, is equal to

$$\int_0^b \int_0^a p_0 \sin\frac{\pi x}{a}\sin\frac{\pi y}{b}\,dx\,dy = \frac{4p_0 ab}{\pi^2} \tag{g}$$

The *reactive forces* at the simply supported *edges* of the plate can act only vertically. Equation 3.23a is applied to obtain the reaction $R_x = V_x$ for the edge $x = a$:

$$R_x = -\frac{p_0}{\pi a(1/a^2 + 1/b^2)^2}\left(\frac{1}{a^2} + \frac{2-\nu}{b^2}\right)\sin\frac{\pi y}{b} \tag{5.8a}$$

Similarly, for the edge $y = b$, setting $R_y = V_y$ from Equation 3.23b, we have

$$R_y = -\frac{p_0}{\pi b(1/a^2 + 1/b^2)^2}\left(\frac{1}{b^2} + \frac{2-\nu}{b^2}\right)\sin\frac{\pi y}{a} \tag{5.8b}$$

The edge reactions thus on the plate also vary sinusoidally (Figure 5.2). The minus sign indicates that they are directed upward as shown in the figure. Conditions of symmetry dictate that the reactions along the support at $x = 0$ and $y = 0$ are identical with those given by Equations 5.8a and 5.8b, respectively.

The sum of distributed reactions on the edges of the plate is

$$F = 2\int_0^a R_y\,dx + 2\int_0^b R_x\,dy$$

This, by substituting Equations 5.8 and integrating, gives

$$F = -\frac{4p_0 ab}{\pi^2} - \frac{8p_0(1-\nu)}{\pi^2 ab(1/a^2 + 1/b^2)^2} \tag{h}$$

It is seen from Equations g and h that the distributed reactions are larger than necessary to compensate for the p loading. Thus, as we already shown in Section 3.9, there will be a concentrated *reaction* $R_c = -F_c = -M_{xy}$ *at each corner* of the plate having a value (Equation 3.24)

$$R_c = \frac{2p_0(1-\nu)}{\pi^2 ab(1/a^2 + 1/b^2)^2} \tag{5.9}$$

The condition that the resultant of the laterally applied and the reactional forces in the plate equal to zero is thus satisfied (Figure 5.2).

Comments: The reason for the corner reactions can be recognized intuitively. When a rectangular plate, supported freely by a rectangular frame, is subjected to center loading, it tends to deform in such a way as to have the *corners rise*, with contact made only at the middle of the sides. To prevent this, the corners must be held down.

EXAMPLE 5.3: Deflection of Plate under Distributed and Concentrated Loads

Find the equation of the elastic surface of a simply supported rectangular plate for two particular cases: (a) the plate is subjected to a load P uniformly distributed over a subregion $4cd$ or so-called *patch load* (Figure 5.3a); (b) the plate carries a (nominal) point load P at $x = x_1$, $y = y_1$ (Figure 5.3b); (c) if the plate is square ($a = b$) and the load P acts at the center, calculate the maximum deflection for $a = 600$ mm, $t = 10$ mm, $E = 70$ GPa, $P = 5$ kN, $\nu = 0.3$.

Solution

a. Since $p(x, y) = P/4cd$, we have, through the use of Equation 5.3,

$$p_{mn} = \frac{P}{abcd}\int_{y_1-d}^{y_1+d}\int_{x_1-c}^{x_1+c} \sin\frac{m\pi x}{a}\sin\frac{m\pi y}{b}\,dx\,dy$$

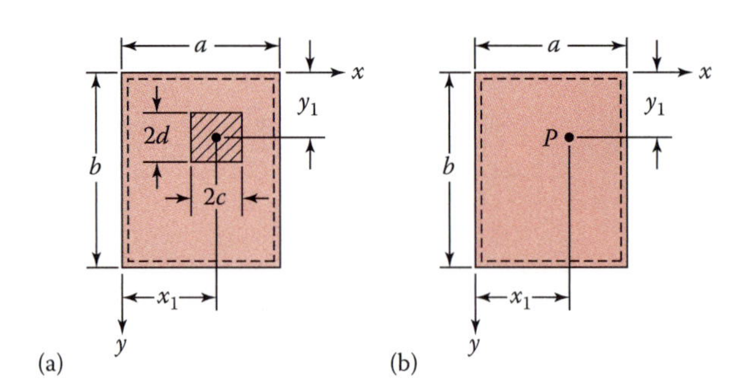

FIGURE 5.3
(a) Patch load; (b) concentrated load.

from which

$$p_{mn} = \frac{4P}{\pi^2 mncd} \sin\frac{m\pi x_1}{a} \sin\frac{n\pi y_1}{b} \sin\frac{n\pi c}{a} \sin\frac{n\pi d}{b} \tag{i}$$

Clearly, for $x_1 = a/2$, $y_1 = b/2$, $c = a/2$, and $d = b/2$, the foregoing reduces to Equation a. Introduction of Equation i into Equation 5.5 gives the required deflection surface.

b. For the case in which c and d are made to approach zero, Equation i appears in the form (Problem 5.11)

$$p_{mn} = \frac{4P}{ab} \sin\frac{m\pi x_1}{a} \sin\frac{n\pi y_1}{b} \tag{5.10}$$

Note that the load P is an idealization of the distributed load $p(x, y)$ concentrated in a very small area of size defined by Equation 4.26. Inserting Equation 5.4 together with Equation 5.10 into Equation 5.1b results in the plate deflection:

$$w = \frac{4P}{\pi^4 Dab} \sum_m^\infty \sum_n^\infty \frac{\sin(m\pi x_1/a)\sin(n\pi y_1/b)}{[(m/a)^2 + (n/b)^2]^2} \sin\frac{m\pi x}{a} \sin\frac{n\pi y}{b} \tag{5.11}$$

Equation 5.11 may be used with the method of superposition (Sections 4.10 and 5.7) to determine the deflection of simply supported rectangular plates under various types of loadings.

c. When the load P is *applied at the center* of the plate ($x_1 = a/2$, $y_1 = b/2$), Equation 5.11 reduces to

$$w = \frac{4P}{\pi^4 Dab} \sum_m^\infty \sum_n^\infty (-1)^{[(m+n)/2]-1} \frac{\sin(m\pi x/a)\sin(n\pi y/b)}{[(m/a)^2 + (n/b)^2]^2} \qquad (m, n = 1, 3, \ldots) \tag{5.12}$$

For a *square* plate ($a = b$), the maximum deflection, which occurs at the center, is from Equation 5.12:

$$w_{\max} = \frac{4Pa^2}{\pi^4 D} \sum_m^\infty \sum_n^\infty \frac{1}{(m^2 + n^2)^2} \qquad (m, n = 1, 3, \ldots)$$

Retaining the first nine terms of this series ($m = 1, n = 1, 3, 5$; $m = 3, n = 1, 3, 5$; $m = 5, n = 1, 3, 5$), we obtain

$$w_{\max} = \frac{4Pa^2}{\pi^4 D}\left[\frac{1}{2^2} + \frac{2}{10^2} + \frac{1}{18^2} + \frac{2}{26^2} + \frac{2}{34^2} + \frac{1}{50^2}\right] = 0.01142\frac{Pa^2}{D}$$

Substituting the given data, we have

$$D = \frac{Et^3}{12(1-\nu^2)} = \frac{70(10)^3}{12(1-0.3^2)} = 6.41\,\text{kN}\cdot\text{m}$$

and

$$w_{\max} = 0.01142\frac{5(0.6)^2}{6.41} = 3.21\,\text{mm}$$

Comment: Note that, the "exact" value is $w_{\max} = 0.01159Pa^2/D$, and the error is thus 1.56%.

EXAMPLE 5.4: Displacement of Plate due to Moment

A simply supported rectangular plate is subjected to a moment M at $x = x_1$, $y = y_1$ (Figure 5.4). Determine an equation for the deflection.

Solution

Observe that the couple $P \cdot e$ is equivalent to M for the case in which e approaches zero (indicated by the dashed lines in the figure). Applying Equation 5.11, we have therefore

$$w = -\frac{4M}{\pi^4 Dab}\sum_m^\infty\sum_n^\infty \frac{\sin(m\pi x/a)\sin(n\pi y/b)}{[(m/a)^2 + (n/b)^2]^2}$$
$$\times \lim_{e\to 0}\left\{\frac{1}{e}\left[\sin\frac{m\pi(x_1 - e)}{a}\sin\frac{n\pi y_1}{b} - \sin\frac{m\pi x_1}{a}\sin\frac{n\pi y_1}{b}\right]\right\}$$

or

$$w = -\frac{4M}{\pi^4 Dab}\sum_m^\infty\sum_n^\infty \frac{\sin(m\pi x/a)\sin(n\pi y/b)}{[(m/a)^2 + (n/b)^2]^2}\frac{\partial}{\partial x_1}\left(\sin\frac{m\pi x_1}{a}\sin\frac{m\pi y_1}{b}\right)$$

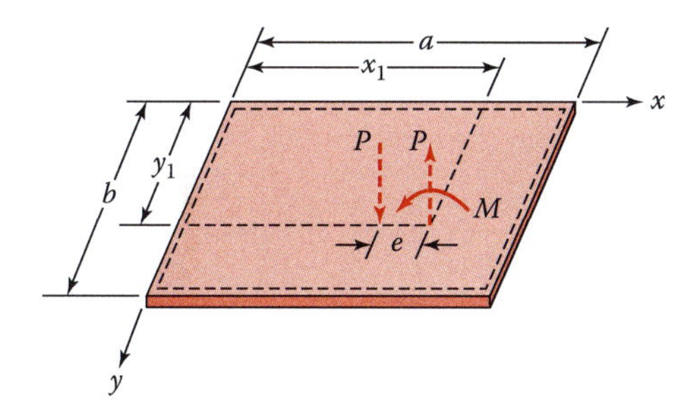

FIGURE 5.4
Concentrated moment at an arbitrary point.

The preceding, on differentiation, provides the following equation for the deflection:

$$w = -\frac{4M}{\pi^3 D a^2 b}\sum_m^\infty \sum_n^\infty \frac{m}{[(m/a)^2 + (n/b)^2]^2}\cos\frac{m\pi x_1}{a}\sin\frac{n\pi y_1}{b}\sin\frac{m\pi x}{a}\sin\frac{n\pi y}{b} \tag{5.13}$$

Substitution of Equation 5.13 into Equation 3.9 leads to the expression for the moments.

5.4 Lévy's Solution for Rectangular Plates

It is seen in the foregoing section that the calculation of bending moments using Equation 5.7 is not very satisfactory because of the slow convergence of the series. An important approach that overcomes this difficulty was developed by M. Lévy [5] in 1900. Another advantage of the *Lévy's solution* as compared with Navier's method is that instead of a double series, we have to deal with a single series (Appendix A). In general, it is easier to perform numerical calculations for a single series than for a double series.

Lévy's method is applicable to the bending of rectangular plates with *particular* boundary conditions on the two opposite sides (say, at $x = 0$ and $x = a$) and *arbitrary* conditions of support on the remaining edges (at $y = \pm b/2$) (Figure 5.5). The total solution consists of the homogeneous solution w_h of Equation 3.18 and the particular solution w_p of Equation 3.17:

$$w = w_h + w_p \tag{a}$$

Inasmuch as $\nabla^4 w_h = 0$ is independent of the loading, a single expression can be derived for w_h that is valid for all rectangular plates having particular boundary conditions on the two opposite sides. Clearly, *for each specific loading $p(x, y)$, a solution w_p must be obtained.*

The homogeneous solution is selected from the following general form:

$$w_h = \sum_{m=1}^{\infty} f_m(y)\begin{Bmatrix}\sin(m\pi x/a)\\ \cos(m\pi x/a)\end{Bmatrix} \tag{5.14}$$

where the function $f_m(y)$ must be obtained such as to fulfill the conditions of the supports at $y = \pm b/2$ and to satisfy Equation 3.18. We proceed with the description of the method by

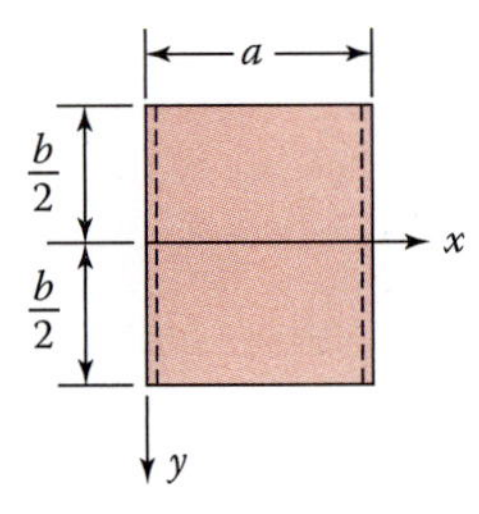

FIGURE 5.5
Usual location of coordinate system of Levy's method.

assuming that *two opposite sides* of the rectangular plate at $x = 0$ and $x = a$ are *simply supported* as shown in Figure 5.5. In this case, Equation 5.14 appears as

$$w_h = \sum_{m=1}^{\infty} f_m(y) \sin \frac{m\pi x}{a} \tag{5.15}$$

Equation 5.15 satisfies the conditions (Equations 3.26b) along the x edges. To complete the solution, we must apply to w the boundary conditions on the *two arbitrary sides* at $y = \pm b/2$.

Substituting Equation 5.15 into Equation 3.18 yields

$$\sum_{m=1}^{\infty} \left[\frac{d^4 f_m}{dy^4} - 2\left(\frac{m\pi}{a}\right)^2 \frac{d^2 f_m}{dy^2} + \left(\frac{m\pi}{a}\right)^4 f_m \right] \sin \frac{m\pi x}{a} = 0$$

In order that this equation be valid for any x,

$$\frac{d^4 f_m}{dy^4} - 2\left(\frac{m\pi}{a}\right)^2 \frac{d^2 f_m}{dy^2} + \left(\frac{m\pi}{a}\right)^4 f_m = 0$$

This is a linear, homogeneous differential equation of fourth order with constant coefficients. Hence, its general solution is (see Problem 5.15)

$$f_m = A'_m e^{m\pi y/a} + B'_m e^{-m\pi y/a} + C'_m y e^{m\pi y/a} + D'_m y e^{-m\pi y/a} \tag{5.16}$$

or, by using hyperbolic functions,

$$f_m = A_m \sinh \frac{m\pi y}{a} + B_m \cosh \frac{m\pi y}{a} + C_m y \sinh \frac{m\pi y}{a} + D_m y \cosh \frac{m\pi y}{a} \tag{5.17}$$

The *homogeneous solution* is therefore

$$\begin{aligned} w_h = \sum_{m=1}^{\infty} \bigg(& A_m \sinh \frac{m\pi y}{a} + B_m \cosh \frac{m\pi y}{a} \\ & + C_m y \sinh \frac{m\pi y}{a} + D_m y \cosh \frac{m\pi y}{a} \bigg) \sin \frac{m\pi x}{a} \end{aligned} \tag{5.18}$$

where A_m, B_m, C_m, and D_m are constants, to be determined later for specified cases.

It is observed that the boundary conditions (Equations 3.26b) along the edges $x = 0$ and $x = a$ are satisfied, if the particular solution is expressed by the following single Fourier series:

$$w_p = \sum_{m=1}^{\infty} k_m(y) \sin \frac{m\pi x}{a} \tag{5.19}$$

where the $k_m(y)$s are functions of y only. Let us also expand $p(x, y)$ in terms of a single Fourier series,

$$p(x,y) = \sum_{m=1}^{\infty} p_m(y) \sin \frac{m\pi x}{a} \tag{5.20}$$

in which

$$p_m(y) = \frac{2}{a} \int_0^a p(x,y) \sin \frac{m\pi x}{a} dx \tag{5.21}$$

Substituting Equations 5.19 and 5.20 into Equation 3.17 and noting the validity of the resulting expression for all values of x between 0 and a, we find that

$$\frac{d^4 k_m}{dy^4} - 2\left(\frac{m\pi}{a}\right)^2 \frac{d^2 k_m}{dy^2} + \left(\frac{m\pi}{a}\right)^4 k_m = \frac{p_m}{D} \tag{b}$$

On determination of a particular solution, k_m, to this ordinary differential equation, we then obtain w_p from Equation 5.19. The method is illustrated by considering the following commonly referred to example.

5.4.1 Simply Supported Rectangular Plate under Uniform Loading

For this case, $p(x, y) = p_0$, for which Equation 5.21 yields, on integration,

$$p_m = \frac{4p_0}{m\pi} \qquad (m = 1, 3, \ldots) \tag{c}$$

Then Equation b becomes

$$\frac{d^4 k_m}{dy^4} - 2\left(\frac{m\pi}{a}\right)^2 \frac{d^2 k_m}{dy^2} + \left(\frac{m\pi}{a}\right)^4 k_m = \frac{4p_0}{m\pi D}$$

The particular solution of the above is $k_m = 4p_0a^4/m^5\pi^5D$. Equation 5.19 is therefore

$$w_p = \frac{4p_0a^4}{\pi^5 D} \sum_{m=1,3,\ldots}^{\infty} \frac{1}{m^5} \sin \frac{m\pi x}{a} \tag{5.22a}$$

This represents the deflection of a uniformly loaded, *simply supported strip* parallel to the x axis and may be rewritten in the following alternative form (see Example 3.1):

$$w_p = \frac{p_0}{24D}(x^4 - 2ax^3 + a^3x) \tag{5.22b}$$

The condition that the plate deflection must be symmetrical with respect to the x axis (i.e., it must have the same values for $+y$ and $-y$ [Figure 5.5]) is satisfied by Equation 5.18 if we let $A_m = D_m = 0$. Then, combining Equations 5.18 and 5.22, we have

$$w = \sum_{m=1,3,\ldots} \left(B_m \cosh\frac{m\pi y}{a} + C_m y \sinh\frac{m\pi y}{a} + \frac{4p_0 a^4}{m^5\pi^5 D} \right) \sin\frac{m\pi x}{a} \tag{5.23}$$

This expression satisfies Equations 3.17 and the simple support restraints at $x = 0$, $x = a$. The remaining edge conditions are

$$w = 0 \qquad \frac{\partial^2 w}{\partial y^2} = 0 \qquad \left(y = \pm\frac{b}{2} \right)$$

Application of the above to w leads to two expressions. These will be satisfied for all values of x when

$$\begin{aligned} &B_m \cosh\alpha_m + C_m \frac{b}{2} \sinh\alpha_m + \frac{4p_0 a^4}{m^5\pi^5 D} = 0 \\ &2\left(\frac{B_m \alpha_m}{b} + C_m \right) \cosh\alpha_m + C_m \alpha_m \sinh\alpha_m = 0 \end{aligned} \tag{d}$$

in which

$$\alpha_m = \frac{m\pi b}{2a}$$

Solution of Equations d gives the unknown constants

$$B_m = -\frac{4p_0 a^4 + m\pi p_0 a^3 b \tanh\alpha_m}{m^5\pi^5 D \cosh\alpha_m} \qquad C_m = \frac{2p_0 a^3}{m^4\pi^4 D \cosh\alpha_m} \tag{e}$$

The deflection surface of the plate (Equation 5.23) may thus be expressed as

$$\begin{aligned} w = \frac{4p_0 a^4}{\pi^5 D} \sum_{m=1,3,\ldots}^{\infty} \frac{1}{m^5} &\left(1 - \frac{\alpha_m \tanh\alpha_m + 2}{2\cosh\alpha_m} \cosh\frac{2\alpha_m y}{b} \right. \\ &\left. + \frac{1}{2\cosh\alpha_m} \frac{m\pi y}{a} \sinh\frac{2\alpha_m y}{b} \right) \sin\frac{m\pi x}{a} \end{aligned} \tag{5.24}$$

The maximum displacement occurs at the center of the plate ($x = a/2$, $y = 0$). That is, from Equation 5.24,

$$w_{\max} = \frac{4p_0 a^4}{\pi^5 D} \sum_{m=1,3,\ldots}^{\infty} \frac{(-1)^{(m-1)/2}}{m^5} \left(1 - \frac{\alpha_m \tanh\alpha_m + 2}{2\cosh\alpha_m} \right)$$

Since

$$\sum_{m=1,3,\ldots}^{\infty} \frac{(-1)^{(m-1)/2}}{m^5} = \frac{5\pi^5}{2^9(3)}$$

we can write the following expression for the maximum deflection of the plate:

$$w_{\max} = \frac{5p_0a^4}{384D} - \frac{4p_0a^4}{\pi^5 D} \sum_{m=1,3,\ldots}^{\infty} \frac{(-1)^{(m-1)/2}}{m^5} \frac{\alpha_m \tanh \alpha_m + 2}{2\cosh \alpha_m} \tag{5.25}$$

The first term above represents the deflection $w_{\max}$ of the middle of a uniformly loaded, simply supported strip. The second term is a very rapidly converging series. For example, in the case of a square plate ($a = b$ and $\alpha_m = m\pi/2$), the maximum displacement is given by

$$w_{\max} = \frac{5p_0a^4}{384D} - \frac{4p_0a^4}{\pi^5 D}(0.68562 - 0.00025 + \cdots) = 0.00406 \frac{p_0a^4}{D}$$

Comment: It is observed that the result obtained, even retaining only the first term of the series in the parentheses, will be accurate to the third significant figure.

Introducing the following notation into Equation 5.25,

$$\delta_1 = \frac{5}{384} - \frac{4}{\pi^5} \sum_{m=1,3,\ldots}^{\infty} \frac{(-1)^{(m-1)/2}}{m^5} \frac{\alpha_m \tanh \alpha_m + 2}{2\cosh \alpha_m} \tag{f}$$

we find the maximum deflection of the plate to be

$$w_{\max} = \delta_1 \frac{p_0a^4}{D} \qquad \left(x = \frac{a}{2}, y = 0\right) \tag{5.26a}$$

Expressions for the moments, edge forces, and stresses of the plate can be derived by following a procedure similar to that described in Section 5.3. The maximum moments in the plate can also be put into the form

$$M_{x,\max} = \delta_2 p_0 a^2 \qquad M_{y,\max} = \delta_3 p_0 a^2 \qquad \left(x = \frac{a}{2}, y = 0\right) \tag{5.26b}$$

Numerical values of the coefficients δ_1, δ_2, and δ_3 are given in Table 5.1 for various aspect ratios b/a of the plate sides (Equations 5.2). It is seen from Table 5.1 that as b/a increases, $w_{\max}$ and $M_{x,\max}$ increase, while $M_{y,\max}$ decreases.

Comments: Note that the maximum deflection and maximum bending moment for a uniformly loaded strip or for a plate bent to a cylindrical surface are determined by making $b/a = \infty$ (Example 3.1b). For $b/a = 4$, the difference between the deflection of the strip and the plate is about 1.5%. However, for $b/a = 5$, this difference is less than 0.5%. The

TABLE 5.1

Constants δ_1, δ_2, and δ_3 for Several Values of the Aspect Ratio

b/a	1.0	2.0	3.0	4.0	5.0	∞
δ_1	0.00406	0.01013	0.01223	0.01282	0.01297	0.01302
δ_2	0.0479	0.1017	0.1189	0.1235	0.1246	0.1250
δ_3	0.0479	0.0464	0.0406	0.0384	0.0375	0.0375

maximum bending moments, for the same ratios, differ 1.2% and 0.3% respectively. We conclude, therefore, that *if the ratio of the sides is large* ($b/a > 4$), then the effect of the short sides is negligible, and the plate may be considered an *infinite strip*.

EXAMPLE 5.5: Deflection and Moment of Plate due to Wind Pressure

A window of a high-rise building is approximated by a rectangular plate with three edges simply supported and one edge clamped. The plate is under uniform wind loading of intensity p_0. Derive an expression for the deflection surface.

Solution

Let the uniformly loaded plate be bounded as shown in Figure 5.6 and assume that the edge $y = 0$ is clamped and that the remaining edges are simply supported. Thus, the *deflection is symmetrical* about the line $x = a/2$, and w may be summed only for odd integers of m. The general solution is obtained by combining Equations 5.18 and 5.22:

$$w = \sum_{m=1,3,\ldots}^{\infty} \left(A_m \sinh\frac{m\pi y}{a} + B_m \cosh\frac{m\pi y}{a} + C_m y \sinh\frac{m\pi y}{a} + D_m y \cosh\frac{m\pi y}{a} + \frac{4p_0 a^4}{m^5\pi^5 D} \right) \sin\frac{m\pi x}{a} \tag{5.27}$$

Boundary conditions are represented by

$$w = 0 \qquad \frac{\partial w}{\partial y} = 0 \qquad (y = 0)$$
$$w = 0 \qquad \frac{\partial^2 w}{\partial y^2} = 0 \qquad (y = b) \tag{g}$$

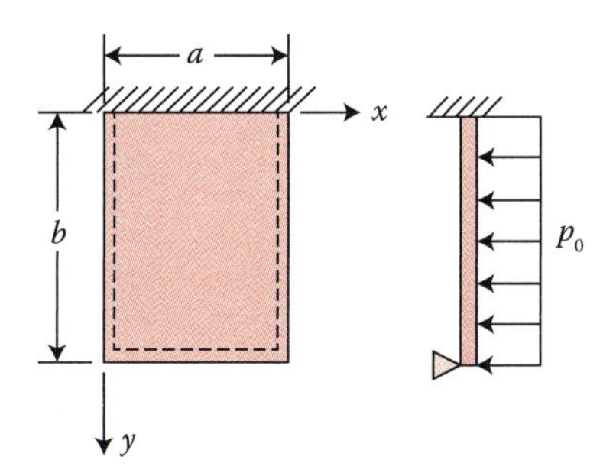

FIGURE 5.6
Plate with simply supported and fixed edges.

Application of Equations g to w leads to values of A_m, B_m, C_m, and D_m:

$$
\begin{aligned}
A_m &= -\frac{a}{m\pi} D_m = \frac{2p_0 a^4}{m^5\pi^5 D} \frac{2\cosh^2\beta_m - 2\cosh\beta_m - \beta_m \sinh\beta_m}{\cosh\beta_m \sinh\beta_m - \beta_m} \\
B_m &= -\frac{4p_0 a^4}{m^5\pi^5 D} \\
C_m &= \frac{2p_0 a^4}{m^5\pi^5 D} \cdot \left(\frac{\beta_m}{b}\right) \cdot \frac{2\sinh\beta_m \cosh\beta_m - \sinh\beta_m - \beta_m \cosh\beta_m}{\sinh\beta_m \cosh\beta_m - \beta_m}
\end{aligned}
\tag{h}
$$

where $\beta_m = m\pi b/a$ When Equations h are inserted into Equation 5.27, an expression for the plate deflection is established.

In the case of a *square plate* ($a = b$), the center deflection and the maximum bending moment are found to be (Problem 5.16)

$$
\begin{aligned}
w &= 0.0028 \frac{p_0 a^4}{D} \qquad \left(x = \frac{a}{2}, y = \frac{b}{2}\right) \\
M_y &= 0.084 p_0 a^2 = M_{\max} \qquad \left(x = \frac{a}{2}, y = 0\right)
\end{aligned}
\tag{i}
$$

Situations involving other combinations of boundary conditions, on the two opposite arbitrary sides of the plate, may be treated similarly as illustrated in the following example.

EXAMPLE 5.6: Displacement of Plate with Various Edge Conditions

The uniform load p_0 acts on a rectangular balcony reinforcement plate with opposite edges $x = 0$ and $x = a$ simply supported, the third edge $y = b$ free, and the fourth edge $y = 0$ clamped (Figure 5.7). Outline the derivation of the expression for the deflection surface w.

Solution

For the situation described, the boundary conditions, Equations 3.25 through 3.27, are

$$
w = 0 \qquad \frac{\partial^2 w}{\partial x^2} = 0 \qquad (x = 0, x = a) \tag{j}
$$

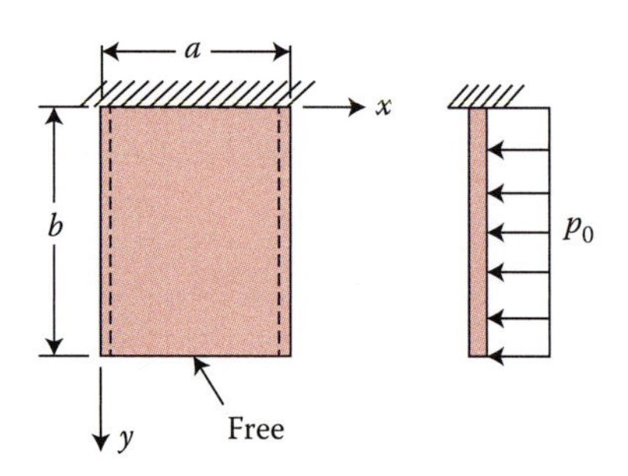

FIGURE 5.7
Plate with free, simply supported, and fixed edges.

$$w = 0 \qquad \frac{\partial^2 w}{\partial y} = 0 \qquad (y = 0) \tag{k}$$

$$\frac{\partial^2 w}{\partial y^2} + \nu \frac{\partial^2 w}{\partial x^2} = 0 \qquad \frac{\partial^3 w}{\partial y^3} + (2 - \nu)\frac{\partial^3 w}{\partial x^2 \partial y} = 0 \qquad (y = b) \tag{l}$$

The particular and the homogeneous solutions are given by Equations 5.22 and 5.18, respectively, both of which satisfy the Equations j. Application of Equations k and l to $w_h + w_p$ leads to definite values of the constants A_m, B_m, C_m, and D_m. The deflection is then obtained by adding Equations 5.18 and 5.22.

EXAMPLE 5.7: Deflection of Plate Strip under Uniform Pressure

Derive an expression for the deflection surface of a very long and narrow rectangular floor panel, or infinite strip, subjected to a uniform load of intensity p_0 (Figure 5.8). Assume that the edges $x = 0$, $x = a$, and $y = 0$ are simply supported.

Solution

The plate deflection can readily be obtained by superposing the solution for an infinite strip w_p given by Equations 5.22 with a solution w_h of Equation 3.18. It is observed from Equation 5.16 that in order for the coefficients of w_h, f_m, and its derivatives to vanish at $y = \infty$, A'_m and C'_m should be equated to zero. Hence, the homogeneous solution w_h can be represented by the following expression:

$$w_h = \sum_{m=1,3,\ldots}^{\infty} \left(B'_m + D'_m y\right) e^{-m\pi y/a} \sin\frac{m\pi x}{a} \tag{5.28}$$

The above, of course, satisfies Equation 3.18.

The boundary conditions of sides $x = 0$ and $x = a$ are fulfilled by Equations 5.22 and 5.28. It remains now to determine B'_m and D'_m in such a manner as to satisfy the boundary conditions on side $y = 0$. Substituting $w = w_h + w_p$ into $w = 0$ and $\alpha^2 w/\alpha y^2 = 0$ and setting $y = 0$, we obtain two equations that after solution yield $B'_m = -4p_0 a^4/\pi^5 D m^5$ and $D'_m = m\pi B'_m/2a$. It then follows that the elastic surface is given by

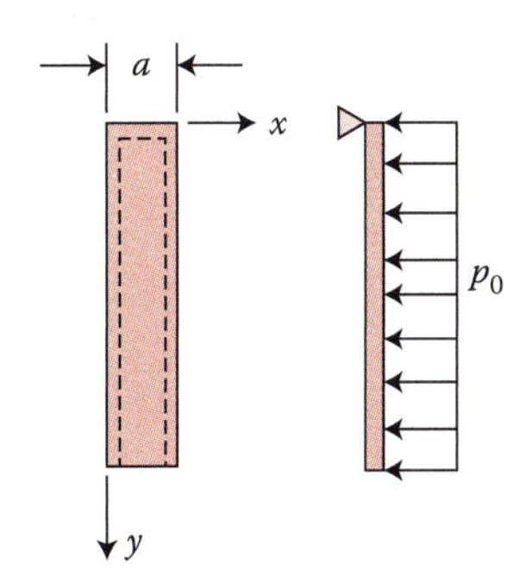

FIGURE 5.8
Plate strip with free and simply supported edges.

$$w = \frac{p_0 a^4}{24D}\left[\frac{x^4}{a^4} - 2\frac{x^3}{a^3} + \frac{x}{a} - \frac{96}{\pi^5}\sum_{m=1,3,\ldots}^{\infty}\frac{1}{m^5}\left(1+\frac{m\pi y}{2a}\right)e^{-m\pi y/a}\sin\frac{m\pi x}{a}\right] \tag{5.29}$$

The corresponding bending moments, along the middle line $x = a/2$ of the plate, are

$$\begin{aligned} M_x &= \frac{p_0 a^2}{2}\left\{\frac{1}{4} - \frac{8}{\pi^3}\sum_m^{\infty}\frac{1}{m^3}\left[1+(1-\nu)\frac{m\pi y}{2a}\right]e^{-m\pi y/a}(-1)^{(m-1)/2}\right\} \\ M_y &= \frac{p_0 a^2}{2}\left\{\frac{\nu}{4} - \frac{8}{\pi^3}\sum_m^{\infty}\frac{1}{m^3}\left[1-(1-\nu)\frac{m\pi y}{2a}\right]e^{-m\pi y/a}(-1)^{(m-1)/2}\right\} \end{aligned} \tag{5.30}$$

These are rapidly convergent series. Taking $m = 1$ and $v = 0.3$, Equations 5.29 and 5.30 yield at $y = \infty$

$$w_{\max} = 0.01302 p_0 a^4/D$$

$$M_{x,\max} = 0.1250 p_0 a^2 \qquad M_y = 0.0375 p_0 a^2$$

as expected (see Table 5.1). Interestingly, M_y attains its *maximum* value of $0.0445 p_0 a^2$ at the distance $y = 0.59a$, at which $\alpha M_y/\alpha y = 0$.

Comment: The cases involving a *clamped edge* at $y = 0$ or a *free edge* at $y = 0$ may be treated in a like manner, applying Equations 3.5 and 3.27, respectively.

EXAMPLE 5.8: Deflection Surface of Plate due to Uniform Load

Determine the expression for the elastic surface of a uniformly loaded rectangular plate with two opposite edges $x = 0$ and $x = a$ simply supported and the other two edges at $y = \pm b/2$ supported by two identical elastic beams that resist bending in vertical planes only and do not resist torsion (Figure 5.9). Denote the flexural rigidity of the supporting beam by *EI*.

Solution

Proceeding as in Section 5.4, we take the deflection surface in the form given by Equation 5.23. The boundary conditions along the edges supported by the beams are $M_y = 0$ and $V_y = EI(\alpha^4 w/\alpha x^4)$. Thus, from Equations 3.9 and 3.22b, we have

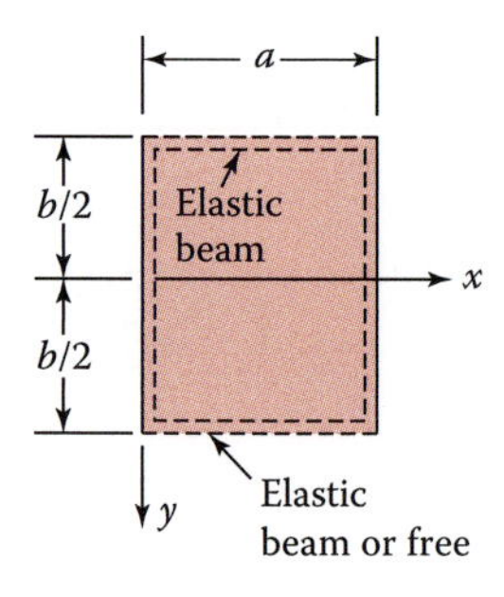

FIGURE 5.9
Variously supported plate.

$$\frac{\partial^2 w}{\partial y^2}+\nu\frac{\partial^2 w}{\partial x^2}=0 \qquad \left(y=\frac{b}{2}\right)$$
$$D\left[\frac{\partial^3 w}{\partial y^3}+(2-\nu)\frac{\partial^3 w}{\partial x^2\partial y}\right]=EI\frac{\partial^4 w}{\partial x^4} \qquad \left(y=\frac{b}{2}\right) \tag{m}$$

Clearly, the latter equation represents the differential equation of the deflection curve of the beam.

Substitution of Equation 5.23 into Equations m yields two expressions. These will be valid for all values of x if

$$\begin{aligned}&B_m(1-\nu)\cosh\alpha_m+C_m\left[\frac{2a}{m\pi}\cosh\alpha_m+\frac{1-\nu}{2}b\sinh\alpha_m\right]\\&=\frac{4\nu p_0 a^4}{m^5\pi^5 D}-B_m[(1-\nu)\sinh\alpha_m+m\pi\lambda\cosh\alpha_m]\\&+C_m\left[(1+\nu)\sinh\alpha_m-\frac{1-\nu}{2}b\cosh\alpha_m-\frac{1}{2}m\pi\lambda b\sinh\alpha_m\right]=\frac{4\lambda}{m^4\pi^4}\end{aligned} \tag{n}$$

where

$$\alpha_m=\frac{m\pi b}{2a} \qquad \lambda=\frac{EI}{aD}$$

Solving the preceding, we obtain

$$B_m=\frac{4p_0a^4}{m^5\pi^5 D}\frac{\nu(1+\nu)\sinh\alpha_m-\nu(1-\nu)(b/2)\cosh\alpha_m-m\pi\lambda[2\cosh\alpha_m+(b/2)\sinh\alpha_m]}{(3+\nu)(1-\nu)\sinh\alpha_m\cosh\alpha_m-(1-\nu)^2(b/2)+2m\pi\lambda\cosh^2\alpha_m}$$
$$C_m=\frac{4}{m^5\pi^5}\frac{\nu(1-\nu)\sinh\alpha_m+m\pi\lambda\cosh\alpha_m}{(3+\nu)(1-\nu)\sin\alpha_m\cosh\alpha_m-(1-\nu)^2(b/2)+2m\pi\lambda\cosh^2\alpha_m}$$

The deflection surface of the plate is determined on substitution of Equations o into Equation 5.23.

Comments: When the supporting *beams are rigid*, we can set $\lambda=\infty$ in Equations o, and B_m and C_m take the identical value as in Section 5.4 for a simply supported plate. Another special case is obtained if we make $\lambda=0$ in Equations o. Then we have the constants B_m and C_m of a plate with two sides simply supported and the other two sides *free*.

5.5 Lévy's Method Applied to Rectangular Plates under Nonuniform Loading

Lévy's approach is now applied to the treatment of bending problems of rectangular plates under nonuniform loading, which is a function of x only. Bounding the plate a shown in Figure 5.5, and assuming that the edges $x=0$ and $x=a$ are simply supported, we can express the loading by the Fourier series,

$$p(x)=\sum_{m=1,2,\ldots}^{\infty}p_m\sin\frac{m\pi x}{a} \tag{5.31}$$

Here

$$p_m = \frac{2}{a}\int_0^a p(x)\sin\frac{m\pi x}{a}dx \tag{5.32}$$

Proceeding as in Section 5.4, we obtain

$$w_p = \frac{a^4}{\pi^4 D}\sum_{m=1,2,\ldots}^{\infty}\frac{p_m}{m^4}\sin\frac{m\pi x}{a} \tag{5.33}$$

The above represents the deflection of a strip under load $p(x)$ and satisfies Equation 3.17 as well as the simple support conditions (Equations 3.26b) at $x = 0$ and $x = a$.

Let us assume that the two arbitrary edges $y = \pm b/2$ are also simply supported. The total deflection expression (Equation 5.23) then becomes

$$w = \sum_{m=1,2,\ldots}^{\infty}\left(B_m\cosh\frac{m\pi y}{a} + C_m y\sinh\frac{m\pi y}{a} + \frac{p_m a^4}{m^4\pi^4 D}\right)\sin\frac{m\pi x}{a} \tag{5.34}$$

where the constants B_m and C_m are to be determined from the conditions at $y = \pm b/2$: $w = 0$, $\alpha^2 w/\alpha y^2 = 0$. Finally, we obtain

$$w = \frac{a^4}{\pi^4 D}\sum_{m=1,2,\ldots}^{\infty}\frac{p_m}{m^4}\left(1 - \frac{2+\alpha_m\tanh\alpha_m}{2\cosh\alpha_m}\cosh\frac{m\pi y}{a} + \frac{(m\pi y/a)\sin(m\pi y/a)}{2\cosh\alpha_m}\right)\sin\frac{m\pi x}{a} \tag{5.35}$$

in which, as before, $\alpha_m = m\pi b/2a$.

Introduction of a given load distribution $p(x)$ into Equation 5.32 yields p_m, following which Equation 5.35 results in the displacements. The moments and stresses are found by applying the usual procedure. Table 5.2 furnishes [3] the values of p_m for various types of load distributions (Problem 5.19).

Consider, as an example, the bending of a *hydrostatically loaded plate* (Figure A of Table 5.2):

$$p_m = \frac{2}{a}\int_0^a \frac{p_0 x}{a}\sin\frac{m\pi x}{a}dx = \frac{2p_0}{m\pi}(-1)^{m+1} \qquad (m = 1,2,\ldots) \tag{a}$$

Equation 5.35, together with Equation a, represents the deflection. Suppose now that the plate is *square* ($a = b$). The deflection occurring at the center ($x = a/2$, $y = 0$) is

$$w = 0.00203\frac{p_0 a^4}{D}$$

TABLE 5.2

Variously Loaded Simply Supported Plates

Geometry	Type of Loading and Expression for p_m
A. 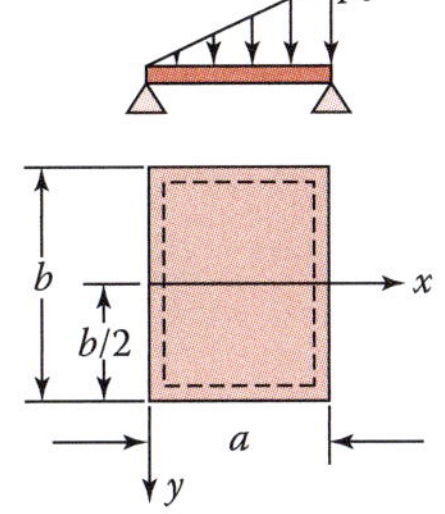	*Hydrostatic loading*: $p = p_0 \dfrac{x}{a}$ $p_m = \dfrac{2p_0}{m\pi}(-1)^{m+1} \qquad (m = 1,2,\ldots)$ *For uniform loading*, p_0 = constant: $p_m = \dfrac{4p_0}{m\pi} \qquad (m = 1,3,\ldots)$
B. 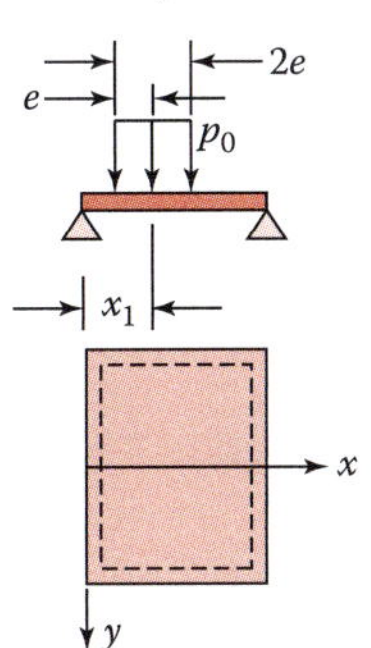	*Uniform load*, from $(x_1 - e)$ to $(x_1 + e)$: $p_m = \dfrac{4p_0}{m\pi} \sin\dfrac{m\pi x_1}{a} \sin\dfrac{m\pi e}{a} \qquad (m = 1,2,\ldots)$
C. 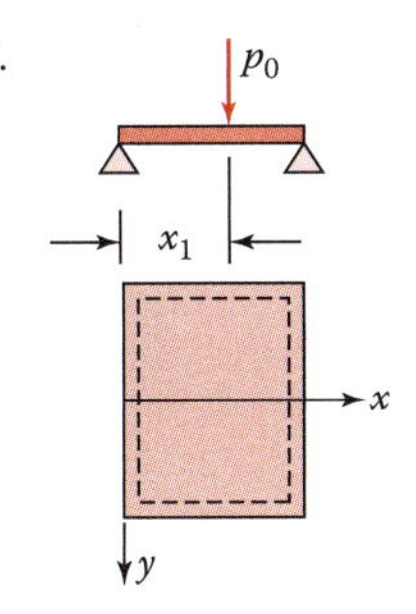	*Line load* p_0 at $x = x_1$: $p_m = \dfrac{2p_0}{a} \sin\dfrac{m\pi x_1}{a} \qquad (m = 1,2,\ldots)$
D. 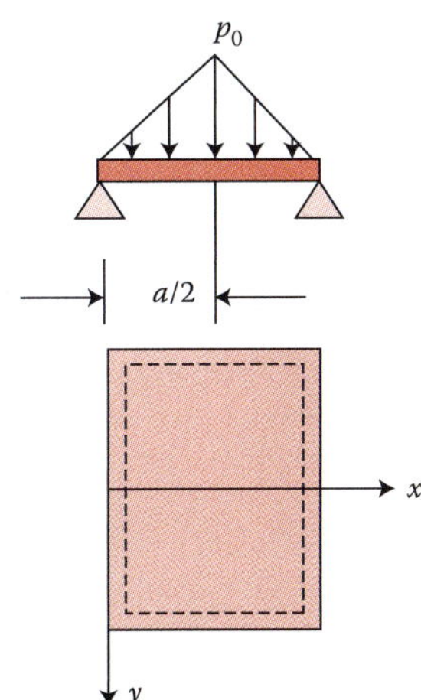	*Triangular loading*: $p = 2p_0 \dfrac{x}{a} \quad$ if $x \leq \dfrac{a}{2}$ $p = 2p_0 \dfrac{a - x}{a} \quad$ if $x \geq \dfrac{a}{2}$ $p_m = \dfrac{8p_0}{\pi^2 m^2}(-1)^{(m+1)/2} \qquad (m = 1,3,\ldots)$

The foregoing result is one-half the displacement of a simply supported rectangular plate under uniform load (Table 5.1).

Of practical importance are problems involving nonuniform *line load* on *long rectangular plates*. Application of Lévy's method to these cases is illustrated in the example that follows.

EXAMPLE 5.9: Deflection of Plate Strip due to Various Loads

A roadway slab of a reinforced concrete highway bridge is approximated by the plate strip with simply supported edges at $x = 0$ and $x = a$ (Figure 5.10). Determine the deflection for three cases: (a) the plate is subjected to a line load $p(x)$ along the x axis (Figure 5.10a); (b) the plate is under a uniformly distributed partial line load (Figure 5.10b); and (c) the plate carries a load P at $x = x_1$, $y = 0$ (Figure 5.10c).

Solution

Because of the plate symmetry about the x axis, only that portion for $y \geq 0$ need be considered. This requires that

$$\frac{\partial w}{\partial y} = 0 \qquad (y = 0) \tag{b}$$

Since each half of the plate will take one-half of the external load, using Equation 3.16b, we have

$$Q_y = -\frac{p(x)}{2} = -D\frac{\partial}{\partial y}(\nabla^2 w) \qquad (y = 0) \tag{c}$$

The minus sign agrees with the convention for shear force discussed in Section 3.4. The homogeneous solution, satisfying Equation 3.18 and boundary conditions at $x = 0$ and $x = a$, is given by Equation 5.28. Inasmuch as there is no pressure $p(x, y)$ distributed over the surface of the plate, the particular solution of Equation 3.17 is zero, and hence $w = w_h$.

a. The Fourier series expansion of the line load $p(x)$ is represented by Equation 5.31. Introducing this expression and Equation 5.28 into Equations b and c results in

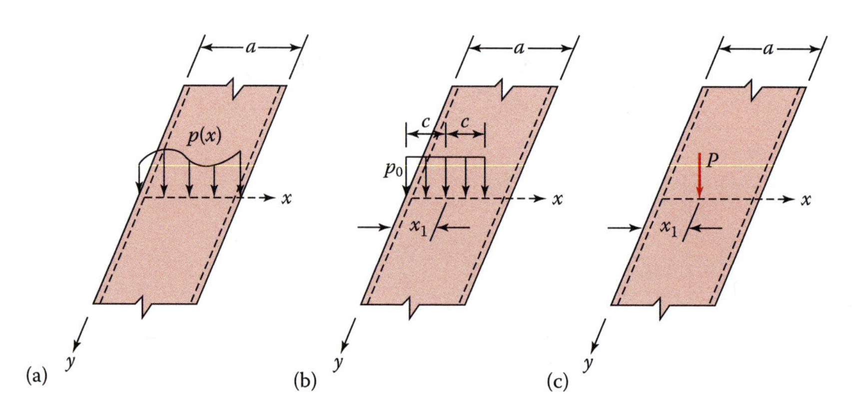

FIGURE 5.10
Simply supported plate strip.

$$D'_m = \frac{a}{m\pi} B'_m \qquad B'_m = \frac{p_m a^3}{4\pi^3 m^3 D}$$

The deflection is therefore

$$w = \frac{a^3}{4\pi^3 D} \sum_{m=1}^{\infty} \frac{p_m}{m^3} \left(1 + \frac{m\pi y}{a}\right) e^{-m\pi y/a} \sin \frac{m\pi x}{a} \tag{5.36}$$

In the particular case of *uniformly distributed line load* of intensity $p(x) = p_0$, we have, by Equation 5.32,

$$p_m = \frac{4p_0}{m\pi} \qquad (m = 1, 3, \ldots)$$

Hence,

$$w = \frac{p_0 a^3}{\pi^4 D} \sum_{m=1}^{\infty} \frac{1}{m^4} \left(1 + \frac{m\pi y}{a}\right) e^{-m\pi y/a} \sin \frac{m\pi x}{a} \tag{5.37}$$

The maximum displacement occurs at the center ($x = a/2$, $y = 0$) of the plate:

$$w_{\max} = \frac{p_0 a^3}{\pi^4 D} \sum_{m=1,3,\ldots}^{\infty} \frac{(-1)^{(m-1)/2}}{m^4} = \frac{5\pi p_0 a^3}{1{,}536 D}$$

b. The coefficients of the Fourier series expansion of the load p_0 are calculated from Equation 5.32:

$$p_m = \frac{4p_0}{\pi} \int_{x_1 - c}^{x_1 + c} \sin \frac{m\pi x}{a} dx = \frac{4p_0}{m\pi} \sin \frac{m\pi x_1}{a} \sin \frac{m\pi c}{a}$$

Equation 5.31 then yields

$$p(x) = \frac{4p_0}{\pi} \sum_{m=1}^{\infty} \frac{1}{m} \sin \frac{m\pi x_1}{a} \sin \frac{m\pi c}{a} \sin \frac{m\pi x}{a}$$

Following a procedure similar to that described in Equation a, we now obtain

$$D'_m = \frac{a}{m\pi} B'_m \quad B'_m = \frac{p_0 a^3}{m^4 \pi^4 D} \sin \frac{m\pi x}{a} \sin \frac{m\pi c}{a}$$

Thus, we arrive at the deflection surface:

$$w = \frac{p_0 a^3}{\pi^4 D} \sum_{m=1}^{\infty} \left(1 + \frac{m\pi y}{a}\right) \frac{e^{-m\pi y/a}}{m^4} \sin \frac{m\pi x_1}{a} \sin \frac{m\pi c}{a} \sin \frac{m\pi x}{a} \tag{5.38}$$

Note that, for $x_1 = a/2$ and $c = a/2$, the foregoing reduces to Equation 5.37, as expected.

c. As is always the case, the load is assumed to act in a very small area of size defined by Equation 4.26. Now, we can substitute $P = 2p_0c$ and sin $(m\pi c/a) \approx m\pi c/a$. Equation 5.38 then becomes

$$w = \frac{Pa^2}{2\pi^3 D}\sum_{m=1}^{\infty}\left(1+\frac{m\pi y}{a}\right)\frac{e^{-m\pi y/a}}{m^3}\sin\frac{m\pi x_1}{a}\sin\frac{m\pi x}{a} \tag{5.39}$$

Comments: Equations 5.36 through 5.39 are valid for $y \geq 0$. In case of a uniform load exerted on a small subrectangle or a patch loading (e.g., wheel load on infinitesimal slab), a solution can be determined by integrating Equation 5.38 for the deflection of a long plate under partial line load [2].

5.6 Rectangular Plates under Distributed Edge Moments

Consider a simply supported rectangular plate subjected to *symmetrically* distributed edge moments at $y = \pm b/2$, described by a Fourier sine series (Figure 5.11):

$$f(x) = \sum_{m=1}^{\infty} M_m \sin\frac{m\pi x}{a} \qquad \left(y = \pm\frac{b}{2}\right) \tag{a}$$

Here M_m represents the unknown set of coefficients:

$$M_m = \frac{2}{a}\int_0^a f(x)\sin\frac{m\pi x}{a}\,dx \tag{b}$$

The boundary conditions are

$$w = 0 \qquad \frac{\partial^2 w}{\partial x^2} = 0 \qquad (x = 0, x = a) \tag{c}$$

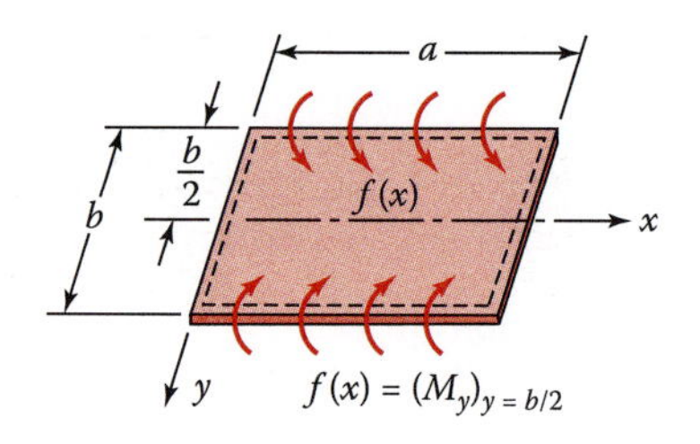

FIGURE 5.11
Plate subjected to moments at two parallel edges.

$$w = 0 \qquad \left(y = \pm\frac{b}{2}\right) \tag{d}$$

$$-D\frac{\partial^2 w}{\partial y^2} = f(x) \qquad \left(y = \pm\frac{b}{2}\right) \tag{e}$$

The solution next proceeds with an assumption of the deflection surface in the form of Equation 5.23 with $p_0 = 0$, except that summation goes over $m = 1, 2, 3, \ldots$:

$$w = \sum_{m=1}^{\infty}\left(B_m \cosh\frac{m\pi y}{a} + C_m y \sinh\frac{m\pi y}{a}\right)\sin\frac{m\pi x}{a} \tag{f}$$

This equation satisfies Equation 3.17 and Equations c as already verified in Section 5.4. Equations d is fulfilled if the terms in parentheses in the above are set equal to zero. As before, setting $\alpha_m = m\pi b/2a$, we have

$$B_m \cosh\alpha_m + C_m \frac{b}{2}\sinh\alpha_m = 0$$

from which

$$B_m = -C_m \frac{b}{2}\tanh\alpha_m$$

Equation f now takes the form

$$w = \sum_{m=1}^{\infty} C_m \left(y \sinh\frac{m\pi y}{a} - \frac{b}{2}\tanh\alpha_m \cosh\frac{m\pi y}{a}\right)\sin\frac{m\pi x}{a} \tag{g}$$

Substitution of Equations g and a into Equation e leads to

$$-2D\sum_{m=1}^{\infty}\frac{m\pi}{a}C_m \cosh\alpha_m \sin\frac{m\pi x}{a} = \sum_{m=1}^{\infty} M_m \sin\frac{m\pi x}{a}$$

It follows that

$$C_m = -\frac{aM_m}{2m\pi D\cosh\alpha_m}$$

The deflection is therefore

$$w = \frac{a}{2\pi D}\sum_{m=1}^{\infty}\frac{\sin(m\pi x/a)}{m\cosh\alpha_m}M_m\left(\frac{b}{2}\tanh\alpha_m \cosh\frac{m\pi y}{a} - y \sinh\frac{m\pi y}{a}\right) \tag{5.40}$$

The moments and the stresses are determined from Equation 5.40 for w.

In the case of *uniformly* distributed moments, we have $f(x) = M_0$, and Equation b gives $M_m = 4M_0/m\pi$. Equation 5.40 then becomes

$$w = \frac{2M_0 a}{\pi^2 D} \sum_{m=1,3,\ldots}^{\infty} \frac{\sin(m\pi x/a)}{m^2 \cosh\alpha_m}\left(\frac{b}{2}\tanh\alpha_m \cosh\frac{m\pi y}{a} - y\sinh\frac{m\pi y}{a}\right) \tag{5.41}$$

For a square plate ($a = b$), deflection and bending moments at the center are determined by the use of Equations 5.41 and 3.9 in the form

$$w = 0.0368\frac{M_0 a^2}{D} \qquad M_x = 0.394\,M_0 \qquad M_y = 0.256\,M_0$$

The displacement occurring along the axis of symmetry is

$$w = \frac{M_0 ab}{\pi^2 D} \sum_{m=1,3,\ldots}^{\infty} \frac{1}{m^2}\frac{\tanh\alpha_m}{\cosh\alpha_m}\sin\frac{m\pi x}{a} \quad (y = 0) \tag{5.42}$$

When $a \gg b$, we can set $\tanh\,\alpha_m \approx a_m$ and $\cosh\,\alpha_m \approx 1$, and the above reduces to

$$w = \frac{M_0 b^2}{2\pi D} \sum_{m=1,3,\ldots}^{\infty} \frac{1}{m}\sin\frac{m\pi x}{a} = \frac{1}{8}\frac{M_0 b^2}{D} \tag{h}$$

Comment: Interestingly, this result is the same as that for the *center deflection* of a strip of length b, subjected to two equal and opposite moments at the ends.

The particular case in which the plate is loaded by an *antimetric* moment distribution $(M_y)_{y=b/2} = -(M_y)_{y=-b/2}$ can be treated similarly by taking the solution of Equation 3.17 in the form of Equation 5.18 and modifying the Equations e as follows:

$$-D\left(\frac{\partial^2 w}{\partial y^2}\right)_{y=b/2} = (M_y)_{y=b/2} \qquad -D\left(\frac{\partial^2 w}{\partial y^2}\right)_{y=b/2} = -(M_y)_{y=-b/2}$$

Furthermore, the general case can be obtained by combination of symmetric and antimetric situations. Solutions for the symmetric (Equation 5.40) and the antimetric moment distributions are useful in dealing with plates with various edge conditions (Section 5.7).

EXAMPLE 5.10: Deflection of Plate under Edge Moment

Determine the expression for the deflection surface of a simply supported rectangular plate subjected to an edge moment $(M_x)_0$ as shown in Figure 5.12.

Solution

We represent the distribution of the bending moment by the Fourier series,

$$(M_x)_0 = \sum_{n=1}^{\infty} M_n \sin\frac{n\pi y}{b} \tag{i}$$

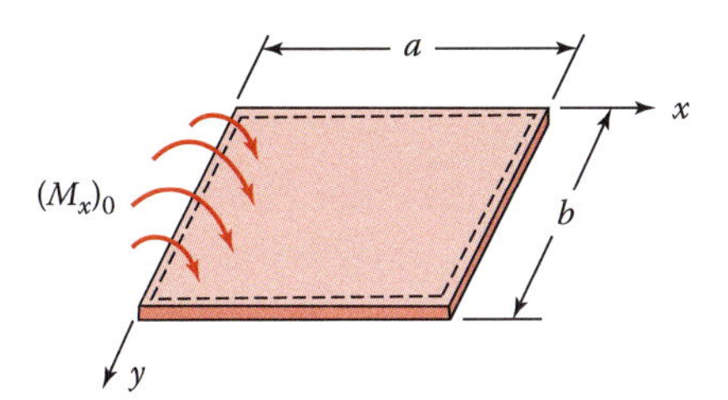

FIGURE 5.12
Simply supported plate with edge moment.

Here

$$M_n = \frac{2}{b}\int_0^b (M_x)_0 \sin\frac{n\pi y}{b}\,dy \tag{j}$$

Since $p = 0$, the solution must satisfy Equation 3.18. Referring to Equation 5.18, we thus take the solution in the form

$$w = \sum_{n=1}^{\infty}\left(A_n \sin\frac{n\pi y}{b} + B_n \cosh\frac{n\pi y}{b} + C_n x \sinh\frac{n\pi x}{b} + D_n x \cosh\frac{n\pi x}{b}\right)\sin\frac{n\pi y}{b} \tag{5.43}$$

in which A_n, B_n, C_n, and D_n are constants.

The boundary conditions are represented by

$$w = 0 \qquad M_X = (M_x)_0 \qquad (x = 0) \tag{k}$$

$$w = 0 \qquad M_x = 0 \qquad (x = a) \tag{l}$$

$$w = 0 \qquad M_x = 0 \qquad (y = 0, y = b) \tag{m}$$

The first of Equations k together with Equation 5.43 gives $B_n = 0$. Since $\sin(n\pi y/b)$ vanishes at $y = 0$ and $y = b$, Equations m are satisfied. The remaining conditions are written as

$$\begin{aligned} -D\left(\frac{\partial^2 w}{\partial x^2} + \nu\frac{\partial^2 w}{\partial y^2}\right)_{x=0} &= (M_x)_0 \\ A_n \sinh\frac{n\pi a}{b} + C_n a \sinh\frac{n\pi a}{b} + D_n a \cosh\frac{n\pi a}{b} &= 0 \\ \left(\frac{\partial^2 w}{\partial x^2} + \nu\frac{\partial^2 w}{\partial y^2}\right)_{x=a} &= 0 \end{aligned} \tag{n}$$

Substituting Equation 5.43 into the preceding gives three linear equations for A_n, C_n, and D_n. Solving for the constants leads to the following expression for the deflection:

$$w = \frac{b}{2\pi D}\sum_{n=1}^{\infty}\frac{M_n \sin(n\pi y/b)}{n \sinh(n\pi a/b)}\left[x \cosh\frac{n\pi(a-x)}{b} - \frac{a \sinh(n\pi x/b)}{\sinh(n\pi a/b)}\right] \tag{5.44}$$

where the coefficients M_n can be calculated from Equation j for each particular case of the moment distribution along the edge $x = 0$.

5.7 Method of Superposition Applied to Bending of Rectangular Plates

The deflection and stress in a rectangular plate with any edge conditions and arbitrary loading can efficiently be determined by the *method of superposition* (Section 4.10). According to this procedure, first a given complex problem is replaced by several simpler situations, each of which can be treated by Navier's or Lévy's approach. The deflections obtained for each replacement plate are then superposed in such a way that the governing equation $\nabla^4 w = p/D$ and the boundary conditions are fulfilled for the original case.

5.7.1 Rectangular Plate with Simple and Fixed Edges under Uniform Load

Consider, for example, the bending of a plate under any lateral load, with one edge clamped and the other edges simply supported (Figure 5.6). The solution begins with the assumption that all edges are simply supported. Then a bending moment along edge $y = 0$ is applied of such a magnitude as to eliminate the rotations due to the lateral load.

The solution procedure for the following problem serves to further illustrate the method.

EXAMPLE 5.11: Analysis of Plate with Simple and Fixed Edges

A rectangular plate has opposite edges at $x = 0$ and $x = a$ simply supported and the other two edges at $y = \pm b/2$ clamped (Figure 5.13a). The plate is subjected to a uniformly distributed load of intensity p_0. We wish to derive an expression for the deflection surface and the moments.

Solution

We shall proceed by superimposing the solutions of each of the two plates illustrated in Figure 5.13b and c. Plate 1 has its edges simply supported and under uniform load p_0. Plate 2 also has all edges simply supported, but in addition the two edges at $y = \pm b/2$ are subjected to uniformly distributed moments yet to be determined.

For plate 1, from Equation 5.24, we have

$$w_1 = \frac{4p_0a^4}{\pi^5 D}\sum_{m=1,3,\ldots}^{\infty}\frac{1}{m^5}\left(1-\frac{\alpha_m\tanh\alpha_m+2}{2\cosh\alpha_m}\cosh\frac{2\alpha_m y}{b} + \frac{1}{2\cosh\alpha_m}\frac{m\pi y}{a}\sinh\frac{2\alpha_m y}{b}\right)\sin\frac{m\pi x}{a} \tag{5.45}$$

The rotation of the bent plate alongside $y = b/2$ is then

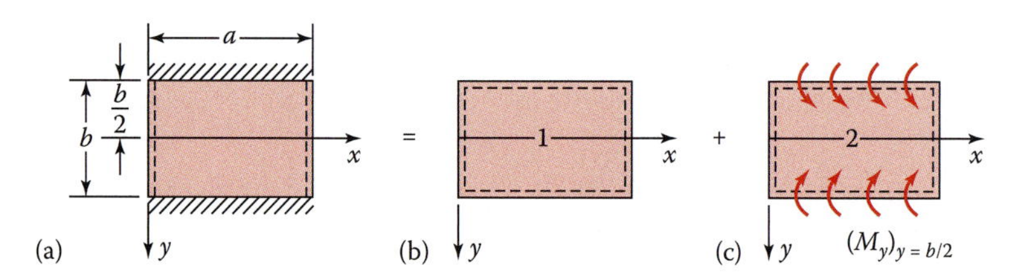

FIGURE 5.13
Method of superposition.

$$\frac{\partial w_1}{\partial y} = \frac{2p_0a^3}{\pi^4 D} \sum_{m=1,3,\ldots}^{\infty} \frac{1}{m^4} [\alpha_m - \tanh\alpha_m(1 + \alpha_m \tanh\alpha_m)] \sin\frac{m\pi x}{a} \tag{a}$$

To prevent this rotation and thus fulfill the actual conditions of the boundary of the initial plate, the following bending moments are applied alongside $y = \pm b/2$ of plate 2:

$$M_y = \sum_{m=1}^{\infty} M_m \sin\frac{m\pi x}{a} \tag{b}$$

wherein the coefficients M_m are obtained so as to make the slope due to these moments equal and opposite to that represented by Equation a. The deflection of plate 2 is given by Equation 5.40, from which, for sides $y = \pm b/2$, we have

$$\frac{\partial w_2}{\partial y} = \frac{a}{2\pi D} \sum_{m=1,3,\ldots}^{\infty} \frac{1}{m} M_m [\tanh\alpha_m(\alpha_m \tanh\alpha_m - 1) - \alpha_m] \sin\frac{m\pi x}{a} \tag{c}$$

The requirement that the slopes for both plates at $y = \pm b/2$ have the same value but areof opposite sign is satisfied if

$$\frac{\partial w_1}{\partial y} = -\frac{\partial w_2}{\partial y} \qquad \left(y = \pm\frac{b}{2}\right)$$

On inserting Equations a and c into the above and solving, we obtain

$$M_m = \frac{4p_0a^2}{m^3\pi^3} \frac{\alpha_m - \tanh\alpha_m(1 + \alpha_m \tanh\alpha_m)}{\alpha_m - \tanh\alpha_m(\alpha_m \tanh\alpha_m - 1)} \tag{5.46}$$

With the expression for M_m determined, the deflection of plate 2 can be obtained by introducing Equation 5.46 into Equation 5.40. Hence,

$$w_2 = -\frac{2p_0a^4}{\pi^5 D} \sum_{m=1,3,\ldots}^{\infty} \frac{\sin(m\pi x/a)}{m^5 \cosh\alpha_m} \frac{\alpha_m - \tanh\alpha_m(1 + \alpha_m \tanh\alpha_m)}{\alpha_m - \tanh\alpha_m(\alpha_m \tanh\alpha_m - 1)} \times\left(\frac{m\pi y}{a}\sinh\frac{m\pi y}{a} - \alpha_m \tanh\alpha_m \cosh\frac{m\pi y}{a}\right) \tag{5.47}$$

The series converges very rapidly, and the first few terms will give a satisfactory result. For instance, when $a = b$, we find that the center deflection given by only the first term in the series is equal to

$$w_2 = 0.00214\frac{p_0a^4}{D} \qquad \left(x = \frac{a}{2}, y = 0\right)$$

The center deflection for plate 1 is (from Table 3.1)

$$w_1 = 0.00406\frac{p_0a^4}{D} \qquad \left(x = \frac{a}{2}, y = 0\right)$$

The maximum displacement, which takes place at the center of a *uniformly loaded square* plate with two simply supported and two fixed edges, is thus

$$w = w_1 - w_2 = 0.00192\frac{p_0 a^4}{D} = w_{\max}$$

The values of bending moments for plate 2 are found by employing the usual procedure, while those for plate 1 are listed in Table 5.1. To ascertain the bending moments for the original plate, the results of the replacement cases are superimposed. It can be shown that the maximum moment occurs at the middle of the fixed edges and for $a = b$ is given by

$$M_y = -0.0697 p_0 a^2 = M_{\max}$$

Deflections and moments at any other points are calculated in a like manner.

5.7.2 Fixed-Edge Rectangular Plate Carries Uniform Load

By employing similar procedures, the solutions for other cases of practical importance may be found. In the case of a *clamped rectangular plate*, for instance, the largest deflection also occurs at the center, and the largest bending moments are again found at the middle of the fixed edges (Figure 5.14).
Their numerical values are as follows [2]:

For $a = b$

$$\begin{aligned} w_0 &= 0.00126\frac{p_0 a^4}{D} = w_{\max} \\ M_x &= M_y = 0.0513 p_0 a^2 = M_{\max} \qquad \left(x = \frac{a}{2}, y = 0\right) \end{aligned} \tag{d}$$

For $b = 2a$

$$\begin{aligned} w_0 &= 0.00254\frac{p_0 a^4}{D} = w_{\max} \\ M_x &= 0.0829 p_0 a^2 = M_{\max} \qquad \left(x = \frac{a}{2}, y = 0\right) \end{aligned} \tag{e}$$

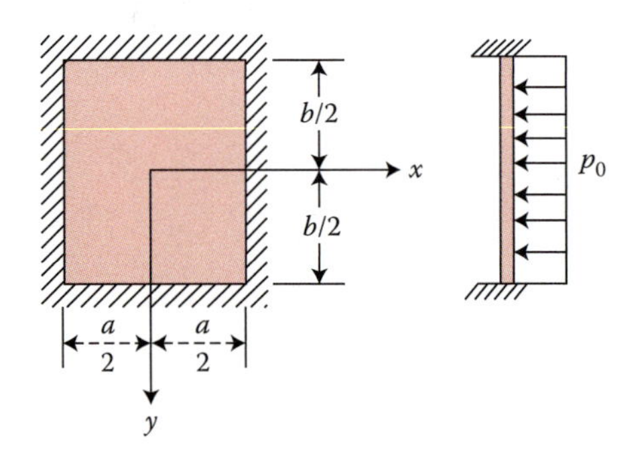

FIGURE 5.14
Clamped plate.

Comments: On comparison of the preceding results for $a = b$ with those values obtained in Examples 5.5 and 5.11, we observe that *as the number of built-in plate edges increases, the deflection and moment produced by the same loading decrease* considerably. A discussion of a number of practical aspects of the edge fixity is found in Section 4.5.

5.8 *The Strip Method

We now present a simple approximate approach, introduced by H. Grashof, for computing deflection and moment in a rectangular plate with arbitrary boundary conditions. In this so-called *strip method,* the plate is assumed to be divided into two systems of strips at right angles to one another, each strip regarded as functioning as a beam (Figure 5.15). The method permits *qualitative analysis* of the plate behavior with ease but is less adequate, in general, in obtaining accurate quantitative results. Note, however, that because this method always gives conservative values for both deflection and moment, it is often employed in practice. A very efficient engineering approach [1] to design of the rectangular floor slabs is also based on the strip method.

Before proceeding to a description of the method, it will prove useful to introduce maximum deflection and moments of a beam with various end conditions. Expressions for such quantities for a beam of length L subjected to a uniform load p, derived from the mechanics of materials, are given in Tables B.7 and B.8.

Consider a rectangular plate under a uniform load p_0 and assume that the plate is divided into strips of spans a and b, carrying the uniform loads p_a and p_b, respectively. The loaded system of beams will be impossible to arrange in such a way as to compose the plate unless the following conditions are met:

$$w_a = w_b \qquad p_0 = p_a + p_b \qquad (x = y = 0) \tag{5.48a, b}$$

In the case of a *plate* with *simply supported* edges, from Table B.7 we have $w_a = 5p_a a^4/384EI$, $w_b = 5p_b b^4/384EI$, and Equation 5.48a yields $p_a a^4 = p_b b^4$. Then Equation 5.48b leads to

$$p_a = p_0 \frac{b^4}{a^4 + b^4} \qquad p_b = p_0 \frac{a^4}{a^4 + b^4} \tag{5.49}$$

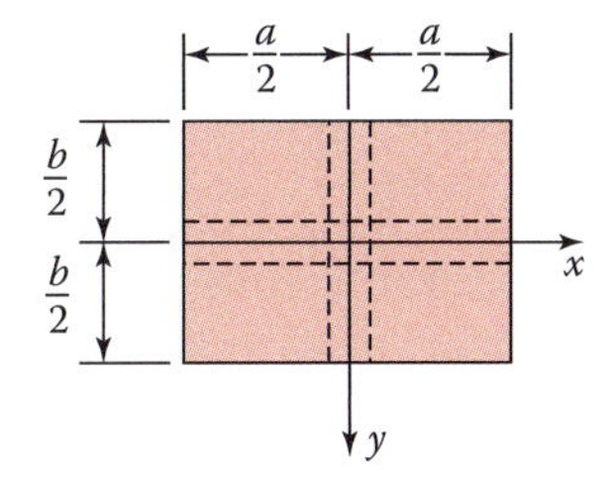

FIGURE 5.15
Equivalent plate strip.

TABLE 5.3

Constants k_1, k_2, and k_3 for Uniformly Loaded Plates

Edge Condition	k_1	k_2	k_3
Simply supported	5/384	1/8	0
Clamped	1/384	1/24	1/12

Equations 5.49 are also valid in the case of a clamped plate since from Table B.8 we again find that

$$p_a a^4 = p_b b^4$$

for $w_a = w_b$.

The deflection and bending moments of the plate are obtained by means of Equations 5.49. Referring to Equations 5.26, we can represent the maximum deflection, occurring at the center of the plate, as follows:

$$w_{\max} = k_1 \frac{p_a a^4}{D} = k_1 \frac{p_0 a^4 b^4}{D(a^4 + b^4)} \tag{5.50}$$

The bending moments at the center are written as

$$M_x = k_2 p_a a^2 = k_2 \frac{p_0 a^2 b^4}{a^4 + b^4} \qquad M_y = k_2 \frac{p_0 a^4 b^2}{a^4 + b^4} \tag{5.51a}$$

and the bending moments at the edges are

$$M_x = k_3 p_a a^2 = k_3 \frac{p_0 a^2 b^4}{a^4 + b^4} \qquad M_y = k_3 \frac{p_0 a^4 b^2}{a^4 + b^4} \tag{5.51b}$$

Here k_1, k_2, and k_3 are constants dependent on the side conditions of the plate. Values for these factors, referring to Tables B.7 and B.8, are listed in Table 5.3. The deflection and moment at any other location in the plate can be expressed in a similar way.

For example, in the case of a uniformly loaded *clamped square plate* ($a = b$), we find from Equation 5.50 and Table 5.3 that

$$w_{\max} = 0.0013 \frac{p_0 a^4}{D} \qquad (x = 0, y = 0)$$

This is accurate to the fourth significant figure as compared with the value given by Equation d in Section 5.7. The results for moments are crude approximations, however.

EXAMPLE 5.12: Displacements and Moments by the Strip Method

A rectangular plate with three edges simply supported and one edge built in is subjected to a uniform load p_0 (Figure 5.16). Determine the deflection at the center and the maximum bending moment of the plate.

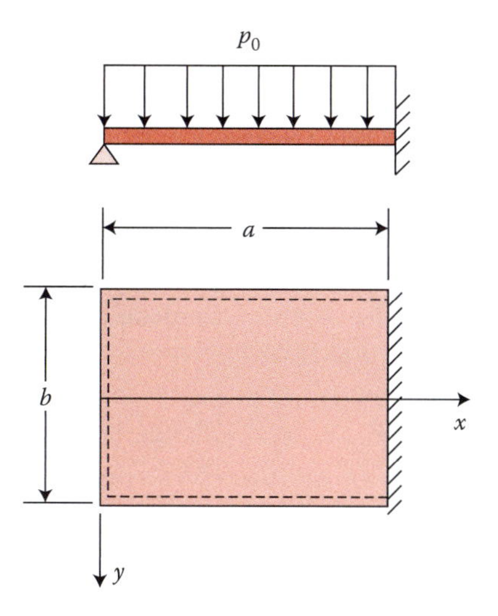

FIGURE 5.16
Plate with various edge conditions.

Solution

The plate is divided into simply supported and clamped simply supported beams. Thus, Equation 5.48a and Tables B.7 and B.8 give

$$\frac{p_a a^4}{192EI} = \frac{5p_b b^4}{384EI}$$

From Equation 5.48b, we then have

$$p_0 = p_0 \frac{5b^4}{2a^4 + 5b^4} \qquad p_b = p_0 \frac{2a^4}{2a^4 + 5b^4}$$

The plate deflection at the center is therefore

$$w = \frac{5}{384} \frac{p_b b^4}{D} = \frac{5}{192} \frac{p_0 a^4 b^4}{(2a^4 + 5b^4)D} \tag{a}$$

We also obtain the following maximum bending moments:

$$M_x = \frac{45}{128} \frac{p_0 a^2 b^4}{2a^4 + 5b^4} \qquad \left(x = \frac{3}{8}L\right)$$
$$M_y = \frac{1}{4} \frac{p_0 a^4 b^2}{2a^4 + 5b^4} \qquad (y = 0) \tag{b}$$

The bending moment at the fixed edge are the following:

$$M_f = \frac{5}{8} \frac{p_0 a^2 b^4}{2a^4 + 5b^4} \qquad (x = a) \tag{c}$$

In the case of a *square plate* ($a = b$), Equations a through c reduce to

$$w = 0.00372 p_0 a^4 / D$$
$$M_x = 0.0502\, p_0 a^2$$
$$M_y = 0.0357\, p_0 a^2$$
$$M_f = 0.0893\, p_0 a^2$$

Comments: The deflection at the center of the plate is 33% greater and the bending moment M_f is 11% greater than the values obtained by the bending theory of plates (Example 5.5).

5.9 *Simply Supported Continuous Rectangular Plates

When a uniform plate extends over a support and has more than one span along its length or width, it is termed *continuous*. In a continuous plate, the several spans may be of varying length. Intermediate supports are provided in the form of beams or columns. The continuous plates, which are often used as *floor slabs*, may be conveniently analyzed by subdividing the plate into individual, simple-span panels. The analysis is based on equilibrium conditions of individual panels and the compatibility of displacements or force at the adjoining edges. Only a continuous plate with a *rigid* intermediate beam is treated in this section; that is, the plate has zero deflection along the axis of the supporting beam. We shall assume that the beam does not prevent rotation of the plate. Hence, the intermediate beam represents a simple support to the plate.

Consider the two-span simply supported continuous plate, half of which is subjected to a uniform load of intensity p_0 (Figure 5.17a). A convenient way of looking at the problem is to draw a FBD of each rectangular panel as shown in Figure 5.17b. The distributed moment along the common edge may be represented by a Fourier series:

$$f(y) = \sum_{m=1,3,\ldots}^{\infty} M_m \sin\frac{m\pi y}{b} \tag{a}$$

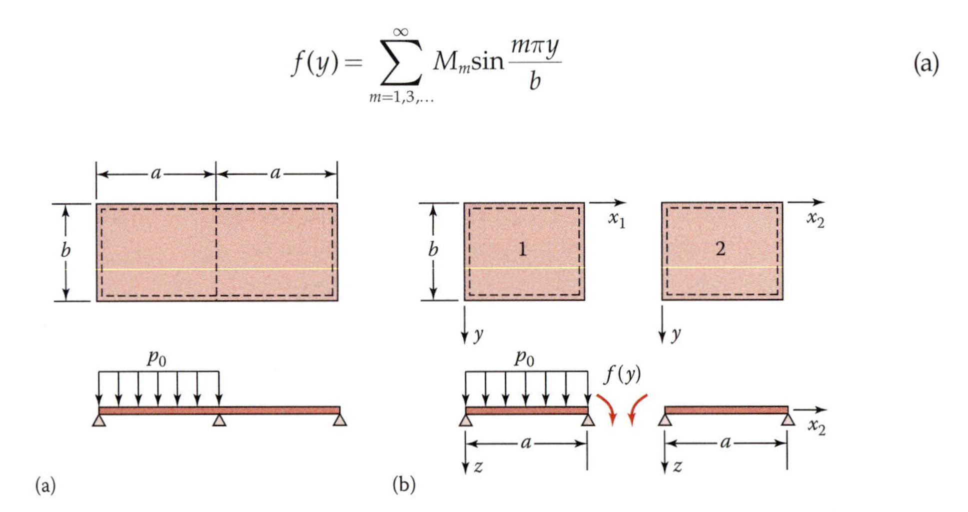

FIGURE 5.17
(a) Partially loaded continuous plate; (b) FBD for panels.

When the set of coefficients M_m is determined, the expressions for the boundary conditions can be handled with ease. The lateral deflection of each simply supported replacement plate may be obtained by application of Lévy's method.

Referring to Figures 5.6 and 5.17b, we conclude from the symmetry in deflections that the general solution for plate 1 is given by Equation 5.27 if y is replaced by x_1, x by y, and a by b. That is,

$$w_1 = \sum_{m=1,3,\ldots}^{\infty} \left(A_m \sinh \lambda_m x_1 + B_m \cosh \lambda_m x_1 + c_m x_1 \sinh \lambda_m x_1 \right.$$
$$\left. + D_m x_1 \cosh \lambda_m x_1 + \frac{4p_0 b^4}{m^5 \pi^5 D}\right) \sin \lambda_m y \tag{5.52a}$$

where

$$\lambda_m = \frac{m\pi}{b} \tag{b}$$

Similarly, for plate 2, setting $p_0 = 0$, the deflection expressed in terms of the coordinates x_2 and y and for a different set of constants is

$$w_2 = \sum_{m=1,3,\ldots}^{\infty} [E_m \sinh \lambda_m x_2 + F_m \cosh \lambda_m x_2$$
$$+ G_m x_2 \sinh \lambda_m x_2 + H_m x_2 \cosh \lambda_m x_2] \sin \lambda_m y \tag{5.52b}$$

The boundary conditions for plate 1 and plate 2 are represented as follows, respectively:

$$\begin{aligned} &w_1 = 0 \qquad \frac{\partial^2 w_1}{\partial x_1^2} = 0 \qquad (x_1 = 0) \\ &w_1 = 0 \qquad -D\frac{\partial^2 w_1}{\partial x_1^2} = \sum_{m=1,3,\ldots}^{\infty} M_m \sin \lambda_m y \qquad (x_1 = a) \end{aligned} \tag{c}$$

and

$$\begin{aligned} &w_2 = 0 \qquad \frac{\partial^2 w_2}{\partial x_2^2} = 0 \qquad (x_2 = 0) \\ &w_2 = 0 \qquad -D\frac{\partial^2 w_2}{\partial x_2^2} = \sum_{m=1,3,\ldots}^{\infty} M_m \sin \lambda_m y \qquad (x_2 = a) \end{aligned} \tag{d}$$

We thus have eight equations (Equations c and d) containing unknown $A_m, .., H_m$, M_m. The required additional equation is obtained by expressing the condition that the slopes must be the same for each panel at the middle support. This continuity requirement is expressed as

$$\frac{\partial w_1}{\partial x_1}\Big|_{x_1=0} = \frac{\partial w_2}{\partial x_1}\Big|_{x_2=0} \tag{e}$$

Introducing Equations 5.52 into the above, we have

$$\begin{aligned} &A_m\lambda_m \cosh\lambda_m a + B_m\lambda_m \sin\lambda_m a + C_m(\sinh\lambda_m a + \lambda_m a\cosh\lambda_m a) \\ &+ D_m(\cosh\lambda_m a + \lambda_m a \sinh\lambda_m a) = E_m\lambda_m + H_m \end{aligned} \tag{f}$$

Application of the edge Equations c and d to Equations 5.52 leads to values for the constants as follows:

$$\begin{aligned} A_m &= \frac{1}{D}\left\{a\frac{\coth\lambda_m a}{\sinh\lambda_m a}\left[\frac{M_m}{2\lambda_m} + \frac{2p_0}{\lambda_m^4 b}(-1+\cosh\lambda_m a)\right]\right. \\ &\quad \left. + \frac{4p_0}{\lambda_m^5 b}\left(\coth\lambda_m a - \frac{\lambda_m a}{2} - \text{csch}\lambda_m a\right)\right\} \\ B_m &= -\frac{4p_0}{\lambda_m^5 Db} \\ C_m &= \frac{2p_0}{\lambda_m^4 Db} \\ D_m &= -\frac{\text{csch}\lambda_m a}{D}\left[\frac{M_m}{2\lambda_m} + \frac{2p_0}{\lambda_m^4 b}(-1+\cosh\lambda_m a)\right] \end{aligned} \tag{g}$$

and

$$\begin{aligned} E_m &= \frac{M_m a}{2\lambda_m D}\left(1-\coth^2\alpha_m a\right) \\ F_m &= 0 \\ G_m &= -\frac{M_m}{2\lambda_m D} \\ H_m &= \frac{M_m}{2\lambda_m D}\coth\lambda_m a \end{aligned} \tag{h}$$

Having Equations g and h available, we obtain, from Equation f, the moment coefficients M_m. Equations 5.52 then give the deflection of the continuous plate from which moments and stresses can also be computed.

The foregoing approach may be extended to include the case of long rectangular plates with many supports, subjected to loading that is symmetric in y. In so doing, an equation similar to that of the three-moment equation of continuous beams is obtained [3]. It is noted that there are situations where the intermediate beams are relatively flexible compared with the flexibility of the plate. The deflections and rotations of the plate are not then taken as zero along the supports but are functions of the bending and torsional stiffnesses of the supporting beams [4].

Comment: The design methods used in connection with continuous plates utilize the solutions derived in the foregoing sections and a number of approximations [2].

5.10 *Rectangular Plates Supported by Intermediate Columns

In this section we consider the bending of a thin continuous plate over many columns. *Flat slabs* exemplify plates of this type. They are peculiar to reinforced-concrete constructions. To attain a simplified expression for the lateral deflection, it is *assumed* that the plate is subjected to a *uniform load* p_0, the column cross sections are so small that their reactions on the plate are regarded as *point loads*, the columns are equally spaced in mutually perpendicular directions, and the dimensions of the plate are large as compared with the column spacing. The foregoing set of assumptions enables one to assume that the bending in all panels, away from the boundary of the plate, is the same. We can therefore restrict our attention to the *bending of one panel alone* and consider it as a uniformly loaded rectangular plate ($a \times b$) supported at the corners by the columns. The origin of coordinates is placed at the center of a panel shown by the shaded area in Figure 5.18a. Clearly, the maximum deflection occurs at the center of the panel.

A solution can be obtained utilizing Lévy's approach (Section 5.4). Accordingly, the deflection may be expressed as a combination of that associated with a strip with uniform load and fixed ends $y = \pm b/2$ and that associated with a rectangular plate. That is, $w_p + w_h$:

$$w = \frac{p_0 b^4}{384D}\left(1 - \frac{4y^2}{b^2}\right)^2 + B_0 + \sum_{m=2,4,\ldots}^{\infty}\left(B_m \cosh\frac{m\pi y}{a} + E_m \frac{m\pi y}{a}\sinh\frac{m\pi y}{a}\right)\cos\frac{m\pi x}{a} \tag{5.53}$$

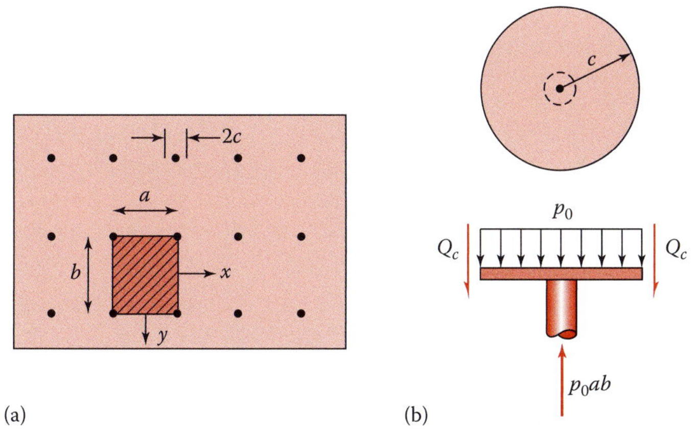

FIGURE 5.18
(a) Continuous plate supported by rows of columns; (b) distribution of support reactions.

where B_0, B_m, and E_m are constants of w_h, yet to be determined. The above satisfies the boundary conditions for the rectangular panel along the x edges,

$$\frac{\partial w}{\partial x}=0 \qquad Q_x=-D\left(\frac{\partial^3 w}{\partial x^3}+\frac{\partial^3 w}{\partial y^2 \partial x}\right)=0 \qquad \left(x=\pm\frac{a}{2}\right) \tag{a}$$

and Equations 3.17 and 3.18.

For purposes of simplifying the analysis, the support forces are regarded as acting over short (infinitesimal) line segments, between $x=a/2-c$ and $x=a/2+c$ or $2c$ (Figure 5.18a). The plate loading is transmitted to the columns by the vertical shear forces. From the conditions of symmetry of the bent panel, we are led to conclude that the slope in the direction of the normal to the boundary and the vertical shear force vanish everywhere on the edges of the panel with the exception of the corner points. Thus,

$$\begin{aligned} &Q_y=0 \qquad \left(0<x<\frac{a}{2}-c\right) \\ &\int_{a/2-c}^{a/2} Q_y\,dx=-\frac{p_0 ab}{4} \end{aligned} \tag{b}$$

The shear force Q_y can be represented by

$$Q_y=C_0+\sum_{m=2,4,\ldots}^{\infty} C_m \cos\frac{m\pi x}{a} \tag{c}$$

Equations c and b, referring to Equation A.4, lead to

$$\begin{aligned} C_0&=-\frac{p_0 b}{2} \\ C_m&=\frac{4}{a}\int_0^{a/2} Q_y \cos\frac{m\pi x}{a}\,dx=-p_0 b(-1)^{m/2} \end{aligned} \tag{d}$$

We are now in a position to represent the boundary conditions for the rectangular panel along the y edges:

$$\begin{aligned} &\frac{\partial w}{\partial y}=0 \qquad \left(y=\pm\frac{b}{2}\right) \\ &Q_y=-D\left(\frac{\partial^3 w}{\partial y^3}+\frac{\partial^3 w}{\partial x^2 \partial y}\right) \\ &\quad=-p_0 b\left[\frac{1}{2}+\sum_{m=2,4,\ldots}^{\infty}(-1)^{m/2}\cos\frac{m\pi x}{a}\right] \qquad \left(y=\pm\frac{b}{2}\right) \end{aligned} \tag{e}$$

Inasmuch as $\alpha w/\alpha y=0$, the second term in the parentheses above is zero. On substituting Equation 5.53 into Equations e, we then determine the constants B_m and E_m. Finally, we find B_0 such that the deflection w is zero at the corners:

$$w = 0 \qquad \left(x = \frac{a}{2}, y = \frac{b}{2}\right) \tag{f}$$

It follows that the values of the constants are

$$\begin{aligned}
B_0 &= \frac{p_0 b^4}{2\pi^3 D}\left(\frac{a}{b}\right)^3 \sum_{m=2,4,\ldots}^{\infty} \frac{1}{m^3}\left(\alpha_m - \frac{\alpha_m + \tanh\alpha_m}{\tanh^2\alpha_m}\right) \\
B_m &= \frac{p_0 b}{2D}(-1)^{m/2}\left(\frac{a}{m\pi}\right)\frac{\alpha_m + \tanh\alpha_m}{\sinh\alpha_m \tanh\alpha_m} \\
E_m &= \frac{p_0 b}{2D}(-1)^{m/2}\left(\frac{a}{m\pi}\right)^3 \frac{1}{\sinh\alpha_m}
\end{aligned} \tag{g}$$

Here $\alpha_m = m\pi b/2a$. Equation 5.53 together with Equations g represent the deflection surface of the panel.

The maximum deflection takes place at the center of the plate ($x = y = 0$) and is

$$\begin{aligned}
w_{\max} &= \frac{p_0 b^4}{384D} - \frac{p_0 a^3 b}{2\pi^3 D}\sum_{m=2,4,\ldots}^{\infty} \frac{(-1)^{m/2}}{m^3}\frac{\alpha_m + \tanh\alpha_m}{\sinh\alpha_m \tanh\alpha_m} \\
&\quad - \frac{p_0 a^3 b}{2\pi^3 D}\sum_{m=2,4,\ldots}^{\infty} \frac{1}{m^3}\left(\alpha_m - \frac{\alpha_m + \tanh\alpha_m}{\tanh^2\alpha_m}\right)
\end{aligned} \tag{5.54}$$

By inserting w from Equation 5.53 together with Equations g into Equations 3.9 and setting $x = y = 0$, we also determine the largest bending moments. The resulting expressions may be put into the following form:

$$w_{\max} = \delta_1 \frac{p_0 b^4}{D} \qquad M_{x,\max} = \delta_2 p_0 b^2 \qquad M_{y,\max} = \delta_3 p_0 b^2 \tag{5.55}$$

Table 5.4 lists values [2] of the constants δ_1, δ_2, and δ_3 for different values of the aspect ratio b/a and $\nu = 0.2$.

Comments: The moments near the columns as calculated by means of Equation 5.53 are much larger than the moments some distance away. The assumption made above, pertaining to support force distribution in an infinitesimal length $2c$, may thus be dispensed with. It can *now* be *assumed* that each support reaction is distributed over a small concentric

TABLE 5.4

Constants δ_1, δ_2, and δ_3 for Several Values of the Aspect Ratio

b/a	1.0	1.2	1.4	1.5	2.0	∞
δ_1	0.00581	0.00428	0.00358	0.00337	0.00292	0.00260
δ_2	0.0331	0.0210	0.0149	0.0131	0.0092	0.0083
δ_3	0.0331	0.0363	0.0384	0.0387	0.0411	0.0417

circular area of the radius c, representing the domain around the ends of the columns. An equivalent circular plate separated from the panel and subjected to uniform load p_0, shear forces Q_c around the periphery, and reactional force p_0ab at the inner fixed boundary is shown in Figure 5.18b. The radius of this plate and the shear force per unit length, as in cases of nearly square panels, are given by [3]

$$c = 0.22a \qquad Q_c = 0.723p_0b - 0.11p_0a \tag{h}$$

Referring to Figure 5.18b, the moment and stress can thus be obtained at the supports (see Section 4.10).

5.11 Rectangular Plates on Elastic Foundation

This section is concerned with the bending of the *rectangular plates on elastic foundation*. As in Section 4.11, the intensity of the foundation reaction q at any point of the plate is assumed proportional to the deflection w at the point. Hence, $q = -kw$, where k is the modulus of the foundation.

The deflection w of a rectangular plate subjected to load p per unit surface area and reaction q must satisfy Equation 3.17:

$$\frac{\partial^4 w}{\partial x^4} + 2\frac{\partial^4 w}{\partial x^2 \partial y^2} + \frac{\partial^4 w}{\partial y^4} = \frac{p - kw}{D} \tag{5.56}$$

and the given boundary conditions.

5.11.1 Simply Supported Plates

Consider the bending of a rectangular plate of sides a and b subjected to arbitrary loading $p(x, y)$ (Figure 5.1a). The plate is resting on an elastic subgrade and simply supported along its edges. The problem is treated by Navier's approach described in Section 5.2. Hence, the load and the deflection are represented by Equations 5.1. To evaluate a_{mn} in Equation 5.1b, we substitute Equations 5.1 into Equation 5.56. It follows that

$$a_{mn} = \frac{p_{mn}}{\pi^4 D[(m/a)^2 + (n/b)^2]^2 + k} \tag{5.57}$$

This value of a_{mn} and Equation 5.1b yields the equation of the plate deflection of the form

$$w = \sum_{m=1}^{\infty}\sum_{n=1}^{\infty} \frac{p_{mn}}{\pi^4 D[(m/a)^2 + (n/b)^2]^2 + k} \sin\frac{m\pi x}{a} \sin\frac{m\pi y}{b} \tag{5.58}$$

where p_{mn} is given by Equation 5.3.

It is seen that for $k = 0$, Equation 5.58 reduces to the solution for the rectangular simply supported plates (Equation 5.5). Expressions for deflection and stress of a rectangular plate resting on an elastic foundation and at the same time simply supported at its edges, under any specific load, may thus easily be determined as illustrated in Section 5.3.

5.11.2 Plates with Arbitrary Boundary Conditions

If any two opposite edges of the plate on an elastic foundation are simply supported, then Lévy's method can be applied advantageously. Example 5.13 illustrates the approach. The method of superposition combined with Lévy's solution may be used for rectangular plates with all possible combinations of simply supported, fixed, and free boundary conditions.

EXAMPLE 5.13: Displacements of Plate on Elastic Foundation

A rectangular plate on an elastic foundation having two opposite edges simply supported and the other two edges clamped is subjected to a uniform load p_0 (Figure 5.13a). Determine the equation of the deflection surface.

Solution

The boundary conditions are

$$w = 0 \qquad \frac{\partial^2 w}{\partial x^2} = 0 \qquad (x = 0, y = a) \tag{a}$$

$$w = 0 \qquad \frac{\partial w}{\partial y} = 0 \qquad \left(y = \pm\frac{b}{2}\right) \tag{b}$$

The differential equation for the deflection, Equation 5.56, becomes

$$\nabla^4 w + \frac{k}{D} w = \frac{p_0}{D} \tag{c}$$

The homogeneous equation

$$\nabla^4 w + \frac{k}{D} w = 0 \tag{d}$$

has the solution of the form given by Equation 5.15:

$$w_h = \sum_{m=1}^{\infty} f_m(y) \sin\frac{m\pi x}{a} \tag{e}$$

This series satisfies Equations a and b. Substitution of Equation e into Equation d results in the differential equation, valid for all values of x:

$$\frac{d^4 f_m}{dy^4} - 2\frac{m^2\pi^2}{a^2}\frac{d^2 f_m}{dy^2} + \left(\frac{m^4\pi^4}{a^4} + \frac{k}{D}\right) f_m = 0 \tag{f}$$

The general solution of this equation is (Problem 5.36)

$$\begin{aligned} f_m = {} & A_m \sinh\beta_m y \sinh\gamma_m y + B_m \sinh\beta_m y \cosh\gamma_m y \\ & + C_m \cosh\beta_m y \sinh\gamma_m y + D_m \cosh\beta_m y \cos\gamma_m y \end{aligned} \tag{g}$$

Here

$$\beta_m^2 = \frac{1}{2}\left(\frac{m^2\pi^2}{a^2} + \sqrt{\frac{m^4\pi^4}{a^4} + \frac{k}{D}}\right)$$
$$\gamma_m^2 = \frac{1}{2}\left(\frac{m^2\pi^2}{a^2} - \sqrt{\frac{m^4\pi^4}{a^4} + \frac{k}{D}}\right) \tag{5.59}$$

and A_m, B_m, C_m, and D_m are constants. From symmetry it can be concluded that f_m in our case is an even function of y. We thus have $B_m = C_m = 0$. The homogeneous solution is therefore

$$w_h = \sum_{m=1}^{\infty} (A_m \sinh\beta_m y \sin\gamma_m y + D_m \cosh\beta_m y \cos\gamma_m y)\sin\frac{m\pi x}{a} \tag{5.60}$$

The particular solution of Equation c is obtained by carrying out the same procedure used in Section 5.4. In so doing, we have

$$w_p = \frac{4p_0}{\pi D}\sum_{m=1}^{\infty} \frac{1}{m\left(\frac{m^4\pi^4}{a^4} + \frac{k}{D}\right)}\sin\frac{m\pi x}{a} \tag{5.61}$$

We observe that for $k = 0$, the preceding equation reduces to Equation 5.22a. Combining Equations 5.60 and 5.61, we arrive at the following expression for the deflection surface:

$$w = \sum_{m=1,3,\ldots}^{\infty} \left[A_m \sinh\beta_m y \sin\gamma_m y + D_m \cosh\beta_m y \cos\gamma_m y + \frac{4p_0}{\pi D}\frac{1}{(m\pi x/a)^4 + (k/D)}\right]\sin\frac{m\pi x}{a} \tag{5.62}$$

The foregoing fulfills the boundary conditions (Equations a). Substitution of Equation 5.62 into Equations b leads to the equations

$$\frac{4p_0}{m\pi D[(m\pi/a)^2 + (k/D)]} + A_m \sinh\frac{\beta_m b}{2}\sin\frac{\gamma_m b}{2} + D_m \cosh\frac{\beta_m b}{2}\cos\frac{\gamma_m b}{2} = 0$$
$$(A_m\beta_m + D_m\gamma_m)\sinh\frac{\beta_m b}{2}\cos\frac{\gamma_m b}{2} + (A_m\gamma_m - D_m\beta_m)\cosh\frac{\beta_m b}{2}\sin\frac{\gamma_m b}{2} = 0 \tag{h}$$

from which the constants A_m and D_m can be found. The solution (Equation 5.62) is thus completed.

5.12 The Ritz Method Applied to Bending of Rectangular Plates

We now apply the Ritz method to the bending of rectangular plates. The strain energy U associated with the bending of a plate is given by Equation 3.44 in Section 3.12. We can represent the work done by the lateral surface loading $p(x, y)$ as

$$W = \iint_A wp\,dx\,dy \tag{5.63}$$

in which A denotes the area of the plate surface. The potential energy $\Pi = U - W$ is therefore

$$\Pi = \frac{D}{2}\iint_A \left\{\left(\frac{\partial^2 w}{\partial x^2} + \frac{\partial^2 w}{\partial y^2}\right)^2 - 2(1-\nu)\left[\frac{\partial^2 w}{\partial x^2}\frac{\partial^2 w}{\partial y^2} - \left(\frac{\partial^2 w}{\partial x \partial y}\right)^2\right]\right\} dx\,dy - \iint wp\,dx\,dy \tag{5.64}$$

For generality and convenience, let us introduce the new *nondimensinal* coordinates α and λ as follows:

$$x = a\alpha \quad \text{and} \quad y = b\lambda \tag{5.65}$$

In doing so, the potential energy, Equation 5.64, can be rewritten in new coordinates in the form

$$\Pi = ab\int_A p_0 w\,d\alpha\,d\lambda - \frac{Db}{2a^3}\int_A \left[\left(\frac{\partial^2 w}{\partial \alpha^2}\right)^2 + \left(\frac{a}{b}\right)^2 + \left(\frac{\partial^2 w}{\partial \lambda^2}\right)^2 + 2(1-\nu)\left(\frac{a}{b}\right)^2\left(\frac{\partial^2 w}{\partial \alpha \partial \lambda}\right)^2 + 2\nu\left(\frac{a}{b}\right)^2\frac{\partial^2 w}{\partial \alpha^2}\frac{\partial^2 w}{\partial \lambda^2}\right] d\alpha\,d\lambda \tag{5.66}$$

The use of this equation is shown in Example 5.15.

The application of the Ritz method is illustrated by considering the bending of a clamped rectangular plate of sides a and b, carrying a uniform load of intensity p_0 (Figure 5.19). The boundary conditions are

$$\begin{aligned} w &= 0 \qquad \frac{\partial w}{\partial x} = 0 \qquad (x = 0, x = a) \\ w &= 0 \qquad \frac{\partial w}{\partial y} = 0 \qquad (y = 0, y = b) \end{aligned} \tag{a}$$

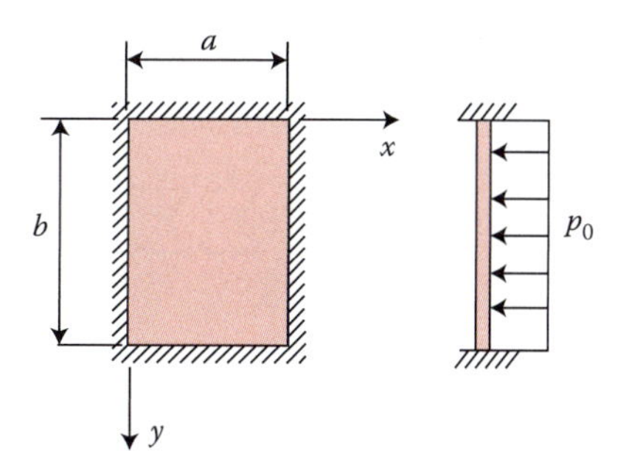

FIGURE 5.19
Clamped plate.

Integration by parts of the last term in Equation 5.64 leads to

$$\begin{aligned}\iint_A \frac{\partial^2 w}{\partial x\,\partial y}\frac{\partial^2 w}{\partial x\,\partial y}dxdy &= \int_S \frac{\partial^2 w}{\partial x\,\partial y}\frac{\partial w}{\partial x}dx - \iint_A \frac{\partial w}{\partial x}\frac{\partial^3 w}{\partial x\,\partial y^2}dx\,dy \\ &= \int_S \frac{\partial^2 w}{\partial x\,\partial y}\frac{\partial w}{\partial x}dx - \int_S \frac{\partial w}{\partial x}\frac{\partial^2 w}{\partial y^2}dy + \iint_A \frac{\partial^2 w}{\partial x^2}\frac{\partial^2 w}{\partial y^2}dx\,dy\end{aligned} \tag{b}$$

According to Equations a, the first two integrals in Equation b become identically zero. Hence,

$$\iint_A \left[\frac{\partial^2 w}{\partial x^2}\frac{\partial^2 w}{\partial y^2} - \left(\frac{\partial^2 w}{\partial x\,\partial y}\right)^2\right]dx\,dy = 0$$

The bending strain energy therefore reduces to

$$U = \frac{D}{2}\iint_A \left(\frac{\partial^2 w}{\partial x^2} + \frac{\partial^2 w}{\partial y^2}\right)^2 dx\,dy \tag{5.67}$$

Assuming a deflection expression of the form

$$w = \sum_{m=1}^{\infty}\sum_{n=1}^{\infty} a_{mn}\left(1 - \cos\frac{2m\pi x}{a}\right)\left(1 - \cos\frac{2n\pi y}{b}\right) \tag{5.68}$$

the boundary conditions (Equations a) are satisfied. Introduction of the above into Equation 5.67 yields

$$\begin{aligned}U = \frac{D}{2}\int_0^b\int_0^a \Bigg\{\sum_{m=1}^{\infty}\sum_{n=1}^{\infty} 4\pi^2 a_{mn}\Bigg[\frac{m^2}{a^2}\cos\frac{2m\pi x}{a}\left(1 - \cos\frac{2n\pi y}{b}\right) \\ + \frac{n^2}{b^2}\cos\frac{2n\pi y}{b}\left(1 - \cos\frac{2m\pi x}{a}\right)\Bigg]\Bigg\}^2 dx\,dy\end{aligned}$$

This leads to

$$\begin{aligned}U = 2\pi^4 abD\Bigg\{\sum_{m=1}^{\infty}\sum_{n=1}^{\infty}\left[3\left(\frac{m}{a}\right)^4 + 3\left(\frac{n}{b}\right)^4 + 2\left(\frac{m}{a}\right)^2\left(\frac{n}{b}\right)^2\right]a_{mn}^2 \\ + \sum_{m=1}^{\infty}\sum_{r=1}^{\infty}\sum_{s=1}^{\infty} 2\left(\frac{m}{a}\right)^4 a_{mr}a_{ms} + \sum_{r=1}^{\infty}\sum_{s=1}^{\infty}\sum_{n=1}^{\infty} 2\left(\frac{n}{b}\right)^4 a_{rn}a_{sn}\Bigg\}\end{aligned} \tag{5.69}$$

which is valid for $r \neq s$. The work done by p_0 is, by application of Equation 5.63,

$$W = p_0\int_0^b\int_0^a\left[\sum_{m=1}^{\infty}\sum_{n=1}^{\infty} a_{mn}\left(1 - \cos\frac{2m\pi x}{a}\right)\left(1 - \cos\frac{2n\pi y}{b}\right)\right]dx\,dy = p_0 ab\sum_{m=1}^{\infty}\sum_{n=1}^{\infty} a_{mn} \tag{5.70}$$

From the minimizing conditions $\partial\Pi/\partial a_{mn} = 0$, it follows that

$$4D\pi^4 ab\left\{\left[3\left(\frac{m}{a}\right)^4 + 3\left(\frac{n}{b}\right)^4 + 2\left(\frac{m}{a}\right)^2\left(\frac{n}{b}\right)^2\right]a_{mn} + \sum_{r=1}^{\infty} 2\left(\frac{m}{a}\right)^4 a_{mr} + \sum_{r=1}^{\infty} 2\left(\frac{n}{b}\right)^4 a_{rn}\right\} - p_0 ab = 0 \tag{c}$$

This is valid for $r \neq n$ and $r \neq m$. Dropping all but the first term a_{11} gives, for Equation c,

$$a_{11} = \frac{p_0 a^4}{4\pi^4 D} \frac{1}{3 + 3(a/b)^4 + 2(a/b)^2}$$

In the case of a square plate ($a = b$), $a_{11} = p_0 a^4/32\pi^4 D$. The maximum deflection takes place at the center of the plate and is obtained by substituting this value of a_{11} into Equation 5.68:

$$w_{\max} = 0.00128 \frac{p_0 a^4}{D}$$

Comments: This is approximately 1.5% greater than the value, given by Equation d of Section 5.7, based on a more elaborate approach. It should be noted that the result obtained, taking only one term of the series, is remarkably accurate. So few terms will not, in general, result in such accuracy in applying the Ritz method.

Let us express the deflection w by retaining seven parameters a_{11}, a_{12}, a_{21}, a_{22}, a_{13}, a_{31}, and a_{33}. Now Equation c results in

$$\left[3 + 3\left(\frac{a}{b}\right)^4 + 2\left(\frac{a}{b}\right)^2\right]a_{11} + 2a_{12} + 2\left(\frac{a}{b}\right)^4 a_{21} + 2a_{13} + 2\left(\frac{a}{b}\right)^4 a_{31} = \frac{p_0 a^4}{4\pi^4 D}$$

$$2a_{11} + \left[3 + 48\left(\frac{a}{b}\right)^4 + 8\left(\frac{a}{b}\right)^2\right]a_{12} + 2a_{13} + 32\left(\frac{a}{b}\right)^2 a_{22} = \frac{p_0 a^4}{4\pi^4 D}$$

$$2\left(\frac{a}{b}\right)^4 a_{11} + \left[48 + 3\left(\frac{a}{b}\right)^4 + 8\left(\frac{a}{b}\right)^2\right]a_{21} + 2\left(\frac{a}{b}\right)^4 a_{31} + 32a_{22} = \frac{p_0 a^4}{4\pi^4 D}$$

$$32a_{21} + 16\left[3 + 3\left(\frac{a}{b}\right)^4 + 2\left(\frac{a}{b}\right)^2\right]a_{22} + 32\left(\frac{a}{b}\right)^4 a_{12} = \frac{p_0 a^4}{4\pi^4 D}$$

$$2a_{11} + 2a_{12} + \left[3 + 243\left(\frac{a}{b}\right)^4 + 18\left(\frac{a}{b}\right)^2\right]a_{31} + 162a_{33} = \frac{p_0 a^4}{4\pi^4 D}$$

$$2\left(\frac{a}{b}\right)^4 a_{11} + 2\left(\frac{a}{b}\right)^4 a_{21} + \left[243 + 3\left(\frac{a}{b}\right)^4 + 18\left(\frac{a}{b}\right)^2\right]a_{31} + 162a_{33} = \frac{p_0 a^4}{4\pi^4 D}$$

$$162\left(\frac{a}{b}\right)^4 a_{13} + 162a_{31} + 81\left[3 + 3\left(\frac{a}{b}\right)^4 + 2\left(\frac{a}{b}\right)^2\right]a_{33} = \frac{p_0 a^4}{4\pi^4 D}$$

Simultaneous solutions of the above for a *square plate* ($a = b$) yield

$$a_{11} = 0.11774p_1 \qquad a_{12} = a_{21} = 0.01184p_1$$
$$a_{22} = 0.00189p_1 \qquad a_{13} = a_{31} = 0.00268p_1$$
$$a_{33} = 0.00020p_1$$

where $p_1 = p_0a^4/4D\pi^4$. On substituting these values into Equation 5.68, the maximum deflection is found to be

$$w_{\max} = 0.00126\frac{p_0a^4}{D}$$

This result is exactly the same value given by Equation d of Section 5.7.

EXAMPLE 5.14: Displacements by the Ritz Method

Determine the deflection of a simply supported rectangular plate subjected to a uniform load p_0 (Figure 5.1).

Solution

The deflection surface of the plate is represented by the series (5.1b). Introducing this series for w, Equations 3.44 and 5.63 become

$$U = \frac{D}{2}\int_0^b\int_0^a\left\{\sum_{m=1}^{\infty}\sum_{n=1}^{\infty}a_{mn}^2\left(\frac{m^2\pi^2}{a^2}+\frac{n^2\pi^2}{b^2}\right)^2\sin^2\frac{m\pi x}{s}\sin^2\frac{n\pi y}{b}\right.$$
$$\left.-2(1-\nu)a_{mn}^2\left[\left(\frac{m\pi}{a}\right)^2\left(\frac{n\pi}{b}\right)^2\left(\sin^2\frac{m\pi x}{a}\sin^2\frac{n\pi y}{b}-\cos^2\frac{m\pi x}{a}\cos^2\frac{n\pi y}{b}\right)\right]\right\}dx\,dy$$

and

$$W = \int_0^b\int_0^a p_0\, a_{mn}\sin\frac{m\pi x}{a}\sin\frac{n\pi y}{b}dx\,dy$$

Because of the orthogonality relations (Equation 5.2), the preceding reduce to, after integration,

$$U = \frac{\pi^4 ab}{8}D\sum_{m=1}^{\infty}\sum_{n=1}^{\infty}a_{mn}^2\left(\frac{m^2}{a^2}+\frac{n^2}{b^2}\right)^2 \tag{5.71}$$

and

$$W = \frac{abp_0}{\pi^2}\sum_{m=1}^{\infty}\sum_{n=1}^{\infty}\frac{a_{mn}^2}{mn}(\cos m\pi - 1)(\cos n\pi - 1) \tag{5.72}$$

The condition $\partial(U - W)/\partial a_{mn} = 0$ therefore gives

$$a_{mn} = \frac{16}{\pi^6 D}\frac{p_0}{mn[(m/a)^2 + (n/b)^2]^2} \qquad (m,n = 1,3,\ldots)$$

Substituting this into Equation 5.1b, we obtain once more the result given by Equation 5.6.

EXAMPLE 5.15: Maximum Deflection of Clamped Plate by the Ritz Method

A rectangular portion ($a \times b$) of a machine room floor has all its edges built in and supports a load P acting at a location $x = x_1$, $y = y_1$ (Figures 5.3 and 5.19). Determine the maximum deflection of the plate.

Solution

From Equation 5.63, the work done by the loading is

$$W = P\sum_{m=1}^{\infty}\sum_{n=1}^{\infty} a_{mn}\left(1 - \cos\frac{2m\pi x_1}{a}\right)\left(1 - \cos\frac{2n\pi y_1}{b}\right) \tag{5.73}$$

The condition $\partial\Pi/\partial a_{mn} = 0$ together with Equations 5.69 and 5.71 now leads to

$$\begin{aligned} &4\pi^4 abD\left\{\left[3\left(\frac{m}{a}\right)^4 + 3\left(\frac{n}{b}\right)^4 + 2\left(\frac{m}{a}\right)^2\left(\frac{n}{b}\right)^2\right]a_{mn} + \sum_{r=1}^{\infty} 2\left(\frac{m}{a}\right)^4 a_{mr}\right. \\ &\left.+\sum_{r=1}^{\infty} 2\left(\frac{n}{b}\right)^4 a_{rn}\right\} - P\left(1 - \cos\frac{2m\pi x_1}{a}\right)\left(1 - \cos\frac{2n\pi y_1}{b}\right) = 0 \end{aligned} \tag{d}$$

valid for $r \neq n$ and $r \neq m$.

Consider, for simplicity, the case of a *square plate* ($a = b$) with P acting at its center and retaining seven parameters, a_{11}, a_{12}, a_{22}, a_{21}, a_{13}, a_{31}, and a_{33}. We then obtain the following values:

$$\begin{aligned} a_{11} &= 0.12662P_1 \qquad & a_{12} = a_{21} &= -0.00601P_1 \\ a_{22} &= 0.00301P_1 \qquad & a_{13} = a_{31} &= 0.00278P_1 \\ & a_{33} = 0.00015P_1 \end{aligned}$$

Here $P_1 = Pa^2/D\pi^4$. The maximum deflection occurring at the center of the plate is, from Equation 5.68, found to be

$$w_{\max} = 0.00543\frac{Pa^2}{D} \tag{e}$$

Comment: This is 3% smaller than the value determined from the solution of Equation 3.17. By retaining more parameters in the series, we expect to improve the result.

EXAMPLE 5.16: Displacements of Rectangular Plate by the Ritz Method

A uniformly loaded rectangular plate has its two opposite edges simply supported, one edge built-in and one edge free (Figure 5.20).
Assumption: Use an approximate solution,

$$w(x, y) = C\left(\frac{x}{a}\right)^2 \sin\frac{\pi y}{b} \tag{5.74}$$

where C is an undetermined coefficient. Apply the Ritz method, to determine: (a) an expression for the deflection surface; (b) the deflection at $x = a$ and $y = b/2$, for $a = b$ and $\nu = 0.3$. What is the value of the deflection, if the plate is made of steel (E = 200 GPa) of dimensions a = 1.8 m, and t = 15 mm, and carries a uniform loading p_0 = 400 kPa?

Solution

The boundary conditions for the plate are

$$w = 0 \qquad (x = 0), \qquad w = 0 \qquad (y = b) \tag{f}$$

$$w = 0, \qquad \frac{\partial w}{\partial x} = 0 \qquad (x = 0)$$

and

$$\frac{\partial^2 w}{\partial x^2} + \frac{\partial^2 w}{\partial y^2} = 0; \qquad \frac{\partial^3 w}{\partial x^3} + (2-\nu)\frac{\partial^3 w}{\partial x \partial y^2} = 0 \qquad (x = a) \tag{g}$$

It can be readily verified that (see Problem 5.43), Equation 5.74 satisfies the geometric conditions (f). As stated in Section 3.13.3, static conditions (g) need notbe fulfilled (Section 3.13.3).

a. From Equations 5.65 and 5.64, the expression for the deflection surface becomes

$$w = C\alpha^2 \sin \pi\lambda \tag{h}$$

Introducing Equation h into Equation 5.66, we have

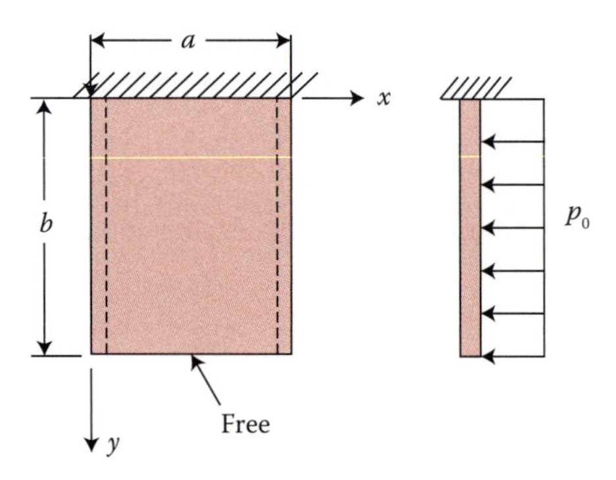

FIGURE 5.20
Rectangular plate with various edge conditions.

$$\Pi = p_0 ab \int_0^1 \int_0^1 C\alpha^2 \sin \pi\alpha \, d\alpha \, d\lambda - \frac{Db}{2a^3} \int_0^1 \int_0^1 [(2C \sin \pi\lambda)^2$$

$$+ \left(\frac{a}{b}\right)^4 (-\pi^2 C\alpha^2 \sin \pi)^2 + 2(1-\nu)\left(\frac{a}{b}\right)^2 (2\pi C\alpha \cos \pi\lambda)^2 \quad \text{(i)}$$

$$+ 2\nu \left(\frac{a}{b}\right)^2 (2C \sin \pi\lambda)(-\pi^2 C\alpha^2 \sin \pi\lambda)] d\alpha d\lambda$$

Hence, $\partial\Pi/\partial C = 0$ leads to the expression:

$$p_0 ab \int_0^1 \int_0^1 \alpha^2 \sin \pi\lambda \, d\alpha \, d\lambda - \frac{Db}{2a^3} \int_0^1 \int_0^1 [8C \sin^2 \pi\lambda$$

$$+ 2\left(\frac{a}{b}\right)^4 \pi^4 C\alpha^4 \sin^2 \pi\lambda + 16(1-\nu)\left(\frac{a}{b}\right)^2 \pi^2 C\alpha^2 \cos^2 \pi\lambda$$

$$- 8\nu \left(\frac{a}{b}\right)^2 C\pi^2 \alpha^2 \sin^2 \pi\lambda] d\alpha \, d\lambda = 0$$

After integration, we can obtain the coefficient as

$$C = \frac{a^4 p_0}{D} \frac{2}{3\pi\left[2 + (\pi^4/10)(a/b)^4 + (4/3)\pi^2(1-\nu)(a/b)^2 - (2/3)\pi^2\nu(a/b)^2\right]} \quad \text{(j)}$$

The expression for the deflection, Equation h, is then

$$w = C\left(\frac{x}{a}\right)^2 \sin \pi \frac{y}{b} \quad (5.75)$$

Here C is given by Equation i.

b. Equation 5.75 together with Equation j lead to the value of the deflection (for $a/b = 1$, $x = a$, and $y = b/2$) as

$$w = 0.01118 \frac{p_0 a^4}{D} \quad \text{(k)}$$

Inserting the given numerical values, we have

$$D = \frac{Et^3}{12(1-\nu^2)} = \frac{200(15^3)}{12(1-0.3^2)} = 61.813 \text{ kN} \cdot \text{m}$$

It follows that

$$w = 0.1118 \frac{400(1.8)^4}{61.813} = 7.6 \text{ mm}$$

Comment: The result given by Equation k differs from the "exact" solution by about 1%.

EXAMPLE 5.17: Natural Frequency of Plate with the Ritz Method

Determine the lowest natural frequency of a rectangular plate clamped on all edges, using the Ritz method. To facilitate the computations, locate the origin of the coordinates at the center of the plate as shown in Figure 5.21.

Solution

We take the deflection of the plate during vibration in the form

$$w = w_0 \cos \omega_n t \tag{1}$$

where w_0 is a function of x and y that determines the mode of vibration and ω_n represents the natural circular frequency of the plate. Substituting Equation 1 into Equations 5.67 and 3.54, expressions for the maximum potential and kinetic energies of vibration are obtained:

$$\begin{aligned} U_{\max} &= \frac{D}{2}\int_{-a}^{a}\int_{-b}^{b}\left(\frac{\partial^2 w_0}{\partial x^2} + \frac{\partial^2 w_0}{\partial y^2}\right)^2 dx\,dy \\ T_{\max} &= \frac{\overline{m}}{2}\omega_n^2 \int_{-a}^{a}\int_{-b}^{b} w_0^2\,dx\,dy \end{aligned} \tag{5.76}$$

Now we assume the function w_0 as follows [10]:

$$w_0 = c_1(x^2 - a^2)^2(y^2 - b^2)^2 \tag{5.77}$$

This satisfies the conditions at the boundary and closely approximates the shape of the first mode of vibration (Figure 5.21). Carrying this into Equations 5.76 and integrating it, we have

$$\begin{aligned} U_{\max} &= c_1^2 D \cdot \frac{2^8}{5^2} a^9 b^9 \left(\frac{1}{a^4} + \frac{2^8}{3^2 \times 7^2}\frac{1}{a^2 b^2} + \frac{1}{b^4}\right) \\ T_{\max} &= \frac{1}{2}\omega_n^2 \overline{m} \cdot c_1^2 \frac{2^4}{5^2} a^9 b^9 \end{aligned}$$

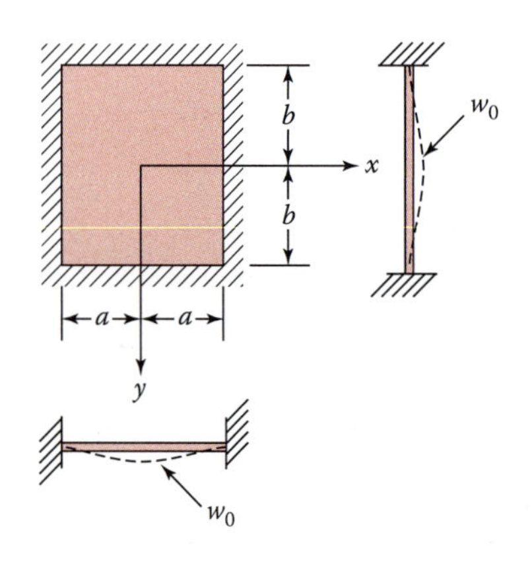

FIGURE 5.21
First mode of vibration of a clamped plate.

It is then necessary only to determine the natural circular frequency in such a manner as to make the total potential energy a minimum: $\partial(U_{max} - T_{max})/\partial c_1 = 0$. In so doing, we have

$$\omega_n = 4\left[2\left(\frac{1}{a^4} + \frac{256}{441a^2b^2} + \frac{1}{b^4}\right)\frac{D}{\bar{m}}\right]^{1/2} \tag{5.78}$$

In the case of a *square plate*, with $a = b$, the foregoing equation gives $\omega_n = (9.09/a^2)/\sqrt{D/\bar{m}}$. The natural frequency of the lowest mode of vibration is therefore

$$f_n = \frac{\omega_n}{2\pi} = \frac{1.446}{a^2}\sqrt{\frac{D}{\bar{m}}} \tag{5.79}$$

This result is about 1% higher than the "exact" value [1].

Comment: Equation 5.79 shows that the frequency is directly proportional to the square root of the flexural rigidity and inversely proportional to the square of the length of the sides of the plate and to the square root of the mass per unit area.

Problems

Sections 5.1 through 5.3

5.1 A structural steel door 1.5 m long, 0.9 m wide, and 15 mm thick is subjected to a uniform pressure p_0. The material properties are $E = 200$ GPa, $\nu = 0.3$, and allowable yield strength $\sigma_{yp} = 240$ MPa. The plate is regarded as simply supported. Determine, using Navier's approach and retaining only the first four terms in the series solution: (a) the limiting value of p_0 that can be applied to the plate without causing permanent deformation; (b) the maximum deflection w that would be produced when p_0 reaches its limiting value.

5.2 A 2 × 2-msquare simply supported steel plate ($E = 200$ GPa, $\nu = 0.3$) of 20-mm thickness carries the loading (in kPa):

$$p = 30\sin\frac{\pi x}{a}\sin\frac{\pi y}{b}$$

Compute: (a) the maximum deflection; (b) the maximum bending stress; (c) corner reactions; (d) the maximum edge reactions.

5.3 Redo Problem 5.2a through c for the case in which the plate is subjected to a uniform loading $p_0 = 30$ kPa.

5.4 A square plate ($a = b$) is simply supported on all edges and is loaded by gravel such that

$$p = p_0 \sin\frac{m\pi x}{a}\sin\frac{n\pi y}{b}$$

(a) Determine the maximum deflection and the corner reaction. (b) Evaluate, by retaining the first two terms of the series solution, the value of p_0 so that the maximum deflection is not to exceed 10 mm for the following data: $E = 165$ GPa, $\nu = 0.25$, $a = 4$ m, and $t = 25$ mm.

5.5 Determine the edge reactions R_x and R_y for a simply supported rectangular plate with concentrated load P at its center (Figure 5.3b). Take $m = n = 1$ and $\nu = 0.3$.

5.6 A simply supported rectangular plate is under the action of hydrostatic pressure expressed by $p = p_0 x/a$, where constant p_0 represents the load intensity along the edge $x = a$ (Figure P5.6). Determine, employing Navier's approach and retaining the first term of the series solution, the equation

$$w = \frac{8p_0}{\pi^6 D} \frac{1}{(1/a^2 + 1/b^2)^2} \sin\frac{\pi x}{a} \sin\frac{\pi y}{b} \tag{P5.6}$$

for the resulting deflection.

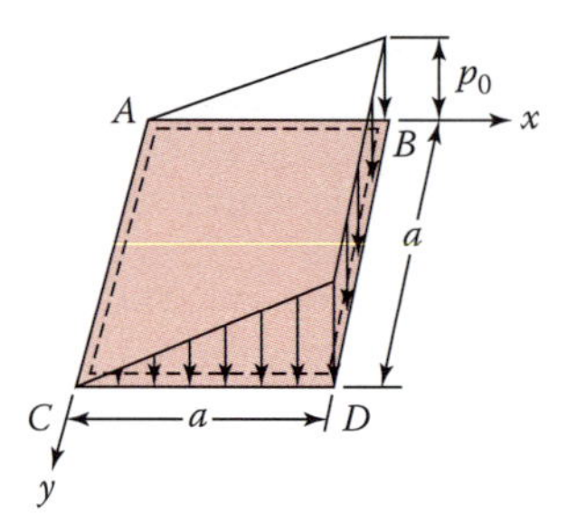

FIGURE P5.6

5.7 Determine the expressions for the moments in the plate described in Problem 5.6 for $a = b$ and $\nu = 0.3$. Calculate: (a) the maximum values of stresses σ_x and σ_y at the center of the plate; (b) the maximum values of stresses τ_{xy}, τ_{yz}, τ_{xz}, and σ_z in the plate.

5.8 A rectangular flat portion of a wind tunnel, simply supported on all edges, is under a uniform pressure p_0 distributed over the subregion shown in Figure 5.3a. Let $b = 2a$, $c = a/4$, $d = a/2$, $x_1 = a/2$, $y_1 = a$, and $\nu = 0.3$. Determine the maximum deflection and maximum stress $\sigma_{x,\max}$ in the plate by retaining the first two terms ($m = 1, n = 1, 3$) of the series solution.

5.9 Determine the corner reaction F_c for a simply supported square plate with a concentrated load P at the center (Figure P5.9). Retain only the first two terms ($m = 1, n = 1, 3$) in the series solution and let $\nu = 1/3$.

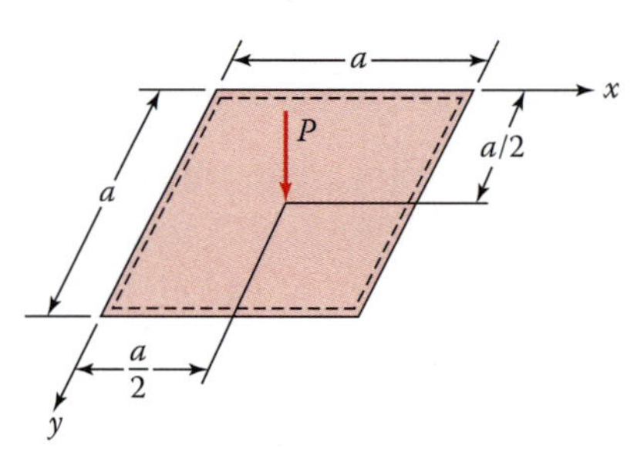

FIGURE P5.9

5.10 A concentrated load P is applied at the center of a simply supported rectangular plate (Figure 5.1a). Determine, if $m = n = 1$, $\nu = 0.3$, and $a = 2b$: (a) the maximum deflection; (b) the maximum stress in the plate.

5.11 Verify the result given by Equation 5.10 by evaluating the limits of $[\sin(m\pi c/a)]/c$ and $[\sin(n\pi y/b)]/d$.

5.12 A simply supported plate is subjected to loads P_1 and P_2 at points $(x = a/2, y = a/4)$ and $(x = a/2, y = 3a/4)$, respectively (Figure 5.1a). Obtain only the first two terms in the series solution and take $a = b$.

Sections 5.4 through 5.7

5.13 Given a rectangular plate simply supported along its edges and subjected to a uniform load of intensity p_0, determine the value of maximum deflection if $b = 2a$. Compare the result with those listed in Table 5.1.

5.14 Determine the expression for shear force Q_x for the plate described in Problem 5.13.

5.15 Verify the results given by Equations 5.16 and 5.17 by assuming a solution of the form $f_m = K_m e^{\lambda_m y}$, where K_m and λ_m are constants.

5.16 Verify the result for w given by Equation i of Example 5.5. Take the first two terms in the series solution.

5.17 Outline the derivation of the expression for the deflection surface of a uniformly loaded plate supported as shown in Figure P5.17.

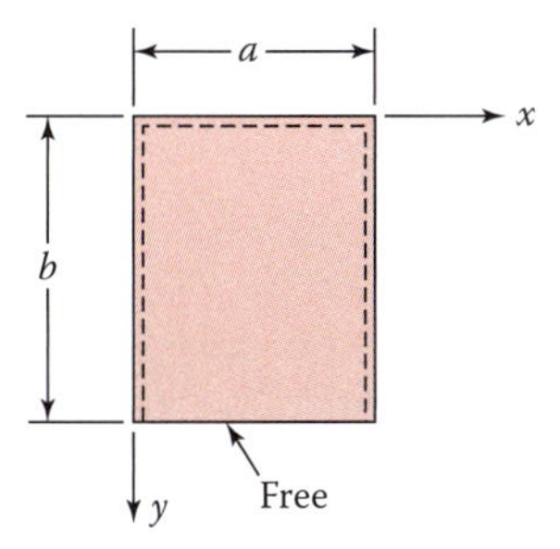

FIGURE P5.17

5.18 Redo Example 5.7 if the edge $y = 0$ of the plate is built in. Assume that all other conditions are the same.

5.19 Verify the results for p_m given in Figures B and C of Table 5.2.

5.20 Determine the equation of the elastic surface for a uniformly loaded plate with two opposite edges simply supported, the third edge free, and the fourth edge clamped (Figure 5.7).

5.21 A water level control structure consists of a vertically positioned simply supported plate. The structure is filled with water up to the upper edge level at $x = 0$. (Figure A of Table 5.2). Show that by taking only the first term of the series solution, the values for the deflection w and the bending moment M_x at the center, for $a = b/3$, are

$$w = 0.00614\frac{p_0 a^4}{D} \qquad M_x = 0.06178 p_0 a^2 \qquad \left(x = \frac{a}{2}, y = 0\right) \tag{P5.21}$$

5.22 Find the equation of the elastic surface for the plate loaded as depicted in Figure C of Table 5.2, if $a = b$ and $x_1 = a/2$.

5.23 Determine the equation of the elastic surface for the plate loaded as shown in Figure B of Table 5.2. Let $x_1 = e = a/2$, $a = b$, and $m = 1$ to compare the result for the maximum deflection with that listed in Table 5.1.

5.24 A square steel plate 10 mm thick is built in at the bottom of a ship drawing 7 m of water. The allowable stress for the plate is 100 MPa. Determine the maximum plate dimension and the maximum deflection. Specific weight of saltwater $g = 10.054$ kN/m³. Take $E = 200$ GPa.

5.25 Determine the equation of the elastic surface of a partially loaded rectangular plate shown in Figure P5.25. Evaluate the deflection at the center of the plate for $a = b$.

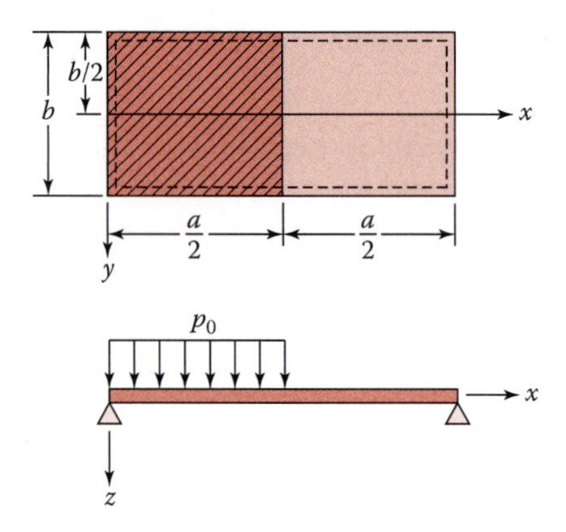

FIGURE P5.25

5.26 A rectangular plate, with two opposite sides $y = \pm b/2$ fixed and the sides $x = 0$ and $x = a$ simply supported, is subjected to a hydrostatic loading as shown in Figure A of Table 5.2. By employing the method of superposition, derive an expression for the reactional moments M_y along the clamped edges $y = \pm b/2$.

Sections 5.8 through 5.12

5.27 Employ the strip method to determine the center deflection and the maximum bending moments for a plate with two opposite edges simply supported and the other two edges clamped (Figure 5.13a). Compare the results with those obtained in Example 5.11.

5.28 Employ the strip method to obtain the approximate center deflection and center bending moment for the simply supported rectangular plate under hydrostatic loading as shown in Figure A of Table 5.2.

5.29 Use the strip method to calculate the maximum deflection and moments for a simply supported plate under uniform loading p_0 (Figure 5.1). Compare the results with those given in Table 5.1, for $b/a = 1$, $b/a = 3$, and $b/a = 5$.

5.30 A *lock-gate* of a reservoir (Figure P5.30a), considered to be a rectangular plate, is subjected to a uniform water pressure p_0 and supported as shown in Figure P5.30b. Determine the maximum stress using the strip method.

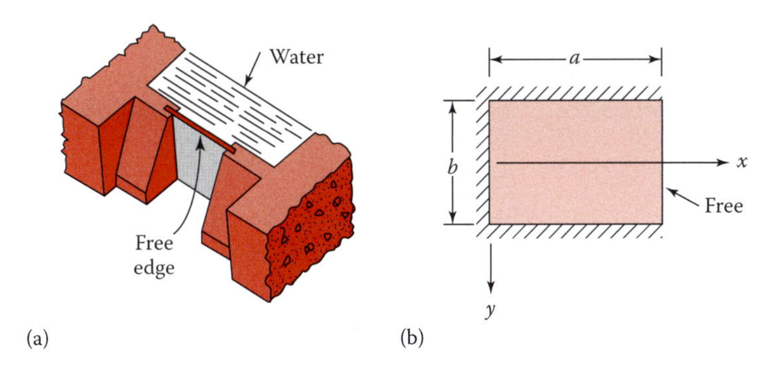

FIGURE P5.30

5.31 Redo Problem 5.30 for the case in which the plate is a square ($a = b$).

5.32 Simply supported silo flooring is loaded as shown in Figure 5.17. Referring to the equations derived in Section 5.9, determine the deflection at the center of plate 1 and plate 2, for $a = b$. Take only the first term of the series solution.

5.33 A uniformly loaded airport platform is supported by equally spaced columns ($a = b$) (Figure 5.18a). Obtain the maximum deflection and maximum bending moment $M_{x,\max}$ in each panel, for $m = 2$. Compare the results with those given in Table 5.4.

5.34 A plate platform under uniform loading p_0 is supported by columns (Figure 5.18a), where $b = 2a$ distance apart. The plate is fabricated of a material of tensile strength σ_u and compressive strength σ_{uc}. Design the platform (find the thickness), selecting a factor of safety n, according to the following failure criteria: (a) the maximum principal stress; (b) Coulomb–Mohr. Use $\sigma_u = 200$ MPa, $\sigma_{uc} = 600$ MPa, $n = 2$, $a = 1$ m, and $p_0 = 2$ MPa.

5.35 Derive the equation of the elastic surface of a simply supported rectangular plate on an elastic foundation (Figure 5.3b) if the plate experiences a central point load P. Use the equilibrium approach of Section 5.11.

5.36 Obtain the maximum stress σ_x in the plate loaded as described in Problem 5.35 by retaining in the series solution the first term only. Given: $b = 4a$ and $\nu = 0.3$.

5.37 A simply supported and uniformly loaded rectangular steel plate rests on an elastic foundation of modulus k. What should be the value of k in order to limit the maximum deflection to 6 mm? Let $E = 200$ GPa, $\nu = 0.3$, $t = 3$ mm, $a = 0.8$ m, $b = 1.4$ m, and $p_0 = 20$ MPa. Retain only the first term of the series solution. The coordinates are placed as shown in Figure 5.1a.

5.38 Verify the result given by Equation g of Section 5.11 by taking a solution of the form $f_m = E_m e^{\lambda_m y}$, where E_m and λ_m are constants.

5.39 Redo Example 5.3b by employing the Ritz method.

5.40 Apply the Ritz method to show that the deflection of the simply supported square plate subjected to a distributed load in the form of a *triangular prism* (Figure P5.40) is expressed by

$$w = \frac{32 p_0 a^4}{\pi^7 D} \sum_{m=1}^{\infty} \sum_{n=1}^{\infty} \frac{(-1)^{(m-1)/2}}{m^2 n (m^2 + n^2)^2} \sin\frac{m\pi x}{a} \sin\frac{n\pi y}{a} \tag{P5.40}$$

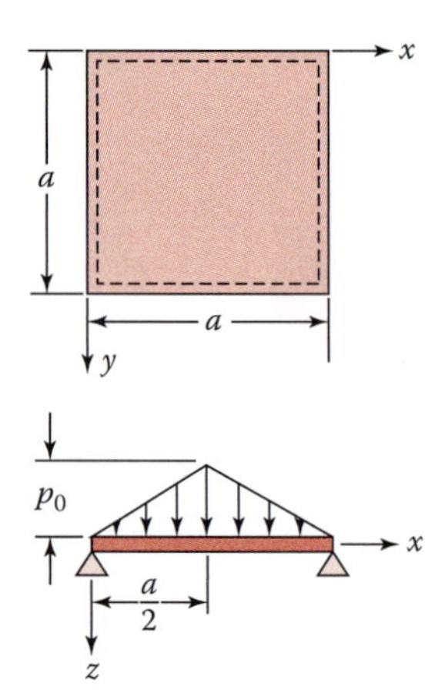

FIGURE P5.40

5.41 Rework Problem 5.35 by employing the Ritz method.

5.42 A uniformly loaded rectangular plate has two opposite sides built in and the other two sides simply supported. Assume a solution of the form

$$w = \sum_{m=1}^{\infty}\sum_{n=1}^{\infty} a_{mn}\left(1-\cos\frac{2m\pi y}{a}\right)\sin\frac{n\pi y}{b} \tag{P5.42}$$

and determine the coefficients a_{mn} employing the Ritz method.

5.43 Verify whether the approximate solution, Equation 5.74, satisfies all edge conditions of the plate depicted in Figure 5.20.

5.44 Assuming that the rectangular plate shown in Figure 5.21 is mounted in a rigid chassis that is subjected to a harmonic vibration with a peak input $g_{in} = 10g$ at a frequency of 62 Hz, calculate the maximum dynamic deflection and stress in the plate. The plate is made of aluminum alloy ($E = 72$ GPa, $\nu = 0.3$, $r = 2.8$ Mg/m^3), having $a = 150$ mm, $b = 300$ mm, and $t = 1$ mm.

5.45 Redo Problem 5.44 for the case in which $a = b = 200$ mm and $f = 50$ Hz.

References

1. R. Szilard, *Theory and Analysis of Plates—Classical and Numerical Methods*, Prentice-Hall, Upper Saddle River, NJ, 2004.
2. S. Timoshenko and S. Woinowsky-Krieger, *Theory of Plates and Shells*, 2nd ed., McGraw-Hill, New York, 1959.
3. W. Flügge (ed.), Handbook of Engineering Mechanics, Chapter 39, McGraw-Hill, New York, 1984.
4. D. McFarland, B. L. Smith, and W. D. Bernhart, *Analysis of Plates*, Spartan Books, Washington, DC, 1972.
5. M. Lévy, Sur l'équibre élastique d'une plaque rectangulaire, *Comptes-Rendus de l'Academie des Sciences, Paris*, 129(1899).

6. A. Nadai, *Die elastichen Platten*, Springer, Berlin, 1925.
7. E. H. Mansfield, *The Bending and Stretching of Plates*, Macmillan, New York, 1964.
8. K. Marguerre and H. T. Woernle, *Elastic Plates*, Blaisdell, Waltham, MA, 1969.
9. W. C. Young, R. C. Budynas, and A. Sadegh, *Roark's Formulas for Stress and Strain*, 8th ed., Chapter 10, McGraw-Hill, New York, 2011.
10. S. Timoshenko, D. H. Young, and W. Weaver Jr., *Vibration Problems in Engineering*, 4th ed., Wiley, Hoboken, NJ, 1974.

6

Plates of Various Geometrical Forms

6.1 Introduction

Structural members designed to resist lateral loading are generally circular or rectangular in shape, but in some situations other geometrical forms are used. As previously discussed, the bending problem of plates of any shape is solved if we derive the displacement function $w(x, y)$, which satisfies $\nabla^4 w = p/D$, and the specified conditions at the boundaries. We now deal with the problem of plate boundaries of geometrical form different from those previously treated.

From the examples worked out so far, it becomes evident that a rigorous determination of the deflection for a plate with more complicated shape is likely to be very difficult. Consequently, if solutions are not readily available, the numerical methods to be discussed in Chapter 7 can be used effectively. In developing approximate formulas for the deflections of polygonal plates, the membrane analogy (Section 3.8) has proved very valuable.

When the uniformity of the cross-sectional area of the plate is interrupted, as in the case of a ship deck or airplane fuselage with holes or windows, a perturbation in stress takes place. Determination of this disturbed stress distribution is of considerable practical importance and is discussed briefly in Section 6.6.

6.2 *Method of Images

Certain problems of plates can be treated by arbitrary extension of the plate and/or introducing fictitious forces to produce the deflection forms sought. Consider, as an example, the bending of an *isosceles right triangular plate* with *simply supported* edges under a concentrated load P acting at arbitrary point $A(x_1, y_1)$. The plate is bounded as shown by solid lines OBC in Figure 6.1.

We begin the solution by using Equation 5.11, derived in Section 5.3. For this purpose, it is assumed that the triangular plate is one-half of a simply supported square plate subjected to concentrated forces P and $-P$ at points A and $A_i[(a - y_1), (a - x_1)]$, respectively (Figure 6.1). Point A_i is the *mirror* or *image point* of A with respect to diagonal BC. Because of the loading described, the square plate deflects in such a way that diagonal BC is a *nodal line* or *fictitious support*. Hence, the deflection of OBC of the square plate is identical with

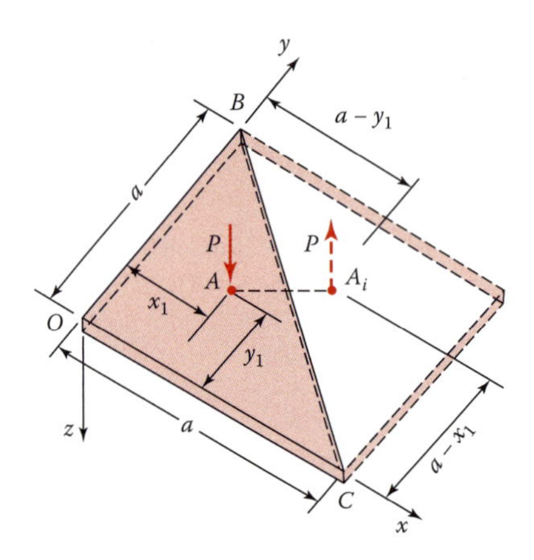

FIGURE 6.1
Simply supported isosceles triangle.

that of a simply supported triangular plate *OBC*. The deflection due to the force *P*, from Equation 5.11, is

$$w_1 = \frac{4Pa^2}{\pi^4 D} \sum_{m=1}^{\infty} \sum_{n=1}^{\infty} \frac{\sin(m\pi x_1/a)\sin(n\pi y_1/a)}{(m^2+n^2)^2} \sin\frac{m\pi x}{a} \sin\frac{n\pi y}{a} \tag{6.1}$$

On substitution of $-P$ for P, $(a - y_1)$ for x_1, and $(a - x_1)$ for y_1 in Equation 6.1, we obtain the deflection due to the force $-P$ at A_i:

$$w_2 = \frac{4Pa^2}{\pi^4 D} \sum_{m=1}^{\infty} \sum_{n=1}^{\infty} (-1)^{m+n} \frac{\sin(m\pi y_1/a)\sin(n\pi x_1/a)}{(m^2+n^2)^2} \sin\frac{m\pi x}{a} \sin\frac{n\pi y}{a} \tag{6.2}$$

The deflection surface of the triangular plate is then

$$w = w_1 + w_2 \tag{6.3}$$

where w_1 and w_2 are given by Equations 6.1 and 6.2, respectively. By applying Equation 6.3 together with the principle of superposition, the deflection may be determined of an isosceles right triangular plate for any kind of loading. The method described above, often referred to as the *method of images*, was introduced by Nadai [1]. Clearly, the foregoing is *not* a general procedure. It is a particular technique that applies to problems similar to that discussed in this section.

In the particular case of a triangular plate under a *uniformly distributed load* of intensity $p_0 = P/dx_1dy_1$, we find, from Equation 6.3, after integration over the area of the plate, that

$$w = \frac{16p_0a^4}{\pi^6 D}\left[\sum_{m=1,3,\ldots}^{\infty} \sum_{n=2,4,\ldots}^{\infty} \frac{n\sin(m\pi x/a)\sin(n\pi y/a)}{m(n^2-m^2)(m^2+n^2)^2} + \sum_{m=2,4,\ldots}^{\infty} \sum_{n=1,3,\ldots}^{\infty} \frac{m\sin(m\pi x/a)\sin(n\pi y/a)}{n(m^2-n^2)(m^2+n^2)^2}\right] \tag{6.4}$$

This series converges very rapidly. With the expression for the deflection of the plate determined, we can obtain the moments and the maximum stresses in the plate from Equations 3.9 and 3.11.

EXAMPLE 6.1: Displacement of Semi-Infinite Plate Strip

A load P acts at $x = x_1$ and $y = y_1$ on a simply supported semi-infinite plate strip (Figure 6.2a). Develop an expression for the deflection surface.

Solution

To use the method of images, we arbitrarily extend the plate strip in the $-y$ direction and apply a second load $-P$ at $x = x_1$, $y = -y_1$ (Figure 6.2b). The line $y = 0$ becomes then a nodal line of the deflection surface of the plate. Therefore, the solution of the problem is obtained by superimposing the deflections produced by both concentrated loads. In so doing, from Equation 5.39, we obtain the deflection surface:

$$w = \frac{Pa^2}{2\pi^3 D}\sum_{m=1}^{\infty}\frac{1}{m^3}\left\{\left[1+\frac{m\pi(y-y_1)}{a}\right]e^{-[m\pi(y_1-y)/a]} - \left[1+\frac{m\pi(y_1+y)}{a}\right]e^{-[m\pi(y_1+y)/a]}\right\}\sin\frac{m\pi x_1}{a}\sin\frac{m\pi x}{a} \tag{6.5}$$

The preceding is valid for $0 \le y \le y_1$. The deflection for the region $y \ge y_1$ is determined in an analogous manner:

$$w = \frac{Pa^2}{2\pi^3 D}\sum_{m=1}^{\infty}\frac{1}{m^3}\left\{\left[1+\frac{m\pi(y-y_1)}{a}\right]e^{-[m\pi(y-y_1)/a]} - \left[1+\frac{m\pi(y+y_1)}{a}\right]e^{-[m\pi(y+y_1)/a]}\right\}\sin\frac{m\pi x_1}{a}\sin\frac{m\pi x}{a} \tag{6.6}$$

Comments: The bending moments can be obtained by inserting Equations 6.5 and 6.6 into Equation 3.9. We note that the resulting moments do not converge very rapidly in the vicinity of the application point of the load P. Thus, it is necessary to express the moments near the point in another (closed) form [2].

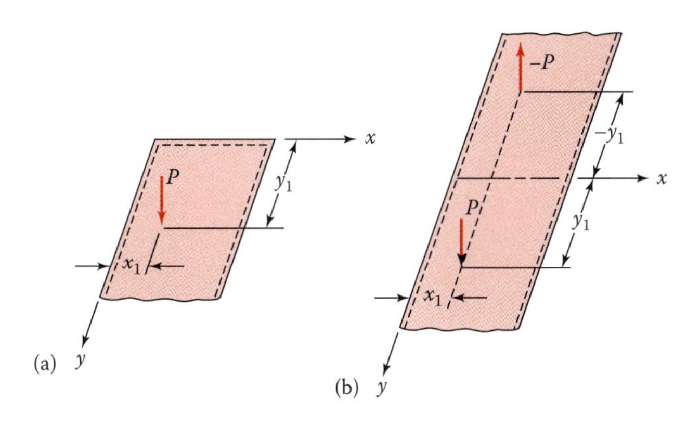

FIGURE 6.2
Simply supported very long plate under concentrated load.

6.3 Equilateral Triangular Plate with Simply Supported Edges

We now treat the case of a simply supported equilateral triangular plate *ABC*, shown in Figure 6.3a. The equations of the boundaries are

$$\begin{aligned} x+\frac{a}{3}&=0 && \text{on } BC \\ \frac{x}{\sqrt{3}}+y-\frac{2a}{3\sqrt{3}}&=0 && \text{on } AC \\ \frac{x}{\sqrt{3}}-y-\frac{2a}{3\sqrt{3}}&=0 && \text{on } AB \end{aligned} \tag{a}$$

Therefore, the function that vanishes at the boundary

$$\begin{aligned} w&=k\left(x+\frac{a}{3}\right)\left(\frac{x}{\sqrt{3}}+y-\frac{2a}{3\sqrt{3}}\right)\left(\frac{x}{\sqrt{3}}-y-\frac{2a}{3\sqrt{3}}\right) \\ &=k\left(\frac{x^3-3xy^2}{3}-\frac{a(x^2+y^2)}{3}+\frac{4a^3}{81}\right) \end{aligned} \tag{b}$$

is the general expression for deflection of an equilateral triangular plate with simply supported edges. Here *k* is determined for a specific case of loading as follows.

6.3.1 Equilateral Triangular Plate under Uniform Moment M_0 along Its Boundary

It can be verified experimentally that the deflection surface of a plate loaded by uniform moment along its boundary and the surface of a uniformly loaded membrane, uniformly stretched over the same triangular boundary, are identical. To arrive at an analytical solution, we introduce Equation b and $p=0$ into Equation 3.21 and set $M=M_0$. It follows that

$$k=\frac{3M_0}{4aD}$$

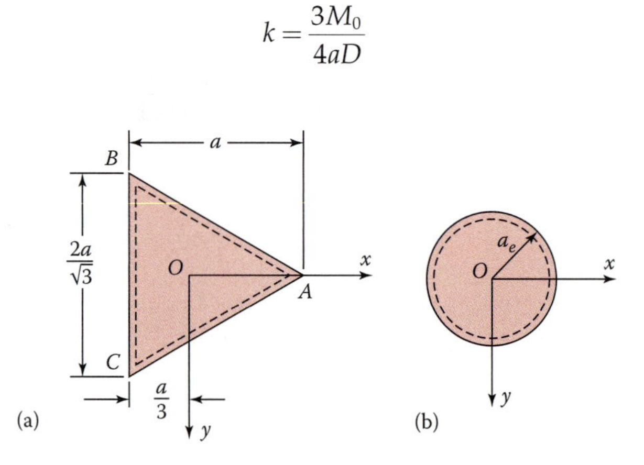

FIGURE 6.3
(a) Equilateral triangular plate; (b) equivalent circular plate.

The deflection is then

$$w = \frac{M_0}{4aD}\left[x^3 - 3xy^2 - a(x^2 + y^2) + \frac{4}{27}a^3\right] \tag{6.7}$$

and the center displacement is

$$w_0 = \frac{M_0 a^2}{27D} \qquad (x = y = 0) \tag{c}$$

The expressions for the moments are determined by means of Equations 6.7 and 3.9 as follows:

$$\begin{aligned} M_x &= \frac{1}{2}M_0\left[1+\nu-(1-\nu)\frac{3x}{a}\right] \\ M_y &= \frac{1}{2}M_0\left[1+\nu+(1-\nu)\frac{3x}{a}\right] \\ M_{xy} &= \frac{3(1-\nu)M_0 y}{2a} \end{aligned} \tag{6.8}$$

The largest moment takes place at the corners and acts on the planes bisecting the angles. For $\nu = 0.3$, it is given by

$$M_{y,\max} = 1.35M_0 \qquad \left(x = \frac{2}{3}a\right) \tag{d}$$

The corresponding stress $\sigma_{y,\max} = 8.10M_0/t^2$.

6.3.2 Equilateral Triangular Plate under Uniform Load p_0

In this case, an expression for deflection of the form

$$w = \frac{p_0}{64aD}\left[x^3 - 3xy^2 - a(x^2 + y^2) + \frac{4}{27}a^3\right]\left(\frac{4}{9}a^2 - x^2 - y^2\right) \tag{6.9}$$

satisfies Equation 3.17 and the conditions of the simply supported edges. Therefore, Equation 6.9 is the solution. By employing Equation 3.9, the bending moments may then be obtained. It can be shown that the center moments are

$$M_x = M_y = (1+\nu)\frac{p_0 a^2}{54} \tag{6.10}$$

The largest moment takes place on the planes bisecting the angles of the triangle. For example, at points along the x axis (Figure 6.3a), for $\nu = 0.3$, we have

$$\begin{aligned} M_{x,\max} &= 0.0248p_0a^2 \qquad (x = -0.062a,\ y = 0) \\ M_{y,\max} &= 0.0259p_0a^2 \qquad (x = 0.129a,\ y = 0) \end{aligned} \tag{e}$$

The maximum stress is given by $\sigma_{y,\max} = 0.155p_0a^2/t^2$.

It is of interest to note that, for *preliminary design purposes*, the plate in Figure 6.3a can be replaced by the plate in Figure 6.3b having the *equivalent radii* [3]:

$$\begin{aligned} a_e &= 0.35a \quad \text{(for uniform load)} \\ a_e &= 0.38a \quad \text{(for concentrated center load)} \end{aligned} \tag{6.11}$$

Hence, the maximum moment and the stress of the plate in Figure 6.3a can be estimated by applying the formulas developed for circular plates in Sections 4.5 and 4.8. Identical approximations can be used for irregular triangular plates, provided that the lengths of the sides do not differ appreciably.

6.4 Elliptical Plates

In this section, consideration is given to the bending of an elliptic plate with semiaxes a and b, subjected to a uniform load of intensity p_0 (Figure 6.4). Two situations are treated in the following paragraphs.

6.4.1 Uniformly Loaded Elliptic Plate with Clamped Edge

The appropriate boundary conditions, referring to Figure 6.4, are

$$w = 0 \qquad \frac{\partial w}{\partial n} = 0 \qquad \left(\text{for } \frac{x^2}{a^2} + \frac{y^2}{b^2} = 1\right) \tag{a}$$

Here n is the normal to the plate boundary. These conditions are satisfied if we assume, for deflection, the expression

$$w = k\left(1 - \frac{x^2}{a^2} - \frac{y^2}{b^2}\right)^2 \tag{b}$$

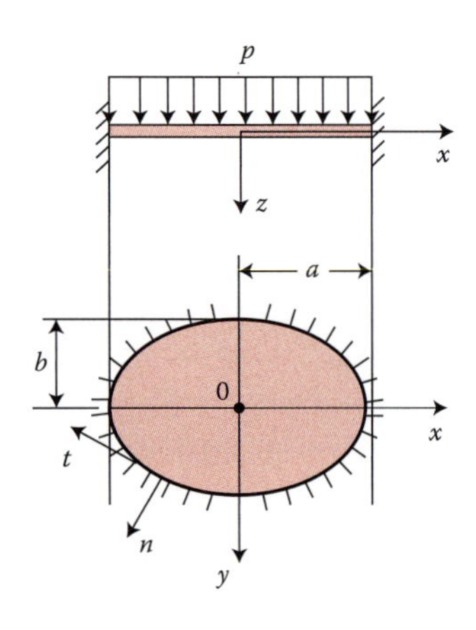

FIGURE 6.4
Elliptical plate with built-in edge.

in which k is a constant. Substituting Equation b into Equation 3.17 and setting $p = p_0$, we have

$$k = \frac{p_0}{8D} \frac{a^4 b^4}{3a^4 + 2a^2b^2 + 3b^4} \tag{c}$$

The required solution is then

$$w = \frac{p_0}{8D} \frac{a^4 b^4}{3a^4 + 2a^2b^2 + 3b^4} \left(1 - \frac{x^2}{a^2} - \frac{y^2}{b^2}\right)^2 \tag{6.12}$$

The maximum deflection occurs at the center of the plate and is given by

$$w_{\max} = \frac{p_0}{8D[(3/b^4) + (2/a^2b^2) + (3/a^4)]} \tag{6.13}$$

By substituting Equation b into Equation 3.9, the bending and twisting moments are obtained

$$\begin{aligned} M_x &= 4Dk\left[\frac{1}{a^2} - \frac{3x^2}{a^4} - \frac{y^2}{a^2b^2} + \nu\left(\frac{1}{b^2} - \frac{3y^2}{b^4} - \frac{x^2}{a^2b^2}\right)\right] \\ M_y &= 4Dk\left[\frac{1}{b^2} - \frac{3y^2}{b^4} - \frac{x^2}{a^2b^2} + \nu\left(\frac{1}{a^2} - \frac{3x^2}{a^4} - \frac{y^2}{a^2b^2}\right)\right] \\ M_{xy} &= -(1-\nu)\frac{8Dk}{a^2b^2}xy \end{aligned} \tag{6.14}$$

At the center ($x = 0$, $y = 0$), the moments are

$$\begin{aligned} M_x &= 4Dk\left(\frac{1}{a^2} + \frac{\nu}{b^2}\right) \\ M_y &= 4Dk\left(\frac{1}{b^2} + \frac{\nu}{a^2}\right) \end{aligned} \tag{d}$$

At the ends of the minor and major axes, we have

$$\begin{aligned} M_x &= -\frac{8Dk}{a^2} \qquad M_y = -\frac{8\nu Dk}{a^2} \qquad (x = \pm a, \quad y = 0) \\ M_x &= -\frac{8\nu Dk}{b^2} \qquad M_y = -\frac{8Dk}{b^2} \qquad (x = 0, \quad y = \pm b) \end{aligned} \tag{6.15}$$

where k is given in Equation c. It is observed from Equation 6.15 that the *maximum* moment occurs at the *extremity of the minor axis*.

TABLE 6.1

Deflections and Moments for Uniformly Loaded Simply Supported Elliptical Plates

a/b	1.0	1.5	2.0	3.0	4.0	∞
$w \cdot (Et^3/p_0b^4)$	0.70	1.26	1.58	1.88	2.02	2.28
$M_x \cdot (1/p_0b^2)$	0.206	0.222	0.210	0.188	0.184	0.150
$M_y \cdot (1/p_0b^2)$	0.206	0.321	0.379	0.433	0.465	0.500

For $a = b$, the solution derived in this section reduces to the results obtained for a circular plate with clamped edge (Section 4.5). In the extreme case, $a \gg b$, Equation 6.13 and the second of Equation d appear as

$$w_{\text{max}} = \frac{p_0(2b)^4}{384D} \qquad M_y = \frac{p_0(2b)^2}{24} \tag{e}$$

Equation e corresponds to the deflection and moment at the center of a fixed-end strip of span $2b$ under uniformly distributed load.

6.4.2 Uniformly Loaded Elliptic Plate with Simply Supported Edge

The boundary conditions, Equations 3.26b, are now expressed in the form:

$$w = 0 \qquad \frac{\partial^2 w}{\partial n^2} = 0 \qquad \left(\text{for } \frac{x^2}{a^2} + \frac{y^2}{b^2} = 1\right) \tag{f}$$

Here, as before, n is a normal to the plate boundary. After routine but somewhat lengthier calculations than in the case of a fixed plate, expressions for deflection and bending moments may be obtained.

Table 6.1 provides, for $\nu = 0.3$, some final numerical results for the deflection and bending moments at the center of the plate. Comparison of these results with those in Table 3.1 show that, *for equal values of a/b*, the values of the *center* deflections and the moments in rectangular and elliptical plates do *not* differ considerably.

6.5 Sector-Shaped Plates

Expressions presented in Section 4.2 for circular plates can also be used for a *plate in the form of a sector* having the straight edges simply supported (Figure 6.5a). The boundary conditions at the circular edges ($r = a$ and $r = b$) are given in Table 4.1. At the edges AC and BD, both the deflection and the tangential bending moment are zero. The *straight edge* restraints, referring to Equations 4.2 and 3.26, are thus

$$w = 0 \qquad \frac{\partial^2 w}{\partial \theta^2} = 0 \qquad (\theta = 0 \text{ and } \theta = \alpha) \tag{a}$$

where, as before, $w = w_h + w_p$.

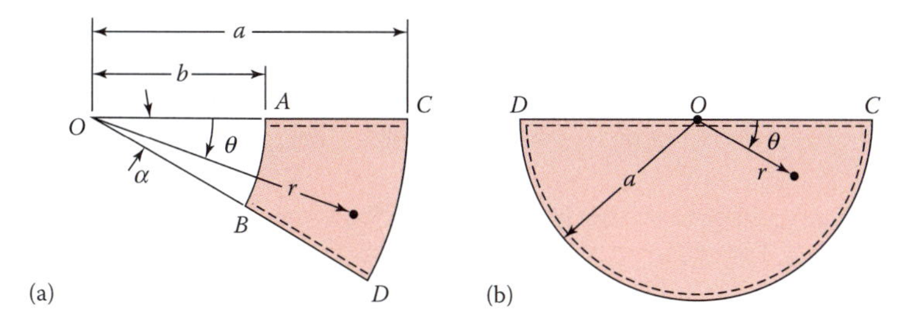

FIGURE 6.5
(a) Sector plate; (b) semicircular plate.

If the plate is under a *uniform loading* p_0, represented by the polar form of sine series given by Equation 5.20,

$$p = \frac{4p_0}{\pi} \sum_{n=1,3\ldots}^{\infty} \frac{1}{n} \sin\frac{n\pi\theta}{\alpha} \tag{b}$$

the particular solution of Equation 4.5 is

$$w_p = \frac{4p_0 r^4}{\pi D} \sum_{n=1,3,\ldots}^{\infty} \frac{1}{n(16 - n^2\pi^2/\alpha^2)(4 - n^2\pi^2/\alpha^2)} \sin\frac{n\pi\theta}{\alpha} \tag{c}$$

As the solution of the homogeneous Equation 4.6, we take only the term of the series (Equation 4.7) that contains sine:

$$w_h = \sum_{n=1}^{\infty} \left(A_n^* r^{n\pi/\alpha} + B_n^* r^{-n\pi/\alpha} + C_n^* r^{2+n\pi/\alpha} + D_n^* r^{2-n\pi/\alpha} \right) \sin\frac{n\pi\theta}{\alpha} \tag{d}$$

Combining Equations c and d, we have the complete expression for the deflection that satisfies the conditions (Equation a). The coefficients $A_n^*, \ldots, D_n^*$ are determined at $r = a$ and $r = b$ of a sector plate, as illustrated in the solution of the sample problem that follows.

EXAMPLE 6.2: Uniformly Loaded Sector Plate

A semicircular aircraft plate of radius a is simply supported along the boundary (Figure 6.5b). The panel is subject to a uniform pressure p_0. (a) Derive an expression for the deflection surface. (b) Calculate the value of the deflection at $r = a/2$ and $\theta = \pi/2$, retaining the first two significant terms of the series solution. *Given*: The plate is made of wrought iron ($G = 190$ GPa, $\nu = 0.3$) with radius $a = 0.4$ m, thickness $t = 8$ mm, and carries $p_0 = 10$ MPa.

Solution

a. We have, for this case, $b = 0$ and $\alpha = \pi$. The values of B_n^* and D_n^* must be taken as zero so that the conditions (*a*) are satisfied along the diameter *CD*. The solution is then

$$w = \sum_{n=1}^{\infty} \left(A_n^* r^n + C_n^* r^{n+2} \right) \sin n\theta + \frac{4p_0 r^4}{\pi D} \sum_{n=1}^{\infty} \frac{1}{n(16 - n^2)(4 - n^2)} \sin n\theta \tag{6.16}$$

The boundary conditions at the circular edge are

$$w=0 \qquad \frac{\partial^2 w}{\partial r^2}+\nu\left(\frac{1}{r}\frac{\partial w}{\partial r}+\frac{1}{r^2}\frac{\partial^2 w}{\partial \theta^2}\right)=0 \qquad (r=a) \tag{6.17}$$

Introducing the series (6.16) into the preceding equation, we have

$$A_n^* = \frac{p_0 a^4 (n+5+\nu)}{a^n n\pi(16-n^2)(2+n)[n+(1/2)(1+\nu)]D}$$

$$C_n^* = -\frac{p_0 a^4 (n+3+\nu)}{a^{n+2} n\pi(4+n)(4-n^2)[n+(1/2)(1+\nu)]D}$$

We thus arrive at the expression for the deflection surface

$$w=\frac{p_0 a^4}{D}\sum_{n=1,3,\ldots}^{\infty}\left\{\frac{4r^4}{a^4}\frac{1}{n\pi(16-n^2)(4-n^2)}+\frac{r^n}{a^n}\frac{n+5+\nu}{n\pi(16-n^2)(2+n)[n+(1/2)(1+\nu)]}\right.$$
$$\left.-\frac{r^{n+2}}{a^{n+2}}\frac{n+3+\nu}{n\pi(4+n)(4-n^2)[n+(1/2)(1+\nu)]}\right\}\sin n\theta \tag{6.18}$$

The expressions for the bending and twisting moments can be obtained from this solution by the use of Equation 4.2.

b. Substituting the given numerical values, we have

$$D=\frac{Et^3}{12(1-\nu^2)}=\frac{190(8^3)}{12(1-0.3^2)}=8.91\ \mathrm{kN\cdot m}$$

and

$$\frac{p_0 a^4}{D}=\frac{(10\times 10^3)(0.4^4)}{8.91}=28.732\ \mathrm{m}^{-1}$$

Equation 6.18 (for $\theta=\pi/2$, $r=a/2$, $\nu=0.3$, $n=1$) gives then (see: Appendix D):

$$w_1=28.732\left\{\frac{4(0.4/2)^4}{0.4^4}\frac{1}{(1)\pi(15)(3)}+\frac{(0.4/2)}{0.4}\frac{6.3}{(1)\pi(15)(3)(1+(1.3/2))}-\frac{(0.4/2)^3}{0.4^3}\frac{4.3}{(1)\pi(5)(3)(1+(1.3/2))}\right\}(1)$$
$$=28.732\{1.768(10^{-3})+0.0135-6.913(10^{-3})\}=0.24\ \mathrm{mm}$$

Similarly, taking the first two significant terms ($n=1, 3$) of the series solution results in $w_{1,3}=w_1+w_3=0.24-8.278(10^{-3})=0.232$ mm.

Following a procedure identical to that employed in the preceding example, we can determine a solution for any sector (of radius a and $b=0$) subtended by the angle α. The resulting expressions for the deflections and bending moments can be put, taking $\nu=0.3$, into the forms

$$w=\delta_1\frac{p_0 a^4}{D} \qquad M_r=\delta_2 p_0 a^2 \qquad M_\theta=\delta_3 p_0 a^2 \tag{6.19}$$

TABLE 6.2
Coefficients δ_1, δ_2, and δ_3 for Various Uniformly Loaded Sector Plates Simply Supported at the Boundary

	$r/a = 1/4$			$r/a = 1/2$			$r/a = 3/4$		
α	δ_1	δ_2	δ_3	δ_1	δ_2	δ_3	δ_1	δ_2	δ_3
$\pi/3$	0.00019	−0.0025	0.00177	0.00080	0.0149	0.0255	0.00092	0.0243	0.0213
$\pi/2$	0.00092	0.0036	0.0319	0.00225	0.0353	0.0352	0.00203	0.0381	0.0286
π	0.00589	0.0692	0.0357	0.00811	0.00868	0.0515	0.00560	0.0617	0.0468

Numerical values of the constants δ_1, δ_2, and δ_3 are furnished in Table 6.2 for various ratios r/a and angle a of a simply supported plate in the form of a sector [2].

The case of a sector-shaped plate, clamped along the circular boundary and simply supported at the straight edges *AC* and *BD*, can be treated in a like manner.

6.6 *Stress Concentration around Holes in a Plate

The bending theory of plates applies only to plates of constant cross-sectional area. If the cross section changes gradually, reasonably accurate results can also be expected. On the other hand, where abrupt variation in the cross section takes place, ordinary bending theory cannot predict the high values of stress that actually exist. The condition referred to occurs in plates with holes, notches, or fillets. In some instances, the stresses in these regions can be analyzed by applying *extended bending theories* that take into account the shear deformation of the plate (Section 3.10). The solutions more often involve considerable difficulty [4–9].

Recall from Section 2.4 that the ratio of the actual maximum stress to the nominal stress in the minimum section of the member is defined as the theoretical or geometric *stress concentration factor K*. Thus,

$$\sigma_{\max} = K\sigma_{nom} \tag{6.20}$$

where σ_{nom} is obtained by applying the familiar flexure formula for beams and plates. As previously noted, stress concentration factors are determined by the theory of elasticity, by finite element computations, from photoelastic models, and by direct strain-gage measurements. These factors are presented in the form of graphs [6], such as shown in Figure 6.6.

It is convenient to examine the stress concentration at the edge of a *small hole in a very large plate* (Example 6.3). The results obtained for this case have been proved applicable to plates of any shape.

EXAMPLE 6.3: Stress Distribution around Hole

A large, thin plate containing a circular hole of radius a and width $2a$, small as compared with the overall dimensions, is subjected to pure bending $M_x = M_0$ and $M_y = 0$ (Figure 6.7). Determine the stress distribution around the hole.

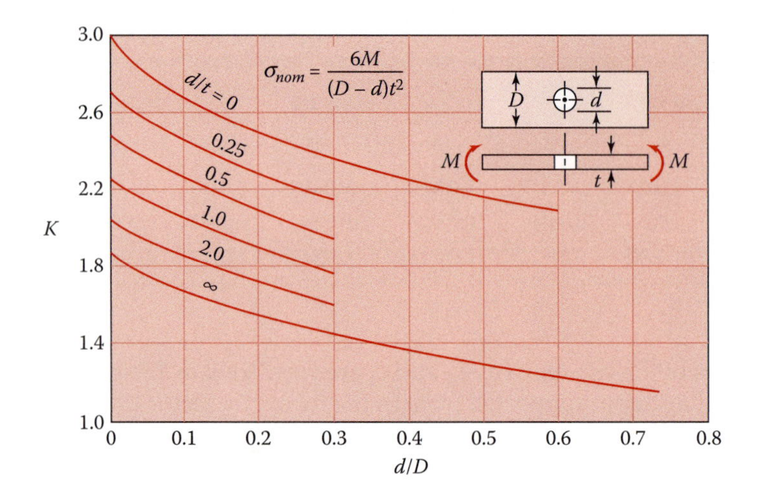

FIGURE 6.6
Stress concentration factors for rectangular plates with hole in bending.

Solution

We place the origin of coordinates at the center of the hole. In the absence of any hole, we have, from polar form of Equation 3.40,

$$w_1 = -\frac{M_0 r^2}{4D(1-\nu^2)}[1-\nu+(1+\nu)\cos 2\theta] \tag{6.21}$$

At $r = a$, the radial moment and effective transverse force due to the M_0, obtained by means of Equations 6.21, 4.2, and 4.4, are

$$\begin{aligned} M_{r1} &= \frac{1}{2}M_0(1+\cos 2\theta) \\ V_{r1} &= -\frac{1}{a}M_0 \cos 2\theta \end{aligned} \tag{6.22}$$

When a hole of radius a is drilled through the plate, we superimpose on the original state of stress an additional state of stress so that the combined radial moments and

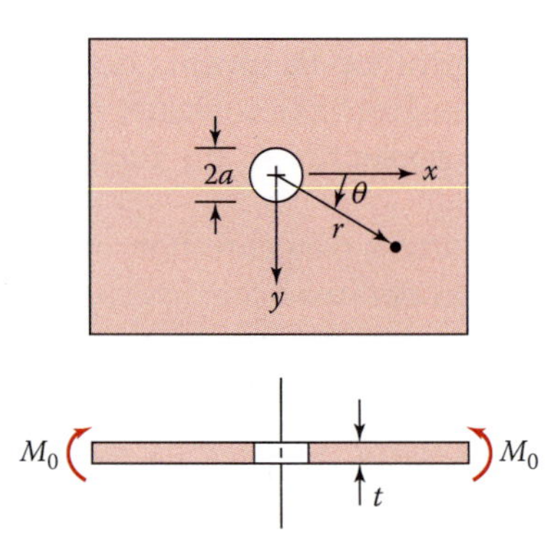

FIGURE 6.7
Bending of a large plate with a hole.

forces are equal to zero at $r = a$, and the superimposed stresses *only* are zero at $r = \infty$. Guided by the expression for w_1, we assume the additional deflection of the form [2]

$$w_2 = -\frac{M_0 a^2}{2D}\left[A \ln r + \left(B - C\frac{a^2}{r^2}\right)\cos 2\theta\right] \tag{a}$$

The foregoing satisfies both conditions stated above and Equation 4.6. The moment and effective transverse force corresponding to Equation a at $r = a$ are determined as follows:

$$\begin{aligned} M_{r2} &= -\frac{1}{2}M_0\{(1-\nu)A + [4\nu B - 6(1-\nu)C]\cos 2\theta\} \\ V_{r2} &= \frac{1}{a}M_0[2(3-\nu)B + 6(1-\nu)C]\cos 2\theta \end{aligned} \tag{b}$$

The conditions along the boundary of the hole are represented by

$$M_{r1} + M_{r2} = 0 \qquad V_{r1} + V_{r2} = 0 \qquad (r = a) \tag{c}$$

Substitution of Equations 6.22 and b into Equation c leads to the values of the constants. The final deflection expression ($w_1 + w_2$) is then obtained. It follows that the tangential moment and shear force at the periphery of the circle ($r = a$) are

$$\begin{aligned} M_\theta &= M_0\left[1 - \frac{2(1+\nu)}{3+\nu}\cos 2\theta\right] \\ Q_\theta &= \frac{4M_0}{(3+\nu)a}\sin 2\theta \end{aligned} \tag{6.23}$$

From Equation 6.23, it is evident that the maximum values of stress resultants occur at $\theta = \pi/2$ and $\theta = \pi/4$, respectively:

$$M_{\theta,\max} = \frac{5+3\nu}{3+\nu}M_0 \qquad Q_{\theta,\max} = \frac{4}{(3+\nu)a}M_0 \tag{6.24}$$

The stress concentration factor is therefore

$$k_b = \frac{5+3\nu}{3+\nu} \tag{6.25}$$

Here reduction in the cross-sectional area due to a small hole has been neglected. For $\nu = 1/3$, Equation 6.25 yields $k_b = 1.80$.

On introducing Equations 6.24 into Equations 3.11 and 3.32 and setting $\nu = 1/3$, the maximum stresses are found to be $\sigma_{\max} = 10.8M_0/t^2$ and $\tau_{\max} = 1.8M_0/at$. The ratio of these stresses is

$$\frac{\tau_{\max}}{\sigma_{\max}} = \frac{t}{6a} \tag{d}$$

When, for example, the hole width is equal to the plate thickness, the above quotient is 1/3. For $2a < t$, the ratio τ_{max}/σ_{max} is larger. The *influence of shear* on the plate deformation is thus *pronounced*.

Comments: A comparison of the results of Figure 6.6 with that of the preceding example shows that the stress concentration factor determined using the *customary thin plate theory* is quite crude. In a like manner, the case of plates with noncircular holes may also be treated.

Problems

Sections 6.1 through 6.3

6.1 A simply supported wing panel in the form of an isosceles right triangle is subjected to a uniform load of intensity p_0 (Figure 6.1). Determine, by retaining only the first term of the series solution, at point $A(x = a/4, y = a/4)$: (*a*) the deflection; (*b*) the bending moment M_x. Take $\nu = 0.3$.

6.2 Resolve Problem 6.1 with the plate subjected to a concentrated load P at point A and $p_0 = 0$. Take first two terms of the series solution and $\nu = 0.3$.

6.3 A simply supported isosceles right triangular plate ($\nu = 1/3$) carries a uniform load of intensity p_0 (Figure 6.1). Determine, by taking the first term of the series solution, at point $A(x_1 = a/3, y_1 = a/2)$: (*a*) the deflection; (*b*) the maximum principal stress.

6.4 Consider an equilateral triangular plate with simply supported edges under a uniform moment M_0 along its boundary (Figure 6.3). Taking $\nu = 1/3$, find: (*a*) the twisting moment on the side AC; (*b*) the concentrated reactions at the corners.

6.5 An equilateral triangular simply supported plate of $a = 240$ mm is under a uniform loading $p_0 = 5$ MPa (Figure 6.3). Calculate: (*a*) the thickness if the allowable stress is not to exceed 100 MPa; (*b*) the center deflection. Take $E = 200$ GPa and $\nu = 0.3$.

6.6 Redo Problem 6.5 For the case in which the plate is fixed at edges. Use the first of Equations 6.11.

6.7 An equilateral triangular steel plate ($E = 200$ GPa, $\nu = 0.3$, $\sigma_{yp} = 240$ MPa) has a thickness of 12 mm. The plate is simply supported (Figure 6.3) and subjected to a uniform pressure $p_0 = 2$ MPa. Based on a factor of safety of 1.5, calculate the minimum side length L_{AB} of the plate and the maximum deflection prior to yielding.

6.8 Resolve Problem 6.7, if the plate is clamped at edges. Apply the second of Equations 6.11.

6.9 A clamped elliptical steel plate is subjected to a uniform pressure $p = 5$ MPa (Figure 6.4). Knowing that $\sigma_{all} = 120$ MPa, $a = 0.3$ m, and $b = 0.2$ m, calculate: (*a*) the thickness; (*b*) the maximum deflection.

6.10 Given a rectangular plate (Figure P6.10a) and an elliptical plate (Figure P6.10b), both simply supported at the edge and under auniform pressure of intensity p_0, find: (*a*) the ratio of the maximum deflection and the maximum bending moment for the rectangular plate to those for the elliptical plate, using $\nu = 0.3$; (*b*) the pressure p_0 that is required to initiate the yielding in each plate, for $c = 220$ mm, $t = 10$ mm, and the tensile yield strength of 270 MPa.

6.11 A semiannular simply supported panel of a flight structure is under a uniform pressure p_0. Determine the deflection of the panel, by retaining only the first term of the series solution, at a point $P(3a/4, \pi/2)$. Compare the result with that given in Table 6.2.

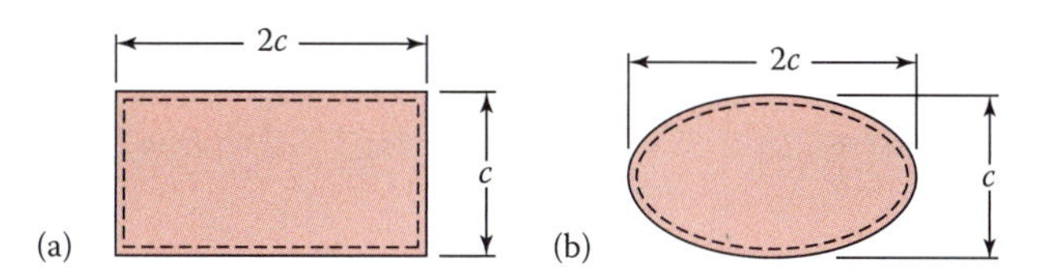

FIGURE P6.10

Sections 6.4 through 6.6

6.12 A manhole cover consists of a cast-iron elliptical plate 1.0 m long, 0.5 m wide, and 0.02 m thick. Determine the maximum uniform loading the plate can carry when the allowable stress is limited to 20 MPa. Use $\nu = 0.3$. Assume that (*a*) the plate is simply supported at the edge and (*b*) the plate is built in at the edge.

6.13 Consider two plates, one having a circular shape of radius c, the other an elliptical shape with semiaxes $a = 2c$ and $b = c$, both clamped at the edge and under a uniform load of intensity p_0. Determine the ratio of the maximum deflection and the maximum bending moment for the elliptical plate to those for the circular plate.

6.14 An elliptical plate built in at the edge is subjected to a linearly varying pressure $p = p_0 x$ (Figure 6.4). Given: An expression for the deflection of the form

$$w = \frac{p_0 x}{24D} \frac{[1-(x/a)^2-(y/b)^2]^2}{(5/a^4)+(1/b^4)+(2/a^2b^2)} \tag{P6.14}$$

Determine whether the above represents a possible solution. By means of Equation P6.14, obtain the bending moment M_x and the stress σ_x(at $x = a$, $y = 0$) for $a = 2b$.

6.15 Redo Problem 6.13 if the edges of the plates are simply supported. Take $\nu = 0.3$.

References

1. A. Nadai, *Die elastichen Platten*, Springer, Berlin, 2014.
2. S. Timoshenko and S. Woinowsky-Krieger, *Theory of Plates and Shells*, 2nd ed., McGraw-Hill, New York, 1959.
3. R. Szilard, *Theory and Analysis of Plates—Classical and Numerical Methods*, Prentice-Hall, Upper Saddle River, NJ, 2004.
4. V. D. Pilkey, *Peterson's Stress Concentration Factors*, 3rd ed., Wiley, Hoboken, NJ, 2008.
5. W. C. Young, R. C. Budynas, and A. Sadegh, *Roark's Formulas for Stress and Strain*, 8th ed., McGraw-Hill, New York, 2011.
6. A. C. Ugural, *Mechanical Design of Machine Components*, 2nd ed., CRC Press, Boca Raton, FL, 2015.
7. J. H. Faupel and F. E. Fisher, *Engineering Design*, 2nd ed., Wiley, New York, 1981.
8. E. Reissner, The effect of transverse shear deformation on the bending of elastic plates, *Transactions ASME*, 67, A62–A77, 1945.
9. H. P. Neuber, *Kerbspannungslehre*, 3rd ed., Springer, Berlin, 1985.

7

Numerical Methods

7.1 Introduction

In the previous chapters, equilibrium and energy methods are employed in the solution of a number of beam- and plate-bending problems. In some cases, however, these analytical solutions are not always possible, and one must resort to approximate *numerical methods*. The use of numerical approaches enables the engineer to expand his or her ability to solve design problems of practical significance. Treated are real shapes and loadings, as distinct from the somewhat limited variety of shapes and loadings amenable to simple analytical solution.

Among the most important of the numerical approaches are the method of *finite differences* (Part A, Section 9.5) and the *finite element method* (FEM) (Part B, Sections 10.6, 13.13, 13.14, Appendix C). Both techniques eventually require the solution of a system of linear algebraic equations. Such calculations are commonly performed by means of a digital computer employing matrix methods. The literature associated with the analysis of beams, plates, and shells by numerical methods is voluminous [1–10].

It is to be noted that the finite difference method is simple, versatile, and suitable for computer and programmable desk calculator use and results in acceptable accuracy for most technical purposes, provided that a relatively fine mesh is used. On the other hand, finite element methods (FEMs) have proved to be extremely powerful and versatile tools for static and dynamic analysis of a wide variety of beam, plate, and shell problems. However, they require the use of electronic digital computers of considerable speed and storage capacity.

Part A: Finite Difference Analysis

7.2 Finite Differences

The method of finite differences replaces the plate differential equation and the expressions defining the boundary conditions with equivalent difference equations. The solution of the bending problem thus reduces to the simultaneous solution of a set of algebraic equations, written for every nodal point within the plate. This section is concerned with the fundamentals of finite differences.

The finite difference expressions may be obtained from the definition of the first derivative with respect to x of a continuous function $y = f(x)$ (Figure 7.1a):

$$\left(\frac{dy}{dx}\right)_n = \lim_{\Delta x \to 0} \frac{y_{n+1} - y_n}{\Delta x}$$

The subscript n denotes any arbitrary point on the curve. Over an increment $\Delta x = h$, the above expression represents an approximation to the derivative given by

$$\left(\frac{dy}{dx}\right)_n \approx \frac{\Delta y_n}{h} = \frac{y_{n+1} - y_n}{h}$$

Here Δy_n is called the *first forward difference* of y at point x_n,

$$\Delta y_n = y_{n+1} - y_n \approx h\left(\frac{dy}{dx}\right)_n \tag{7.1}$$

The *first backward difference* at n, ∇y_n, is

$$\nabla y_n = y_n - y_{n-1} \approx h\left(\frac{dy}{dx}\right)_n \tag{7.2}$$

The central differences contain nodal points symmetrically located with respect to x_n and often result in *more accurate* approximations than forward and backward differences.

The *first central difference* δy_n is thus

$$\delta y_n = \frac{1}{2}(y_{n+1} - y_{n-1}) \approx h\left(\frac{dy}{dx}\right)_n \tag{7.3}$$

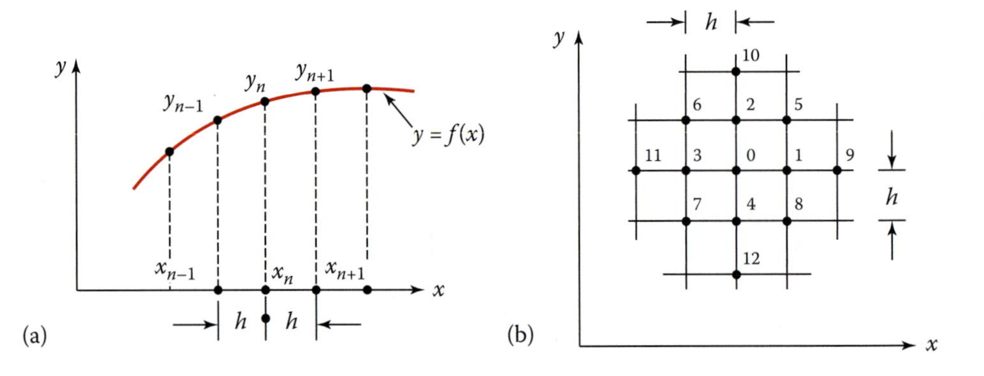

FIGURE 7.1
(a) Finite difference approximation of $f(x)$; (b) rectangular boundary divided into a square mesh.

A procedure identical to that used above will yield the higher-order derivatives. We shall hereafter consider only the central differences. The second derivative can be written using the difference representation of the first derivative

$$h^2\left(\frac{d^2y}{dx^2}\right)_n \approx \Delta(\nabla y_n) = \nabla(\Delta y_n) = \delta^2 y_n$$

The *second central difference* at x_n, after substitution of Equations 7.1 and 7.2 into the above, is

$$\begin{aligned}\delta^2 y_n &= \Delta y_n - \Delta y_{n-1} = (y_{n+1} - y_n) - (y_n - y_{n-1}) \\ &= y_{n+1} - 2y_n + y_{n-1} \approx h^2\left(\frac{d^2y}{dx^2}\right)_n\end{aligned} \tag{7.4}$$

The *third* and the *fourth central differences* are

$$\begin{aligned}\delta^3 y_n &= \delta(\delta^2 y_n) = \delta y_{n+1} - 2\delta y_n + \delta y_{n-1} \\ &= \frac{1}{2}(y_{n+2} - y_n) - (y_{n+1} - y_{n-1}) + \frac{1}{2}(y_n - y_{n-2}) \\ &= \frac{1}{2}(y_{n+2} - 2y_{n+1} + 2y_{n-1} - y_{n-2}) \approx h^3\left(\frac{d^3y}{dx^3}\right)_n\end{aligned} \tag{a}$$

and

$$\begin{aligned}\delta^4 y_n &= \delta^2(\delta^2 y_n) = \delta^2 y_{n+1} - 2\delta^2 y_n + \delta^2 y_{n-1} \\ &= (y_{n+2} - 2y_{n+1} + y_n) - 2(y_{n+1} - 2y_n + y_{n-1}) + (y_n - 2y_{n-1} + y_{n-2}) \\ &= y_{n+2} - 4y_{n+1} + 6y_n - 4y_{n-1} + y_{n-2} \approx h^4\left(\frac{d^4y}{dx^4}\right)_n\end{aligned} \tag{b}$$

Hereafter, the term *finite differences* will be used to refer to central differences.

We now discuss the case of the function $w\,(x, y)$ of two variables. For the purpose of illustration, consider a rectangular plate. By taking $\Delta x = \Delta y = h$, the plate is divided into a square *mesh* (Figure 7.1b). Equations 7.3 and 7.4 lead to

$$\frac{\partial w}{\partial x} \approx \frac{1}{h}\delta_x w \qquad \frac{\partial w}{\partial y} \approx \frac{1}{h}\delta_y w$$

$$\frac{\partial^2 w}{\partial x^2} \approx \frac{1}{h^2}\delta_x^2 w \qquad \frac{\partial^2 w}{\partial y^2} \approx \frac{1}{h^2}\delta_y^2 w \qquad \frac{\partial^2 w}{\partial x \partial y} \approx \frac{1}{h}\delta_x\left(\frac{\partial w}{\partial y}\right)$$

Here the subscripts x and y designate the directions in which the differences are taken. The above expressions, based on the definition of the partial derivatives, are written at a point 0 as follows:

$$\begin{aligned}\frac{\partial w}{\partial x} &\approx \frac{1}{2h}[w(x+h,y)-w(x-h,y)] = \frac{1}{2h}(w_1 - w_3)\\ \frac{\partial w}{\partial y} &= \frac{1}{2h}(w_2 - w_4)\end{aligned} \tag{c}$$

and

$$\begin{aligned}\frac{\partial^2 w}{\partial x^2} &\approx \frac{1}{h^2}[w(x+h,y)-2w(x,y)+w(x-h,y)] = \frac{1}{h^2}(w_1 - 2w_0 + w_3)\\ \frac{\partial^2 w}{\partial y^2} &\approx \frac{1}{h^2}(w - 2w_0 + w_4)\\ \frac{\partial^2 w}{\partial x \partial y} &\approx \frac{1}{h^2}\delta_x(\delta_y w) = \frac{1}{2^2}(\delta_x w_2 - \delta_x w_4) = \frac{1}{4h^2}(w_5 - w_6 + w_7 - w_8)\end{aligned} \tag{d}$$

The finite difference approximation of the Laplace operator at the point 0 is thus

$$\nabla^2 w = \frac{1}{h^2}(w_1 + w_2 + w_3 + w_4 - 4w_0) \tag{7.5}$$

The formulas for approximating the higher-order partial derivatives are developed by applying Equations a and b in the x and y directions, respectively:

$$\begin{aligned}\frac{\partial^3 w}{\partial x^3} &\approx \frac{1}{h^3}\delta_x^3 w = \frac{1}{2h^3}(w_9 - 2w_1 + 2w_3 - w_{11})\\ \frac{\partial^4 w}{\partial x^4} &\approx \frac{1}{h^4}\delta_x^2 w = \frac{1}{h^4}(w_9 - 4w_1 + 6w_0 - 4w_3 + w_{11})\\ \frac{\partial^3 w}{\partial y^3} &\approx \frac{1}{h^3}\delta_y^3 w = \frac{1}{2h^3}(w_{10} - 2w_2 + 2w_4 - w_{12})\\ \frac{\partial^4 w}{\partial y^4} &\approx \frac{1}{h^4}\delta_y^4 w = \frac{1}{h^4}(w_{10} - 4w_2 + 6w_0 - 4w_4 + w_{12})\end{aligned} \tag{e}$$

The mixed derivatives are also represented by

$$\begin{aligned}\frac{\partial^3 w}{\partial x \partial y^2} &\approx \frac{1}{b^3}\delta_x\left(\delta_y^2 w\right) = \frac{1}{b^3}(\delta_x w_2 - 2\delta_x w_0 + \delta_x w_4)\\ &= \frac{1}{2b^3}(w_5 - w_6 - 2w_1 + 2w_3 + w_8 - w_7)\\ \frac{\partial^3 w}{\partial x^2 \partial y} &\approx \frac{1}{b^3}\delta_y\left(\delta_x^2 w\right) = \frac{1}{2b^3}(w_5 + w_6 - 2w_2 + 2w_4 - w_8 - w_7)\\ \frac{\partial^4 w}{\partial x^2 \partial y^2} &\approx \frac{1}{b^4}\delta_x^2\left(\delta_y^2 w\right) = \frac{1}{b^4}[w_5 + w_6 + w_7 + w_8 + 4w_0 - 2(w_1 + w_2 + w_3 + w_4)]\end{aligned} \tag{f}$$

Having available the various derivatives as difference approximations, we can readily obtain the finite difference equivalent of the plate equations. For reference purposes, some useful finite difference operators are represented in the form of *coefficient patterns* in Table 7.1. It is noted that the *center point* in each case is the node about which each operator is written.

Difference operators in Cartesian coordinates x and y are well adapted to the solution of problems involving rectangular domains. When the plate has a curved or irregular boundary, special operators must be used at nodal points adjacent to the boundary (Section 7.4). One of the non-Cartesian meshes most commonly employed to cover the polygonal and irregular boundaries is the *triangular mesh* (Section 7.6). If the plate shape is a parallelogram, it may often be more accurate and convenient to use coordinates parallel to the edges of the plate, *skew coordinates*. The *polar mesh* is used in connection with shapes having some degree of axisymmetry (Section 7.5). The finite difference operators in any coordinate set are developed through transformation of the equations that relate the x and y coordinates to that set. In all cases, the procedure for determining the deflections and the moments is the same as described in the following section.

TABLE 7.1

Coefficient Patterns for Some Finite Difference Operators

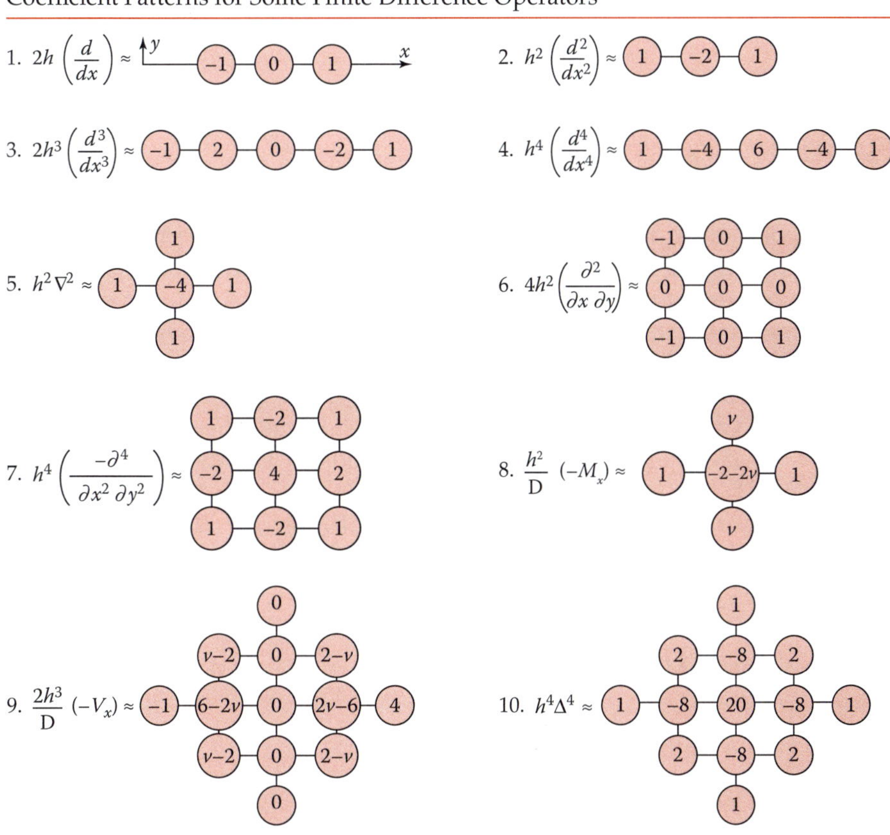

7.3 Solution of the Finite Difference Equations

We are now in a position to transform the differential equation of the bent plate into an algebraic equation. Let us write this equation for an interior point, such as 0 in Figure 7.1b. Referring to the operator ∇^4 of Table 7.1, we find that the difference equation corresponding to Equation 3.17 is

$$\begin{aligned}&[w_9+w_{10}+w_{11}+w_{12}+2(w_5+w_6+w_7+w_8)\\&\quad-8(w_1+w_2+w_3+w_4)+20w_0]\frac{1}{h^4}=\frac{p_0}{D}\end{aligned} \tag{7.6}$$

Similar expressions are written for every nodal point within the plate. At the same time, the boundary conditions must also be converted into finite difference form. The set of difference equations is then solved to yield the deflection.

As an alternate approach to the plate-bending problem, it is shown in Section 3.8 that Equations 3.17 can be replaced by two second-order equations (Equations 3.20 and 3.21). The application of the operator ∇^2 of Table 7.1 to the latter equations at point 0 leads to

$$(M_1+M_2+M_3+M_4-4M_0)\frac{1}{h^2}=-p_0 \tag{7.7}$$

$$(w_1+w_2+w_3+w_4-4w_0)\frac{1}{h^2}=-\frac{M_0}{D} \tag{7.8}$$

Identical equations are written for the remaining nodes within the plate. Solution of the problem now requires the determination of those values of M and w satisfying the system of algebraic equations and the given boundary conditions. In the case of a *simply supported* plate, M and w are zero at the boundary, and we can solve the first set of equations *independently* of the second system to find all the values of M within the boundary. The second set is then solved. For plates with edges *clamped or free* or for *mixed* boundary conditions, it is necessary to solve both sets *simultaneously*. Note that for plates with mixed boundary conditions, the values of M may be different at the edges. The deflection w may then be obtained more advantageously by direct application of Equation 7.6 instead of from Equations 7.7 and 7.8.

With the values of M and w available at all nodes, we can derive expressions for the moments and shear forces from Equations 3.9 and b of Section 3.8. These are, at point 0,

$$\begin{aligned}M_x&=\frac{D}{h^2}[2w_0-w_1-w_3+\nu(2w_0-w_2-w_4)]\\M_y&=\frac{D}{h^2}[2w_0-w_2-w_4+\nu(2w_0-w_1-w_3)]\\M_{xy}&=\frac{D(1-\nu)}{4h^2}(-w_5+w_6+w_7+w_8)\end{aligned} \tag{7.9}$$

and

$$Q_x = \frac{1}{2h}(M_1 - M_3) \qquad Q_y = \frac{1}{2h}(M_2 - M_4) \tag{7.10}$$

The stresses are now readily determined through the application of Equations 3.11 and 3.32.

7.3.1 Load Representation

The factors serving to complicate the analysis of real problems may generally be reduced to irregularities in the shape of the member and nonuniformity in the applied load. Situations involving irregular boundaries are treated in Section 7.4. By replacing the actual configuration of load with suitable approximations, frequently very little sacrifice in accuracy is encountered.

Consider the case of a nonuniformly loaded plate (or beam). There are a number of ways of determining an equivalent load. A simple approach to this problem is to regard the load as *stepped;* that is, to replace the loading with a series of uniformly distributed loads. Figure 7.2a shows such a representation in the xz plane of the plate. Note that this method yields good results only if the mesh width h is relatively small and the variation of the load in terms of load intensity between mesh points is not excessive. *Concentrated* loads may also be represented by stepped distribution in a similar manner.

A better accuracy, however, is obtained by the use of an equivalent loading that *varies linearly* from nodal point to nodal point, thus avoiding abrupt changes in load (Figure 7.2b). In any method, the *nodal loads* attributable to variable loading may readily be determined by evaluating the static resultants. By *straight-line* approximation (Figure 7.2b) for square mesh, it can be shown [8] that the value of the nodal load at the node 0 is

$$(p)_0 = \frac{1}{36}[p_5 + p_6 + p_7 + p_8 + 4(p_1 + p_2 + p_3 + p_4) + 16p_0] \tag{7.11}$$

Identical expressions are written for every mesh points within the plate. The coefficient pattern of nodal load is given in Figure 7.3.

The finite difference method can best be illustrated by reference to numerical examples.

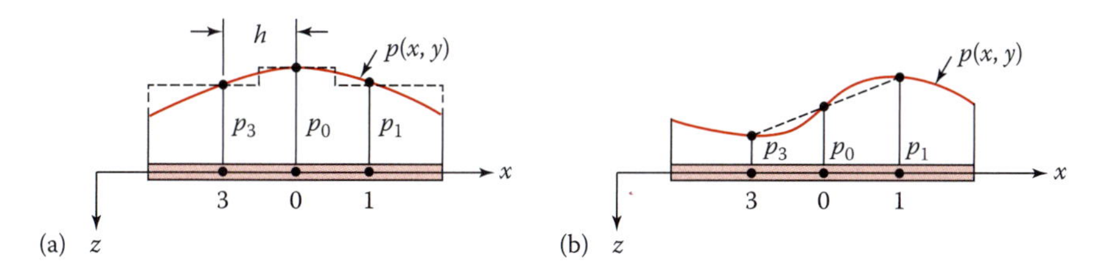

FIGURE 7.2
Distributed load: (a) Stepped representation; (b) straight-line representation.

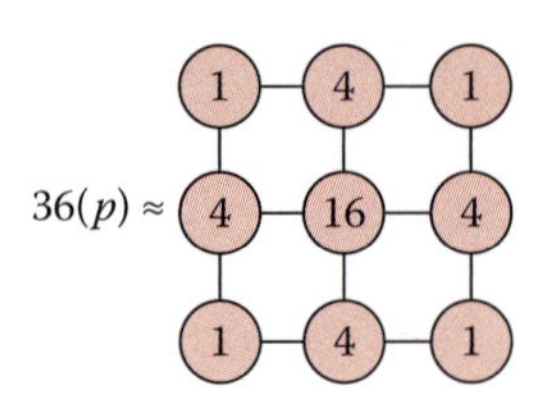

FIGURE 7.3
Coefficient pattern of straight-line averaging of distributed load.

EXAMPLE 7.1: Analysis of Plate under Constant Load

Use finite difference techniques to analyze the bending of a square plate ($a \times a$) with simply supported edges subjected to a uniformly distributed load of p_0.

Solution

The domain is divided into a number of small squares, 16, for example. We thus have $h = a/4$. In labeling nodal points, it is important to take into account any conditions of symmetry that may exist. This has been done in Figure 7.4a. It is seen that only one-eighth of the surface need then be considered, this indicated in the figure by darker shaded area. Form the boundary conditions at points located on the boundary, M and w are zero. Application of the operator ∇^2 of Table 7.1 to $\nabla^2 M = -p$ at the nodes 1, 2, and 3 results in

$$2M_2 - 4M_1 = -p_0h^2$$
$$2M_1 + M_3 - 4M_2 = -p_0h^2$$
$$4M_2 - 4M_3 = -p_0h^2$$

Simultaneous solution yields

$$M_1 = \frac{11}{16}p_0h^2 \qquad M_2 = \frac{7}{8}p_0h^2 \qquad M_3 = \frac{9}{8}p_0h^2 \tag{a}$$

Similarly, $\Delta^2 w = -M/D$ together with Equations a leads to

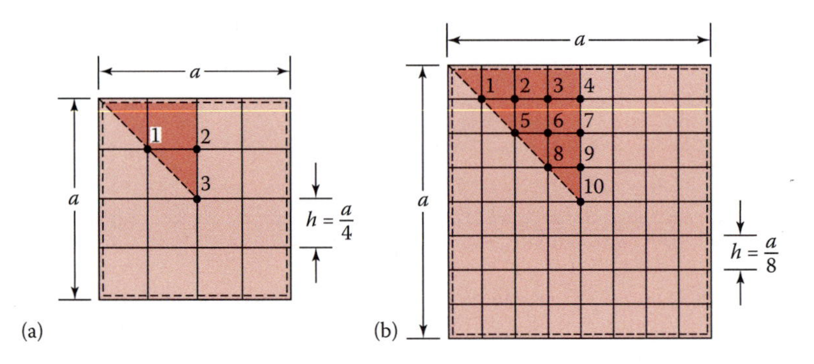

FIGURE 7.4
Simply supported plate with uniform load.

$$2w_2 - 4w_1 = -\frac{11p_0h^4}{16D}$$

$$2w_1 + w_3 - 4w_2 = -\frac{7p_0h^4}{8D}$$

$$4w_2 - 4w_3 = -\frac{9p_0h^4}{8D}$$

From the preceding, after setting $h = a/4$, we obtain

$$w_1 = 0.00214\frac{p_0a^4}{D} \qquad w_2 = 0.00293\frac{p_0a^4}{D} \qquad w_3 = 0.00403\frac{p_0a^4}{D}$$

Deflection w_3 at the center of the plate is 0.79% less than the "exact" value given in Example 5.1. The bending moment at the center of the plate is, from Equations 3.19 and a for $\nu = 0.3$,

$$M_x = M_y = (1+\nu)\frac{M_3}{2} = 0.0457p_0a^2$$

This value is approximately 4.5% less than the "exact" value, $0.0479p_0a^2$.

By means of a finer network, we expect to improve the results. Let us subdivide the domain into 64 small squares, each with $h = a/8$. Taking into account the symmetry, we number the nodal points as shown in Figure 7.4b. The values of M and w are zero on the boundary. Writing the finite difference equations at points 1 through 10, we have 20 simultaneous equations for the 20 unknowns M and w at the inner nodal points. Solving these, we obtain

$$\begin{aligned}
M_1 &= 0.01778p_0a^2 & M_6 &= 0.05377p_0a^2\\
M_2 &= 0.02774p_0a^2 & M_7 &= 0.05664p_0a^2\\
M_3 &= 0.03291p_0a^2 & M_8 &= 0.06523p_0a^2\\
M_4 &= 0.03452p_0a^2 & M_9 &= 0.06888p_0a^2\\
M_5 &= 0.04466p_0a^2 & M_{10} &= 0.07278p_0a^2
\end{aligned}$$

and

$$\begin{aligned}
w_1 &= 0.000663p_0a^4/D & w_6 &= 0.002733p_0a^4/D\\
w_2 &= 0.001186p_0a^4/D & w_7 &= 0.002937p_0a^4/D\\
w_3 &= 0.001515p_0a^4/D & w_8 &= 0.003507p_0a^4/D\\
w_4 &= 0.001627p_0a^4/D & w_9 &= 0.003770p_0a^4/D\\
w_5 &= 0.002134p_0a^4/D & w_6 &= 0.004055p_0a^4/D
\end{aligned}$$

The deflection at the center w_{10} is now 0.12% less than the exact value. For the center, $M_x = M_y = 0.0473p_0a^2$, which is 1.22% less than the "exact" value. Thus, for the situation described, a small number of subdivisions of the plate yields results of acceptable accuracy for practical applications. On the basis of the results for $h = a/4$ and $a/8$, a still better approximation can be obtained by applying extrapolation approaches or by reducing the net size h further.

To calculate the moments at any other location, we apply finite difference form of Equations 3.9. For example, at point 7, for $v = 0.3$, we have

$$(M_x)_7 = D[2w_7 - 2w_6 + \nu(2w_7 - w_4 - w_9)]\frac{64}{a^2} = 0.0353p_0a^2$$

$$(M_y)_7 = D[2w_7 - w_4 - w_9 + \nu(2w_7 - 2w_6)]\frac{64}{a^2} = 0.0384p_0a^2$$

In a like manner, the twisting moment at point 6 is found to be

$$(M_{xy})_6 = D(1-\nu)(w_2 - w_4 - w_6 + w_9)\frac{16}{a^2} = 0.00668p_0a^2$$

We can obtain the shear forces at node 6, referring to Equation 7.10, as follows:

$$(Q_x) = (M_7 - M_5)\frac{4}{a} = 0.0479p_0a$$

$$(Q_y) = (M_3 - M_8)\frac{4}{a} = 0.1293p_0a$$

The support reactions are computed similarly through the use of Equations 3.23 and 3.24.

EXAMPLE 7.2: Deflection of Plate Carrying Variable Load

A simply supported square plate is subjected to a distributed load in the form of a *triangular prism* (Figure 7.5). (a) Use a *stepped load* representation (Figure 7.2a) to ascertain the plate deflections. Take $h = a/4$. (b) Calculate center deflection for the given data: $a = 0.3$ m, $t = 12$ mm, $E = 190$ GPa, $\nu = 0.3$, $p_0 = 10$ MPa.

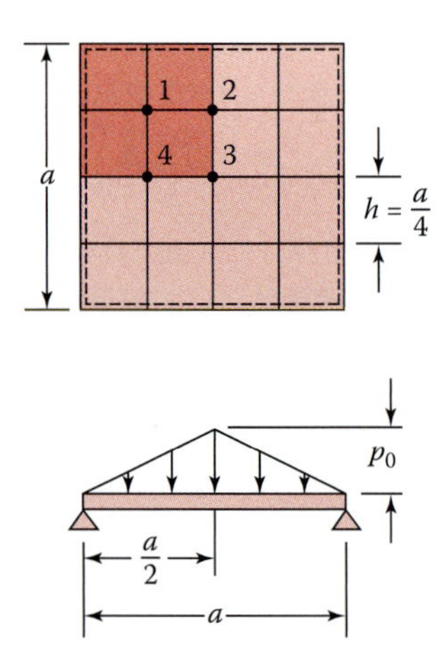

FIGURE 7.5
Simply supported plate with triangular prism loading.

Solution

For this case, the nodal forces are $p_1 = p_4 = p_0/2$ and $p_2 = p_3 = p_0$.

a. Because of symmetry, only a quarter of the section need be considered. As before, applying the operator ∇^2 of Table 7.1 to $\nabla^2 M = -p$ at points 1, 2, 3, and 4, we have

$$M_2 + M_4 - 4M_1 = -\frac{1}{2}p_0h^2$$
$$2M_1 + M_3 - 4M_2 = -p_0h^2$$
$$2M_2 + 2M_4 - 4M_3 = -p_0h^2$$
$$2M_1 + M_3 - 4M_4 = -\frac{1}{2}p_0h^2$$

By solving these equations, we find that

$$M_1 = \frac{7}{16}p_0h^2 \qquad M_2 = \frac{11}{16}p_0h^2$$
$$M_3 = \frac{7}{8}p_0h^2 \qquad M_4 = \frac{9}{16}p_0h^2$$

In a like manner, the use of $\nabla^2 w = -M/D$ and the preceding values gives

$$w_2 + w_4 - 4w_1 = -\frac{7}{16}p_0h^4$$
$$2w_1 + w_3 - 4w_2 = -\frac{11}{6}p_0h^4$$
$$2w_2 + 2w_4 - 4w_3 = -\frac{7}{8}p_0h^4$$
$$2w_1 + w_3 - 4w_4 = -\frac{9}{16}p_0h^4$$

from which, after setting $h = a/4$, it follows that

$$w_1 = 0.00148\frac{p_0a^4}{D} \qquad w_2 = 0.00214\frac{p_0a^4}{D}$$
$$w_3 = 0.00293\frac{p_0a^4}{D} \qquad w_4 = 0.00202\frac{p_0a^4}{D}$$

The center deflection w_3 is approximately 11.5% more than the "exact" value [10], $0.00263p_0a^4/D$.

Interestingly, with *straight-line* representation (Figure 7.2b), applying the operator of Figure 7.3 at nodes 1, 2, 3, and 4, respectively, we obtain

$$(p)_1 = (p)_4 = \frac{1}{2}p_0 \qquad (p)_2 = (p)_3 = \frac{5}{16}p_0$$

Use of these values of the nodal forces for the same mesh size (Problem 7.5) leads to $w_3 = 0.00262p_0a^4/D$.

b. Introducing the given numerical values into Equation 3.10, we obtain

$$D = \frac{Et^3}{12(1-\nu^2)} = \frac{190(12^3)}{12(0.91)} = 30.066 \text{ kN} \cdot \text{m}$$

The value of center deflection is therefore

$$w_3 = 0.00293 \frac{p_0 a^4}{D} = 0.00293 \frac{10(10^3)(0.3^4)}{30.066} = 7.89 \text{ mm}$$

EXAMPLE 7.3: Deflection and Moment at Various Points of Plate

Determine the deflection and moment at various points of a clamped-edge square plate of sides a under a uniformly distributed load p_0 per unit surface area. Take $h = a/4$ and employ two approaches of solution: (a) application of Equations 3.20 and 3.21; (b) application of Equation 3.17.

Solution

The boundary conditions are (Figure 7.6)

$$w = 0 \qquad \frac{\partial w}{\partial x} = \frac{\partial w}{\partial y} = 0 \qquad \left(x = y = \pm\frac{a}{2}\right) \tag{b}$$

That is, the deflection and slopes are zero at the points on the border. Referring to Equation 7.3, for the first central difference, we obtain at the boundary

$$2hw'_n = w_{n+1} - w_{n-1} = 0 \tag{c}$$

where the prime denotes the first derivative with respect to x or y. Equations b and c indicate that the values of w (and M) at the nodes immediately outside the boundary are equal to the values of w (and M) at the points immediately inside the boundary along the same normal. Because of symmetry, only the triangular darker shaded area is considered, with the corresponding labeling of nodes shown in Figure 7.6.

a. Applying the operator ∇^2 of Table 7.1 to $\nabla^2 M = -p_0$ at points 1, 2, and 3, respectively, we have

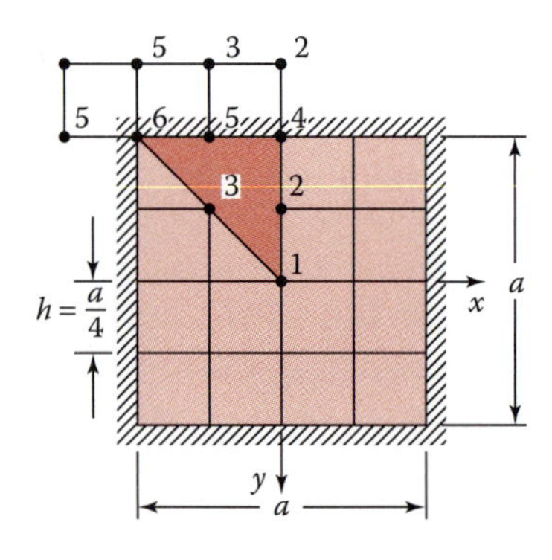

FIGURE 7.6
Clamped plate with uniform load.

$$
\begin{aligned}
4M_2 - 4M_1 &= -p_0h^2 \\
2M_3 + M_1 + M_4 - 4M_2 &= -p_0h^2 \\
2M_5 + 2M_2 - 4M_3 &= -p_0h^2
\end{aligned}
$$

and, referring to $\nabla^2 w = -M/D$,

$$
\begin{aligned}
4w_2 - 4w_1 &= -M_1h^2/D \\
2w_3 + w_1 - 4w_2 &= -M_2h^2/D \\
2w_2 - 4w_3 &= -M_3h^2/D \\
2w_2 &= -M_4h^2/D \\
2w_3 &= -M_5h^2/D \\
0 &= M_6
\end{aligned}
$$

Solving the above expressions simultaneously, substituting $h = a/4$, we obtain

$$
\begin{aligned}
&M_1 = 0.03792p_0a^2 \qquad && M_1 = 0.00180p_0a^4/D \\
&M_2 = 0.02230p_0a^2 && M_1 = 0.00121p_0a^4/D \\
&M_3 = 0.01369p_0a^2 && M_1 = 0.00082p_0a^4/D \\
&M_4 = 0.03862p_0a^2 && \\
&M_5 = 0.026159p_0a^2 && \\
&M_6 = 0 &&
\end{aligned}
\tag{d}
$$

It follows from Equation 3.19 that, for $\nu = 0.3$,

$$
\begin{aligned}
(M_x)_1 &= (M_y)_1 = 0.02465p_0a^2 \\
(M_x)_4 &= (M_y)_4 = -0.02510p_0a^2
\end{aligned}
$$

Comments: Compared with the series solution for this problem, the calculated deflection w_1 is 43% higher, the numerical value of moments at point 1 is about 6% higher, while at point 4 it is about 25% lower. Similar calculations conducted for $n = 8$ result in a center deflection $w_1 = 0.00143p_0a^4/D$, about 13% higher than the series solution.

b. Application of the operator ∇^4 of Table 7.1 to $\nabla^4 w = p_0/D$ at nodes 1, 2, and 3, respectively, leads to

$$
\begin{aligned}
-32w_2 + 8w_3 + 20w_1 &= \frac{p_0}{D}\left(\frac{a}{4}\right)^4 \\
-8w_1 - 16w_3 + 26w_2 &= \frac{p_0}{D}\left(\frac{a}{4}\right)^4 \\
2w_1 - 16w_2 + 24w_3 &= \frac{p_0}{D}\left(\frac{a}{4}\right)^4
\end{aligned}
$$

Solution of the above set yields values of w_1, w_2, and w_3 equal to those given by Equations d. The accuracy may be improved by increasing the number of nodes or by applying an extrapolation procedure.

EXAMPLE 7.4: Deflection of Variously Supported Plate

A stockroom floor, half of which is carrying a uniform loading, is approximated by a continuous plate with opposite edges $y = \pm a/2$ clamped and remaining edges and midspan simply supported (Figure 7.7a). We wish to obtain deflections at points 1, 2, and 3 (Figure 7.7b).

Solution

The boundary conditions are

$$\begin{aligned} &w = 0 & & (x = 0, x = a, x = 2a) \\ & & \frac{\partial^2 w}{\partial x^2} = 0 & \quad (x = 0, x = 2a) \\ &w = 0 & \frac{\partial w}{\partial y} = 0 & \quad \left(y = \frac{a}{2}, y = -\frac{a}{2}\right) \end{aligned} \tag{e}$$

Taking into account symmetry of the solution about the x axis and satisfying the finite difference expressions (Section 7.2) corresponding to Equations e, we shall number the nodes as shown in the figure. The operator ∇^4 (Table 7.1) is applied to $\nabla^4 w = p_0/D$ at points 1, 2, 3, and 4, respectively, to yield

$$\begin{aligned} 13w_1 - 6w_2 + w_4 &= p_0 h^4/D \\ 12w_2 - 6w_1 &= p_0 h^4/D \\ 12w_3 - 6w_4 &= 0 \\ 12w_4 - 6w_3 + w_1 &= 0 \end{aligned}$$

Simultaneous solution of the foregoing, after setting $h = a/3$, results in

$$\begin{aligned} w_1 &= 0.00187 p_0 a^2/D & \qquad w_3 &= -0.0009 p_0 a^4/D \\ w_2 &= 0.00196 p_0 a^4/D & \qquad w_4 &= -0.00019 p_0 a^4/D \end{aligned}$$

By decreasing the size of the mesh, the accuracy of the solution can be improved.

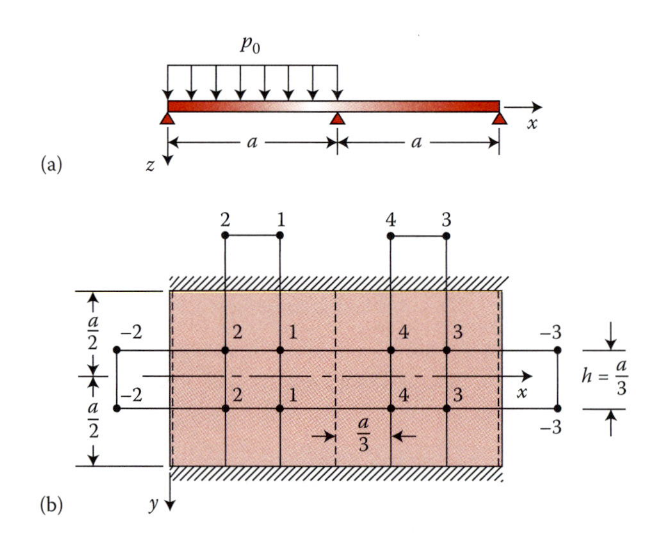

FIGURE 7.7
Continuous plate with partial load.

EXAMPLE 7.5: Analysis of Plate with Various Supports

Consider the case of a uniformly loaded rectangular plate with two adjacent edges simply supported, the third edge free, and the fourth edge built in (Figure 7.8a). Use the finite difference method with $h = a/4$ to obtain w at various points.

Solution

The domain is divided into 12 squares of sides $a/4$ (Figure 7.8b). At the nodes of the simply supported and the built-in edges, we have $w = 0$. The finite difference formula corresponding to Equation 3.17 is applied to all interior nodes and to nodes 16 and 24 on the free edge, where the deflections are to be found. This application gives *eight* equations involving *36* nodes, indicated in the mesh on the figure. The boundary conditions for each edge, replaced by the finite difference forms, are listed in Table 7.2. It is seen from the table that all edge conditions are represented by a total of *28* expressions. We thus have a total of 36 independent equations in terms of the 36 values of the deflections corresponding to each node.

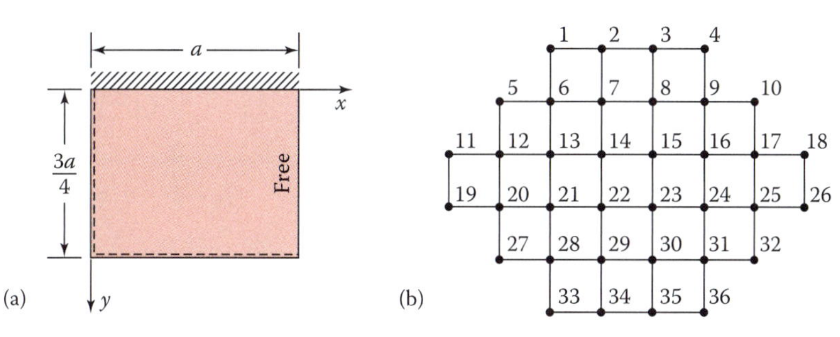

FIGURE 7.8
Variously supported plate with uniform load.

TABLE 7.2
Boundary Conditions for the Plate Shown in Figure 7.8

Edge(s)	Conditions
A. Clamped at $y = 0$	$w = 0$: $w_5 = w_6 = w_7 = w_8 = w_9 = 0$ $\partial w/\partial y = 0$: $w_{13} = w_1$, $w_{14} = w_2$, $w_{15} = w_3$, $w_{16} = w_4$ Continuity of zero curvature ($\partial^2 w/\partial x^2 = 0$) along x axis: $w_8 = -w_{10} = 0$
B. Simply supported at $x = 0$ and $y = 3a/4$	$w = 0$: $w_{12} = w_{20} = w_{27} = w_{28} = w_{29} = w_{30} = w_{31} = 0$ $\partial^2 w/\partial x^2 = 0$: $w_{13} = -w_{11}$, $w_{21} = -w_{19}$ $\partial^2 w/\partial y^2 = 0$: $w_{21} = -w_{33}$, $w_{22} = -w_{34}$, $w_{23} = -w_{35}$, $w_{24} = -w_{36}$ Continuity of zero curvature ($\partial^2 w/\partial x^2 = 0$) along edge $y = 3a/4$: $w_{30} = -w_{32} = 0$
C. Free at $x = a$	$V_x = 0$, at nodes 16 and 24, respectively: $w_{18} - w_{14} + (\nu - 2)(w_8 - w_{25} + w_{23} - w_{10}) + (6 - 2\nu)(w_{15} - w_{17}) = 0$ $w_{26} - w_{22} + (\nu - 2)(w_{15} + w_{30} - w_{17} - w_{32}) + (6 - 2\nu)(w_{23} - w_{25}) = 0$ $M_x = 0$, at nodes 16 and 24, respectively: $w_{15} + w_{17} + \nu(w_9 + w_{24}) - 2(1 + \nu)w_{16} = 0$ $w_{23} + w_{25} + \nu(w_{16} + w_{31}) - 2(1 + \nu)w_{24} = 0$

Writing the finite difference expressions for Equation 3.17 by applying the operator ∇^4 of Table 7.1 at nodes 13, 14, 15, 16, 21, 22, 23, and 24 and inserting the ws given in cases A and B of Table 7.2, we have eight simultaneous equations containing 12 unknown values of w at nodal points 13 through 18 and 21 through 26. Case C of Table 7.2 provides four more equations in terms of the same unknown values of w. The resulting 12 independent expressions are represented in the following matrix form (for $\nu = 0.3$):

$$
\begin{bmatrix}
20 & -8 & 1 & 0 & 0 & 0 & -8 & 2 & 0 & 0 & 0 & 0 \\
-8 & 21 & -8 & 1 & 0 & 0 & 2 & -8 & 2 & 0 & 0 & 0 \\
1 & -8 & 21 & -8 & 1 & 0 & 0 & 2 & -8 & 2 & 0 & 0 \\
0 & 1 & -8 & 21 & -8 & 1 & 0 & 0 & 2 & -8 & 2 & 0 \\
-8 & 2 & 0 & 0 & 0 & 0 & 18 & -8 & 1 & 0 & 0 & 0 \\
2 & -8 & 2 & 0 & 0 & 0 & -8 & 19 & -8 & 1 & 0 & 0 \\
0 & 2 & -8 & 2 & 0 & 0 & 1 & -8 & 19 & -8 & 1 & 0 \\
0 & 0 & 2 & -8 & 2 & 0 & 0 & 1 & -8 & 19 & -8 & 1 \\
0 & -1 & 5.4 & 0 & -5.4 & 1 & 0 & 0 & -1.7 & 0 & 1.7 & 0 \\
0 & 0 & -1.7 & 0 & 1.7 & 0 & 0 & -1 & 5.4 & 0 & -5.4 & 1 \\
0 & 0 & 1 & -2.6 & 1 & 0 & 0 & 0 & 0 & 0.3 & 0 & 0 \\
0 & 0 & 0 & 0.3 & 0 & 0 & 0 & 0 & 1 & -2.6 & 1 & 0
\end{bmatrix}
\begin{Bmatrix}
w_{13} \\ w_{14} \\ w_{15} \\ w_{16} \\ w_{17} \\ w_{18} \\ w_{21} \\ w_{22} \\ w_{23} \\ w_{24} \\ w_{25} \\ w_{26}
\end{Bmatrix}
= \frac{p_0 h^4}{D}
\begin{Bmatrix}
1 \\ 1 \\ 1 \\ 1 \\ 1 \\ 1 \\ 1 \\ 1 \\ 0 \\ 0 \\ 0 \\ 0
\end{Bmatrix}
$$

The above equations are solved to yield

$$
\begin{aligned}
&w_{13} = 0.25819N && w_{14} = 0.38943N && w_{15} = 0.45037N \\
&w_{16} = 0.51951N && w_{17} = 0.70839N && w_{18} = 1.07433N \\
&w_{21} = 0.30383N && w_{22} = 0.46598N && w_{23} = 0.54558N \\
&w_{24} = 0.63989N && w_{25} = 0.96226N && w_{26} = 2.27756N
\end{aligned}
$$

where $N = p_0 h^4/D$ and $h = a/4$. From these values, we determine the bending moment at any node in the plate as illustrated in the previous examples.

7.4 *Plates with Curved Boundaries

We now treat the bending problems of *simply supported plates* having *curved* or *irregular boundaries*. Dividing a portion of such a boundary into a square mesh (Figure 7.9a) shows that the ∇^2 operator (Table 7.1) does not apply to point 0 because of the unequal lengths of arms 01, 02, 03, and 04. When at least one of the arms is not equal to h, the operator pattern is referred to as an *irregular star*. There are available various approaches to such situations. The one that will be discussed assumes that it is required to develop an irregular star using the actual boundary points, rather than those falling "outside," associated with the continued regular mesh.

It is assumed that in the region immediately surrounding the node 0, the expression for w can be approximated by

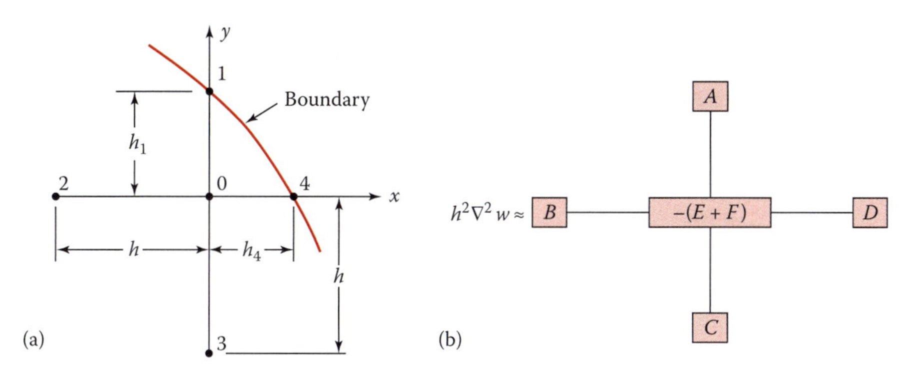

FIGURE 7.9
(a) Curved boundary; (b) coefficient pattern of the finite difference operator for $\nabla^2 w$ for curved boundary.

$$w(x, y) = w_0 + a_1 x + a_2 y + a_3 x^2 + a_4 y^2 \tag{a}$$

This equation, referring to Figure 7.9a, gives the following expressions for w at points 1, 2, 3, and 4:

$$\begin{aligned} w_1 &= w_0 + a_2 h_1 + a_4 h_1^2 \\ w_2 &= w_0 - a_1 h + a_3 h^2 \\ w_3 &= w_0 - a_2 h + a_4 h^2 \\ w_4 &= w_0 + w_1 h_4 + a_3 h_4^2 \end{aligned}$$

from which

$$\begin{aligned} a_1 &= \frac{b^2(w_4 - w_0) + b_4^2(w_0 - w_2)}{bb_4(b + b_4)} \qquad & a_2 &= \frac{b^2(w_1 - w_0) + b^2(w_0 - w_3)}{bb_1(b + b_1)} \\ a_3 &= \frac{b(w_4 - w_0) - b_4(w_0 - w_2)}{bb_4(b + b_4)} \qquad & a_4 &= \frac{b(w_1 - w_0) + b_1(w_0 - w_3)}{bb_1(b + b_1)} \end{aligned} \tag{b}$$

At point 0 ($x = 0$, $y = 0$), Equation a gives

$$\left(\frac{\partial^2 w}{\partial x^2}\right)_0 = 2a_3 \qquad \left(\frac{\partial^2 w}{\partial y^2}\right)_0 = 2a_4 \tag{c}$$

Next, we introduce Equations b into Equations c. Adding the resulting expressions, the operator $\nabla^2 w$ is obtained in the form

$$h^2\left(\frac{\partial^2 w}{\partial x^2} + \frac{\partial^2 w}{\partial y^2}\right)_0 \approx \frac{2w_1}{\alpha_1(1+\alpha_1)} + \frac{2w_2}{1+\alpha_4} + \frac{2w_3}{1+\alpha_1} + \frac{2w_4}{\alpha_4(1+\alpha_4)} - \left(\frac{2}{\alpha_1} + \frac{2}{\alpha_4}\right) w_0 \tag{7.12a}$$

in which $\alpha_i = h_i/h$, $i = 1, 4$. For convenience, the above may be written as

$$h^2(\nabla^2 w)_0 \approx Aw_1 + Bw_2 + Cw_3 + Dw_4 - (E+F)w_0 \tag{7.12b}$$

The pattern associated with Equations 7.12 is shown in Figure 7.9b.

The case where all the arms h_i, $i = 1, 2, 3, 4$, are smaller than h may also be treated by following a procedure identical with that described in this section. In so doing, we obtain

$$h^2(\nabla^2 w)_0 \approx \frac{2w_1}{\alpha_1(\alpha_1+\alpha_3)} + \frac{2w_2}{\alpha_2(\alpha_2+\alpha_4)} + \frac{2w_3}{\alpha_3(\alpha_1+\alpha_3)} + \frac{2w_4}{\alpha_4(\alpha_2+\alpha_4)} - \left(\frac{2}{\alpha_1\alpha_3} + \frac{2}{\alpha_2\alpha_4}\right)w_0 \tag{7.13}$$

The finite difference equivalent forms of the Laplace operator, Equations 7.12 and 7.13, may be applied in the same manner as the standard form given by Equation 7.5.

For situations involving other than simply supported plates with curved boundaries, a somewhat different approach for approximating $\nabla^2 w$ is used. In any case, the development of the finite difference expression that replaces $\nabla^2 w$ near the curved boundary is not simple.

EXAMPLE 7.6: Elliptical Plate Analysis

Determine the deflection and the bending moments at the center of a simply supported elliptical steel plate under uniform load of intensity p_0 (Figure 7.10). Let $a = 0.15$ m, $b = 0.1$ m, and $h = 0.05$ m. Compare the results with those given in Table 6.1.

Solution

Because of the symmetry, it is sufficient to find a solution in one-fourth of the entire domain. The equation of the ellipse together with the given data yields $h_1 = 0.044$ m, $h_2 = 0.0245$ m, and $h_3 = 0.03$ m. At nodes 1, 2, 3, and 4 the standard finite difference expression applies, while at 5 and 6 we employ the operator of Figure 7.9b. Note that $w = 0$ and $M = 0$ at the boundary. We can thus write six finite difference equations corresponding to Equation 3.20, presented in matrix form as follows:

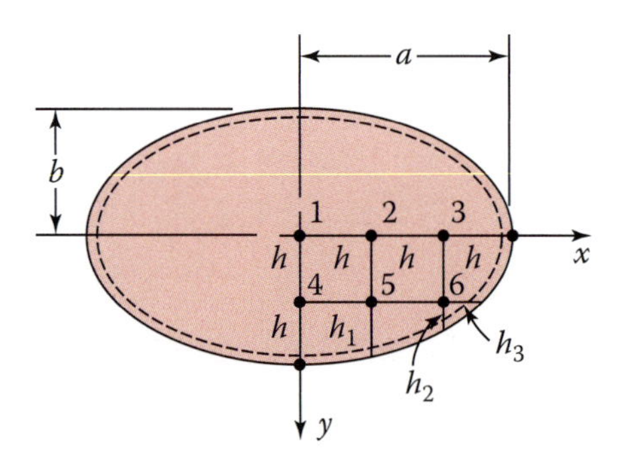

FIGURE 7.10
Simply supported elliptical plate with uniform load.

$$\begin{bmatrix} -4 & 2 & 0 & 2 & 0 & 0 \\ 1 & -4 & 1 & 0 & 2 & 0 \\ 0 & 1 & -4 & 0 & 0 & 2 \\ 1 & 0 & 0 & -4 & 2 & 0 \\ 0 & 1.06 & 0 & 1 & -4.26 & 1 \\ 0 & 0 & 1.34 & 0 & 1.25 & -7.41 \end{bmatrix} \begin{Bmatrix} M_1 \\ M_2 \\ M_3 \\ M_4 \\ M_5 \\ M_6 \end{Bmatrix} = -p_0 h^2 \begin{Bmatrix} 1 \\ 1 \\ 1 \\ 1 \\ 1 \\ 1 \end{Bmatrix} \tag{d}$$

which, when solved, yields

$$\begin{aligned} M_1 &= 1.384 p_0 h^2 \quad M_2 = 1.230 p_0 h^2 \quad M_3 = 0.769 p_0 h^2 \\ M_4 &= 1.038 p_0 h^2 \quad M_5 = 0.884 p_0 h^2 \quad M_6 = 0.423 p_0 h^2 \end{aligned} \tag{e}$$

Similarly, application of Equation 3.21 at nodes 1 through 6, together with the values of moment given by Equations e, results in six equations from which

$$\begin{aligned} w_1 &= 1.520 \frac{p_0 h^4}{D} \qquad w_2 = 1.282 \frac{p_0 h^4}{D} \qquad w_3 = 0.674 \frac{p_0 h^4}{D} \\ w_4 &= 1.066 \frac{p_0 h^4}{D} \qquad w_5 = 0.853 \frac{p_0 h^4}{D} \qquad w_6 = 0.323 \frac{p_0 h^4}{D} \end{aligned} \tag{f}$$

The bending moments at the center, by means of Equations f and 7.9, are therefore

$$\begin{aligned} M_x &= \frac{D}{h_2}[2w_1 - 2w_2 + \nu(2w_1 - 2w_4)] = 0.187 p_0 b^2 0 \\ M_y &= \frac{D}{h^2}[2w_1 - 2w_4 + \nu(2w_1 - 2w_2)] = 0.263 p_0 b^2 = M_{\max} \end{aligned} \tag{g}$$

The value of M_{max} is 18% less than the result listed in Table 6.1. The maximum deflection occurs at the center, and from Equations f, with $\nu = 0.3$, we obtain

$$w_1 = 1.038 \frac{p_0 b^4}{Et^3} \tag{h}$$

Comment: This is 17.6% less than the value furnished in Table 6.1.

7.5 *The Polar Mesh

It is convenient to employ a *polar mesh* (Figure 7.11a) to cover the domains of the circular plates. The Laplacian, in terms of the polar coordinates r and θ, is, from Section 4.2:

$$\nabla^2 w = \frac{\partial^2 w}{\partial r^2} + \frac{1}{r}\frac{\partial w}{\partial r} + \frac{1}{r^2}\frac{\partial^2 w}{\partial \theta^2} \tag{4.1}$$

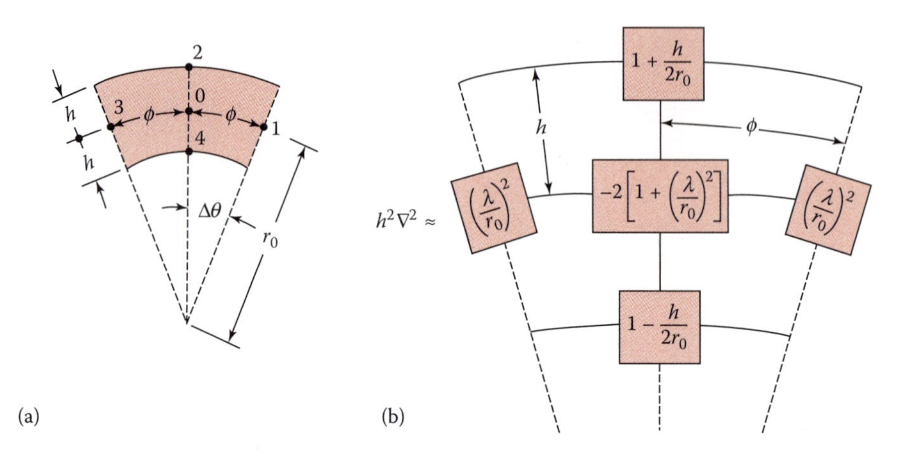

FIGURE 7.11
(a) Polar mesh; (b) coefficient pattern of the finite difference operator $\nabla^2 w$ in polar coordinates.

Using the notation of Figure 7.11a and by means of Equations 7.3 and 7.4, we have

$$\frac{\partial^2 w}{\partial r^2} \approx \frac{1}{h^2}(w_2 - 2w_0 + w_4) \qquad \frac{\partial w}{\partial r} \approx \frac{1}{2h}(w_2 - w_4)$$

$$\frac{\partial^2 w}{\partial \theta^2} \approx \frac{1}{\phi^2}(w_1 - 2w_0 + w_3)$$

where

$$h = \Delta r \qquad \phi = \Delta\theta$$

The pattern for ∇^2 in polar coordinates is presented by Figure 7.11b, in which $\lambda = h/\phi$.

Other operator pattern may be developed in a like manner. Referring to the operator of Figure 7.11b, the governing equations (Equations 3.20 and 3.21) are replaced by the finite differences in terms of the nodal values of a polar mesh.

7.6 *The Triangular Mesh

Some plates have boundary configurations that can be covered with ease by the use of a *triangular mesh* shown in Figure 7.12a. In this case, it is necessary to express the governing plate equations in terms of the *triangular coordinates*, q_1, q_2, and q_3. Referring to Figure 7.12b, the equations relating the Cartesian coordinates to the triangular coordinates are

$$\begin{aligned} x &= q_1 + q_2\cos\alpha + q_3\cos\beta \\ y &= q_2\sin\alpha + q_3\sin\beta \end{aligned} \tag{7.14}$$

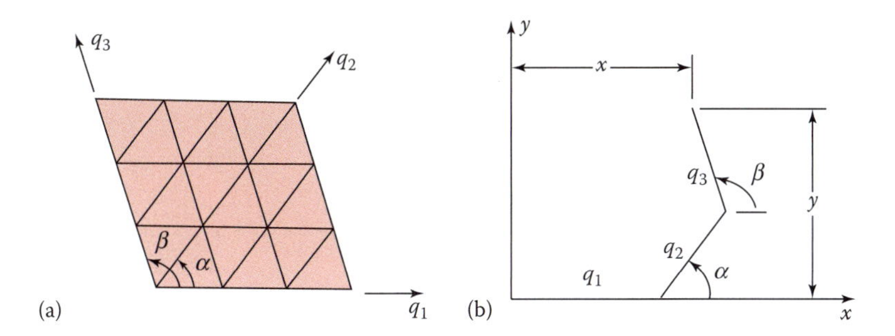

FIGURE 7.12
(a) Triangular mesh; (b) triangular coordinates.

in which α and β are the angles between q_1 and q_2 and between q_3 and q_1, respectively. The following partial derivatives are obtained from the above:

$$\begin{aligned} &\frac{\partial x}{\partial q_1}=1 \qquad &&\frac{\partial x}{\partial q_2}=\cos\alpha \qquad &&\frac{\partial x}{\partial q_3}=\cos\beta \\ &\frac{\partial y}{\partial q_1}=0 &&\frac{\partial y}{\partial q_2}=\sin\alpha &&\frac{\partial y}{\partial q_3}=\sin\beta \end{aligned} \tag{a}$$

When an expression for $w(x, y)$ is given, Equations 7.14 and a can be employed to define the corresponding expression $w(q_1, q_2, q_3)$ and its derivatives with respect to the triangular coordinates. The chain rule, together with Equations a, yields

$$\begin{aligned} \frac{\partial w}{\partial q_1}&=\frac{\partial w}{\partial x}\frac{\partial x}{\partial q_1}+\frac{\partial w}{\partial y}\frac{\partial y}{\partial q_1}=\frac{\partial w}{\partial x} \\ \frac{\partial w}{\partial q_2}&=\frac{\partial w}{\partial x}\cos\alpha+\frac{\partial w}{\partial y}\sin\alpha \\ \frac{\partial w}{\partial q_3}&=\frac{\partial w}{\partial x}\cos\beta+\frac{\partial w}{\partial y}\sin\beta \end{aligned} \tag{b}$$

The second partial derivatives of the deflection, written in the matrix from, are then (Problem 7.18)

$$\begin{Bmatrix} \partial^2 w/\partial q_1^2 \\ \partial^2 w/\partial q_2^2 \\ \partial^2 w/\partial q_3^2 \end{Bmatrix} = \begin{bmatrix} 1 & 0 & 0 \\ \cos^2\alpha & 2\sin\alpha\cos\alpha & \sin^2\alpha \\ \cos^2\beta & 2\sin\beta\cos\beta & \sin^2\beta \end{bmatrix} \begin{Bmatrix} \partial^2 w/\partial x^2 \\ \partial^2 w/\partial x\partial y \\ \partial^2 w/\partial y^2 \end{Bmatrix} \tag{c}$$

On introducing the first of Equations c into the second and the third and eliminating $\partial^2 w/\partial x\, \partial y$, we obtain

$$\frac{\partial^2 w}{\partial y^2}=R\left[2\frac{\partial^2 w}{\partial q_1^2}\cos\alpha\cos\beta\sin(\beta-\alpha)-\frac{\partial^2 w}{\partial q_2^2}\sin 2\beta+\frac{\partial^2 w}{\partial q_3^2}\sin 2\alpha\right]$$

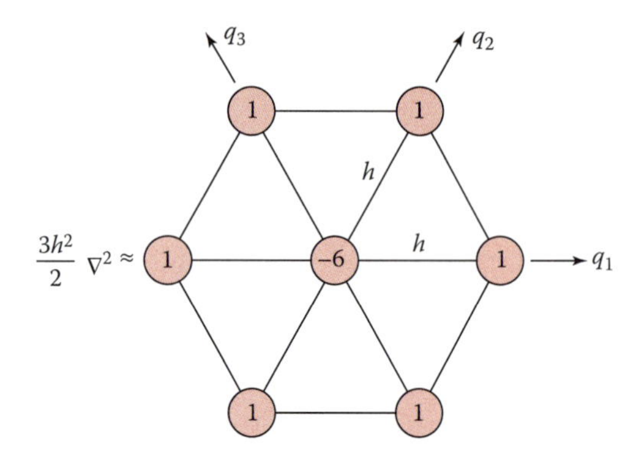

FIGURE 7.13
Coefficient pattern of the finite difference operator $\nabla^2 w$ for an equilateral triangular mesh.

and thus

$$\begin{aligned}\nabla^2 w &= \frac{\partial^2 w}{\partial x^2} + \frac{\partial^2 w}{\partial y^2} = \frac{\partial^2 w}{\partial q_1^2} + \frac{\partial^2 w}{\partial q_2^2} \\ &= R\left[\frac{\partial^2 w}{\partial q_1^2}\sin 2(\beta-\alpha) - \frac{\partial^2 w}{\partial q_2^2}\sin 2\beta + \frac{\partial^2 w}{\partial q_3^2}\sin 2\alpha\right]\end{aligned} \tag{7.15}$$

where $R = (1/2)[\sin\alpha \sin(\beta-\alpha)]^{-1}$.

In the case of a commonly used equilateral triangular mesh, where $\alpha = 60°$ and $\beta = 120°$, Equation 7.15 reduces to

$$\nabla^2 w = \frac{2}{3}\left(\frac{\partial^2 w}{\partial q_1^2} + \frac{\partial^2 w}{\partial q_2^2} + \frac{\partial^2 w}{\partial q_3^2}\right) \tag{7.16}$$

The finite difference operator pattern for the above, obtained by application of the operator for the second derivative (Table 7.1) in the q_1, q_2, q_3 directions, is given by Figure 7.13. Other finite difference operators may be developed similarly. The finite difference equivalent of Equations 3.20 and 3.21 can then be readily written in terms of the triangular coordinates.

EXAMPLE 7.7: Center Deflection of Skew Plate

Determine the center deflection of a simply supported skew plate under uniform loading of intensity p_0 (Figure 7.14). Use $h = a/4$.

Solution

The domain is divided into 32 small triangles. In labeling nodal points, the condition of symmetry is taken into account as shown in the figure. Note that $M = 0$ and $w = 0$ at the boundary. Operator ∇^2 of Figure 7.13 is applied to Equations 3.20 and 3.21 at nodal points 1, 2, 3, and 4, resulting in the following two sets of expressions:

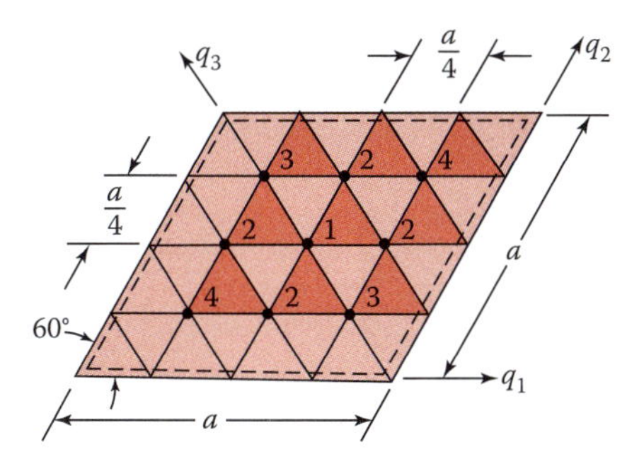

FIGURE 7.14
Skew plate with uniform load.

$$\begin{bmatrix} -6 & 4 & 2 & 0 \\ 1 & -5 & 1 & 1 \\ 1 & 2 & -6 & 0 \\ 0 & 2 & 0 & -6 \end{bmatrix} \begin{Bmatrix} M_1 \\ M_2 \\ M_3 \\ M_4 \end{Bmatrix} = -\frac{3}{2} p_0 h^2 \begin{Bmatrix} 1 \\ 1 \\ 1 \\ 1 \end{Bmatrix} \tag{d}$$

and

$$\begin{bmatrix} -6 & 4 & 2 & 0 \\ 1 & -5 & 1 & 1 \\ 1 & 2 & -6 & 0 \\ 0 & 2 & 0 & -6 \end{bmatrix} \begin{Bmatrix} w_1 \\ w_2 \\ w_3 \\ w_4 \end{Bmatrix} = -\frac{3}{2} \frac{h^2}{D} \begin{Bmatrix} M_1 \\ M_2 \\ M_3 \\ M_4 \end{Bmatrix} \tag{e}$$

from the above, $w_1 = 0.00283 p_0 a^4/D$.

Comments: By increasing the number of nodes, we expect to improve the result. For example, selecting $h = a/8$, it can be shown that the center deflection $w_1 = 0.00262 p_0 a^4/D$. The "exact" coefficient for the solution is 0.00256.

Part B: Finite Element Analysis

7.7 The FEM

The powerful FEM was developed in the 1960s simultaneously with the widespread use of digital computers and the increasing emphasis on numerical methods. The solution is obtained *without* use of the governing differential equations. This approach permits the complete automation of all procedures. The FEM may be viewed as an approximate Ritz method combined with a variational principle applied to continuum mechanics. It results in calculation of stress and deflection in a plate with a degree of ease and precision never before possible.

In the FEM, a structural member is *discretized* into a finite number of elements (usually triangular or rectangular in shape), connected at their nodes and along hypothetic interelement boundaries. Hence, equilibrium and compatibility must be satisfied at each

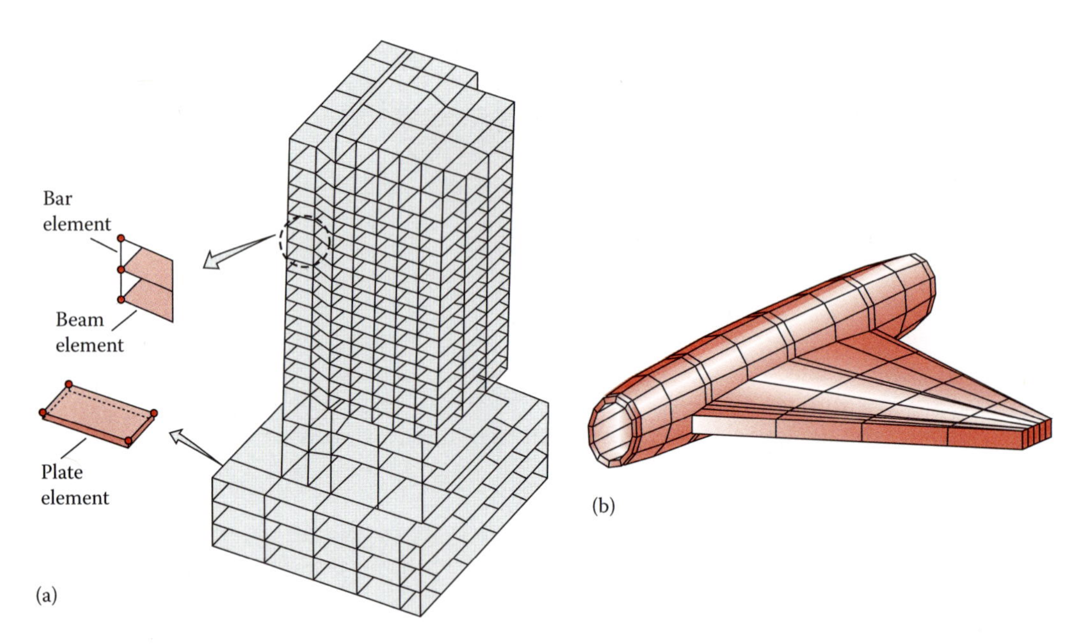

FIGURE 7.15
Finite element models of two common structures [2,10]: (a) Multistory building; (b) wing of an aircraft.

node and along the boundaries between elements. Figure 7.15 shows, as an example, how a multistory hotel building is modeled using bar, beam, column, and plate elements. Interestingly, this model can be used for static and dynamic analyses [2]. There are a number of FEMs. We discuss here only the commonly used finite displacement approach, wherein the governing set of algebraic equations is expressed in terms of unknown nodal displacements (see *Appendix C*).

The application of fundamental concepts of the FEM has already been extended to practical problems in most engineering fields. These include thick plates and shells, geometric and material nonlinearities, plasticity, vibration, viscoelasticity and viscoplasticity, fracture, laminated plates and shells, buckling, thermal stresses, dynamic response, aero- and hydroelastic analysis of structural systems, and so on. Most of the foregoing developments have now been well *coded into commercial programs*. The FEM offers numerous advantages, including

1. Structural shape of components that can easily be described
2. Ability to deal with discontinuities
3. Ability to handle composite and anisotropic materials
4. Ease of dealing with dynamic and thermal loadings
5. Ability to treat combined load conditions
6. Ability to handle nonlinear structural problems
7. Capacity for complete automation

In Sections 7.10 through 7.12, three of the *simplest* and earliest (beam, triangular, and rectangular) finite elements are derived to demonstrate the basic formulative method.

The treatment given here is very brief. There are various types of finite elements that lead to improved solutions. The reader wishing to pursue this subject further should consult the specialized literature, such as those listed in the references.

7.8 Properties of a 2D Finite Element

To begin with, a number of basic quantities relevant to individual *finite elements* of an isotropic plate is defined. The derivations are based on the assumptions of *small deflection* theory, described in Section 3.3. Illustrated in Chapter 8 is the determination of stiffness matrices for plate elements with nonisotropic properties, after derivation of the stress–strain relations for the orthotropic materials. The plate, in general, may have any nonuniform shape and loading.

Consider the thin plate of Figure 7.16, which is replaced by an assembly of triangular finite elements indicated by the dashed lines. The properties belonging to a discrete element will be designated by *e*.

7.8.1 Displacement Matrix

The nodal displacements $\{\delta\}_e$ are related to the displacements within the element by means of a *displacement function* $\{w\}_e$. The latter is expressed in the following general form:

$$\{w\}_e = [P]\{\delta\}_e \tag{7.17}$$

where the braces indicate a *column* matrix and the matrix $[P]$ is a function of position, to be later determined for a specific element. This matrix is often referred to as *shape function*. It is, of course, desirable that a displacement function be chosen such that the true displacement field is as closely represented as possible.

7.8.2 Strain, Stress, and Elasticity Matrices

Referring to Equations 3.3, we define, for commonality in finite element analysis of all types of problems, a *generalized "strain"–displacement* matrix of the form

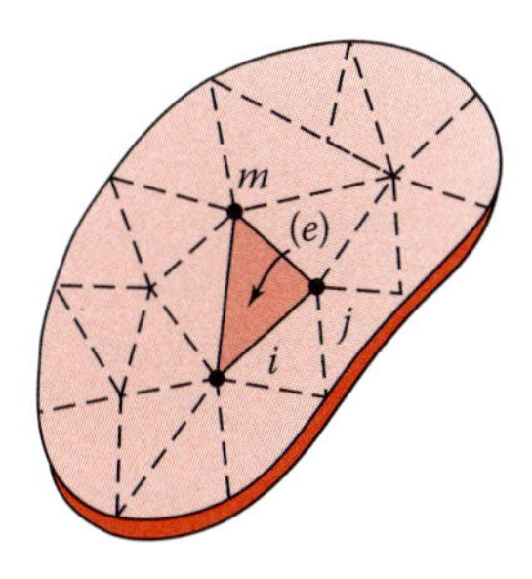

FIGURE 7.16
Plate divided into finite elements.

$$\begin{Bmatrix} \varepsilon_x \\ \varepsilon_y \\ \gamma_{xy} \end{Bmatrix}_e = \left\{ -\frac{\partial^2 w}{\partial x^2}, -\frac{\partial^2 w}{\partial y^2}, -2\frac{\partial^2 w}{\partial x \partial y} \right\} \tag{7.18a}$$

or

$$\{\varepsilon\}_e = [B]\{\delta\}_e \tag{7.18b}$$

in which $[B]$ is also yet to be determined. The *stress–generalized "strain"* relationship, from Equations 3.7, is as follows:

$$\begin{Bmatrix} \sigma_x \\ \sigma_y \\ \tau_{xy} \end{Bmatrix}_e = \frac{Ez}{1-\nu^2} \begin{bmatrix} 1 & \nu & 0 \\ \nu & 1 & 0 \\ 0 & 0 & (1-\nu)/2 \end{bmatrix} \{\varepsilon\}_e \tag{7.19a}$$

Concisely,

$$\{\sigma\}_e = z[D^*]\{\varepsilon\}_e \tag{7.19b}$$

Moments are connected to stresses by Equations 3.8a:

$$\{M\}_e = \begin{Bmatrix} M_x \\ M_y \\ M_{xy} \end{Bmatrix}_e = \int_{-t/2}^{t/2} z\{\sigma\}_e dz \tag{7.20}$$

Substitution of Equations 7.19 into the above yields the following moment-generalized "strain" relations:

$$\{M\}_e = \left(\int_{-t/2}^{t/2} z^2 [D^*] dz \right) \{\varepsilon\}_e \tag{7.21a}$$

or

$$\{M\}_e = [D]\{\varepsilon\}_e \tag{7.21b}$$

The *elasticity matrix* for an isotropic plate is therefore

$$[D] = \frac{t^3}{12}[D^*] = \frac{Et^3}{12(1-\nu^2)} \begin{bmatrix} 1 & \nu & 0 \\ \nu & 1 & 0 \\ 0 & 0 & (1-\nu)/2 \end{bmatrix} \tag{7.22}$$

The stress–strain relations, and thus elasticity matrix, differ for anisotropic materials (Section 8.8).

Due to many causes (e.g., shrinkage, temperature changes), some *initial strains* may be produced in the plate. In the case of a transversely loaded and heated plate, for instance, the stress and moment matrices become, respectively,

$$\{\sigma\}_e = z[D^*](\{\varepsilon\} - \{\varepsilon_0\})_e$$

and

$$\{M\}_e = [D](\{\varepsilon\} - \{\varepsilon_0\})_e$$

wherein $\{\varepsilon_0\}_e$ is the thermal strain matrix. The thermal stress problems are treated in Chapter 11.

7.9 General Formulation of the FEM

A convenient approach for derivation of the finite element governing expressions and characteristics is based on the principle of potential energy. The variation in the potential energy $\Delta\Pi$ of the entire plate shown in Figure 7.16, from Equation 3.52, is

$$\begin{aligned}\Delta\Pi = &\sum_1^n \iint_A (M_x\Delta\varepsilon_x + M_y\Delta\varepsilon_y + 2M_{xy}\Delta\gamma_{xy})dx\,dy \\ &-\sum_1^n \iint_A (p\Delta w)dx\,dy = 0\end{aligned} \tag{7.23a}$$

where n, A, and p represent the number of uniform thickness elements making up the plate, surface area of an element, and the lateral load per unit surface area, respectively. Equation 7.23a may be rewritten as follows:

$$\sum_1^n \iint_A \left(\{\Delta\varepsilon\}_e^T\{M\}_e - p\Delta w\right)dx\,dy = 0 \tag{7.23b}$$

in which superscript T denotes the transpose of a matrix, Introduction of Equations 7.17, 7.18, and 7.21 into Equation 7.23b yields

$$\sum_1^n \{\Delta\delta\}_e^T([K]_e\{\delta\}_e - \{Q\}_e) = 0 \tag{a}$$

The element *stiffness* matrix $[k]_e$ equals

$$[k]_e = \iint_A [B]^T [D][B] dx\, dy \tag{7.24}$$

The element *nodal force* matrix $\{Q\}_e$, due to initial strain and transverse load, is

$$\{Q\}_e = \iint_A [B]^T [D]\{\varepsilon_0\} dx\, dy + \iint_A [P]^T p\, dx\, dy \tag{7.25}$$

Since the changes in $\{\delta\}_e$ are independent and arbitrary, Equation a leads to the expression

$$[k]_e \{\delta\}_e = \{Q\}_e \tag{7.26}$$

for the element nodal force equilibrium.

We now assemble Equation a to obtain

$$\{\Delta\delta\}^T ([K]\{\delta\} - \{Q\}) = 0 \tag{b}$$

The above must be valid for all $\{\Delta\delta\}$. This yields the following *governing equations* for the entire plate:

$$[K]\{\delta\} = \{Q\} \tag{7.27}$$

where

$$[K] = \sum_1^n [k]_e \qquad \{Q\} = \sum_1^n \{Q\}_e \tag{7.28}$$

We observe that the *plate stiffness matrix* $[K]$ and the *plate nodal force matrix* $[Q]$ are determined by superposition of all the elements' stiffness and nodal force matrices, respectively.

The *general procedure* for solving a plate- (beam- or shell-) bending problem by the FEM is summarized as follows:

1. Determine $[k]_e$ from Equation 7.24 in terms of the given element properties. Generate $[K] = \Sigma[k]_e$.
2. Determine $\{Q\}_e$ from Equation 7.25 in terms of the applied loading. Generate $\{Q\} = \Sigma\{Q\}_e$.
3. Determine the nodal displacements from Equation 7.27 by satisfying the boundary conditions: $[\delta] = [K]^{-1}\{Q\}$.

Then determine element moment from $\{M\}_e = [D]\{\delta\}_e$ and the element stress from Equations 7.19 or 3.11.

The finite element analysis, from start to finish, may be outlined [11] as in Figure 7.17. The first step is to *define the analysis problem* as clearly as possible (e.g., static, dynamic, whether the solution will be 2D or 3D) in order to choose a finite element model and the

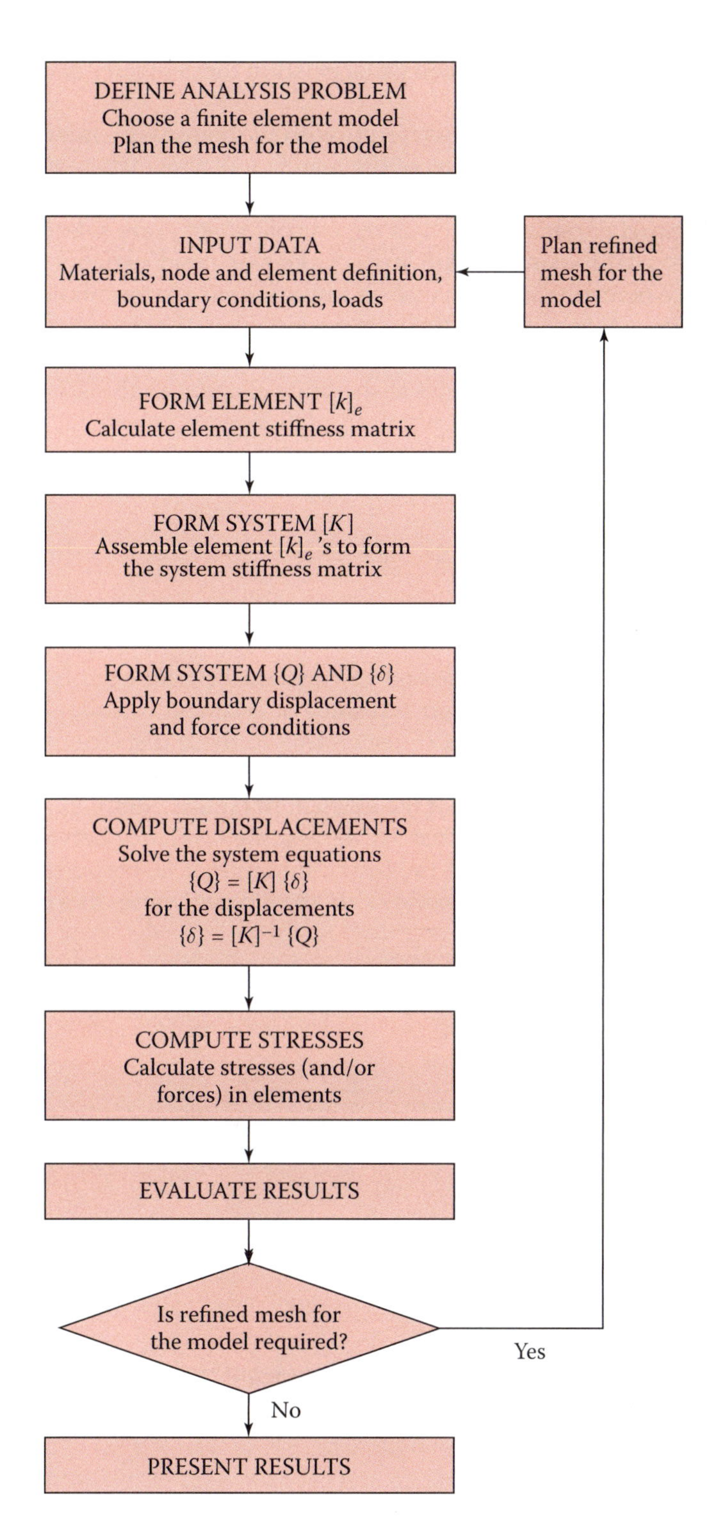

FIGURE 7.17
The finite element analysis block diagram.

corresponding mesh. Exploring symmetry that exists in the geometry and loading reduces the number of computations.

Comparison of displacement and stress results with the boundary conditions will provide a good check of the analysis. *Evaluation* of the results will show where to refine the mesh to begin the convergence to an accurate solution. Regions within the model with the *high stress* values and rapid variation as well as regions of *low stress* are selected for *refinement*. *Reducing element size* in regions of high stress values or rapid variation provides refinement. Convergence of results of the analysis is important to assure validity of the analysis.

The foregoing procedure will be better understood when the characteristics of a certain element are derived. Formulation of the governing equations of beam elements, which shows how the physical behavior relates to the finite element analysis, is presented in the next section. Two of the most commonly used plate-bending elements, each requiring a different type of general displacement function, are discussed in Sections 7.11 and 7.12.

7.10 Beam Element

Consider an initially straight *beam element* of constant flexural rigidity EI and length L (Figure 7.18a). The transverse deflection and slope at each end or node of the element are designated by w and θ, respectively. Corresponding to these displacements, a transverse shear force F_z and a bending moment M act at each node. The foregoing quantities are shown in the deflected configuration of the element in Figure 7.18b.

In the formulation of the stiffness matrix, we assume *no loading* between nodes.* Thus, the expression of the elastic curve of the beam, Equation 2.33, becomes $d^4w/dx^4 = 0$. We assume that the deflection w at any point within the element varies as a cubic polynomial function of x:

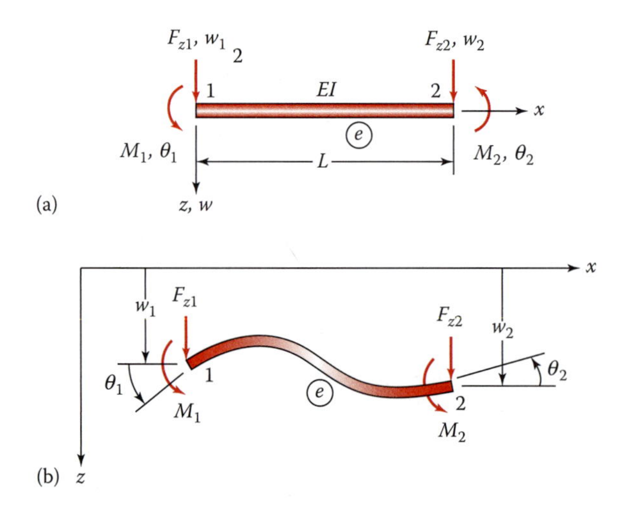

FIGURE 7.18
Beam element: (a) Undeformed state; (b) deformed state.

* In the elements where there is a distributed load, the equivalent nodal load components are used [12].

$$w = a_1 + a_2x + a_3x^2 + a_4x^3 \tag{a}$$

where the constants a are found from the conditions at the ends. The stiffness matrix may then be obtained by applying Equation 7.24.

However, for a beam element, the required relationship between the nodal forces and displacements may be obtained directly from well-known slope–deflection equations and by using the method of superposition. In so doing, it can be shown that

$$\begin{aligned} M_1 &= \frac{6EI}{L^2}w_1 + \frac{4EI}{L}\theta_1 - \frac{6EI}{L^2}w_2 + \frac{2EI}{L}\theta_2 \\ M_2 &= \frac{6EI}{L^2}w_1 + \frac{2EI}{L}\theta_1 - \frac{6EI}{L^2}w_2 + \frac{4EI}{L}\theta_2 \end{aligned} \tag{b}$$

Applying equations of statics, we have

$$\begin{aligned} F_{z1} &= -F_{z2} = \frac{1}{L}(M_1 + M_2) \\ &= \frac{12EI}{L^3}w_1 + \frac{6EI}{L^2}\theta_1 - \frac{12EI}{L^3}w_2 + \frac{6EI}{L^2}\theta_2 \end{aligned} \tag{c}$$

The nodal force–displacement relations, in the matrix form, are therefore

$$\begin{Bmatrix} F_{z1} \\ M_1 \\ F_{z2} \\ M_2 \end{Bmatrix} = \frac{EI}{L^3}\begin{bmatrix} 12 & 6L & -12 & 6L \\ 6L & 4L^2 & -6L & 2L^2 \\ -12 & -6L & 12 & -6L \\ 6L & 2L^2 & -6L & 4L^2 \end{bmatrix}\begin{Bmatrix} w_1 \\ \theta_1 \\ w_2 \\ \theta_2 \end{Bmatrix} \tag{7.29a}$$

Symbolically, this may be written as

$$\{Q\}_e = [k]_e\{\delta\}_e \tag{7.29b}$$

where Q represents the force F and moment M components.

Equations 7.29 define the *stiffness matrix* $[k]_e$ *for a beam element* lying along a local coordinate axis x. With the stiffness matrix developed, solution of problems involving beam elements proceeds as illustrated in Example 7.9.

7.10.1 Methods of Assemblage of the $[k]_e$s

As noted in Section C2.1, to carry out proper summation, we can expand the $[k]_e$s for each element to the order of the system stiffness matrix by adding rows and columns of zeros. However, for the problems involving a large number of elements, it becomes tedious to apply this method. Alternatively, we can *label* the *columns and rows* of $[k]_e$ according to the *displacement components* associated with it (Example 7.9). In so doing, the system matrix $[K]$ is obtained simply by adding terms from the individual stiffness matrix into their corresponding location in $[K]$.

EXAMPLE 7.8: Displacements of Cantilever Beam

Figure 7.19a illustrates a cantilever beam of length L and flexural rigidity EI under a uniformly distributed load of intensity p with reactions also noted by the dashed lines. *Find:* The vertical deflection and rotation at the free end.

SOLUTION

Only one finite element is used to represent the entire beam. Note that, the distributed load is replaced by the equivalent forces and moments as shown in Figure 2.19b (see case 4, Table B.8). The boundary conditions are $\upsilon_1 = 0$ and $\theta_1 = 0$. Observe that

$$F_{1y} = F_{2y} = -\frac{pL}{2} \qquad M_2 = -M_1 = \frac{pL^2}{2}$$

Then, *force–displacement relations*, Equation 7.29a take the form

$$\begin{Bmatrix} -\dfrac{pL}{2} \\ \dfrac{pL^2}{12} \end{Bmatrix} = \frac{EI}{L^3}\begin{bmatrix} 12 & -6L \\ -6L & 4L \end{bmatrix}\begin{Bmatrix} \upsilon_2 \\ \theta_2 \end{Bmatrix}$$

Solving for the *displacements*, we have

$$\begin{Bmatrix} \upsilon_2 \\ \theta_2 \end{Bmatrix} = \frac{L}{6EI}\begin{bmatrix} 2L^2 & 3L \\ 3L & 6 \end{bmatrix}\begin{Bmatrix} -\dfrac{pL}{2} \\ \dfrac{pL^2}{12} \end{Bmatrix}$$

or

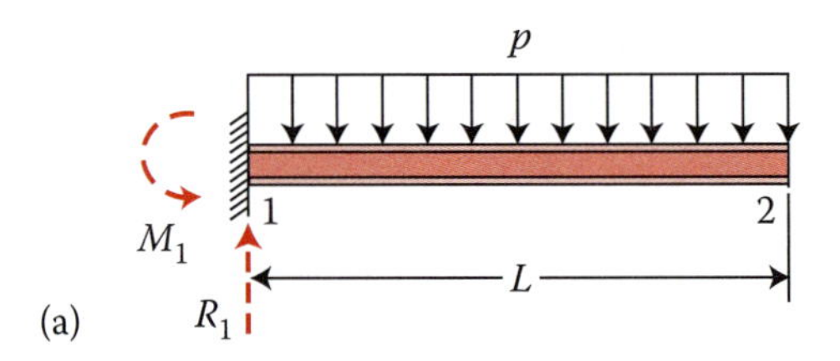

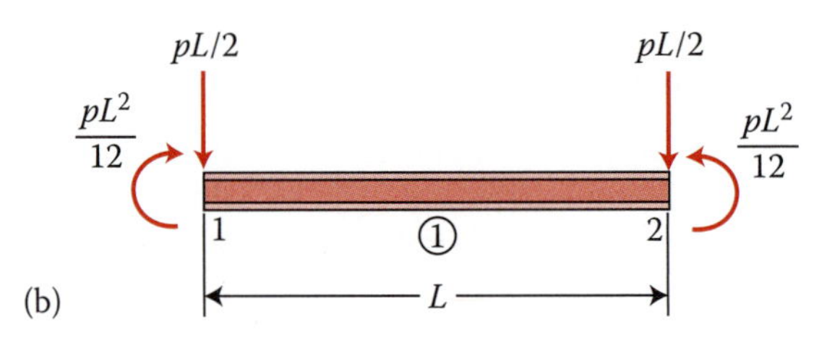

FIGURE 7.19
Uniformly loaded cantilever beam.

$$\begin{Bmatrix} v_2 \\ \theta_2 \end{Bmatrix} = \begin{Bmatrix} -\dfrac{pL^4}{8EI} \\ -\dfrac{pL^3}{6EI} \end{Bmatrix}$$

Comment: The minus signs indicate a downward deflection and a clockwise rotation at right end, node 2, respectively.

EXAMPLE 7.9: Analysis of Statically Indeterminate Beam

A propped cantilever beam of flexural rigidity EI carries an end load P (Figure 7.20a). Calculate, by discreditizing the beam into elements with nodes 1, 2, and 3, as shown in the figure: (a) the nodal displacements; (b) the nodal forces and moments.

Solution

Through the use of Equations 7.29, we obtain

$$[k]_1 = \frac{EI}{L^3}\begin{bmatrix} 12 & 6L & -12 & 6L \\ & 4L & -6L & 2L^2 \\ & & 12 & -6L \\ \text{Symmetric} & & & 4L^2 \end{bmatrix} \begin{matrix} w_1 \\ \theta_1 \\ w_2 \\ \theta_2 \end{matrix}$$

(columns: w_1, θ_1, w_2, θ_2)

and

$$[k]_2 = \frac{EI}{L^3}\begin{bmatrix} 12 & 6L & -12 & 6L \\ & 4L & -6L & 2L^2 \\ & & 12 & -6L \\ \text{Symmetric} & & & 4L^2 \end{bmatrix} \begin{matrix} w_2 \\ \theta_2 \\ w_3 \\ \theta_3 \end{matrix}$$

(columns: w_2, θ_2, w_3, θ_3)

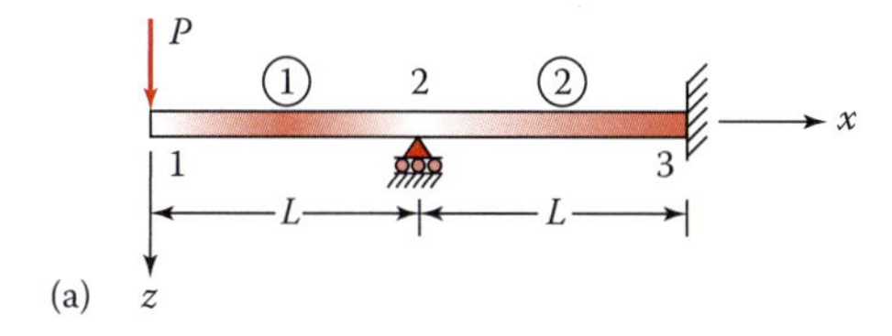

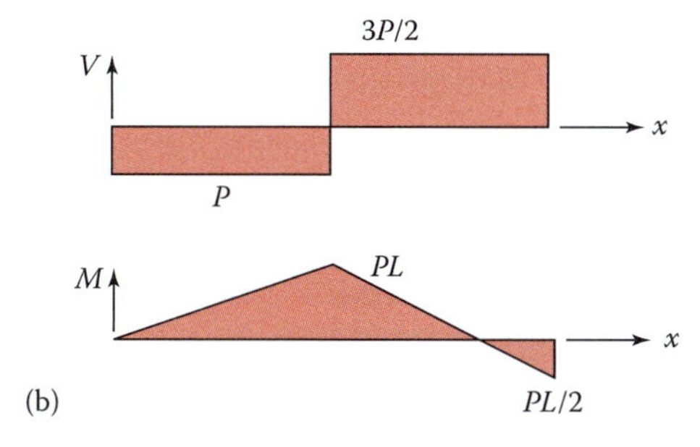

FIGURE 7.20
Continuous beam with end load.

a. The stiffness matrix of the beam can now be assembled: $[K] = [k]_1 + [k]_2$. The governing expressions for the beam are thus

$$\begin{Bmatrix} F_{z1} \\ M_1 \\ F_{z2} \\ M_2 \\ F_{z3} \\ M_3 \end{Bmatrix} = \frac{EI}{L^3} \begin{bmatrix} 12 & 6L & -12 & 6L & 0 & 0 \\ & 4L^2 & -6L & 2L^2 & 0 & 0 \\ & & 24 & 0 & -12 & 6L \\ & & & 8L^2 & -6L & 2L^2 \\ & & & & 12 & -6L \\ \text{Symmetric} & & & & & 4L^2 \end{bmatrix} \begin{Bmatrix} w_1 \\ \theta_1 \\ w_2 \\ \theta_2 \\ w_3 \\ \theta_3 \end{Bmatrix} \quad (7.30a)$$

(columns: w_1, θ_1, w_2, θ_2, w_3, θ_3)

or

$$\{Q\} = [K]\{\delta\} \quad (7.30b)$$

The boundary conditions are $w_2 = 0$, $\theta_3 = 0$, and $w_3 = 0$. The foregoing equations become, after partitioning the first, second, and fourth, associated with the unknown displacements:

$$\begin{Bmatrix} P \\ 0 \\ 0 \end{Bmatrix} = \frac{EI}{L^3} \begin{bmatrix} 12 & 6L & 6L \\ 6L & 4L^2 & 2L^2 \\ 6L & 2L^2 & 8L^2 \end{bmatrix} \begin{Bmatrix} w_1 \\ \theta_1 \\ \theta_2 \end{Bmatrix}$$

Solving, we obtain

$$w_1 = \frac{7PL^3}{12EI} \qquad \theta_1 = -\frac{3PL^2}{4EI} \qquad \theta_2 = -\frac{PL^2}{4EI}$$

b. Carrying Equations d together with $w_2 = w_3 = 0$ and $\theta_3 = 0$ into Equation 7.30a, after multiplying, nodal forces and moments are obtained as

$$F_{z1} = P \qquad M_1 = 0 \qquad F_{z2} = -\frac{5}{2}P$$

$$M_2 = PL \qquad F_{z3} = \frac{3}{2}P \qquad M_3 = -\frac{1}{2}PL$$

The shear V and moment M diagrams are shown in Figure 7.20b.

EXAMPLE 7.10: Analysis of Statically Indeterminate Stepped Beam

Figure 7.21a portrays a propped cantilever beam of flexural rigidity EI and $2EI$ for the parts 1–2 and 2–3, respectively, carries a concentrated load P at point 2. *Find:* (a) The nodal displacements; (b) The nodal forces and moments.

Solution

The beam is discretized into elements 1 and 2 with nodes 1, 2, and 3 as depicted in Figure 7.21a. There are a total of six displacement components for the beam before the

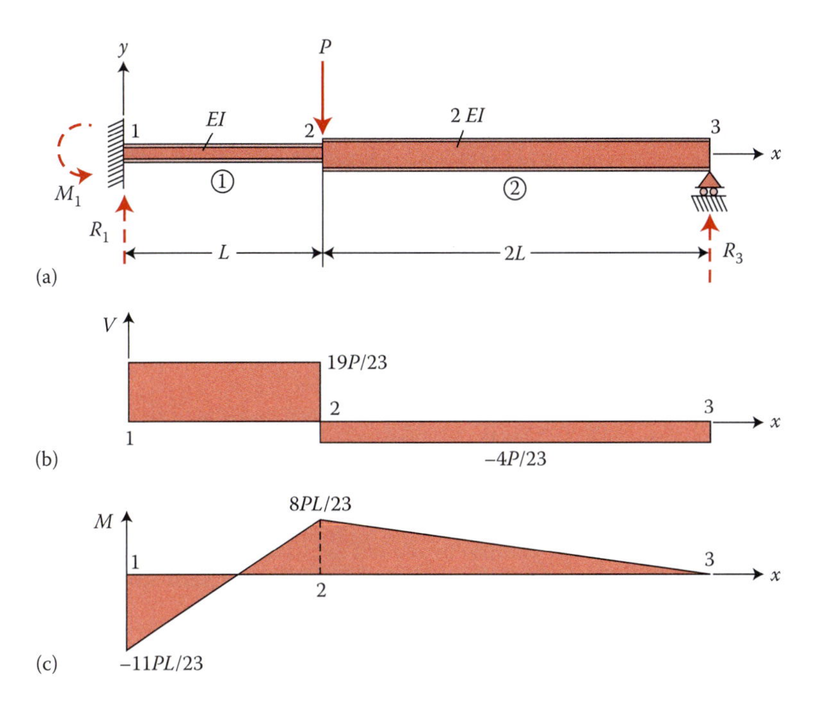

FIGURE 7.21
Propped up cantilever beam with concentrated load.

boundary conditions applied. Thus, the order of the system stiffness matrix must be 6×6. From Equation 7.29, the stiffness matrix for the *element 1*, with $(EI/L^3)_1 = EI/L^3$ and $L_1 = L$:

$$
\begin{array}{c} \\ [k]_1 = \dfrac{EI}{L^3} \end{array}
\begin{array}{c}
\begin{array}{cccccc} v_1 & \theta_1 & v_2 & \theta_2 & v_3 & \theta_3 \end{array} \\
\begin{bmatrix}
12 & 6L & -12 & 6L & 0 & 0 \\
6L & 4L^2 & -6L & 2L^2 & 0 & 0 \\
-12 & -6L & 12 & -6L & 0 & 0 \\
6L & 2L^2 & -6L & 4L^2 & 0 & 0 \\
0 & 0 & 0 & 0 & 0 & 0 \\
0 & 0 & 0 & 0 & 0 & 0
\end{bmatrix}
\end{array}
\begin{array}{c} \\ v_1 \\ \theta_1 \\ v_2 \\ \theta_2 \\ v_3 \\ \theta_3 \end{array}
\quad \text{(c)}
$$

Likewise, for *element 2*, with $(EI/L^3)_2 = EI/4L^3$, and $L_2 = 2L$, we have after rearrangement:

$$
\begin{array}{c} \\ [k]_2 = \dfrac{EI}{L^3} \end{array}
\begin{array}{c}
\begin{array}{cccccc} v_1 & \theta_1 & v_2 & \theta_2 & v_3 & \theta_3 \end{array} \\
\begin{bmatrix}
0 & 0 & 0 & 0 & 0 & 0 \\
0 & 0 & 0 & 0 & 0 & 0 \\
0 & 0 & 3 & 3L & -3 & 3L \\
0 & 0 & 3L & 4L^2 & -3L & 2L^2 \\
0 & 0 & -3 & -3L & 3 & -3L \\
0 & 0 & 3L & -2L^2 & -3L & 4L^4
\end{bmatrix}
\end{array}
\begin{array}{c} \\ v_1 \\ \theta_1 \\ v_2 \\ \theta_2 \\ v_3 \\ \theta_3 \end{array}
$$

Note, in these expressions, the nodal displacements are indicated with associativity of the rows and columns of the member stiffness matrices; in accordance with, rows and columns of zeros are added.

a. The *system stiffness matrix* of the beam can now be superimposed $[K] = [k]_1 + [k]_2$. The governing equations for the beam, with $F_{2y} = -P$ (Figure 7.21b), are

$$\begin{Bmatrix} R_1 \\ M_1 \\ -P \\ 0 \\ R_3 \\ 0 \end{Bmatrix} = \frac{EI}{L^3}\begin{bmatrix} 12 & 6L & -12 & 6L & 0 & 0 \\ 6L & 4L^2 & -6L & 2L^2 & 0 & 0 \\ -12 & -6L & 15 & -3L & 3 & 3L \\ 6L & 2L^2 & -3L & 8L^2 & -3L & 2L^2 \\ 0 & 0 & -3 & -3L & 3 & -3L \\ 0 & 0 & 3L & 2L^2 & -3L & 4L^2 \end{bmatrix}\begin{Bmatrix} v_1 \\ \theta_1 \\ v_2 \\ \theta_2 \\ v_3 \\ \theta_3 \end{Bmatrix}$$

The *boundary conditions* are $v_1 = 0$, $\theta_1 = 0$, and $v_3 = 0$. Multiplying out the foregoing relationship corresponding to the unknown displacements, we have

$$\begin{Bmatrix} -P \\ 0 \\ 0 \end{Bmatrix} = \frac{EI}{L^3}\begin{bmatrix} 15 & -3L & 3L \\ -3L & 8L^2 & 2L^2 \\ 3L & 2L^2 & 4L^2 \end{bmatrix}\begin{Bmatrix} v_2 \\ \theta_2 \\ \theta_3 \end{Bmatrix} \tag{d}$$

and

$$\begin{Bmatrix} R_1 \\ M_1 \\ R_3 \end{Bmatrix} = \frac{EI}{L^3}\begin{bmatrix} -12 & 6L & 0 \\ -6L & 2L^2 & 0 \\ -3 & -3L & -3L \end{bmatrix}\begin{Bmatrix} v_2 \\ \theta_2 \\ \theta_3 \end{Bmatrix} \tag{e}$$

Solution of Equation d by the method of matrix inversion, results in *deflection and slopes*,

$$\begin{Bmatrix} v_2 \\ \theta_2 \\ \theta_3 \end{Bmatrix} = \frac{L^2}{276EI}\begin{bmatrix} 28L & 18 & -30 \\ 18 & 51/L & -39/L \\ -30 & -39/L & 111/L \end{bmatrix}\begin{Bmatrix} -P \\ 0 \\ 0 \end{Bmatrix} = \frac{PL^2}{276EI}\begin{Bmatrix} -28L \\ -18 \\ 30 \end{Bmatrix} \tag{f}$$

The minus sign means a downward deflection at node 2 and clockwise rotation of left end 1; the positive sign indicates a counterclockwise rotation at right end 3 (Figure 7.18), as cognized intuitively.

b. Substituting the displacements found into Equation e, after multiplying and simplifying, *nodal forces* and *moments* are

$$\begin{Bmatrix} R_1 \\ M_1 \\ R_3 \end{Bmatrix} = \frac{P}{276L}\begin{bmatrix} -12 & 6L & 0 \\ -6L & 2L^2 & 0 \\ -3 & -3L & -3L \end{bmatrix}\begin{Bmatrix} -28L \\ -18 \\ 30 \end{Bmatrix} = \frac{1}{23}\begin{Bmatrix} 19P \\ 11PL \\ 4P \end{Bmatrix}$$

Comments: Often, it is necessary to compute the nodal forces and moments associated with *each element* to analyze the entire structure. For the situation under consideration, it may easily be observed from a FBD of element 2 that $M_2 = R_3(2L) = 8PL/23$. Consequently, we sketch the shear and moment diagrams for the beam as shown in Figure 7.21b and c, respectively.

7.11 Triangular Finite Element

The triangular element can easily accommodate irregular boundaries and can be graduated in size to permit small elements in regions of stress concentration. Because of this, it is used extensively in the finite element approach. Consider as the finite element model a *triangular plate element ijm* coinciding with the xy plane (Figure 7.22a). Note the *counterclockwise* numbering convention of the nodes. Each nodal displacement of the element has three components: a deflection in the z direction (w), a rotation about the x axis (θ_x), and a rotation about the y axis (θ_y). Rotations are related to the slopes as follows:

$$\theta_x = \frac{\partial w}{\partial y} \qquad \theta_y = \frac{\partial w}{\partial x} \tag{7.31}$$

The positive directions of the rotations are determined by the *right-hand rule* as shown in the figure.

7.11.1 Displacement Function

The nodal displacement matrix for the element is

$$\{\delta\}_e = \begin{Bmatrix} \delta_i \\ \delta_j \\ \delta_m \end{Bmatrix} = \{w_i, \theta_{xi}, \theta_{yi}, w_j, \theta_{xj}, \theta_{yj}, w_m, \theta_{xm}, \theta_{ym}\} \tag{7.32}$$

The displacement function, defining the displacement at any point within the element *ijm*, is chosen to be a modified third-order polynomial of the form

$$w_e = a_1 + a_2 x + a_3 y + a_4 x^2 + a_5 xy + a_6 y^2 + a_7 x^3 + a_8(x^2 y + xy^2) + a_9 y^3 \tag{7.33}$$

leading to a reasonably simple theoretical development. Observe that the number of terms in the above is identical to the number of nodal displacements of the element.

This function preserves the *continuity of displacements* but not the slopes along the element surfaces. For practical engineering purposes, however, in most cases the accuracy of the solution based on Equation 7.33 is acceptable. A displacement function of 18 order,

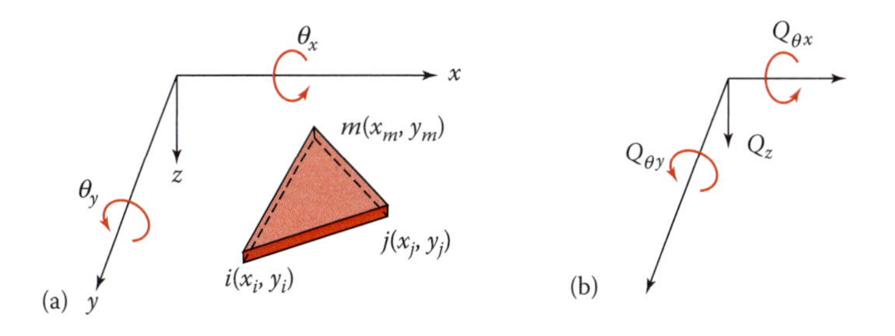

FIGURE 7.22
(a) Triangular finite element, and (b) its external nodal forces.

corresponding to a triangle with six nodes, leads to improved results, but the analysis becomes more involved than described here; hence, it is considered to be beyond the scope of this introductory treatise.

When coefficients a_1 through a_9 are known, Equation 7.33 will provide the displacement at all locations in the plate. The nodal displacements can be written as follows:

$$\begin{Bmatrix} w_i \\ \theta_{xi} \\ \theta_{yi} \\ w_j \\ \theta_{xj} \\ \theta_{yi} \\ w_m \\ \theta_{xm} \\ \theta_{ym} \end{Bmatrix} = \begin{bmatrix} 1 & x_i & y_i & x_i^2 & x_i y_i & y_i^2 & x_i^3 & (x_i^2 y_i + x_i y_i^2) & y_i^3 \\ 0 & 0 & 1 & 0 & x_i & 2y_i & 0 & (x_i^2 + 2x_i y_i) & 3y_i^2 \\ 0 & 1 & 0 & 2x_i & y_i & 0 & 3x_i^2 & (2x_i y_i + y_i^2) & 0 \\ 1 & x_j & y_j & x_j^2 & x_j y_j & y_j^2 & x_j^3 & (x_j^2 y_j + x_j y_j^2) & y_j^3 \\ 0 & 0 & 1 & 0 & x_j & 2y_j & 0 & (x_j^2 + 2x_j y_j) & 3y_j^2 \\ 0 & 1 & 0 & 2x_j & y_j & 0 & 3x_j^2 & (2x_j y_j + y_j^2) & 0 \\ 1 & x_m & y_m & x_m^2 & x_m y_m & y_m^2 & x_m^3 & (x_m^2 y_m + x_m y_m^3) & y_m^3 \\ 0 & 0 & 1 & 0 & x_m & 2y_m & 0 & (x_m^2 + 2x_m y_m) & 3y_m^2 \\ 0 & 1 & 0 & 2x_m & y_m & 0 & 3x_m^2 & (2x_m y_m + y_m^2) & 0 \end{bmatrix} \begin{Bmatrix} a_1 \\ a_2 \\ a_3 \\ a_4 \\ a_5 \\ a_6 \\ a_7 \\ a_8 \\ a_9 \end{Bmatrix} \quad (7.34a)$$

or, concisely,

$$\{\delta\}_e = [C]\{a\} \quad (7.34b)$$

From the foregoing, the solution for the unknown constants is

$$\{a\} = [C]^{-1}\{\delta\}_e \quad (7.35)$$

Equations 7.34 show that matrix [C] is dependent on the coordinate dimensions of the nodal points. The displacement function may now be written in the form of Equation 7.17:

$$\{w\}_e = [P]\{\delta\}_e = [L][C]^{-1}\{\delta\}_e \quad (7.36)$$

where

$$[L] = [1, x, y, x^2, xy, y^2, x^3, x^2y + xy^2, y^3] \quad (7.37)$$

Substitution of Equation 7.33 into Equations 7.18 gives

$$\{\varepsilon\}_e = \begin{bmatrix} 0 & 0 & 0 & -2 & 0 & 0 & -6x & -2y & 0 \\ 0 & 0 & 0 & 0 & 0 & -2 & 0 & -2x & -6y \\ 0 & 0 & 0 & 0 & -2 & 0 & 0 & -4(x+y) & 0 \end{bmatrix} \{a_1, a_2, \ldots, a_9\} \quad (7.38a)$$

or

$$\{\varepsilon\}_e = [H]\{a\} \tag{7.38b}$$

We can determine the generalized "strain"–displacement matrix, by introducing Equation 7.35 into Equations 7.38:

$$\{\varepsilon\}_e = [B]\{\delta\}_e = [H][C]^{-1}\{\delta\}_e$$

We thus have

$$[B] = [H][C]^{-1} \tag{7.39}$$

7.11.2 The Stiffness Matrix

On substituting $[B]$ from Equation 7.39 into Equation 7.24, we have

$$[k]_e = [[C]^{-1}]^T \left(\iint [H]^T [D][H] dx\, dy \right) [C]^{-1} \tag{7.40}$$

Here the matrices $[H]$, $[D]$, and $[C]$ are given by Equations 7.38, 7.22, and 7.34, respectively. After multiplying the matrices under the integral sign, the integrations can be performed to determine the element stiffness matrix.

7.11.3 External Nodal Forces

The nodal forces due to transverse surface loading may be obtained from Equation 7.25 or by physical intuition. The element nodal force matrix is represented by

$$\{Q\}_e = \begin{Bmatrix} Q_i \\ Q_j \\ Q_m \end{Bmatrix} = \{Q_{zi}, Q_{\theta xi}, Q_{\theta yi}, Q_{zj}, Q_{\theta xj}, Q_{\theta yj}, Q_{zm}, Q_{\theta xm}, Q_{\theta ym}\} \tag{7.41}$$

where Q_z, Q_{qx}, and $Q_{\theta y}$ denote the lateral force in the z direction, the moment about the x axis, and the moment about the y axis (Figure 7.20b), respectively.

The determination of the element nodal forces is demonstrated by the following longhand solution.

EXAMPLE 7.11: Nodal Forces of Plate Element

The element 123, shown in Figure 7.23, represents a portion of a thin elastic plate that is under uniform loading of intensity p_0. Determine the nodal force matrix. Assume that the weight of element is negligible.

Solution

Referring to the figure (where $i = 1, j = 2, m = 3$), matrix $[C]$, defined by Equations 7.34, is obtained as

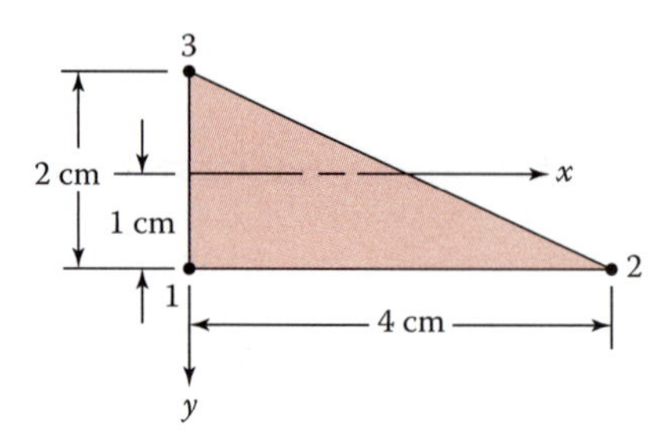

FIGURE 7.23
Triangular finite element.

$$[C]=\begin{bmatrix} 1 & 0 & 1 & 0 & 0 & 1 & 0 & 0 & 1 \\ 0 & 0 & 1 & 0 & 0 & 2 & 0 & 0 & 3 \\ 0 & 1 & 0 & 0 & 1 & 0 & 0 & 1 & 0 \\ 1 & 4 & 1 & 16 & 4 & 1 & 64 & 20 & 1 \\ 0 & 0 & 1 & 0 & 4 & 2 & 0 & 24 & 3 \\ 0 & 1 & 0 & 8 & 1 & 0 & 48 & 9 & 0 \\ 1 & 0 & -1 & 0 & 0 & 1 & 0 & 0 & -1 \\ 0 & 0 & 1 & 0 & 0 & -2 & 0 & 0 & 3 \\ 0 & 1 & 0 & 0 & -1 & 0 & 0 & 1 & 0 \end{bmatrix}$$

From the above we have

$$[[C]^{-1}]^T=\begin{bmatrix} 0.5 & 0 & 0.75 & -0.187 & 0 & 0 & 0.031 & 0 & -0.25 \\ -0.25 & 0.042 & -0.25 & 0.042 & 0 & 0.25 & 0 & -0.042 & 0.25 \\ 0 & 0.583 & 0 & -0.417 & 0.5 & 0 & 0.062 & -0.083 & 0 \\ 0 & 0 & 0 & 0.187 & 0 & 0 & -0.031 & 0 & 0 \\ 0 & -0.042 & 0 & -0.042 & 0 & 0 & 0 & 0.042 & 0 \\ 0 & 0 & 0 & -0.25 & 0 & 0 & 0.062 & 0 & 0 \\ 0.5 & 0 & -0.75 & 0 & 0 & 0 & 0 & 0 & 0.25 \\ 0.25 & 0 & -0.25 & 0 & 0 & -0.25 & 0 & 0 & 0.25 \\ 0 & 0.417 & 0 & -0.083 & -0.5 & 0 & 0 & 0.083 & 0 \end{bmatrix}$$

The nodal forces due to uniform surface loading p_0 are given by Equations 7.25 and 7.36:

$$\{Q\}_e=[[C]^{-1}]^T p_0 \iint [L]^T\,dx\,dy \tag{a}$$

Substituting Equation 7.37 and the limits of integrations into Equation a, we have

$$\{Q\}_e=[[C]^{-1}]^T p_0 \int_{-1}^{1}\int_{0}^{2y+2} \{1,x,y,x^2,xy,y^2,x^3,x^2y+xy^2,y^3\}\,dx\,dy$$

This expression is readily integrated and then multiplied by $[[C]^{-1}]^T$ to yield

$$\{Q\}_e = p_0\{1.60, -0.49, 0.89, 1.20, -0.31, -1.07, 1.20, 0.53, 0.71\}$$

The element stiffness matrix $[k]_e$ may be determined similarly (see Problem 7.27).

It is clear that the FEM, even in the simplest of cases, requires considerable algebra. For any problem of practical significance, the digital computer must be employed to perform the necessary matrix algebra. Compared with other methods, however, the finite element approach offers a distinct advantage in the treatment of plates having irregular shapes, nonuniform load, and isotropic or anisotropic materials.

7.12 Rectangular Finite Element

Let us now consider the *rectangular element* shown in Figure 7.24a. To ensure at least an approximate fulfillment of the continuity of slopes, three nodal displacement components described in Section 7.11 are taken into account.

7.12.1 Displacement Function

The element nodal displacement matrix is represented by

$$\{\delta\}_e = \{w_i, \theta_{xi}, \theta_{yi}, w_j, \theta_{xj}, \theta_{yj}, w_m, \theta_{xm}, \theta_{ym}, w_n, \theta_{xn}, \theta_{yn}\} \tag{7.42}$$

The following polynomial expression for the displacement of the element *ijmn* is selected [1]:

$$\begin{aligned} w_e = a_1 + a_2x + a_3y + a_4x^2 + a_5xy + a_6y^2 + a_7x^3 + a_8x^2y \\ + a_9xy^2 + a_{10}y^3 + a_{11}x^3y + a_{12}xy^3 \end{aligned} \tag{7.43}$$

Nodal displacements, on introduction of Equations 7.43 and 7.31 into Equation 7.42, are next found. In concise form, these are

$$\{\delta\}_e = [C]\{a\} \tag{7.44}$$

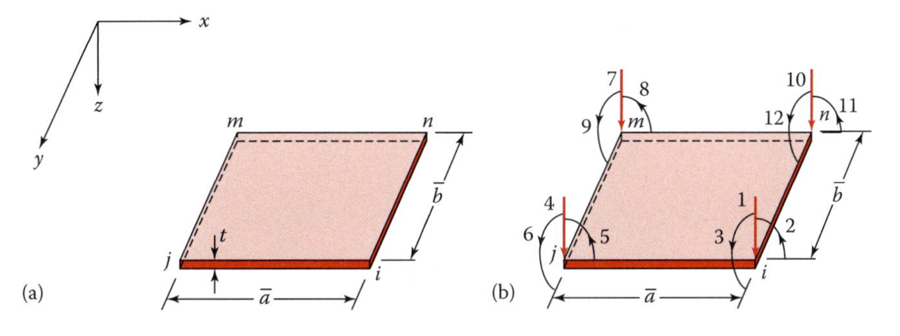

FIGURE 7.24
(a) Rectangular finite element; (b) kinematic freedom of nodal points.

in which [C], a 12 × 12 matrix, depends on the nodal coordinates as in Equations 7.34. Inversion of the above provides the values of the unknown coefficients $a_1, \ldots a_{12}$:

$$\{a\} = [C]^{-1}\{\delta\}_e \tag{7.45}$$

The displacement function is expressed by Equation 7.36. However, the matrix [L] in that equation now has the form

$$[L] = [1, x, y, x^2, xy, y^2, x^3, x^2y, xy^2, y^3, x^3y, xy^3]$$

Inserting Equation 7.44 into Equations 7.18, we have

$$\{\varepsilon\}_e = \begin{bmatrix} 0 & 0 & 0 & -2 & 0 & 0 & -6x & -2y & 0 & 0 & -6xy & 0 \\ 0 & 0 & 0 & 0 & 0 & -2 & 0 & 0 & -2x & -6y & 0 & -6xy \\ 0 & 0 & 0 & 0 & -2 & 0 & 0 & -4x & -4y & 0 & -6x^2 & -6y^2 \end{bmatrix} \{a_1, \ldots, a_{12}\} \tag{7.46a}$$

or

$$\{\varepsilon\}_e = [H]\{a\} \tag{7.46b}$$

As before, the generalized "strain"–displacement matrix is found on inserting Equation 7.45 into Equations 7.46.

7.12.2 The Stiffness Matrix

The element stiffness matrix $[k]_e$ is obtained by introducing matrices [H], [D], and [C] given by Equations 7.46, 7.22, and 7.44 into Equation 7.40. It follows that

$$[k]_e = \frac{Et^3}{108\bar{a}\bar{b}(1-\nu^2)}[R]\left\{[k_1]+[k_2]+\nu[k_3]+\frac{1-\nu}{2}[k_4]\right\}[R] \tag{7.47}$$

Explicit expressions of bending-stiffness coefficients $[k_1]$ to $[k_4]$ and the matrix [R] are given in Table 7.3 [1,8]. These coefficients in general provide rapid convergence and satisfactory accuracy. It is to be noted that the *numbering system* in the table refers to Figure 7.24b.

7.12.3 External Nodal Forces

The element nodal force matrix is

$$\{Q\}_e = \begin{Bmatrix} Q_i \\ Q_j \\ Q_m \\ Q_n \end{Bmatrix}$$

where the Qs are defined in Section 7.11.

TABLE 7.3

Bending Stiffness Coefficients in Equations 7.47 and 8.45 for a Rectangular Plate Element

$[k_1] = \epsilon^2$ (Symmetric)

	1	2	3	4	5	6	7	8	9	10	11	12
1	60											
2	30	20										
3	0	0	0									
4	−60	−30	0	60								
5	30	10	0	−30	20							
6	0	0	0	0	0	0						
7	−30	−15	0	30	−15	0	60					
8	15	5	0	−15	10	0	−30	20				
9	0	0	0	0	0	0	0	0	0			
10	30	15	0	−30	15	0	−60	30	0	60		
11	15	10	0	−15	5	0	−30	10	0	30	20	
12	0	0	0	0	0	0	0	0	0	0	0	0

$[k_2] = \epsilon^{-2}$ (Symmetric)

	1	2	3	4	5	6	7	8	9	10	11	12
1	60	0										
2	0	0										
3	−30	0	20									
4	30	0	−15	60								
5	0	0	0	0	0							
6	−15	0	10	−30	0	20						
7	−30	0	15	−60	0	30	60					
8	0	0	0	0	0	0	0	0				
9	−15	0	5	−30	0	10	30	0	20			
10	−60	0	30	−30	0	15	30	0	15	60		
11	0	0	0	0	0	0	0	0	0	0	0	
12	−30	0	10	−15	0	0	15	0	10	30	0	20

$[k_3] =$ (Symmetric)

	1	2	3	4	5	6	7	8	9	10	11	12
1	30											
2	15	0										
3	−15	−15	0									
4	−30	0	15	30								
5	0	0	0	−15	0							
6	15	0	0	−15	15	0						
7	30	0	0	−30	15	0	30					
8	0	0	0	15	0	0	−15	0				
9	0	0	0	0	0	0	15	−15	0			
10	−30	−15	0	30	0	0	−30	0	−15	30		
11	−15	0	0	0	0	0	0	0	0	15	0	
12	0	0	0	0	0	0	−15	0	0	15	15	0

(Continued)

TABLE 7.3 (*Continued*)

Bending Stiffness Coefficients in Equations 7.47 and 8.45 for a Rectangular Plate Element

$[k_4] =$ (Symmetric)

	1	2	3	4	5	6	7	8	9	10	11	12
1												
2	6											
3	−6	0										
4		−6	6									
5	6	−2	0	−6								
6	6	0		−6	0							
7		6	−6		6	6						
8	−6	2	0	6		0	−6					
9	6	0	2	−6	0	−2	6	0				
10		−6	6		−6	−6		6	−6			
11	−6		0	6	2	0	−6	−2	0	6		
12	6	0	−2	6	0	2	−6	0		6	0	

$$[R] = \begin{bmatrix} [r] & [0] & [0] & [0] \\ [0] & [r] & [0] & [0] \\ [0] & [0] & [r] & [0] \\ [0] & [0] & [0] & [r] \end{bmatrix} \quad \text{where} \quad [r] = \begin{bmatrix} 1 & 0 & 0 \\ 0 & \bar{b} & 0 \\ 0 & 0 & \bar{a} \end{bmatrix}$$

Note: For *numbering system* and sign convention, see Figure 7.24; $\epsilon = \bar{b}/\bar{a}$.

The unknown displacements, strains, and stresses may now be calculated employing the general procedure of Section 7.9.

EXAMPLE 7.12: Analysis of Plate with Simple and Fixed Edges

Consider a square plate of sides a with two opposite edges $x = 0$ and $x = a$ simply supported and the remaining edges clamped (Figure 7.25a). Compute the maximum value of w if the plate is subjected to a uniformly distributed load of intensity p_0. Take $a = 2$ m and $v = 0.3$.

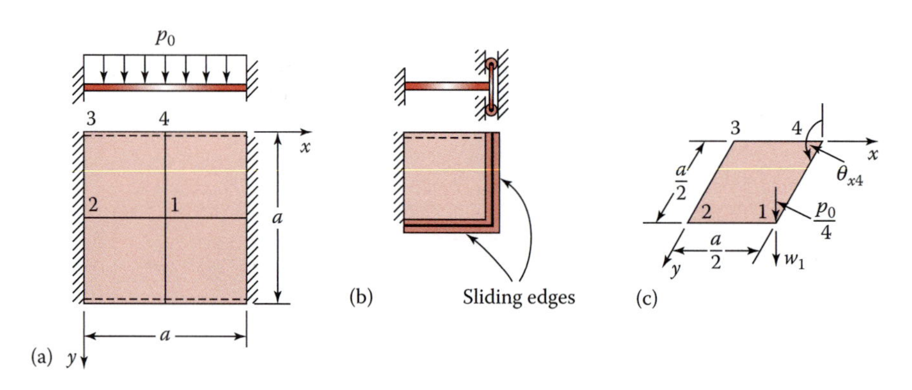

FIGURE 7.25

Finite element representation of a plate problem: (a) Actual plate; (b) substitute plate; (c) nodal force and displacements.

Solution

Symmetry in deflection dictates that only one quarter-plate need be analyzed, provided that *sliding-edge* conditions (Equation 3.28) are introduced along the lines of symmetry. The *substitute plate* is shown in Figure 7.25b.

For the sake of simplicity in calculations, we employ only one element per quarter-plate and lump the load at node 1. The boundary constraints permit only a lateral displacement w_1 at node 1 and a rotation θ_{x4} at node 4 (Figure 7.25c). The displacement matrix is

$$\{\delta\}_e = \{w_1, 0, 0, 0, 0, 0, 0, 0, 0, 0, 0, \theta_{x4}\}$$

The stiffness matrix (Table 7.3) reduces to

$$[k]_e = \frac{Et^3}{180(1-0.09)} \begin{bmatrix} (60+60+0.3\times 30+0.35\times 84) & \text{Symmetric} \\ (0-30+0-0.35\times 6) & (0+20+0+0.35\times 8) \end{bmatrix} \begin{matrix} 1 \\ 12 \end{matrix}$$

(columns labelled 1 and 12)

The force–displacement relationship, Equation 7.26, is thus

$$\begin{Bmatrix} p_0/4 \\ 0 \end{Bmatrix}_e = \frac{Et^3}{163.8} \begin{bmatrix} 158.4 & -32.1 \\ -32.1 & 22.8 \end{bmatrix} \begin{Bmatrix} w_1 \\ \theta_{x4} \end{Bmatrix}_e \quad \text{(a)}$$

From the foregoing, we obtain

$$w_1 = w_{\max} = 0.3617 \frac{p_0}{Et^3}$$

The "exact" solution of this problem (see Example 5.11) is $0.3355p_0/Et^3$.

When the quarter-plate is divided into 4, 16, and 25 elements, applying Equation 7.27 the results are, respectively [8],

$$w_{\max} = 0.3512p_0/Et^3$$
$$w_{\max} = 0.3397p_0/Et^3$$
$$w_{\max} = 0.3378p_0/Et^3$$

It is apparent that the accuracy of the solution increases as the mesh is refined.

We note that, in the case of a uniformly loaded square plate with *all edges clamped*, $\theta_{x4} = 0$ in Figure 7.25c. Then Equation a reduces to

$$\frac{p_0}{4} = \frac{Et^3}{163.8}(158.4)w_1$$

or

$$w_1 = w_{\max} = 0.2585 \frac{p_0}{Et^3}$$

Comment: This deflection is 17.4% more than the "exact" value given in Section 5.7.

Problems

Sections 7.1 through 7.3

7.1 Verify that the effective shear forces are represented, as finite difference approximations at point 0 (Figure 7.1b), in the form

$$V_x = -\frac{D}{2h^3}[w_9 - w_{11} - 2(3-\nu)(w_1 - w_3) + (2-\nu)(w_5 - w_6 - w_7 + w_8)]$$
$$V_y = -\frac{D}{2h^3}[w_{10} - w_{12} - 2(3-\nu)(w_2 - w_4) + (2-\nu)(w_5 + w_6 - w_7 - w_8)] \quad \text{(P7.1)}$$

Referring to Table 7.1, check the correctness of the result for V_x.

7.2 Compute the deflection at points 1 through 5 for the beam and loading shown in Figure P7.2.

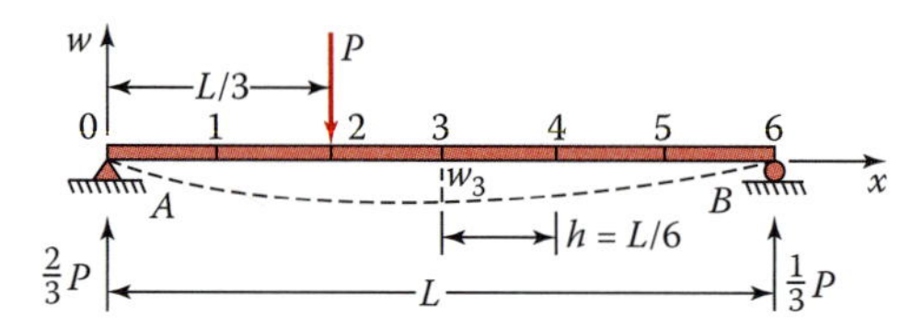

FIGURE P7.2

7.3 Determine the deflection of the free end of the cantilever beam loaded as shown in Figure P7.3.

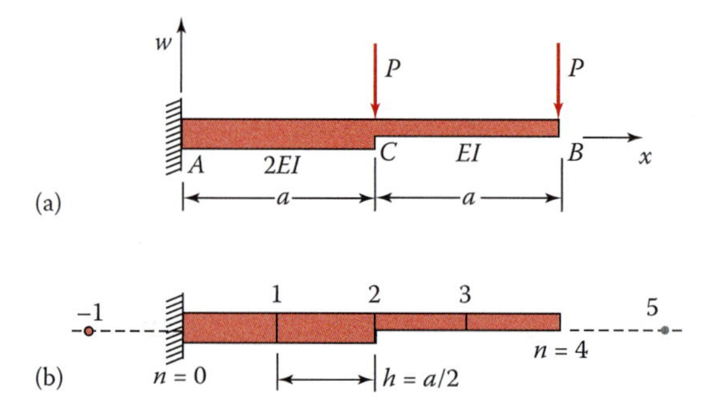

FIGURE P7.3

7.4 Using Table 7.1, calculate the midspan deflection and slope at support A of the beam loaded as shown in Figure P7.4.

7.5 A simply supported beam of length L is subjected to a distributed load that varies linearly as shown in Figure P7.5. Determine the deflection at the center, using the finite difference method. Take $n = 4$. Compare the result with the "exact" value $w_{max} = 5p_0L^4/768EI$.

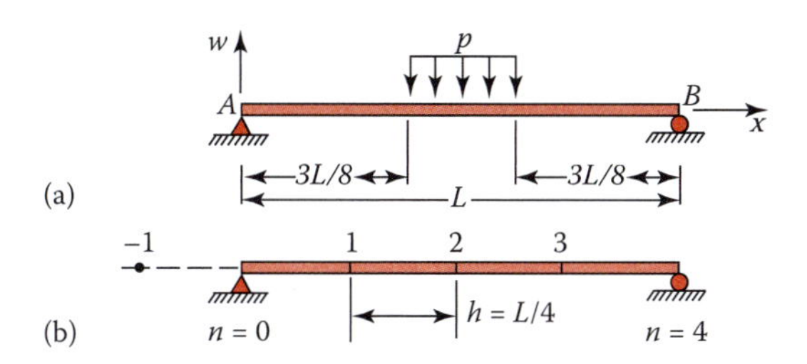

FIGURE P7.4

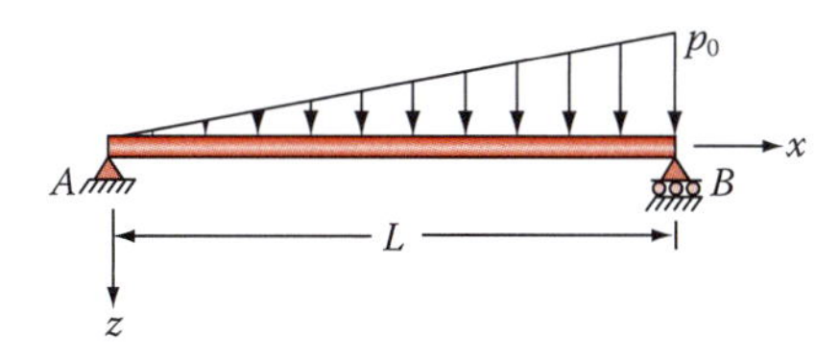

FIGURE P7.5

7.6 Determine the maximum deflection $w_{\max}$ and bending stress in a simply supported rectangular plate of sides a and $b = 1.5a$ that is subjected to a uniform load of intensity p_0 (Figure P7.6). Employ the finite difference method, taking $h = a/4$ and $\nu = 0.3$. Compare the results obtained with the "exact" values: $w_{\max} = 0.00772p_0a^4/D$ and $\sigma_{y,\max} = 0.4872p_0(a/t)^2$.

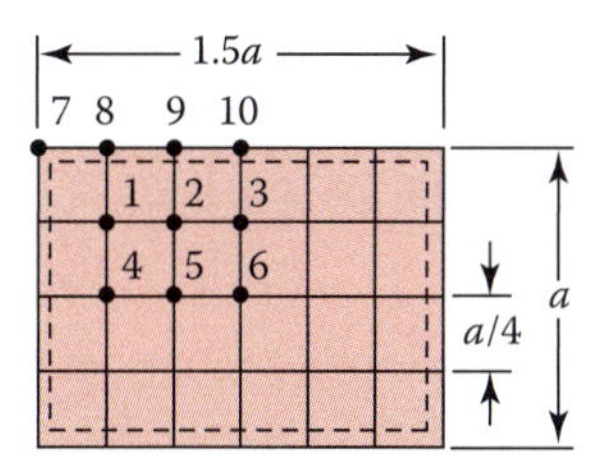

FIGURE P7.6

7.7 Redo Problem 7.6 for the case of a rectangular plate built in at the edges, the "exact" values are $w_{\max} = 0.0022p_0a^4/D$ and $\sigma_{y,\max} = 0.4542p_0(a/t)^2$.

7.8 Resolve Example 7.2 by approximating the nodal forces in accordance with a straight-line representation (Figure 7.2b).

7.9 Calculate the maximum principal stress at the nodal point 22 for the uniformly loaded plate shown in Figure 7.8. Use $\nu = 0.3$.

7.10 For the uniformly loaded plate shown in Figure 7.8, determine the maximum principal strain at the nodal point 24 for $a = 2$ m, $t = 20$ mm, $p_0 = 15$ kPa, $E = 200$ GPa, and $\nu = 0.3$.

7.11 Determine the deflections at the nodal points 1 through 6 for a uniformly loaded plate supported as shown in Figure P7.11. Apply the finite difference method.

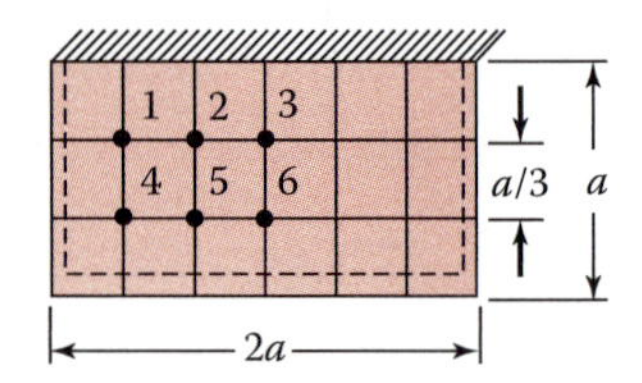

FIGURE P7.11

Sections 7.4 through 7.6

7.12 through 7.17 Each of the variously shaped plates shown in the Figures P7.12 through P7.17 is simply supported at all edges and carries a uniform loading of intensity p_0. Determine the deflection w at the nodal points labeled on the mesh configurations describing the plates.

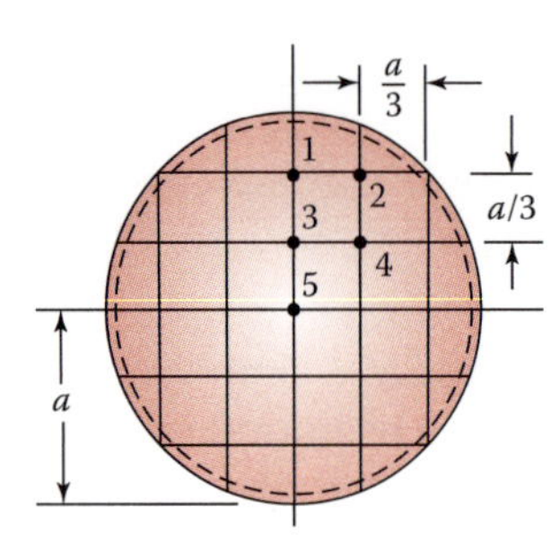

FIGURE P7.12

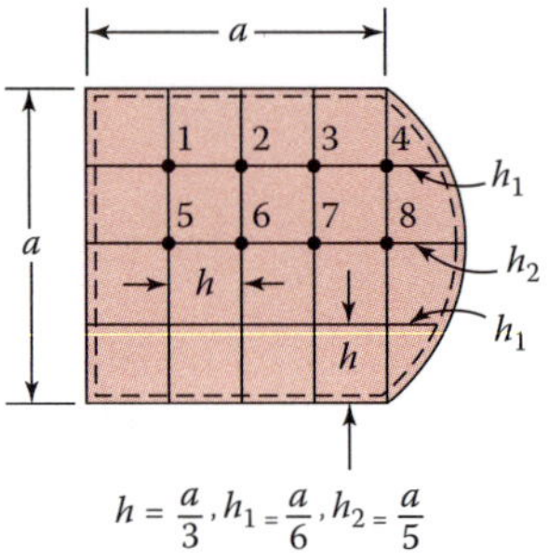

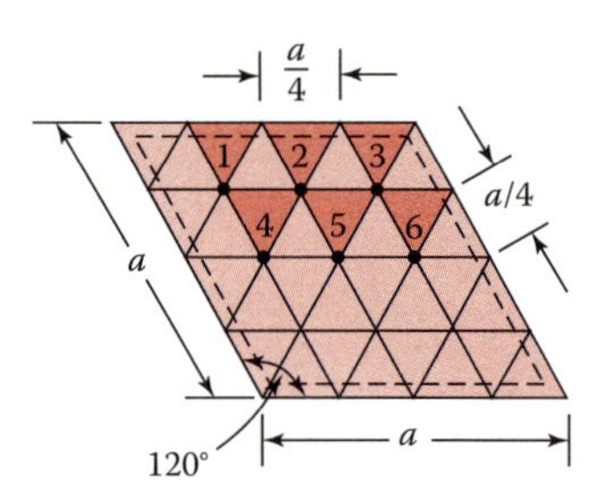

FIGURE P7.13

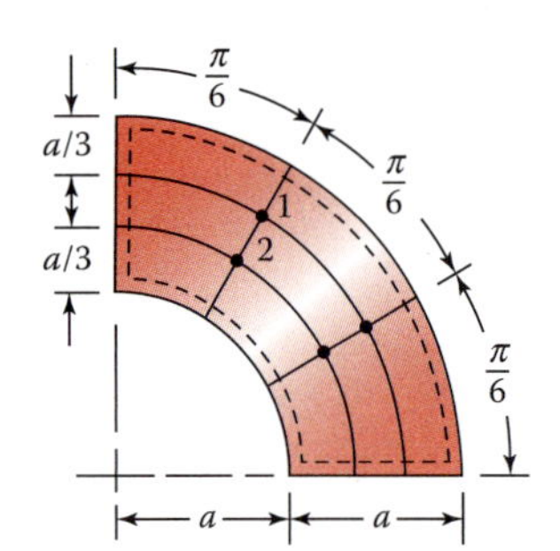

FIGURE P7.14

FIGURE P7.15

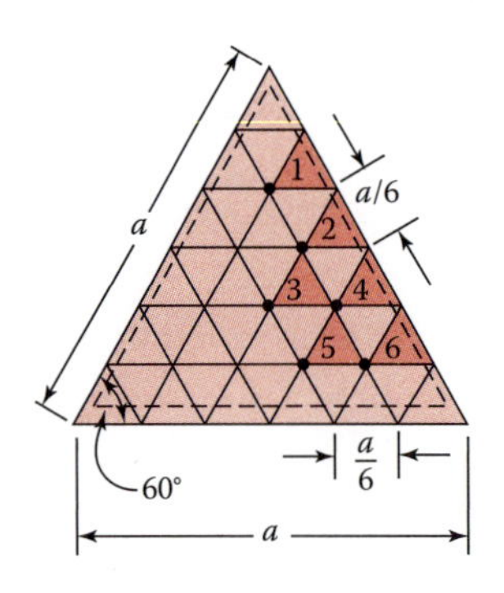

FIGURE P7.16

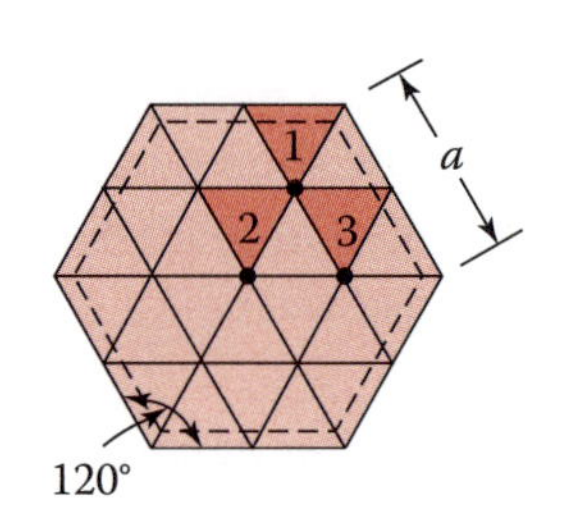

FIGURE P7.17

7.18 Verify the results given by Equations c of Section 7.6 using the chain rule.

7.19 Redo Problem 7.14 if the inner edge is built in and the remaining edges are simply supported.

7.20 For the beam of constant flexural rigidity EI shown in Figure P7.20, determine: (a) the stiffness matrix for each element; (b) the system stiffness matrix.

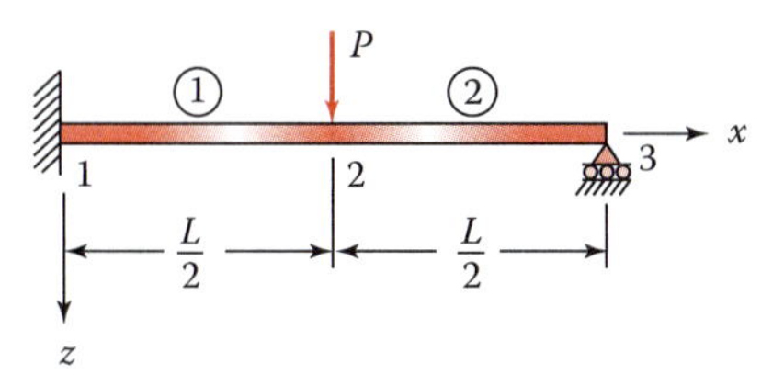

FIGURE P7.20

Sections 7.7 through 7.12

7.21 Figure P7.21 shows a cantilever of constant flexural rigidity EI under a concentrated load P at its free end. Find: (a) The deflection v_1 and angle of rotation θ_1 at the free end; (b) The reactions R_2 and M_2 at the fixed end.

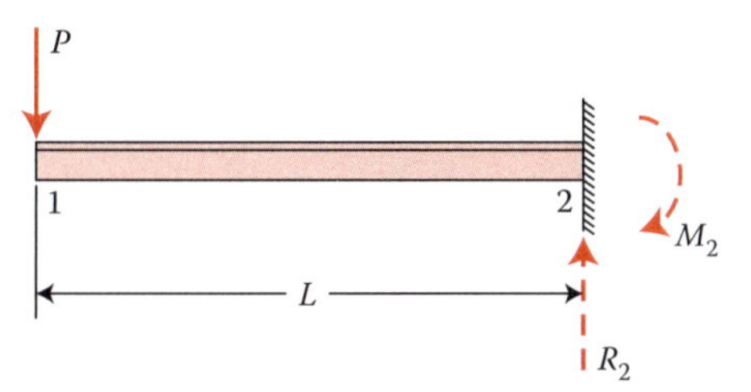

FIGURE P7.21

7.22 A simple beam 1–3 of length L and flexural rigidity EI carries a uniformly distributed load of intensity p as depicted in Figure P7.22. Replace the applied load with the equivalent nodal loads (see Table B.8), to determine the deflection of the beam at midpoint 2.

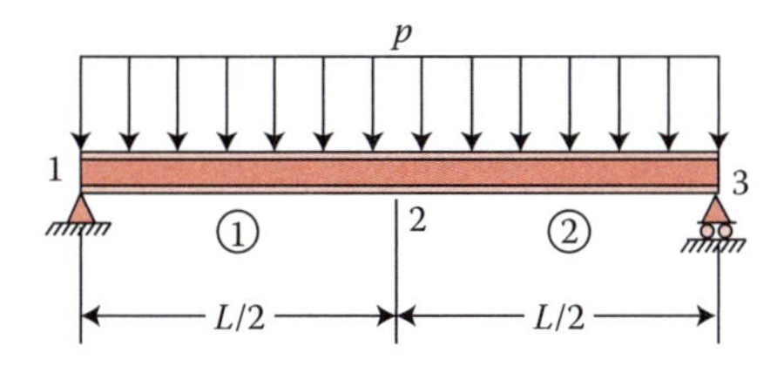

FIGURE P7.22

7.23 A cantilever beam 1–2 of length L and constant flexural rigidity EI supports a concentrated load P at the midspan (Figure P7.23). Replace the applied load with the equivalent nodal loads acting at each end of the beam (see Table B.8), to find the vertical deflection v_2 and angle of rotation θ_2 at the free end.

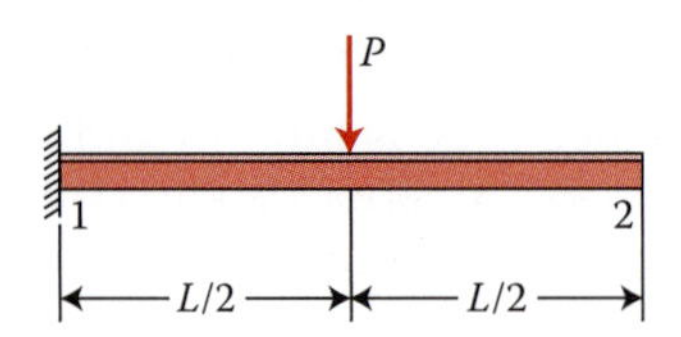

FIGURE P7.23

7.24 Figure P7.24 portrays a propped cantilever beam with flexural rigidities *EI* and *EI*/4 for the parts 1–2 and 2–3, respectively, carrying concentrated load *P* and moment 3*PL* at point 2. *Find*: The displacements υ_2, θ_2, and θ_3. *Given:* $L = 1.5$ m, $P = 30$ kN, $E = 210$ GPa, $I = 10 \times 10^6$ mm^4.

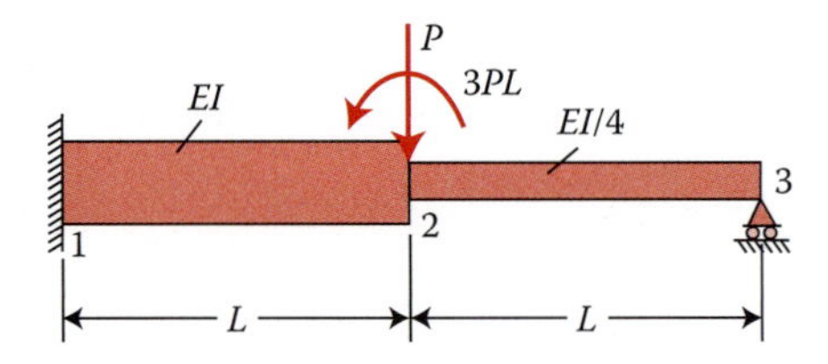

FIGURE P7.24

7.25 A propped cantilever beam of flexural rigidity *EI* supports a vertical load *P* at its midspan, as shown in Figure P7.25. *Find*: (a) The stiffness matrix for each element; (b) The system stiffness matrix; (c) The nodal displacements; (d) The forces and moments at the ends of each member; (e) The shear and moment diagrams. *Given:* $P = 40$ kN, $L = 2$ m, $EI = 1.3 \times 10^6$ N·m^2.

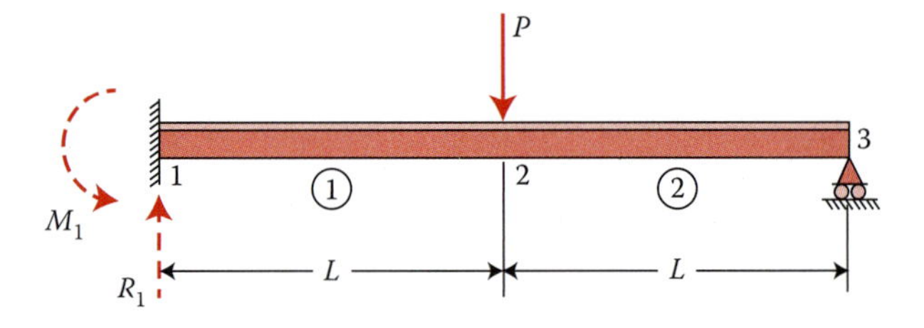

FIGURE P7.25

7.26 Figure P7.26 illustrates a stepped beam with fixed ends under a concentrated center load *P* that produces a vertical deflection Δ at the midpoint 2. Find, in terms of *EI*, *L*, and Δ, as required: (a) The load *P*; (b) The slope at point 2.

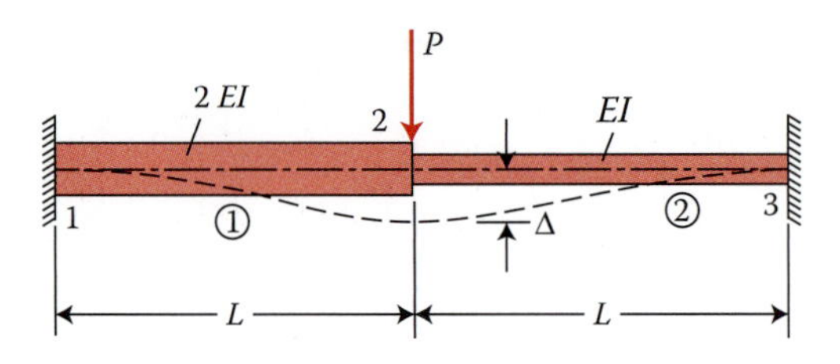

FIGURE P7.26

7.27 Develop the stiffness matrix of the finite element, described in Example 7.11, in terms of E and ν.

7.28 Derive the matrix [C] given by Equation 7.44 for the rectangular element of sides a and b (Figure 7.24) by locating the origin of xyz at the node m.

7.29 Calculate maximum deflection of a simply supported and uniformly loaded 2×2 m-square plate, using only one element per quarter-plate. Compare the result with that given in Table 5.1.

7.30 Redo Problem 7.29 for the case in which the plate is subjected to a concentrated load P at its center and $p_0 = 0$.

References

1. O. C. Zienkiewitcz and R. I. Taylor, *The Finite Element Method*, 6th ed., Elsevier, Oxford, UK, 2005.
2. T. Y. Yang, *Finite Element Structural Analysis*, Prentice-Hall, Upper Saddle River, NJ, 1986.
3. R. D. Cook et al., *Concepts and Applications of Finite Element Analysis*, 4th ed., Wiley, Hoboken, NJ, 2002.
4. K. J. Bathe, *Finite Element Procedures in Engineering Analysis*, Prentice-Hall, Upper Saddle River, NJ, 1996.
5. E. Hinton and D. R. J. Owen, *Finite Element Software for Plates and Shells*, Pineridge Press, Swansea, UK, 1984.
6. M. Bernadou, *Finite Element Methods for Thin Shell Problems*, Wiley, Chichester, UK, 1996.
7. A. C. Ugural and S. K. Fenster, *Advanced Mechanics of Materials and Applied Elasticity*, 5th ed., Prentice-Hall, Upper Saddle River, NJ, 2013.
8. R. Szilard, *Theory and Analysis of Plates—Classical and Numerical Methods*, Prentice-Hall, Upper Saddle River, NJ, 2004.
9. E. Kreyszic, *Advanced Engineering Mathematics*, 10th ed., Wiley, Hoboken, NJ, 2011.
10. S. Timoshenko and S. Woinowsky-Krieger, *Theory of Plates and Shells*, 2nd ed., McGraw-Hill, New York, 1959.
11. A. C. Ugural, *Mechanical Design of Machine Components*, CRC Press, Boca Raton, FL, 2015.
12. D. L. Logan, *First Course in the Finite Element Method*, 5th ed., Cengage Learning, Stanford, CT, 2011.

8

Anisotropic Plates

8.1 Introduction

The plates analyzed thus far have been assumed to be composed of a single homogeneous and isotropic material. However, plates of anisotropic materials have important applications because of their exceptionally high bending stiffness. A nonisotropic or *anisotropic* material displays direction-dependent properties. Simplest among them are those in which the material properties differ in *two* mutually perpendicular directions. A material so described is *orthotropic*.

Some *wood* materials may be modeled by orthotropic properties provided that local effects are ignored. A number of manufactured materials are approximated as orthotropic. Examples include corrugated and rolled metal sheet, fillers in sandwich plate construction, plywood, fiber-reinforced composites, reinforced concrete, and gridwork. The latter consists of two systems of equally spaced parallel ribs (beams), mutually perpendicular and attached rigidly at the point of intersection.

As already discussed in Section 1.14, a material having two or more distinct constituents is considered a *composite*. Usually, components consist of high-strength reinforcement material (e.g., steel, glass, fiber) embedded in a surrounding material (e.g., resin, concrete, nylon) termed *matrix*. Thus, a composite material has a relatively large strength-to-weight ratio as well as other desirable characteristics. Classical structural materials are more and more substituted by composite materials. For instance, orthotropic single or composite material plates and shells are widely used—in a great many space applications (Figure 8.1 depicts [1] an example.), in buildings, in pressure vessels, and in engine components, such as clutches and pump diaphragms.

Presented in this chapter are the fundamental equations for the small-deflection theory of bending of thin orthotropic, laminated composite, and sandwich plates. Orthotropic properties of several commonly employed materials are discussed, as are applications to various orthotropic plates. Topics on plates made from two or more different isotropic and anisotropic materials and on numerical methods are also included. The technical literature contains an abundance of specialized information on orthotropic and composite plates [2–10].

8.2 Basic Relationships

Solution of the bending problem of orthotropic plate requires reformulation of Hooke's law. Equation 3.6 now assumes the following form:

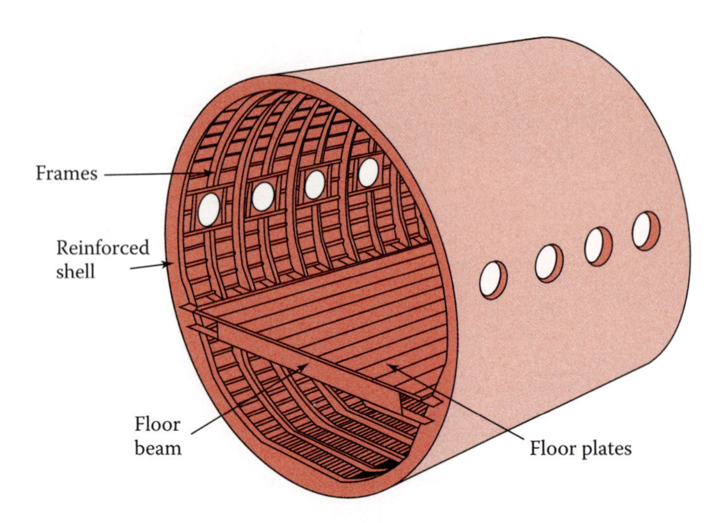

FIGURE 8.1
Orthotropic members of a typical flight structure.

$$\begin{aligned}\sigma_x &= E_x\varepsilon_x + E_{xy}\varepsilon_y \\ \sigma_y &= E_y\varepsilon_y + E_{xy}\varepsilon_x \\ \tau_{xy} &= G\gamma_{xy}\end{aligned} \tag{8.1}$$

in which the four moduli, E_x, E_y, E_{xy}, and G, are all independent of one another. In particular, Equation 1.35 is *no longer applicable*, as this relation involving E, ν, and G relies on material isotropy. An alternate representation of Equations 8.1 is

$$\begin{aligned}\sigma_x &= \frac{E'_x}{1-\nu_x\nu_y}(\varepsilon_x + \nu_y\varepsilon_y) \\ \sigma_y &= \frac{E'_y}{1-\nu_x\nu_y}(\varepsilon_y + \nu_x\varepsilon_x) \\ \tau_{xy} &= G\gamma_{xy}\end{aligned} \tag{8.2}$$

Here ν_x, ν_y, and E'_x, E'_y are the effective Poisson's ratios and effective moduli of elasticity, respectively. Subscripts x and y relate to the directions. The shear modulus of elasticity G is the *same* for both isotropic and orthotropic materials. Clearly, the two sets of elastic constants in Equations 8.1 and 8.2 are connected by

$$\begin{aligned}E_x &= \frac{E'_x}{1-\nu_x\nu_y} \\ E_y &= \frac{E'_y}{1-\nu_x\nu_y} \\ E_{xy} &= \frac{E'_x\nu_y}{1-\nu_x\nu_y} = \frac{E'_y\nu_x}{1-\nu_x\nu_y}\end{aligned} \tag{8.3}$$

The strain–displacement relations (Equations 3.3) are based on purely geometrical considerations and are *unchanged* for orthotropic plates. The same is true for the conditions of equilibrium of Section 3.7. The stresses are obtained by introducing Equations 3.3 into Equation 8.1:

$$\begin{aligned}\sigma_x &= -z\left(E_x\frac{\partial^2 w}{\partial x^2}+E_{xy}\frac{\partial^2 w}{\partial y^2}\right)\\ \sigma_y &= -z\left(E_y\frac{\partial^2 w}{\partial y^2}+E_{xy}\frac{\partial^2 w}{\partial x^2}\right)\\ \tau_{xy} &= -2Gz\frac{\partial^2 w}{\partial x\partial y}\end{aligned} \tag{8.4}$$

The formulas for the moments, by inserting the foregoing into Equations 3.8 and integrating the resulting expression, are found to be

$$\begin{aligned}M_x &= -\left(D_x\frac{\partial^2 w}{\partial x^2}+D_{xy}\frac{\partial^2 w}{\partial y^2}\right)\\ M_y &= -\left(D_y\frac{\partial^2 w}{\partial y^2}+D_{xy}\frac{\partial^2 w}{\partial x^2}\right)\\ M_{xy} &= -2G_{xy}\frac{\partial^2 w}{\partial x\partial y}\end{aligned} \tag{8.5}$$

where

$$D_x=\frac{t^3E_x}{12}\qquad D_y=\frac{t^3E_y}{12}\qquad D_{xy}=\frac{t^3E_{xy}}{12}\qquad G_{xy}=\frac{t^3G}{12} \tag{8.6}$$

The expressions for D_x, D_y, D_{xy}, and G_{xy} represent the *flexural rigidities* and the *torsional rigidity* of an orthotropic plate, respectively. We can obtain the vertical shear forces in the plate by substituting Equations 8.5 into Equations b and c of Section 3.7:

$$\begin{aligned}Q_x &= -\frac{\partial}{\partial x}\left(D_x\frac{\partial^2 w}{\partial x^2}+H\frac{\partial^2 w}{\partial y^2}\right)\\ Q_y &= -\frac{\partial}{\partial y}\left(D_y\frac{\partial^2 w}{\partial y^2}+H\frac{\partial^2 w}{\partial x^2}\right)\end{aligned} \tag{8.7}$$

where

$$H=D_{xy}+2G_{xy} \tag{8.8}$$

The governing *differential equation of deflection* for an orthotropic plate, through the use of Equations 3.15 and 8.5, is expressed in the form

$$D_x \frac{\partial^4 w}{\partial x^4} + 2H \frac{\partial^4 w}{\partial x^2 \partial y^2} + D_y \frac{\partial^4 w}{\partial y^4} = p \tag{8.9}$$

which is solved for w, on satisfying the given boundary conditions, as illustrated in Sections 8.4 and 8.5.

8.3 Determination of Rigidities

Discussed briefly in this section is the determination of the rigidities of orthotropic materials used in plate structures. Practical considerations often lead to assumptions, with regard to material properties, resulting in approximate expressions for the elastic constants. The accuracy of these approximations is generally the most significant factor in the orthotropic plate problem.

The orthotropic plate moduli and Poisson's ratios

$$E'_x, E'_y, \nu_x, \nu_y, G \tag{8.10}$$

are obtained by tension and shear tests, as in the case of isotropic materials. The plate rigidities are calculated from Equations 8.3, 8.6, and 8.8:

$$\begin{aligned}
D_x &= \frac{t^3 E'_x}{12(1-\nu_x\nu_y)} \qquad & G_{xy} &= \frac{t^3 G}{12} \\
D_y &= \frac{t^3 E'_y}{12(1-\nu_x\nu_y)} \qquad & H &= D_{xy} + 2G_{xy} \\
D_{xy} &= \frac{t^3 E'_x \nu_y}{12(1-\nu_x\nu_y)} = \frac{t^3 E'_y \nu_x}{12(1-\nu_x\nu_y)} & &
\end{aligned} \tag{8.11}$$

When it is *not possible*, however, to determine the constants of Equation 8.10 experimentally, we resort to approximations derived using analytical techniques. The latter approaches consist of constructing an orthotropic plate with elastic properties equal to the average properties of components of the original plate. Such a plate is termed an *equivalent* or *transformed orthotropic plate*. For example, in the case of a plate reinforced by ribs, the bending stiffness of the ribs and the plating are combined and taken to be uniform across the replacement model. Subsequently, the constants of Equation 8.10 are approximated with Equations 8.11, giving the rigidities. For reference purposes, Table 8.1 presents the rigidities for some commonly encountered cases [4–7].

It is noted that for the case in which $E'_x = E'_y = E$ (and hence $\nu_x = \nu_y = \nu$), Equations 8.3 become

$$E_x = E_y = \frac{E}{1-\nu^2} \qquad E_{xy} = \frac{\nu E}{1-\nu^2} \tag{8.12}$$

TABLE 8.1

Various Orthotropic Plates

Geometry	Rigidities
A. Reinforced-concrete slab with x- and y-directed reinforcement steel bars 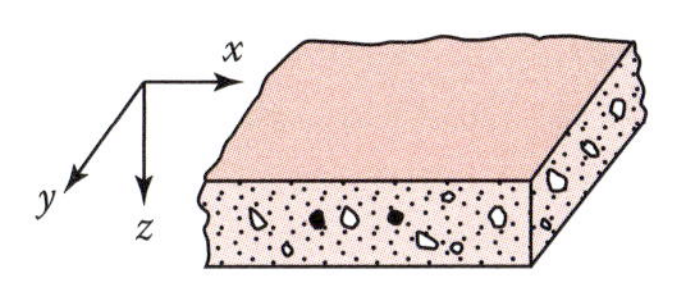 	$D_x = \frac{E_c}{1-\nu_c^2}\left[I_{cx} + \left(\frac{E_s}{E_c}-1\right)I_{sx}\right]$ $D_y = \frac{E_c}{1-\nu_c^2}\left[I_{cy} + \left(\frac{E_s}{E_c}-1\right)I_{sy}\right]$ $G_{xy} = \frac{1-\nu_c}{2}\sqrt{D_xD_y} \quad H = \sqrt{D_xD_y} \quad D_{xy} = \nu_c\sqrt{D_xD_y}$ ν_c: Poisson's ratio for concrete E_c, E_s: Elastic modulus of concrete and steel, respectively $I_{cx}(I_{sx})$, $I_{cy}(I_{sy})$: Moment of inertia of the slab (steel bars) about neutral axis in the section x = constant and y = constant, respectively
B. Plate reinforced by equidistant stiffeners 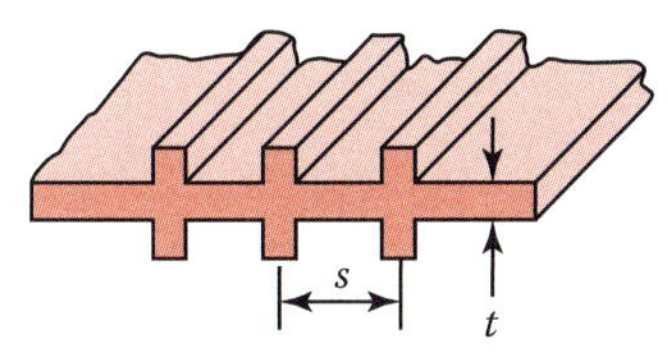 	$D_x = H = \frac{Et^3}{12(1-\nu^2)} \qquad D_y = \frac{Et^3}{12(1-\nu^2)} + \frac{E'I}{s}$ E, E': Elastic modulus of plating and stiffeners, respectively ν: Poisson's ratio of plating s: Spacing between centerlines of stiffeners I: Moment of inertia of the stiffener cross section with respect to midplane of plating
C. Plate reinforced by a set of equidistant ribs 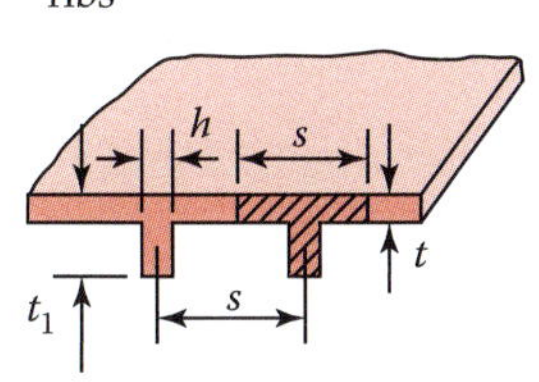 	$D_x = \frac{Est^3}{12\left[s-h+h(t/t_1)^3\right]} \qquad D_y = \frac{EI}{s}$ $H = 2G'_{xy} + \frac{C}{s} \qquad D_{xy} = 0$ C: Torsional rigidity of one rib I: Moment of inertia about neutral axis of a T-section of width s (shown as shaded) G'_{xy} : Torsional rigidity of the plating E: Elastic modulus of the plating
D. Corrugated plate 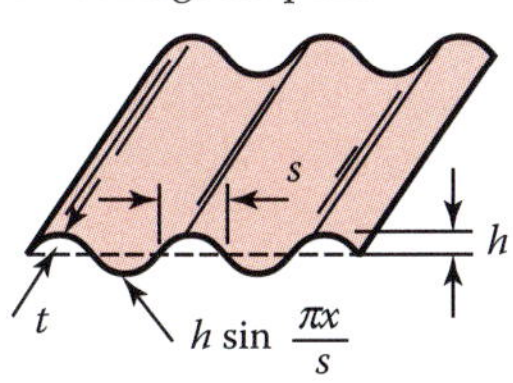 	$D_x = \frac{s}{\lambda}\frac{Et^3}{12(1-\nu^2)}, \quad D_y = EI, \quad H = \frac{\lambda}{a}\frac{Et^3}{12(1+\nu)}, \quad D_{xy} = 0$ Where $\lambda = s\left(1+\frac{\pi^2h^2}{4s^2}\right) \quad I = 0.5h^2t\left[1-\frac{0.81}{1+2.5(h/2s)^2}\right]$

Consequently

$$D_x = D_y = \frac{Et^3}{12(1-\nu^2)} \qquad G_{xy} = \frac{Et^3}{24(1+\nu)} \qquad H = \frac{Et^3}{12(1-\nu^2)} = D \tag{8.13}$$

It follows that Equation 8.9, as expected, reduces to that of an isotropic plate given by Equation 3.17.

8.4 Rectangular Orthotropic Plates

The general procedure for the determination of the deflection and stress in *rectangular orthotropic plates* is identical to that employed for isotropic plates. In this section, we shall first apply Navier's method to treat the case of a simply supported orthotropic plate. Then the problem of an orthotropic plate with mixed support conditions is discussed with the approach proposed by Lévy. There is also a brief discussion on the calculation of the deflection of an orthotropic plate by the use of the method of finite differences.

8.4.1 Application of Navier's Method

Consider a simply supported rectangular orthotropic plate under a nonuniform load $p(x, y)$. We take the coordinate axes as shown in Figure 8.2 and represent the load and deflection in the form of double trigonometric series.

Substitution of Equations 5.1 (see Section 5.2) into Equation 8.9 then gives

$$\sum_{m=1}^{\infty}\sum_{n=1}^{\infty}\left\{a_{mn}\left(\frac{m^4\pi^4}{a^4}D_x+2H\frac{m^2n^2\pi^4}{a^2b^2}+\frac{n^4\pi^4}{b^4}D_y\right)-p_{mn}\right\}\sin\frac{m\pi x}{a}\sin\frac{n\pi y}{b}=0$$

Inasmuch as the above must be valid for all x and y, it follows that the terms in the brackets must be zero, leading to

$$a_{mn}=\frac{p_{mn}}{(m^4\pi^4/a^4)D_x+2H(m^2n^2\pi^4/a^2b^2)+(n^4\pi^4/b^4)D_y} \tag{8.14}$$

The expression of the plate-deflection surface, by substitution of Equations 8.14 and 5.3 into Equation 5.1b, is therefore

$$w=\frac{4}{ab}\sum_{m=1}^{\infty}\sum_{n=1}^{\infty}\int_0^b\int_0^a\frac{p(x,y)\sin(m\pi x/a)\sin(n\pi y/b)dxdy}{(m^4\pi^4/a^4)D_x+2H(m^2n^2\pi^4/a^2b^2)+(n^4\pi^4/b^4)D_y}$$
$$\times\sin\frac{m\pi x}{a}\sin\frac{n\pi y}{b} \tag{8.15}$$

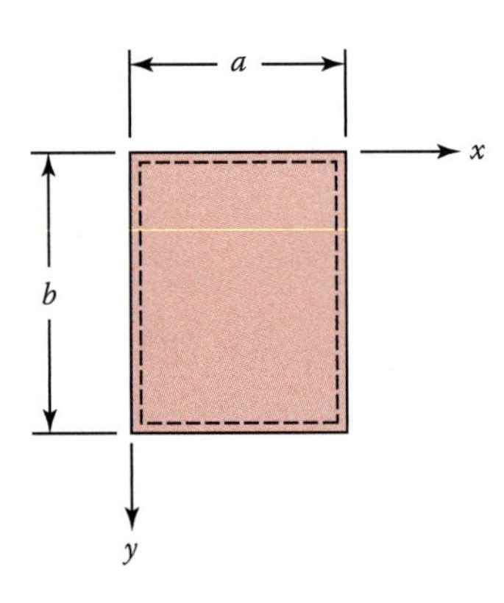

FIGURE 8.2
Simply supported orthotropic rectangular plate.

For an isotropic material, from Equations 8.12 and 8.13, $D_x = D_y = H = D$, and the above coincides with Equation 5.5.

EXAMPLE 8.1: Deflection of Reinforced-Concrete Plate

A reinforced-concrete rectangular plate simply supported on all edges is under a uniformly distributed load p_o (Figure 8.2). Determine: (a) the equation of the deflection surface; (b) the deflection and bending moments at the center.

Solution

a. Referring to Section 5.3, we can readily obtain from Equation 8.15

$$w = \frac{16p_0}{\pi^6} \sum_m^\infty \sum_n^\infty \frac{\sin(m\pi x/a)\sin(n\pi y/b)}{mn[(m^4/a^4)D_x + 2H(m^2n^2/a^2b^2) + (n^4/b^4)D_y]} \tag{8.16}$$

When the material is isotropic, the above reduces to Equation 5.6. For a *reinforced-concrete plate,* from Table 8.1, we have $H = \sqrt{D_x D_y}$. Based on the notation

$$a_1 = a\sqrt[4]{\frac{D}{D_x}} \quad b_1 = b\sqrt[4]{\frac{D}{D_y}} \tag{a}$$

Equation 8.16 becomes

$$w = \frac{16p_0}{D\pi^6} \sum_m^\infty \sum_n^\infty \frac{\sin(m\pi x/a)\sin(n\pi y/b)}{mn\left(m^2/a_1^2 + n^2/b_1^2\right)^2} \quad (m,n = 1,3,\ldots) \tag{b}$$

which is of the same from as Equation 5.6. We are led to conclude that the center deflection of the reinforced-concrete plate ($a \times b$) having rigidities D_x, D_y is equal to that of an isotropic plate ($a_1 \times b_1$) of rigidity D.

b. Having the expression for the deflection of the plate available, we can obtain the bending moments from Equations 8.5 and the stresses by applying the relationships of Section 3.5. The resulting expressions for the deflection and the bending moments at the center can be written in the form

$$\begin{aligned} w &= \delta_1 \frac{p_0 b}{D_y} \\ M_x &= \left(\delta_2 + \delta_3 \frac{E_{xy}}{E_x}\sqrt{\frac{D_x}{D_y}}\right)\frac{p_0 a^2}{\gamma} \\ M_y &= \left(\delta_2 + \delta_3 \frac{E_{xy}}{E_y}\sqrt{\frac{D_y}{D_x}}\right)p_0 b^2 \end{aligned} \tag{8.17}$$

where

$$\gamma = \frac{a}{b}\sqrt[4]{\frac{D_y}{D_x}}$$

TABLE 8.2

Coefficients δ_1, δ_2, and δ_3 in Equations 8.17

γ	1.0	1.5	2.0	3.0	4.0	5.0	∞
δ_1	0.00407	0.00772	0.01013	0.01223	0.01282	0.01297	0.01302
δ_2	0.0368	0.0280	0.0174	0.0055	0.0015	0.0004	0
δ_3	0.0368	0.0728	0.0964	0.1172	0.1230	0.1245	0.1250

Numerical values [2] of the coefficients δ_1, δ_2, and δ_3 are furnished in Table 8.2.

8.4.2 Application of Lévy's Method

We now consider the bending of an orthotropic rectangular plate simply supported on the two opposite edges at $x = 0$ and $x = a$ and arbitrarily supported on the other two opposite sides at $y = \pm b/2$ (Figure 8.3). Let the plate be subjected to a nonuniform loading which is a function of x only.

The homogeneous form of Equation 8.9 is

$$D_x \frac{\partial^4 w_h}{\partial x^4} + 2H \frac{\partial^4 w_h}{\partial x^2 \partial y^2} + D_y \frac{\partial^4 w_h}{\partial y^4} = 0 \tag{8.18}$$

A solution of the foregoing equation, consistent with Lévy's technique (discussed in Sections 5.4 and 5.5), is given by Equation 5.15. Carrying this into Equation 8.18, we obtain

$$D_x \left(\frac{m\pi}{a} \right)^4 f_m - 2H \left(\frac{m\pi}{a} \right)^2 \frac{d^2 f_m}{dy^2} + D_y \frac{d^4 f_m}{dy^4} = 0 \tag{c}$$

The roots of the corresponding characteristic equation are

$$\lambda_{1,2,3,4} = \pm \frac{m\pi}{a} \sqrt{\frac{1}{D_y} \left(H \pm \sqrt{H^2 - D_x D_y} \right)} \tag{d}$$

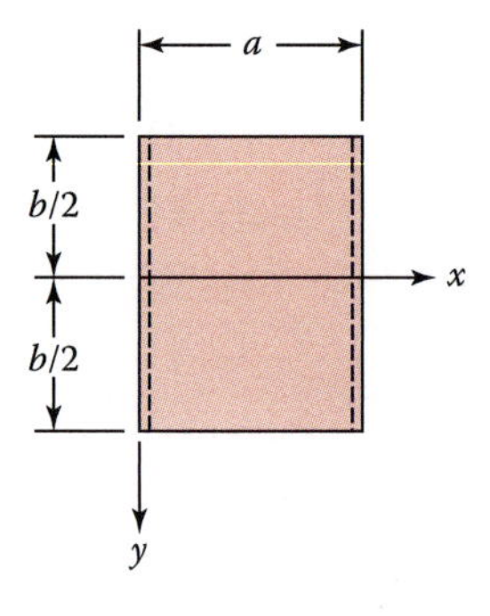

FIGURE 8.3
Orthotropic rectangular plate with mixed edge conditions.

The homogeneous solution is therefore

$$w_h = \sum_{m=1}^{\infty} \left(c_1 e^{\lambda_1 y} + c_2 e^{\lambda_2 y} + c_3 e^{\lambda_3 y} + c_4 e^{\lambda_4 y} \right) \sin \frac{m\pi x}{a} \tag{8.19}$$

where c_1, c_2, c_3, and c_4 are constants.

The loading, expanded in terms of sine series, is defined by Equation 5.31. Substituting Equations 5.15 and 5.31 into Equation 8.9 and noting the validity of the resulting expression for all values of x, we have

$$D_x \left(\frac{m\pi}{a} \right)^4 k_m - 2H \left(\frac{m\pi}{a} \right)^2 \frac{d^2 k_m}{dy^2} + D_y \frac{d^4 k_m}{dy^4} = p_m$$

It follows that $k_m = (p_m/D_x)(a/m\pi)^4$. Equation 5.19 becomes

$$w_p = \sum_{m=1}^{\infty} \frac{p_m}{D_x} \left(\frac{a}{m\pi} \right)^4 \sin \frac{m\pi x}{a} \tag{8.20}$$

The general expression for the deflection surface is thus

$$w = \sum_{m=1}^{\infty} \left[c_1 e^{\lambda_1 y} + c_2 e^{\lambda_2 y} + c_3 e^{\lambda_3 y} + c_4 e^{\lambda_4 y} + \frac{p_m}{D_x} \left(\frac{a}{m\pi} \right)^4 \right] \sin \frac{m\pi x}{a} \tag{8.21}$$

In the preceding p_m is given by Equation 5.21. The solution given by Equation 8.19 satisfies Equation 8.9 and simple support restraints at $x = 0$ and $x = a$. The coefficients c_1, c_2, c_3, and c_4 can easily by calculated for the particular conditions on the edges at $y = \pm b/2$ (Figure 8.2), as already demonstrated in Chapter 5.

8.4.3 Application of the Finite Difference Method

Numerical solutions of the deflection of orthotropic rectangular plates are determined by following the basic procedures of Chapter 7, as will be illustrated in Examples 8.2 and 8.6. Referring to Table 7.1, the pertinent coefficient pattern for finite difference expression of Equation 8.9 of the orthotropic plate may readily be obtained. This is given in Figure 8.4 for the case of evenly distributed nodes, that is, $\Delta x = \Delta y = h$.

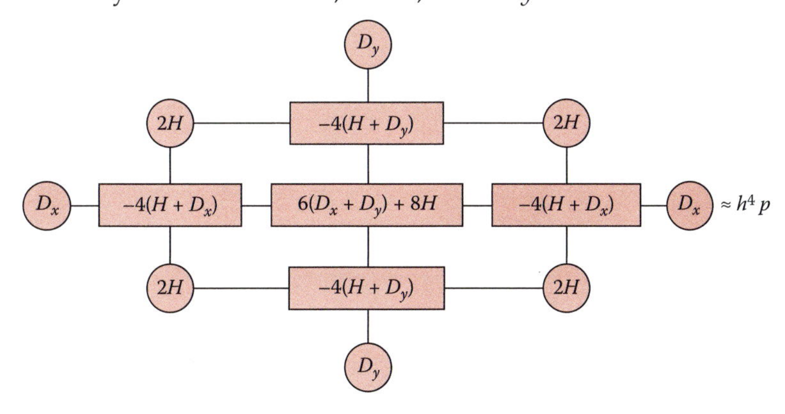

FIGURE 8.4
Coefficient pattern for finite difference expressions of orthotropic equation.

EXAMPLE 8.2: Deflection by Finite Difference Method of Orthotropic Plate

A square orthotropic plate is subjected to a uniform loading of intensity p_0. Assume that the plate edges are clamped and parallel with the principal directions of orthotropy. Determine the deflection, using the finite difference approach, by dividing the domain into equal nets with $h = a/4$. Take $D_x = D_0$, $D_y = 0.5D_0$, $H = 1.2148D_0$, and $\nu_x = \nu_y = 0.3$.

Solution

For this case, the governing expression for deflection, Equation 8.9, appears as

$$\frac{\partial^4 w}{\partial x^4} + 2.4296\frac{\partial^4 w}{\partial x^2 \partial y^2} + 0.5\frac{\partial^4 w}{\partial y^4} = \frac{p_0}{D_0} \tag{e}$$

Considerations of symmetry indicate that only one-quarter (darker shaded portion) of the plate need be analyzed (Figure 8.5). The conditions that the slopes vanish at the edges are satisfied by numbering the nodes located outside the plate surface as shown in the figure (see Example 7.2). The values of w are zero on the boundary.

Applying Figure 8.4 at the nodes 1, 2, 3, and 4, we obtain

$$\begin{bmatrix} 18.718 & -17.718 & -13.718 & 9.718 \\ -8.859 & 20.718 & 4.859 & -13.718 \\ -6.859 & 4.859 & 19.718 & -17.718 \\ 2.429 & -6.859 & -8.859 & 21.718 \end{bmatrix} \begin{Bmatrix} w_1 \\ w_2 \\ w_3 \\ w_4 \end{Bmatrix} = \frac{p_0 h^4}{D_0} \begin{Bmatrix} 1 \\ 1 \\ 1 \\ 1 \end{Bmatrix} \tag{f}$$

The simultaneous solution of Equations f results in

$$w_1 = 0.5334\frac{p_0 h^4}{D_0} \qquad w_2 = 0.3549\frac{p_0 h^4}{D_0}$$

$$w_3 = 0.3746\frac{p_0 h^4}{D_0} \qquad w_4 = 0.2513\frac{p_0 h^4}{D_0}$$

Comments: The center deflection, $w_1 = 0.0021p_0a^4/D_0$, is about 24% more than the "exact" value [2] of $0.00156p_0a^4/D_0$. By decreasing the size of the mesh increment, the accuracy of the solution can be improved.

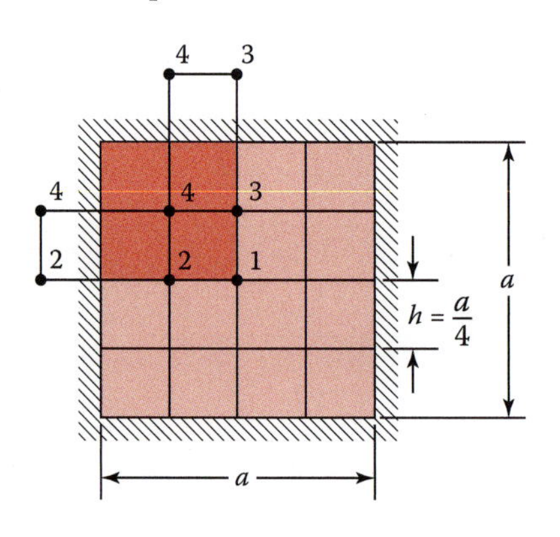

FIGURE 8.5
Clamped orthotropic plate with uniform loading.

8.5 Elliptic and Circular Orthotropic Plates

Consider an *elliptic orthotropic plate* with semiaxes a and b, clamped at the edge and subjected to the uniformly distributed load p_0 (Figure 6.4). Assume that the principal axes of the ellipse and the principal directions of the orthotropic material are parallel. The solution procedure follows a pattern similar to that described in Section 6.4. Thus, we let

$$w = k\left(1 - \frac{x^2}{a^2} - \frac{y^2}{b^2}\right)^2 \tag{a}$$

in which k is a constant to be determined. Substitution of the above into Equation 8.9 leads to an expression that is satisfied when

$$k = \frac{p_0}{8}\frac{a^4 b^4}{3b^4 D_x + 2a^2 b^2 H + 3a^4 D_y} \tag{b}$$

The expression describing the deflected surface of the plate is then

$$w = \frac{p_0}{8}\frac{a^4 b^4}{3b^4 D_x + 2a^2 b^2 H + 3a^4 D_y}\left(1 - \frac{x^2}{a^2} - \frac{y^2}{b^2}\right)^2 \tag{8.22}$$

This equation satisfies the boundary conditions for an elliptic plate with fixed edge, presented by Equations a in Section 6.4. The maximum deflection occurs at the center of the plate and is given by

$$w_{\max} = \frac{p_0 a^4 b^4}{24b^4 D_x + 16a^2 b^2 H + 24a^4 D_y} \tag{8.23}$$

As anticipated, in the case of an isotropic plate, Equations 8.22 and 8.23 reduce to Equations 6.12 and 6.13, respectively. Expressions for the moments may then be obtained from Equations 8.5.

The result obtained above for an elliptic plate may readily be reduced to the case of a circular plate by setting $b = a$. For a built-in-edge *orthotropic circular plate* of radius a under uniform load, we have, from Equation 8.22,

$$w = \frac{p_0}{64D_1}(a^2 - r^2)^2 \tag{8.24}$$

where

$$r = \sqrt{x^2 + y^2} \qquad D_1 = \frac{1}{8}(3D_x + 2H + 3D_y) \tag{c}$$

When $D_x = D_y = H = D$, Equation 8.24 is identical with Equation 4.19, the deflection formula for an isotropic circular plate.

The bending moments and the twisting moment are calculated by means of Equation 8.24, which, when introduced into Equations 8.5, yields

$$M_x = \frac{p_0}{16D_1}\left[(D_x + D_{xy})(a^2 - r^2) - 2\left(D_x x^2 + D_{xy} y^2\right)\right]$$
$$M_y = \frac{p_0}{16D_1}\left[(D_y + D_{xy})(a^2 - r^2) - 2\left(D_y y^2 + D_{xy} x^2\right)\right] \tag{8.25}$$
$$M_{xy} = \frac{p_0}{4D_1} D_{xy} xy$$

The stresses are then determined through application of the formulas given in Section 3.5.

8.6 Deflection by the Energy Method

As already pointed out in Section 3.12, the *energy methods* usually offer considerable ease of solution compared with equilibrium approaches to the analysis of orthotropic plates. Now Equation 3.44 for strain energy becomes (Problem 8.14)

$$U = \frac{1}{2}\int_0^b\int_0^a \left[D_x\left(\frac{\partial^2 w}{\partial x^2}\right)^2 + 2D_{xy}\frac{\partial^2 w}{\partial x^2}\frac{\partial^2 w}{\partial y^2} + D_y\left(\frac{\partial^2 w}{\partial y^2}\right)^2 + 4G_{xy}\left(\frac{\partial^2 w}{\partial x \partial y}\right)^2\right] dxdy \tag{8.26}$$

while the rest of the procedure remains the same as in the case of the isotropic plates. Application of the Ritz method for determination of deflection in orthotropic plates is illustrated in the examples to follow.

EXAMPLE 8.3: Deflection by Energy Method of Simple Orthotropic Plate

A simply supported rectangular plate is subjected to a lateral loading $p(x, y)$ distributed over its surface (Figure 8.2). Derive an expression for deflection w.

Solution

The deflection can be represented by the sine series defined by Equation 5.1b. Substitution of this series into Equation 8.26 gives, after integrating and letting $H = D_{xy} + 2G_{xy}$,

$$U = \frac{ab}{8}\sum_{m=1}^{\infty}\sum_{n=1}^{\infty} a_{mn}^2 \pi^4 \left[D_x\left(\frac{m}{a}\right)^4 + 2H\left(\frac{mn}{ab}\right)^2 + D_y\left(\frac{n}{b}\right)^4\right] \tag{8.27}$$

The work done by the load, from Equation 5.63, is

$$W = \int_0^b\int_0^a p(x,y) a_{mn} \sin\frac{m\pi x}{a}\sin\frac{n\pi y}{b} dxdy \tag{8.28}$$

The potential energy $\Pi = U - W$ may be written in the form

$$\Pi = U - W = \frac{ab}{8}\sum_{m=1}^{\infty}\sum_{n=1}^{\infty}\left\{a_{mn}^2\pi^4\left[D_x\left(\frac{m}{a}\right)^4 + 2H\left(\frac{mn}{ab}\right)^2 + D_y\left(\frac{n}{b}\right)^4\right] - 2a_{mn}p_{mn}\right\}$$

where p_{mn} is defined by Equation 5.3. From the conditions $\partial\Pi/\partial a_{mn} = 0$, we obtain once more the result given by Equation 8.14. The deflection is thus identical to that ascertained in Section 8.4.

EXAMPLE 8.4: Deflection by Energy Method of Clamped Orthotropic Plate

Determine an expression for the deflection surface of a clamped orthotropic plate under uniform load p_0 (Figure 8.6).

Solution

The boundary conditions are

$$\begin{aligned} w &= 0 \qquad \frac{\partial w}{\partial x} = 0 \qquad \left(x = \pm\frac{a}{2}\right) \\ w &= 0 \qquad \frac{\partial w}{\partial y} = 0 \qquad \left(y = \pm\frac{b}{2}\right) \end{aligned} \tag{a}$$

Assume a solution of the form

$$w = k\left(x^2 - \frac{a^2}{4}\right)^2\left(y^2 - \frac{b^2}{4}\right)^2 \tag{b}$$

wherein k is a constant. Note that this choice satisfies the foregoing boundary restraints. The corresponding strain energy, by Equation 8.26, is

$$U = \frac{a^5b^5}{225}\left[\frac{k^2}{49}(7D_xb^4 + 4Ha^2b^2 + 7D_ya^4)\right]$$

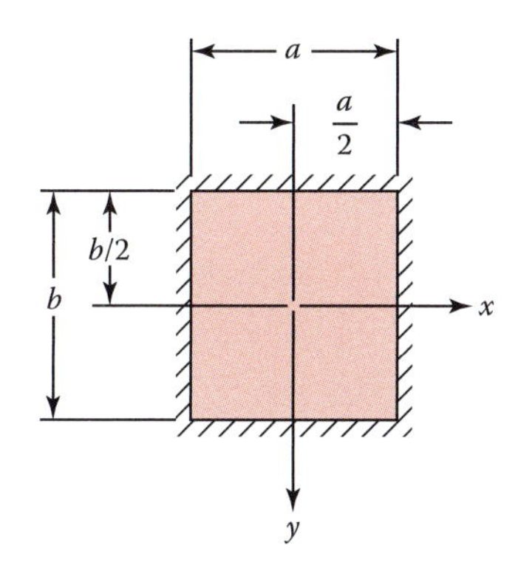

FIGURE 8.6
Clamped orthotropic plate with uniform loading.

Carrying Equation b and $p = p_0$ into Equation 5.63, we have

$$W = \int_0^b \int_0^a p_0 k \left(x^2 - \frac{a^2}{4}\right)^2 \left(y^2 - \frac{b^2}{4}\right)^2 dxdy = kp_0 \frac{a^5 b^5}{900}$$

The minimizing condition, $\partial \Pi \partial k = 0$, yields

$$k = \frac{49}{8} \frac{p_0}{7D_x b^4 + 4Ha^2 b^2 + 7D_y a^4}$$

The deflection is then

$$w = \frac{49p_0}{8} \frac{(x^2 - a^2/4)^2 (y^2 - b^2/4)^2}{7D_x b^4 + 4Ha^2 b^2 + 7D_y a^4} \tag{8.29}$$

The maximum deflection occurs at the center of the plate and is given by

$$w_{\max} = \frac{0.003418 p_0 a^4}{D_x + 0.5714 H(a/b)^2 + D_y (a/b)^2} \tag{c}$$

In the particular case of an *isotropic square plate*, $a = b$, $D_x = D_y = H = D$, and the solution becomes

$$w_{\max} = \frac{0.003418 p_0 a^4}{D[1 + 0.5714(a/b)^2 + (a/b)^2]} = 0.00133 \frac{p_0 a^4}{D}$$

Note that, retaining the first term of the series solution (see Section 5.12), we have $w_{\max} = 0.00128 p_0 a^4 / D$.

EXAMPLE 8.5: Maximum Deflection by Energy Method of Triangular Orthotropic Plate

A clamped plate in the form of a right triangle carries a uniform load p (Figure 8.7). Determine the maximum deflection if the plate is made of an orthotropic sheet in such a way that the principal directions x and y of the orthotropy are parallel to the side a and b.

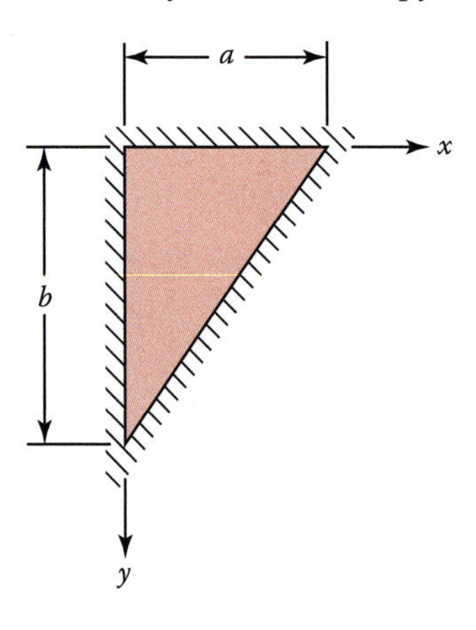

FIGURE 8.7
Clamped triangular orthotropic plate under uniform loading.

Solution

The deflection surface can be approximated by the expression

$$w = kx^2y^2\left(1 - \frac{x}{a} - \frac{y}{b}\right)^2 \tag{d}$$

that stratifies the edge conditions. Substituting the preceding into Equations 8.26 and 5.63 and setting $p = p_0$, after integrating we obtain U and W.

The requirement that $\Pi = U - W$ have minimum value $\partial\Pi/\partial k = 0$ leads to the value of k. Then the deflection is given in the form

$$w = \frac{0.3125p_0(a/b)^2}{D_x + H(a/b)^2 + D_y(a/b)^2}x^2y^2\left(1 - \frac{x}{a} - \frac{y}{b}\right)^2 \tag{8.30}$$

The maximum deflection takes place at the center ($x = a/3$, $y = b/3$) and is found to be

$$w_{\max} = \frac{0.000429p_0a^4}{D_x + H(a/b)^2 + D_y(a/b)^2} \tag{e}$$

Comment: Various orthotropic triangular plates subject to different end conditions may also be treated in a manner analogous to that given in this example [2].

8.7 *Plates of Isotropic Multilayers

Structures composed of an arbitrary number of bounded layers (e.g., aircraft and marine windshields and portions of space vehicles) can often be satisfactorily approximated by considering a *laminated plate* (Figure 1.17). Generally, in these structural assemblies, each layer may possess a different thickness, orientation of the principal axes, and anisotropic properties (see Section 8.10). We shall treat here only the plates consisting of *isotropic* layers.

Multilayered plates are fabricated such that they act as single-layer materials. The layers cannot slip over each other, and the displacements remain continuous across the bond; the strains vary linearly over the thickness. Thus, strain–displacement relations (Equations 3.3), for the kth layer (Figure 8.8), are expressed as follows:

$$\varepsilon_x^{(k)} = -z_k\frac{\partial^2 w}{\partial x^2} \qquad \varepsilon_y^{(k)} = -z_k\frac{\partial^2 w}{\partial y^2} \qquad \gamma_{xy}^{(k)} = -2z_k\frac{\partial^2 w}{\partial x \partial y} \tag{8.31}$$

Hooke's law (Equations 3.6) now appears as

$$\begin{aligned} \sigma_x^{(k)} &= \frac{E_k}{1-\nu_k^2}\left[\varepsilon_x^{(k)} + \nu_k\varepsilon_y^{(k)}\right] \\ \sigma_y^{(k)} &= \frac{E_k}{1-\nu_k^2}\left[\varepsilon_y^{(k)} + \nu_k\varepsilon_x^{(k)}\right] \\ \tau_{xy}^{(k)} &= \frac{E_k}{2(1+\nu_k)}\gamma_{xy}^{(k)} \end{aligned} \tag{8.32}$$

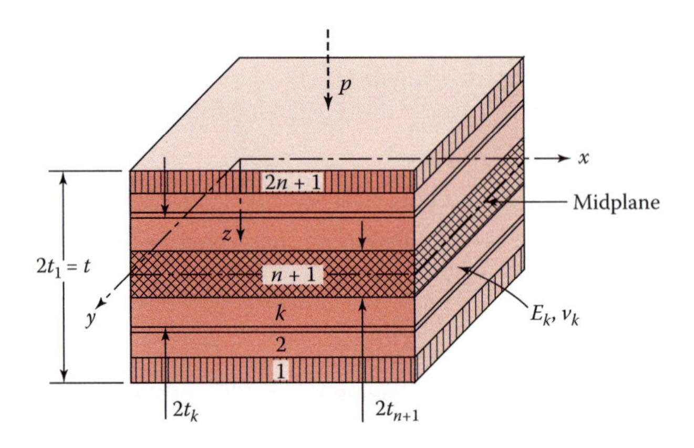

FIGURE 8.8
Symmetrically constructed layered plate.

Substituting strains defined by Equations 8.31 into the above, integrating over each layer, and summing the results, we obtain the stress resultants acting at the midplane:

$$\begin{Bmatrix} M_x \\ M_y \\ M_{xy} \end{Bmatrix} = \sum_k \int_{z_{k-1}}^{z_k} \begin{Bmatrix} \sigma_x \\ \sigma_y \\ \tau_{xy} \end{Bmatrix}_{(k)} z dz \tag{8.33}$$

Stresses defined by Equations 3.7, for the kth layer, are

$$\begin{aligned}
\sigma_x^{(k)} &= -z_k \frac{E_k}{1-\nu_k^2}\left(\frac{\partial^2 w}{\partial x^2} + \nu_k \frac{\partial^2 w}{\partial y^2}\right) \\
\sigma_y^{(k)} &= -z_k \frac{E_k}{1-\nu_k^2}\left(\frac{\partial^2 w}{\partial y^2} + \nu_k \frac{\partial^2 w}{\partial x^2}\right) \\
\tau_{xy}^{(k)} &= -z_k \frac{E_k}{1+\nu_k} \frac{\partial^2 w}{\partial x \partial y}
\end{aligned} \tag{8.34}$$

The general method of deriving the governing equation for multilayered plates follows a pattern identical with that described in Chapter 3. It can be shown that [7] the differential equation (Equations 3.17) now becomes

$$\nabla^4 w = \frac{p}{D_t} \tag{8.35}$$

where D_t is the *transformed flexural rigidity* of laminated plates. Layered plates of a symmetric structure about the midplane are of practical significance. For a plate of $2n + 1$ *symmetrical isotropic layers* (Figure 8.8), the transformed flexural rigidity D_t and Poisson's ratio ν_t are given by [2,9]:

$$D_t = \frac{2}{3}\left[\sum_{k=1}^{n} \frac{E_k}{1-\nu_k^2}\left(t_k^3 - t_{k+1}^3\right) + \frac{E_{n+1}t_{n+1}^3}{1-\nu_{n+1}^2}\right]$$
$$\nu_t = \frac{2}{3D_t}\left[\sum_{k=1}^{n} \frac{E_k\nu_k}{1-\nu_k^2}\left(t_k^3 - t_{k+1}^3\right) + \frac{E_{n+1}\nu_{n+1}t_{n+1}^3}{1-\nu_{n+1}^2}\right] \quad (8.36)$$

If boundary conditions, transverse load p, and the isotropic material properties of each layer are known, Equation 8.35 may be solved for $w(x, y)$. The stress components in the kth layer may then be computed from Equations 8.34.

We observe that, on introduction of transformed rigidity and Poisson's ratio, solution of a multilayered-plate problem reduces to that of a corresponding homogeneous plate. All analytical and numerical techniques are thus equally applicable to homogeneous and laminated plates.

8.8 The Finite Element Solution

In Sections 8.4 through 8.6, solutions of orthotropic plate problems were limited to simple cases in which there was uniformity of structural geometry and loading. In this section, the finite element approach of Chapter 7 is applied for computation of displacement and stress in an *orthotropic plate* of *arbitrary shape* and thickness, subjected to *nonuniform loads*.

For plates made of any nonisotropic material, it is necessary to rederive the elasticity matrix $[D]$. When the principal directions of orthotropy are parallel to the directions of the x and y coordinates, the stress–generalized "strain" relationship is given by Equations 8.2. Written in the matrix form, they are as follows:

$$\begin{Bmatrix} \sigma_x \\ \sigma_y \\ \sigma_{xy} \end{Bmatrix}_e = z\begin{bmatrix} \dfrac{E'_x}{1-\nu_x\nu_y} & \dfrac{\nu_y E'_x}{1-\nu_x\nu_y} & 0 \\ \dfrac{\nu_x E'_y}{1-\nu_x\nu_y} & \dfrac{E'_y}{1-\nu_x\nu_y} & 0 \\ 0 & 0 & G \end{bmatrix}\begin{Bmatrix} \varepsilon_x \\ \varepsilon_y \\ \gamma_{xy} \end{Bmatrix}_e \quad (8.37a)$$

or, succinctly,

$$\{\sigma\}_e = z[D^*]\{\varepsilon\}_e \quad (8.37b)$$

The *elasticity matrix*, from Equation 7.22, is therefore

$$[D] = \frac{t^3}{12}\begin{bmatrix} \dfrac{E'_x}{1-\nu_x\nu_y} & \dfrac{\nu_y E'_x}{1-\nu_x\nu_y} & 0 \\ \dfrac{\nu_x E'_y}{1-\nu_x\nu_y} & \dfrac{E'_y}{1-\nu_x\nu_y} & 0 \\ 0 & 0 & G \end{bmatrix} \quad (8.38)$$

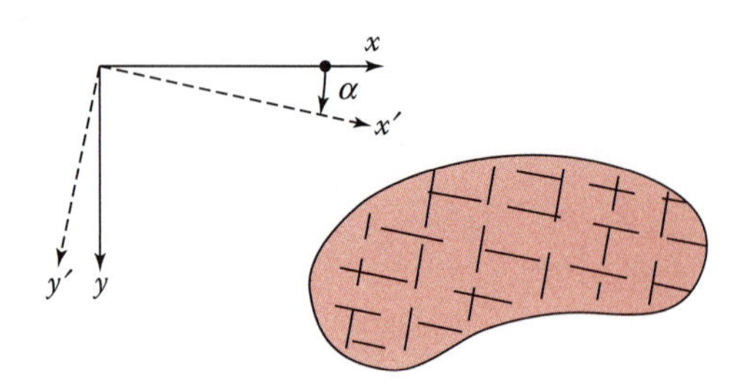

FIGURE 8.9
Plate with principal directions (x', y') of orthotropy.

The principal directions of orthotropy usually do *not coincide* with the x and y directions, however. Let us consider a plate in which x' and y' represent the principal directions of the material oriented at angle a with the reference coordinate axes (Figure 8.9). The stress and generalized "strain," in the directions of these coordinates, are related by

$$\begin{Bmatrix} \sigma_{x'} \\ \sigma_{y'} \\ \tau_{x'y'} \end{Bmatrix}_e = z \begin{bmatrix} \dfrac{E'_{x'}}{1-\nu_{x'}\nu_{y'}} & \dfrac{\nu_{y'}E'_{x'}}{1-\nu_{x'}\nu_{y'}} & 0 \\ \dfrac{\nu_{x'}E'_{y'}}{1-\nu_{x'}\nu_{y'}} & \dfrac{E'_{y'}}{1-\nu_{x'}\nu_{y'}} & 0 \\ 0 & 0 & G \end{bmatrix} \begin{Bmatrix} \varepsilon_{x'} \\ \varepsilon_{y'} \\ \gamma_{x'y'} \end{Bmatrix}_e \tag{8.39a}$$

or

$$\{\sigma'\}_e = z[D^{*'}]\{\varepsilon'\}_e \tag{8.39b}$$

where $E'_{x'}$, $E'_{y'}$, $\nu_{x'}$, and $\nu_{y'}$ are the material properties referred to the x' and y' directions.

Equations for transformation of the strain components ε_x, ε_y, γ_{xy} at any point of the plate, referring to Equations 1.25, are written in the following matrix form:

$$\begin{Bmatrix} \varepsilon_{x'} \\ \varepsilon_{y'} \\ \gamma_{x'y'} \end{Bmatrix} = \begin{bmatrix} \cos^2\alpha & \sin^2\alpha & \sin\alpha\cos\alpha \\ \sin^2\alpha & \cos^2\alpha & -\sin\alpha\cos\alpha \\ -2\sin\alpha\cos\alpha & 2\sin\alpha\cos\alpha & \cos^2\alpha - \sin^2\alpha \end{bmatrix} \begin{Bmatrix} \varepsilon_x \\ \varepsilon_y \\ \gamma_{xy} \end{Bmatrix} \tag{8.40a}$$

Concisely,

$$\{\varepsilon'\} = [T]\{\varepsilon\} \tag{8.40b}$$

where $[T]$ is called the *strain transformation matrix*. Similarly, the transformation relating stress components in x, y, z to those in x', y', z' is written as

$$\{\sigma\}_e = [T]^T\{\sigma'\}_e \tag{8.41}$$

On introducing Equations 8.39 together with Equations 8.40 into the above, we obtain

$$\{\sigma\}_e = z[T]^T[D^{*'}][T]\{\varepsilon\} = z[D^*]\{\varepsilon\}_e \tag{8.42}$$

in which

$$[D^*] = [T]^T[D^{*'}][T] \tag{8.43}$$

We thus have, from Equations 7.22 and 8.43, the expression

$$[D] = \frac{t^3}{12}[T]^T[D^{*'}][T] \tag{8.44}$$

for the *elasticity matrix* of the orthotropic plate in which the principal *directions* of orthotropy are *not oriented* along the x and y axes.

With Equations 8.38 or 8.44, explicit expressions of stiffness matrices for orthotropic plate elements may be evaluated as outlined in the preceding chapter. In the case of a *rectangular*, orthotropic plate element, we obtain

$$[k]_e = \frac{1}{15\bar{a}b}[R]\{D_x[k_1] + D_y[k_2] + D_{xy}[k_3] + G_{xy}[k_4]\}[R] \tag{8.45}$$

The coefficients $[k_1]$ through $[k_4]$ and $[R]$ are listed in Table 7.3. For any particular orthotropic material, the appropriate values of the rigidities D_x, D_y, D_{xy}, and G_{xy} (Table 8.1) are specified.

The process of arriving at solutions for the orthotropic plates is identical to that described in Sections 7.9, 7.11, and 7.12.

EXAMPLE 8.6: Rework Example 8.2 with Different Data

Reconsider the orthotropic plate described in Example 8.2. Calculate the deflection, using the finite element method. Let $D_x = D_0$, $D_y = 3D_0$, $D_{xy} = 0.4D_0$, $G_{xy} = 0.5D_0$, $a = 2$ m, and divide the domain into four elements.

Solution

The boundary conditions permit only a lateral displacement W_1 at the center at which the concentrated nodal force is equal to $P_0/4$ (see Figure 7.25c). On following a procedure similar to that described in Example 7.12, from Equation 8.35 and Table 7.3 together with the given data, we readily obtain

$$\frac{p_0}{4} = \frac{D_0}{15}[60 + 3\times 60 + 0.4\times 30 + 0.5\times 84]w_1$$

The foregoing yields $w_1 = w_{\max} = 0.0193p_0/D_0$.

8.9 A Typical Layered Orthotropic Plate

Generally, analysis of laminated composite structures requires: a knowledge of anisotropic elasticity, structural theories (i.e., kinematics of deformation) of laminates; analytical and computational methods to obtain the solution of the governing equations; and failure criteria to predict the modes of failures and to find failure loads. Presented in this section are the basic equations for a commonly used plate of bonded *three-layer orthotropic materials*. We consider a *cross-ply* laminate where the fibers of the midlayer and the two outer layers are oriented along the x and y directions, respectively, as shown in Figure 8.10. It is assumed that all layers are of the same type of material (e.g., as in plywood).

Hooke's law is modified for an orthotropic single material in Section 8.2. The strain–displacement relations (Equations 3.3), the equilibrium conditions of Section 3.7, and the definitions (Equation 3.8a) of the moments may be employed without any change. The stress–displacement relationship (Equations 8.4) is expressed for the inner and the outer layers, respectively, as

$$\begin{Bmatrix} \sigma_x \\ \sigma_y \\ \tau_{xy} \end{Bmatrix}_i = -z \begin{bmatrix} E_x & E_{xy} & 0 \\ E_{xy} & E_y & 0 \\ 0 & 0 & G \end{bmatrix}_i \begin{bmatrix} \dfrac{\partial^2 w}{\partial x^2} \\ \dfrac{\partial^2 w}{\partial y^2} \\ 2\dfrac{\partial^2 w}{\partial x \partial y} \end{bmatrix} \tag{8.46a}$$

and

$$\begin{Bmatrix} \sigma_x \\ \sigma_y \\ \tau_{xy} \end{Bmatrix}_o = -z \begin{bmatrix} E_y & E_{xy} & 0 \\ E_{xy} & E_x & 0 \\ 0 & 0 & G \end{bmatrix}_o \begin{bmatrix} \dfrac{\partial^2 w}{\partial x^2} \\ \dfrac{\partial^2 w}{\partial y^2} \\ 2\dfrac{\partial^2 w}{\partial x \partial y} \end{bmatrix} \tag{8.46b}$$

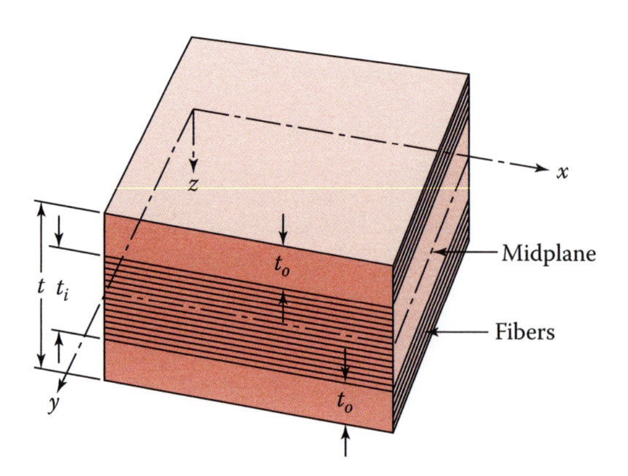

FIGURE 8.10
A three-cross-ply composite plate.

In the preceding, E_x, E_y, and E_{xy} are defined by Equations 8.3.

Introducing the preceding expressions into Equation 3.8a, with reference to Figure 8.9, we have

$$M_x = \int_{-t_i/2}^{t_i/2} \sigma_x^{(i)} z dz + 2\int_{t_i/2}^{t/2} \sigma_x^{(o)} z dz$$

$$= -\frac{1}{12}\left[E_x t_i^3 + E_y\left(t^3 - t_i^3\right)\right]\frac{\partial^2 w}{\partial x^2} - \frac{1}{12}E_{xy}t^3\frac{\partial^2 w}{\partial y^2}$$

Similarly, equations for the other moment components are derived. In so doing, altogether we have

$$\begin{aligned} M_x &= -\left(D_{11}\frac{\partial^2 w}{\partial x^2} + D_{12}\frac{\partial^2 w}{\partial y^2}\right) \\ M_y &= -\left(D_{22}\frac{\partial^2 w}{\partial y^2} + D_{12}\frac{\partial^2 w}{\partial x^2}\right) \\ M_{xy} &= -2D_{33}\frac{\partial^2 w}{\partial x \partial y} \end{aligned} \tag{8.47}$$

Here the plate properties are given by the following:

Flexural rigidities

$$\begin{aligned} D_{11} &= \frac{1}{12}\left[E_x t_i^3 + E_y\left(t^3 - t_i^3\right)\right] \\ D_{22} &= \frac{1}{12}\left[E_y t_i^3 + E_x\left(t^3 - t_i^3\right)\right] \\ D_{12} &= \frac{1}{12}E_{xy}t^3 \end{aligned} \tag{8.48a}$$

Torsional rigidity

$$D_{33} = \frac{1}{12}Gt^3 \tag{8.48b}$$

Finally, the new elastic law (Equations 8.47) is introduced into equations of equilibrium of Section 3.6. Then the *governing equation of deflection* for the plate that consists of three orthotropic layers, referring to Section 8.2, is expressed in the form

$$D_{11}\frac{\partial^4 w}{\partial x^4} + 2H_1\frac{\partial^4 w}{\partial x^2 \partial y^2} + D_{22}\frac{\partial^4 w}{\partial y^4} = p \tag{8.49}$$

where

$$H_1 = D_{12} + 2D_{33} \tag{8.50}$$

Equation 8.49 is solved for w, on fulfilling the prescribed edge conditions, as illustrated before.

In the particular case of *isotrophy*, we have E_x, E_y, and E_{xy} given by Equation 8.12. Hence,

$$D_{11} = D_{22} = D \qquad D_{12} = \nu D \qquad D_{33} = \frac{Et^3}{24(1+\nu)}$$

and Equations 8.47 and 8.49 reduce to the previous Equations 3.9 and 3.17. When $t_i = t$ (i.e., $t_o = 0$), another special case is obtained. We then arrive at the expression for a solid board of orthotropic plate, Equations 8.5, 8.6, and 8.9.

Comments: It is observed that the theory of anisotropic plates is *not* more involved than that of isotropic plates. The only additional difficulty is encountered in determining the plate rigidities, discussed in Section 8.3 and in the next section.

8.10 Laminated Composite Plates

We now discuss the bending of *laminated plates* consisting of two or more *orthotropic layers* bonded together at *various orientations*. Each layer may be of single or composite material. The basic expression derived here are the generalized forms of that given in the preceding section.

Consider a plate of n orthotropic arbitrarily oriented layers (Figure 8.11). As before, the strains are defined by Equations 8.31. Stress–displacement relations for the kth layer can be written from Equation 8.42 as follows:

$$\begin{Bmatrix} \sigma_x \\ \sigma_y \\ \tau_{xy} \end{Bmatrix}_k = -z[D^*]_k \begin{Bmatrix} \dfrac{\partial^2 w}{\partial x^2} \\ \dfrac{\partial^2 w}{\partial y^2} \\ 2\dfrac{\partial^2 w}{\partial x \partial y} \end{Bmatrix} = z[D^*]_k \begin{Bmatrix} \varepsilon_x \\ \varepsilon_y \\ \gamma_{xy} \end{Bmatrix} \tag{8.51}$$

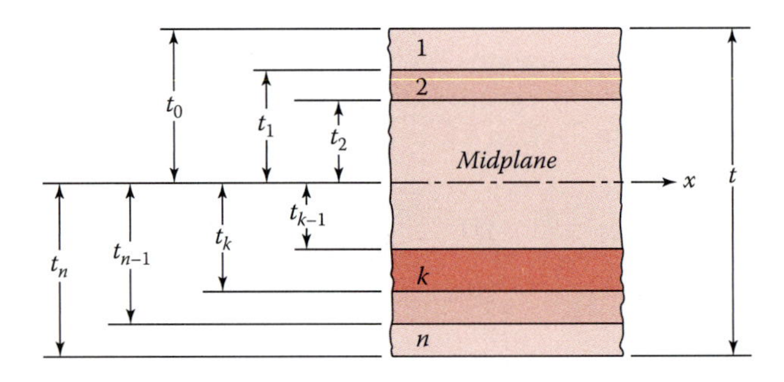

FIGURE 8.11
An n arbitrarily oriented orthotropic composite plate.

The terms $D_{ij}^*(i,j=1,2,3)$ of the *lamina rigidity matrix* $[D^*]$ are determined by introducing $[T]$ and its transpose from Equation 8.40 into Equation 8.43. In so doing, we obtain

$$[D^*]=\begin{bmatrix} D_{11}^* & D_{12}^* & D_{13}^* \\ D_{12}^* & D_{22}^* & D_{23}^* \\ D_{13}^* & D_{23}^* & D_{33}^* \end{bmatrix} \tag{8.52}$$

where

$$\begin{aligned}
D_{11}^* &= E_{x'}\cos^4\alpha + E_{y'}\sin^4\alpha + 2(E_{x'y'}+2G)\sin^2\alpha\cos^2\alpha \\
D_{22}^* &= E_{x'}\sin^4\alpha + E_{y'}\cos^4\alpha + 2(E_{x'y'}+2G)\sin^2\alpha\cos^2\alpha \\
D_{12}^* &= (E_{x'}+E_{y'}-4G)\sin^2\alpha\cos^2\alpha + E_{x'y'}(\sin^4\alpha+\cos^4\alpha) \\
D_{33}^* &= (E_{x'}+E_{y'}-2E_{x'y'}-2G)\sin^2\alpha\cos^2\alpha + G(\sin^4\alpha+\cos^4\alpha) \\
D_{13}^* &= (E_{x'}-E_{x'y'}-2G)\sin\alpha\cos^3\alpha - (E_{y'}-E_{x'y'}-2G)\sin^3\alpha\cos\alpha \\
D_{23}^* &= (E_{x'}-E_{x'y'}-2G)\sin^3\alpha\cos\alpha - (E_{y'}-E_{x'y'}-2G)\sin\alpha\cos^3\alpha
\end{aligned} \tag{8.53}$$

The quantities $E_{x'}$, $E_{y'}$, and $E_{x'y'}$ are referred to the principal directions (x', y') of the material. The α is the angle between these axes and the reference coordinates (x, y) of the lamina (Figure 8.9).

We note that the rigidity matrix now contains six different terms. However, it can be verified that D_{13}^* and D_{23}^* are *not* independent but linear combinations of the remaining terms [4]. Thus, the behavior of the orthotropic layer is governed by the basic rigidities D_{11}^*, D_{22}^*, D_{12}^*, and D_{33}^*.

Substitution of Equation 8.51 into Equation 3.8a results in

$$\begin{Bmatrix} M_x \\ M_y \\ M_{xy} \end{Bmatrix} = -\sum_{k=1}\int_{t_{k-1}}^{t_k} \begin{bmatrix} D_{11}^* & D_{12}^* & D_{13}^* \\ D_{12}^* & D_{22}^* & D_{23}^* \\ D_{13}^* & D_{23}^* & D_{33}^* \end{bmatrix}_k \begin{Bmatrix} \dfrac{\partial^2 w}{\partial x^2} \\ \dfrac{\partial^2 w}{\partial y^2} \\ 2\dfrac{\partial^2 w}{\partial x\partial y} \end{Bmatrix} z^2 dz \tag{8.54}$$

The equations for the bending and the twisting moments acting at the midplane of the plate are therefore

$$\begin{Bmatrix} M_x \\ M_y \\ M_{xy} \end{Bmatrix} = -\begin{bmatrix} D_{11} & D_{12} & D_{13} \\ D_{12} & D_{22} & D_{23} \\ D_{13} & D_{23} & D_{33} \end{bmatrix} \begin{Bmatrix} \dfrac{\partial^2 w}{\partial x^2} \\ \dfrac{\partial^2 w}{\partial y^2} \\ 2\dfrac{\partial^2 w}{\partial x\partial y} \end{Bmatrix} \tag{8.55}$$

where the terms of the lamina rigidity matrix [D] are

$$D_{ij} = \frac{1}{3}\left(D_{ij}^*\right)_k \left(t_k^3 - t_{k-1}^3\right) \qquad (i, j = 1, 2, 3) \tag{8.56}$$

For the case of the plate shown in Figure 8.9, the preceding equations are reduced to Equations 8.48.

Now, on following the procedure of Section 8.8, the governing differential equation for the deflection of the multilayered composite plate may be obtained. This expression can be written in the form [5] similar to that given in Equation 8.49.

We observe from Equations 8.53 and 8.56 that D_{13} and D_{23} *become zero* for *cross-ply* laminate where all the laminae are oriented at 0° and 90° (Section 8.9). For any other value of α, we have $D_{13} \neq 0$ and $D_{23} \neq 0$. However, if the plate is constructed by laying down alternate laminae at equal positive and negative angles, the D_{13} and D_{23} terms do become small, particularly if the number of layers is large.

Comparison of the plate rigidities D_{ij} is illustrated in the solution of the following sample problems.

EXAMPLE 8.7: Rigidity Matrix of Two-Layer Plate

Consider a two-layer plate having the play orientations of 0° and 30° with the laminate axes (Figure 8.12). The bottom ply (2) is a 0° layer with a thickness of 6 mm, whereas the 30° top lamina (1) is 4 mm thick. The elastic constants, referred to the principal directions of orthotropy, are the same for the layers:

$$E'_{x'} = 14.4 \text{ GPa} \qquad E'_{y'} = 3.6 \text{ GPa}$$
$$G = 5 \text{ GPa} \qquad \nu_{x'} = 0.4 \qquad \nu_{y'} = 0.1$$

Calculate the plate rigidity matrix.

Solution

Substituting the given constants into Equations 8.39, we obtain (in GPa) for the two laminae

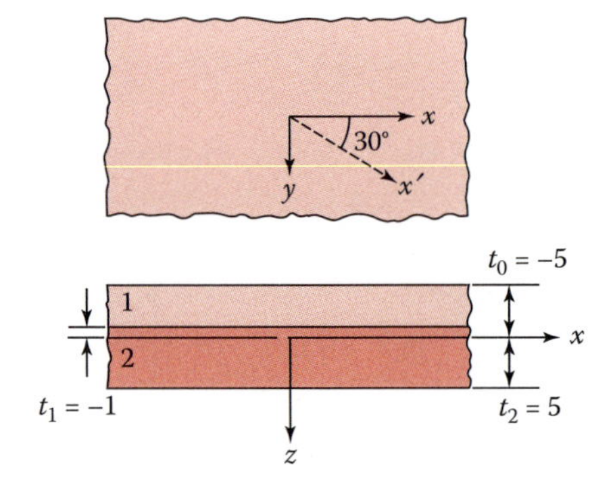

FIGURE 8.12
Two layer plate.

$$[D^{*\prime}] = \begin{bmatrix} \dfrac{14.4}{1-0.04} & \dfrac{0.1(14.4)}{1-0.04} & 0 \\ \dfrac{0.4(3.6)}{1-0.04} & \dfrac{3.6}{1-0.04} & 0 \\ 0 & 0 & 5 \end{bmatrix} = \begin{bmatrix} 15 & 1.5 & 0 \\ 1.5 & 3.75 & 0 \\ 0 & 0 & 5 \end{bmatrix} \tag{a}$$

For layer 2, we have $\alpha = 0°$, and Equations 8.53 lead to

$$[D^*]_2 = [D^{*\prime}] \tag{b}$$

Similarly, for layer 1, substituting the terms of $[D^{*\prime}]$ and $\alpha = 30°$ into Equations 8.53, we have

$$D^*_{11} = 15\cos^4 30° + 3.75\sin^4 30° + 2(1.5 + 2\times 5)\sin^2 30°\cos^2 30° = 12.984$$
$$D^*_{22} = 15\sin^4 30° + 3.75\cos^4 30° + 2(1.5 + 2\times 5)\sin^2 30°\cos^2 30° = 7.359$$
$$D^*_{12} = (15 + 3.75 - 4\times 5)\sin^2 30°\cos^2 30° + 1.5(\sin^4 30° + \cos^4 30°) = 0.703$$
$$D^*_{33} = (15 + 3.75 - 2\times 5)\sin^2 30°\cos^2 30° + 5(\sin^4 30° + \cos^4 30°) = 4.203$$
$$D^*_{13} = (15 - 1.5 - 2\times 5)\sin 30°\cos^3 30° - (3.75 - 1.5 - 2\times 5)\sin^3 30°\cos 30° = 1.975$$
$$D^*_{23} = (15 - 1.5 - 2\times 5)\sin^3 30°\cos 30° - (3.75 - 1.5 - 2\times 5)\sin 30°\cos^3 30° = 2.895$$

The rigidity matrix (in GPa) of layer 1 referred to the x and y axes is thus

$$[D^*]_1 = \begin{bmatrix} 12.984 & 0.703 & 1.975 \\ 0.703 & 7.359 & 2.895 \\ 1.975 & 2.895 & 4.203 \end{bmatrix} \tag{c}$$

The formulas (Equation 8.56), with reference to Figure 8.12, lead to

$$\begin{aligned} D_{ij} &= \frac{1}{3}\sum_{k=1}^{2}\left(D^*_{ij}\right)_k\left(t_k^3 - t_{k-1}^3\right) \\ &= \frac{1}{3}\left(D^*_{ij}\right)_1[(-1)^3 - (-5)^3] + \frac{1}{3}\left(D^*_{ij}\right)_2[(5)^3 - (-1)^3] \\ &= \frac{124}{3}\left(D^*_{ij}\right)_1 + 42\left(D^*_{ij}\right)_2 \end{aligned}$$

Then, carrying the results given by Equations b and c into the preceding, we have the lamina rigidity matrix (in N · m):

$$[D] = \frac{124}{3}\begin{bmatrix} 12.984 & 0.703 & 1.975 \\ 0.703 & 7.359 & 2.895 \\ 1.975 & 2.895 & 4.203 \end{bmatrix} + 42\begin{bmatrix} 15 & 1.5 & 0 \\ 1.5 & 3.75 & 0 \\ 0 & 0 & 5 \end{bmatrix} = \begin{bmatrix} 1166.672 & 92.057 & 81.633 \\ 92.057 & 461.672 & 119.660 \\ 81.633 & 119.660 & 383.724 \end{bmatrix} \tag{d}$$

The moments in terms of the curvatures of the composite plate are found on introducing Equation d into Equation 8.55.

EXAMPLE 8.8: Rigidity Matrix of Three-Layer Plate

Find the rigidity matrix D for a three-ply laminate having the top and the bottom layers 4-mm thick and oriented at 45° to the plate axes, while the 6-mm thick midlayer is oriented at 0° (Figure 8.13). Assume that each lamina has the identical properties as in the preceding example.

Solution

Equation b applies to layer 2. Substituting Equation a together with $\alpha = 45°$ into Equations 8.53 gives (in GPa)

$$D_{11}^* = 15\cos^4 45° + 3.75\sin^4 45° + 2(1.5+10)\sin^2 45°\cos^2 45° = 10.437$$

$$D_{22}^* = 15\sin^4 45° + 3.75\cos^4 45° + 2(1.5+10)\sin^2 45°\cos^2 45° = 10.437$$

$$D_{12}^* = (15 + 3.75 - 20)\sin^2 45°\cos^2 45° + 1.5(\sin^4 45° + \cos^4 45°) = 0.437$$

$$D_{33}^* = (15 + 3.75 - 3 - 10)\sin^2 45°\cos^2 45° + 5(\sin^4 45° + \cos^4 45°) = 3.937$$

$$D_{13}^* = (15 - 1.5 - 10)\sin 45°\cos^3 45° - (3.75 - 1.5 - 10)\sin^3 45°\cos 45° = 2.812$$

$$D_{23}^* = (15 - 1.5 - 10)\sin^3 45°\cos 45° - (3.75 - 1.5 - 10)\sin 45°\cos^3 45 = 2.812$$

For layers 1 and 3, we thus have

$$[D^*]_1 = [D^*]_3 = \begin{bmatrix} 10.437 & 0.437 & 2.812 \\ 0.437 & 10.437 & 2.812 \\ 2.812 & 2.812 & 3.937 \end{bmatrix} \tag{e}$$

Equation 8.56 is applied to Figure 8.13 to give, after letting $\left(D_{ij}^*\right)_1 = \left(D_{ij}^*\right)_3$,

$$\begin{aligned} D_{ij} &= \frac{1}{3}\sum_{k=1}^{3}\left(D_{ij}^*\right)_k\left(t_k^3 - t_{k-1}^3\right) \\ &= \frac{1}{3}\left(D_{ij}^*\right)_1[(-3)^3 - (-7)^3] + \frac{1}{3}\left(D_{ij}^*\right)_2[(3)^3 - (-3)^3] \\ &\quad + \frac{1}{3}\left(D_{ij}^*\right)_3[(7)^3 - (3)^3] = \frac{632}{3}\left(D_{ij}^*\right)_1 + 18\left(D_{ij}^*\right)_2 \end{aligned}$$

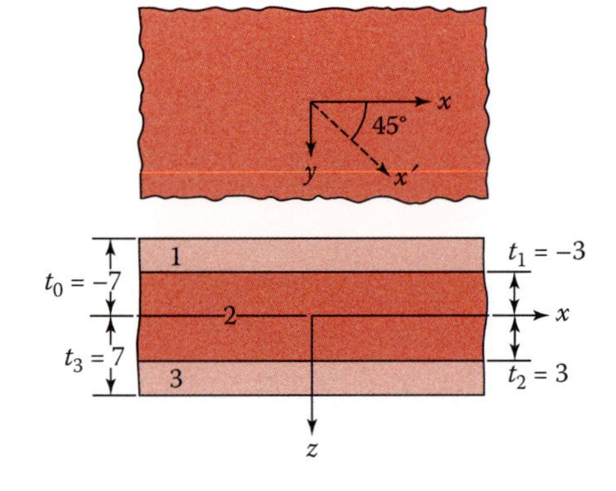

FIGURE 8.13
Three layer plate.

Substituting Equations a and e into the foregoing, we obtain (in N · m)

$$[D] = \frac{632}{3}\begin{bmatrix} 10.437 & 0.437 & 2.812 \\ 0.437 & 10.1437 & 2.812 \\ 2.812 & 2.812 & 3.937 \end{bmatrix} + 18\begin{bmatrix} 15 & 1.5 & 0 \\ 1.5 & 3.75 & 0 \\ 0 & 0 & 5 \end{bmatrix}$$

$$= \begin{bmatrix} 2468.8 & 119.2 & 592.5 \\ 119.2 & 2266.3 & 592.5 \\ 592.5 & 592.5 & 919.5 \end{bmatrix}$$

Comments: It is noted that each layered composite is unique in its properties and hence must be distinctively described. However, this is inconvenient for plates having a large number of layers. In practice, a concise identification of a laminate is facilitated through the use of the standard *laminate code* [5].

8.11 Sandwich and Honeycomb Plates

Sandwich plates are often employed in aerospace structures, shipbuilding, and building structures. There are a variety of sandwich construction, where basic idea is to separate two thin flat sheets (called *faces* or *skins*) of strong material (*metal* or *fiber* composites) by a low density comparatively weak core material. Usually, the core material is hardened foam or has a honeycomb (Figure 8.14a) or corrugated construction. Figure 8.14b shows a fabric sandwich construction, of which honeycomb or plastic foam is a special case. This has been very common for applications that require high strength–weight ratio combined with rigidity. Interestingly, sandwich plates act somewhat like I-beams. In the former, facings correspond to the web of the latter which carry the shear and helps to prevent buckling and wrinkling of the faces.

Figure 8.15 depicts an a by b rectangular sandwich plate of thickness t. The midplane of the plate lies in the xy plane. The faces are each of thickness h and the core is of

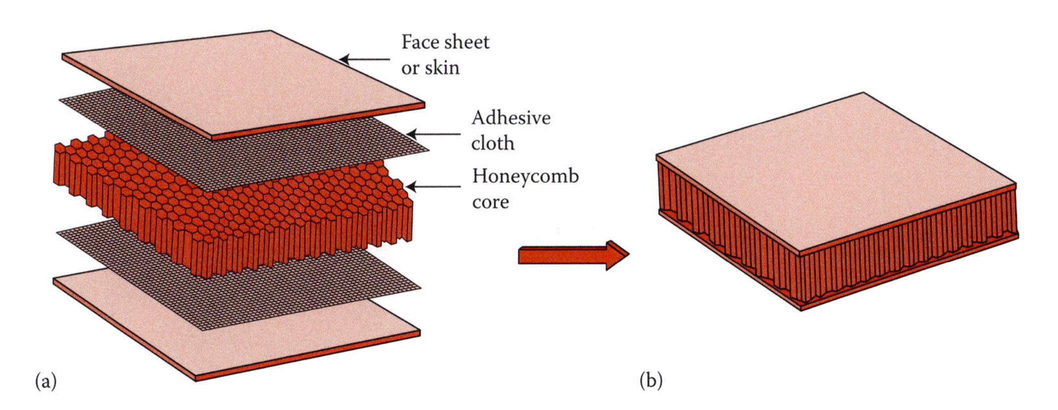

FIGURE 8.14
Sandwich material with honeycomb core: (a) Typical construction; (b) sandwich panel.

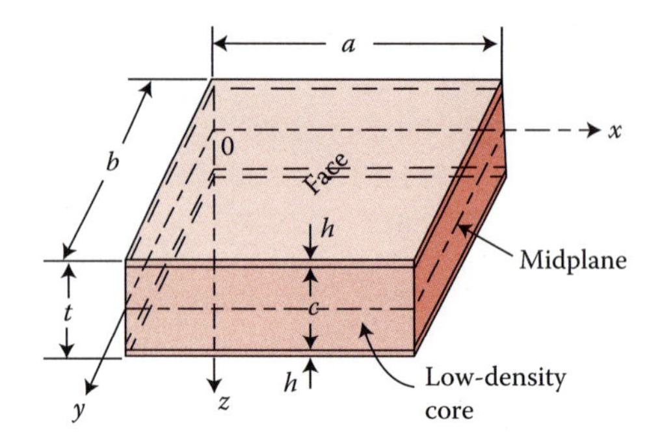

FIGURE 8.15
Rectangular sandwich plate.

thickness c. The *small deflection theory of sandwich plates* is based upon the assumptions that include

a. Faces are thin compared to the thickness of the core c, that is, $h \ll c$.
b. The faces carry all the bending and twisting moments.
c. The core makes no contribution to the bending stiffness.
d. The transverse shear stiffness of the core must be included in all sandwich calculations [9].

In the following section, only preliminary design applications of this type structure are discussed briefly.

8.11.1 Design of Sandwich-Type Beams and Plates

The stress and strain formulas of the mechanics of materials (see Section I) can be applied, in designing of sandwich-type beams and plates, provided that the correct values of the cross-sectional properties are used. Let us refer again Figure 8.15. The moment of inertia, can be obtained through the use of the parallel-axis theorem ($I = I_1 + Ad^2$) as

$$I = 2\left(\frac{1}{12}bh^3\right) + 2bh\left(\frac{c+h}{2}\right)^2 \tag{a}$$

Since the face thickness h is small, neglecting its powers, the foregoing reduces to

$$I = bhc\left(h + \frac{c}{2}\right) \tag{8.57}$$

By assumption (c), Equation 3.10, becomes

$$D = \frac{E}{1-\nu^2}\left(\int_{-t/2}^{t/2} t^2 dt - \int_{-c/2}^{c/2} c^2 dc\right)$$

After integrating, the flexural rigidity of the plate takes the form

$$D = \frac{Et^3}{12(1-\nu^2)}(t^3 - c^3) \tag{8.58}$$

Here E represents the modulus of elasticity of the facing material, ν its Poisson's ratio, and t and c are defined in Figure 8.15. Often in sandwich construction $c/t > 5$, and hence Equation 8.58 can be rewritten as

$$D = \frac{Eh(t+c)^2}{8(1-\nu^2)} \tag{8.59}$$

Having the section properties developed above, it is necessary to have material properties available. The modulus of elasticity of a facing material is usually known. In case for which the core material is *isotropic* and with the modulus of elasticity given, the shear modulus of elasticity can be found from the equation

$$G = \frac{E}{2(1+\nu)} \tag{1.35}$$

However, frequently it is desirable to know the shear modulus of elasticity of the core material. This can be found, in some cases, by experimental techniques or using empirical formulas [10].

Problems

Sections 8.1 through 8.5

8.1 A plate is reinforced by single equidistant stiffeners (Table 8.1). Compute the rigidities. The plate and stiffeners are made of steel with E = 200 GPa, ν = 0.3, s = 200 mm, t = 20 mm, and $I = 12 \times 10^{-7}$ m^4.

8.2 Determine the rigidities of an orthotropic steel bridge deck that may be approximated as a steel plate reinforced by a set of equidistant steel ribs (Table 8.1). Assume the following properties: $t = h = 10$ mm, $t_1 = 30$ mm, $s = 100$ mm, $\nu = 0.3$, and $E = 210$ GPa. Torsional rigidity of one rib is $C = JG = 0.246h^3(t_1 - t)G$.

8.3 Calculate the rigidities of a concrete slab with x- and y-directed reinforcement steel bars (Table 8.1). Use $I_{cx} = 1.5I_{cy} = 30 \times 10^8$ mm^4, $I_{sx} = 2I_{sy} = 2 \times 10^8$ mm^4, $E_s = 7E_c = 210$ GPa, and $\nu_c = 0.15$.

8.4 A rectangular building floor slab made of a reinforced-concrete material is subjected to a concentrated center load P (Figure 5.3b). Determine expressions for: (a) the deflection surface; (b) the bending moment M_x. The plate edges can be assumed to be simply supported. Take $b = 2a$, $m = n = 1$, $I_{sx} = I_{cx}/2$, $I_{sy} = I_{cy}/2$, $t = 0.2$ m, $E_s = 200$ GPa, $\nu_c = 0.15$, and $E_c = 21.4$ GPa.

8.5 Determine the value of the largest defection in the plate described in Problem 8.4. If $a = b$ retain the first two terms of the series solution.

8.6 A simply supported square plate of sides a is subjected to a uniform load p_0 (Figure 8.2). The plate is constructed from the material described in Problem 8.1. What should be the value of p_0 for an allowable deflection $w_{max} = 1$ mm? Retain only the first term of the series solution.

8.7 Determine, by taking $n = m = 1$, the center deflection of a simply supported square plate uniformly loaded by p_0 Assume that the plate is constructed of the material described in Problem 8.2.

8.8 A steel clamped manhole cover, subjected to uniform load p_0 consists of a flat plate reinforced by equidistant steel stiffeners and is elliptical in form (Figure 6.3). The material properties are given in Problem 8.1, and $a = 2b = 4$ m. Compute the maximum displacement w, assuming that the principal x and y axes of the ellipse and the material coincide.

8.9 Redo Problem 8.8. for a circular plate, $a = b$.

8.10 Derive expressions for the bending moments of an orthotropic elliptical plate with a built-in edge.

Sections 8.6 through 8.10

8.11 Determine the largest value of the deflection of a clamped rectangular plate (Figure 8.6). Assume that the plate is made of the material described in Problem 8.2. Take $b = 1.2a = 2$ m and $p_0 = 100$ kPa.

8.12 Compute the maximum deflection of a clamped right triangular plate (Figure 8.7). The plate is constructed of the reinforced equidistant stiffeners described in Problem 8.1. Let $b = 2a = 1$ m and $p_0 = 250$ MPa.

8.13 Rework Problem 8.12 for the case in which the clamped right triangular plate is made of the reinforced concrete described in Problem 8.3 and $p_0 = 500$ kPa.

8.14 Show that Equation 3.44 for strain energy appears in the form given by Equation 8.26 in the case of orthotropic plates.

8.15 Derive an expression for the deflection of an orthotropic clamped rectangular plate under a uniform load p_0 (Figure 5.19). Assume, by setting $m\pi/a = \alpha_m$ and $2n\pi/b = \beta_n$,

$$w = \sum_m^\infty \sum_n^\infty a_{mn}(1 - \cos\alpha_m x)(1 - \cos\beta_n y)$$

Use the Ritz method by retaining the first term of the series solution. Find the maximum deflection if $a = b$.

8.16 A 5-mm-thick large plate is fabricated of an orthotropic material having the properties

$$E'_{x'} = 2E'_{y'} = 2.2G = 13.6\,\text{GPa} \quad \nu_{x'} = 2\nu_{y'} = 0.2$$

The angle between the principal directions of the material (x', y') and the reference axes (x, y) is $\alpha = 30°$ (Figure 8.9). Determine the elasticity matrix of the plate.

8.17 Consider a five-layer plate with each layer having the same thickness. The properties of the layers are as shown in Figure P8.17. Verify that the transformed rigidity of the plate is given by

$$D_t = \frac{t^3}{1500}\left(\frac{99E_1}{1-\nu_1^2} + \frac{26E_2}{1-\nu_2^2}\right) \tag{P8.17}$$

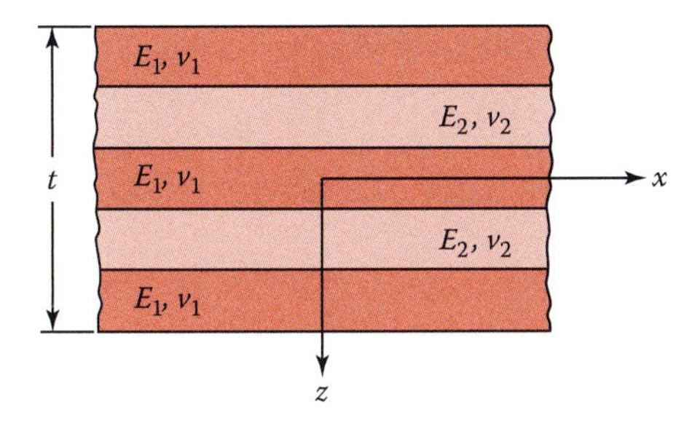

FIGURE P8.17

8.18 A three-ply laminate has the top and bottom layers oriented at 45° to the plate axes, while the midlayer is oriented at 0° (Figure 8.13). Assume that each ply is 4 mm thick and has the stiffness matrix (in GPa).

$$[D^{*\prime}] = \begin{bmatrix} 30 & 1 & 0 \\ 1 & 3 & 0 \\ 0 & 0 & 1 \end{bmatrix}$$

Calculate the plate curvatures, strains, and stresses in the laminate if $M_x = 25$ kN, $M_y = 0$, and $M_{xy} = 0$.

8.19 Determine the rigidity matrix D of a three-ply plate. The top and bottom layers are 3 mm and 4 mm thick and oriented at 45° and −45° to the plate axis, respectively, while the 5-mm-thick midlayer is oriented at 0°. Assume that each lamina has the same properties as the lamina considered in Example 8.8.

8.20 Resolve Problem 8.19 if all laminae are oriented to the plate axis: (a) at 45°; (b) at 30°.

References

1. A. C. Ugural, *Mechanics of Materials*, Wiley, Hoboken, NJ, 2008.
2. S. G. Lekhnitskii, *Anisotropic Plates*, 2nd ed., Gordon and Breach, New York, 1968.
3. J. R. Vinson, *The Behavior of Sandwich Structures of Isotropic and Composite Materials*, Technomic, Lancaster, PA, 1999.
4. M. S. Troitsky, *Stiffened Plates—Bending, Stability, and Vibration*, Elsevier, New York, 1976.
5. B. D. Agarwal and L. J. Broutman, *Analysis and Performance of Fiber Composites*, Wiley, New York, 1990.

6. S. Timoshenko and S. Woinowsky-Krieger, *Theory of Plates and Shells*, 2nd ed., McGraw-Hill, New York, 1959.
7. H. Altenbach and G. I. Mikhasev (eds.), *Shell and Membrane Theories in Mechanics and Biology*, Springer, Switzerland, 2015.
8. J. N. Reddy and A. Miravete, *Practical Analysis of Composite Laminates*, 2nd ed., CRC Press, Boca Raton, FL, 2005.
9. E. Ventsel and T. Krauthammer, *Thin Walled Plates and Shells: Theory and Applications*, CRC Press, Boca Raton, FL, 2011.
10. J. H. Fauppel and F. E. Fisher, *Engineering Design*, 2nd ed., Wiley, Hoboken, NJ, 1981.

9

Plates under Combined Loads

9.1 Introduction

The classical stress analysis relations of the small deformation theory of plates resulting from lateral loading have been developed in the preceding chapters. Attention will now be directed to situations in which *combined lateral and in-plane* or *direct forces* act at a plate section. The latter forces are also referred to as *membrane forces*. These forces may be applied directly at the plate edges, or they may arise as a result of temperature changes (Chapter 11).

To begin, the governing differential equations are modified to include the simultaneous action of the combined loading. This is followed by consideration of buckling stresses caused by in-plane compression, pure shear, and biaxial compression on application of equilibrium, energy, and finite difference methods, respectively. The problem of plates with small initial curvature under the action of combined forces is next discussed. The chapter concludes with consideration of a plate bent into a simple surface of practical importance.

9.2 Governing Equation for the Deflection Surface

The midplane is *strained* subsequent to combined loading, and assumption 2 of Section 3.3 is no longer valid. However, w is still regarded as *small* so that the remaining suppositions of Section 3.3 hold and yet large enough so that the products of the in-plane forces or their derivatives and the derivatives of w are of the same order of magnitude as the derivatives of the shear forces (Q_x and Q_y). Thus, as before, the stress resultants are given by Equations 3.9 and 3.16.

Consider a plate element of sides dx and dy under the action of direct forces N_x, N_y and $N_{xy} = N_{yx}$, which are functions of x and y only. Assume the body forces to be negligible. The top and front views of such an element are shown in Figure 9.1a and b, respectively. The other resultants due to a lateral load, which also act on the element, are shown in Figure 3.6. Referring to Figure 9.1, from the equilibrium of $N_x dy$ forces, we obtain

$$\left(N_x + \frac{\partial N_x}{\partial x}dx\right)dy\cos\beta' - N_x dy\cos\beta \tag{a}$$

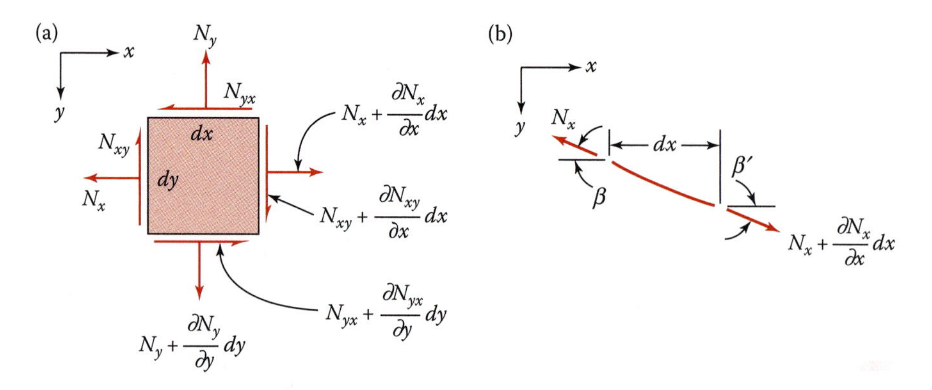

FIGURE 9.1
Forces on the midplane of a plate element: (a) Top view; (b) front view.

in which $\beta' = \beta + (\partial\beta/\partial x)dx$. Writing

$$\cos\beta = (1-\sin^2\beta)^{1/2} = 1 - \frac{1}{2}\sin^2\beta + \cdots = 1 - \frac{\beta^2}{2} + \cdots$$

and noting that for β small, $\beta^2/2 \ll 1$ and $\cos\beta \approx 1$, and that likewise, $\cos\beta' \approx 1$, Equation a reduces to $(\partial N_x/\partial x)dx\,dy$. The sum of the x components of $N_{xy}dx$ is treated in a similar way. The condition $\sum F_x = 0$ then leads to

$$\frac{\partial N_x}{\partial x} + \frac{\partial N_{xy}}{\partial y} = 0 \tag{9.1}$$

Furthermore, the condition $\sum F_y = 0$ results in

$$\frac{\partial N_{xy}}{\partial x} + \frac{\partial N_y}{\partial y} = 0 \tag{9.2}$$

To describe equilibrium in the z direction, it is necessary to consider the z components of the in-plane forces acting at each edges of the element. The z component of the normal forces acting on the x edges is equal to

$$-N_x dy \sin\beta + \left(N_x + \frac{\partial N_x}{\partial x}dx\right)dy \sin\beta' \tag{b}$$

Inasmuch as β and β' are small, $\sin\beta \approx \beta \approx \partial w/\partial x$ and $\sin\beta' \approx \beta'$, and hence

$$\beta' \approx \beta + \frac{\partial\beta}{\partial x}dx = \frac{\partial w}{\partial x} + \frac{\partial^2 w}{\partial x^2}dx$$

Neglecting higher-order terms, Equation b is therefore

$$-N_x dy\frac{\partial w}{\partial x}+\left(N_x+\frac{\partial N_x}{\partial x}dx\right)dy\left(\frac{\partial w}{\partial x}+\frac{\partial^2 w}{\partial x^2}dx\right)$$
$$=N_x\frac{\partial^2 w}{dx^2}dxdy+\frac{\partial N_x}{\partial x}\frac{\partial w}{\partial x}dxdy$$

Similarly, the z component of the normal forces N_y is obtained as

$$N_y\frac{\partial^2 w}{\partial y^2}dxdy+\frac{\partial N_y}{\partial y}\frac{\partial w}{\partial y}dxdy$$

The z components of the *shear forces* N_{xy} *on the* x *edges* of the element are determined as follows. The slope of the deflection surface in the y direction on the x edges equals $\partial w/\partial y$ and $\partial w/\partial y + (\partial^2 w/\partial x\partial y)\, dx$. The z-directed component of the shear forces, $-N_{xy}dy$ and $[N_{xy} + (\partial N_{xy}/\partial x)dx]dy$, is then

$$N_{xy}\frac{\partial^2 w}{\partial x\partial y}dx\,dy+\frac{\partial N_{xy}}{\partial x}\frac{\partial w}{\partial y}dx\,dy$$

An expression identical to the above is found for the z projection of *shear forces* N_{yx} acting *on the* y *edges*:

$$N_{yx}\frac{\partial^2 w}{\partial x\partial y}dxdy+\frac{\partial N_{yx}}{\partial y}\frac{\partial w}{\partial x}dxdy$$

For the forces in Figures 9.1 and 3.6, from $\sum F_z = 0$, we thus have

$$\frac{\partial Q_x}{\partial x}+\frac{\partial Q_y}{\partial y}+p+N_x\frac{\partial^2 w}{\partial x^2}+N_y\frac{\partial^2 w}{\partial y^2}+2N_{xy}\frac{\partial^2 w}{\partial x\partial y}$$
$$+\left(\frac{\partial N_x}{\partial x}+\frac{\partial N_{yx}}{\partial y}\right)\frac{\partial w}{\partial x}+\left(\frac{\partial N_{xy}}{\partial x}+\frac{\partial N_y}{\partial y}\right)\frac{\partial w}{\partial y}=0 \qquad \text{(c)}$$

It is observed from Equations 9.1 and 9.2 that the terms within the parentheses in the above expression vanish. As the direct forces do not result in any moment along the edges of the element, Equations b and c of Section 3.7 and hence Equations 3.16 are unchanged.

Introduction of Equations 3.16 into Equation c yields

$$\frac{\partial^4 w}{\partial x^4}+2\frac{\partial^4 w}{\partial x^2\partial y^2}+\frac{\partial^4 w}{\partial y^4}=\frac{1}{D}\left(p+N_x\frac{\partial^2 w}{\partial x^2}+N_y\frac{\partial^2 w}{\partial y^2}+2N_{xy}\frac{\partial^2 w}{\partial x\partial y}\right) \qquad (9.3)$$

where the flexural rigidity of the plate D is defined by Equation 3.10. Equations 9.1 through 9.3 are the *governing differential equations* for a thin plate, subjected to *combined* lateral and in-plane forces. It is observed that Equations 3.17 is now replaced by Equation 9.3 to determine the deflection surface of the plate. Either Navier's or Lévy's method may be applied to obtain a solution.

EXAMPLE 9.1: Deflection of Plate under Combined Loading

A rectangular plate with simply supported edges is subject to the action of combined uniform lateral load p_0 and uniform tension N (Figure 9.2). Derive the equation of the deflection surface.

Solution

In this particular case, $N_x = N =$ constant and $N_y = N_{xy} = 0$, and hence Equations 9.1 and 9.2 are identically satisfied. The lateral load p_0 can be represented by the following (Section 5.3):

$$p = \frac{16p_0}{\pi^2} \sum_m^{\infty} \sum_n^{\infty} \frac{1}{mn} \sin\frac{m\pi x}{a} \sin\frac{n\pi y}{b} \qquad (m, n = 1, 3, \ldots)$$

Inserting the above in Equation 9.3, we have

$$\frac{\partial^4 w}{\partial x^4} + 2\frac{\partial^4 w}{\partial x^2 \partial y^2} + \frac{\partial^4 w}{\partial y^4} - \frac{N}{D}\frac{\partial^2 w}{\partial x^2} = \frac{16p_0}{\pi^2 D} \sum_m^{\infty} \sum_n^{\infty} \frac{1}{mn} \sin\frac{m\pi x}{a} \sin\frac{n\pi y}{b} \tag{d}$$

The conditions at the simply supported edges, expressed by Equations a of Section 5.2, are satisfied by assuming a deflection of the form given by Equation 5.1b. When this is introduced into Equation d, we obtain

$$a_{mn} = \frac{16p_0}{\pi^6 Dmn\left[\left(\dfrac{m^2}{a^2} + \dfrac{n^2}{b^2}\right) + \dfrac{N}{D}\left(\dfrac{m}{\pi a}\right)^2\right]} \qquad (m, n = 1, 3, \ldots)$$

The deflection is thus

$$w = \frac{16p_0}{\pi^6 D} \sum_m^{\infty} \sum_n^{\infty} \frac{\sin(m\pi x/a)\sin(n\pi y/b)}{mn\left[\left(\dfrac{m^2}{a^2} + \dfrac{n^2}{b^2}\right)^2 + \dfrac{N}{D}\left(\dfrac{m}{\pi a}\right)^2\right]} \tag{9.4}$$

Comment: On comparison of Equations 5.6 and 9.4, we are led to conclude that the presence of a tensile (compressive) force decreases (increases) the plate deflection.

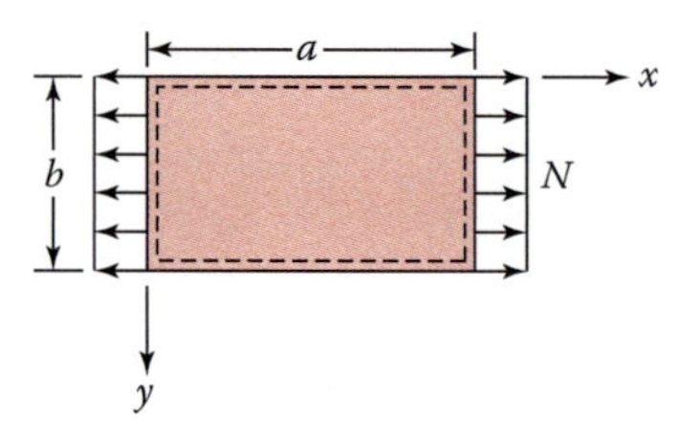

FIGURE 9.2
Simple plate with combined surface and in-plane tensile loads.

9.3 Buckling of Plates

When a plate is compressed in its midplane, it becomes unstable and begins to *buckle* at a certain *critical value* of the in-plane force. Buckling of plates is qualitatively similar to column buckling [1–3]. However, a buckling analysis of the former case is not performed as readily as for the latter. *Classical buckling* problems, by the so-called *equilibrium method* and the conditions that result in the lowest *eigenvalue*, or the *actual buckling load*, are not at all obvious in many situations [4–12]. This is especially true in plates having other than simply supported edges. Often in these cases, the energy method (Section 9.4) is used to good advantage to obtain the approximate buckling loads.

Thin plates or sheets, although quite capable of carrying tensile loadings, are poor in resisting compression. Usually, buckling or wrinkling phenomena observed in compressed plates (and shells) take place rather suddenly and are very dangerous. Fortunately, there is close correlation between theory and experimental data concerned with buckling of plates subjected to various types of loads and edge conditions.

Note that the behavior of flat plates after buckling is of considerable interest. The *post-buckling* analysis of plates is usually difficult since it is basically a nonlinear problem. Slightly curved plates, under simultaneous actions of in-plane compressive forces and lateral loads, exhibit a third type of instability behavior termed *snap-through buckling*. This is characterized by reversal of deflections produced by the nonlinear relationship between the buckling load and displacements.

When plates are under simultaneous actions of in-plane compressive forces and shear forces, combined with lateral bending, buckling occurs at lower load magnitudes than when those forces act individually [6]. Illustrations of the compressed plate behavior, using the equilibrium method, are presented in the following two examples. Clearly, in the absence of lateral loading, we take $p = 0$ in Equation 9.3.

EXAMPLE 9.2: Critical Load and Stress in Plate under Uniaxial Compression

A rectangular plate, with sides ($y = 0$ and $y = b$) free and edges ($x = 0$ and $x = a$) simply supported, is subjected to uniaxial in-plane compression forces N (Figure 9.3). Determine the critical load and stress.

Solution

A plate strip parallel to the x axis may be considered to be a pin-ended column that deflects in the xz plane. It is recalled from Section 3.4 that flexural rigidity EI of a narrow beam (or column) is replaced by $EI/(1 - v^2)$ in the case of a wide beam or plate. On this basis, Euler's formula for the plate shown in Figure 9.3, from Equation 2.5, is

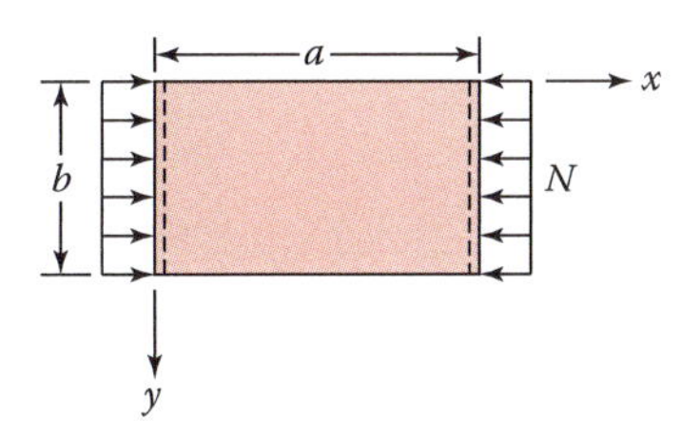

FIGURE 9.3
Plate with simply supported and free edges with in-plane compression.

$$N_{cr} = \frac{\pi^2 EI}{(1-\nu^2)a^2} \tag{9.5}$$

Here $I = bt^3/12$, and t represents the thickness of the plate. The critical stress, N_{cr}/bt, is thus

$$\sigma_{cr} = \frac{\pi^2 E}{12(1-\nu^2)}\left(\frac{t}{a}\right)^2 \tag{9.6}$$

EXAMPLE 9.3: Buckling Analysis of Plate under Uniaxial Compression

A simply supported rectangular plate is subjected to uniaxial in-plane compression forces N (Figure 9.4). Determine the buckling load and stress.

Solution

For this case, $N_x = -N =$ constant and $N_y = N_{xy} = 0$, and hence Equations 9.1 and 9.2 are satisfied identically. The governing equation for the displacement becomes

$$D\nabla^4 w + N\frac{\partial^2 w}{\partial x^2} = 0 \tag{9.7}$$

We assume the solution in the form

$$w = \sum_m^\infty \sum_n^\infty a_{mn} \sin\frac{m\pi x}{a}\sin\frac{n\pi y}{b} \qquad (m,n = 1,2,\ldots) \tag{5.1b}$$

which, when substituted into Equation 9.7, leads to

$$\sum_m^\infty \sum_n^\infty \left[D\pi^4\left(\frac{m^2}{a^2}+\frac{n^2}{b^2}\right)^2 - N\pi^2\frac{m^2}{a^2}\right] a_{mn}\sin\frac{m\pi x}{a}\sin\frac{n\pi y}{b} = 0$$

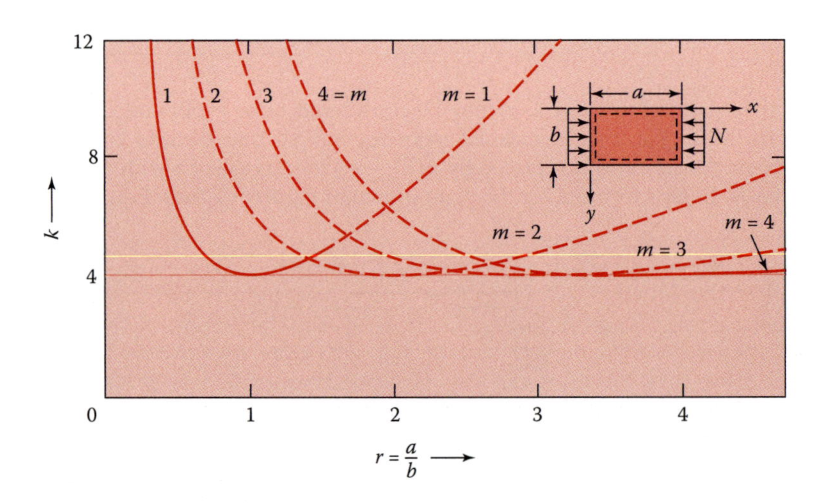

FIGURE 9.4
Variation of the buckling load factor k with aspect ratio r for various values of m.

The nontrivial solution is

$$\pi^4 D\left(\frac{m^2}{a^2}+\frac{n^2}{b^2}\right)^2 - N\pi^2\frac{m^2}{a^2}=0$$

from which

$$N=\frac{\pi^2 a^2 D}{m^2}\left(\frac{m^2}{a^2}+\frac{n^2}{b^2}\right)^2=\frac{\pi^2 D}{b^2}\left(\frac{mb}{a}+\frac{n^2 a}{mb}\right)^2 \tag{9.8}$$

Clearly, when N attains the value given by the right-hand side of Equation 9.8, we have $a_{mn} \neq 0$, and hence $w \neq 0$, indicating plate buckling. It is observed from Equation 9.8 that the minimum value of N occurs when $n = 1$. Thus, when the simply supported plate buckles, the *buckling mode* given by Equation 9.8 can be only *one half-sine wave*, sin $(\pi y/b)$, across the span, while several half-waves in the direction of compression can occur. The resulting expression is thus

$$N_{cr}=\frac{\pi^2 D}{b^2}\left(\frac{m}{r}+\frac{r}{m}\right)^2=k\frac{\pi^2 D}{b^2} \tag{9.9}$$

for the critical load. Here $k = [(m/r) + (r/m)]^2$, and *aspect ratio* $r = a/b$.

To ascertain the aspect ratio r at which the critical load is a minimum, we set

$$\frac{dN_{cr}}{dr}=\frac{2\pi^2 D}{b^2}\left(\frac{m}{r}+\frac{r}{m}\right)\left(-\frac{m^2}{r^2}+1\right)=0$$

from which $m/r = 1$. This provides the following *minimum* value of the critical load:

$$N_{cr}=\frac{4\pi^2 D}{b^2} \tag{9.10}$$

The corresponding critical stress, N_{cr}/t, is given by

$$\sigma_{cr}=\frac{\pi^2 E}{3(1-\nu^2)}\left(\frac{t}{b}\right)^2 \tag{9.11}$$

The variations of the buckling load factor k as functions of aspect ratio r for $m = 1, 2, 3, 4$ are plotted in Figure 9.4. Obviously, for a specific m, the magnitude of k depends on r only. Referring to the figure, the magnitude of N_{cr} and the number of half-waves m for any value of the aspect ratio r can readily be found. In the case of $r = 1.5$, for instance, from Figure 9.4, $k = 4.34$ and $m = 2$. The corresponding critical load is $N_{cr} = 4.34\pi^2 D/b^2$, under which the plate will buckle into two half-waves in the direction of the loading, as shown in Figure 9.5.

It is also observed from Figure 9.4 that a plate m times as long as it is wide will buckle in m half-sine waves. Thus, a *long plate* ($b \ll a$) with simply supported edges under a uniaxial compression *tends to buckle* into a number of *square cells* of side dimension b; its critical load for all practical purposes is given by Equation 9.10.

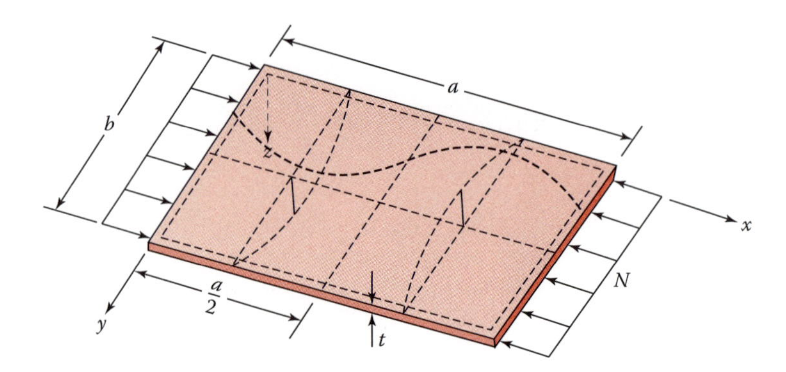

FIGURE 9.5
Deflection of the plate into half-sine curves of $m = 2$ and $n = 1$.

EXAMPLE 9.4: Buckling Load of Plate with Biaxial Compression

Develop an expression for the buckling load of A simply supported rectangular plate carries uniformly compressed in two directions (Figure 9.6). (a) Develop an expression for the buckling load; (b) What is the critical load, for the case in which a square plate ($a = b$) with $N_x = N_y = -N$?

Solution

For the situation described, compressive loads are $-N_x$ and $-N_y$ and shear force is $N_{xy} = 0$. So, Equation 9.3 reduces to

$$\frac{\partial^4 w}{\partial x^4} + 2\frac{\partial^4 w}{\partial x^2 \partial y^2} + \frac{\partial^4 w}{\partial y^4} + \frac{1}{D}\left(N_x \frac{\partial^2 w}{\partial x^2} + N_y \frac{\partial^2 w}{\partial y^2}\right) = 0 \tag{9.12}$$

a. We use w in the form of the Expression 5.1b, which fulfills the simply supported boundary conditions. Inserting this into Equation 9.12 leads to

$$a_{mn}\left[\left(\frac{m^2}{a^2} + \frac{n^2}{b^2}\right)^2 - \frac{1}{\pi^2 D}\left(N_x \frac{m^2}{a^2} + N_y \frac{n^2}{b^2}\right)\right] = 0 \tag{9.13}$$

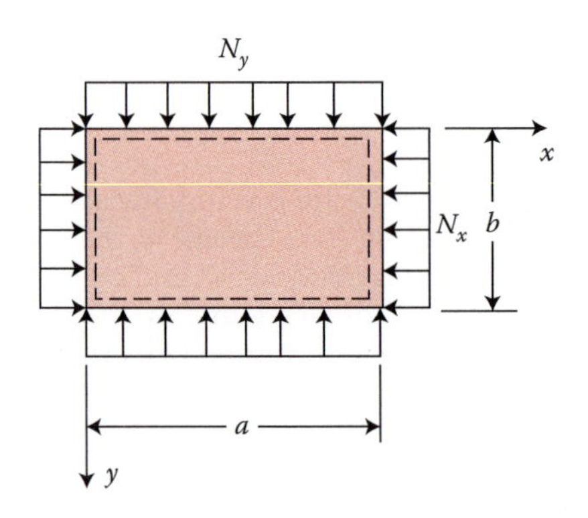

FIGURE 9.6
Simply supported plate subjected to biaxial compression.

A nontrivial solution equals

$$N_x\left(\frac{m}{a}\right)^2+N_y\left(\frac{n}{b}\right)^2=D\pi^2\left[\left(\frac{m}{a}\right)^2+\left(\frac{n}{b}\right)^2\right]^2 \tag{9.14}$$

b. Through the use of Equation 9.14 with $N_x=N_y=N$, we obtain

$$N=\frac{\pi^2 D}{b^2}\left[n^2+\left(\frac{mb}{a}\right)^2\right]$$

It is seen from the above expression that the minimum value of N occurs for $m=n=1$. It follows that

$$N_{cr}=\frac{\pi^2 D}{b^2}\left[1+\left(\frac{b}{a}\right)^2\right] \tag{9.15}$$

This equation results in, for the square plate ($a=b$),

$$N_{cr}=2\frac{\pi^2 D}{b^2} \tag{9.16}$$

Comment: A comparison of Equations 9.16 and 9.10 shows that, when the square plate is under uniform biaxial compression forces N, the critical value of these loads is *half* of the critical load of the square plate compressed by uniaxial load only.

9.4 Application of the Energy Method

The principle of minimum potential energy may be employed to analyze plates under the action of lateral and in-plane loading. We shall first develop expressions for the midsurface strains. Work done by the direct forces may then be evaluated and the *energy method* applied readily to any particular problem.

Consider element $dx\,dy$ representing a point at the *midplane* of a plate (Figure 9.7a).

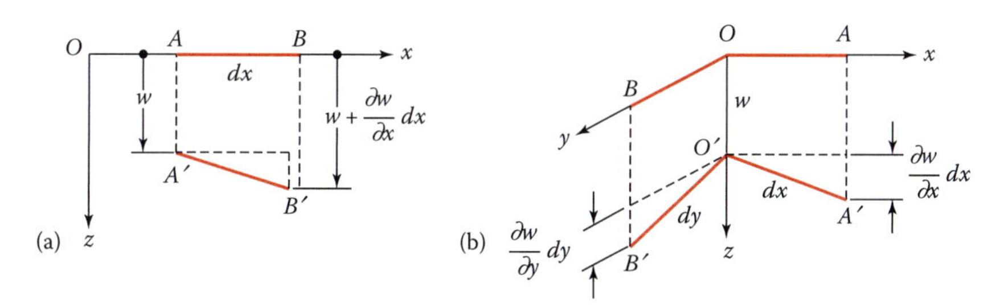

FIGURE 9.7
Deformation due to the displacement w: (a) Normal strain; (b) shear strain.

Subsequent to the bending, linear element AB is displaced to become $A'B'$. Inasmuch as mid-plane stressing does *not* occur, original element length dx remains unchanged, and its horizontal projection due to the displacement w is

$$\left[dx^2 - \left(\frac{\partial w}{\partial x}dx\right)^2\right]^{1/2} = dx - \frac{1}{2}\left(\frac{\partial w}{\partial x}\right)^2 dx + \cdots \tag{a}$$

The *midplane* displacement per unit length (to a second approximation) in the x direction is therefore

$$\varepsilon_x = \frac{1}{2}\left(\frac{\partial w}{\partial x}\right)^2 \tag{9.17a}$$

Similarly, unit displacement of the midplane in the y direction equals

$$\varepsilon_y = \frac{1}{2}\left(\frac{\partial w}{\partial y}\right)^2 \tag{9.17b}$$

For the purpose of determining shear strains associated with plate bending, consider now two infinitesimal linear elements OA and OB in the x and y directions (Figure 9.7b). Because of z displacement w, these elements move to the positions $O'A'$ and $O'B'$, having direction cosines l_1, m_1, n_1, and l_2, m_2, n_2, respectively, given by

$$l_1 = \frac{[dx^2 - ((\partial w/\partial x)dx)^2]^{1/2}}{dx} \approx 1 - \frac{1}{2}\left(\frac{\partial w}{\partial x}\right)^2 \qquad m_1 = 0 \qquad n_1 = \frac{(\partial w/\partial x)dx}{\partial x} = \frac{\partial w}{\partial x} \tag{b}$$

and

$$l_2 = 0 \qquad m_2 \approx 1 - \frac{1}{2}\left(\frac{\partial w}{\partial y}\right)^2 \qquad n_2 = \frac{\partial w}{\partial y} \tag{c}$$

Consider the following:

$$\gamma_{xy} = \frac{\pi}{2} - \sphericalangle A'O'B' = \sin\left(\frac{\pi}{2} - \sphericalangle A'O'B'\right) = \cos\sphericalangle A'O'B' = l_1 l_2 + m_1 m_2 + n_1 n_2$$

This expression, on substitution of Equations b and c, yields the midplane shear strain

$$\gamma_{xy} = \frac{\partial w}{\partial x}\frac{\partial w}{\partial y} \tag{9.17c}$$

Note that in previous discussions of plate bending, the strains given by Equations 9.17 were always omitted. They are now taken into account since their products with direct forces may be of the same order of magnitude as the strain energy of bending. However, these strains are considered very small in comparison with

$$\varepsilon_x = \frac{1}{Et}(N_x - \nu N_y) \qquad \varepsilon_y = \frac{1}{Et}(N_y - \nu N_x) \qquad \gamma_{xy} = \frac{N_{xy}}{Gt} \tag{9.18}$$

caused by in-plane forces. It is further assumed that *direct forces are applied first* (*before lateral loads*) *and that they remain unchanged during plate bending*. The latter assumption is widely used in the *classical treatment of plates* as well as beams and shells.

The work done by the direct forces, due to displacement w only, then equals

$$W = \frac{1}{2}\iint_A \left[N_x \left(\frac{\partial w}{\partial x}\right)^2 + N_y \left(\frac{\partial w}{\partial w}\right)^2 + 2N_{xy}\frac{\partial w}{\partial x}\frac{\partial w}{\partial y} \right] dx\,dy \tag{9.19}$$

where A is the area of the plate. The plate-strain energy due to bending is given by

$$U = \frac{1}{2}\iint_A D\left\{ \left(\frac{\partial^2 w}{\partial x^2} + \frac{\partial^2 w}{\partial y^2}\right)^2 - 2(1-\nu)\left[\frac{\partial^2 w}{\partial x^2}\frac{\partial^2 w}{\partial y^2} - \left(\frac{\partial^2 w}{\partial x \partial y}\right)^2\right]\right\} dx\,dy \tag{3.44}$$

Application of the basic expressions derived in the preceding paragraphs is illustrated in the discussion of a plate-buckling problem that follows. The plate is taken to be subjected to *constant direct forces* (N_x, N_y, N_{xy}) *during bending just before it buckles*. This means that the magnitude of these in-plane forces is just equal to their critical values. Subsequently, it is assumed that the plate undergoes some small disturbances and that buckling takes place. At the time of transition from one to the other of these equilibrium forms, no energy is gained or lost. Hence, the work done by the direct forces must be equal to the bending-strain energy stored in the plate. That is, buckling occurs if $U = W$. Thus, we have the expression

$$\Pi = U - W = 0 \tag{3.49}$$

for the potential energy of the plate.

EXAMPLE 9.5: Critical Load of Plate Subjected to Varying Axial Compression

What is the buckling load for a plate with three edges $x = 0$, $x = a$, and $y = 0$ simply supported and one edge free at $y = b$ (Figure 9.8)? The plate carries linearly distributed compressive forces N_x.

Solution

An approximate expression for the deflection surface of the plate is given by

$$w = \sin\frac{n\pi x}{a}\sum_{i=1}^{N} C_i y^i \tag{9.20}$$

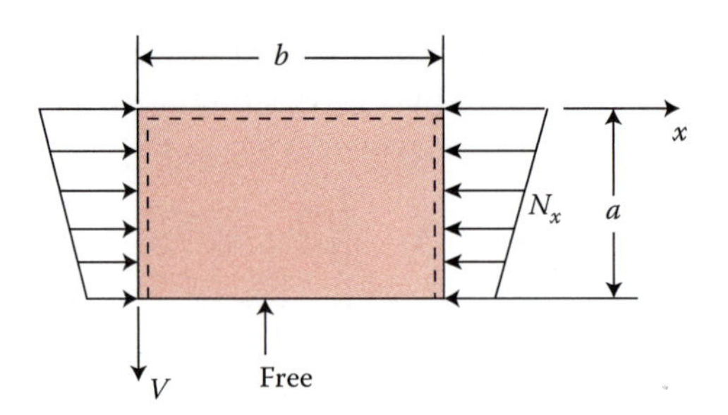

FIGURE 9.8
Plate with free and simply supported edges under linearly varying compression.

The foregoing satisfies the geometric boundary conditions: $w = 0$ at $x = 0$ and $x = a$, and $w = 0$ at $y = 0$. Taking only one term ($i = 1$) in the series, Equation 9.20, we have

$$\frac{\partial y}{\partial x} = C_1 y\left(\frac{n\pi}{a}\cos\frac{n\pi x}{a}\right) \qquad \frac{\partial^2 w}{\partial x^2} = -C_1 y\left(\frac{n\pi}{a}\right)^2 \sin\frac{n\pi x}{a}$$

$$\frac{\partial^2 w}{\partial x \partial y} = C_1\left(\frac{n\pi}{a}\cos\frac{n\pi x}{a}\right) \qquad \frac{\partial^2 w}{\partial y^2} = 0$$

The loading may be expressed as follows:

$$N_x = -N\left\{1 + \left(\alpha\frac{y}{b}\right)\right\} \qquad N_y = N_{xy} = 0$$

in which $\alpha > 0$ denotes some fixed parameter.

Introducing the foregoing into Equation 3.49, after integrating, leads to

$$\Pi = C_1^2\frac{ab}{4}\left\{D\left[\frac{b^2}{3}\left(\frac{n\pi}{a}\right)^4 + 2(1-\nu)\left(\frac{n\pi}{a}\right)^2\right] - N\left(\frac{n\pi}{a}\right)^2 b^2\left(\frac{1}{3} + \frac{\alpha}{4}\right)\right\}$$

From the minimizing condition $\partial\Pi/\partial C_1 = 0$, it follows that

$$N_n = \frac{n^2\pi^2 D}{b^2}\left[\frac{b^2/a^2 + 6(1-\nu)/n^2\pi^2}{1 + 3\alpha/4}\right]$$

Retaining only the first term ($n = 1$) of this solution, the buckling load is found to be

$$N_{cr} = \frac{\pi^2 D}{b^2}\left[\frac{b^2/a^2 + 6(1-\nu)/\pi^2}{1 + 3\alpha/4}\right] \tag{9.21}$$

Comments: By changing the values of α, we can obtain various particular cases of loading. For example, by taking $\alpha = 0$ we have the case of a uniformly distributed compressive force, as shown in Figure 9.3.

EXAMPLE 9.6: Critical Load due to Pure Shearing forces

Determine the buckling load of a simply supported plate under the action of uniform shearing forces $N_{xy} = S$, $N_x = N_y = 0$, as shown in Figure 9.9.

Solution

Assume that the deflection surface of the plate is described by an expression of the form

$$w = \sum_{m}^{\infty}\sum_{n}^{\infty} a_{mn} \sin\frac{m\pi x}{a}\sin\frac{n\pi y}{b} \tag{d}$$

Clearly, the foregoing satisfies the boundary conditions at the simply supported plate edges. The work done by S during the buckling of the plate is, from Equation 9.14,

$$W = S\int_0^b\int_0^a \frac{\partial w}{\partial x}\frac{\partial w}{\partial y}dxdy$$

Inserting Equation d into this equation and observing that

$$\int_0^a \sin\frac{m\pi x}{a}\cos\frac{p\pi x}{a}dx = \begin{cases} 0 & (m \pm p \text{ is an even number}) \\ \dfrac{2a}{\pi}\dfrac{m}{m^2 - p^2} & (m \pm p \text{ is an odd number}) \end{cases}$$

we obtain

$$W = 4S\sum_{m}^{\infty}\sum_{n}^{\infty}\sum_{p}^{\infty}\sum_{q}^{\infty} a_{mn}a_{pq}\frac{mnpq}{(m^2 - p^2)(q^2 - n^2)} \tag{e}$$

wherein $m \pm p$ and $n \pm q$ are odd numbers.

The strain energy associated with the bending of the buckled plate is given by

$$U = \frac{\pi^4 ab}{8}D\sum_{m}^{\infty}\sum_{n}^{\infty} a_{mn}^2\left(\frac{m^2}{a^2} + \frac{n^2}{b^2}\right)^2 \tag{5.71}$$

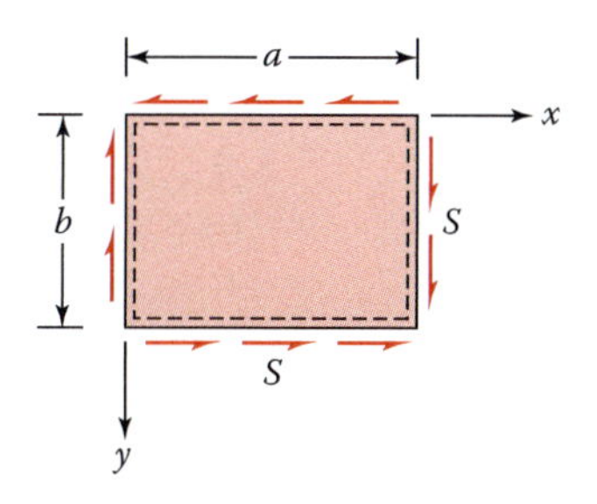

FIGURE 9.9
Simply supported rectangular plate with shear load.

Equation 3.49 now becomes

$$\Pi = \frac{D\pi^4 ab}{8}\sum_m^\infty\sum_n^\infty a_{mn}^2\left(\frac{m^2}{a^2}+\frac{n^2}{b^2}\right)^2 - 4S\sum_m^\infty\sum_n^\infty\sum_p^\infty\sum_q^\infty a_{mn}a_{pq}\frac{mnpq}{(m^2-p^2)(q^2-n^2)} = 0 \tag{f}$$

To ascertain the critical value of the shearing forces, the coefficient a_{mn} is to be found such that S is a minimum. This, it can be demonstrated (Section 3.12), is equivalent to requiring that Π be a minimum. On application of Equation 3.53, we thus have

$$\frac{\pi^2 Dab}{4}a_{mn}\left(\frac{m^2}{a^2}+\frac{n^2}{b^2}\right)^2 - 8S\sum_p^\infty\sum_q^\infty a_{pq}\frac{mnpq}{(m^2-p^2)(q^2-n^2)} = 0 \tag{9.22}$$

in which p and q must be such that $m \pm p$, $n \pm q$ are odd numbers. On the basis of the notation,

$$r = \frac{a}{b} \qquad \lambda = \frac{\pi^2 D}{32rb^2S} \tag{9.23}$$

Equation 9.22 assumes the following convenient form:

$$\lambda a_{mn}\frac{(m^2+n^2r^2)^2}{r^2} - \sum_p^\infty\sum_q^\infty a_{pq}\frac{mnpq}{(m^2-p^2)(q^2-n^2)} = 0 \tag{9.24}$$

We thus have a system of linear equation in a_{mn}, of which an approximate solution may be found by retaining a finite number of parameters a_{mn}.

If, for example, only two parameters, a_{11} and a_{22}, are kept, we have, from Equation 9.24,

$$\frac{\lambda(1+r^2)^2}{r^2}a_{11} + \frac{4}{9}a_{22} = 0$$
$$\frac{16\lambda(1+r^2)^2}{r^2}a_{22} + \frac{4}{9}a_{11} = 0$$

These equations have the following determinant:

$$\begin{vmatrix} \dfrac{\lambda(1+r^2)^2}{r^2} & \dfrac{4}{9} \\ \dfrac{4}{9} & \dfrac{16\lambda(1+r^2)^2}{r^2} \end{vmatrix} = 0$$

with the solution

$$\lambda = \pm\frac{1}{9}\frac{r^2}{(1+r^2)^2} \tag{g}$$

The magnitude of the critical load, from Equation 9.23, is thus

$$S_{cr} = \pm \frac{9\pi^4 D}{32b^2} \frac{(1+r^2)^2}{r^3} \tag{9.25}$$

Comment: This result differs by approximately 15% from the "exact" solution [4] for a square plate ($r = 1$). Accuracy may be improved by retaining additional parameters.

EXAMPLE 9.7: Critical Load of Stiffened Plate

A simply supported rectangular plate, stiffened by one longitudinal rib of negligible torsional rigidity, is subjected to uniaxial compression forces N as shown in Figure 9.10. Determine the critical buckling load by taking the first term of the series solution.

Solution

The surface of the buckled, stiffened plate is represented by

$$w = a_{11} \sin \frac{\pi x}{a} \sin \frac{\pi y}{b} \tag{h}$$

The potential energy of the plate, in the absence of the rib, from Equations 3.44, 3.49, and 5.71, is

$$(\Pi)_p = \frac{\pi^4 ab}{8} D a_{11}^2 \left(\frac{1}{a^2} + \frac{1}{b^2} \right)^2 - \frac{N}{2} \int_0^b \int_0^a \left(\frac{\partial w}{\partial x} \right)^2 dxdy \tag{i}$$

Since in-plane pressure acting on the edges of the plate equals N/t, the in-plane load acting on the rib is thus $(N/t)\, A_s$. Here A_s denotes the cross-sectional area of the rib. Then, the potential energy of the stiffener can be expressed as

$$(\Pi)_s = \frac{EI_s}{2} \int_0^a \left(\frac{\partial^2 w}{\partial x^2} \right)_{y=b/2} dx - \frac{NA_s}{t} \int_0^a \left(\frac{\partial w}{\partial x} \right)^2_{y=b/2} dx \tag{j}$$

in which I_s represents the moment of inertia of the stiffener cross section with respect to the midplane of the plate.

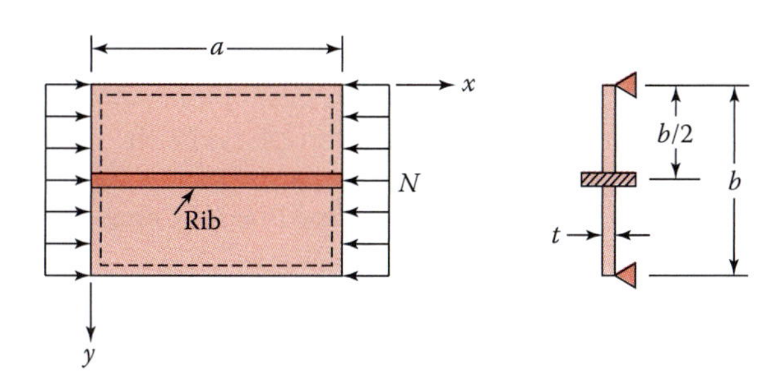

FIGURE 9.10
Simply supported stiffened rectangular plate.

The potential energy of the stiffened plate, $\Pi = \Pi_p + \Pi_s$, after we substitute Equation h into Equations i and j and let

$$\lambda = \frac{Nb^2}{\pi^2 D} \tag{k}$$

takes the form

$$\Pi = \frac{\pi^4 bD}{8a^3}\left[a_{11}^2\left(\frac{1}{a^2}+\frac{1}{b^2}\right)^2 + \frac{2EI_s}{bD}\left(a_{11}\sin\frac{\pi}{2}\right)^2\right] - \frac{\pi^4 D\lambda}{8ab}\left[a_{11}^2 + \frac{2A_s}{bt}\left(a_{11}\sin\frac{\pi}{2}\right)^2\right]$$

Through the use of the condition $\partial\Pi/\partial a_{11} = 0$, we have

$$\lambda = \frac{(1+r^2)^2 + 2\alpha}{(1+2\beta)r^2}$$

where

$$r = \frac{a}{b} \qquad \alpha = \frac{EI_s}{bD} \qquad \beta = \frac{A_s}{bt} \tag{l}$$

The critical load, from Equation k, is therefore

$$N_{cr} = \frac{\pi^2 D}{b^2}\frac{(1+r^2)^2 + 2\alpha}{(1+2\beta)r^2} \tag{9.26}$$

Comment: The preceding equation yields very accurate results for longer plates ($r > 2$). In the case of shorter plates, the first two significant terms of the series solution (Equation d) must be retained [4].

9.5 *The Finite Difference Solution

The preceding sections have been concerned with rectangular plates subjected to uniform loads. The determination of the buckling load by means of analytical methods may be quite tedious and difficult, as was observed. To enable the stress analyst to cope with the numerous compressed plates of practical importance, numerical techniques must be relied on. The following examples apply to the *method of finite difference* (Chapter 7) to the governing differential equations of a plate under in-plane loads.

EXAMPLE 9.8: Buckling Load of Plate Subjected Biaxial Compression

Compute the buckling load of a simply supported square plate under uniform compressive loading per unit of boundary length (Figure 9.9a).

Solution

The plate is subjected to constant in-plane loads $N_x = N_y = -N$ and $p = N_{xy} = 0$. On substitution, Equation 9.3 reduces to

$$\nabla^4 w + \frac{N}{D}\nabla^2 w = 0 \tag{9.27}$$

Along the edges, the conditions are described by

$$\begin{aligned} w &= 0 \qquad \frac{\partial^2 w}{\partial x^2} = 0 \qquad \left(x = \pm\frac{a}{2}\right) \\ w &= 0 \qquad \frac{\partial^2 w}{\partial y^2} = 0 \qquad \left(y = \pm\frac{a}{2}\right) \end{aligned} \tag{a}$$

Hence,

$$\nabla^2 w = 0 \qquad \text{(on the boundary)}$$

and the problem is reduced to finding the solution of

$$\nabla^2 M + \frac{N}{D}M = 0 \tag{9.28a}$$

$$\nabla^2 w = M \tag{9.28b}$$

under the condition that $w = 0$ and $M = 0$ on the boundary. For $w = 0$ along the boundary, it is observed that Equation 9.28b, for the trivial solution $M = 0$ also leads to the trivial solution $w = 0$ It is concluded therefore that treating the problem requires the solution of Equation 9.28a only.

The finite difference expression corresponding to Equation 9.28a, referring to Figure 9.11b, is written at node 0 as follows:

$$M_1 + M_2 + M_3 + M_4 + \left(\frac{Na^2}{n^2 D} - 4\right)M_0 = 0 \tag{9.29}$$

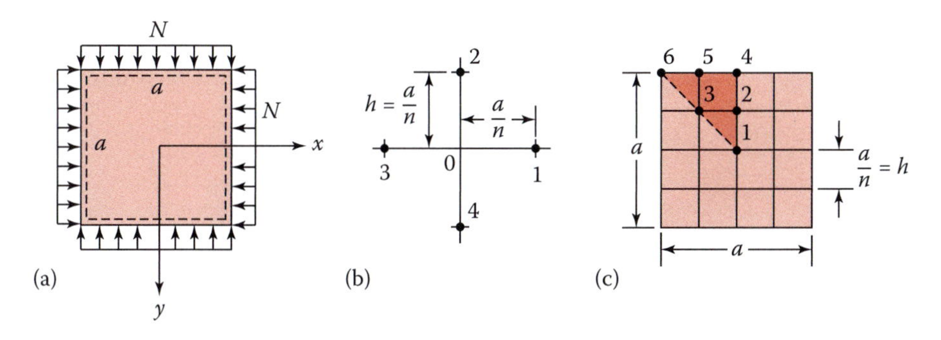

FIGURE 9.11
Simply supported plate with compression loading.

where n represents the number of divisions of the sides. For convenience, in the computations we shall denote

$$\lambda = \frac{Na^2}{n^2D} - 4 = \frac{K}{n^2} - 4 \tag{9.30}$$

with

$$K = \frac{Na^2}{D} \tag{9.31}$$

The plate is now divided into a number of small squares (e.g., 16). Because of conditions of symmetry, we need only treat one-eighth of the plate, shown by the shaded area in Figure 9.11c. On applying Equation 9.29 at nodes 1, 2, and 3 with $M_4 = M_5 = M_6 = 0$ along the boundary, three equations can be written. These are represented in the following matrix form:

$$\begin{bmatrix} \lambda & 4 & 0 \\ 1 & \lambda & 2 \\ 0 & 2 & \lambda \end{bmatrix} \begin{Bmatrix} M_1 \\ M_2 \\ M_3 \end{Bmatrix} = 0$$

The above leads to a nontrivial solution if the determinant is set equal to zero. In so doing, we obtain $\lambda = -2.8284$, and hence $K = 18.75$. The buckling load, from Equation 9.31, is thus

$$N_{cr} = 18.75 \frac{D}{a^2}$$

Comment: The result is smaller than the "exact" solution by 5.1% [4]. By increasing the number of subdivisions, the accuracy may be improved.

EXAMPLE 9.9: Critical Load by Finite Difference Method

A portion of a missile launcher support fixture, approximated by a square plate with opposite edges $y = \pm a/2$ built in and the other sides simply supported (Figure 9.12a), carries uniform compressive loads N. Determine the critical buckling load.

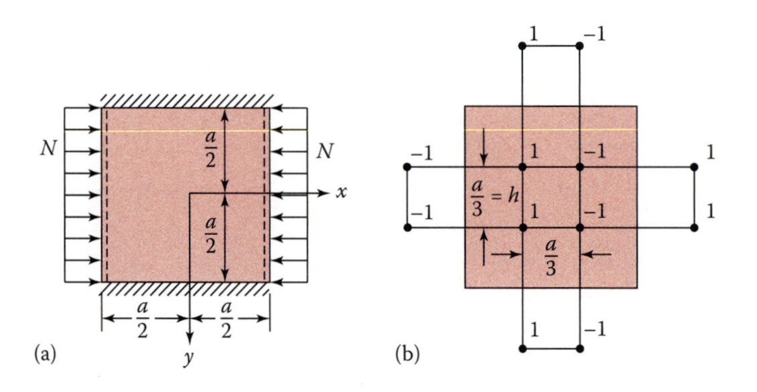

FIGURE 9.12
Variously supported plate under biaxial compression.

Solution

We have $N_x = -N$, and $N_y = N_{xy} = p = 0$. Equation 9.3 then becomes

$$\nabla^4 w + \frac{N}{D}\frac{\partial^2 w}{\partial x^2} = 0 \tag{9.32}$$

The boundary conditions are

$$\begin{aligned} &w = 0 \qquad \frac{\partial^2 w}{\partial x^2} = 0 \qquad \left(x = \pm\frac{a}{2}\right) \\ &w = 0 \qquad \frac{\partial w}{\partial y} = 0 \qquad \left(y = \pm\frac{a}{2}\right) \end{aligned} \tag{b}$$

As before, employing $h = a/n$ and $K = Na^2/D$, we obtain the coefficient pattern for the finite difference expression of Equation 9.32 (Figure 9.13).

Proceeding with the finite difference solution, the plate is subdivided into nine small squares. *Antisymmetrical deflections in the direction of the compression result in a smaller buckling load.* The finite difference equivalents of Equations b are fulfilled by numbering the nodes located outside the boundary as shown in Figure 9.12b. Note that the values of w are zero along the boundary. On applying Figure 9.13 at point 1, we have

$$\begin{aligned} &(-w_1 + w_1 + 0 + 0) + 2(0 + 0 + 0 - w_1) - 8(0 + 0 + w_1 - w_1) \\ &\qquad + 20w_1 + \frac{K}{9}(0 - 2w_1 - w_1) = 0 \end{aligned}$$

from which

$$w_1\left(18 - \frac{K}{3}\right) = 0$$

or $K = 54 = 5.471\pi^2$. Thus,

$$N_{cr} = \frac{5.471\pi^2 D}{a^2} \tag{c}$$

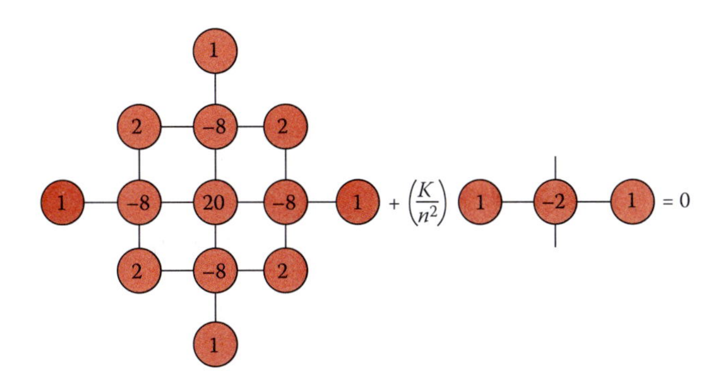

FIGURE 9.13
Coefficient pattern of finite difference expression of Equation 9.32.

It can be shown that using $n = 4$ and the antisymmetrical deflections, the critical load is (Problem 9.10)

$$N_{cr} = \frac{6.193\pi^2 D}{a^2} \tag{d}$$

for the square plate under consideration.

9.6 Plates with Small Initial Curvature

For a plate with an *initial curvature* subjected to the action of *lateral load only*, the governing equation (Equations 3.17) still holds, provided that at any point the magnitude of the *initial* deflection w_0 is *small* compared with the plate thickness t. The *total* deflection w is obtained by the superposition of w_0 and deflection w_1 due to the lateral load. Here w_1 is determined by solving Equations 3.17 as in the case of flat plates.

However, as might be anticipated on physical grounds, the load-carrying capacity and deformation of a plate under *in-plane and lateral loading* are significantly affected by *any* initial curvature. To take into account the extent of this influence, it can be shown that Equation 9.3 can be modified as follows [4]:

$$\nabla^4 w_1 = \frac{1}{D}\left(p + N_x \frac{\partial^2 w}{\partial x^2} + N_y \frac{\partial^2 w}{\partial y^2} + 2N_{xy} \frac{\partial^2 w}{\partial x \partial y}\right) \tag{9.33}$$

where $w = w_0 + w_1$. This is the *governing differential equation* for deflection of thin plates with *small initial curvature*. Note that the left- and the right-hand sides of the above depend on the *change* in the curvature and the *total* curvature of the plate, respectively.

On comparison of Equations 3.17 and 9.33, it is observed that the influence of an initial curvature on the deflection is identical with that of a fictitious *equivalent lateral load*:

$$N_x \frac{\partial^2 w_0}{\partial x^2} + N_y \frac{\partial^2 w_0}{\partial y^2} + 2N_{xy} \frac{\partial^2 w_0}{\partial x \partial y}$$

We conclude, therefore, that a plate experiences bending under direct forces *only* if it has an initial curvature.

Consider, as an example, a simply supported plate for which the *unloaded shape* is described by

$$w_0 = \sum_m^{\infty} \sum_n^{\infty} a_{mn} \sin\frac{m\pi x}{a} \sin\frac{n\pi y}{b} \tag{a}$$

Assume that the plate edges at $x = 0$ and $x = a$ are subjected to uniform compressive forces $N_x = -N$ (Figure 9.14a). We observe that Figure 9.14b depicts the initial deflection for $m = n = 1$.

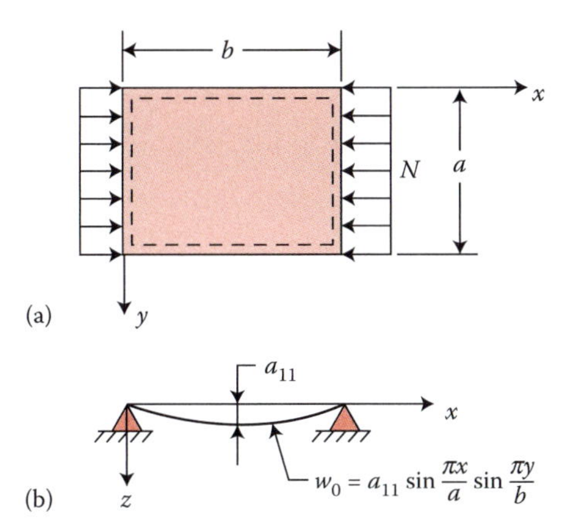

FIGURE 9.14
Simply supported plate subjected to: (a) In-plane compression; (b) initial deflection.

The differential equation (Equation 9.33), with Equation a introduced, is

$$\nabla^4 w_1 = \frac{1}{D}\left(N\frac{\pi^2}{a^2}\sum_m^{\infty}\sum_n^{\infty} a_{mn}m^2 \sin\frac{m\pi x}{a}\sin\frac{n\pi y}{b} - N\frac{\partial^2 w_1}{\partial x^2}\right) \tag{b}$$

When the series solution

$$w_1 = \sum_m^{\infty}\sum_n^{\infty} b_{mn}\sin\frac{m\pi x}{a}\sin\frac{n\pi y}{b} \tag{c}$$

is inserted into Equation b, it is found that

$$b_{mn} = \frac{a_{mn}N}{(\pi^2 D/a^2)[m+(n^2a^2/mb^2)]^2 - N} \tag{d}$$

The solution of Equation b is determined by substituting the above into Equation c. The total deflection is then

$$w = w_0 + w_1 = \sum_m^{\infty}\sum_n^{\infty}\frac{a_{mn}}{1-\alpha}\sin\frac{m\pi x}{a}\sin\frac{n\pi y}{b} \tag{9.34}$$

where

$$\alpha = \frac{N}{(\pi^2 D/a^2)[m+(n^2a^2/mb^2)]^2} \tag{9.35}$$

For $m = n = 1$, the maximum deflection of the plate, which occurs at the center (Figure 9.14b), is given by

$$w_{\max} = \frac{a_{11}}{1-\alpha} \tag{9.36}$$

in which now

$$\alpha = \frac{N}{(\pi^2 D/a^2)[1+(a^2/b^2)]^2}$$

Note that the deflection expression derived above is analogous to the relation for initially curved columns [2].

From Equation d, it is seen that b_{mn} (and thus w_1) increases with increasing N. When N reaches the critical value given by Equation 9.8, the denominator of Equation d vanishes, and w_1 grows without limit. This means that the plate buckles, as described in the alternate manner in Section 9.3.

To write the equation for a plate subjected to uniform *tensile forces* (Figure 9.2), it is required only to *change the sign* of N in the equations of the foregoing example. By following an approach similar to that described above, the deflection of an initially curved plate under simultaneous action of in-plane forces N_x, N_y, and N_{xy} may also be readily obtained.

9.7 *Bending to a Cylindrical Surface

Assume that a *long* rectangular plate of uniform thickness t is bent into a *cylindrical surface* with its generating line parallel to the y axis. For this case, $w = w(x)$, and the *governing equation for deflection*, Equation 9.3, reduces to

$$D\frac{d^4w}{dx^4} - N_x\frac{d^2w}{dx^2} = p \tag{9.37}$$

When axial force N_x is zero, the foregoing agrees with Equation 3.34. Observe that the calculation of the plate deflection simplifies to the solution of Equation 9.37, which is of the same form as the differential equation for deflection of beams under the action of lateral and axial forces.

If plate edges are *not free* to move horizontally or *are immovable*, a tension in the plate is produced *depending on* the magnitude of *lateral deflection* w. The problem then becomes complicated. The tensile forces in the plate carry part of the lateral loading through membrane action.

A typical case in which bending to a cylindrical surface occurs is illustrated below.

EXAMPLE 9.10: Analysis of Tie-Rod

A rectangular plate, the length of which is large in comparison with its width, is subjected to a uniform loading of intensity p_0. The longitudinal edges of the plate are free to rotate but otherwise immovable. (a) Develop expressions for lateral deflection and moment. (b) Calculate the lateral deflection and stresses for $L = 1.2$ m, $t = 10$ mm, $p_0 = 10$ kPa, $E = 200$ GPa, and $\nu = 0.3$.

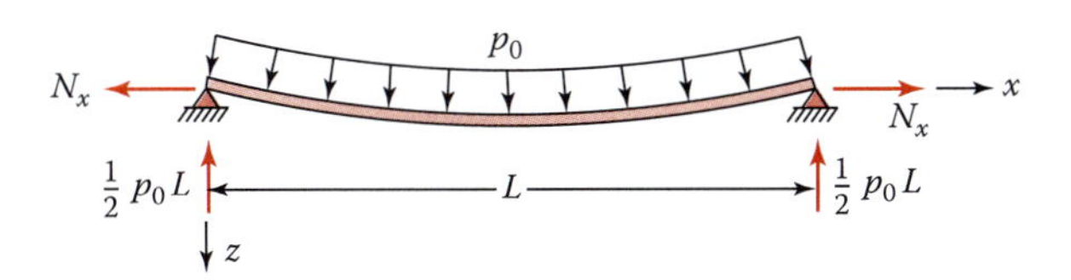

FIGURE 9.15
Tie-rod with uniform load.

Solution

A strip of unit width removed from a plate of this type will be in the same condition as a laterally and axially loaded beam or so-called *tie-rod* (Figure 9.15). The value of the axial tensile forces $N_x = N$ is such that horizontal movement of the edges is prevented.

a. The bending moment at any section x of the strip is described by

$$M(x) = \frac{1}{2}p_0Lx - \frac{1}{2}p_0x^2 - Nw$$

In terms of this moment, Equation 9.37 may be written

$$D\frac{d^2w}{dx^2} = -M(x) \tag{9.38}$$

We observe that differentiating Equation 9.38 twice results in Equation 9.37. Employing the notation

$$\lambda^2 = \frac{NL^2}{4D} \tag{9.39}$$

the solution of Equation 9.38 appears as follows:

$$w = C_1\sinh\frac{2\lambda x}{L} + C_2\cosh\frac{2\lambda x}{L} + \frac{p_0L^3x}{8\lambda^2D} - \frac{p_0L^2x^2}{8\lambda^2D} - \frac{p_0L^4}{16\lambda^4D} \tag{a}$$

The deflections vanish at the ends, $w = 0$ at $x = 0$ and $x = L$, of the strip, and thus

$$C_1 = \frac{p_0L^4}{16\lambda^4D}\frac{1-\cosh2\lambda}{\sinh2\lambda} \quad C_2 = \frac{p_0L^4}{16\lambda^4D} \tag{b}$$

Using the identities

$$\cosh 2\lambda = \cosh^2\lambda + \sinh^2\lambda$$

$$\sinh 2\lambda = 2\sinh\lambda\cosh\lambda$$

$$\cosh^2\lambda = 1 + \sinh^2\lambda$$

and substituting Equations b into Equation a, we obtain w in convenient form. It follows that the deflection curve of the strip is

$$w = \frac{p_0 L^4}{16\lambda^4 D}\left\{\frac{\cosh\lambda[1-(2x/L)]}{\cosh\lambda} - 1\right\} + \frac{p_0 L^2 x}{8\lambda^2 D}(L-x) \tag{9.40}$$

The maximum displacement occurs at midspan ($x = L/2$):

$$w_{\max} = \frac{5p_0 L^4}{384D} f_1(\lambda) \tag{9.41}$$

where

$$f_1(\lambda) = \frac{24}{5\lambda^4}\left(\operatorname{sech}\lambda - 1 + \frac{\lambda^2}{2}\right)$$

From Equation 9.40, we find that at ($x = L/2$):

$$M_{\max} = -D\frac{d^2 w}{dx^2} = \frac{p_0 L^2}{8} f_2(\lambda) \tag{9.42}$$

In the preceding, we have

$$f_2(\lambda) = \frac{2}{\lambda^2}(1 - \operatorname{sech}\lambda)$$

Interestingly, if there were no tensile reactions at the ends of the strip, the maximum deflection and moment would be

$$w_{\max} = \frac{5p_0 L^4}{384D} \qquad M_{\max} = \frac{1}{8} p_0 L^2 \tag{c}$$

The effects of N on displacement and moment are given by f_1 (λ) and f_2 (λ), which diminish rapidly with increasing λ.

Observe that the displacement and moment depend on the quantity λ and hence the axial forces, as defined by Equation 9.39. To determine N, we must consider the deformations. The extension of the strip produced by the axial tensile forces is, from Equation 9.17a,

$$\frac{1}{2}\int_0^L \left(\frac{dw}{dx}\right)^2 dx$$

During bending, the lateral contraction of the strip in the plane of the plate is assumed to be zero: $\varepsilon_y = 0$. Hooke's law,

$$\varepsilon_x = \frac{1}{Et}(N_x - \nu N_y) \qquad 0 = \frac{1}{Et}(N_y - \nu N_x)$$

yields $N_y = vN_x$ and $\varepsilon_x = N(1 - v^2)/Et$. It follows that

$$\frac{N(1-\nu^2)L}{Et} = \frac{1}{2}\int_0^L \left(\frac{dw}{dx}\right)^2 dx \tag{d}$$

A good *approximation* for N is found by selecting a deflection curve of the form

$$w = \frac{a_1}{1+\alpha}\sin\frac{\pi x}{L} \tag{e}$$

in which a_1 designates the midspan deflection produced by the *lateral load only*. The term $a_1/(1 + \alpha)$ is given by Equation 9.36, wherein α is replaced by $-\alpha$, as the axial forces in the present case are tensile. The quantity α in Equation 9.35, with $m = 1$, $n = 0$, $b = 1$, and $a = L$, becomes

$$\alpha = \frac{NL^2}{\pi^2 D} \tag{f}$$

Introducing Equation e into Equation d and *integrating*, we have

$$\frac{NL(1-\nu^2)}{Et} = \frac{\pi^2 a_1^2}{4L(1+\alpha)^2}$$

Finally, inserting Equations f and 3.10 into the above, we find that

$$\alpha(1+\alpha)^2 = \frac{3a_1^2}{t^2} \tag{9.43}$$

In *any* particular case, the preceding is solved for α, and the quantity λ, from Equations 9.39 and f, is then

$$\lambda^2 = \frac{NL^2}{4D} = \frac{\pi^2\alpha}{4} \tag{g}$$

For the present case, from Equations c,

$$a_1 = \frac{5p_0L^4}{384D}$$

and Equation 9.43 becomes

$$\alpha(1+\alpha)^2 = \frac{3}{t^2}\left(\frac{5p_0L^4}{384D}\right)^2 \tag{9.44}$$

b. Substituting the given data into Equation 3.10, we have

$$D = 200\times10^9(0.01)^3/12(1-0.09) = 1.8315\times10^4$$

Equation 9.44 then yields

$$\alpha(1+\alpha)^2 = \frac{3}{(0.01)^2}\left[\frac{5\times 70{,}000(1.2)^4}{384\times 1.83\times 10^4}\right]^2 = 319.94$$

from which

$$\alpha = 6.1895 \qquad \text{and} \qquad \lambda = \frac{\pi}{2}\sqrt{\alpha} = 3.9079$$

The in-plane tensile stress is now readily calculated as follows:

$$\sigma_{xp} = \frac{N}{t} = \frac{4\lambda^2 D}{L^2 t} = \frac{E\lambda^2}{3(1-v^2)}\left(\frac{t}{L}\right)^2$$
$$= \frac{200(10^9)(3.9079)^2}{3(1-0.09)}\left(\frac{0.01}{1.2}\right)^2 = 77.69 \text{ MPa}$$

The maximum bending moment is given by Equation 9.42 and the corresponding maximum bending stress equals

$$\sigma_{xb} = \frac{6M_{\max}}{t^2} = \frac{3}{4}p_0\left(\frac{L}{t}\right)^2 f_2(\lambda)$$
$$= \frac{3}{4}(70{,}000)\left(\frac{1.2}{0.01}\right)^2 \frac{2[1-\text{sech}\,(3.9079)]}{(3.9079)^2} = 95.03 \text{ MPa}$$

The maximum stress in the plate is therefore

$$\sigma_{\max} = \sigma_{xp} + \sigma_{xb} = 77.69 + 95.03 = 172.72 \text{ MPa}$$

Comment: Plates having other end conditions may be treated similarly (see Problem 9.16).

Problems

Sections 9.1 through 9.3

9.1 An aircraft wing panel is approximated as a rectangular plate under a uniform pressure p_0 and uniform tension N (Figure 9.1). Assume that the panel is made of an aluminum alloy 6061-T6 (Table B.3). Let $a = 2b = 500$ mm, $p_0 = 40$ kPa, $N = 20$ kPa, $t = 10$ mm, and $\nu = 0.3$. Compute the value of maximum deflection by taking the first term ($m = 1, n = 1$) of Equation 9.4.

9.2 A simply supported rectangular plate is subjected to a uniform axial compression N (Figure 9.3). The plate is constructed of high-strength, ASTM-A243 steel with $b = 200$ mm, $t = 4$ mm, $\nu = 0.29$, and $E = 200$ GPa (Table B.3). Calculate the critical load N_{cr} and buckling stress.

9.3 A steel ship bulkhead is a rectangular plate of length a and width b. The plate, assumed to be simply supported, is subject to uniform lateral pressure p_0 and uniform tensile forces N along the four edges. Derive a general expression for the deflection sur-face and determine the maximum deflection w and maximum stress σ_x for $a = b$, $m = 1$, and $n = 1$.

9.4 A simply supported rectangular plate carries uniform tensile force N along sides $x = 0$ and $x = a$ (Figure 9.2) and hydrostatic surface pressure described by

$$p(x, y) = p_0 \frac{x}{a}$$

Derive an expression for the deformed surface.

9.5 A structural component in the interior of a spacecraft consists of a square plate of sides a and thickness t. The plate may be approximated as simply supported on all edges and subjected to uniform biaxial compression N. Determine the buckling stress by using Equation 9.3.

9.6 Determine the buckling load of the plate under biaxial compression loads (Figure 9.6), for the case in which $N_x = N$ and $N_y = \alpha N$. Here the quantity $\alpha > 0$ represents some fixed known parameter.

9.7 If there are body forces acting in the midplane of the plate, show that the governing differential equation for deflection becomes

$$\nabla^4 w = \frac{1}{D}\left(p + N_x \frac{\partial^2 w}{\partial x^2} + N_y \frac{\partial^2 w}{\partial y^2} + 2N_{xy} \frac{\partial^2 w}{\partial x \partial y} - F_x \frac{\partial w}{\partial x} - F_y \frac{\partial w}{\partial y}\right) \tag{P9.7}$$

Here F_x and F_y denote the x- and y-directed body forces per unit area of the mid-plane of the plate.

Sections 9.4 through 9.7

9.8 Compute the buckling load S_{cr} of a simply supported steel plate of sides $a = 2b$ in pure shear (Figure 9.9). Let $a = 300$ mm, $t = 6$ mm, $E = 210$ GPa, and $\nu = 0.29$.

9.9 A simply supported square plate is under the action of a lateral load P at its center C and a uniform in-plane tension N, as shown in Figure P9.9. Derive the equation of the deflection surface, using the energy method and by retaining the first term of the series solution.

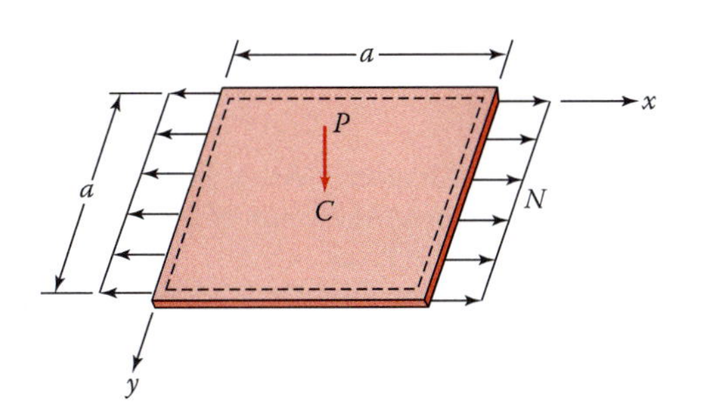

FIGURE P9.9

9.10 Redo Problem 9.5 employing the energy method.

9.11 Rework Example 9.3 employing the energy approach.

9.12 A simply supported square plate is subjected to uniform compression loads $N_x = -N$ applied to two opposite sides. Determine the buckling load, using the method of finite differences with $n = 3$.

9.13 Verify the result given by Equation d of Example 9.9.

9.14 Apply the finite difference approach to obtain the buckling load for a clamped square plate under uniform biaxial pressure N per unit of boundary length. Use $n = 3$.

9.15 A simply supported rectangular plate with an initial deflection defined by

$$w_0 = a_0 \sin\frac{\pi x}{a} \sin\frac{\pi y}{b}$$

is subject to uniform biaxial tensile forces N. Determine the maximum displacement for $a = b$.

9.16 A long rectangular plate with edges clamped carries a uniform load of intensity p_0. An elemental strip of unit plate width, bent into a cylindrical surface, is similar to that shown in Figure 9.15, except that now the ends are fixed, and bending moments M_0 thus occur there. Demonstrate that the differential equation of the deflection curve is expressed by

$$D\frac{d^2w}{dx^2} - Nw = -\frac{p_0 Lx}{2} + \frac{p_0 x^2}{2} - M_0 \tag{P9.16a}$$

Obtain the following solution:

$$\begin{aligned} w = &-\frac{p_0 L^4}{16\lambda^3 D}\sinh\frac{2\lambda x}{L} + \frac{p_0 L^4}{16\lambda^3 D}\coth\lambda\cosh\frac{2\lambda x}{L} \\ &+\frac{p_0 L^3 x}{8\lambda^2 D} - \frac{p_0 L^2 x^2}{8\lambda^2 D} - \frac{p_0 L^4}{16\lambda^3 D}\coth\lambda \end{aligned} \tag{P9.16b}$$

for the resulting displacement of the plate. Here λ, defined by Equation 9.39, is calculated as in Section 9.7.

References

1. A. C. Ugural, *Mechanics of Materials*, Chapter 9, Wiley, Hoboken, NJ, 2008.
2. A. C. Ugural and S. K. Fenster, *Advanced Mechanics of Materials and Applied Elasticity*, 5th ed., Chapter 11, Prentice-Hall, Upper Saddle River, NJ, 2012.
3. G. J. Simitses, *Elastic Stability of Structures*, Prentice-Hall, Upper Saddle River, NJ, 1976.
4. S. Timoshenko and J. M. Gere, *Theory of Elastic Stability*, 2nd ed., McGraw-Hill, New York, 1961.
5. B. Aalami and D. G. Williams, *Thin Plate Design for Transverse Loading*, Wiley, New York, 1975.
6. R. Szilard, *Theory and Analysis of Plates—Classical and Numerical Methods*, Chapter 6, Prentice-Hall, Upper Saddle River, NJ, 2004.

7. D. J. Peery and J. J. Azar, *Aircraft Structures*, 2nd ed., Chapter 11, McGraw-Hill, New York, 1982.
8. J. H. Faupel and F. E. Fisher, *Engineering Design*, 2nd ed., Wiley, New York, 1981.
9. J. N. Reddy, *Theory and Analysis of Plates and Shells*, 2nd ed., CRC Press, Boca Raton, FL, 2007.
10. G. Gerard and H. Becker, *Handbook of Structural Stability*, NACA TN 3782, National Advisory Committee for Aeronautics, Washington, DC, 1975.
11. W. Flügge, *Handbook of Engineering Mechanics*, McGraw-Hill, New York, 1984.
12. D. O. Brush and B. O. Almroth, *Buckling of Bars, Plates, and Shells*, McGraw-Hill, New York, 1975.

10

Large Deflection of Plates

10.1 Introduction

In previous sections of the text, relatively small plate deflection is assumed ($w < t$). In some applications of thin plates, however, the maximum deflection is equal to or larger than the plate thickness. Because of these *large* displacements ($w \geq t$) the midplane stretches and hence the in-plane tensile stresses developed within the plate stiffen and add considerable load resistance to it that is not predicted by the small-deflection bending theory. For such situations, an extended plate theory must be employed, accounting for the effects of large deflections. The *large-deflection theory* of plates assumes that the deflections are no longer small in comparison with the thickness but are nevertheless small compared with the remaining plate dimensions.

The large-displacement behavior for plates of simple form is illustrated in Section 10.2, principally to give some idea of the additional load-carrying action. The behavior described is also generally valid for plates of any other shape. Section 10.3 describes the differences between the small- and large-deflection theories. The general analytical solution of plate problems is formulated in Section 10.4. This is followed by the application of the energy method to the solution of problems involving large plate deflections. Section 10.6 presents a numerical treatment for the bending of plates experiencing large deflections.

10.2 Plate Behavior When Deflections are Large

In discussing large-deflection behavior, we must distinguish the cases of midplane deformation into "developable" or "nondevelopable" surfaces. A developable surface completely recovers its original flat shape and dimensions; developable behavior implies the absence of any deformation. Cylinders and cones have developable surfaces, while a sphere or a saddle is a nondevelopable surface. When a plate bends into a cylindrical geometry, for example, the tension in the midplane can be produced only if the end supports are immovable (Section 9.7). Thus, as mentioned in Section 3.3, the limitation $w_{max} < t$ does not hold in cases where a simply supported plate bends into a developable surface, and the classical formulas are valid until yielding impends or $w_{max} \rightarrow a$, where a is the smaller span length of the plate. It is recalled from Section 3.4 that a developable surface has a zero Gaussian curvature.

For plates of ordinary proportions, the edge conditions have a pronounced effect on the magnitude of the direct tensile stresses that may be developed within them. In the case of

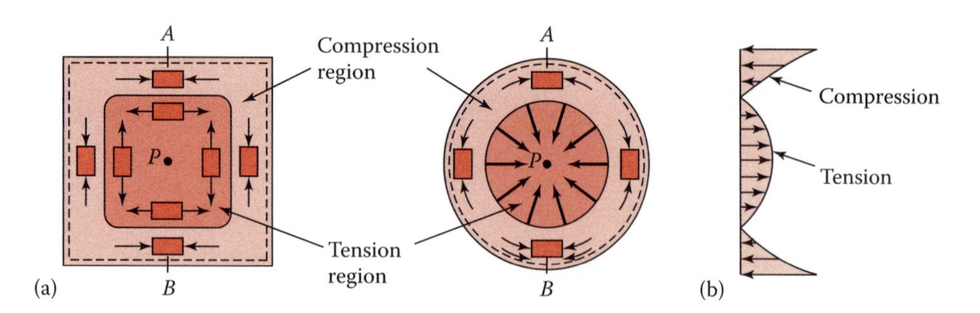

FIGURE 10.1
(a) Plates with large deflections; (b) significance of the direct tension.

a simply supported plate undergoing a general large displacement, the plate edges will be *free* of stresses in the direction normal to the boundary, while all other points within the plate will not be stress free. The tensile stresses, which increase with the distance from an edge, are caused by compressive stresses in a tangential direction near the plate edges. In practice, the latter stresses often produce wrinkling or buckling near the edges of a simply supported plate. On the other hand, normal and tangential stresses may both occur at the edges of a built-in plate. Clearly, only when the plate edges are clamped and fixed are the direct tensile stresses utilized fully by carrying some of the lateral loading.

Consider the behavior of a square or circular simply supported plate under a concentrated center load P (Figure 10.1a). While the plate is being subjected to an increasing load, it goes through the pure bending, small-displacement range to the large-displacement range. At this stage, the relationship between P and the deflection w is *no longer linear* (Section 10.3) because of the *change in plate geometry*. At higher loads, a central tensile and outer compressive stress region develops. The distribution of stress at a section through points A and B of the plate is illustrated by the sketch shown in Figure 10.1b.

10.3 Comparison of Small- and Large-Deflection Theories

It is now demonstrated that all relationships derived thus far for plates bending into nondevelopable surfaces are true in general *only* if the maximum deflection is small in comparison with the thickness of the plate. For this purpose, we shall discuss the solutions of the circular plate problem based on approximate and exact approaches. The general behavior for circular plates when deflections are large also applies to rectangular, elliptical, or other shapes of plates. Data for rectangular plates similar to those introduced here for circular plates are given in the technical literature [1–8].

10.3.1 An Approximate Method for the Circular Plates

According to this approach, which is in good agreement with the exact theory of the circular plate, the bending solution and the *membrane or very thin plate* solution are treated separately. The partial loads carried by the membrane and the bending actions in the plate are then added and equated to the actual plate loading.

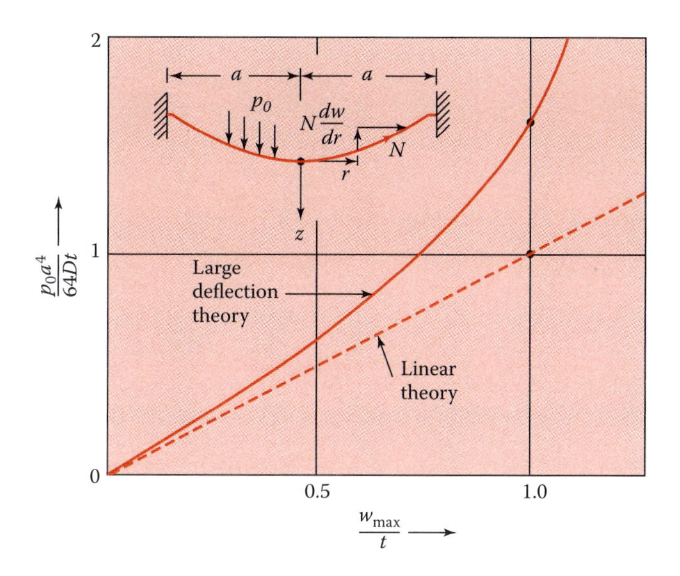

FIGURE 10.2
Comparison of large-deflection and linear (small-displacement) theories for circular plates based on an approximate approach.

As an example, consider the case of a *clamped-edge circular plate* subjected to a uniform load of intensity p_0 (Figure 10.2). The *bending solution* for the maximum deflection occurring in the center ($r = 0$) is, from Equation 4.19,

$$w_{max} = \frac{p_0 a^4}{64D} \quad \text{or} \quad p_0 = \frac{64D}{a^4} w_{max} \tag{10.1}$$

To derive the *membrane solution*, refer to Figure 10.2, where N denotes the *constant* tensile force per unit length. The static equilibrium of vertical forces is expressed by $2\pi r N(dw/dr) = p_0 \pi r^2$, $dw/dr = p_0 r/2N$, from which integration of the latter expression for the slope at $r = a$ leads to

$$w_{max} = \frac{p_0 a^2}{4N} \tag{a}$$

Determination of the value of N in Equation a requires consideration of the midplane deformations.

The radial elongation produced by the deflection w is found from Equation 9.12a as follows:

$$\frac{1}{2}\int_0^a \left(\frac{dw}{dr}\right)^2 dr = \frac{1}{2}\int_0^a \left(\frac{p_0 r}{2N}\right)^2 dr = \frac{p_0^2 a^3}{24N^2}$$

The strain is therefore

$$\varepsilon = \frac{p_0^2 a^2}{24N^2} \tag{b}$$

This strain is the *same* in all directions. Hence, $(1 - \nu)N/t = E\varepsilon$ together with Equation b yields

$$N^3 = \frac{E}{1-\nu}\frac{p_0^2 a^2 t}{24} \tag{c}$$

Finally, elimination of N between Equations a and c results in the membrane solution:

$$p_0 = \frac{8}{3}\frac{E}{1-\nu}\frac{t}{a}\left(\frac{w_{max}}{a}\right)^3 \tag{10.2}$$

The load actually carried by the plate equals the sum of the partial loads resisted by the bending and the membrane actions. On application of Equations 10.1 and 10.2, we have

$$p_0 = \frac{64D}{a^3}\left(\frac{w_{max}}{a}\right) + \frac{8}{3}\frac{E}{1-\nu}\frac{t}{a}\left(\frac{w_{max}}{a}\right)^3 \tag{d}$$

An alternate form of Equation d, obtained by taking $\nu = 0.3$, is

$$\frac{p_0 a^4}{64Dt} = \frac{w_{max}}{t}\left[1 + 0.65\left(\frac{w_{max}}{t}\right)^2\right] \tag{10.3}$$

Clearly, the first and the second bracketed terms represent the bending and the membrane solutions, respectively.

To illustrate the variation of load and deflection for a uniformly loaded circular plate with clamped edge, w_{max}/t and $p_0 a^4/64Dt$ are plotted in Figure 10.2. We observe that the *linear theory*, which neglects the membrane action, is *satisfactory for* $w_{max} < t/2$ and that larger deflections produce greater error. When $w_{max} = t$, for example, there is a 65%t error in the load according to the bending theory alone. The experimental data agrees well with the result given by *large-deflection theory* that considers both the bending and the membrane actions.

In Section 10.5, the preceding problem is discussed by the energy method, which sheds further light on the stretching of the midsurface.

EXAMPLE 10.1: Large Deflection of Fixed Edge Plate with Uniform Loading

A circular steel plate ($E = 200$ GPa, $\nu = 0.3$) 5 mm thick and 0.5 m in diameter is clamped along the edge and subjected to a uniformly distributed load of 160 kPa. Determine the maximum deflection.

Solution

If the linear theory held, the deflection (and stress) would be directly proportional to the load. The center deflection would then be

$$w_{\max} = \frac{p_0 a^4}{64D} = \frac{160{,}000(0.25)^4 \times 12(1-0.09)}{64\times 200\times 10^9 (0.005)^3} = 4.3 \text{ mm}$$

which is much more than half the thickness. Thus, the ordinary theory does *not* lead to a result of acceptable accuracy.

We use Equation 10.3, which takes into account both membrane and bending stresses:

$$0.0043 = w_{\max}\left[1 + 0.65\left(\frac{w_{\max}}{0.005}\right)^2\right]$$

From this expression, by trial and error, the correct value of deflection is found to be $w_{\max} = 3.3$ mm.

Comment: Stress at any point in the plate may then be obtained by the superposition of the bending stress (Section 4.5) and the membrane stress (N/t), where N is calculated from Equation a.

10.3.2 Exact Solution for the Circular Plate Problem

To obtain a *more satisfactory solution* of the problem of large deflection of a uniformly loaded circular plate, it is necessary to solve the governing equations of equilibrium by a series method. In so doing, the maximum deflections (for $v = 0.3$) and stresses in a *circular plate with clamped edge can* be represented graphically as shown in Figure 10.3 [1,7]. If the deflection of the plate is found from Figure 10.3a, the corresponding stress can be obtained by using Figure 10.3b or vice versa (see Problem 10.3).

Similarly, the maximum deflections (for $v = 0.3$) and stresses occurring at the center of a *simply supported circular plate* are represented in Figure 10.4. The coordinates of these curves have the same meaning as those of Figure 10.3 for a plate whose edge is fixed. In each case, the maximum stress is the *sum* of the values of the *bending stress* ($6M/t^2$) and the *membrane stress* (N/t). For purposes of comparison, Figures 10.3a and 10.4a also include straight dashed lines showing the results from the linear theory. It is seen that the *error* of the linear theory *increases* as the load and deflection increase.

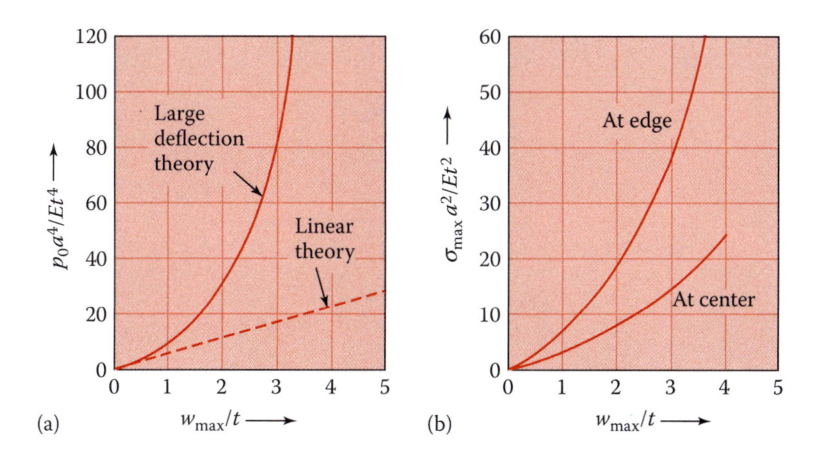

FIGURE 10.3
Maximum displacements and stresses in clamped circular plates having large deflections.

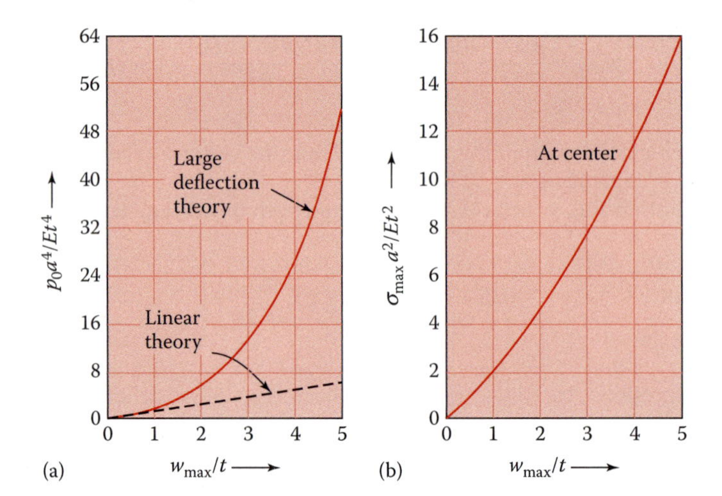

FIGURE 10.4
Maximum displacements and stresses in simply supported circular plates having large deflections.

10.4 General Equations for Large Deflections of Plates

A modified form of Equation 9.3 may be employed in the analysis of deformation in a plate under transverse loading, resulting in relatively large elastic deflection. It is important to note that the in-plane forces N_x, N_y, and N_{xy} *do not, as before, depend only* on the external loading acting in the xy plane. They are now affected also by the *stretching of the midplane* of the plate produced by large deflections *due to bending*. For a thin plate element, the x and y equilibria of direct forces are expressed by Equations 9.1 and 9.2. To ascertain the values of N_x, N_y, and N_{xy} in Equations 9.1 and 9.2, a third expression will be developed from the midplane strain–displacement relations as follows.

When the stretching of the midplane occurs *during the bending* of the plate, any point in the *midplane* experiences x-, y-, and z-directed displacements, $u_0 = u$, $v_0 = v$, and w, respectively (Section 3.4). Then resultant strain components are found by combining Equations 3.1 and 9.17. Thus, the following are the *general* midplane strain–displacement relations due to bending and stretching of the plate:

$$\begin{aligned}
\varepsilon_x &= \frac{\partial u}{\partial x} + \frac{1}{2}\left(\frac{\partial w}{\partial x}\right)^2 \\
\varepsilon_y &= \frac{\partial v}{\partial y} + \frac{1}{2}\left(\frac{\partial w}{\partial y}\right)^2 \\
\gamma_{xy} &= \frac{\partial v}{\partial x} + \frac{\partial u}{\partial y} + \frac{\partial w}{\partial x}\frac{\partial w}{\partial y}
\end{aligned} \tag{10.4}$$

These equations are the *simplified* nonlinear case of *small strains* and *large rotations* that may be characterized by *small finite deformations*. The displacements are still nonlinear, but they must be small so that the range of deformations remains elastic and no yielding occurs. For applications to thin structures such as beams, plates, and shells, new simplifications have been proposed recently in the nonlinear elasticity theory [8,9].

As the strains are evidently not independent of one another, by differentiating ε_x twice with respect to y, ε_y twice with respect to x, and γ_{xy} with respect to x and y, the following familiar relationship is obtained:

$$\frac{\partial^2 \varepsilon_x}{\partial y^2} + \frac{\partial^2 \varepsilon_y}{\partial x^2} - \frac{\partial^2 \gamma_{xy}}{\partial x \partial y} = \left(\frac{\partial^2 w}{\partial x \partial y}\right)^2 - \frac{\partial^2 w}{\partial x^2}\frac{\partial^2 w}{\partial y^2} \tag{a}$$

According to Hooke's law,

$$\varepsilon_x = \frac{1}{Et}(N_x - \nu N_y) \qquad \varepsilon_y = \frac{1}{Et}(N_y - \nu N_x) \qquad \gamma_{xy} = \frac{N_{xy}}{Gt} \tag{b}$$

On introduction of Equations b into Equation a, we thus have the third equation in terms of N_x, N_y, and N_{xy}. This expression, as well as Equations 9.1 and 9.2, is identically satisfied by the *stress function* $\Phi(x, y)$, related to the in-plane forces as follows:

$$N_x = t\frac{\partial^2 \Phi}{\partial y^2} \qquad N_y = t\frac{\partial^2 \Phi}{\partial x^2} \qquad N_{xy} = -t\frac{\partial^2 \Phi}{\partial x \partial y} \tag{c}$$

Substitution of the foregoing into Equations b yields

$$\begin{aligned} \varepsilon_x &= \frac{1}{E}\left(\frac{\partial^2 \Phi}{\partial y^2} - \nu\frac{\partial^2 \Phi}{\partial x^2}\right) \\ \varepsilon_y &= \frac{1}{E}\left(\frac{\partial^2 \Phi}{\partial x^2} - \nu\frac{\partial^2 \Phi}{\partial y^2}\right) \\ \gamma_{xy} &= -\frac{2(1+\nu)}{E}\frac{\partial^2 \Phi}{\partial x \partial y} \end{aligned} \tag{d}$$

Inserting Equations d into Equation a results in

$$\frac{\partial^4 \Phi}{\partial x^4} + 2\frac{\partial^4 \Phi}{\partial x^2 y^2} + \frac{\partial^4 \Phi}{\partial y^4} = E\left[\left(\frac{\partial^2 w}{\partial x \partial y}\right)^2 - \frac{\partial^2 w}{\partial x^2}\frac{\partial^2 w}{\partial y^2}\right] \tag{10.5}$$

and introduction of Equations c into Equation 9.3 leads to

$$\frac{\partial^4 w}{\partial x^4} + 2\frac{\partial^4 w}{\partial x^2 \partial y^2} + \frac{\partial^4 w}{\partial y^4} = \frac{t}{D}\left[\frac{p}{t} + \frac{\partial^2 \Phi}{\partial y^2}\frac{\partial^2 w}{\partial x^2} + \frac{\partial^2 \Phi}{\partial x^2}\frac{\partial^2 w}{\partial y^2} - 2\frac{\partial^2 \Phi}{\partial x \partial y}\frac{\partial^2 w}{\partial x \partial y}\right] \tag{10.6}$$

Equations 10.5 and 10.6 are the *governing differential equations for large deflections* of thin plates. Determination of Φ and w requires the solution of these equations, which must, of course, satisfy the boundary conditions. Once the stress function is known, the midplane stresses are obtained through the use of Equations c. Knowing the deflection w, on application of Equations 3.7, 3.30, and 3.29, as in the case of small deflection, the normal and shear stresses, respectively, are determined.

Equations 10.5 and 10.6 were introduced by *von Kármán* in 1910, for whom they are named [10]. Unfortunately, where realistic problems are concerned, solving these coupled, nonlinear, partial differential equations may be a formidable task. Exact solutions of von Kármán equations have been achieved only for very few cases [11]. A number of approximate solutions have been determined for the uniformly loaded plates of simple regular shapes [1,2,6]. Only as a result of recent progress in the development of numerical approaches has the general problem of plates been treated satisfactorily [12–14]. We shall discuss the powerful finite element solution of large deflections of thin plates in Section 10.6.

Comments: In concluding this discussion, consider the bending of the plate into a cylindrical surface (Section 9.7). If the generating line of the cylindrical surface is parallel to the y axis, $w = w(x)$, $\partial^2\Phi/\partial x^2$, and $\partial^2\Phi/\partial y^2$ are constants. Hence, Equation 10.5 is satisfied identically, and Equation 10.6 reduces to Equation 9.37. This particular case of bending has already been treated in Examples 3.1 and 9.8.

10.5 Deflections by the Energy Method

In an investigation of the large deflection of thin plates, the energy method frequently can be employed to good advantage. The strain energy of the plate consists of the energy of bending and the energy due to straining of the midsurface. The first part of this energy is defined by Equation 3.44.

The strain energy associated with the stretching of the midplane of the plate or the *membrane strain energy*, with reference to Section 3.13, is given by

$$U_m = \frac{1}{2}\iint (N_x\varepsilon_x + N_y\varepsilon_y + N_{xy}\gamma_{xy})dx\,dy \tag{10.7a}$$

This becomes, by the use of Equations b for Section 10.4,

$$U_m = \frac{Et}{2(1-\nu^2)}\iint\left[\varepsilon_x^2 + \varepsilon_y^2 + 2\nu\varepsilon_x\varepsilon_y + \frac{1}{2}(1-\nu)\gamma_{xy}^2\right]dx\,dy \tag{10.7b}$$

By substitution of the values of the strains from Equations 10.4 into the preceding equation, the strain energy is obtained in the form

$$\begin{aligned} U_{nt} = \frac{Et}{2(1-\nu^2)}\int\int\Bigg\{&\left(\frac{\partial u}{\partial x}\right)^2 + \frac{\partial u}{\partial x}\left(\frac{\partial w}{\partial x}\right)^2 + \left(\frac{\partial v}{\partial y}\right)^2 + \frac{\partial v}{\partial y}\left(\frac{\partial w}{\partial x}\right)^2 \\ &+ \frac{1}{4}\left[\left(\frac{\partial w}{\partial x}\right)^2 + \left(\frac{\partial w}{\partial y}\right)^2\right]^2 + 2\nu\left[\frac{\partial u}{\partial x}\frac{\partial v}{\partial y} + \frac{1}{2}\frac{\partial v}{\partial y}\left(\frac{\partial w}{\partial x}\right)^2 + \frac{1}{2}\frac{\partial u}{\partial x}\left(\frac{\partial w}{\partial y}\right)^2\right] \\ &+ \frac{1-\nu}{2}\left[\left(\frac{\partial u}{\partial y}\right)^2 + 2\frac{\partial u}{\partial y}\frac{\partial v}{\partial x} + \left(\frac{\partial v}{\partial x}\right)^2 + 2\frac{\partial u}{\partial y}\frac{\partial w}{\partial x}\frac{\partial w}{\partial y} + 2\frac{\partial v}{\partial x}\frac{\partial w}{\partial x}\frac{\partial w}{\partial y}\right]\Bigg\}dx\,dy \end{aligned} \tag{10.8}$$

The energy method provides a simple approach to the determination of the large deflections of plates, as illustrated in the solution of the following numerical sample problems.

EXAMPLE 10.2: Analysis of Membrane by Energy Method

What are the deflection and stress at the center of a very thin square plate of sides $2b$ due to a uniform load p_0 per unit surface area (Figure 10.5)? The displacements u, v, and w are taken to be zero at the edges. The plate under consideration may be regarded as flexible membrane having no resistance to bending action.

Solution

Assuming that the origin of coordinates x, y is placed at the center of the plate, the displacement may be represented by

$$\begin{aligned} w &= a_0 \cos\frac{\pi x}{2b}\cos\frac{\pi y}{2b} \\ u &= c_0 \sin\frac{\pi x}{b}\cos\frac{\pi y}{2b} \\ v &= c_0 \sin\frac{\pi y}{b}\cos\frac{\pi x}{2b} \end{aligned} \tag{a}$$

where a_0 and c_0 are unknown coefficients. Equations a satisfy the boundary conditions. It is also observed that the requirement of symmetry of deformation is fulfilled by u, v, and w.

Introducing Equations a into Equation 10.8 and integrating, we find that for $\nu = 0.25$,

$$U_m = \frac{Et}{7.5}\left[\frac{5\pi^4}{64}\frac{a_0^4}{b^2} - \frac{17\pi^2}{6}\frac{c_0 a_0^2}{b} + c_0^2\left(\frac{35\pi^2}{4} + \frac{80}{9}\right)\right] \tag{b}$$

The work done by the uniformly distributed load is

$$W = \int_{-b}^{b}\int_{-b}^{b} p_0 a_0 \cos\frac{\pi x}{2b}\cos\frac{\pi y}{2b}\,dx\,dy \tag{c}$$

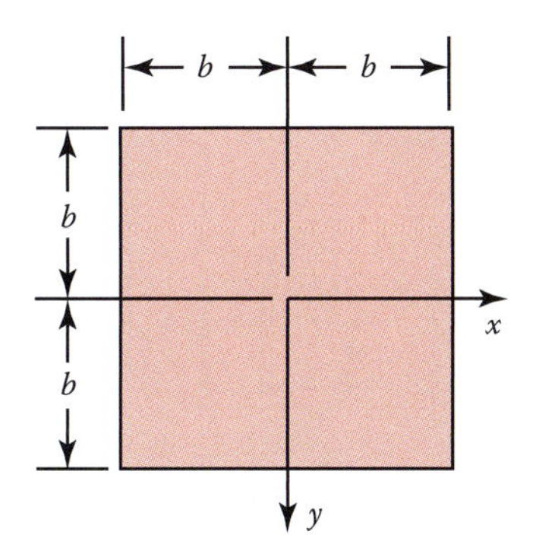

FIGURE 10.5
Uniformly loaded very thin square plate or membrane.

Potential energy $\Pi = U - W$. The conditions $\partial\Pi/\partial a_0 = 0$ and $\partial\Pi/\partial c_0 = 0$ yield two equations, the solution of which results in the following values for the coefficients:

$$a_0 = 0.818b\sqrt[3]{\frac{p_0 b}{Et}} \qquad c_0 = 0.147\frac{a_0^2}{b} \tag{10.9}$$

The maximum lateral deflection occurs at the center and is given by $w_{max} = 0.818b\sqrt[3]{p_0 b/Et}$. On applying Equations 10.4 together with Equations a, the tensile strain at $x = y = 0$ is found to be $\varepsilon_x = \varepsilon_y = \pi c_0/b = 0.462a_0^2/b^2$. The associated tensile stress is

$$\sigma = \frac{E}{1-\nu}\left(0.462\frac{a_0^2}{b^2}\right) = 0.412\sqrt[3]{\frac{p_0^2 E b^2}{t^2}} \tag{10.10}$$

Thus, we observe that the deflection and stress are not proportional to the intensity of the load but vary as the 1/3 power and the 2/3 power of that intensity, respectively.

EXAMPLE 10.3: Large Deflection Analysis of Plate by Energy Method

Derive an expression for the maximum deflection of the clamped circular plate subjected to a uniform load p_0 (Figure 10.6) by using the energy method.

Solution

The edge conditions, $w = 0$ and $dw/dr = 0$ at $r = 0$, are satisfied by

$$w = w_{max}\left(1 - \frac{r^2}{a^2}\right)^2 \tag{d}$$

Here w_{max} represents the maximum deflection occurring in the center of the plate. Equation d is actually an approximation to the true surface since it disregards the midsurface stretching. The associated strain energy of bending, from Equation 4.41, is

$$U_b = \frac{32\pi}{3}\frac{w_{max}^2}{a^2}D \tag{e}$$

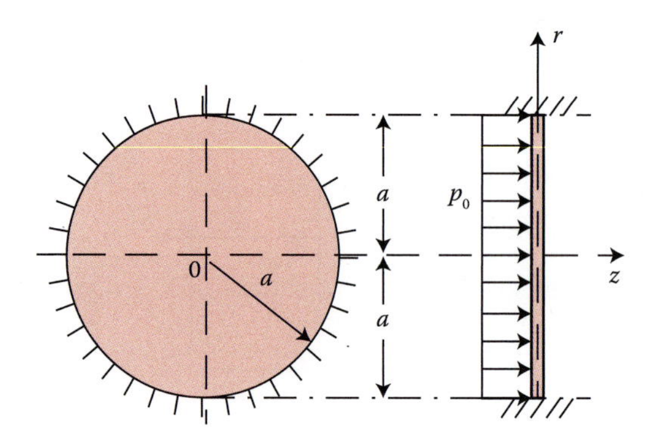

FIGURE 10.6
Clamped edge plate under uniform pressure.

The radial displacement u vanishes at the center and at the clamped boundary of the plate. These conditions are fulfilled by the series

$$u = r(a-r)\left(c_1 + c_2 r + c_3 r^2 + \Delta\right) \tag{10.11}$$

where $c_1, c_2, c_3, \ldots$ are unknown constants.

Consider now only the first two terms of the preceding series:

$$u = r(a-r)(c_1 + c_2 r) \tag{f}$$

The strain energy corresponding to the stretching of the midsurface, carrying Equations d and f into Equation P10.10b, is

$$U_m = \frac{\pi Eta^2}{1-\nu^2}\left(0.250c_1a^2 + 0.1167c_2^2a^4 + 0.300c_1c_2a^3 - 0.06768c_1\frac{w_{max}^2}{a} + 0.05456c_2w_{max}^2 + 0.30528\frac{w_{max}^2}{a^4}\right) \tag{g}$$

Note that in the foregoing parentheses, ν is taken as 0.3. The conditions $\partial U_m/\partial c_1 = 0$ and $\partial U_m/\partial c_2 = 0$ that make U_m a minimum lead to

$$c_1 = 1.185\frac{w_{max}^2}{a^3} \qquad c_2 = -1.75\frac{w_{max}^2}{a^4}$$

Inserting these into Equation g, we have

$$U_m = 2.59\pi D\frac{w_{max}^4}{a^2t^2} \tag{h}$$

The total energy is obtained by combining Equations e and h as

$$U = \frac{32\pi D}{3}\frac{w_{max}^2}{a^2}\left(1 + 0.244\frac{w_{max}^2}{t^2}\right) \tag{i}$$

in which the second term is the energy associated with the straining of the midsurface of the plate. It is clear that, if $w_{max} < t$, the strain energy is mainly due to bending. The work done by p_0 is given by

$$W = 2\pi\int_0^a wp_0r\,dr = 2\pi p_0w_{max}\int_0^a\left(1-\frac{r^2}{a^2}\right)^2 r\,dr \tag{j}$$

The potential energy function, $\Pi = U - W$, is obtained by combining Equations i and j and integrating. Then, application of the minimizing condition, $\partial\Pi/\partial w_{max} = 0$, yields an expression for the maximum deflection of the form

$$\frac{p_0a^4}{64D} = w_{max}\left[1 + 0.488\left(\frac{w_{max}}{t}\right)^2\right] \tag{10.12}$$

For example, taking $w_{\max} = t/2$, the foregoing gives $w_{\max} = 0.89\ p_0a^4/64D$. This value is approximately 3.5% higher than the more accurate solution $0.86\ p_0a^4/64D$ found from Equation 10.6. An improved approximation results by retaining more terms of the series (Equation 10.11).

Comments: Observe that, in the case of *very thin plates*, the deflection $w_{\max}$ may become many times greater than t. In such situations, the resistance of the plate to bending can be omitted, and the plate can be considered as a flexible membrane. That is, the first term on the right-hand side of Equation 10.12, being small in comparison with the second term, is neglected. It follows that

$$\frac{p_0a^4}{64Dt} = 0.488\left(\frac{w_{\max}}{t}\right)^3 \tag{10.13}$$

and the center deflection becomes, for $\nu = 0.3$, $w_{\max} = 0.704a\sqrt[3]{p_0a/Et}$.

The energy method can similarly be applied in the case of large deflections of plates having other boundary conditions and shapes.

10.6 The Finite Element Solution

The FEM (Chapter 7) is here applied to determining the large displacements and stresses in a plate of arbitrary shape under general loading. The behavior of plates experiencing large deflections is illustrated in Section 10.2. Clearly, we must now extend the formulations of Section 7.9 to include the effect of midplane plate strains and corresponding in-plane stresses.

To begin with, assume an initial system of in-plane forces applied to the plate and regard these as constants during bending (Section 9.4). On this basis, in-plane strains of an element will be represented, from Equations 10.4, as

$$\{\bar{\varepsilon}\}_e = \left\{\frac{1}{2}\left(\frac{\partial w}{\partial x}\right)^2, \frac{1}{2}\left(\frac{\partial w}{\partial y}\right)^2, \frac{\partial w}{\partial x}\frac{\partial w}{\partial y}\right\} \tag{10.14a}$$

The "generalized" bending-strain–displacement relations are given by Equations 7.18:

$$\{\varepsilon\}_e = \left\{-\frac{\partial^2 w}{\partial x^2}, -\frac{\partial^2 w}{\partial y^2}, -2\frac{\partial^2 w}{\partial x \partial y}\right\} \tag{10.14b}$$

The stress resultants are thus composed of *constant* direct forces and moment

$$\begin{Bmatrix} N_x \\ N_y \\ N_{xy} \end{Bmatrix}_e = t\begin{Bmatrix} \bar{\sigma}_x \\ \bar{\sigma}_y \\ \bar{\tau}_{xy} \end{Bmatrix}_e \qquad \begin{Bmatrix} M_x \\ M_y \\ M_{xy} \end{Bmatrix}_e = \int_{-t/2}^{t/2} \begin{Bmatrix} \sigma_x \\ \sigma_y \\ \tau_{xy} \end{Bmatrix}_e z\,dz \tag{10.15a}$$

or

$$\{N\}_e = t\{\bar{\sigma}\}_e \qquad \{M\}_e = \int_{-t/2}^{t/2} \{\sigma\}_e \, z \, dz \tag{10.15b}$$

The bending stresses {s} and the "generalized" bending strains {e} are related by Equation 7.19. In-plane stresses $\{\bar{\sigma}\}$ and in-plane strains $\{\bar{\varepsilon}\}$ are connected by Hooke's law:

$$\{\bar{\sigma}\}_e = z[D^*]\{\bar{\varepsilon}\}_e \tag{10.16}$$

In the preceding,

$$[D^*] = \frac{E}{1-\nu^2}\begin{bmatrix} 1 & \nu & 0 \\ \nu & 1 & 0 \\ 0 & 0 & (1-\nu)/2 \end{bmatrix} \tag{10.17}$$

represent the *elasticity matrix for plane stress.*

Inasmuch as the direct and bending strains are taken to be *independent*, Equation 3.51 for the potential energy of plates is

$$\Pi = \frac{1}{2}\iint_A \{\varepsilon\}_e^T \{M\}_e \, dx \, dy + \frac{t}{2}\iint_A \begin{Bmatrix} \dfrac{\partial w}{\partial x} \\ \dfrac{\partial w}{\partial y} \end{Bmatrix}^T [\bar{\sigma}]_e \begin{Bmatrix} \dfrac{\partial w}{\partial x} \\ \dfrac{\partial w}{\partial y} \end{Bmatrix} dx \, dy$$

$$-\iint_A (pw) dx \, dy \tag{10.18}$$

where

$$[\bar{\sigma}]_e = \begin{bmatrix} \bar{\sigma}_x & \bar{\tau}_{xy} \\ \bar{\tau}_{xy} & \bar{\sigma}_y \end{bmatrix}_e \tag{10.19}$$

We shall proceed with the discussion by considering the case of a specific element.

10.6.1 Rectangular Finite Element

The bending properties of this simple element are developed in Section 7.12. We now treat its in-plane deformation properties. The derivatives (slopes) of w are, from Equation 7.43,

$$\begin{Bmatrix} \dfrac{\partial w}{\partial x} \\ \dfrac{\partial w}{\partial y} \end{Bmatrix}_e = \begin{bmatrix} 0 & 1 & 0 & 2x & y & 0 & 3x^2 & 2xy & y^2 & 0 & 3x^2y & y^3 \\ 0 & 0 & 1 & 0 & x & 2y & 0 & x^2 & 2xy & 3y^2 & x^3 & 3xy^2 \end{bmatrix} \{a_1, \ldots, a_{12}\} \tag{10.20a}$$

or

$$\{\theta\}_e = [S]\{a\} \tag{10.20b}$$

Applying Equation 7.45, we have

$$\{\theta\}_e = [S][C]^{-1}\{\delta\}_e = [G]\{\delta\}_e \tag{10.21}$$

in which

$$[G] = [S][C]^{-1} \tag{10.22}$$

It is observed that $[G]$ is a matrix defined only in terms of the coordinates.

The potential-energy expression, on introducing Equations 10.21 into Equation 10.18, is therefore

$$\begin{aligned} \Pi = \frac{1}{2}\iint_A \{\varepsilon\}_e^T\{M\}_e\,dx\,dy + \frac{t}{2}\{\delta\}_e^T\left(\iint_A [G]^T[\bar{\sigma}][G]dx\,dy\right)\{\delta\}_e \\ -\iint_A (pw)dx\,dy \end{aligned} \tag{10.23}$$

As in Section 7.9, it can be demonstrated readily that the principle of potential energy now yields

$$[Q\}_e = [k]_e\{\delta\}_e + [k_G]_e\{\delta\}_e = [k_T]_e\{\delta\}_e \tag{10.24}$$

These are *modified equations of equilibrium* of nodal forces. The new term

$$[k_G]_e = t[C^{-1}]^T\left(\iint_A [S]^T[\bar{\sigma}][S]dx\,dy\right)[C^{-1}] \tag{10.25}$$

is known as the *initial stress* or *geometric stress* matrix. Appropriate definitions of the small-displacement stiffness matrix $[k]_e$ and nodal force matrix $\{Q\}_e$ are given by Equations 7.24 and 7.25, respectively. The $[K_T]_e$ is termed the *total stiffness matrix* of the element. Equation 10.25 may be expressed in the following convenient form:

$$[k_G]_e = \bar{\sigma}_x[k_{Gx}]_e + \bar{\sigma}_y[kGy]_e + \bar{\tau}_{xy}[k_{Gxy}]_e \tag{10.26}$$

wherein $[k_{Gx}]_e$, for instance, designates

$$[k_{Gx}]_e = t[C^{-1}]^T \left(\iint_A [S]^T \begin{bmatrix} 1 & 0 \\ 0 & 0 \end{bmatrix} [S] dx\, dy \right) [C^{-1}] \tag{10.27}$$

Relationships for other components of the geometric matrix, $[k_{Gy}]_e$ and $[k_{Gxy}]_e$, may be written in a like manner.

The *generalized procedure* for solving a plate *large-deflection problem* is summarized as follows:

1. Assume the initial direct stresses $\{\bar{\sigma}\}$ to be zero. Apply the procedure (steps 1 through 3) of Section 7.9 to obtain $[k]_e$, $\{Q\}_e$, and the hence small-deflection solution for the nodal displacements $\{\delta\}_e$.
2. Compute the slopes at some representative point within the element (e.g., at the centroid) from Equation 10.21: $\{\theta\}_e = [G]\{\delta\}_e$.
3. Compute the in-plane strains $\{\bar{\varepsilon}\}_e$ from Equations 10.14.
4. Compute the in-plane stresses from Equation 10.16: $\{\bar{\sigma}\}_e = z[D^*]\{\bar{\varepsilon}\}_e$.
5. Compute the geometrical stiffness $[k_G]_e$ from Equation 10.26 in terms of the given element properties.
6. Compute the element total stiffness matrix from Equation 10.24: $[k_T]_e = [k]_e + [k_G]_e$.
7. Repeat steps 1 through 4, each time with a new $[k_T]_e$ found by applying steps 5 and 6 until satisfactory convergence of the in-plane stresses $\{\bar{\sigma}\}_e$ s is attained.

Comments: The results determined by applying the finite element method to the large deflection of thin plates agrees well with analytical solutions [13]. It is interesting to note that some classical buckling problems of plates may also be treated by the approach described in this section.

Problems

Sections 10.1 through 10.6

10.1 Calculate the maximum deflection and the maximum stress in the plate of Example 10.1 for uniformly distributed loads: (a) 50 kPa; (b) 200 kPa.

10.2 Verify that, for axisymmetrically bent thin plates, the strain–large-displacement relations are

$$\varepsilon_r = \frac{du}{dr} + \frac{1}{2}\left(\frac{\partial w}{dr}\right)^2 \qquad \varepsilon_\theta = \frac{u}{r} \tag{P10.2}$$

Here u and w are the r- and z-directed displacements, respectively.

10.3 A 1-m-diameter and 10-mm-thick circular plate ($E = 70$ GPa, $\sigma_{yp} = 280$ MPa) with clamped edge is under a uniform load p_0. Determine, using Figure 10.3: (a) the limiting value of p_0 that can be applied to the plate without causing permanent deformation; (b) the maximum deflection corresponding to three-quarters of this allowable load.

10.4 Redo Problem 10.3 for the case in which the plate is simply supported. Use $a = 0.6$ m and $t = 12$ mm.

10.5 Apply Hooke's law and Equations P10.2 to derive the following stress–displacement relations for axisymmetrically loaded plates:

$$\sigma_r = \frac{E}{1-\nu^2}\left[\frac{du}{dr} + \frac{1}{2}\left(\frac{dw}{dr}\right)^2 + \nu\frac{u}{r}\right]$$
$$\sigma_\theta = \frac{E}{1-\nu^2}\left[\frac{u}{r} + \nu\frac{du}{dr} + \frac{\nu}{2}\left(\frac{dw}{dr}\right)^2\right] \tag{P10.5}$$

10.6 Very thin and square plate described in Example 10.2 is made of cold-rolled brass (Table B.3) and subjected to a uniform pressure $p_0 = 15$ kPa. Compute the value of maximum deflection and associated tensile stress. The dimensions are $b = 200$ mm and $t = 3$ mm.

10.7 What are the values of thickness and stress in the plate of Problem 10.6 for the case in which the maximum deflection is limited to 4 mm?

10.8 Find the value of maximum deflection of the very thin circular, clamped plate considered in Example 10.3. The plate is made of a high-strength, ASTM-A242 steel (Table B.3) and is under a uniform pressure $p_0 = 20$ kPa. Take $a = 250$ mm, $t = 5$ mm, and $\nu = 0.3$.

10.9 Show that the system of Equations 10.5 and 10.6, for axisymmetrically bent plates, assumes the form

$$\nabla^4\Phi = -\frac{E}{2}L(w,w)$$
$$\nabla^4 w = \frac{t}{D}L(w,\Phi) + \frac{p}{D} \tag{P10.9}$$

where

$$L(w,\Phi) = \frac{1}{r}\frac{d\Phi}{dr}\frac{d^2w}{dr^2} + \frac{1}{r}\frac{dw}{dr}\frac{d^2\Phi}{dr^2}$$

and $L(w, w)$ is found on replacement of w by Φ in the above expression.

10.10 Verify that the strain energy, due to the stretching of the midplane, is expressed by

$$U_m = \frac{\pi Et}{1-\nu^2}\int_0^a \left(\varepsilon_r^2 + \varepsilon_\theta^2 + 2\nu\varepsilon_r\varepsilon_\theta\right) r\,dr \tag{P10.10a}$$

or

$$U_m = \frac{\pi Et}{1-\nu^2}\int_0^a\left[\left(\frac{du}{dr}\right)^2+\frac{du}{dr}\left(\frac{dw}{dr}\right)^2+\frac{u^2}{r^2}+\frac{2\nu u}{r}\frac{du}{dr}+\frac{\nu u}{r}\left(\frac{dw}{dr}\right)^2+\frac{1}{4}\left(\frac{dw}{dr}\right)^4\right]r\,dr \tag{P10.10b}$$

for a thin axisymmetrically loaded circular plate.

References

1. S. Timoshenko and S. Woinowsky-Krieger, *Theory of Plates and Shells*, 2nd ed., McGraw-Hill, New York, 1959.
2. R. Szilard, *Theory and Analysis of Plates*, Chapter 3, Prentice-Hall, Upper Saddle River, NJ, 2004.
3. K. Marguerre and H. T. Woernle, *Elastic Plates*, Blaisdell, Waltham, MA, 1970.
4. J. Prescott, *Applied Elasticity*, Dover, New York, 1946.
5. E. H. Mansfield, *Bending and Stretching of Plates*, 2nd ed., Cambridge University Press, Cambridge, 1989.
6. W. C. Young, R. C. Budynas, and A. M. Sadegh, *Roark's Formulas for Stress and Strain*, 8th ed., McGraw-Hill, New York, 2011.
7. A. P. Boresi and J. R. Schmidth, *Advanced Mechanics of Materials*, 6th ed., Wiley, Hoboken, NJ, 2003.
8. V. V. Novozhilov, *Foundations of the Nonlinear Theory of Elasticity*, Graylock Press, Rochester, NY, 1953.
9. J. N. Reddy, *Theory and Analysis of Elastic Plates and Shells*, 2nd ed., CRC Press, Boca Raton, FL, 2007.
10. T. von Kármán, Festigkeitsprobleme in Machinenbau, *Encl. der Math. Wiss.*, 4, 1910, 348–351.
11. S. Levy, *Bending of rectangular plates with large deflections*, NACA Report 737, National Advisory Committee for Aeronautics, Washington, DC, 1942.
12. J. T. Oden, *Finite Elements of Nonlinear Continua*, McGraw-Hill, New York, 1972.
13. O. C. Zienkiewitcz and R. I. Taylor, *The Finite Element Method*, 6th ed., Elsevier, Oxford, UK, 2005.
14. B. Aalami and D. G. Williams, *Thin Plate Design for Transverse Loading*, Wiley, Hoboken, NJ, 1975.

11

Thermal Stresses in Plates

11.1 Introduction

Solution for the deflection and stress in plates subjected to temperature variation requires reformulation of the stress–strain relationship. This is accomplished by superposition of the strain attributable to stress and that due to temperature. For homogeneous isotropic materials, a *change in temperature* $\Delta T = T - T_0$ produces uniform linear strain in every direction. Here T and T_0 are the final and the initial temperatures, respectively. Recall from Section 1.9 that the *thermal strains* are expressed as

$$\varepsilon_t = \alpha(\Delta T) \tag{11.1}$$

In the preceding α, an experimentally determined material property, is called the coefficient of thermal expansion. Over a moderate temperature change, α remains reasonably constant. In SI units, α is expressed in meters per degree Celsius. Coefficients of thermal expansion for common materials are listed in Table B.3. For isotropic materials, a change in temperature produces no shear strains, that is, $\gamma_t = 0$. In this text, the modulus of elasticity E and coefficient of thermal expansion α are treated as constants over the temperature ranges involved.

Stresses due to the restriction of thermally induced expansion or contraction of a body are termed *thermal stresses* (Section 2.3). When a free plate is heated *uniformly*, there are produced normal strains but *no* thermal stresses. If, however, the plate experiences a *nonuniform* temperature field, if the displacements are prevented from occurring freely because of the restrictions placed on the boundary even with a uniform temperature, or if the material displays *anisotropy* even with uniform heating, thermal stresses will occur.

11.2 Stress, Strain, and Displacement Relations

The total x and y strains ε_x and ε_y are obtained by adding to the thermal strains the *strains due to stress* resulting from external forces:

$$\varepsilon_x = \frac{1}{E}(\sigma_x - \nu\sigma_y) + \alpha(\Delta T)$$
$$\varepsilon_y = \frac{1}{E}(\sigma_y - \nu\sigma_x) + \alpha(\Delta T) \tag{11.2}$$
$$\gamma_{xy} = \frac{\tau_{xy}}{G}$$

From Equations 11.2, the total stress components are given by

$$\sigma_x = \frac{E}{1-\nu^2}[\varepsilon_x + \nu\varepsilon_y - (1+\nu)\alpha(\Delta T)]$$
$$\sigma_y = \frac{E}{1-\nu^2}[\varepsilon_y + \nu\varepsilon_x - (1+\nu)\alpha(\Delta T)] \tag{11.3}$$
$$\tau_{xy} = G\gamma_{xy}$$

An increase in temperature ΔT is algebraically positive.

In deriving the strain–displacement relationships, a state of plane stress is assumed. It is further assumed that straight lines, initially normal to the midsurface, remain straight and normal to that surface after heating. Thus, the assumptions of Section 3.3, with the exception of assumption 2 in that section, apply. Substituting Equation b of Section 3.4 into Equations 3.1a through 3.1c and denoting $u_0 = u$ and $v_0 = v$, we obtain the following expressions for the strains in terms of displacements:

$$\varepsilon_x = \frac{\partial u}{\partial x} - z\frac{\partial^2 w}{\partial x^2}$$
$$\varepsilon_y = \frac{\partial v}{\partial y} - z\frac{\partial^2 w}{\partial y^2} \tag{11.4}$$
$$\gamma_{xy} = \left(\frac{\partial u}{\partial y} + \frac{\partial v}{\partial x}\right) - 2z\frac{\partial^2 w}{\partial x\,dy}$$

In the foregoing, the first terms represent the strain components in the midplane of the plate, and $w(x, y)$ represents the transverse deflection.

11.3 Stress Resultants

The stresses distributed over the thickness of the plate result in *in-plane forces* and *moments* per unit length as shown in Figures 9.1 and 3.6. The latter quantities are given by Equation 3.8a. The in-plane force components are represented by

$$\begin{Bmatrix} N_x \\ N_y \\ N_{xy} \end{Bmatrix} = \int_{-t/2}^{t/2} \begin{Bmatrix} \sigma_x \\ \sigma_y \\ \tau_{xy} \end{Bmatrix} dz \tag{11.5}$$

Introducing Equations 11.3 and 11.4 into Equations 3.8a and 11.5, we obtain the *stress resultants*

$$
\begin{aligned}
N_x &= \frac{Et}{1-\nu^2}\left(\frac{\partial u}{\partial x}+\nu\frac{\partial v}{\partial y}\right)-\frac{N^*}{1-\nu} \\
N_y &= \frac{Et}{1-\nu^2}\left(\frac{\partial v}{\partial y}+\nu\frac{\partial u}{\partial x}\right)-\frac{N^*}{1-\nu} \\
N_{xy} &= \frac{Et}{2(1+\nu)}\left(\frac{\partial u}{\partial y}+\frac{\partial v}{\partial x}\right) \\
M_x &= -D\left(\frac{\partial^2 w}{\partial x^2}+\nu\frac{\partial^2 w}{\partial y^2}\right)-\frac{M^*}{1-\nu} \\
M_y &= -D\left(\frac{\partial^2 w}{\partial y^2}+\nu\frac{\partial^2 w}{\partial x^2}\right)-\frac{M^*}{1-\nu} \\
M_{xy} &= -(1-\nu)D\frac{\partial^2 w}{\partial x\partial y}
\end{aligned}
\tag{11.6}
$$

Here the quantities

$$
N^* = \alpha E\int_{-t/2}^{t/2}(\Delta T)dz \qquad M^* = \alpha E\int_{-t/2}^{t/2}(\Delta T)z\,dz \tag{11.7}
$$

are termed the *thermal stress resultants.*

The components of stress may now be determined by the substitution of Equations 11.4 into Equations 11.3 and elimination of the displacement derivatives through the use of Equations 11.6:

$$
\begin{aligned}
\sigma_x &= \frac{1}{t}\left(N_x+\frac{N^*}{1-\nu}\right)+\frac{12z}{t^3}\left(M_x+\frac{M^*}{1-\nu}\right)-\frac{E\alpha(\Delta T)}{1-\nu} \\
\sigma_y &= \frac{1}{t}\left(N_y+\frac{N^*}{1-\nu}\right)+\frac{12z}{t^3}\left(M_y+\frac{M^*}{1-\nu}\right)-\frac{E\alpha(\Delta T)}{1-\nu} \\
\tau_{xy} &= \frac{1}{t}N_{xy}+\frac{12z}{t^3}M_{xy}
\end{aligned}
\tag{11.8}
$$

Equations 11.8 permit the direct calculation of the stress components for a plate of any cross section subject to an arbitrary temperature distribution. To derive the stress equations in polar coordinates, one need only replace subscripts x by r and y by θ in these expressions.

The first and second terms of Equations 11.8 are due to the *in-plane forces* and *bending moments*, respectively. The third term represents *in-plane stress* in the case of uniform heating and *bending stress* in the case of nonuniform heating. An example of the latter is a plate with its upper surface heated and lower surface cooled. At some location (x, y), there will develop compressive stress $-\alpha E(\Delta T)/(1-\nu)$ in the upper half and tensile stress $+\alpha E(\Delta T)/(1-\nu)$ in the lower half.

Comment: It is observed that Equations 11.8 are of the same form as those associated with beams under *compound* loading.

11.4 The Governing Differential Equations

For the *governing equation for deflection w*, the procedure described in Section 3.8 and the modified moment resultants of Equations 11.6 are combined to yield

$$D\nabla^4 w = p - \frac{1}{1-\nu}\nabla^2 M^* \tag{11.9}$$

For the case of nonuniformly heated free plate, that is, $p = 0$, Equation 11.9 reduces to

$$D\nabla^4 w = -\frac{1}{1-\nu}\nabla^2 M^* \tag{11.10}$$

It is usual to denote

$$p^* = -\frac{1}{1-\nu}\nabla^2 M^* \tag{11.11}$$

where p^* is termed the *equivalent transverse load*. We observe from Equations 11.9 and 11.10 that it is possible to superimpose the deflections due to the temperature alone with those due to transverse loads alone.

The 2D equilibrium and compatibility equations (in the xy plane) are employed to obtain the forces N as shown in the outline that follows, similar to those employed in Section 10.4. The differential equations of equilibrium in the plane of the plate are given by Equations 9.1 and 9.2. These expressions are identically satisfied by the stress function $\Phi\,(x, y)$, related to the force resultants as follows:

$$N_x = \frac{\partial^2 \Phi}{\partial y^2} \qquad N_{xy} = -\frac{\partial^2 \Phi}{\partial x \partial y} \qquad N_y = \frac{\partial^2 \Phi}{\partial x^2} \tag{11.12}$$

For a *plate of constant thickness* and negligible weight, the equation of compatibility is

$$\frac{\partial^2}{\partial y^2}(N_x - \nu N_y + N^*) + \frac{\partial^2}{\partial x^2}(N_y - \nu N_x + N^*) = 2(1+\nu)\frac{\partial^2 N_{xy}}{\partial x \partial y} \tag{a}$$

Substitution of Equations 11.12 into the above yields

$$\nabla^4 \Phi + \nabla^2 N^* = 0 \tag{11.13}$$

What has been accomplished is the formulation of a plane stress problem in *thermoelasticity* in such a way as to require the solutions of a single partial differential equation (which must, of course, fulfill the boundary conditions).

In summary, the solution of a thermoelastic-plate problem requires the solution of Equation 11.13 for the midplane forces and Equation 11.9 for the deflection w and corresponding moments. These *two solutions may be obtained independently* of one another and the stresses due to each added if the total stresses are required. The second problem, the determination of the transverse deflection, is the *direct concern* of plate theory. The first problem is a concern of the theory of elasticity and is *not* treated in this text.

The *boundary conditions* at edge $x = a$ of a rectangular plate (Figure 3.8), Equations 3.25 through 3.27 are now expressed as shown in Table 11.1. We observe from the table that

TABLE 11.1

Boundary Conditions for Rectangular Plates under Thermal Loading

Edge	Clamped	Simply Supported	Free
At $x = a$	$w = 0$	$w = 0$	$\frac{\partial^2 w}{\partial x^2} = -\frac{M^*}{(1-\nu)D}$
	$\frac{\partial w}{\partial x} = 0$	$\frac{\partial^2 w}{\partial x^2} = -\frac{M^*}{(1-\nu)D}$	$D\left[\frac{\partial^3 w}{\partial x^3} + (2-\nu)\frac{\partial^3 w}{\partial x \partial y^2}\right] + \frac{1}{1-\nu}\frac{\partial M^*}{\partial x} = 0$

the boundary conditions for a simply supported and a free-edge thermoelastic plate are nonhomogeneous. Other kinds of boundary conditions may also be obtained by employing the stress resultants derived in this chapter and the procedure presented in Section 3.9.

In the sections to follow, several methods are described, useful in the solution of *thermal bending problems* of elastic thin plates. A number of references are available for those seeking a more thorough treatment [1–8]. *Design data* and other useful information dealing with thermal stresses are presented in the papers of Goodier [9] and others [10].

11.5 Simply Supported Rectangular Plate Subject to an Arbitrary Temperature Distribution

This section deals with the deflection and stress due to nonuniform heating of a simply supported rectangular plate (Figure 11.1). The approach introduced may be extended to all polygonal simply supported plates. The boundary conditions are represented by the following (Table 11.1):

$$\begin{array}{lll} w = 0 & M_x = 0 & (x = 0, x = a) \\ w = 0 & M_y = 0 & (y = 0, y = b) \end{array} \tag{a}$$

From Equations 11.6 and a, the following expression applies to the boundary:

$$D\nabla^2 w = -\frac{M^*}{1-\nu} \tag{b}$$

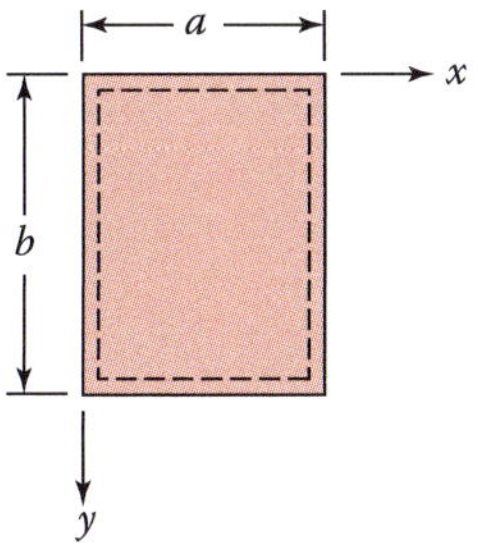

FIGURE 11.1
Simply supported plate subjected to nonuniform heating.

The governing equation, Equation 11.10, is equivalent to

$$D\nabla^2 w + \frac{M^*}{1-\nu} = f(x,y) \qquad \nabla^2 f = 0 \tag{c, d}$$

Equations b and c result in $f = 0$, which is an appropriate solution of Equation d.

The problem at hand, as in Section 9.5, is thus represented by a second-order differential equation,

$$D\nabla^2 w = -\frac{M^*}{1-\nu} \tag{11.14}$$

which must of course satisfy the boundary conditions

$$w = 0 \qquad (x = 0, a;\ y = 0, b) \tag{11.15}$$

The solution is obtained by the application of Fourier series for load (moment) and deflection:

$$M^* = \sum_{m=1}^{\infty}\sum_{n=1}^{\infty} p_{mn} \sin\frac{m\pi x}{a} \sin\frac{n\pi y}{b} \tag{11.16a}$$

$$w = \sum_{m=1}^{\infty}\sum_{n=1}^{\infty} a_{mn} \sin\frac{m\pi x}{a} \sin\frac{n\pi y}{b} \tag{11.16b}$$

The coefficients p_{mn} are, from Equation 5.3,

$$p_{mn} = \frac{4}{ab}\int_0^b\int_0^a M^* \sin\frac{m\pi x}{a} \sin\frac{n\pi y}{b}\, dx\, dy \tag{11.17}$$

Substitution of Equations 11.16a, 11.17, and 11.16b into Equation 11.14 leads to

$$a_{mn} = \frac{1}{(1-\nu)\pi^2 D} \frac{p_{mn}}{(m/a)^2 + (n/b)^2} \tag{11.18}$$

The deflection w corresponding to the thermal loading $M^*(x, y)$ has thus been determined.

11.6 Simply Supported Rectangular Plate with Temperature Distribution Varying over the Thickness

The solution of a *simply supported plate* subjected to *nonuniform heating* such that the temperature varies through the thickness only, $T(z)$, can readily be obtained from the results of Section 11.5. In this case, the thermal loading M^* is constant, and Equation 11.17, after integration, leads to

$$p_{mn} = \frac{4M^*}{\pi^2 mn}[1-(-1)^m][1-(-1)^n] \tag{a}$$

Substitution of Equations a and 11.18 into Equation 11.16b yields the following expression for deflection:

$$w = \frac{16M^*}{(1-\nu)D\pi^4}\sum_m^\infty \sum_n^\infty \frac{\sin(m\pi x/a)\sin(n\pi y/b)}{mn[(m/a)^2+(n/b)^2]} \qquad (m,n=1,3,\ldots) \tag{11.19}$$

The bending moments and stresses in the plate may now be calculated from Equations 11.6 and 11.8. As already noted in Example 5.1, while the expression for deflection (Equation 11.19) converges very rapidly, the relationship for moments does not.

An *alternate solution* [1] of Equations 11.14 and 11.15, more suitable to the computation of moments, may be obtained by the use of a *simple series* for w and M^* rather than the double series as before (Section 5.4). That is, the solution proceeds with an assumption of the deflection surface in the form of Equation 5.15:

$$w = \sum_{m=1}^\infty f_m(y)\sin\frac{m\pi x}{a} \tag{b}$$

Here the function $f_m(y)$ must be obtained such as to satisfy the boundary conditions (Figure 11.1). Thus,

$$f_m = 0 \qquad (y=0, y=b) \tag{c}$$

The *constant* moment M^* may be represented as in Section 5.6 by a Fourier series,

$$M^* = \sum_{m=1}^\infty p_m \sin\frac{m\pi x}{a} \tag{d}$$

where

$$p_m = \frac{2}{a}\int_0^a M^* \sin\frac{m\pi x}{a}dx = \frac{4M^*}{m\pi}$$

Finally, carrying Equations b and d into Equation 11.14, we obtain

$$\frac{d^2 f_m}{dy^2} - \left(\frac{m\pi}{a}\right)^2 f_m = -\frac{4M^*}{(1-\nu)Dm\pi} \qquad (m=1,3,\ldots) \tag{e}$$

The solution of the preceding equation that satisfies Equation c may be expressed as

$$w = \frac{4M^*}{(1-\nu)D\pi} \sum_{m=1}^{\infty} \frac{1}{m\alpha_m^2} \left[1 - \frac{\sinh \alpha_m y + \sinh \alpha_m (b-y)}{\sinh \alpha_m y} \right] \sin \frac{m\pi x}{a} \tag{11.20}$$

in which $\alpha_m = m\pi/a$. Having the expression for the deflection of the plate available, we can obtain the moments in the usual manner.

11.7 Analogy between Thermal and Isothermal Plate Problems

We now demonstrate that an analogy exists between the thermal and isothermal plate-bending problems, serving as a basis of a convenient procedure to determine the deflection. The analogy is *complete only for the determination of deflection*. The *thermal stresses are ascertained by adding* $-\alpha E(\Delta T)/(1-\nu)$ to the stress components σ_x and σ_y of the isothermal solution.

11.7.1 Plates with Clamped Edges

The problem of the bending of built-in plates as a result of nonuniform thermal load requires the solution of Equation 11.10 together with the specified boundary conditions given in Table 11.1. Note that the boundary conditions for a clamped edge do not involve explicitly the temperature. Thus, it is observed from a comparison of Equations 11.9 and 11.10 that the solution sought is identical to that for the same shaped clamped plate subject to the equivalent transverse load p^*. The *thermal problem* is therefore *reduced* to an *isothermal* one, and the results and techniques of the latter case are valid for the problem under consideration. Table 11.2 provides a list of some examples.

In the case of a plate of *arbitrary shape* undergoing thermal variation *through the thickness only*, we have $\nabla^2 M^* = 0$ and $p^* = 0$. According to the analogy, $w = 0$, and the corresponding stresses, from Equations 11.8, are

$$\begin{aligned}
\sigma_x &= \frac{1}{t}\left(N_x + \frac{N^*}{1-\nu}\right) - \frac{\alpha E(\Delta T)}{1-\nu} \\
\sigma_y &= \frac{1}{t}\left(N_y + \frac{N^*}{1-\nu}\right) - \frac{\alpha E(\Delta T)}{1-\nu} \\
\tau_{xy} &= \frac{N_{xy}}{t}
\end{aligned} \tag{11.21}$$

TABLE 11.2

Plates with Clamped Edge(s)

Geometry	Loading (p^*)	Solution
Rectangular	Uniform	Sections 5.7 and 5.12
Circular (solid)	Uniform	Section 4.5
	Radial	Section 11.8
Annular	Radial	Section 11.8
	Uniform	Table 4.3

In Equations 11.21, the first and the second term represent the *plane*- and the *bending*-stress components, respectively.

11.7.2 Plates with Simply Supported or Free Edges

An analogy also exists between heated and unheated plates with other than clamped supports. In this case a modification of the edge conditions is required inasmuch as they contain the temperature. At a simply supported edge of the analogous isothermal plate, $w = 0$ as before, but a bending moment $M^*/(1 - \nu)$ must be assumed to apply. In a like manner, at a free edge of the analogous unheated plate, a force equal to $(\partial M^*/\partial x)/(1 - \nu)$ must be applied. It is thus observed that a thermal solution can always be determined by *superposition of various isothermal solutions.*

Consider, for example, the bending caused by a nonuniform temperature distribution of a simply supported plate. The deflection of the plate is determined by adding the deflection of an unheated, simply supported plate subject to the surface load p^* to the deflection of an unheated plate carrying no transverse load but subject to the moment $M^*/(1 - \nu)$ acting at its edges.

Comments: The foregoing analogy is also useful in the experimental analysis of elastic heated plates. This is because it may be easier to test a plate at constant temperature with given transverse and edge loadings and then to impose on it arbitrary temperature distributions.

EXAMPLE 11.1: Thermal Analysis of Circular Plate

An aircraft window, which can be represented approximately as a simply supported circular plate, is subjected to uniform temperature T_1 and uniform temperature T_2 at the lower and the upper surfaces, respectively (Figure 11.2a). Determine the deflection and bending stress if the plate is free of stress at 0°C. Assume that the temperature through the thickness varies linearly and that $T_1 > T_2$.

Solution

The plate of Figure 11.2a is replaced by the plates shown in Figure 11.2b and c. The temperature difference between the faces is $T_1 - T_2$, and that between either face and the midsurface is $\Delta T = (1/2)(T_1 + T_2)$. Since for the present case M^* is a function of z only, Equation 11.11 gives an equivalent loading $p^* = 0$. The thermal stress resultant is, from Equation 11.7,

$$M^* = \alpha E \int_{-t/2}^{t/2} \left[\frac{1}{2}(T_1 + T_2) + \frac{1}{2}(T_1 - T_2)\frac{z}{t/2} \right] z\,dz = \frac{\alpha E t^2}{12}(T_1 - T_2) \tag{11.22}$$

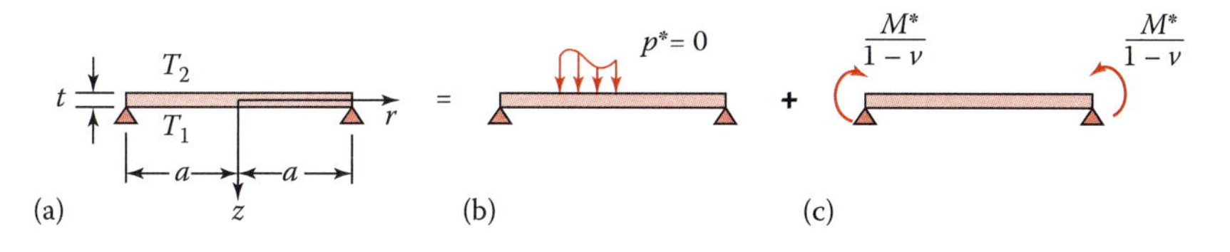

FIGURE 11.2
Simply supported circular plate with a temperature differential.

For the *plate of* Figure 11.2b, the bending stress at the faces is

$$\sigma_r = \sigma_\theta = -\frac{\alpha E(\Delta T)}{1-\nu} = -\frac{\alpha E}{2(1-\nu)}(T_1 - T_2) \tag{a}$$

Also, there is no deflection of this plate ($w = 0$).

In the case of the plate in Figure 11.2c, introducing Equation 11.22 into Equation 4.33 and setting $b = 0$, $M_1 = 0$, and $M_2 = M^*/(1 - \nu)$, we have

$$\begin{aligned} w &= \frac{\alpha(a^2 - r^2)}{2t}(T_1 - T_2) \\ M_r = M_\theta &= \frac{M^*}{1-\nu} = \frac{\alpha E t^2}{12(1-\nu)}(T_1 - T_2) \end{aligned} \tag{11.23a, b}$$

The bending stresses at the faces of the plate, on introduction of Equations 11.22 and 11.23b into Equation 11.8, are as follows:

$$\sigma_r = \sigma_\theta = \frac{\alpha E}{2(1-\nu)}(T_1 - T_2) \tag{b}$$

The resultant stress in the *original plate* is obtained by addition of the stresses given by Equations a and b:

$$\sigma_r = \sigma_\theta = 0$$

This is the result expected. Equation 11.23a leads to the relationship

$$w_{\max} = \frac{\alpha a^2}{2t}(T_1 - T_2) \tag{11.24}$$

for the maximum deflection of the original plate.

11.8 Axisymmetrically Heated Circular Plates

Consider the bending of an axisymmetrically heated circular plate having simply supported or clamped edge conditions and subjected to temperatures varying with the r and z coordinates, $T(r, z)$, such that the equivalent transverse load $p^* = p^*(r)$. The expressions for *moments*, Equations 4.9, for the situation described become

$$\begin{aligned} M_r &= -D\left(\frac{d^2w}{dr^2} + \frac{\nu}{r}\frac{dw}{dr}\right) - \frac{M^*}{1-\nu} \\ M_\theta &= -D\left(\frac{1}{r}\frac{dw}{dr} + \nu\frac{d^2w}{dr^2}\right) - \frac{M^*}{1-\nu} \end{aligned} \tag{11.25}$$

The plate deflection must satisfy the *differential equation* (Equation 4.10b),

$$\frac{1}{r}\frac{d}{dr}\left\{r\frac{d}{dr}\left[\frac{1}{r}\frac{d}{dr}\left(r\frac{dw}{dr}\right)\right]\right\}=\frac{p^*}{D} \tag{11.26}$$

with the *boundary conditions*,

$$\begin{aligned} &w=0 \quad \frac{dw}{dr}=0 \\ &w=0 \quad D\left(\frac{d^2w}{dr^2}+\frac{\nu}{r}\frac{dw}{dr}\right)+\frac{M^*}{1-\nu}=0 \end{aligned} \tag{11.27}$$

Equations 11.27 refer to *clamped* and *simply supported edges*, respectively. They are derived by employing a procedure identical with that described in Section 3.9 together with Equations 11.25.

The *general solution* of Equation 11.26, referring to Equation 4.13, may be expressed as

$$w=c_1\ln\frac{r}{b}+c_2r^2\ln\frac{r}{b}+c_3(r^2-b^2)+c_4+w_p \tag{11.28}$$

where the *c*s are determined from the boundary conditions. Note that inner radius $b=0$ for solid plates. The outer radius of the plate is designated a (Figure 4.8). The *particular solution*, denoted by w_p in Equation 11.28, can be determined by four successive integrations of Equation 11.26:

$$w_p=\int_b^r\frac{1}{r_4}\int_b^{r_4}r_3\int_b^{r_3}\frac{1}{r_2}\int_b^{r_2}\frac{r_1p^*}{D}dr_1\,dr_2\,dr_3\,dr_4 \tag{11.29}$$

In the case of an *annular plate* under a temperature distribution such that $p^*=$ constant, Equation 11.29 appears as follows (Problem 11.14):

$$w_p=\frac{p^*}{64D}\left[r^4-5b^4-4b^2(b^2+2r^2)\ln\frac{r}{b}+4b^2r^2\right] \tag{11.30}$$

When $b=0$, Equation 11.30 yields the particular solution of a solid plate.

For the sake of simplicity in the representation of the results, let

$$w_p^{(n)}=\frac{d^nw_p}{dr^n} \qquad (r=a) \tag{11.31}$$

The values of constants are listed [1] in terms of the foregoing notation in Table 11.3 (see also Problem 11.11). In this table, M_a^* represents the value of the thermal stress resultant at $r=a$. Given a temperature distribution *T*(*r*, *z*), the deflection and moment in a solid or annular plate with simply supported or built-in edges can thus be obtained from Equations 11.28 and 11.25. Plates with other edge conditions can similarly be treated.

TABLE 11.3

Constants c_1 through c_4 in Equation 11.28

Geometry	Constants in Equation 11.28
a. Solid plate (clamped at edge $r = a$)	$c_1 = c_2 = 0$ $c_3 = -\frac{1}{2a} w_p^{(1)} \quad c_4 = \frac{a}{2} w_p^{(1)} - w_p^{(0)}$
b. Solid plate (supported at edge $r = a$)	$c_1 = c_2 = 0$ $c_3 = -\frac{1}{2(1+\nu)} \left[\frac{M_a^*}{(1-\nu)D} + w_p^{(2)} + \frac{\nu}{a} w_p^{(1)} \right]$ $c_4 = \frac{a^2}{2(1+\nu)} \left[\frac{M_a^*}{(1-\nu)D} + w_p^{(2)} + \frac{\nu}{a} w_p^{(1)} \right] - w_p^{(0)}$
c. Annular plate (clamped at inner $r = b$ and outer $r = a$ edges) 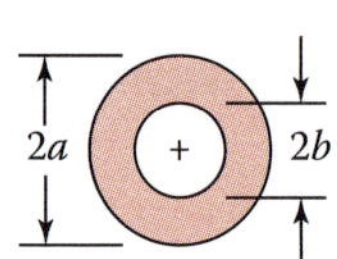	$c_1 = \frac{b^2 a w_p^{(1)} [a^2 - b^2 - 2a^2 \ln(a/b)] + 4a^2 b^2 w_p^{(0)} \ln(a/b)}{(a^2 - b^2)^2 - 4a^2 b^2 \ln^2(a/b)}$ $c_2 = \frac{-a w_p^{(1)} [a^2 - b^2 - 2b^2 \ln(a/b)] + 2(a^2 - b^2) w_p^{(0)}}{(a^2 - b^2)^2 - 4a^2 b^2 \ln^2(a/b)}$ $c_3 = \frac{a w_p^{(1)} [(a^2 - b^2)] \ln(a/b) + w_p^{(0)} [b^2 - a^2 - 2a^2 \ln(a/b)]}{(a^2 - b^2)^2 - 4a^2 b^2 \ln^2(a/b)}$ $c_4 = 0$

EXAMPLE 11.2: Thermal Deflection of Simple Circular Plate

Redo the problem discussed in Example 11.1 using the relationships developed in this section.

Solution

As $p^* = 0$, we have, from Equation 11.31, $w_p^{(n)} = 0$ for all n. Referring to row B of Table 11.3,

$$c_1 = c_2 = 0 \qquad -c_3 = \frac{c_4}{a^2} = \frac{M_a^*}{2(1-\nu^2)D} = \frac{6M^*}{Et^3}$$

Here M_a^* is replaced by M^* because the thermal loading does not vary with radius. Equation 11.28 then yields

$$w = \frac{6M^*}{Et^3}(a^2 - r^2)$$

This expression with Equation 11.22 results in a solution of w, identical to that found in Example 11.1. On inserting Equations 11.23a and 11.22 into Equations 11.25, we find that $M_r = M_\theta = 0$. Applying Equations 11.8, we again obtain $\sigma_r = \sigma_\theta = 0$.

EXAMPLE 11.3: Thermal Stresses in Clamped Plate

Determine the deflections and stresses in a circular plate with clamped edge for the following cases: (a) the plate is stress free at temperature T_0 and uniformly heated to a temperature T_1 (Figure 11.3a); (b) the plate experiences a steady-state, linear temperature variation between its faces (Figure 11.3b). Assume that the plate is free of stress at 0°C and that $T_1 > T_2$.

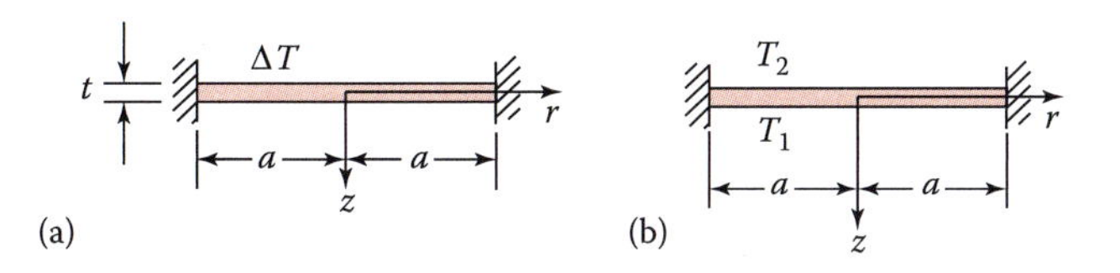

FIGURE 11.3
Fixed circular plate with a temperature differential.

Solution

a. The uniform temperature differential is $\Delta T = T_1 - T_0$. We have $p^* = 0$, and thus $w_p^{(n)} = 0$ for all n. Row A of Table 11.3 then gives

$$c_1 = c_2 = c_3 = c_4 = 0$$

Hence, Equation 11.28 yields

$$w = 0 \tag{a}$$

The thermal stress resultant is

$$M^* = \alpha E(\Delta T) \int_{-t/2}^{t/2} z\,dz = 0$$

Equations 11.25 then lead to $M_r = M_\theta = 0$. The magnitude of the bending stresses at the plate surfaces, from Equations 11.8, is therefore

$$\sigma_r = \sigma_\theta = \frac{E\alpha(\Delta T)}{1-\nu} \tag{11.32}$$

b. We now have $\Delta T = (T_1 - T_2)/2$ (Figure 11.3b). The displacement $w = 0$, as before. The thermal stress resultant M^* is defined by Equation 11.22, and the bending moments are found from Equations 11.25:

$$M_r = M_\theta = -\frac{\alpha E t^2}{12(1-\nu)}(T_1 - T_2)$$

By means of Equations 11.8, we have

$$\sigma_r = \sigma_\theta = \frac{E\alpha(\Delta T)}{1-\nu} = \frac{E\alpha}{2(1-\nu)}(T_1 - T_2) \tag{11.33}$$

for the magnitude of the bending stress at the upper and the lower faces of the plate.

Comment: It is noted that inasmuch as $p^* = 0$, Equations 11.32 and 11.33 may readily be found without any calculation by the use of analogy between heated and isothermal clamped plates.

Problems

Sections 11.1 through 11.6

11.1 An aluminum airplane wing panel is assumed to be stress free at 20°C. After a time at cruising speed, the temperatures on the heated (upper) and the cooled (lower) surfaces are 54°C and 24°C, respectively. Compute the thermal stress resultants for a linear temperature transition. Assume that the panel edges are clamped. Use $E = 70$ GPa, $\alpha = 23.2 \times 10^{-6}/°C$, $\nu = 0.3$, and $t = 6$ mm.

11.2 Determine the values of stresses σ_x and σ_y and at the lower and middle surfaces of the plate considered in Problem 11.1.

11.3 Derive expressions for the bending moments in a simply supported square plate of sides a with temperature distribution linearly varying over the thickness only.

11.4 Determine the maximum deflection w and the maximum moment M_x in a simply supported square plate of sides a, subject to a temperature field Az^3, where A is a constant. Retain only the first term of the series solutions. Let $\nu = 0.3$.

11.5 Compute the values of the maximum deflection w and maximum moment M_x for the simply supported plate described in Problem 11.4. Assume that the plate is made of a structural, ASTM-A36 steel (Table B.3), $t = 5$ mm, and $a = 120$ mm.

11.6 Redo Problem 11.4 for the case in which the plate is subjected to a temperature field $Az^2 + Bz$, where A and B are constants.

11.7 A plate of any shape is clamped along its boundaries and is loaded by temperature $\Delta T = 2T_1 z/t$. Determine the maximum stress, moment, deflection, and strain. Use $\nu = 1/3$.

Sections 11.7 and 11.8

11.8 Compute the value of the maximum deflection for the plate described in Example 11.1. Take $\alpha = 80(10^{-6})/°C$, $a = 220$ mm, $t = 8$ mm, and $T_1 - T_2 = 20°C$.

11.9 Consider a built-in edge, square plate of sides a, under an equivalent transverse thermal loading p^* (x, y) (Figure 5.19). Employ the analogy between isothermal and thermal problems and apply the Ritz method, taking $m = n = 1$ to obtain an expression for the deflection surface w.

11.10 A square plate is simply supported at $x = 0$ and $x = a$. The remaining edges at $y = \pm a/2$ are clamped (Figure 5.13a). The plate is subjected to a temperature distribution Ay^2z^3, where A is a constant. Derive an approach to the evaluation of the center displacement w. Use the analogy with the isothermal solutions given in Sections 5.6 and 5.7.

11.11 Verify the results given in Cases A and B of Table 11.3.

11.12 Determine the deflection w and the stress σ_r at the center of a simply supported circular plate experiencing a temperature field Azr^3, where A is a constant.

11.13 Resolve Problem 11.12 for a plate clamped on all edges.

11.14 Determine the displacement w at $r = 2a$ of a hollow plate having inner ($r = a$) and outer ($r = 3a$) edges built in. Assume a temperature distribution Bz^5r^2, where B is a constant.

References

1. B. A. Boley and J. H. Weiner, *Theory of Thermal Stresses*, Dover, Mineola, New York, 2011.
2. D. J. Johns, *Thermal Stress Analysis*, Pergamon, New York, 1965.
3. W. Nowacki, *Thermoelasticity*, 2nd ed., English translation by H. Zorski, Pergamon, New York, 1986.
4. B. E. Gatewood, *Thermal Stresses*, McGraw-Hill, New York, 1957.
5. D. Burgreen, *Elements of Thermal Analysis*, C.P. Press, New York, 1971.
6. R. Szilard, *Theory and Analysis of Plates*, Prentice-Hall, Upper Saddle River, NJ, 2004.
7. J. R. Vinson, *Structural Mechanics: The Behavior of Plates and Shells*, Section 3.6, Wiley, New York, 1974.
8. E. Ventzel and T. Krouthammer, *Thin Plates and Shells: Theory, Analaysis, and Applications*, CRC Press, Boca Raton, FL, 2001.
9. J. N. Goodier, Thermal stresses, In: *Design Data and Methods*, pp. 74–77, ASME, New York, 1953.
10. W. C. Young, R. C. Budynas, and A. M. Sadegh, *Roark's Formulas for Stress and Strain*, 8th ed., McGraw-Hill, New York, 2011.

Section III

Shells

Sydney Opera House. This modern building is comprised of three groups of interlocking shells, which roof two main performance halls and a restaurant. Shell elements, typically curved, are assembled to make many large engineering structures. We will show that the load-resisting action of a shell differs from that of other structural forms. The applied load is resisted predominantly by the in-plane stressing of the shell.

12

Membrane Stresses in Shells

12.1 Introduction

Until now, our concern was with the analysis of beams and thin flat plates. We now extend the discussion to *curved surface structures* termed *thin shells*. Examples of shells include pressure vessels, airplane wings and fuselages, pipes, ship and submarine hulls, the exterior of rockets, missiles, automobile tires, incandescent lamps, caps, roof domes, chimneys, cooling towers, factory or car sheds, and a variety of containers. Each of these has walls that are curved. Inasmuch as a *curved plate* can be viewed as a portion of a shell, the general equations for thin shells are also applicable to curved plates.

We shall limit our discussions (except Sections 13.13 and 13.14) to shells of constant thickness, small in comparison with the other two dimensions. As in the treatment of plates, the plane bisecting the shell thickness is called the *midsurface*. To describe the shape of a shell, we need only know the geometry of the midsurface and the thickness of the shell at each point. Shells of technical significance are often defined as *thin* when the ratio of thickness t to radius curvature r is equal to or less than 1/20. For thin shells of practical importance, this ratio may be 1/1000 or smaller.

The major *thin-shell structures* include concrete shell structures, often cast as a dome, bridge, or saddle roof; lattice shell structures, so-called gridshell structures, usually in the form of a dome or hyperboloid structure; membrane or fabric structures and other tensile structures, cable domes, and pneumatic structures. In practice, cylindrical shells are often selected for covering massive amount of spaces, such as the roofs of garages, supermarkets, power stations, and warehouses. The thin-walled shell structures have more strength with respect to their self-weight and high stiffness, they are gaining in popularity in areas where there is risk of earthquake and hurricane. Thus, it is necessary to analyze the shell structures to ensure the best and safe design method for the future, to avoid the potential accidents.

Membrane stresses in typical shell structures in the forms of sphere, cylinder, cone, ellipsoid, toroid, hyperbolic paraboloid, and multisphere are discussed in the examples and a case study that follow.

12.2 Theories and General Behavior of Shells

The analysis of shell-like structures often embraces *two* distinct, *commonly applied theories*. The first of these, the *membrane theory*, usually applies to a rather large part of the entire shell. A membrane, either flat or curved, is identified as a body of the same shape as a plate or shell but incapable of conveying moments or shear forces. In other words, a membrane is a 2D analog of a flexible string with the exception that it can resist compression.

The second, the *bending theory* or *general theory*, includes the effects of bending. Thus, it permits the treatment of discontinuities in the stress distribution taking place in a limited region in the vicinity of a load or structural discontinuity. However, information relative to shell *membrane stresses* is usually of much greater practical significance than the knowledge of the *bending stresses*. The former are also far simpler to calculate.

For thin shells having no abrupt changes in thickness, slope, or curvature, the membrane stresses are uniform throughout the wall thickness. The bending theory generally comprises a membrane solution, corrected in those areas in which discontinuity effects are pronounced. The goal is thus not the improvement of the membrane solution but rather the analysis of stresses and strains due to the edge forces or concentrated loadings, which cannot be accomplished by membrane theory only.

It is important to note that membrane forces are independent of bending and are completely defined by the conditions of static equilibrium. As no material properties are used in the derivation of these forces, the *membrane theory applies to all shells* made of *any material* (e.g., metal, fabric, reinforced concrete, sandwich shell, soap film, gridwork shell, and plywood). Various relationships developed for bending theory in the following chapters, excluding Sections 15.11 and 15.12, are restricted to homogeneous, elastic, isotropic shells. A number of references in the analysis and design of shell-like structures are available for those seeking a more thorough treatment [1–16].

The *basic* kinematic *assumptions* associated with the deformation of a thin shell as used in small-deflection analysis are as follows:

1. The ratio of shell thickness to radius of curvature of the midsurface is small in comparison with unity.
2. Deflections are small compared with shell thickness.
3. Plane sections through a shell, taken normal to the midsurface, remain plane and become normal to the *deformed midsurface* after the shell is subjected to bending. This hypothesis implies that strains γ_{xz} and γ_{yz} are negligible. Normal strain ε_z due to transverse loading may also be omitted.
4. The z-directed normal stress σ_z is negligible.

With the exception of midsurface deformation, these hypotheses are familiar to the reader since they have their equivalent counterparts in the beam- and plate-bending theories. In this chapter, we consider only shells and loadings for which bending stresses are negligibly small. Applications are also presented of the governing equations of the membrane theory to specific practical cases.

12.3 Load Resistance Action of a Shell

The common deformational behavior of beams, thin plates, and thin shells is illustrated by the unified set of assumptions (Sections 2.7, 3.3, and 12.2). The load-carrying mechanisms of these members do not resemble one another, however. The *load-resisting*

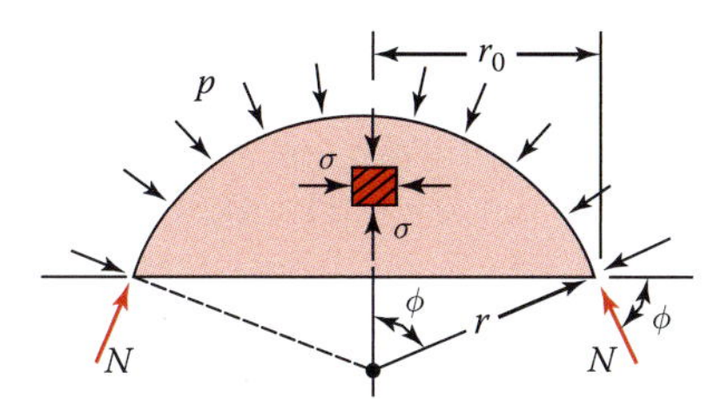

FIGURE 12.1
Truncated spherical shell subjected to uniform pressure.

action of a shell differs from that of other structural forms is underscored by noting the extraordinary capacity of an eggshell or an electric lightbulb to withstand normal forces, this despite their thinness and fragility. (A hen's egg has a radius along the axis of revolution $r = 20$ mm and a thickness $t = 0.4$ mm; thus, $t/r = 1/50$.) The above behavior contrasts markedly with similar materials in plate or beam configurations under lateral loading. A shell, being curved, can develop in-plane forces (thrusts) to form the primary resistance action in addition to those forces and moments existing in a plate or beam.

To describe the phenomenon, consider a part of a spherical shell of radius r and thickness t, subjected to a uniform pressure of intensity p (Figure 12.1). The condition that the sum of vertical forces be zero is expressed as $2\pi r_0 N \sin\phi - p\pi r_0^2 = 0$ or

$$N = \frac{pr_0}{2\sin\phi} = \frac{pr}{2}$$

in which N is the in-plane force per unit of circumference. This relationship is valid anywhere in the shell, as N is seen not to vary with ϕ. We note that, in contrast to the case of plates, load is sustained by the midsurface.

It is next demonstrated that the bending stresses play an insignificant role in the load-resisting action. Based on the symmetry of the shell and the loading, the state of stress at a point, represented by an infinitesimal element, is as shown in Figure 12.1. The compressive *direct stress* has the form

$$\sigma = -\frac{pr}{2t} \tag{a}$$

The stress normal to the midsurface is negligible, and thus the direct strain, from Hooke's law, is

$$\varepsilon = \frac{1}{E}(\sigma - \nu\sigma) = -(1-\nu)\frac{pr}{2Et} \tag{b}$$

Associated with this strain, the reduced circumference is $2\pi r' = 2\pi(r + r\varepsilon)$, or $r' = r(1 + \varepsilon)$. The *variation in curvature* χ (chi) is thus

$$\chi = \frac{1}{r'} - \frac{1}{r} = \frac{1}{r}\left(\frac{1}{1+\varepsilon} - 1\right) = -\frac{\varepsilon}{r}\left(\frac{1}{1+\varepsilon}\right) = -\frac{\varepsilon}{r}(1 - \varepsilon + \varepsilon^2 - \cdots)$$

Neglecting higher-order terms because of their small magnitude and introducing Equation b, the above expression becomes

$$\chi = -\frac{\varepsilon}{r} = \frac{(1-\nu)p}{2Et} \tag{c}$$

A relationship for the shell-bending moment is derived from the plate formulas. For the spherical shell under consideration, we have $\chi = \chi_x = \chi_y$. Thus, Equations 3.9 and c lead to

$$M = -D(\chi_x + \nu\chi_y) = -D(1-\nu^2)\frac{p}{2Et} = -\frac{pt^2}{24}$$

Hence, the *bending stress* is given by

$$\sigma_b = \frac{6M}{t^2} = -\frac{p}{4} \tag{d}$$

The ratio of the direct stress to the bending stress is

$$\frac{\sigma}{\sigma_b} = \frac{2r}{t} \tag{e}$$

Comment: It is observed that the direct or membrane stress is very much larger than the bending stress as $(t/2r) \ll 1$. We are led to conclude therefore that the *applied load is resisted predominantly by the in-plane stressing* of the shell.

The foregoing discussion relates to the simplest shell configuration. However, the conclusions drawn with respect to the basic action apply to any geometry and loading at locations away from the edges or points of application of concentrated load. Should there be asymmetries in load or shape, shearing stresses as well as the membrane and bending stresses will be present. Treated in greater detail in the sections that follow is the state of membrane stress in shells of various geometry.

It is noted that thin shells may be vulnerable to local *buckling* under compression stresses. The problem of shell *instability* is treated in Chapter 15. For reference purposes, we now introduce the following relation [1] useful in the prediction of the *critical stress* σ_{cr} for *local buckling* of a thin shell:

$$\sigma_{cr} = k\frac{Et}{r} \tag{f}$$

Here E is the modulus of elasticity, and the value of constant k can be determined by rational analysis, such as described in Sections 15.13 and 15.14, and incorporated with an empirical factor in order to relate theoretical values to actual test data. The value $k \approx 0.25$ is often used.

Consider, for example, a concrete or masonry shell for which $E = 20$ GPa and $t/r = 1/500$. We have

$$\sigma_{cr} = 0.25 \times 20 \times 10^9 \left(\frac{1}{500}\right) = 10\,\text{MPa}$$

This, compared with the ultimate strength of the concrete of 21 MPa, demonstrates the importance of buckling analysis in predicting allowable load.

Shell structures should thus be checked for the possibility of buckling in compressed areas as well as for yielding or facture in those sections subjected to tensile forces (Example 12.2b).

12.4 Geometry of Shells of Revolution

The geometry of a shell is completely defined if we know the shape of its midsurface and its thickness at all points. Consider a particular type of shell described by a *surface of revolution* (Figure 12.2). Examples include the sphere, cylinder, and cone. The midsurface of a shell of revolution is generated by rotation of a so-called *meridian curve* about an axis lying in the plane of the curve. Figure 12.2 shows that a point on the shell is conveniently located by coordinates θ, ϕ, r_0 and that elemental surface *ABCD* is defined by two *meridians* and two *parallel circles* or *parallels*.

The planes associated with the *principal radii of curvature* r_1 and r_2 at any point on the midsurface of the shell are the *meridian plane* and the *parallel plane* at the point in question, respectively. The corresponding principal *curvatures* of the meridian and parallel planes at any point may be denoted by κ_1 and κ_2, so that $\kappa_1 = 1/r_1$ and $\kappa_2 = 1/r_2$ (see Section 3.4). Observe from the figure that the radii of curvature r_1 and r_2 are thus related to sides *CD* and *AC*. The principal radius r_2 generates the shell surface in the direction perpendicular to the direction of the tangent to the meridian curve. The two radii r_0 and r_2 are related inasmuch as $r_0 = r_2 \sin\phi$ (Figure 12.2). It follows that lengths of the curvilinear shell element are

$$L_{AC} = r_0\, d\theta = r_2 \sin\phi\, d\theta \qquad L_{CD} = r_1\, d\phi \tag{a}$$

It is assumed in the above description that the principal radii of curvature of the shell r_1 and r_2 are known constants. In the case of a radius of curvature that varies from point

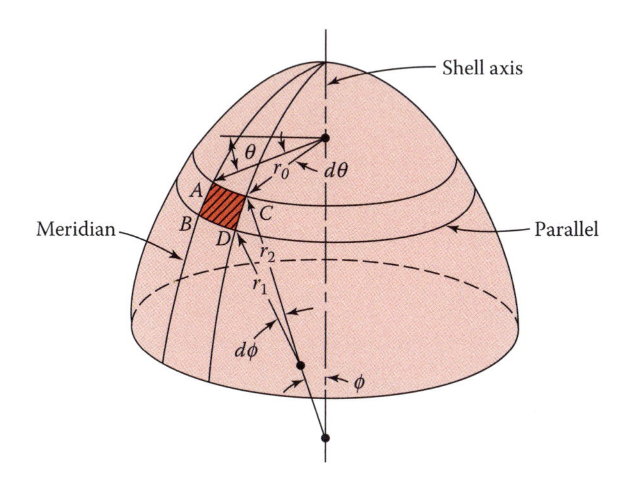

FIGURE 12.2
A shell of revolution. *Note:* Meridian and parallel planes are associate with the principal radii of curvature r_1 and r_2, respectively.

to point, the radii are computed by applying the equation that defines the shell shape, along with various relationships of differential geometry of a surface [2], as illustrated in Example 12.5.

12.5 Symmetrically Loaded Shells of Revolution

In axisymmetrical problems involving shells of revolution, no shear forces exist, and there are only two unknown membrane forces, or membrane stress resultants, per unit length, N_θ and N_ϕ. The governing equations for the so-called *hoop force* N_θ and the *meridional force* N_ϕ are derived from two equilibrium conditions. Figure 12.3a and b shows two different views of the element *ABCD* cut from the shell of Figure 12.2. Prescribed by the condition of symmetry, the membrane forces and the loading display no variation with θ. The externally applied forces per unit surface area are represented by the components p_y and p_z in the y and z directions, respectively.

Description of the *z equilibrium* requires that the z components of the loading as well as of the forces acting on each edge of the element be considered. The z-directed distributed load carried on the surface area of the element is

$$p_z r_0 r_1 \, d\theta \, d\phi$$

The force acting on the top edge of the element equals $N_\phi r_0 d\theta$. *Neglecting higher terms*, the force on the bottom edge is also $N_\phi r_0 d\theta$. The z-directed component at each edge is then $N_\phi r_0 \, d\theta \sin(d\phi/2)$. This force is nearly equal to $N_\phi r_0 \, d\theta(d\phi/2)$, yielding the following magnitude of the resultant for both edges:

$$N_\phi r_0 \, d\theta \, d\phi$$

As the cross-sectional area along each of the two sides of the element is $r_1 \, d\phi$, the force on these areas is $N_\theta r_1 \, d\phi$. The resultant in the direction of the radius of the parallel

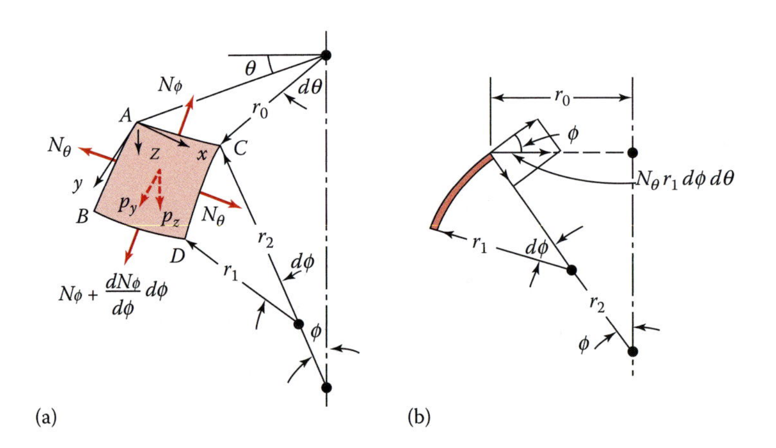

FIGURE 12.3
Diagrams for analysis of symmetrically loaded shells of revolution: (a and b) Membrane forces and surface loads on a shell element.

plane for both such forces is $N_\theta r_1\, d\phi\, d\theta$, producing the following component in the z direction:

$$N_\theta r_1\, d\phi\, d\theta \sin\phi$$

For the forces considered above, from $\Sigma F_z = 0$, we have

$$N_\phi r_0 + N_\theta r_1 \sin\phi + p_z r_0 r_1 = 0 \tag{a}$$

This expression may be converted to simpler form by dividing by $r_0 r_1$ and replacing r_0 by r_2 sin ϕ. By so doing, one of the basic relations for the axisymmetrically loaded shell is found as follows:

$$\frac{N_\phi}{r_1} + \frac{N_\theta}{r_2} = -p_z \tag{12.1a}$$

The *equilibrium* of forces in the direction of the meridional tangent, that is, in the *y direction,* is expressed as

$$\frac{d}{d\phi}(N_\phi r_0) d\phi\, d\theta - N_\theta r_1 d\phi\, d\theta \cos\phi + p_y r_1\, d\phi\, r_0\, d\theta = 0 \tag{b}$$

The first term represents the sum of normal forces acting on edges *AC* and *BD,* while the third term is the loading component. The second term of Equation b is the component in the y direction of the radial resultant force $N_\theta r_1\, d\phi\, d\theta$ acting on faces *AB* and *CD.* Dividing Equation b by $d\theta\, d\phi$, the equation of equilibrium of the y-directed forces is now

$$\frac{d}{d\phi}(N_\phi r_0) - N_\theta r_1 \cos\phi = -p_y r_1 r_0 \tag{12.2}$$

It is noted that an equation of equilibrium that can be used *instead of* Equation 12.2 follows readily by isolating part of the shell intercepted by angle ϕ (Figure 12.4a). Here force F represents the *resultant of all external loading* applied to this free body. Recall that from conditions of symmetry, forces N_ϕ are constant around the edge. Equilibrium of the

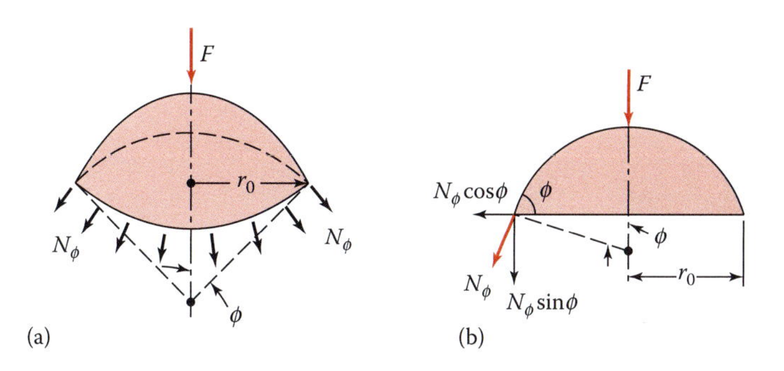

FIGURE 12.4
(a) Meridian forces and resultant loading (F) acting on a truncated shell defined by the angle ϕ; (b) components the meridional force N_ϕ.

vertical forces (see Figure 12.4b) is therefore described by $2\pi r_0 N_\phi \sin\phi + F = 0$, and it follows that

$$N_\phi = -\frac{F}{2\pi r_0 \sin\phi} \tag{12.1b}$$

We verify below that Equation 12.1b is an *alternative form* of Equation 12.2. Substitution of N_θ from Equation 12.1a into Equation 12.2 and multiplication of the resulting expression by $\sin\phi$ leads to

$$\frac{d}{d\phi}(r_0 N_\phi)\sin\phi + r_0 N_\phi \cos\phi = -r_1 r_2 p_z \cos\phi \sin\phi - r_1 r_2 p_y \sin^2\phi$$

Clearly, the left-hand side of the foregoing equation may be written as

$$\frac{d}{d\phi}(r_0 N_\phi \sin\phi) = \frac{d}{d\phi}(r_2 N_\phi \sin^2\phi)$$

and force N_ϕ determined through integration:

$$N_\phi = -\frac{1}{r_2 \sin^2\phi}\left[\int r_1 r_2 (p_z \cos\phi + p_y \sin\phi)\sin\phi \, d\phi + c\right] \tag{c}$$

In the preceding, constant c represents the effects of the loads that may be applied to a shell element (Figure 12.4a). Thus, introduction of $2\pi c = F$, $p_z = p_y = 0$, and $r_0 = r_2 \sin\phi$ into Equation c results in the value of N_ϕ, defined by Equation 12.1b.

Equations 12.1 are sufficient to determine the membrane forces N_θ and N_ϕ from which the stresses are readily obtained. Negative algebraic results indicate compressive stresses.

Comments: Because of the freedom of motion in the z direction, for the axisymmetrically loaded shells of revolution considered, strains are produced such as to ensure consistency with the field of stress and compatibility with one another [3]. The action cited demonstrates the basic difference between the problem of a shell membrane and one of *plane stress*. In the latter case, a compatibility equation is required. However, it is clear that when a shell is subject to the action of concentrated surface loadings or is constrained at its boundaries, membrane theory cannot everywhere fulfill the conditions of deformation. The complete solution is obtained only by application of bending theory.

12.6 Some Typical Cases of Shells of Revolution

The membrane stresses in any particular axisymmetrically loaded shell in the form of a surface of revolution may be determined from the governing expressions of equilibrium developed in Section 12.5. Treated in the following paragraphs are several common structural members. As pointed out previously, the principal radii of curvature r_1 and r_2 at any point on the midsurface of the shell denote the meridian plane and parallel plane, respectively.

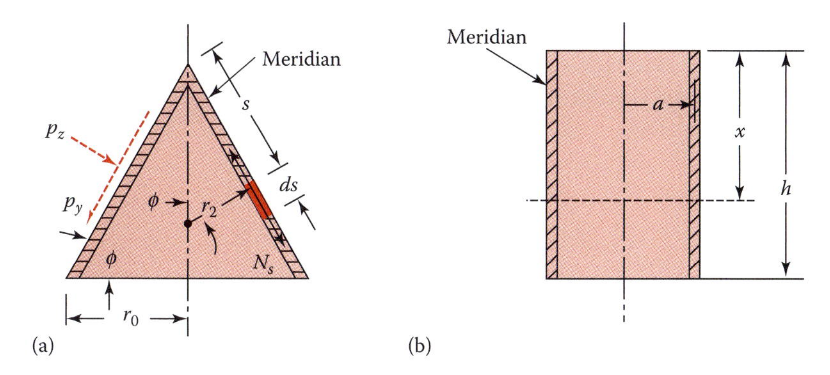

FIGURE 12.5
(a) Conical shell; (b) vertical cylindrical shell.

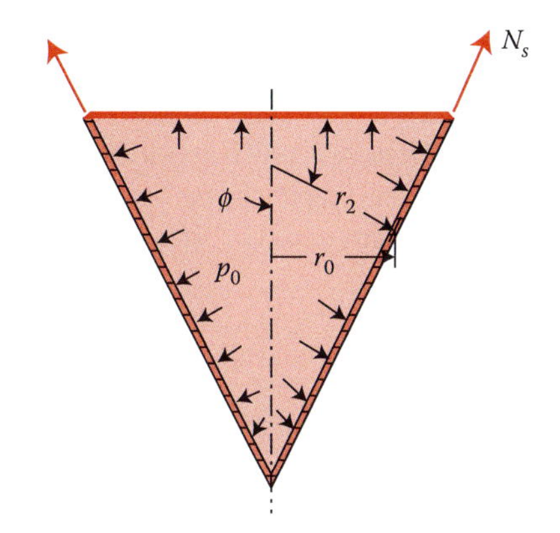

FIGURE 12.6
Conical pressure vessel.

It is interesting to note that for the circular cylinder and the cone, the meridian is a straight line and hence $\kappa_1 = 0$ (Figure 12.5). Thus, these are shells of zero Gaussian curvatures. For ellipsoids of revolution, the sphere (Figure 12.6) or paraboloids of revolution (Figure 12.9), both the principal curvatures are in the same direction, and hence these surfaces have positive Gaussian curvatures: They are called *synclastic*.

In the case of the hyperboloid of revolution, the curvatures of the meridian and the parallel sections are in opposite directions; that is, r_1 and r_2 lie on opposite sides of the surface at the point. Therefore, this shell has a negative Gaussian curvature and is termed anticlastic (see Figure 3.12b). For the toroidal shell (Figure 12.8), the Gaussian curvature changes from positive to negative as we move along the surface.

12.6.1 Spherical Shell

For spherical shells we can set the *mean* radius $a = r_1 = r_2$. Then Equations 12.1 appear in the form

$$N_\phi + N_\theta = -p_z a$$
$$N_\phi = -\frac{F}{2\pi a \sin^2\phi} \tag{12.3}$$

The simplest case is that of a spherical shell subjected to *constant internal gas pressure* p, like a balloon. We now have $p = -p_z$, $\phi = 90°$, and $F = -\pi a^2 p$. Inasmuch as any section through the center results in the identical free body, $N_\phi = N_\theta = N$. The stress, from Equations 12.3, is therefore

$$\sigma = \frac{N}{t} = \frac{pa}{2t} \tag{12.4}$$

where t is the thickness of the shell. The expansion of the sphere, by Hooke's law, is then

$$\delta_{sphere} = \frac{a}{Et}(N - \nu N) = \frac{pa^2}{2Et}(1-\nu) \tag{12.5}$$

12.6.2 Conical Shell

In this typical case (Figure 12.5a), angle ϕ is the constant ($r_1 = \infty$) and can no longer serve as a coordinate on the meridian. Instead, we introduce coordinate s, the distance of a point of the midsurface, *usually measured from the vertex*, along the generator. Accordingly, $d\phi/ds = 1/r_1$ or $r_1 = \infty$, as physically appreciated. Differentiating this slope–curvature relationship with respect to s results in

$$\frac{d}{d\phi} = r_1 \frac{d}{ds} \tag{a}$$

Also,

$$r_0 = s\cos\phi \qquad r_2 = s\cot\phi \qquad N_\phi = N_s \tag{b}$$

These relationships, when introduced into Equations 12.2 and 12.1a, lead to

$$\begin{gathered} \frac{d}{ds}(N_s s) - N_\theta = -p_y s \\ N_\theta = -p_z s\cot\phi = -\frac{p_z r_0}{\sin\phi} \end{gathered} \tag{12.6a, b}$$

where r_0 is the mean radius at the base. Obviously, load components p_y and p_z are in the s and radial directions, respectively. The sum of Equations 12.6 yields

$$\frac{d}{ds}(N_s s) = -(p_y + p_z\cot\phi)s$$

The meridional force, on integration of the above expression, is

$$N_s = -\frac{1}{s}\int (p_y + p_z\cot\phi)s\,ds \tag{12.7}$$

An alternate form of Equation 12.6a may be obtained from Equation 12.1b and Equation b. The membrane forces are then

$$N_s = -\frac{F}{2\pi r_0 \sin\phi}$$
$$N_\theta = -\frac{p_z r_0}{\sin\phi} \tag{12.8}$$

It is observed that given an external load distribution, hoop and meridional stresses can be computed *independently*.

In case of a *conical vessel under internal pressure,* we have $p = -p_z$ and $F = \pi r_0^2\ p$. Equations 12.8 result in

$$N_s = \frac{pr_0}{2\sin\phi} \qquad N_\theta = \frac{pr_0}{\sin\phi} \tag{c}$$

The expansion of a conical vessel can then be found to be as

$$\delta_{cone} = \frac{r_0}{Et}(N_\theta - \nu N_s) = \frac{r_0^2 p}{2Et\sin\phi}(2-\nu) \tag{d}$$

12.6.3 Circular Cylindrical Shell

To obtain the stress resultants in a circular cylindrical shell (Figure 12.5b), we can begin with the cone equations, setting $\phi = \pi/2$, $p_z = p_r$ and mean radius $a = r_0 =$ constant. Hence, Equations 12.8 become

$$N_s = N_x = -\frac{F}{2\pi a} \qquad N_\theta = -p_r a \tag{12.9}$$

in which x is measured in the axial direction.

For a closed-end cylindrical vessel under *constant internal pressure*, $p = -p_r$ and $F = -\pi a^2 p$. Equations 12.9 then yield the following axial and hoop stresses:

$$\sigma_x = \frac{pa}{2t} \qquad \sigma_\theta = \frac{pa}{t} \tag{12.10}$$

Note that the tangential stresses in the cylinder are twice the magnitude of the tangential stresses in the sphere. From Hooke's law, the extension of the radius of the cylinder under the action of the stresses given above is

$$\delta_{cylinder} = \frac{a}{E}(\sigma_\theta - \nu\sigma_x) = \frac{pa^2}{2Et}(2-\nu) \tag{12.11}$$

Solutions of various other cases of practical significance may be obtained by employing a procedure similar to that described in the foregoing paragraphs, as demonstrated in Examples 12.1 through 12.8.

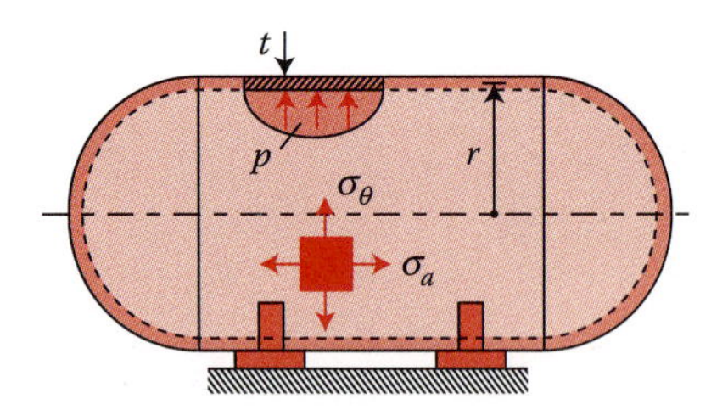

FIGURE 12.7
Cylindrical pressure vessel with hemispherical ends.

EXAMPLE 12.1: Stress and Deflection of Compressed Air Tank

A steel cylindrical tube having hemispherical ends or heads, supported by two cradles as depicted in Figure 12.7, contains internal air at a pressure of p. *Find*: (a) The stresses in the vessel if each portion has the same radius r and the thickness t; (b) Expansion of the cylinder. *Given*: $r = 0.6$ m, $t = 12$ mm, $p = 2$ MPa, $E = 200$ GPa, $\nu = 0.3$. *Assumptions:* The cradles act as simple supports. The weight of the tank may be omitted.

Solution

a. Axial stress in the cylinder and the tangential stress in the spherical ends are identical. Through the use of Equation 12.4:

$$\sigma_a = \sigma = \frac{pr}{2t} = \frac{2(10^6)(0.6)}{2(0.012)} = 50 \text{ MPa}$$

The tangential stress in the cylinder, by Equation 12.10, equals

$$\sigma_\theta = \frac{pr}{2t} = 2\sigma_a = 100 \text{ MPa}$$

b. Substitution of the given data into Equation 12.11 gives cylinder deflection

$$\sigma_{cylinder} = \frac{pr^2}{2Et}(1-\nu) = \frac{2(10^6)(0.6)^2(1.7)}{2(200\times 10^9)(0.012)} = 0.26 \text{ mm}$$

Comment: The maximum stress that occurs on the surface of the tank ($\sigma_r = p$) is negligibly small in comparison to the tangential and axial stresses. The state of stress in the wall of the thin-walled vessel may thus be assumed to be *biaxial.*

EXAMPLE 12.2: Analysis of Dome Carrying its Own Weight

Consider a simply supported *covered market dome* of radius a and thickness t, carrying only its own weight p per unit area. (a) Determine the stresses for a dome of half-spherical geometry (Figure 12.8a). (b) Assume that the hemi spherical dome is constructed of 70-mm-thick concrete of unit weight 23 kN/m^3 and span $2a = 56$ m. Apply the maximum principal stress theory to evaluate the shell's ability to resist failure by fracture. The ultimate compressive strength or crushing strength of concrete $\sigma_{uc} = 21$ MPa, and $E = 20$ GPa. Also check the possibility of local buckling. (c) Determine the stresses in a dome which is a truncated half-sphere (Figure 12.8b).

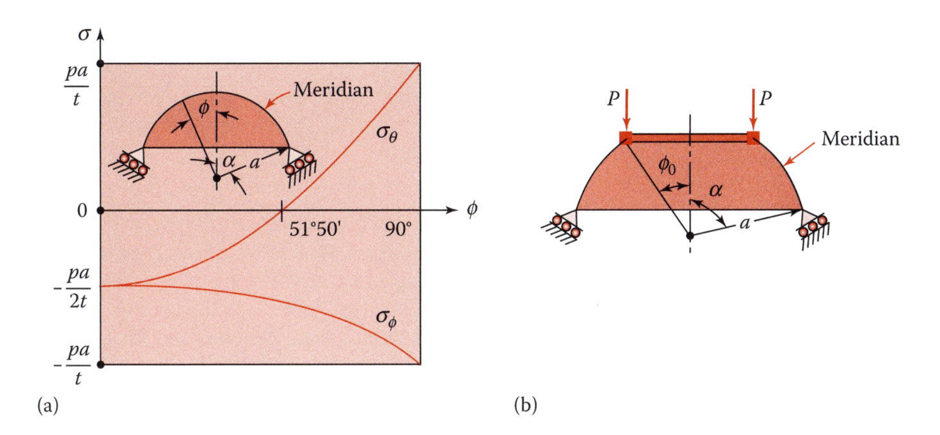

FIGURE 12.8
(a) Hemispherical dome; (b) truncated hemispherical dome.

Solution

The *components of the dome weight* are

$$p_x = 0 \qquad p_y = p\sin\phi \qquad p_z = p\cos\phi \tag{12.12}$$

a. Referring to Figure 12.8a, the weight of that part of the done subtended by ϕ is

$$F = \int_0^{\phi} p\,2\pi a\sin\phi\; a\,d\phi = 2\pi a^2 p(1-\cos\phi) \tag{a}$$

Introduction into Equations 12.3 of p_z and F given by Equations 12.12 and a and division of the results by t yields the membrane stresses:

$$\begin{aligned} \sigma_\phi &= -\frac{ap(1-\cos\phi)}{t\sin^2\phi} = -\frac{ap}{t(1+\cos\phi)} \\ \sigma_\theta &= -\frac{ap}{t}\left(\cos\phi - \frac{1}{1+\cos\phi}\right) \end{aligned} \tag{12.13}$$

These stresses are plotted for a *hemisphere* in Figure 12.8a. Clearly, σ_ϕ is always compressive; its value increases with ϕ from $-pa/2t$ at the crown to pa/t at the edge. The *sign* of σ_θ, on the other hand, *changes* with the value of ϕ. The second of the above equations yields $\sigma_\theta = 0$ for $\phi = 51°50'$. When ϕ is smaller than this value, σ_θ is compressive. For $\phi > 51°50'$, σ_θ is tensile, as shown in the figure.

b. The maximum compressive stress in the dome is $\sigma_\phi = pa/t = (0.023t)a/t = 0.023 \times 28 = 0.644$ MPa. Note that no failure occurs, as $|\sigma_\phi| < |\sigma_{uc}|$. Clearly, even for large domes, *the stress level due to self-weight is far from the limit stress of the material, at least in compression.* Note also that concrete is weak in tension and that a different conclusion may emerge from consideration of failure due to direct tensile forces. If the tensile strength of the material is smaller than 0.644 MPa, an assessment of tensile reinforcement will be required to ensure satisfactory design.

On application of Equation f of Section 12.3, the stress level at which local buckling occurs in the dome is found to be

$$\sigma_{cr} = 0.25(20 \times 10^9)\left(\frac{7}{2800}\right) = 12.5 \text{ MPa}$$

It is observed that there is no possibility of local buckling, as $\sigma_\phi < \sigma_{cr}$.

c. Most domes are not closed at the upper portion and have a lantern, a small tower for lighting, and ventilation. In this case, a reinforcing ring is used to support the upper structure as shown in Figure 12.8b. Let $2\phi_0$ be the angle corresponding to the opening and P the vertical load per unit length acting on the reinforcement ring. The resultant of the total load on that portion of the dome subtended by the angle ϕ, Equation a, is then

$$\begin{aligned} F &= 2\pi \int_{\phi_0}^{\phi} a^2 p \sin\phi \, d\phi + P\, 2\pi a \sin\phi_0 \\ &= 2\pi a^2 p(\cos\phi_0 - \cos\phi) + 2\pi P a \sin\phi_0 \end{aligned} \tag{b}$$

Using Equations 12.3, we obtain

$$\begin{aligned} \sigma_\phi &= -\frac{ap}{t}\frac{\cos\phi_0 - \cos\phi}{\sin^2\phi} - \frac{P}{t}\frac{\sin\phi_0}{\sin^2\phi} \\ \sigma_\theta &= \frac{ap}{t}\left(\frac{\cos\phi_0 - \cos\phi}{\sin^2\phi} - \cos\phi\right) + \frac{P}{t}\frac{\sin\phi_0}{\sin^2\phi} \end{aligned} \tag{12.14}$$

for the hoop and the meridional stresses.

Note that the circumferential strain in the dome under the action of membrane stresses may be computed from Hooke's law:

$$\varepsilon_\theta = \frac{1}{E}(\sigma_\theta - \nu\sigma_\phi) \tag{c}$$

This strain contributes to a change $\varepsilon_\theta a \sin\alpha$ in edge radius. The simple support shown in the figure, free to move as the shell deforms under loading, ensures that no bending is produced in the neighborhood of the edge.

EXAMPLE 12.3: Force Analysis in Edge Supported Truncated Cone

Consider a planetarium dome that may be approximated as an edge-supported *truncated cone*. Derive expressions for the hoop and meridional forces for two conditions of loading: (a) the shell carries its own weight p per unit area (Figure 12.9a); (b) the shell carries a snow load assumed to be uniformly distributed over the plan (Figure 12.9b).

Solution

a. Referring to Figure 12.7a,

$$r_2 = s\cot\phi \qquad r_0 = s\cos\phi \qquad p_z = p\cos\phi \tag{d}$$

The weight of that part of the cone defined by s–s_0 is determined from

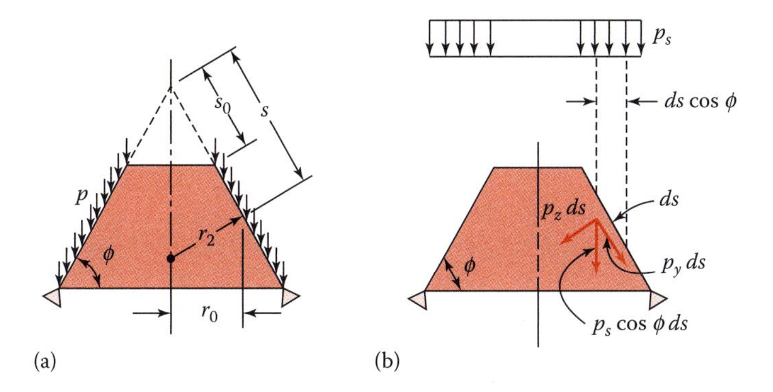

FIGURE 12.9
Truncated cone under, (a) its own weight, (b) snow load.

$$F = \int_{s_0}^{s} p\,2\pi r_2 \sin\phi\, ds = 2\pi p \int_{s_0}^{s} s \cot\phi \sin\phi\, ds$$

or

$$F = \pi p \cos\phi (s^2 - s_0^2) + c$$

As no force acts at the top edge, $c = 0$. Now the s- and θ-directed stress resultants can readily be obtained. Substituting F and Equations d into Equations 12.8 gives

$$\begin{aligned} N_s &= -\frac{p}{2s}\frac{s^2 - s_0^2}{\sin\theta} \\ N_\theta &= -ps\frac{\cos^2\phi}{\sin\phi} \end{aligned} \tag{12.15}$$

b. To analyze the components of the *snow load* p_s, we use a sketch of the forces acting on a midsurface element *ds* (Figure 12.9b). Referring to this figure, we have

$$p_x = 0 \qquad p_y = p_s \sin\phi\cos\phi \qquad p_z = p_s \cos^2\phi \tag{12.16}$$

From Equations 12.8, d, and 12.16, we have

$$N_\theta = -p_s s \frac{\cos^3\phi}{\sin\phi} \tag{12.17a}$$

Similarly, Equations 12.7, d, and 12.16 yield

$$\begin{aligned} N_s &= -\frac{1}{s}\int \left(p_s \sin\phi\cos\phi + p_s\cos^2\phi\cot\phi\right) s\, ds + \frac{c}{s} \\ &= \frac{p_s s}{2}\cot\phi + \frac{c}{s} \end{aligned}$$

The condition that $N_s = 0$ at $s = s_0$ leads to $c = (1/2)p_s s_0^2 \cot\phi$. Hence,

$$N_s = -\frac{p_s}{2s}\left(s^2 - s_0^2\right)\cot\phi \tag{12.17b}$$

On dividing Equations 12.17 by t, the membrane stresses are obtained.

Comments: It is interesting to note that the *three typical loads* (weight per unit surface area, snow load per plan area, and wind load per surface area) are of the *same order.* For ordinary structures these might approximate 1500–2000 Pa.

EXAMPLE 12.4: Stresses in Pressurized Toroidal Shell

A shell in the shape of a *torus* or *doughnut* of circular cross section is subjected to internal pressure p (Figure 12.10). Determine membrane stresses σ_ϕ and σ_θ.

Solution

Consider the portion of the shell defined by ϕ. The vertical equilibrium of forces leads to

$$2\pi r_0 \cdot N_\phi \sin\phi = \pi p(r_0^2 - b^2)$$

or

$$N_\phi = \frac{p(r_0^2 - b^2)}{2r_0 \sin\phi} = \frac{pa(r_0 + b)}{2r_0}$$

Introducing N_ϕ into Equation 12.1a, setting $p_z = -p$ and $r_1 = a$, we obtain

$$N_\theta = \frac{pr_2(r_0 - b)}{2r_0} = \frac{pa}{2}$$

The stresses are then

$$\sigma_\phi = \frac{pa(r_0 + b)}{2r_0 t} \qquad \sigma_\theta = \frac{pa}{2t} \tag{12.18}$$

Comment: Observe that σ_θ is constant throughout the shell from the condition of symmetry.

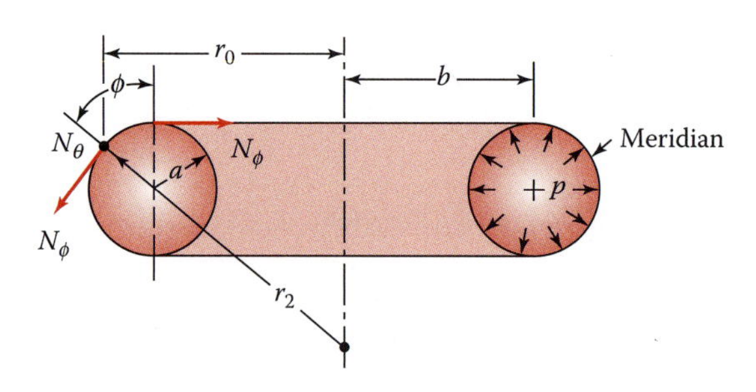

FIGURE 12.10
Toroidal shell.

EXAMPLE 12.5: Stresses in Pressurized Ellipsoidal Shell

Figure 12.11 represents the end enclosure of a cylindrical vessel in the form of a half *ellipsoid* of semiaxes a and b. Determine the membrane stresses resulting from an internal steam pressure p.

Solution

Expressions for the principal radii of curvature r_1 and r_2 will be required. The equation of the ellipse $b^2x^2 + a^2y^2 = a^2b^2$ leads $y = \pm b\sqrt{a^2 - x^2}/a$. The magnitude of the derivatives of this expression are [17]

$$y' = \frac{bx}{a\sqrt{a^2 - x^2}} = \frac{b^2x}{a^2y} \qquad y'' = \frac{b^4}{a^2y^3} \tag{e}$$

Referring to the figure,

$$\tan\phi = y' = \frac{x}{h} \qquad r_2 = \sqrt{h^2 + x^2} \tag{f}$$

From the first of Equations e and f, $h = a\sqrt{a^2 - x^2}/b$, which, when substituted into the second of Equations f, yields r_2. Introduction of Equations e into the familiar expression for the curvature, $[1 + (y')^2]^{3/2}/y''$, gives the radius r_1. Thus,

$$r_1 = \frac{(a^4y^2 + b^4x^2)^{3/2}}{a^4b^4} \qquad r_2 = \frac{(a^4y^2 + b^4x^2)^{1/2}}{b^2} \tag{12.19}$$

The load resultant is represented by $F = \pi p r_2^2 \sin^2\phi$. The membrane forces can then be determined from Equations 12.1 in terms of the principal curvatures. It follows that

$$\sigma_\phi = \frac{pr_2}{2t} \qquad \sigma_\theta = \frac{p}{t}\left(r_2 - \frac{r_2^2}{2r_1}\right) \tag{12.20}$$

At the *crown* (top of the shell), $r_1 = r_2 = a^2/b$, and Equations 12.20 reduce to

$$\sigma_\phi = \sigma_\theta = \frac{pa^2}{2bt}$$

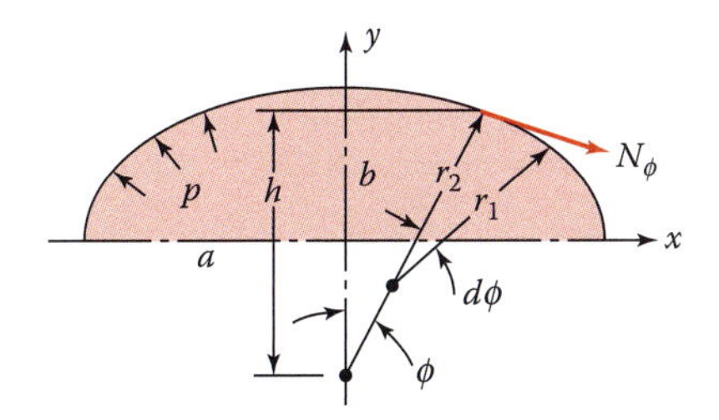

FIGURE 12.11
Ellipsoidal shell.

At the *equator* (base of the shell), $r_1 = b^2/a$ and $r_2 = a$, and Equations 12.20 appear as

$$\sigma_\phi = \frac{pa}{2t} \qquad \sigma_\theta = \frac{pa}{t}\left(1 - \frac{a^2}{2b^2}\right)$$

Comments: We note that the hoop stress σ_θ becomes compressive for $a^2 > 2b^2$. Clearly, the meridian stresses σ_ϕ are always tensile. A ratio $a/b = 1$, the case of a sphere, yields the lowest stress.

EXAMPLE 12.6: Design of Parabolic Pressure Vessel

A parabolic shell, closed at the top by a thick plate, is subjected to internal pressure p, as depicted in Figure 12.12a. What is the minimum allowable thickness t_{all} of the shell at level A–A? *Given:* $p = 1.5$ kPa. *Design decision*: The parabola is expressed by $y = x^2/100$, where x and y are in millimeters. *Assumption*: At section A–A the membrane stress is limited to $\sigma_{all} = 120$ MPa.

Solution

The force resultant for part of the vessel *below* plane A–A (Figure 12.12b) is denoted by the force F. At level A–A, we have $y = 625$ mm. It follows that $x = r_0 = \sqrt{100(625)} = 250$ mm and $dy/dx = x/50 = 5$. By geometry, we have

$$r_2 = 250\left(\frac{\sqrt{26}}{5}\right) = 255 \text{ mm}$$

Then familiar curvature equation results in

$$r_1 = \frac{[1 + (dy/dx)^2]^{1.5}}{d^2y/dx^2} = \frac{(1 + 5^2)^{1.5}}{1/50} = 6{,}629 \text{ mm} = 6.629 \text{ m}$$

The membrane forces per unit length at level A–A can now be found by using Equation 12.3 as follows:

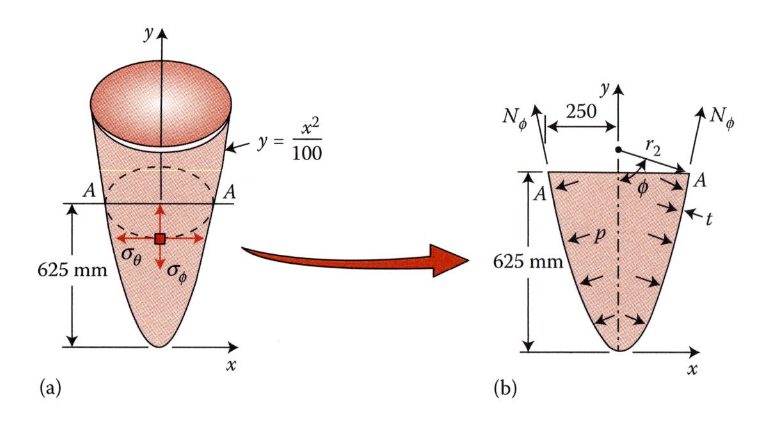

FIGURE 12.12
(a) Parabolic shell with closed at top under internal pressure; (b) FBD of lower portion of shell.

$$N_\phi = -\frac{F}{2r_0 \sin\phi} = -\frac{-p\pi r_0^2}{2\pi r_0(5/\sqrt{26})} = \frac{\sqrt{26}pr_0}{10}$$
$$= \frac{\sqrt{26}(1500)(0.25)}{10} = 191.2\ \text{N/m}$$

and

$$\frac{N_\phi}{r_1} + \frac{N_\theta}{r_2} = -p_z; \qquad \frac{191.2}{6.629} + \frac{N_\theta}{0.255} = 1,500$$

Solving, $N_\theta = 375.1$ kN. Since $N_\theta > N_\phi$, we obtain

$$t_{all} = \frac{N_\theta}{\sigma_{\text{all}}} = \frac{375.1}{120} = 3.13\ \text{mm}$$

Comment: At level *A–A*, the minimum allowable thickness should be 3.2 mm.

EXAMPLE 12.7: Stresses in Conical Shells

Finding the membrane stresses in a thin metal *container* of *conical shape*, supported from the top. Consider two specific cases: (a) the shell is subjected to an internal pressure p; (b) the shell is filled with a liquid of specific weight γ (Figure 12.13).

Solution

For such a tank, we have

$$\phi = \frac{\pi}{2} - \alpha \qquad r_0 = y\tan\alpha \tag{g}$$

a. Equations 12.8 with $F = -p\pi r_0^2$, after dividing by the thickness t, then become

$$\sigma_s = \frac{pr_0}{2t\cos\alpha} \qquad \sigma_\theta = \frac{pr_0}{t\cos\alpha} \tag{12.21}$$

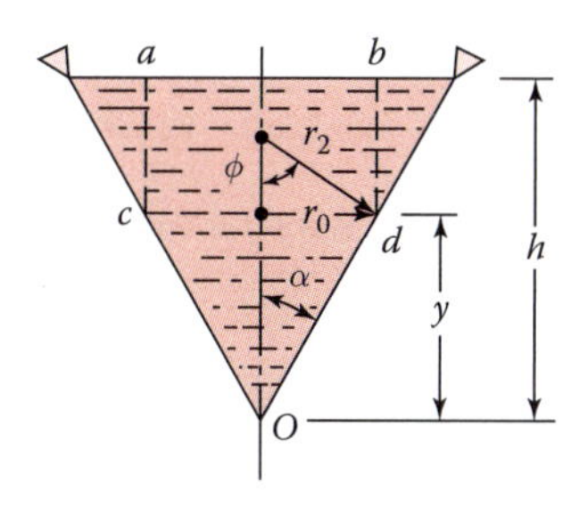

FIGURE 12.13
Conical vessel supported at top.

b. According to the familiar laws of hydrostatics, the pressure at any point in the shell equals *the weight of a column of unit cross-sectional area of the liquid* at that point. At any arbitrary level y, the *pressure* is therefore

$$p = -p_z = \gamma(h - y) \tag{h}$$

Substituting Equations g and h into the second of Equations 12.8, after division by t, we have

$$\sigma_\theta = \frac{\gamma(h-y)y}{t}\frac{\tan\alpha}{\cos\alpha} \tag{12.22a}$$

Differentiating with respect to y and equating to zero reveals that the maximum value of the above stress occurs at $y = h/2$ and is given by

$$\sigma_{\theta,\max} = \frac{\gamma h^2}{4t}\frac{\tan\alpha}{\cos\alpha}$$

The load is equal to the weight of the liquid of volume *acOdb*. That is,

$$F = -\pi\gamma y^2\left(h - y + \frac{1}{3}y\right)\tan^2\alpha$$

Introducing this value into the first of Equations 12.8 and dividing the resulting expression by t leads to

$$\sigma_s = \frac{\gamma(h - 2y/3)y}{2t}\frac{\tan\alpha}{\cos\alpha} \tag{12.22b}$$

The maximum value of this stress, $\sigma_{s,\max} = 3h^2\gamma \tan\alpha/16t\cos\alpha$, occurs at $y = 3h/4$.

EXAMPLE 12.8: Stress Analysis of Spherical Tank Filled to Capacity

Determine the membrane forces in a *spherical storage tank* filled with liquid of specific weight γ and supported on a cylindrical pipe (Figure 12.14a).

Solution

The loading is expressed as

$$p = -p_z = \gamma a(1 - \cos\phi)$$

Because of this pressure, the resultant force F for the portion intercepted by ϕ is

$$\begin{aligned} F &= -2\pi a^2 \int_0^\phi \gamma a(1-\cos\phi)\sin\phi\cos\phi\, d\phi \\ &= -2\pi a^3\gamma\left[\frac{1}{6} - \frac{1}{2}\cos^2\phi\left(1 - \frac{2}{3}\cos\phi\right)\right] \end{aligned}$$

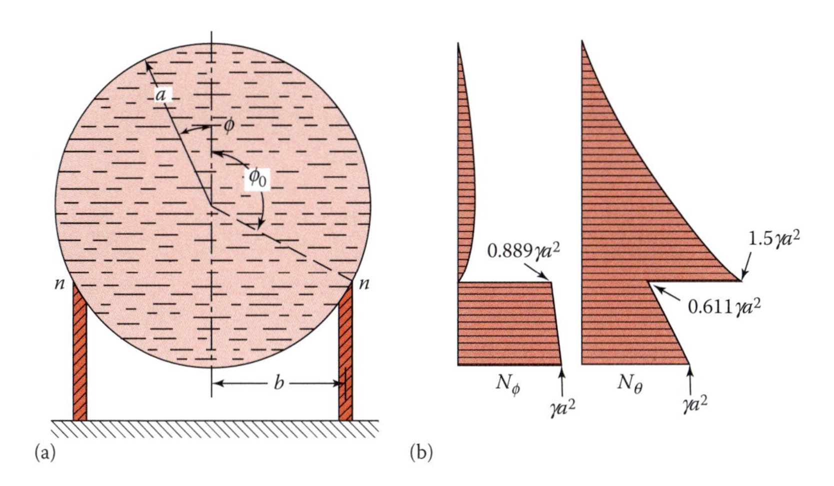

FIGURE 12.14
(a) Tank supported on a circular pipe; (b) variation of membrane forces.

Inserting the above into Equations 10.3,

$$\begin{aligned} N_\phi &= \frac{\gamma a^2}{6\sin^2\phi}[1-\cos^2\phi(3-2\cos\phi)] = \frac{\gamma a^2}{6}\left(1-\frac{2\cos^2\phi}{1+\cos\phi}\right) \\ N_\theta &= \frac{\gamma a^2}{6}\left(5-6\cos\phi+\frac{2\cos^2\phi}{1+\cos\phi}\right) \end{aligned} \tag{12.23}$$

Equations 12.23 are valid for $\phi < \phi_0$.

In determining F for $\phi > \phi_0$, the sum of the vertical support reactions, which is equal to the total weight of the liquid ($4\gamma\pi a^3/3$), must also be taken into account in addition to the internal pressure loading. That is,

$$F = -\frac{4}{3}\pi a^3\gamma - 2\pi a^3\gamma\left[\frac{1}{6}-\frac{1}{2}\cos^2\phi\left(1-\frac{2}{3}\cos\phi\right)\right]$$

Equations 12.3 now yield

$$\begin{aligned} N_\phi &= \frac{\gamma a^2}{6}\left(5+\frac{2\cos^2\phi}{1-\cos\phi}\right) \\ N_\theta &= \frac{\gamma a^2}{6}\left(1-6\cos\phi-\frac{2\cos^2\phi}{1-\cos\phi}\right) \end{aligned} \tag{12.24}$$

Comments: From Equations 12.23 and 12.24, it is observed that both forces N_ϕ and N_θ change values abruptly at the support ($\phi = \phi_0$). This is illustrated for a support position at $\phi = 120°$ in Figure 12.14b. A discontinuity in N_θ means a discontinuity of the deformation of the parallel circles on the immediate sides of *nn*. Thus, the deformation associated with the membrane solution is *not compatible* with the continuity of the structure at support *nn*.

CASE STUDY 12.1 Analysis of Multisphere Vessels

The preceding analysis of a number of thin-walled shells demonstrates that a *sphere* is an *optimum* shape for an internally pressurized closed vessel. It is well known that a spherical vessel is very efficient type of structure since it has uniformly stressed membrane. In addition, a sphere has the largest volume per surface area with a minimum thickness, and thus lowest material weight and cost. However, when requirements exceed those practicable for a single sphere, *multiple intersecting spheres* with n intersections (such as tandem configuration portrayed in Figure 12.15, for $n = 3$) can be used. This kind of vessel gives longitudinal dimensions nearly that of a cylinder though retaining most of the advantages of the sphere.

Given: A typical multiple-spherical vessel of thickness t and radius a is subjected to internal pressure p (Figure 12.15).

Assumption: In the vessel shown the bulkhead (or reinforcement element) attachment B–B is such that the uniform membrane stress in the shell is not disturbed. That is bending (see Section 14.11) is not produced in the intersections. The bulkhead consists of a concentric ring for access to all segments of the vessels.

Find:

a. The radial expansion of the intersecting circle.
b. The ratio of length (L) to radius (a) for a vessel with n intersections.
c. Outline the determination of the bulkhead geometry.

SOLUTION

a. The membrane stress in a sphere is given by

$$\sigma = \frac{pa}{2t} \tag{i}$$

and corresponding unit elongation at the intersection of $pa(1-\nu)/2Et$. The radial growth of the intersecting circle, with radius $r_0 = a \sin \phi$, equals

$$\sigma = pa^2(1-\nu)\sin\phi/2Et \tag{j}$$

where ϕ denotes the intersection angle.

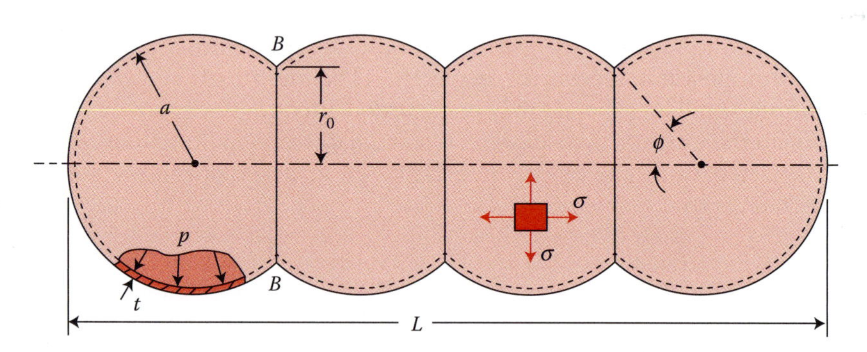

FIGURE 12.15
Intersecting spheres with internal pressure.

b. The required L/a may readily be found, referring to Figure 12.15:

$$L/a = 2(n\cos\phi + 1) \tag{k}$$

c. Enclosed volume and the weight of a single sphere are, respectively,

$$V = 4\pi a^3/3 \quad \text{and} \quad W = 4\pi a^2 t\gamma$$

in which γ is the specific weight. The preceding results in terms of the membrane stress, is as follows:

$$(W/V)_{sphere} = 3p\,\gamma/2\sigma \tag{l}$$

Similarly, the ratio $(W/V)_{intersecting}$ of enclosed weight to volume ratio of a multiple-sphere with n intersections is expressed [12]. Then, the optimum geometry for a solid bulkhead is analyzed from $(W/V)_{intersecting}/(W/V)_{sphere}$. Cross-sectional area of the bulkhead may be approximated by [14],

$$A = 2ta\,n\cos\phi\sin\phi/(1-\nu) \tag{m}$$

Comments: Multiple-spherical shells have a practical application in the economical design of vessels for extremely high pressures. In addition, they are the basic construction employed for deep-diving submarines which must have both minimum weight for buoyancy and maximum strength for pressure.

12.7 Axially Symmetric Deformation

We now discus the displacements in *symmetrically loaded* shells of revolution by considering an element AB of length $r_1\,d\phi$ of the meridian in an unstrained shell. Let the displacements in the direction of the tangent to the meridian and in the direction normal to the midsurface be denoted by v and w, respectively (Figure 12.16). After straining, AB is displaced to position A′B′.

In the analysis that follows, the small deformation approximation is employed, and higher-order infinitesimal terms are neglected. The deformation experienced by an element of infinitesimal length $r_1\,d\phi$ may be regarded as composed of an increase in length $(dv/d\phi)\,d\phi$ due to the tangential displacements and a decrease in length $w\,d\phi$ produced by the radial displacement w. The meridional strain ε_ϕ, the total deformation per unit length of the element AB, is thus

$$\varepsilon_\phi = \frac{1}{r_1}\frac{dv}{d\phi} - \frac{w}{r_1} \tag{12.25a}$$

The deformation of an element of a parallel circle may be treated in a like manner. It can be shown that the increase in radius r_0 of the circle, produced by the displacements v

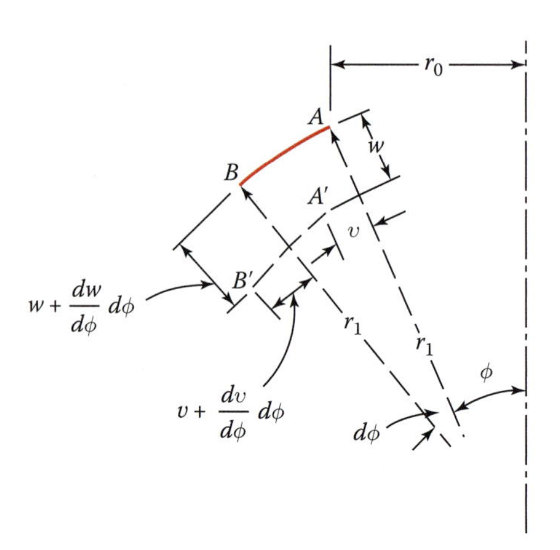

FIGURE 12.16
Displacements of midsurface.

and w, is $v \cos \phi - w \sin \phi$. Inasmuch as the circumference of the parallel circle expands in direct proportion to its radius,

$$\varepsilon_\theta = \frac{1}{r_0}(v \cos\phi - w \sin\phi) \tag{a}$$

Recalling that $r_0 = r_2 \sin \phi$, we write the hoop strain as follows:

$$\varepsilon_\theta = \frac{1}{r_2}(v \cot\phi - w) \tag{12.25b}$$

Elimination of w from Equations 12.25 leads to the following differential equation for v:

$$\frac{dv}{d\phi} - v \cot\phi = r_1 \varepsilon_\phi - r_2 \varepsilon_\theta \tag{b}$$

The strains are related to the membrane stresses by Hooke's law:

$$\varepsilon_\phi = \frac{1}{E}(\sigma_\phi - \nu\sigma_\theta) \qquad \varepsilon_\theta = \frac{1}{E}(\sigma_\theta - \nu\sigma_\phi) \tag{c}$$

Introduction of the above into Equation b gives

$$\frac{dv}{d\phi} - v \cot\phi = \frac{1}{E}[\sigma_\phi(r_1 + \nu r_2) - \sigma_\theta(r_2 + \nu r_1)] \tag{d}$$

We observe that the symmetric deformations of a shell of revolution may be determined by integrating Equation d when the membrane stresses are known. Next, we let

$$\frac{dv}{d\phi} - v \cot\phi = f(\phi)$$

This equation has the solution

$$v=\left[\int\frac{f(\phi)}{\sin\phi}d\phi+c\right]\sin\phi \tag{e}$$

The constant of integration c is determined from a boundary condition. Once v has been found, we can readily obtain w from Equation 12.25b.

EXAMPLE 12.9: Displacements of Spherical Dome due to Its Own Weight

Determine the displacements of the *spherical roof dome* supporting its own weight (Figure 12.8a).

Solution

For the half-sphere under consideration $r_1 = r_2 = a$, and stresses σ_ϕ and σ_θ are given by Equations 12.13. Equation d is therefore

$$\frac{dv}{d\phi}-v\cot\phi=\frac{a^2p(1+\nu)}{Et}\left(\cos\phi-\frac{2}{1+\cos\phi}\right)=f(\phi)$$

Inserting this expression into Equation e, we obtain

$$v=\frac{a^2p(1+\nu)}{Et}\left[\sin\phi\ln(1+\cos\phi)-\frac{\sin\phi}{1+\cos\phi}\right]+c\sin\phi \tag{f}$$

It is necessary to choose c such that $v = 0$ at $\phi = \alpha$ (Figure 12.5a). It follows that

$$c=\frac{a^2p(1+\nu)}{Et}\left[\frac{1}{1+\cos\alpha}-\ln(1+\cos\alpha)\right] \tag{g}$$

On substituting this value of c into Equation f, deflection v is obtained, and Equation 12.25b then yields w. It is noted that if the *support* displacement w is to be determined, we need not employ Equation f, as $v = 0$ there; the second of Equations c and Equation 12.25b directly give the solution.

12.8 Asymmetrically Loaded Shells of Revolution

In the bending of a shell of revolution under *unsymmetrical loading*, not only do normal forces N_ϕ and N_θ act on the sides of an element, but shearing forces $N_{\theta\phi}$ and $N_{\phi\theta}$ do as well (Figure 12.17). The moment equilibrium requires that $N_{\theta\phi} = N_{\phi\theta}$, *as is always the case for a thin shell* (Section 13.2). The surface load, referred to the unit area of the midsurface, has components p_x, p_y, and p_z.

The x-directed forces are as follows. The force

$$\frac{\partial N_\theta}{\partial\theta}r_1 d\theta\, d\phi \tag{a}$$

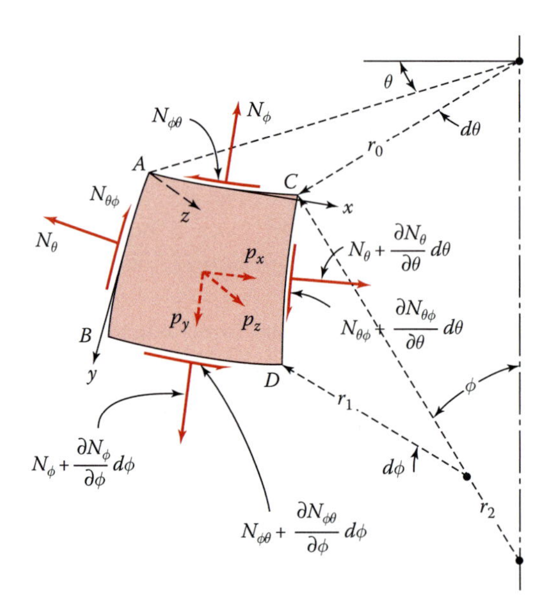

FIGURE 12.17
Element with membrane forces and surface loads.

is *due to the variation of* N_θ. *Horizontal components of the forces* $N_{\theta\phi}\, r_1\, d\phi$ acting on the faces *AB* and *CD* of the element make an angle $d\theta$ and thus have the following resultant in the x direction:

$$N_{\theta\phi}\, r_1\, d\phi \cos\phi\, d\theta \tag{b}$$

The difference of the shearing forces acting on faces *AC* and *BD* of the element is expressed as

$$N_{\theta\phi}\frac{dr_0}{d\phi}d\phi\, d\theta + \frac{\partial N_{\theta\phi}}{\partial\phi} r_0\, d\phi\, d\theta = \frac{\partial}{\partial\phi}(r_0 N_{\theta\phi})d\theta\, d\phi \tag{c}$$

The component of the external force is

$$p_x r_0 r_1\, d\theta\, d\phi \tag{d}$$

The x equilibrium condition thus reads

$$\frac{\partial}{\partial\phi}(r_0 N_{\theta\phi}) + \frac{\partial N_\theta}{\partial\theta} r_1 + N_{\theta\phi} r_1 \cos\phi + p_x r_0 r_1 = 0 \tag{12.26a}$$

To the expression governing the y equilibrium of the symmetrically loaded case (Section 12.4), we must add the force

$$\frac{\partial N_{\theta\phi}}{\partial\theta} r_1 d\theta\, d\phi$$

produced by the difference in the shearing forces acting on the faces *AB* and *CD* of the element. Inasmuch as the projection of the shearing forces on the z axis vanishes, Equation

12.1a remains valid for the present case as well. The equilibrium of y- and z-directed forces is therefore satisfied by the expressions

$$\frac{\partial}{\partial \phi}(N_{\phi} r_0) + \frac{\partial N_{\theta\phi}}{\partial \theta} r_1 - N_{\theta} r_1 \cos\phi + p_y r_1 r_0 = 0 \quad (12.26b)$$

$$\frac{N_{\phi}}{r_1} + \frac{N_{\theta}}{r_2} = -p_z \quad (12.26c)$$

Equations 12.26 permit determination of the membrane forces in a shell of revolution with nonsymmetrical loading that may, in general, vary with θ and ϕ. Such a case is discussed in the next section.

Comment: We note that the governing equations of equilibrium for the spherical, conical, and cylindrical shells may readily be deduced from Equations 12.26 on following a procedure identical to that described in Section 12.6.

12.9 *Shells of Revolution under Wind Loading

It is usual to represent dynamic loading such as wind and earthquake effects by statically equivalent or pseudostatic loading adequate for purposes of design. The *wind load* on shells is composed of pressure on the wind side and suction on the leeward side. Only the *load component acting perpendicular to the midsurface* p_z is considered important. Components p_x and p_y are due to friction forces and are of negligible magnitude. Assuming for the sake of simplicity that the wind acts in the direction of the meridian plane $\theta = 0$, the components of wind pressure are as follows:

$$p_x = 0 \qquad p_y = 0 \qquad p_z = p \sin\phi \cos\theta \quad (12.27)$$

In the above, p represents the *static* wind pressure intensity. For purposes of illustration, Figure 12.18 shows the distribution of the static-design wind load on a spherical dome. This distribution should be regarded as a rough approximation [1,2].

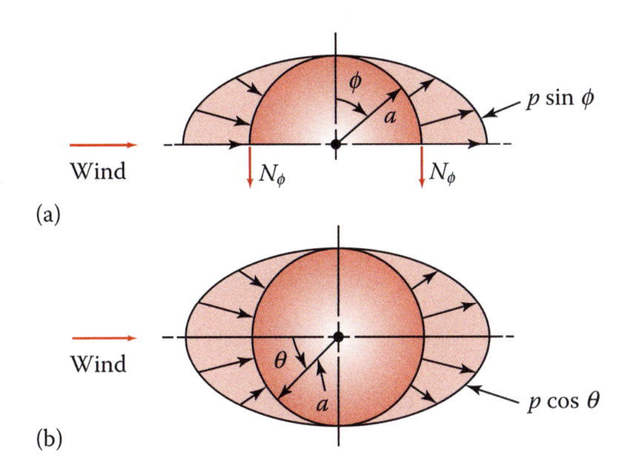

FIGURE 12.18
Hemispherical shell under wind pressure: (a) Front view; (b) top view.

Proceeding with the solution, we substitute Equations 12.27 into 12.26 to obtain

$$\begin{aligned}
&\frac{\partial}{\partial\phi}(r_0 N_\phi)+\frac{\partial N_{\theta\phi}}{\partial\theta}r_1 - N_\theta r_1\cos\phi = 0\\
&\frac{\partial}{\partial\phi}(r_0 N_{\theta\phi})+\frac{\partial N_\theta}{\partial\theta}r_1 + N_{\theta\phi} r_1\cos\phi = 0\\
&N_\phi r_0 + N_\theta r_1 \sin\phi = -p r_0 r_1 \sin\phi\cos\theta
\end{aligned} \tag{a}$$

The third expression, when substituted into the first and second, eliminates N_θ. The *equations of equilibrium* for shells of revolution *under* the action of *wind pressure* are then

$$\begin{aligned}
&\frac{\partial N_\phi}{\partial\phi}+\left(\frac{1}{r_0}\frac{dr_0}{d\phi}+\cot\phi\right)N_\phi+\frac{r_1}{r_0}\frac{\partial N_{\theta\phi}}{\partial\theta}=-pr_1\cos\phi\cos\theta\\
&\frac{\partial N_{\theta\phi}}{\partial\phi}+\left(\frac{1}{r_0}\frac{dr_0}{d\phi}+\frac{r_1}{r_2}\cot\phi\right)N_{\theta\phi}-\frac{1}{\sin\phi}\frac{\partial N_\phi}{\partial\theta}=-pr_1\sin\theta\\
&N_\theta = -pr_0\cos\theta-\frac{N_\phi r_0}{r_1\sin\phi}
\end{aligned} \tag{12.28a–c}$$

Illustrated in the solution of the following problem is the determination of membrane stresses.

EXAMPLE 12.10: Stresses in a Circular Cone under Wind Pressure

Consider a *mushroom-like shelter*, a shell having the shape of a circular cone, supported by a column at the vertex (Figure 12.19). Find the hoop, meridian, and shear stresses if the shell is submitted to a wind pressure described by Equations 12.27.

Solution

Referring to Figure 12.15,

$$r_1=\infty \qquad r_2 = s\tan\alpha \qquad ds = r_1\, d\phi \qquad r_0 = s\sin\alpha \tag{b}$$

from which

$$\frac{d}{d\phi}=r_1\frac{d}{ds} \qquad \frac{dr_0}{ds}=\sin\alpha \tag{c}$$

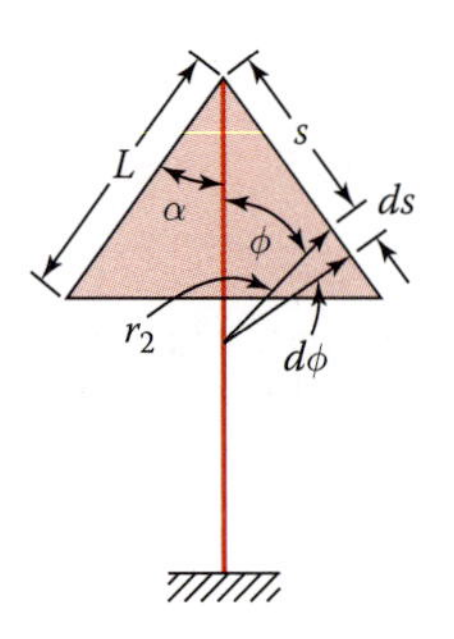

FIGURE 12.19
Circular cone supported by a column.

In addition, we replace N_ϕ by N_s and $N_{\theta\phi}$ by $N_{\theta s}$. On introducing these together with Equations 12.27 into Equation 12.28b, we obtain, after integration,

$$N_{\theta s} = -\frac{1}{s^2}\left(\frac{ps^3}{3} + c\right)\sin\theta \tag{d}$$

The condition that the edge $s = L$ of the shell is free is fulfilled by setting $c = -pL^3/3$ in the above expression. We thus have

$$\tau_{\theta s} = \frac{N_{\theta s}}{t} = \frac{p}{3t}\frac{L^3 - s^3}{s^2}\sin\theta \tag{12.29a}$$

After substituting Equations b, c, and 12.29a, we find that Equation 12.28a becomes

$$\frac{\partial N_s}{\partial s} + \frac{N_s}{s} = -\left(\frac{p}{3}\frac{L^3 - s^3}{s^3\sin\alpha} + p\sin\alpha\right)\cos\theta$$

Integration of the above leads to

$$\sigma_s = \frac{N_s}{t} = \frac{p\cos\theta}{t\sin\alpha}\left(\frac{L^3 - s^3}{3s^2} - \frac{L^2 - s^2}{2s}\cos^2\alpha\right) \tag{12.29b}$$

Equation 12.28c then results in the hoop stress:

$$\sigma_\theta = \frac{N_\theta}{t} = -\frac{ps\sin\alpha\cos\theta}{t} \tag{12.29c}$$

Comments: It is seen from Equations 12.29a and 12.29b that meridian and shear stresses grow without limit at the top ($s = 0$), as expected at a point support. It can be shown that the vertical resultant of forces N_s and $N_{\theta s}$ transmitted in a parallel circle approaches the total load of the shell when $s = s_1$. Here s_1 is a particular finite length. To avoid infinite stresses, the conical shell must be assumed to be fastened to the column along a circle corresponding to s_1.

12.10 Cylindrical Shells of General Shape

A cylindrical shell is formed by moving a straight line, the *generator*, parallel to its initial direction along a closed path. Depicted in Figure 12.20a is a *cylindrical shell of arbitrary cross section*. A shell element is bordered by two adjacent generators and two planes normal to the axial axis x, spaced dx apart. The element so described is located by coordinates x and θ.

Assume a *nonuniform loading* to act on the shell. Then, a FBD of a membrane element contains the forces shown in Figure 12.20b. The x and θ components of the externally applied forces per unit area are labeled p_x and p_θ and are indicated to act in the directions

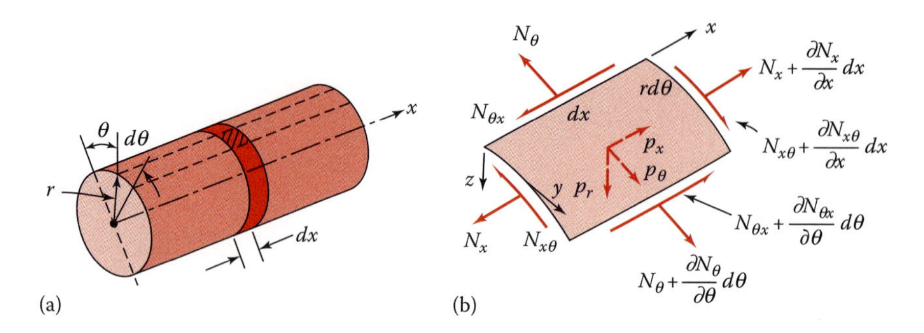

FIGURE 12.20
(a) Cylindrical shell of general shape; (b) membrane forces and surface loads on shell element.

of increasing x and θ. The radial or normal component of the loading p_r acts in the positive, inward direction. The equilibrium of forces in the x, θ, and r directions is represented by

$$\frac{\partial N_x}{\partial x} dx\, r\, d\theta + \frac{\partial N_{x\theta}}{\partial \phi} d\theta\, dx + p_x\, dx\, r\, d\theta = 0$$
$$\frac{\partial N_\theta}{\partial \theta} d\theta\, dx + \frac{\partial N_{x\theta}}{\partial x} dx\, r\, d\theta + p_\theta\, dx\, r\, d\theta = 0$$
$$N_\theta\, dx\, d\theta + p_r\, dx\, r\, d\theta = 0$$

Dividing by the differential quantities, we obtain the equations of equilibrium of a cylindrical shell. It follows that

$$\begin{aligned} N_\theta &= -p_r r \\ \frac{\partial N_{x\theta}}{\partial x} + \frac{1}{r}\frac{\partial N_\theta}{\partial \theta} &= -p_\theta \\ \frac{\partial N_x}{\partial x} + \frac{1}{r}\frac{\partial N_{x\theta}}{\partial \theta} &= -p_x \end{aligned} \tag{12.30a–c}$$

As already mentioned in Section 12.7, the above expressions could be obtained directly from Equations 12.26. It is observed that they are simple in structure and may be solved one by one.

For a prescribed loading, N_θ is readily found from Equation 12.30a. Subsequently, $N_{x\theta}$ and N_x are determined by integrating Equations 12.30b and 12.30c. In so doing, we have

$$\begin{aligned} N_\theta &= -p_r r \\ N_{x\theta} &= -\int\left(p_\theta + \frac{1}{r}\frac{\partial N_\theta}{\partial \theta}\right)dx + f_1(\theta) \\ N_x &= -\int\left(p_x + \frac{1}{r}\frac{\partial N_{x\theta}}{\partial \theta}\right)dx + f_2(\theta) \end{aligned} \tag{12.31}$$

Here $f_1(\theta)$ and $f_2(\theta)$ are arbitrary functions of integration to be evaluated on the basis of the edge conditions. These functions arise because of the integration of partial derivatives.

EXAMPLE 12.11: Forces in Cylinder filled with Liquid

A long, horizontal, cylindrical *conduit* is supposed as shown in Figure 12.21 and filled to capacity with a liquid of specific weight γ. Determine the membrane forces under two assumptions: (a) there is free spanning, with expansion joints at both ends; (b) both ends are rigidly fixed.

Solution

At an arbitrary level defined by angle θ, the pressure is $\gamma a(1 - \cos\theta)$, and the external forces are thus

$$p_r = -\gamma a(1-\cos\theta) \qquad p_\theta = p_x = 0 \tag{a}$$

where the minus sign indicates the outward direction. Substituting the above into Equations 12.31, we have

$$\begin{aligned} N_\theta &= \gamma a^2(1-\cos\theta) \\ N_{x\theta} &= -\int \gamma a \sin\theta dx + f_1(\theta) = -\gamma a x \sin\theta + f_1(\theta) \\ N_x &= \int \gamma x \cos\theta dx - \frac{1}{a}\int \frac{df_1}{d\theta} dx + f_2(\theta) \\ &= \frac{\gamma x^2}{2}\cos\theta - \frac{x}{a}\frac{df_1}{d\theta} + f_2(\theta) \end{aligned} \tag{b}$$

a. The conditions for the simply supported edges are represented by

$$N_x = 0 \qquad \left(x = \pm\frac{L}{2}\right) \tag{c}$$

Introduction of Equations b into Equation c yields

$$\begin{aligned} 0 &= \frac{\gamma L^2}{8}\cos\theta - \frac{L}{2a}\frac{df_1}{d\theta} + f_2(\theta) \\ 0 &= \frac{\gamma L^2}{8}\cos\theta + \frac{L}{2a}\frac{df_1}{d\theta} + f_2(\theta) \end{aligned}$$

Adding and subtracting the foregoing expressions provide, respectively,

$$f_2(\theta) = -\frac{\gamma L^2}{8}\cos\theta \tag{d}$$

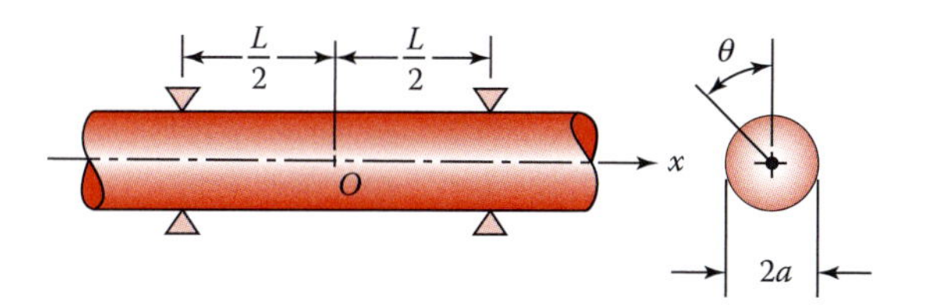

FIGURE 12.21
Circular cylinder filled with a liquid.

$$\frac{df_1}{d\theta} = 0 \quad \text{or} \quad f_1(\theta) = 0 + c \tag{e}$$

From the second of Equations b, we observe that c in Equation e represents the value of the uniform load N_x at $x = 0$. This load is not present because the pipe is free of torque; thus, $c = 0$. The solution, then, from Equations b and d is written as follows:

$$\begin{aligned} N_\theta &= \gamma a^2(1 - \cos\theta) \\ N_{x\theta} &= -\gamma a x \sin\theta \\ N_x &= -\frac{\gamma}{8}(L^2 - 4x^2)\cos\theta \end{aligned} \tag{12.32a–c}$$

Note that the shear $N_x\theta$ and the normal N_x forces, respectively, represent identical span-wise distributions with the shear force and the bending moment of a *uniformly loaded beam.*

Through the application of Hooke's law,

$$u = \frac{1}{Et}\int_{-L/2}^{L/2}(N_x - \nu N_\theta)dx \tag{f}$$

On introducing the membrane forces, Equations 12.32, the above expression provides the axial deformation.

b. In this case, no change occurs in the length of the generator:

$$\int_{-L/2}^{L/2}(N_x - \nu N_\theta)dx = 0 \tag{g}$$

where

$$\begin{aligned} N_x &= -\frac{\gamma}{8}(L^2 - 4x^2)\cos\theta + f_2(\theta) \\ N_\theta &= \gamma a^2(1 - \cos\theta) \end{aligned} \tag{h}$$

Introduction of Equations h into Equation g yields

$$f_2(\theta) = \nu\gamma a^2(1 - \cos\theta) + \frac{\gamma L^2}{12}\cos\theta$$

Hence,

$$N_x = \nu\gamma a^2(1 - \cos\theta) + \frac{\gamma}{2}\left(x^2 - \frac{L^2}{12}\right)\cos\theta \tag{12.33}$$

We find that the circumferential strain, $\varepsilon_\theta = (N_\theta - \nu N_x)/Et$, is not zero. Because clamped edges inhibit any such deformations at the ends, some bending of the pipe occurs near the supports. The membrane solution, Equations 12.32a, 12.32b, and 12.33, thus will agree very well with measurement at distances approximately a from the supports. The detail of the distribution of support-reaction forces is obtained by application of the bending theory.

Sections of cylindrical shells are often used as *roofing structures*. In such cases, the shells may be either simply supported at the ends or cantilevered out. To ascertain the membrane stresses in these *open shells*, so-called *curved panels* or *barrel vaults* (Equations 12.31) can again be employed, as shown in the example that follows.

EXAMPLE 12.12 Stress Distribution in Curved Panel

A cantilever shell of length L, radius a, and half-angle α carries its own weight p (Figure 12.22). Determine the distribution of the stresses in the member.

Solution

By Equations 12.12, we have

$$p_r = p\cos\theta \qquad p_\theta = p\sin\theta \qquad p_x = 0$$

Introducing the preceding into Equations 12.31,

$$\begin{aligned} N_\theta &= -pa\cos\theta \\ N_{x\theta} &= -2px\sin\theta + f_1(\theta) \\ N_x &= -\frac{px^2}{a}\cos\theta - \frac{x}{a}\frac{df_1}{d\theta} + f_2(\theta) \end{aligned} \tag{i}$$

The boundary conditions are

$$N_{x\theta} = 0 \qquad N_x = 0 \qquad (x = L) \tag{j}$$

Substitution of Equations i into Equation j gives

$$f_1(\theta) = 2pL\sin\theta \qquad f_2(\theta) = \frac{pL^2}{a}\cos\theta$$

Equations i, after division by t, result in

$$\begin{aligned} \sigma_\theta &= -\frac{pa}{t}\cos\theta \\ \tau_{x\theta} &= \frac{2p}{t}(L-x)\sin\theta \\ \sigma_x &= \frac{p}{at}(L-x)^2\cos\theta \end{aligned} \tag{12.34}$$

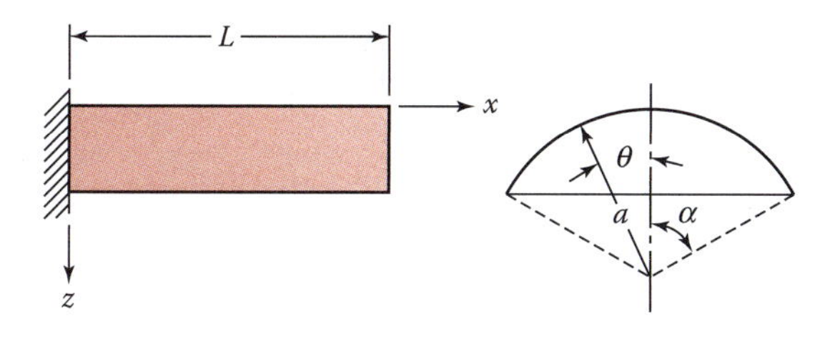

FIGURE 12.22
Cantilever shell in the form of curved panel.

Comments: Observe that σ_θ and $\tau_{x\theta}$ do not vanish along the edges $\theta = \pm\alpha$, as they should for free edges. Also, σ_x is always tensile, while it should be compressive along free edges. Thus, in the case of an open type of shell, the stresses as predicted by the membrane theory are *unsatisfactory*. The true elastic behavior of such a shell can be determined only by considering membrane stresses together with bending stresses, as will be discussed in Section 15.7.

12.11 *Folded Structures

A *folded structure*, also called *folded plate*, acts basically in the same manner as a barrel vault. Typical folded structures consist of an assembly of *flat plate strips*, as depicted in Figure 12.23. Like cylindrical shells, these prismatic members must be properly supported.

Membrane theory of folded structures is subject to severe limitations, resulting from incompatibilities in the deformations pertinent to the membrane forces. Thus, a rather lengthy bending theory is absolutely necessary for a realistic stress analysis. The *folded plate theories* are discussed in a number of references [3,6].

The use of each theory of folded structures demands a numerical accuracy that can be achieved only by a computer. Such a treatment of these structures is beyond the scope of this text. Obviously, the possibilities of analysis of folded plate and shell structures by the finite element method are enormous, as will be discussed in Section 13.13.

12.12 *Shell of General Form

So far, only two special types of shells have been examined: the shells of revolution and cylindrical shells. In this section, first we develop the equations of equilibrium for the *shell of general form*. Then, a particular shell that does not fall into the groups considered before is treated in Example 12.13.

The surface of a shell, in the general case, may be defined by the equation $z = f(x, y)$. A typical element *ABCD* of the midsurface of such a shell and its projection on the *xy*

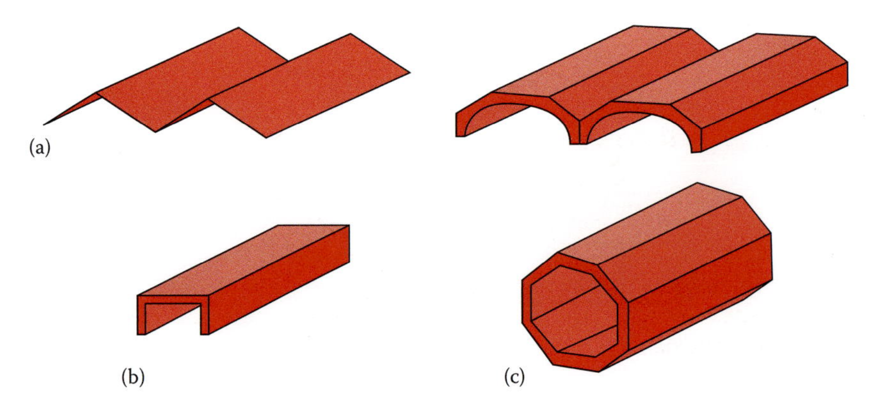

FIGURE 12.23
Folded structures: (a) Roof; (b) bridge; (c) octagonal tube.

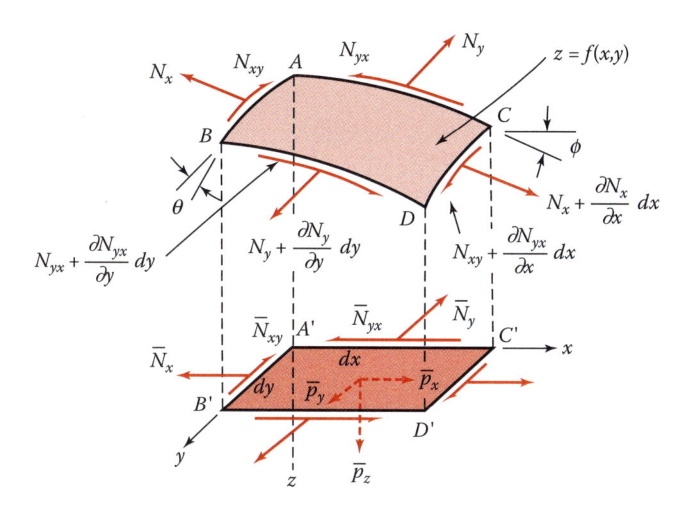

FIGURE 12.24
A shell of general shape $z = f(x, y)$ and its projection on the xy plane.

plane $A'B'C'D$ are shown in Figure 12.24. The inclination of the element to the horizontal is measured by the angles ϕ and θ.

The components of applied load per unit area in the xy plane are denoted by $\bar{p}_x, \bar{p}_y$, and $\bar{p}_z$. The shell element is considered to be in a membrane state of stress under the forces N_x, N_y, and N_{xy} having projections in the xy plane $\bar{N}_x, \bar{N}_y$, and $\bar{N}_{xy}$, respectively. Referring to Figure 12.24, we can write (Problem 12.29)

$$N_x = \bar{N}_x \frac{\cos\theta}{\cos\phi} \qquad N_y = \bar{N}_y \frac{\cos\phi}{\cos\theta} \qquad N_{xy} = \bar{N}_{xy} \tag{12.35}$$

where $\tan \phi = \partial z/\partial x$ and $\tan \theta = \partial z/\partial y$.

On following a procedure similar to that described in Section 9.2, the conditions of the equilibrium of forces in the x and y directions are obtained in the form

$$\begin{aligned} \frac{\partial \bar{N}_x}{\partial x} + \frac{\partial \bar{N}_{xy}}{\partial y} + \bar{p}_x = 0 \\ \frac{\partial \bar{N}_{xy}}{\partial x} + \frac{\partial \bar{N}_y}{\partial y} + \bar{p}_y = 0 \end{aligned} \tag{12.36a, b}$$

In a like manner, $\Sigma F_z = 0$ leads to

$$\frac{\partial}{\partial x}\left(\bar{N}_x \frac{\partial z}{\partial x} + \bar{N}_{xy} \frac{\partial z}{\partial y}\right) + \frac{\partial}{\partial y}\left(\bar{N}_y \frac{\partial z}{\partial y} + \bar{N}_{xy} \frac{\partial z}{\partial x}\right) + \bar{p}_z = 0 \tag{a}$$

Performing the differentiation as indicated in the preceding and taking into account Equations 12.36a, b, we have

$$\bar{N}_x \frac{\partial^2 z}{\partial x^2} + 2\bar{N}_{xy} \frac{\partial^2 z}{\partial x \partial y} + \bar{N}_y \frac{\partial^2 z}{\partial y^2} = -\bar{p}_z + \bar{p}_x \frac{\partial z}{\partial x} + \bar{p}_y \frac{\partial z}{\partial y} \tag{12.36c}$$

Hence, provided that a solution of Equations 12.36 may be obtained for the projected forces, the actual forces may then be readily determined from the relationships (Equations 12.35).

In most cases, it is advantageous to introduce a stress function $\Phi(x, y)$ that reduces Equations 12.36 to one second-order equation as follows. By analogy with plate problems (see Section 10.4), in order to satisfy Equations 12.36a and 12.36b, we let

$$\bar{N}_x = \frac{\partial^2 \Phi}{\partial y^2} - \int \bar{p}_x \, dx \qquad \bar{N}_y = \frac{\partial^2 \Phi}{\partial x^2} - \int \bar{p}_y \, dy \qquad \bar{N}_{xy} = -\frac{\partial^2 \Phi}{\partial x \partial y} \tag{12.37}$$

Then, Equation 12.36c becomes

$$\begin{aligned} &\frac{\partial^2 \Phi}{\partial x^2}\frac{\partial^2 z}{\partial y^2} - 2\frac{\partial^2 \Phi}{\partial x \partial y}\frac{\partial^2 z}{\partial x \partial y} + \frac{\partial^2 \Phi}{\partial y^2}\frac{\partial^2 z}{\partial x^2} = -\bar{p}_z + \bar{p}_x\frac{\partial z}{\partial x} + \bar{p}_y\frac{\partial z}{\partial y} + \frac{\partial^2 z}{\partial x^2} \\ &\int \bar{p}_x \, dx + \frac{\partial^2 z}{\partial y^2}\int \bar{p}_y \, dy \end{aligned} \tag{12.38}$$

The solution of the problem is thus reduced to the determination of stress function Φ.

EXAMPLE 12.13: Force Analysis in Hyperbolic Shell Carrying Two Typical Loads

Determine the membrane forces in a *hyperbolic paraboloid* shell with a midsurface defined by the expression

$$z = -\frac{xy}{c} \tag{b}$$

where $c = a^2/h$ (Figure 12.25) for two particular cases: (a) the shell is subjected to a vertical load p_s per unit of its horizontal projection (e.g., snow load); (b) the shell carries its own weight p per unit surface area. Assume that edge beams (or shear diaphragms) are provided to resist shear forces along the four edges. These members are taken to be incapable of resisting normal forces:

$$\begin{aligned} N_x = \bar{N}_x = 0 \qquad (x = 0) \\ N_y = \bar{N}_y = 0 \qquad (y = 0) \end{aligned} \tag{c}$$

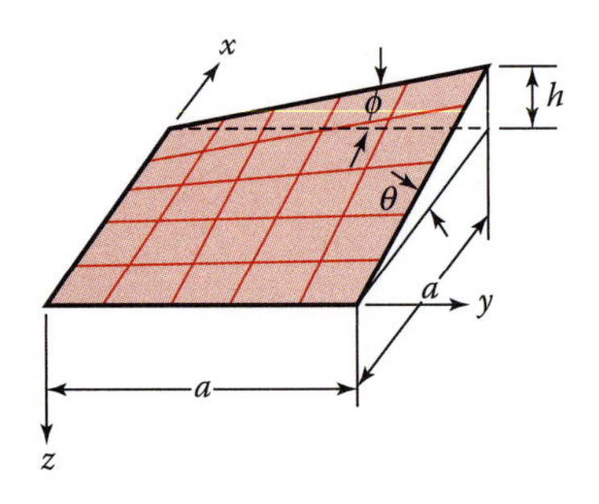

FIGURE 12.25
Hyperbolic paraboloid shell.

Solution

Partial derivatives of Equation b are

$$\frac{\partial z}{\partial x} = -\frac{y}{c} \qquad \frac{\partial z}{\partial y} = -\frac{x}{c} \quad \frac{\partial^2 z}{\partial x^2} = \frac{\partial^2 z}{\partial y^2} = 0 \qquad \frac{\partial^2 z}{\partial x \partial y} = -\frac{1}{c} \tag{d}$$

Introducing the foregoing into Equation 12.38, we obtain

$$\frac{\partial^2 \Phi}{\partial x \partial y} = -\frac{1}{2} c\bar{p}_z$$

Integration of this relation with respect to x and y gives

$$\Phi = -\frac{1}{2}\bar{p}_z cxy + f_1(x) + f_2(y) \tag{e}$$

a. For this case, we have $p = p_s$, and Equations 12.37 become

$$\bar{N}_x = \frac{d^2 f_2}{dy^2} \qquad \bar{N}_y = \frac{d^2 f_1}{dx^2} \qquad \bar{N}_{xy} = \frac{1}{2} p_s c \tag{f}$$

We observe from the preceding that the boundary conditions (Equations c) are fulfilled if $d^2f_2/dy^2 = 0$ and $d^2f_1/dx^2 = 0$. Hence, Equations 12.35 give

$$N_{xy} = \frac{1}{2} p_s c = \frac{1}{2h} p_s a^2 \qquad N_x = N_y = 0 \tag{12.39}$$

indicating that the entire shell is in a state of *pure shear*.

b. The *unit* surface area of the shell is defined by [17]

$$\sqrt{\left(\frac{\partial z}{\partial x}\right)^2 + \left(\frac{\partial z}{\partial y}\right)^2 + 1} = \frac{1}{c}\sqrt{x^2 + y^2 + c^2} \tag{g}$$

It follows that

$$\bar{p}_z = \frac{p}{c}\sqrt{x^2 + y^2 + c^2} \tag{h}$$

The shear force, from Equations 12.37, e, and 12.35, is therefore

$$N_{xy} = \frac{p}{2}\sqrt{x^2 + y^2 + c^2} \tag{12.40a}$$

Carrying this into Equation 12.36a, after integrating, we have

$$\bar{N}_x = -\frac{py}{2}\ln\left[x + \sqrt{x^2 + y^2 + c^2}\right] + f(y)$$

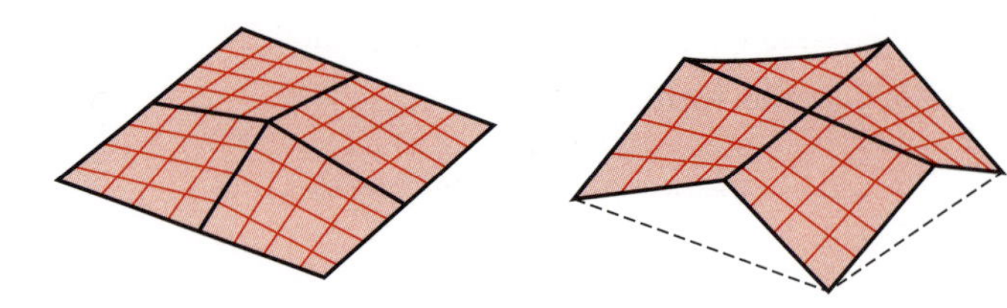

FIGURE 12.26
Typical roof shells.

In a similar manner, Equation 12.36b results in

$$\bar{N}_y = -\frac{px}{2}\ln\left[y + \sqrt{x^2 + y^2 + c^2}\right] + f(x)$$

Satisfying the conditions (Equations c), we obtain

$$f(y) = \frac{py}{2}\ln\sqrt{y^2 + c^2} \qquad f(x) = \frac{px}{2}\ln\sqrt{x^2 + c^2}$$

Thus, the normal forces, through the use of Equations 12.35, are

$$\begin{aligned} N_x &= -\frac{py}{2}\frac{\cos\theta}{\cos\phi}\ln\frac{x + \sqrt{x^2 + y^2 + c^2}}{\sqrt{y^2 + c^2}} \\ N_y &= -\frac{px}{2}\frac{\cos\phi}{\cos\theta}\ln\frac{y + \sqrt{x^2 + y^2 + c^2}}{\sqrt{x^2 + c^2}} \end{aligned} \qquad (12.40b, c)$$

where $\tan\phi = x/c$ and $\tan\theta = y/c$.

Comments: Because of the ease with which the hyperbolic paraboloid surface may be generated from a series of straight lines parallel to the x and y axes, *roofs* are often formed by the combination of four shells of the type shown in Figure 12.25. Figure 12.26 depicts two typical cases. When all four panels are uniformly loaded, edge beams resist axial forces alone. These structures are, however, precariously balanced; when the panels carry different loads, simple equilibrium is impossible [10].

12.13 *Breakdown of Elastic Action in Shells

From a design point of view, it is clear that for practical purposes it is necessary to obtain the proper dimensions of a member that can sustain a prescribed loading without suffering failure by breakdown of elastic action. As already alluded to (Section 2.11), the dimensions that should be assigned to a loaded element depend on the failure theory held concerning the cause of the yielding or fracture. The former will be illustrated in the examples to follow for two particular shell structures.

EXAMPLE 12.14: Design of Cylindrical Pressure Vessel

A circular cylindrical vessel with closed ends is subjected to internal pressure p. The tube is fabricated of a material having a yield point stress σ_{yp}. Determine the required wall thickness t remote from the ends, according to two yield criteria discussed in Section 2.11.

Solution

The hoop and the axial shell stresses are given by Equations 12.10:

$$\sigma_1 = \sigma_\theta = \frac{pa}{t} \qquad \sigma_2 = \sigma_x = \frac{pa}{2t} \tag{a}$$

Maximum shear stress theory. As σ_1 and σ_2 are of the same sign, we obtain, from Equations 2.42b and a,

$$t = \frac{pa}{\sigma_{yp}} \tag{b}$$

Maximum energy of distortion theory. Equations 2.43 and a yield

$$t = \frac{\sqrt{3}}{2}\frac{pa}{\sigma_{yp}} \tag{c}$$

EXAMPLE 12.15: Design of Conical Tank

A conical tank is constructed of steel of yield point σ_{yp} in tension and $(3/2)\sigma_{yp}$ in compression (Figure 12.13). The tank is filled with a liquid of specific weight γ and is edge supported. Determine the proper wall thickness if based on a factor of safety n, using the maximum energy of distortion theory.

Solution

Expressions for the hoop and longitudinal stresses, from Equations 12.22, are

$$\sigma_1 = \sigma_\theta = \gamma(h-y)y\frac{\tan\alpha}{t\cos\alpha} \qquad \sigma_2 = \sigma_s = \gamma\left(h - \frac{2}{3}y\right)y\frac{\tan\alpha}{2t\cos\alpha} \tag{d}$$

The largest values of the principal stresses are given by

$$\begin{aligned}\sigma_{1,\max} &= \frac{\gamma h^2}{4t}\frac{\tan\alpha}{\cos\alpha} \qquad \left(y = \frac{h}{2}\right)\\ \sigma_{2,\max} &= \frac{3\gamma h^2}{16t}\frac{\tan\alpha}{\cos\alpha} \qquad \left(y = \frac{3h}{4}\right)\end{aligned} \tag{e}$$

Equations e demonstrate that the principal stresses assume their maximum values at different locations. The location at which the combined principal stresses is critical is ascertained as follows. First, Equations d are substituted into Equations 2.43:

$$\left[\gamma(h-y)y\frac{\tan\alpha}{t\cos\alpha}\right]^2 + \left[\gamma\left(h-\frac{2}{3}y\right)y\frac{\tan\alpha}{2t\cos\alpha}\right]^2$$

$$-\left[\gamma(h-y)y\frac{\tan\alpha}{t\cos\alpha}\right]\left[\gamma\left(h-\frac{2}{3}y\right)y\frac{\tan\alpha}{2t\cos\alpha}\right]=\left(\frac{\sigma_{yp}}{n}\right)^2 \tag{f}$$

Next, the derivative of the foregoing with respect to y is set equal to zero to yield

$$y=0.52h \tag{g}$$

Substitution of Equation g into Equation f results in the required wall thickness:

$$t=0.225\frac{\gamma h^2 n}{\sigma_{yp}}\frac{\tan\alpha}{\cos\alpha} \tag{h}$$

Problems

Sections 12.1 through 12.7

12.1 A hemispherical roof dome is subjected to a snow load p_s (Figure 12.8a). Develop the expressions

$$\sigma_\phi=-\frac{p_s a}{2t} \qquad \sigma_\theta=-\frac{p_s a}{2t}\cos 2\phi \tag{P12.1}$$

for the meridional and hoop stresses.

12.2 A simply supported hemispherical roof dome is made of cold rolled (510) bronze (Table B.3) and carries a snow load of p_s. Calculate the magnitudes for: (a) the meridional and hoop stresses; (b) the critical stress for local buckling. *Data*: $E = 110$ GPa and $\sigma_{yp} = 520$ MPa (Table B.3), $a = 9$ m, $t = 30$ mm, $p_s = 2$ kPa. *Assumption*: Stresses due to its own weight of the shell may be neglected.

12.3 An observation dome of a pressurized aircraft is of ellipsoidal shape (Figure 12.11). It is constructed of 6-mm-thick plastic material. Determine the limiting value of the pressure differential the shell can resist given a maximum stress of 14 MPa. The lengths of the semiaxes are $a = 0.15$ m and $b = 0.12$ m.

12.4 Redo Problem 12.3 for the case in which the dome is of spherical shape ($a = b = 0.15$ m) and can resist a maximum stress of 28 MPa.

12.5 A conical aluminum container of thickness 3 mm, apex angle $\alpha = 45°$, and height $h = 3$ m is filled with water (Figure 12.13). Taking $\nu = 0.3$, compute the locations measured vertically above the apex for which (a) the hoop strain is zero and (b) the hoop strain is maximum.

12.6 An edge-supported hemispherical container is filled with a liquid of specific weight γ (Figure P12.6).

a. Demonstrate that the membrane stresses are given by

$$\sigma_\phi=\frac{\gamma a^2}{3t}\left(1+\frac{\cos^2\phi}{1+\cos\phi}\right) \qquad \sigma_\theta=\frac{\gamma a^2}{3t}\left(2\cos\phi-\frac{1}{1+\cos\phi}\right) \tag{P12.6}$$

b. Determine the radial (w) and the circumferential (v) deformations of the shell. *Note*: The volume of the portion mOn of the shell is $V = (1/3)\pi h^2(3a - h)$.

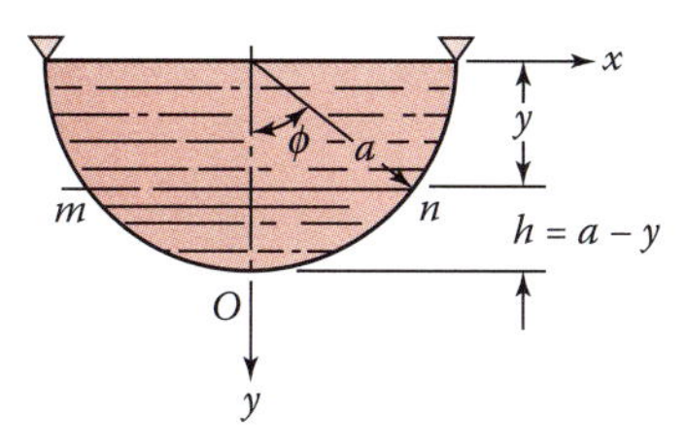

FIGURE P12.6

12.7 A vertical cylindrical steel tank of radius a and height h having suitable roof and bottom head is completely filled with a liquid of density γ and is subjected to an additional internal pressure of p (Figure 12.5b). Let $a = 8$ m, $h = 15$ m, $p = 100$ kPa, and $\gamma = 10$ kN/m^3. Calculate the wall thickness, for an allowable stress of 140 MPa: (a) at $x = h/2$; (b) at $x = 5\,h/8$; (c) of the roof, approximating it as a clamped thin circular plate.

12.8 Determine the maximum stress at point A of a football of uniform skin thickness $t = 3$ mm if the internal pressure is 90 kPa (Figure P12.8).

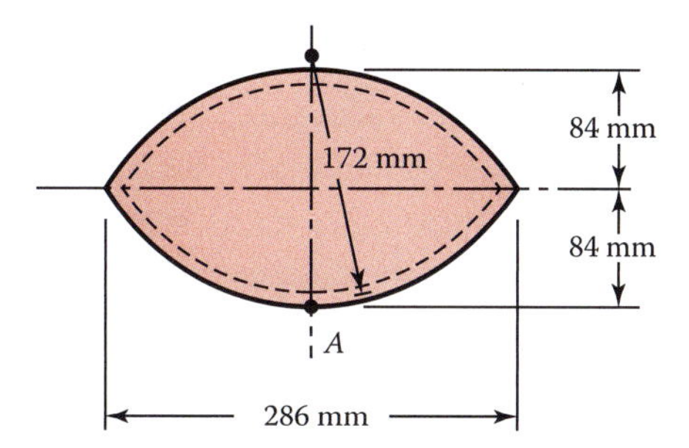

FIGURE P12.8

12.9 A toroidal pressure vessel of outer and inner diameters 3 and 2 m, respectively, is to be used to store a gas pressure of 1.6 MPa (Figure 12.10). Calculate the required minimum thickness of the vessel if the allowable stress is limited to 200 MPa.

12.10 A spherical tank of 20-m mean diameter is filled with water ($\gamma = 9.81$ kN/m^3) and supported along the line nn (Figure 12.14). If the tank is made of steel with an allowable stress of 125 MPa, calculate the minimum required wall thickness.

12.11 Calculate the maximum required wall thickness at $\phi = 90°$ for the spherical tank, supported and loaded as described in Problem 12.10.

12.12 A cylindrical tank has a suspended conical bottom as shown in Figure P12.12. Derive the following expressions for the membrane forces in the bottom part:

$$N_\theta = \gamma(H + h - y)t\frac{\tan\alpha}{\cos\alpha}$$

$$N_s = \frac{\gamma}{2}\left(H + h - \frac{2}{3}y\right)y\frac{\tan\alpha}{\cos\alpha} \qquad \text{(P12.12)}$$

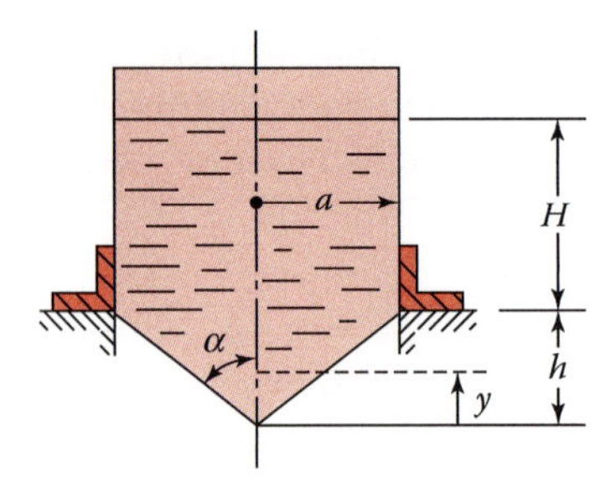

FIGURE P12.12

12.13 A supported truncated conical shell carries an upper edge load p_e (Figure P12.13). Derive the following expressions for the membrane stresses:

$$\sigma_s = -p_e \frac{s_0}{st} \frac{1}{\sin\phi} \qquad \sigma_\theta = \tau_{s\theta} = 0 \tag{P12.13}$$

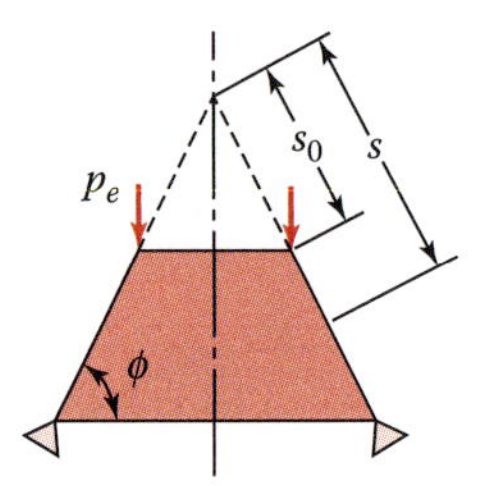

FIGURE P12.13

12.14 The *compound tank* shown in Figure P12.14 consists of a conical shell with a spherical bottom. The tank is filled to a level h_1 with a liquid of specific weight γ. Show that the hoop and the meridional stresses in the conical part are given by

$$\begin{aligned} \sigma_\theta &= \frac{\gamma}{t} y(h_1 - y) \frac{\tan\alpha}{\cos\alpha} \\ \sigma_s &= \frac{\gamma}{6t}\left[(3h_1 - 2y)y \frac{\sin\alpha}{\cos^2\alpha} + \frac{a^2(2a - h_2)\cos^2\alpha}{y\sin\alpha} + \frac{a^3\cos^2\alpha}{y}\right] \end{aligned} \tag{P12.14}$$

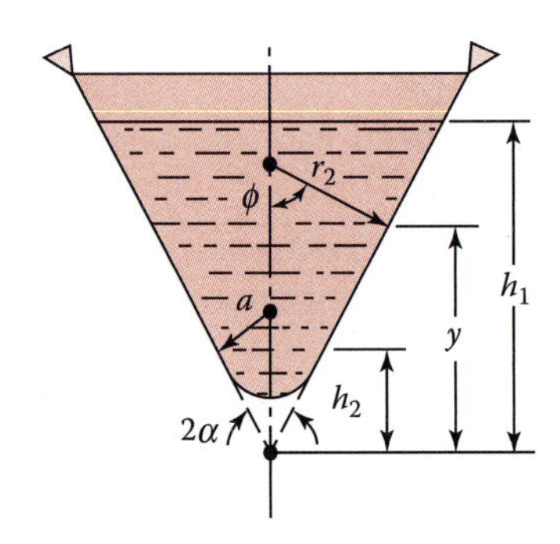

FIGURE P12.14

The solution for the bottom part is governed by Equations P12.6. Note that at the juncture of the two parts, a ring must be provided to resist the difference in the horizontal components of the meridional forces N_s in the cone and the N_ϕ in the sphere.

12.15 A compound tank of uniform thickness t consists of a cylindrical shell and a hemispherical bottom (Figure P12.15). The tank is filled with a liquid of specific weight γ up to a level H. Determine the stresses σ_ϕ and σ_θ at section mn in terms of ϕ, a, H, t, and γ, as needed. Note that the volume of the portion mOn of the shell is $V = (1/3)\pi h^2(3a - h)$.

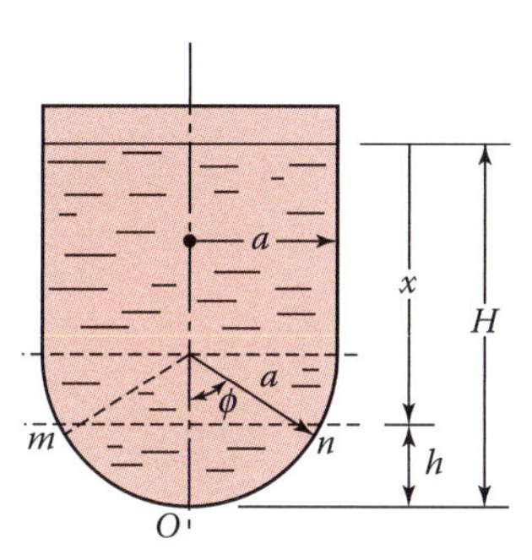

FIGURE P12.15

12.16 A 5-mm-thick shell in the form of a paraboloid and closed by a thick plate at top (Figure P12.16) is subjected to an internal pressure of $p_0 = 5$ MPa. Determine the membrane stresses σ_ϕ and σ_θ at the level $h = 500$ mm. Note that the expression of the generating parabola is $y = x^2/5$, with x and y in millimeters.

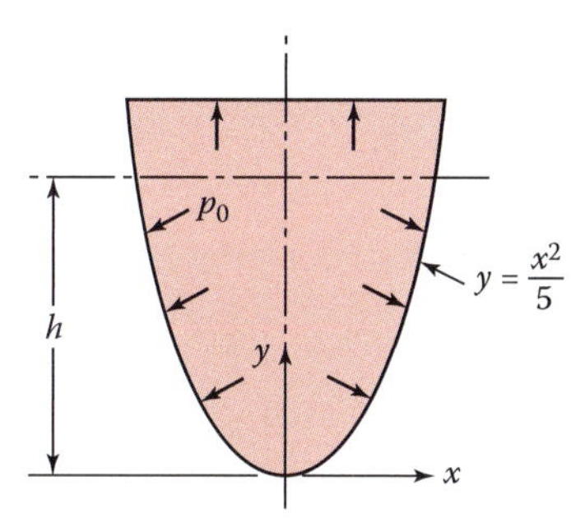

FIGURE P12.16

12.17 The mushroom-like shelter shown in Figure 12.19 is assumed to support only its own weight per unit surface area p. Derive the following expressions for the membrane forces:

$$N_\theta = -ps\sin\alpha\tan\alpha \qquad N_s = -\frac{p}{2}\frac{L^2 - s^2}{s\cos\alpha} \tag{P12.17}$$

12.18 Verify that, in the spherical tank described in Example 12.8, the maximum shear stress is represented by

$$\tau_{\max} = \frac{\gamma a^2}{6t}\left(1 + \frac{1}{1+\cos\phi}\right)(1-\cos\phi) \tag{P12.18}$$

12.19 Figure P12.19 shows a compound tank comprised of a cylindrical shell and a spherical bottom. The tank is filled to a level h with a liquid of specific weight γ. Derive the following expressions for the membrane stresses in the bottom part:

$$\begin{aligned}\sigma_\theta &= \frac{\gamma a^2}{6t}\left(\frac{3h}{a} + 5 - 6\cos\phi + \frac{2\cos^2\phi}{1+\cos\phi}\right)\\ \sigma_\phi &= -\frac{\gamma a^2}{6t}\left(\frac{3h}{a} + 1 - \frac{2\cos^2\phi}{1+\cos\phi}\right)\end{aligned} \tag{P12.19}$$

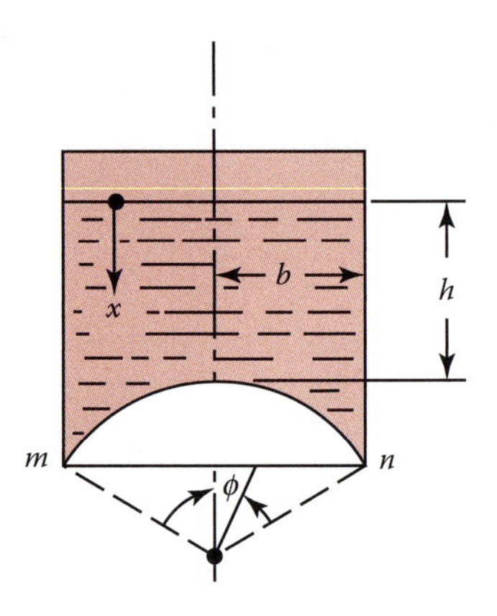

FIGURE P12.19

12.20 Determine the radial (w) and the circumferential (v) deformations in the spherical tank described in Example 12.8.

Sections 12.8 through 12.13

12.21 For the tank of Problem 12.19, using Equations 12.31, verify that the membrane forces in the cylindrical part are

$$\sigma_\theta = \frac{\gamma bx}{t} \qquad \sigma_x = c \qquad \tau_{xy} = 0 \tag{P12.21}$$

Here the constant c of the axial stress may be produced by the weight of a roof, for example. What minimum wall thickness is required for the tank if $a = h = 2b = 10$ m, γ 14 kN/m³ and the allowable stress is limited to 100 MPa? *Note*: Stresses in the bottom part are given by Equations P12.19.

12.22 A pipeline in the form of an *open semicircular channel* is filled with a liquid of specific weight γ (Figure P12.22). Derive the following expressions for the membrane forces:

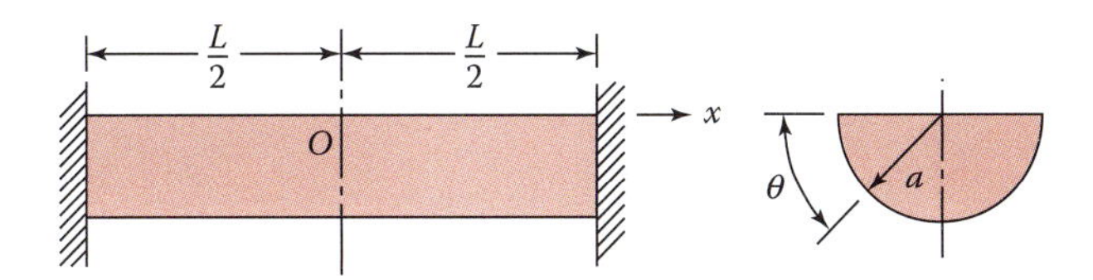

FIGURE P12.22

$$N_\theta = \gamma a^2 \sin\theta \quad N_{x\theta} = \gamma a x \cos\theta$$
$$N_x = \frac{\gamma}{2}\left(\frac{L^2}{12} - x^2\right)\sin\theta + \upsilon N_\theta \tag{P12.22}$$

12.23 The roof of an aircraft hangar, supporting its own weight p, may be approximated as a semicircular cross-sectional cylinder or so-called *barrel vault* (Figure P12.23). At both ends of the shell, it is assumed that the conditions of simple support prevail. Develop the following expressions for the membrane stresses:

$$\sigma_\theta = -\frac{pa}{t}\cos\theta \qquad \tau_{x\theta} = -\frac{2px}{t}\sin\theta$$
$$\sigma_x = -\frac{p}{4at}(L^2 - 4x^2)\cos\theta \tag{P12.23}$$

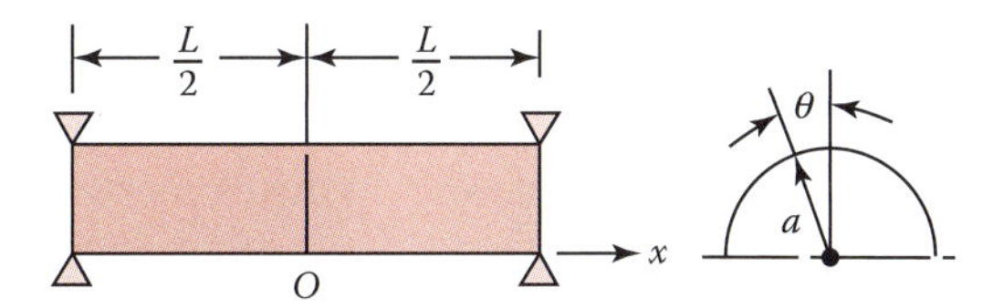

FIGURE P12.23

12.24 Redo Problem 12.23 if the ends of the semicircular cylinder are fixed.

12.25 A horizontal circular pipe of radius a, length L, and thickness t is filled with a gas at a constant pressure p. Determine the membrane stresses in the pipe if the ends are assumed to be built in.

12.26 A simply supported, horizontal, circular cylinder of radius a (Figure 12.21), thickness t, and length L carries its own weight p. Derive the following expressions for the membrane stresses:

$$\sigma_\theta = -\frac{pa}{t}\cos\theta \qquad \tau_{x\theta} = -\frac{2px}{t}\sin\theta$$
$$\sigma_x = -\frac{p}{4at}(L^2 - 4x^2)\cos\theta \tag{P12.26}$$

12.27 Redo Problem 12.26 for the case of rigidly built-in cylinder ends. Verify that now the membrane stresses are represented by

$$\sigma_\theta = -\frac{pa}{t}\cos\theta \qquad \tau_{x\theta} = -\frac{2px}{t}\sin\theta$$
$$\sigma_x = \frac{p}{t}\left(\frac{x^2}{a} - \frac{L^2}{12a} - \nu a\right)\cos\theta \tag{P12.27}$$

12.28 Consider a thin-walled circular pipe of mean radius a, loaded as a cantilever (Figure P12.28). Derive the following expressions for the membrane stresses:

$$\sigma_\theta = 0 \qquad \tau_{x\theta} = -\frac{P}{\pi a t}\sin\theta$$
$$\sigma_x = \frac{Px}{\pi a^2 t}\cos\theta \tag{P12.28}$$

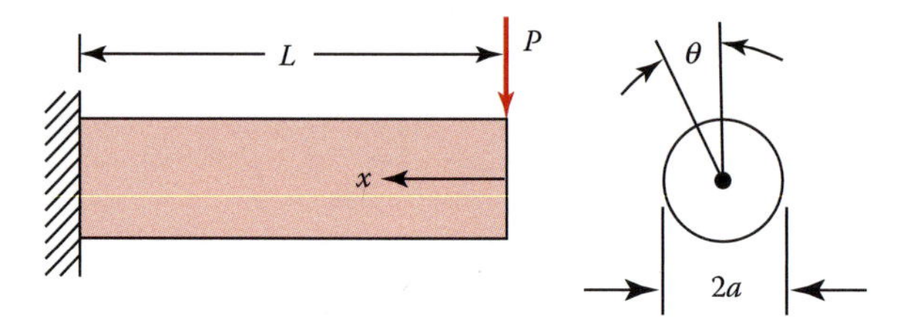

FIGURE P12.28

12.29 Verify the relationship (Equations 12.35) by evaluating the x- and y-directed horizontal components of the forces acting on the element and its projection shown in Figure 12.24.

12.30 A closed-end cylinder, 240 mm in diameter and 5 mm thick, is made of high-strength, ASTM-A242 steel (Table B.3). What is the allowable pressure the shell can carry on the basis of a safety factor $n = 2.5$? Apply the two yielding theories of failure discussed in Section 2.11.

12.31 A circular toroidal shell is subjected to internal pressure $p = 1.5$ MPa (Figure 12.10). Assume that the yield stress $\sigma_{yp} = 200$ MPa. The dimensions of the torus are $a = 0.06$ m, $b = 0.4$ m, and $t = 1$ mm. What is the factor to safety, assuming failure to occur in accordance with the energy of distortion theory?

References

1. J. Heyman, *Equilibrium of Shell Structure*, Oxford University Press, New York, 1977.
2. P. L. Gould, *Static Analysis of Shells*, Lexington Books, New York, 1977.
3. W. Flügge, *Stresses in Shells*, 2nd ed., Springer, Berlin, 1990.
4. E. H. Baker, L. Kovalevsky, and F. L. Rish, *Structural Analysis of Shells*, Krieger, Melbourne, FL, 1981.
5. P. Seide, *Elastic Deformation of Thin Shells*, Noordhoff International, Leyden, 1975.
6. J. E. Gibson, *Thin Shells: Computing and Theory*, Pergamon, New York, 1980.

7. A. Pflüger, *Elementary Statics of Shells*, 2nd ed., McGraw-Hill, New York, 1961.
8. H. Kraus, *Thin Elastic Shells*, Wiley, Hoboken, NJ, 2002.
9. S. Timoshenko and S. Woinowsky-Krieger, *Theory of Plates and Shells*, 2nd ed., Chapter 14, McGraw-Hill, New York, 1959.
10. W. Flügge (ed.), *Handbook of Engineering Mechanics*, Sections 40.1 through 40.4, McGraw-Hill, New York, 1984.
11. J. R. Vinson, *Structural Mechanics: The Behavior of Plates and Shells*, Chapter 7, Wiley, New York, 1974.
12. J. H. Faupel and F. E. Fisher, *Engineering Design*, 2nd ed., Wiley, Hoboken, NJ, 1981.
13. W. C. Young, R. C. Budynas, and A. M. Sadegh, *Roark's Formulas for Stress and Strain*, 8th ed., McGraw-Hill, New York, 2011.
14. J. F. Harvey, *Theory and Design of Pressure Vessels*, 2nd ed., Van Nostrand Reinhold Company, New York, 1974.
15. A. E. H. Love, *A Treatise on the Mathematical Theory of Elasticity*, Dover, New York, 1944.
16. L. H. Donnell, *Beams, Plates, and Shells*, McGraw-Hill, New York, 1976.
17. E. Kreyszig, *Advanced Engineering Mathematics*, 10th ed., Wiley, Hoboken, NJ, 2011.

13

Bending Stresses in Shells

13.1 Introduction

It was observed in Chapter 12 that membrane theory cannot, in all instances, provide solutions compatible with the actual conditions of deformation. This theory also fails to predict the state of stress at the boundaries and in certain other areas of the shell. These shortcomings are avoided by application of bending theory, considering membrane forces, shear forces, and moments to act on the shell structure.

To develop the governing differential equations for the *midsurface* displacements u, v, and w, which define the geometry or kinematics of deformation of a shell, we proceed as in the case of plates. We shall begin by deriving the basic relationship between the stress resultants and the deformations for shells of general shape. The relationships for the stresses and strain energy under an arbitrary loading are developed in Sections 13.4 and 13.5.

The complete *bending theory* is mathematically intricate, and the first solutions involving shell-bending stresses date back to only 1920 [1–13]. With the exception of Section 13.13, we shall, in this chapter, limit consideration to the most significant practical case involving rotationally symmetrical loading. Stress analysis of cylindrical shells under general loads is postponed to Chapter 15, after applications to various common structural members are presented in Chapter 14.

13.2 Shell Stress Resultants

In deriving an expression for the stress resultants, that is, the resultant forces and moments representing the internal stresses, consider an infinitesimal element (Figure 13.1a). This element is defined by two pairs of planes, normal to the midsurface of the shell. The origin of a Cartesian coordinate system is located at a corner of the element, as shown, with the x and y axes tangent to the lines of principal curvature and z perpendicular to the midsurface.

Because of shell curvature, the arc lengths of an element located a distance z from the midsurface are not simply ds_x and ds_y, the lengths measured on the midsurface, but rather

$$\frac{ds_x(r_x - z)}{r_x} = \left(1 - \frac{z}{r_x}\right)ds_x \qquad \frac{ds_y(r_y - z)}{r_y} = \left(1 - \frac{z}{r_y}\right)ds_y \tag{a}$$

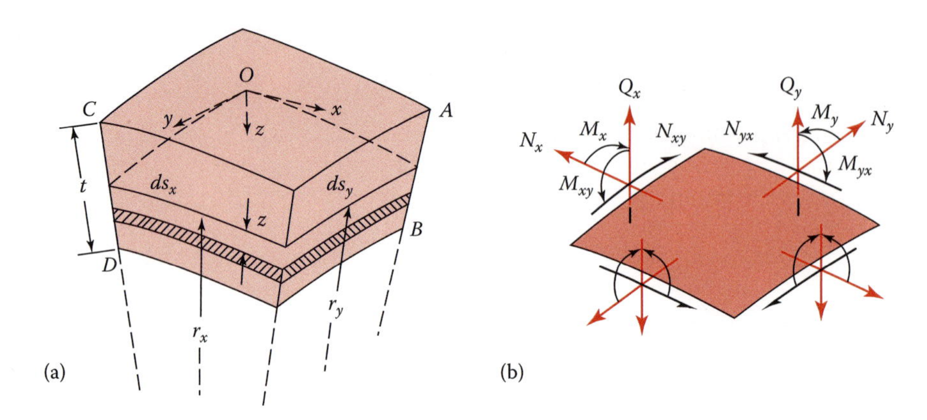

FIGURE 13.1
(a) Shell elements; (b) stress resultants on shell element.

Here r_x and r_y are the *radii of principal curvatures*, in the xz and yz planes, respectively.

The stresses acting on the plane faces of the element are σ_x, σ_y, τ_{xy}, τ_{xz}, and τ_{yz}. Letting N_x represent the resultant normal force acting on plane face yz per unit length and using the true arc length given above, we have

$$N_x\, ds_y = \int_{-t/2}^{t/2} \sigma_x \left(1 - \frac{z}{r_y}\right) ds_y\, dz$$

Observe that $1-(z/r_y)ds_y$ represents the length of arc lying in yz plane of the element located a distance z from midsurface. Canceling arbitrary length ds_y, this becomes

$$N_x = \int_{-t/2}^{t/2} \sigma_x \left(1 - \frac{z}{r_y}\right) dz = \int_{-t/2}^{t/2} \sigma_x (1 - z\kappa_y)\, dz$$

Expressions for the remaining *stress resultants* per unit length are derived in a similar manner. The complete set shown in Figure 13.1b is thus

$$\begin{Bmatrix} N_x \\ N_y \\ N_{xy} \\ N_{yx} \\ Q_x \\ Q_y \end{Bmatrix} = \int_{-t/2}^{t/2} \begin{Bmatrix} \sigma_x(1-z\kappa_y) \\ \sigma_y(1-z\kappa_x) \\ \tau_{xy}(1-z\kappa_y) \\ \tau_{yx}(1-z\kappa_x) \\ \tau_{xz}(1-z\kappa_y) \\ \tau_{yz}(1-z\kappa_x) \end{Bmatrix} dz \qquad \begin{Bmatrix} M_x \\ M_y \\ M_{xy} \\ M_{yx} \end{Bmatrix} = \int_{-t/2}^{t/2} \begin{Bmatrix} \sigma_x(1-z\kappa_y) \\ \sigma_y(1-z\kappa_x) \\ \tau_{xy}(1-z\kappa_y) \\ \tau_{yx}(1-z\kappa_x) \end{Bmatrix} zdz \tag{13.1}$$

The sign convention is the same as in the treatment of plates.

It may be concluded that even through $\tau_{xy} = \tau_{yx}$, shearing forces N_{xy} and N_{yx} are not generally equal, nor are twisting moments M_{xy} and M_{yx}. This is because in general $r_x \neq r_y$. For *thin shells*, however, for which we are concerned, t is small relative to r_x and r_y, and consequently *z/r_y may be neglected in comparison with unity.* On this basis, $N_{xy} = N_{yx}$ and $M_{xy} = M_{yx}$. The stress resultants are thus described by the *same* expressions that apply to thin plates.

13.3 Force, Moment, and Displacement Relations

To relate the stress resultants to the shell deformation, σ_x, σ_y, and τ_{xy} must be evaluated in terms of strains. According to our assumption, the *z*-directed stress is neglected, $\sigma_z = 0$. Hooke's law is then written as

$$\begin{aligned}\sigma_x &= \frac{E}{1-\nu^2}(\varepsilon_x + \nu\varepsilon_y)\\ \sigma_y &= \frac{E}{1-\nu^2}(\varepsilon_y + \nu\varepsilon_x)\\ \tau_{xy} &= \gamma_{xy} G\end{aligned} \tag{a}$$

Let us first determine the strains appearing in the above expressions. Consider the deformed shell element of Figure 13.2, nothing that, by assumption 3 of Section 12.2, sides *mn* and *m′n′* are straight lines. Accordingly, the normal and shearing strains, ε_z, γ_{xz}, γ_{yz} are neglected. The figure shows the midsurface stretched and side *mn* rotated with respect to its original configuration. The unit elongation ε_x of a fiber of length l_f, located in the *xz* plane a distance *z* from the midsurface, is given by

$$\varepsilon_x = \frac{\Delta l_f}{l_f} \tag{b}$$

In the foregoing, Δl_f is the elongation experienced by l_f. Referring to Figure 13.2,

$$l_f = ds_x\left(1 - \frac{z}{r_x}\right) \qquad \Delta l_f = ds_x(1+\varepsilon_{x0})\left(1 - \frac{z}{r'_x}\right) - l_f \tag{c}$$

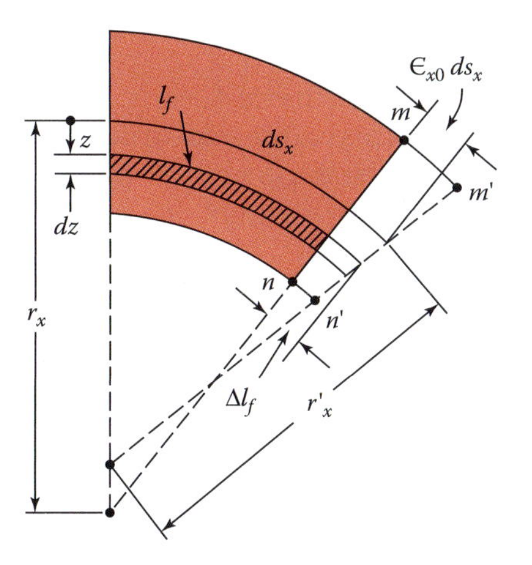

FIGURE 13.2
Deformations of midsurface.

where ε_{x0} represents the x-directed *midsurface* unit deformation, r_x' the radius of curvature after deformation, and ds_x the length of the midsurface fiber. Substituting Equations c into Equation b, we have

$$\varepsilon_x = \frac{\varepsilon_{x0}}{1-(z/r_x)} - \frac{z}{1-(z/r_x)}\left[\frac{1}{(1-\varepsilon_{x0})r_x'} - \frac{1}{r_x}\right]$$

in which r_x is the curvature prior to deformation. This is because for the case under analysis, $t \ll r_x$, z/r_x may be omitted. In addition, it can be demonstrated that the *influence of* ε_{x0} *on curvature* is negligible. Introducing the foregoing considerations, the above expression becomes

$$\varepsilon_x = \varepsilon_{x0} - z\left(\frac{1}{r_x'} - \frac{1}{r_x}\right) = \varepsilon_{x0} - z\chi_x \tag{13.2a}$$

where χ represents the *change of curvature* of the midsurface. The unit elongation at any distance normal to the midsurface is thus related to the midsurface stretch and the change in curvature associated with deformation. For the y direction, a similar expression is obtained:

$$\varepsilon_y = \varepsilon_{y0} - z\left(\frac{1}{r_y'} - \frac{1}{r_y}\right) = \varepsilon_{y0} - z\chi_y \tag{13.2b}$$

The nomenclature parallels that used in connection with ε_x.

The distribution of shear strain γ_{xy} is next evaluated. Let γ_{xy0} denote the shearing strain of the midsurface. Due to the rotation of edge AB relative to Oz about the x axis (Figure 13.1a) and γ_{xy0} and referring to the third of Equations 3.3a for plates, we have

$$\gamma_{xy} = \gamma_{xy0} - 2z\chi_{xy} \tag{13.2c}$$

Here χ_{xy} designates the *twist* of the midsurface. Clearly, it represents the effect of the rotation of the shell elements about a normal to the midsurface. Equations 13.2 indicate that the *strain* in a shell *varies linearly* across the thickness t.

On substitution of Equations 13.2 into a, we obtain

$$\begin{aligned}
\sigma_x &= \frac{E}{1-\nu^2}[\varepsilon_{x0} + \nu\varepsilon_{y0} - z(\chi_x + \nu\chi_y)] \\
\sigma_y &= \frac{E}{1-\nu^2}[\varepsilon_{y0} + \nu\varepsilon_{x0} - z(\chi_y + \nu\chi_x)] \\
\tau_{xy} &= (\gamma_{xy0} - 2z\chi_{xy})G
\end{aligned} \tag{13.3}$$

Finally, when Equation 13.3 is introduced into Equation 13.1, neglecting terms z/r_x and z/r_y as before, the *stress resultants* become

$$
\begin{aligned}
N_x &= \frac{Et}{1-\nu^2}(\varepsilon_{x0}+\nu\varepsilon_{y0}) & N_y &= \frac{Et}{1-\nu^2}(\varepsilon_{y0}+\nu\varepsilon_{x0}) \\
M_x &= -D(\chi_x+\nu\chi_y) & M_y &= -D(\chi_y+\nu\chi_x) \\
N_{xy} &= N_{yx} = \frac{\gamma_{xy0}Et}{2(1+\nu)} & M_{xy} &= M_{yx} = -D(1-\nu)\chi_{xy}
\end{aligned}
\tag{13.4}
$$

In the preceding, $D = Et^3/12(1 - \nu^2)$ defines the *flexural rigidity* of the shell, the same as for a plate. Equations 13.4 are the *constitutive equations* for shells.

Comment: Should the actual conditions be such as to permit bending to be neglected, the analysis of stress is vastly simplified, as M_x, M_y, and $M_{xy} = M_{yx}$ now vanish. What remains are the membrane forces N_x, N_y, and $N_{xy} = N_{yx}$.

13.4 Compound Stresses in a Shell

We are now in a position to express the *compound stresses* in a shell produced by the forces and moments. For this purpose, we substitute the strains and deformations of Equations 13.4 into Equations 13.3 with the result that

$$
\begin{aligned}
\sigma_x &= \frac{N_x}{t}+\frac{12M_xz}{t^3} & \sigma_y &= \frac{N_y}{t}+\frac{12M_yz}{t^3} \\
\tau_{xy} &= \frac{N_{xy}}{t}+\frac{12M_{xy}z}{t^3}
\end{aligned}
\tag{13.5}
$$

The first terms above clearly describe *membrane stress*, and the second terms describe *bending stress*. We observe that distribution of the stress components σ_x, σ_y, and τ_{xy} within the shell is *linear*.

It can be verified, as for plates or beams, that the vertical *shear stresses* are governed by a *parabolic* distribution:

$$
\tau_{xz} = \frac{3Q_x}{2t}\left(1-\frac{4z^2}{t^2}\right) \qquad \tau_{yz} = \frac{3Q_y}{2t}\left(1-\frac{4z^2}{t^2}\right)
\tag{13.6}
$$

Their values, as in the case of plates, are small in comparison with the other plane-stress components.

The fundamental stress relationships are thus *identical* for beams, plates, and shells. In Chapter 12, methods for determining membrane forces were discussed for shells of various shapes. Cases involving bending moments are treated in the sections that follow and in Chapters 14 and 15. Knowing the stress resultants, we can readily compute the stress at any point within a shell through the application of Equations 13.5 and 13.6.

13.5 Strain Energy in the Bending and Stretching of Shells

Equations 3.42 through 3.44, on application of the appropriate stresses and strains, lead to a strain-energy expression for the shells. As in the bending of plates, we assume that the transverse shearing strains (γ_{xz}, γ_{yz}) and the normal stress (σ_z) vanish. The *components* of strain energy of a deformed shell are the *bending-strain energy* U_b and the *membrane-strain energy* U_m. That is,

$$U = U_b + U_m \tag{a}$$

The bending-strain energy, on replacing the curvatures κ_x, κ_y, κ_{xy} in Equation 3.44 by the changes in curvature χ_x, χ_y, χ_{xy}, is found to be

$$U_b = \frac{1}{2}D\iint_A \left[(\chi_x + \chi_y)^2 - 2(1-\nu)\left(\chi_x\chi_y - \chi_{xy}^2\right)\right]dx\,dy \tag{13.7}$$

where A represents the surface area of the shell.

The membrane energy is associated with midsurface stretching produced by the in-plane forces and is given by

$$U_m = \frac{1}{2}\iint_A (N_x\varepsilon_{x0} + N_y\varepsilon_{y0} + N_{xy}\gamma_{xy0})dx\,dy \tag{13.8a}$$

Introduction of Equations 13.4 into the above expression leads to the following form involving the strains and elastic constants:

$$U_m = \frac{Et}{2(1-\nu^2)}\iint_A \left[(\varepsilon_{x0} + \varepsilon_{y0})^2 - 2(1-\nu)\left(\varepsilon_{x0}\varepsilon_{y0} - \frac{1}{4}\gamma_{xy0}^2\right)\right]dx\,dy \tag{13.8b}$$

Equations 13.7 and 13.8 permit the energy to be evaluated readily for a number of commonly encountered shells of regular shape and regular loading. The strain energy plays an important role in treating the bending and buckling problems of shells (Chapter 15).

13.6 Axisymmetrically Loaded Circular Cylindrical Shells

Pipes, tanks, boilers, and various other vessels under internal pressure exemplify the *axisymmetrically loaded cylindrical shell*. Because of symmetry, an element cut from a cylinder or radius a will have only the stress resultants show in Figure 13.3: N_θ, M_θ, N_x, and Q_x acting on it N_θ, M_θ, N_x. Furthermore, the circumferential force and moment, N_θ and M_θ, do not vary with θ. The circumferential displacement v thus vanishes, and we need consider only the x and z displacements, u and w.

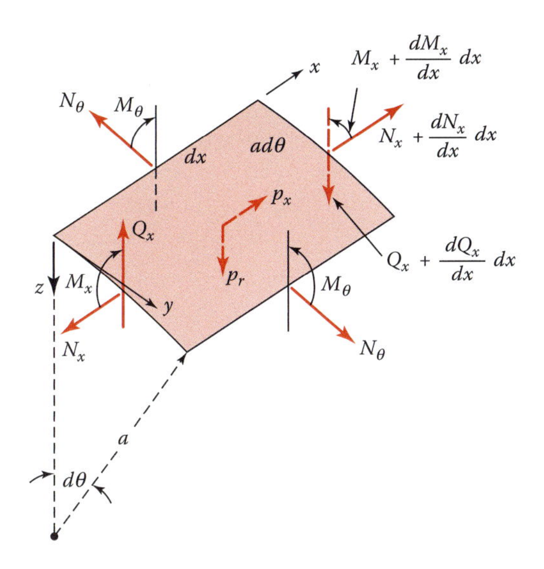

FIGURE 13.3
Stress resultants on element of axisymmetrically loaded circular cylindrical shell.

Subject to the foregoing simplifications, only three of the six equilibrium equations of the shell element remain to be satisfied. Suppose also that the external loading is as shown in Figure 13.3. *Equilibrium of forces* in the x (axial) and z (radial) directions now requires that

$$\begin{aligned} &\frac{dN_x}{dx} dx\, a\, d\theta + px\, a\, d\theta\, dx = 0 \\ &\frac{dQ_x}{dx} dx\, a\, d\theta + N_\theta\, dx\, d\theta + p_r\, dx\, a\, d\theta = 0 \end{aligned} \tag{a}$$

Equilibrium of moments about the y axis is governed by

$$\frac{dN_x}{dx} dx\, a\, d\theta - Q_x\, a\, d\theta\, dx = 0 \tag{b}$$

Equations a and b are, through cancellation of like terms, rewritten as

$$\begin{aligned} &\frac{dN_x}{dx} + p_x = 0 \\ &\frac{dQ_x}{dx} + \frac{1}{a} N_\theta + p_r = 0 \\ &\frac{dM_x}{dx} - Q_x = 0 \end{aligned} \tag{13.9a–c}$$

It is interesting to note that Equation 13.9c is a statement of the basic beam relationship: The shearing force is the first derivative of the bending moment. From Equation 13.9a, the axial force N_x is

$$N_x = -\int p_x\, dx + c \tag{c}$$

where c is a constant of integration. Clearly, the unknown quantities Q_x, N_θ, and M_x cannot be determined from Equations 13.9b and 13.9c, alone, and it is therefore necessary to examine the midsurface displacements.

Because $v = 0$, the *strain-displacement relations* are, from symmetry,

$$\varepsilon_x = \frac{du}{dx} \qquad \varepsilon_\theta = \frac{(a-w)\,d\theta - a\,d\theta}{a\,d\theta} = -\frac{w}{a} \tag{d}$$

Applying Hooke's law, we have

$$N_x = \frac{Et}{1-\nu^2}(\varepsilon_x + \nu\varepsilon_\theta) = \frac{Et}{1-\nu^2}\left(\frac{du}{dx} - \nu\frac{w}{a}\right)$$

from which

$$\frac{du}{dx} = \frac{1-\nu^2}{Et} \qquad N_x + \nu\frac{w}{a} \tag{13.10a}$$

Then, from Hooke's law and Equations d, the circumferential force is found to be

$$N_\theta = \frac{Et}{1-\nu^2}(\varepsilon_\theta + \nu\varepsilon_x) = -\frac{Et}{1-\nu^2}\left(\frac{w}{a} - \nu\frac{du}{dx}\right) \tag{13.11}$$

The *bending moment displacement relations* are the same as for a plate bent into a cylindrical surface. That is, because $d^2w/dy^2 = 0$,

$$M_x = -D\frac{d^2w}{dx^2} \qquad M_\theta = \nu M_x \tag{13.12}$$

where D is the *flexural rigidity* of the shell, given by Equation 3.10. Employing Equations 13.9b and 13.9c and eliminating Q_x, the following is obtained:

$$\frac{d^2M_x}{dx^2} + \frac{1}{a}N_\theta + p_r = 0$$

Finally, when the preceding expression is combined with Equations 13.10a, 13.11, and 13.12, we have

$$\frac{d^2}{dx^2}\left(D\frac{d^2w}{dx^2}\right) + \frac{Et}{a^2}w - \frac{\nu N_x}{a} - p_r = 0 \tag{13.13a}$$

For a shell of *constant thickness*, Equation 13.13a becomes

$$D\frac{d^4w}{dx^4} + \frac{Et}{a^2}w - \frac{\nu N_x}{a} - p_r = 0 \tag{13.13b}$$

A more convenient form of this expression is

$$\frac{d^4w}{dx^4} + 4\beta^4 w - \frac{\nu N_x}{aD} = \frac{p_r}{D} \tag{13.10b}$$

In the preceding, we have

$$\beta^4 = \frac{Et}{4a^2D} = \frac{3(1-\nu^2)}{a^2t^2} \tag{13.14}$$

where *geometric parameter* β has the dimension of L^{-1}, the reciprocal of length. Equations 13.10b or 13.13a represent the *governing displacement conditions* for a symmetrically loaded circular cylindrical shell. When an *axial load does not exist*, $N_x = 0$, and Equations 13.10 simplify to

$$\begin{aligned} \frac{du}{dx} &= \nu \frac{w}{a} \\ \frac{d^4w}{dx^4} + 4\beta^4 w &= \frac{p_r}{D} \end{aligned} \tag{13.15a, b}$$

The first of these, on integration, directly yields u. Equation 13.15b is an ordinary differential equation with constant coefficients. It also represents [13] the equation of a beam of flexural rigidity D, resting on an elastic foundation and subject to loading p_r. The homogeneous solution of Equation 13.15b is given by

$$w_h = c_1 e^{m_1 x} + c_2 e^{m_2 x} + c_3 e^{m_3 x} + c_4 e^{m_4 x}$$

Here c_1, c_2, c_3, and c_4 are constants, and m_1, m_2, m_3, and m_4 are the roots of the expression

$$m^4 + 4\beta^4 = 0$$

This equation may be written, by addition and subtraction of $4m^2\beta^2$, as $(m^2 + 2\beta^2)^2 - 4m^2\beta^2 = 0$. Hence, $m^2 + 2\beta^2 = \pm m\beta$. We thus have

$$m = \pm\beta(1 \pm i)$$

It follows that

$$w_h = e^{-\beta x}\left(c_1 e^{i\beta x} + c_2 e^{-i\beta x}\right) + e^{\beta x}\left(c_3 e^{i\beta x} + c_4 e^{-i\beta x}\right)$$

Let $f(x)$ represent the particular solution w_p. It is noted that the results of membrane theorycan always be considered as the particular solutions of the equations of bending theory (Section 14.4). The general solution of Equation 13.15b may therefore be written as

$$w = e^{-\beta x}(C_1 \cos \beta x + C_2 \sin \beta x) + e^{\beta x}(C_3 \cos \beta x + C_4 \sin \beta x) + f(x) \tag{13.16}$$

where C_1, C_2, C_3, and C_4 are arbitrary constants of integration, determined on the basis of the appropriate boundary conditions. The next section serves to illustrate application of the theory.

13.7 A Typical Case of the Axisymmetrically Loaded Cylindrical Shell

This section deals with the bending problem of a cylinder with length very large compared with its diameter, the so-called *infinite cylinder*, subjected to a load p uniformly distributed along a circular section (Figure 13.4). Inasmuch as there is no pressure p_r distributed over the surface of the shell and $N_x = 0$, we set $f(x) = 0$ in Equation 13.16. The solution may be written as follows:

$$w = e^{-\beta x}(C_1 \cos \beta x + C_2 \sin \beta x) + e^{\beta x}(C_3 \cos \beta x + C_4 \sin \beta x) \tag{13.17}$$

Because of shell symmetry, boundary conditions for the right half are deduced from the fact that as $x \to \infty$, the deflection and all derivatives of w with respect to x must vanish. The conditions are fulfilled if $C_3 = C_4 = 0$ in Equation 13.17. We thus have

$$w = e^{-\beta x}(C_1 \cos \beta x + C_2 \sin \beta x) \tag{13.18}$$

In as much as $N_x = 0$, Equations 13.10a and 13.11 yield

$$N_\theta = -\frac{Etw}{a} \tag{13.19}$$

From Equations 13.12 and 13.9c,

$$M_x = -D\frac{d^2w}{dx^2} \qquad M_\theta = -\nu D\frac{d^2w}{dx^2} \qquad Q_x = \frac{dM_x}{dx} = -D\frac{d^3w}{dx^3} \tag{13.20}$$

The conditions applicable immediately to the right of the load are

$$Q_x = -D\frac{d^3w}{dx^3} = -\frac{P}{2} \qquad \frac{dw}{dx} = 0 \tag{a}$$

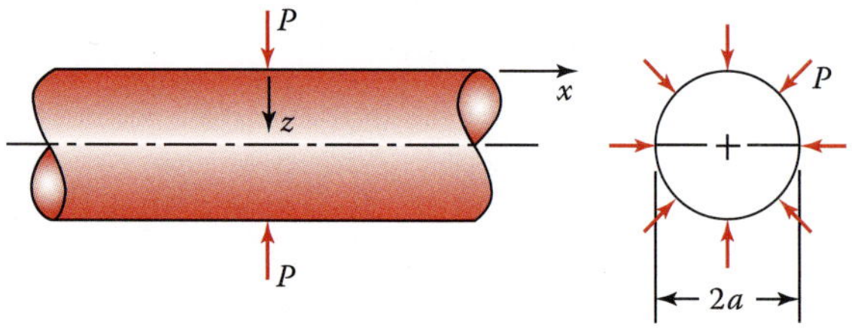

FIGURE 13.4
Long (infinite) cylinder carrying a uniform line load P.

These describe the respective requirements that each half of the cylinder carries one-half the external load and that the slope vanishes at the center due to the symmetry. Introducing Equations a into Equation 13.18 and setting $x = 0$,

$$C_1 = C_2 = \frac{P}{8\beta^3 D}$$

The *displacement* is therefore

$$w = \frac{Pe^{-\beta x}}{8\beta^3 D}(\sin\beta x + \cos\beta x) \tag{13.21}$$

This result may be expressed as

$$w = \frac{Pe^{-\beta x}}{8\beta^3 D}\left[\sqrt{2}\sin\left(\beta x + \frac{\pi}{4}\right)\right]$$

We observe that the deflection attenuates with distance as an exponentially damped sine wave of wavelength, for $\nu = 0.3, 2\pi/\beta \approx 4.89\sqrt{at}$.

The following *notations* are used to more conveniently represent the expressions for deflection and stress resultants:

$$\begin{aligned}
f_1(\beta x) &= e^{-\beta x}(\cos\beta x + \sin\beta x) \\
f_2(\beta x) &= e^{-\beta x}\sin\beta x = -\frac{1}{2\beta}f_1' \\
f_3(\beta x) &= e^{-\beta x}(\cos\beta x - \sin\beta x) = \frac{1}{\beta}f_2' = -\frac{1}{2\beta^2}f_1'' \\
f_4(\beta x) &= e^{-\beta x}\cos\beta x = -\frac{1}{2\beta}f_3' = -\frac{1}{2\beta^2}f_2'' = \frac{1}{4\beta^3}f_1''' \\
f_1(\beta x) &= -\frac{1}{\beta}f_4'
\end{aligned} \tag{13.22}$$

Table 13.1 furnishes numerical values of these functions for various values of βx. The term βx is *dimensionless* and is usually thought of as *expressed in radians*. Substituting Equation 13.21 into Equations 13.19 and 13.20 gives

$$\begin{aligned}
w &= \frac{P}{8\beta^3 D}f_1(\beta x) \\
N_\theta &= -\frac{EtP}{8\beta^3 Da}f_1(\beta x) \\
M_x &= \frac{P}{4\beta}f_3(\beta x) \\
M_\theta &= \frac{\nu P}{4\beta}f_3(\beta x) \\
Q_x &= -\frac{P}{2}f_4(\beta x)
\end{aligned} \tag{13.23}$$

TABLE 13.1

Selected Values of the Functions Defined by Equations 13.22

βx	$f_1(\beta x)$	$f_2(\beta x)$	$f_3(\beta x)$	$f_4(\beta x)$	βx	$f_1(\beta x)$	$f_2(\beta x)$	$f_3(\beta x)$	$f_4(\beta x)$
0.0	1.000	0.000	1.000	1.000	3.0	−0.042	0.007	−0.056	−0.049
0.2	0.965	0.163	0.640	0.802	3.2	−0.043	−0.002	−0.038	−0.041
0.4	0.878	0.261	0.356	0.617	3.4	−0.041	−0.009	−0.024	−0.032
0.6	0.763	0.310	0.143	0.453	3.6	−0.037	−0.012	−0.012	−0.024
0.8	0.635	0.322	−0.009	0.313	3.8	−0.031	−0.014	−0.004	−0.018
1.0	0.508	0.310	−0.111	0.199	4.0	−0.026	−0.014	0.002	−0.012
1.2	0.390	0.281	−0.172	0.109	4.2	−0.020	−0.013	−0.006	−0.007
1.4	0.285	0.243	−0.201	0.042	4.4	−0.016	−0.012	0.008	−0.004
1.6	0.196	0.202	−0.208	−0.006	4.6	−0.011	−0.010	0.009	−0.001
1.8	0.123	0.161	−0.199	−0.038	4.8	−0.008	−0.008	0.009	0.001
2.0	0.067	0.123	−0.179	−0.056	5.0	−0.005	−0.007	0.008	0.002
2.2	0.024	0.090	−0.155	−0.065	5.5	0.000	−0.003	0.006	0.003
2.4	−0.006	0.061	−0.128	−0.067	6.0	0.002	−0.001	0.003	0.002
2.6	−0.025	0.038	−0.102	−0.064	6.5	0.002	0.000	0.001	0.001
2.8	−0.037	0.020	−0.078	−0.057	7.0	0.001	0.001	0.000	0.001

These expressions are valid for $x \geq 0$. For the left half of the cylinder, we take x in the opposite direction to that shown in Figure 13.4. The maximum deflection and moment occur at $x = 0$, found from Equations 13.23 to be

$$w_{\max} = \frac{P}{8\beta^3 D} = \frac{Pa^2\beta}{2Et} \qquad M_{\max} = \frac{P}{4\beta} \tag{13.24}$$

The largest values of bending stress are found at $x = 0$ and $z = t/2$, determined by applying Equations 13.4 and 13.23:

$$\sigma_{x,\max} = \frac{3P}{2\beta t^2} \qquad \sigma_{\theta,\max} = \frac{P\beta}{2}\left(-\frac{a}{t} + \frac{3\nu}{\beta^2 t^2}\right) \tag{b}$$

The foregoing are the *maximum axial* and *circumferential stresses* in the cylinder, respectively.

Referring to Table 13.1, it is observed that each quantity in Equations 13.22 and hence in Equations 13.23 decreases with increasing βx. Because of this, in most engineering applications, the *effect of the concentrated loads* may be *neglected*at locations for which $x > (\pi/\beta)$. Therefore, it is concluded that bending is of a local character. A shell of length $L = 2\pi/\beta$, loaded at midlength, will experience maximum deflection and bending moment nearly identical with those associated with a *long shell.*

Comment: Application of Equations 13.23 together with the principle of superposition permits the determination of deflection and stress resultants in long cylinders under any other kind of loading.

EXAMPLE 13.1: Deflection of Cylinder under Line Loading

A very long cylinder of radius a is subjected to a uniform loading p over L of its length (Figure 13.5). Derive an expression for the deflection at an arbitrary point O within length L.

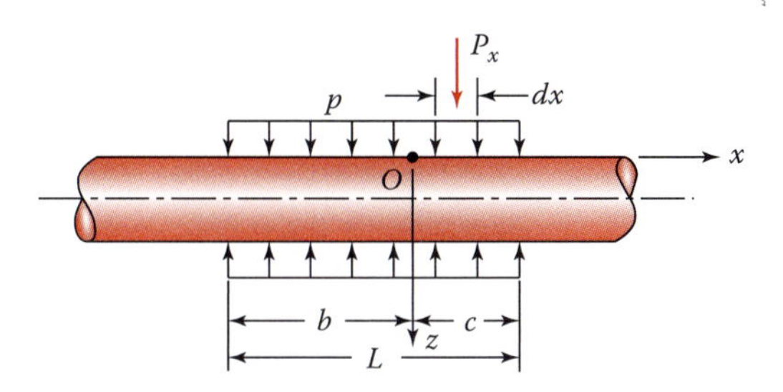

FIGURE 13.5
Cylinder with partial loading.

Solution

Through the use of the first of Equations 13.23, the deflection Δw at point O due to load $P_x = p\,dx$ is

$$\Delta w = \frac{pdx}{8\beta^3 D} f_1(\beta x)$$

The displacement at point O produced by the entire load is then

$$w = \int_0^b \frac{p}{8\beta^3 D} f_1(\beta x)dx + \int_0^c \frac{p}{8\beta^3 D} f_1(\beta x)dx$$

Inserting into the above $f_1(\beta x)$ from Equations 11.22, we obtain after integration

$$w = \frac{pa^2}{2Et}[2 - e^{-\beta b}\cos(\beta b) - e^{-\beta c}\cos(\beta c)]$$

or

$$w = \frac{pa^2}{2Et}[2 - f_4(\beta b) - f_4(\beta c)] \tag{c}$$

The maximum deflection of the cylinder occurs at midlength of the distributed load at the point at which $b = c$. Note that *if b and c are large*, the values of $f_4(\beta b)$ and $f_4(\beta c)$ are quite small, and the deflection will approximately equal pa^2/Et.

Comment: In a like manner, applying the last four of Equations 13.23, we can obtain the expressions for the stress resultants at O.

13.8 Shells of Revolution under Axisymmetrical Loads

Let us consider a body in the *general form of a shell of revolution* subjected to rotationally symmetrical loads. The sphere, cone, and circular cylinder (Section 13.6) are typical simple geometries in this category. To begin with, we define the state of stress resultant acting at a point of such shells, represented by an infinitesimal element in Figure 13.6. Conditions of symmetry dictate that only the resultants Q_ϕ, M_θ, M_ϕ, N_θ, and N_ϕ exist and that the

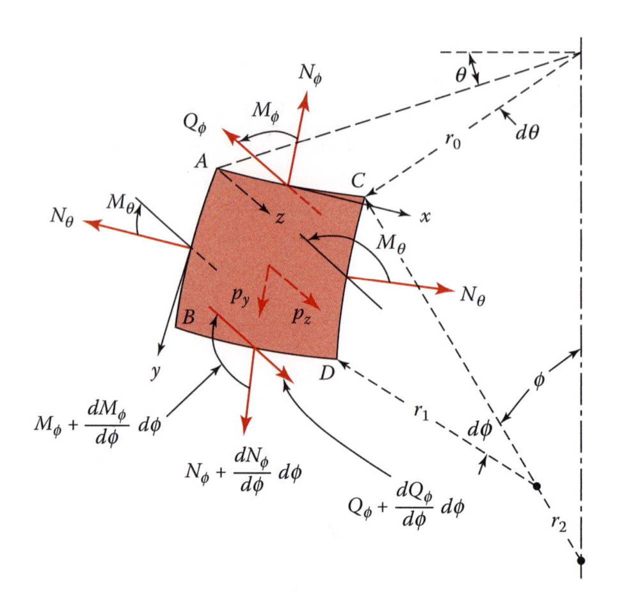

FIGURE 13.6
Stress resultants on axisymmetrically loaded shell of revolution.

normal forces N_θ and the bending moments M_θ cannot vary with θ. The notations for the radii of curvature and the angular orientation are identical with those of membrane theory (Figure 12.2).

The development of the equilibrium equations of shell element *ABCD* proceeds in a manner similar to that described in Section 12.4. The condition that *summation of the y-directed forces* be equal to zero is fulfilled by

$$\frac{d}{d\phi}(N_\phi r_0 d\theta)d\phi - N_\theta r_1\, d\theta\, d\phi \cos\phi - Q_\phi\, r_0\, d\theta\, d\phi + p_y r_1\, d\phi\, r_0 d\theta = 0 \tag{a}$$

The first two and the last terms of Equation a are already specified by Equation 12.2. The third term is due to the shear forces $Q_\phi\, r_0\, d_\theta$ on faces *AC* and *BD* of the element. These faces form an angle d_ϕ with one another.

The condition of *equilibrium in the z direction* may readily be obtained by adding the increment of the shear forces $Q_\phi\, r_0\, d\theta$ to Equation a of Section 12.5. That is,

$$N_\phi r_0 d\theta\, d\phi + N_\theta r_1 d\phi\, d\theta \sin\phi + \frac{d}{d\phi}(Q_\phi r_0\, d\theta)d\phi + p_z r_1\, d\phi r_0\, d\theta = 0 \tag{b}$$

Finally, we write the expression

$$\frac{d}{d\phi}(M_\phi\, r_0 d\theta)d\phi - Q_\phi\, r_0 d\theta\, r_1 d\phi - M_\theta\, r_1 d\phi \cos\phi\, d\theta = 0 \tag{c}$$

for the *equilibrium of the forces around the x axis.* The terms of Equation c are described as follows: The first is the increment of the bending moments $M_\phi\, r_0\, d\theta$; the second represents the moment of the shear forces $Q_\phi\, r_0\, d\theta$; and the third is the resultant of the moments $M_\theta\, r_1\, d\phi$. Note that the two moment *vectors* $M_\theta\, r_1\, d\phi$ acting on the side faces AB and CD of the element are not parallel. Their horizontal components $M_\theta\, r_1\, d\phi \cos\phi$ form an angle $d\theta$ with one another and thus have the resultant expressed by the last terms.

Dropping the factor $d\theta d\phi$ in Equations a through c, common to all terms, we obtain the *equation of equilibrium for axisymmetrical shells in bending*:

$$\begin{aligned}
&\frac{d}{d\phi}(N_\phi r_0) - N_\theta r_1 \cos\phi - r_0 Q_\phi + r_0 r_1 p_y = 0 \\
&N_\phi r_0 + N_\theta r_1 \sin\phi + \frac{d(Q_\phi r_0)}{d\phi} + p_z r_1 r_0 = 0 \\
&\frac{d}{d\phi}(M_\phi r_0) - M_\phi r_1 \cos\phi - Q_\phi r_1 r_0 = 0
\end{aligned} \tag{13.25}$$

The governing equations for the common shells of revolution subjected to axisymmetrical loads may be derived from the foregoing expressions.

13.8.1 Conical Shells

For this case (Figure 12.4a), it is observed in Section 12.5 that ϕ = constant ($r_1 = \infty$). Thus,

$$r_2 = s\cot\phi \qquad r_1\, d\phi = ds \qquad N_\phi = N_s \qquad M_\phi = M_s \qquad Q_\phi = Q_s$$

Employing these, the equations of equilibrium (Equations 13.25) assume the form

$$\begin{aligned}
&\frac{d}{ds}(N_s s) - N_\theta = -p_y s \\
&N_\theta + \frac{d}{ds}(Q_s s)\cot\phi = -p_z s\cot\phi \\
&\frac{d}{ds}(M_s s) - Q_s s + M_\theta = 0
\end{aligned} \tag{13.26}$$

13.8.2 Spherical Shells

Denoting the radius of the midsurface of the shellby a, we have $r_1 = r_2 = a$ and $r_0 = a \sin\phi$. The equilibrium conditions (Equations 13.25) then simplify to

$$\begin{aligned}
&\frac{d}{d\phi}(N_\phi \sin\phi) - N_\theta \cos\phi - Q_\phi \sin\phi = -p_y a \sin\phi \\
&N_\phi \sin\phi + N_\theta \sin\phi + \frac{d}{d\phi}(Q_\phi \sin\phi) = -p_z a \sin\phi \\
&\frac{d}{d\phi}(M_\phi \sin\phi) - M_\theta \cos\phi - a Q_\phi \sin\phi = 0
\end{aligned} \tag{13.27}$$

13.8.3 Cylindrical Shells

In this case (Figure 13.3), we can use Equation 13.26 for the cone, letting $s = x = r_2 \tan \phi$, $\phi = \pi/2$, and $r_2 = a$. By so doing, we obtain expressions that are identical with Equations 13.9 of Section 13.6.

Interestingly, the first expression of Equations 13.26 agrees with the corresponding expression of membrane theory. It is noted that by canceling the terms involving shear forces and moments, we reduce Equations 13.26, 13.27, and 13.9 to the conditions of membrane theory of conical, spherical, and cylindrical shells, respectively.

13.9 Governing Equations for Axisymmetrical Displacements

In the preceding section, it is observed that three equilibrium conditions (Equations 13.25) of an *axisymmetrically loaded shell of revolution* contain five unknown stress resultants: N_ϕ, N_θ, Q_ϕ, M_ϕ, and M_θ. To reduce the number of unknowns to three, relationships involving the forces (N_ϕ, N_θ) the moments (M_ϕ, M_θ) and the displacement components (v, w) are developed in the paragraphs that follow.

The membrane strains and the displacements at a point of the midsurface are connected by Equations 12.25:

$$\varepsilon_\phi = \frac{1}{r_1}\frac{dv}{d\phi} - \frac{w}{r_1} \qquad \varepsilon_\theta = \frac{v}{r_2}\cot\phi - \frac{w}{r_2} \tag{13.28}$$

The force–resultant strain relations (Equations 13.4) then lead to

$$\begin{aligned} N_\phi &= \frac{Et}{1-\nu^2}\left[\frac{1}{r_1}\left(\frac{dv}{d\phi} - w\right) + \frac{\nu}{r_2}(v\cot\phi - w)\right] \\ N_\theta &= \frac{Et}{1-\nu^2}\left[\frac{1}{r_2}(v\cot\phi - w) + \frac{\nu}{r_1}\left(\frac{dv}{d\phi} - w\right)\right] \end{aligned} \tag{13.29}$$

Identical expressions for M_ϕ and M_θ can be obtained by considering the variations in curvature of a shell element (Figure 13.6). For this purpose, we examine the meridional section of the shell element (Figure 13.7). The rotations of the tangent of the top face AC consist of a rotation with respect to a perpendicular to the meridian plane by an amount v/r_1 due to the displacement v of point A to point A' (Figure 13.7a) and a rotation about the same axis by $dw/(r_1\, d\phi)$ produced by the additional displacement of point B with respect to point A (Figure 13.7b). Denoting the *total rotation of the upper edge AC* by V, we thus have

$$V = \frac{v}{r_1} + \frac{dw}{r_1\, d\phi} \tag{13.30}$$

The top and bottom faces of the element initially make an angle $d\theta$ with one another. The *rotation of the bottom face BD* is therefore

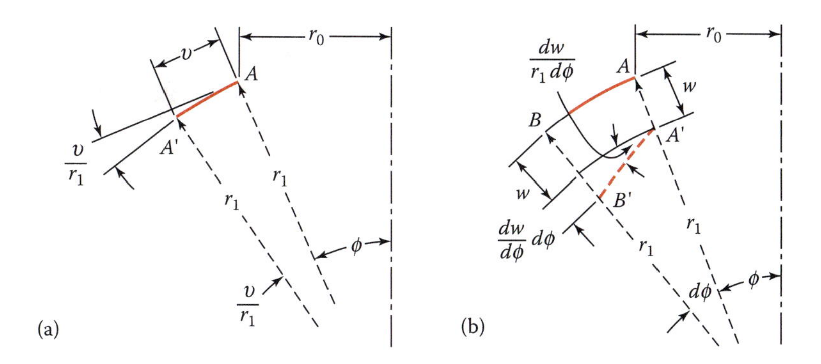

FIGURE 13.7
Variation in curvature of element of axisymmetrically loaded shell.

$$\frac{v}{r_1}+\frac{dw}{r_1\,d\phi}+\frac{d}{d\phi}\left(\frac{v}{r_1}+\frac{dw}{r_1\,d\phi}\right)d\phi$$

The variation of curvature of the meridian, the angular variation divided by the length r_1 $d\phi$ of the arc, is thus

$$\chi_\phi=\frac{1}{r_1}\frac{d}{d\phi}\left(\frac{v}{r_1}+\frac{dw}{r_1\,d\phi}\right) \tag{13.31a}$$

It is observed that, due to the symmetry of deformation, each of the lateral edges *AB* and *CD* of the shell element also rotates in its meridian plane by an angle defined by Equation 13.30. It may be verified that the *unit normal* to the right face of a shell element has a y-directed component equal to $-d\phi$ cosϕ. Thus, the rotation of face *CD* in its own plane has a component with respect to the y axis given by

$$-\left(\frac{v}{r_1}+\frac{dw}{r_1\,d\phi}\right)\cos\phi\,d\theta$$

Dividing this rotation by length r_0 $d\theta$, we have the change of curvature:

$$\chi_\theta=\left(\frac{v}{r_1}+\frac{dw}{r_1\,d\phi}\right)\frac{\cos\phi}{r_0}=\left(\frac{v}{r_1}+\frac{dw}{r_1\,d\phi}\right)\frac{\cot\phi}{r_2} \tag{13.31b}$$

Finally, inserting Equations 13.31 into Equations 13.4, we obtain the *moment–displacement relations*:

$$\begin{aligned}M_\phi&=-D\left[\frac{1}{r_1}\frac{d}{d\phi}\left(\frac{v}{r_1}+\frac{dw}{r_1 d\phi}\right)+\frac{\nu}{r_2}\left(\frac{v}{r_1}+\frac{dw}{r_1 d\phi}\right)\cot\phi\right]\\M_\theta&=-D\left[\left(\frac{v}{r_1}+\frac{dw}{r_1 d\phi}\right)\frac{\cot\phi}{r_2}+\frac{\nu}{r_1}\frac{d}{d\phi}\left(\frac{v}{r_1}+\frac{dw}{r_1 d\phi}\right)\right]\end{aligned} \tag{13.32}$$

Now Equations 13.25 together with Equations 13.29 and 13.32 lead to three expressions in three unknowns: υ, w, and Q_ϕ. Furthermore, by using the first of the resulting three equations, the shear force Q_ϕ can readily be eliminated in the last two expressions. The expressions (Equations 13.25) are thus reduced to two equations in two unknowns: υ and w. These *governing equations for displacements*, usually transformed into *new variables* [1] are employed to treat the shell-bending problem.

Applications of the equations derived in the preceding paragraphs are presented in the sections to follow.

13.10 Spherical Shells under Axisymmetrical Load

Consideration is now given to the bending of *spherical shells of constant thickness* subjected to a uniform normal pressure p (Figure 13.8).

When the radius of the midsurface is a, we have $r_1 = r_2 = a$ and $r_0 = a \sin \phi$.

The components of the loading are as follows:

$$p_y = 0 \qquad p_z = p \tag{a}$$

The *conditions of equilibrium* (Equations 13.27) are then

$$\begin{aligned} &\frac{d}{d\phi}(N_\phi \sin\phi) - N_\theta \cos\phi - Q_\phi \sin\phi = 0 \\ &N_\phi \sin\phi + N_\theta \sin\phi + \frac{d}{d\phi}(Q_\phi \sin\phi) + pa \sin\phi = 0 \\ &\frac{d}{d\phi}(M_\phi \sin\phi) - M_\theta \cos\phi - Q_\phi a \sin\phi = 0 \end{aligned} \tag{13.33a–c}$$

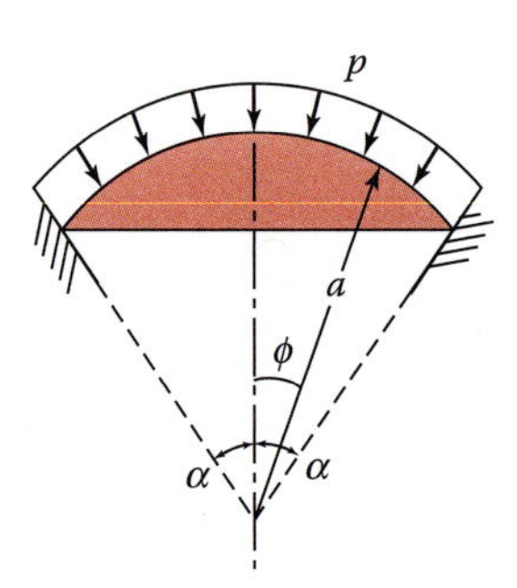

FIGURE 13.8
Spherical dome with fixed edge.

The elastic law, Equations 13.29 and 13.32, together with Equation 13.30, becomes

$$
\begin{aligned}
N_\phi &= \frac{Et}{a(1-\nu^2)}\left[\left(\frac{dv}{d\phi} - w\right) + \nu(v\cos\phi - w)\right] \\
N_\theta &= \frac{Et}{a(1-\nu^2)}\left[(v\cot\phi - w) + \nu\left(\frac{dv}{d\phi} - w\right)\right] \\
M_\phi &= -\frac{D}{a}\left[\frac{dV}{d\phi} + \nu V\cot\phi\right] \\
M_\theta &= -\frac{D}{a}\left[V\cot\phi + \nu\frac{dV}{d\phi}\right]
\end{aligned}
\tag{13.34a–d}
$$

The *foregoing expressions* may be *reduced* to a pair of equations for Q_ϕ and V as follows. Substituting M_ϕ and M_θ from Equations 13.34 into Equation 13.33c and dividing the resulting expression by sin ϕ yields the first equation:

$$\frac{d^2V}{d\phi^2} + \frac{dV}{d\phi}\cot\phi - V(\cot^2\phi + \nu) = -\frac{a^2Q_\phi}{D} \tag{13.35}$$

A second equation necessarily contains Equations 13.33a, 13.33b, 13.34a, and 13.34b. Employing the former two and eliminating Q_ϕ, we obtain an expression that may be written as

$$\frac{d}{d\phi}\left(N_\phi\sin^2\phi\right) + \frac{d}{d\phi}(Q_\phi\sin\phi\cos\phi) + \frac{d}{d\phi}\left(\frac{1}{2}pa\sin^2\phi\right) = 0$$

Then, meridional force N_ϕ is determined through integrating the preceding equation:

$$N_\phi = -Q_\phi\cot\phi - \frac{1}{2}pa \tag{b}$$

Here the constant of integration is ignored. Carrying Equation b into Equation 13.33b, we have

$$N_\theta = -\frac{dQ_\phi}{d\phi} - \frac{pa}{2} \tag{c}$$

The meridional force N_ϕ and hoop force N_θ are thus represented in terms of the shearing force Q_ϕ. Next, from Equations 13.34a and 13.34b, we readily obtain

$$\frac{dv}{d\phi} - w = \frac{a}{Et}(N_\phi - \nu N_\theta) \tag{d}$$

and

$$v\cot\phi - w = \frac{a}{Et}(N_\phi - \nu N_\theta) \tag{e}$$

Eliminating w from these expressions gives

$$\frac{dv}{d\phi} - v\cot\phi = \frac{a(1+\nu)}{Et}(N_\phi - N_\theta) \tag{f}$$

Differentiation of Equation e results in

$$\frac{dv}{d\phi}\cot\phi - \frac{v}{\sin^2\phi} - \frac{dw}{d\phi} = \frac{a}{Et}\left(\frac{dN_\theta}{d\phi} - \nu\frac{dN_\phi}{d\phi}\right) \tag{g}$$

The derivative $dv/d\phi$ is eliminated by using Equations f and g to yield

$$v + \frac{dw}{d\phi} = aV = \frac{a}{Et}\left[(1+\nu)(N_\phi - N_\theta)\cot\phi - \frac{dN_\theta}{d\phi} - \nu\frac{dN_\phi}{d\phi}\right]$$

Finally, substituting Equations b and c into the preceding, we arrive at the following equation:

$$\frac{d^2Q_\phi}{d\phi^2} + \cot\phi\frac{dQ_\phi}{d\phi} - Q_\phi(\cot^2\phi - \nu) = EtV \tag{13.36}$$

Comments: Thus, the *problem of bending* of a spherical shell under uniform pressure is *reduced* to the solution of Equations 13.35 and 13.36 for Q_ϕ and V. However, laborious computations are required for this elimination method. The approximate theory, to be taken up in Section 13.12, is often preferred in engineering practice.

13.11 Comparison of Bending and Membrane Stresses

The equations for the stress resultants (Equations 13.29 and 13.32) may be used to gauge the accuracy of the membrane analysis discussed in Chapter 12. To determine the bending moments, which were omitted in the membrane theory, the expressions for the displacements developed in Section 12.6 are introduced into Equations 13.32. The bending stresses are then obtained from Equations 13.5. On *comparison of the stress magnitudes* so determined with those of the membrane stresses, conclusions may be drawn with respect to the accuracy of the membrane theory, as in the following example.

EXAMPLE 13.2: Bending and Membrane Stresses in Spherical Dome

Consider the spherical dome supporting its own weight described in Example 12.2. Assume that the supports are as shown in Figure 12.8a. (a) Determine the bending stress; (b) compare the bending and membrane stresses.

Solution

a. The displacements v and w, from Equations f and 12.25b of Section 12.6 together with Equations 12.13, are

$$v = \frac{a^2 p(1+\nu)}{Et}\left(\frac{1}{1+\cos\alpha} - \frac{1}{1+\cos\phi} + \ln\frac{1+\cos\phi}{1+\cos\alpha}\right)\sin\phi$$
$$w = v\cot\phi - \frac{a^2 p}{Et}\left(\frac{1+\nu}{1+\cos\phi} - \cos\phi\right) \tag{a}$$

Introduction of the above into Equations 13.32 yields the following expressions for the moments:

$$M_\theta = M_\phi = \frac{pt^2}{12}\frac{2+\nu}{1-\nu}\cos\phi \tag{b}$$

The magnitude of the bending stress σ_b at the surface of the shell is then

$$\sigma_b = \frac{p}{2}\frac{2+\nu}{1-\nu}\cos\phi \tag{c}$$

b. The value of membrane stress σ from Equations 12.13 is

$$\sigma = \frac{ap}{t(1+\cos\phi)}$$

The ratio of the bending stress to the membrane stress,

$$\frac{\sigma_b}{\sigma} = \frac{2+\nu}{2(1-\nu)}\frac{t}{a}(1+\cos\phi)\cos\phi \tag{d}$$

has a maximum value at the top of the shell ($\phi = 0$). For $\nu = 0.3$,

$$\left(\frac{\sigma_b}{\sigma}\right)_{\max} = 3.29\frac{t}{a} \tag{e}$$

For a thin shell ($a > 20t$), the above ratio is small. It is thus seen that membrane theory provides a result of sufficient accuracy.

The values of the forces N_θ and $N\phi$ may be determined by introducing Equation b into Equation 13.27. However, it can be verified that (see Problem 13.13) these closer approximations for the forces will differ little from the results of Equations 12.13.

Comments: We are led to conclude from the foregoing example that, for thin shells, the stresses (and the displacements) as ascertained from the *membrane theory* have values of *acceptable accuracy*. The results will of course be inaccurate if expansion of the shell edge is prevented by a support. In the latter case, the forces exerted by the support on the shell produce bending in the vicinity of the edge (Section 13.12). The local stresses caused by concentrated forces discussed in Section 13.7 will be treated in considerable detail in Chapters 14 and 15.

13.12 *Simplified Theory of Spherical Shells under Axisymmetrical Load

Elaborate investigations show that the *dominant terms* in Equations 13.35 and 13.36 are those involving *higher derivatives* [1,2]. Based on this premise, we may exclude V and Q_ϕ (and their first derivatives) in Equations 13.35 and 13.36, respectively. Thus,

$$\begin{aligned} &\frac{d^2V}{d\phi^2}+\frac{1}{D}Q_\phi a^2=0 \\ &\frac{d^2Q_\phi}{d\phi^2}-EtV=0 \end{aligned} \tag{13.37a, b}$$

From the preceding, eliminating V, we obtain the *simple governing equation*:

$$\frac{d^4Q_\phi}{d\phi^4}+4\lambda^4 Q_\phi=0 \tag{13.38}$$

where

$$\lambda^4=\frac{Eta^2}{4D}=3(1-\nu^2)\left(\frac{a}{t}\right)^2 \tag{13.39}$$

Note that the form of the left-hand side of Equation 13.38 is the same as that of Equation 13.15b, which is used in the treatment of the axisymmetrically loaded circular cylindrical shells. The general solution of the foregoing is

$$Q_\phi=e^{\lambda\phi}(C_1\cos\lambda\phi+C_2\sin\lambda\phi)+e^{-\lambda\phi}(C_3\cos\lambda\phi+C_4\sin\lambda\phi) \tag{a}$$

in which C_1 through C_4 are arbitrary constants.

Recall that the bending of cylindrical shells is of local character (Section 13.7). This conclusion is also valid in the case of spherical shells. The distributed edge forces must therefore damp out toward the interior of a continuous shell (Figure 13.8). If follows that we need to retain in Equation a only the first two terms, which decrease as the angle ϕ decreases. Therefore, the solution reduces to

$$Q_\phi=e^{\lambda\phi}(C_1\cos\lambda\phi+C_2\sin\lambda\phi) \tag{13.40}$$

Now we use Equations b and c of Section 13.10 to obtain

$$N_\theta=-\lambda e^{\lambda\phi}[(C_1+C_2)\cos\lambda\phi+(C_2-C_1)\sin\lambda\phi]-\frac{pa}{2} \tag{13.41a}$$

and

$$N_\phi = -e^{\lambda\phi}[C_1 \cos\lambda\phi + C_2 \sin\lambda\phi]\cot\phi - \frac{pa}{2} \tag{13.41b}$$

Note that λ, defined by Equation 13.39, for thin shells ($a/t > 20$) is large compared to unity. Hence, of the terms contained *within the brackets* in the preceding, only those *multiplied by* λ will be *significant* [5]. Neglecting the other terms within the brackets in Equation 13.41b, we thus have

$$N_\phi = -\frac{1}{2}pa \tag{13.41c}$$

Equations 13.37b and 13.40 yield the following expression for the angle of rotation:

$$V = \frac{2\lambda^2}{Et}e^{\lambda\phi}(C_2 \cos\lambda\phi - C_1 \sin\lambda\phi) \tag{13.42}$$

From Equations 13.34c and 13.34d, on retaining only (higher) derivatives, we have

$$\begin{aligned} M_\phi &= -\frac{2D\lambda^3}{Eta}e^{\lambda\phi}[(C_2 - C_1)\cos\lambda\phi - (C_1 + C_2)\sin\lambda\phi] \\ M_\theta &= \nu M_\phi \end{aligned} \tag{13.43a, b}$$

The *displacement* δ in the planes of the parallel circles is often important. This can readily be obtained from Equation 12.25b of Section 12.6 as

$$\delta = a\sin\phi\ \varepsilon_\phi = (v\cot\phi - w)\sin\phi$$

or by employing Equation e of Section 13.10:

$$\delta = \frac{a\sin\phi}{Et}(N_\theta - \nu N_\phi)$$

Substitution of Equations 13.41b and 13.41c into this equation gives

$$\delta = -\frac{a\sin\phi}{Et}\left\{e^{\lambda\phi}[\lambda(C_1 + C_2)\cos\lambda\phi + \lambda(C_2 - C_1)\sin\lambda\phi] + \frac{pa}{2}(1-\nu)\right\} \tag{13.44}$$

By the use of Equations 13.40 through 13.44, various particular cases can be easily treated. The results obtained this way have very satisfactory accuracy.

EXAMPLE 13.3: Force and Moment in Spherical Dome with Built-in Edge

Determine the stress resultants in the spherical shell with clamped edge and submitted to the action of a uniform normal pressure p (Figure 13.8).

Solution

The boundary conditions

$$V = 0 \qquad \delta = 0 \qquad (\phi = \alpha)$$

yield the respective equations:

$$\frac{2\lambda^2}{Et} e^{\lambda\alpha}(C_2 \cos\lambda\alpha - C_1 \sin\lambda\alpha) = 0$$
$$\left\{\lambda e^{\lambda\alpha}[(C_1 + C_2)\cos\lambda\alpha + (C_2 - C_1)\sin\lambda\alpha] + \frac{pa}{2}(1-\nu)\right\} = 0$$

The foregoing leads to

$$C_1 = -\frac{pa}{2\lambda}(1-\nu)e^{-\lambda\alpha}\cos\lambda\alpha$$
$$C_2 = C_1 \tan\lambda\alpha$$

We then obtain the following equations in place of Equations 13.40 through 13.43:

$$\begin{aligned}
Q_\phi &= -\frac{pa}{2\lambda}(1-\nu)e^{\lambda(\phi-\alpha)}\cos\lambda(\alpha-\phi) \\
N_\phi &= -\frac{pa}{2} \\
N_\theta &= \frac{pa}{2}(1-\nu)e^{\lambda(\phi-\alpha)}[\cos\lambda(\alpha-\phi) + \sin\lambda(\alpha-\phi)] - \frac{pa}{2} \\
M_\phi &= \frac{pa^2(1-\nu)}{4\lambda^2}e^{\lambda(\phi-\alpha)}[\sin\lambda(\alpha-\phi) - \cos\lambda(\alpha-\phi)] \\
M_\theta &= \nu M_\phi
\end{aligned} \tag{13.45}$$

Comment: Interestingly, the shear force (Q_ϕ) and the bending moments (M_ϕ, M_θ) damp out in moving from the shell edge (at $\phi = \alpha$) to the crown (at $\phi = 0$).

EXAMPLE 13.4: Force and Moment in Dome

Redo the preceding example for the case in which the edge of the spherical shell is simply supported (Figure 13.9).

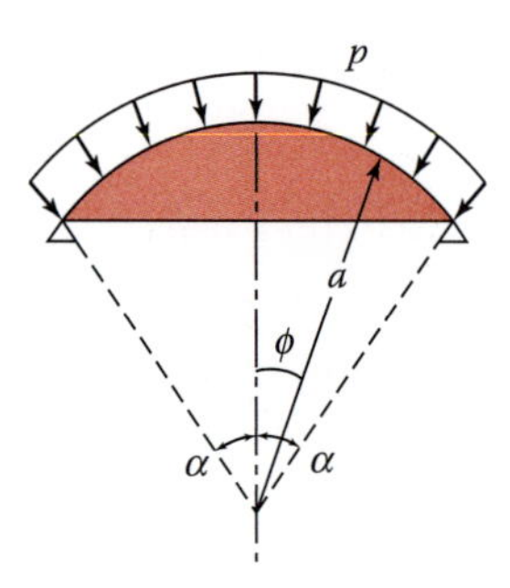

FIGURE 13.9
Simply supported spherical dome under pressure.

Solution

For this situation, the boundary conditions

$$M_\phi = 0 \qquad \delta = 0 \qquad (\phi = \alpha)$$

when introduced into Equations 13.43a and 13.44 lead to

$$C_2 - C_1 - (C_2 + C_1)\tan\lambda\alpha = 0$$
$$(C_1 + C_2)\cos\lambda\alpha + (C_2 - C_1)\sin\lambda\alpha + \frac{pa(1-\nu)}{2\lambda}e^{-\lambda\alpha} = 0$$

Solution of these equations gives the unknown constants

$$C_1 = -\frac{pa(1-\nu)}{4\lambda}e^{-\lambda\alpha}(\cos\lambda\alpha - \sin\lambda\alpha)$$
$$C_2 = -\frac{pa(1-\nu)}{4\lambda}e^{-\lambda\alpha}(\cos\lambda\alpha + \sin\lambda\alpha)$$

The stress resultants may thus be expressed in the form

$$\begin{aligned}
Q_\phi &= -\frac{pa(1-\nu)}{4\lambda}e^{\lambda(\phi-\alpha)}[\cos\lambda(\alpha-\phi) - \sin\lambda\alpha(\alpha-\phi)] \\
N_\phi &= -\frac{pa}{2} \\
N_\theta &= \frac{pa(1-\nu)}{2}e^{\lambda(\phi-\alpha)}\cos\lambda(\alpha-\phi) - \frac{pa}{2} \\
M_\phi &= \frac{pa^2(1-\nu)}{4\lambda^2}e^{\lambda(\phi-\alpha)}\sin\lambda(\alpha-\phi) \\
M_\theta &= \nu M_\phi
\end{aligned} \tag{13.46}$$

The values of the stresses produced by the foregoing forces and moments are determined by application of Equations 13.5 and 13.6.

13.13 The Finite Element Representations of Shells of General Shape

The factors that complicate an analysis of shell problems may generally be reduced to irregularities in the shape or thickness of the shell and nonuniformity of the applied load. By replacing the actual geometry of the structure and the load configuration with suitable finite element approximations (Section 7.7), very little sacrifice in accuracy is encountered.

Consider the case of a *shell of variable thickness* and general *arbitrary shape*. There are a number of ways of obtaining an equivalent shell that will not significantly compromise the elastic response. For example, we can replace the actual shell with a series of *curved* or *flat* (straight) triangular elements or *finite elements* of other form, attached at their edges and corners. Whatever the true load configuration may be, it is then reduced to a series of concentrated or distributed forces applied to each finite element.

When a shell of revolution is subjected to nonuniform load, the usual finite element approach is to replace a shell element with two flat elements, one subjected to direct force resultants and the other to moment resultants. The applied load may be converted to uniform or concentrated forces also acting on the replacement elements. In-plane and bending effects may then be analyzed separately and superposed. Hence, a shell element may be developed as a combination of a *membrane* element and a *plate* element of the same shape. The shell is thus idealized as an assemblage of flat elements.

Curved elements have been proposed to secure an improved approximation of shells, but analysis employing them is more complex than is the case using straight elements. In the general treatment of axisymmetrically loaded shells given in the next section, the latter elements are considered.

13.14 The Finite Element Solution of Axisymmetrically Loaded Shells

Recent developments open the way to adequate models for representing the behavior of shells. Some shells, for example, can be portrayed by flat elements of rectangular or quadrilateral shape, with good stiffness matrices available for such elements. The following derivation of the *properties of the finite element* is one of the simplest [8,9].

An axisymmetrically loaded shell may be represented by a series of *conical frustra* (Figure 13.10a). Each element is thus a ring generated by the *straight line* segment between two parallel circles or "nodes," say i and j (Figure 13.10b). The *thickness may vary* from element to element. As before, the displacement of a point in the midsurface is specified by two components v and w in the meridional and normal directions, respectively. Referring to Figure 13.10b, the "strain"–displacement relations, Equations 13.28 and 13.31, simplify to

$$\{\varepsilon\} = \begin{Bmatrix} \varepsilon_s \\ \varepsilon_\theta \\ \chi_s \\ \chi_\theta \end{Bmatrix} = \begin{Bmatrix} dv/ds \\ (w\cos\phi + v\sin\phi)/r \\ -d^2w/ds^2 \\ -(dw/ds)\sin\phi/r \end{Bmatrix} \tag{13.47}$$

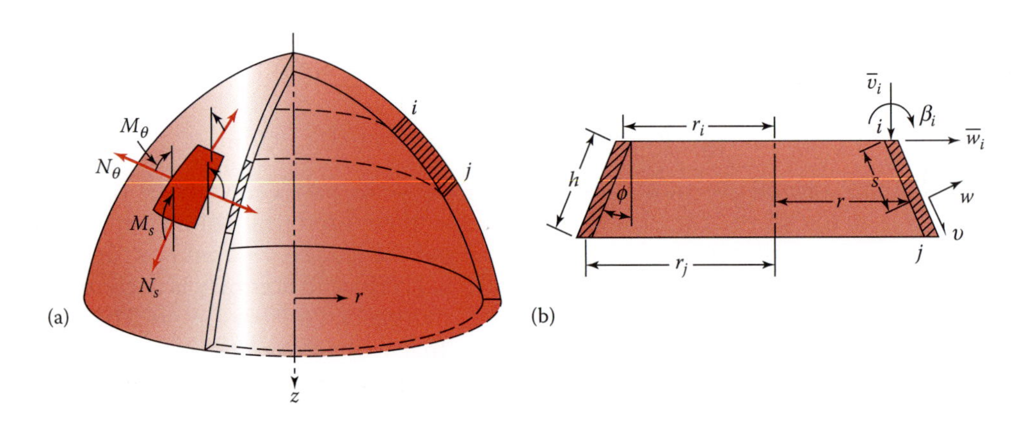

FIGURE 13.10
(a) Conical frustra; (b) finite element in the form of a ring.

The stress–resultant strain relations, Equations 13.29 and 13.32, are then

$$\begin{Bmatrix} N_s \\ N_\theta \\ M_s \\ M_\theta \end{Bmatrix} = \frac{Et}{1-\nu^2}\begin{bmatrix} 1 & \nu & 0 & 0 \\ \nu & 1 & 0 & 0 \\ 0 & 0 & t^2/12 & \nu t^2/12 \\ 0 & 0 & \nu t^2/12 & t^2/12 \end{bmatrix}\begin{Bmatrix} \varepsilon_s \\ \varepsilon_\theta \\ \chi_s \\ \chi_\theta \end{Bmatrix} \tag{13.48a}$$

or

$$\{N_s, N_\theta, M_s, M_\theta\} = [D]\{\varepsilon\} \tag{13.48b}$$

Here $[D]$ is the elasticity matrix for the isotropic axisymmetrically loaded shell.

Three displacements are chosen at each node (Figure 13.10b). The element *nodal displacement* matrix is thus

$$\{\delta\}_e = \begin{Bmatrix} \delta_i \\ \delta_j \end{Bmatrix} = \begin{Bmatrix} \bar{v}_i, & \bar{w}_i, & \beta_i \\ \bar{v}_j, & \bar{w}_j, & \beta_j \end{Bmatrix} \tag{13.49}$$

where $\bar{v}$, $\bar{w}$, and β represent the axial movement, radial movement, and rotation, respectively. The *displacements within* the element, expressed in standard form, are

$$\{f\} = \begin{Bmatrix} v \\ w \end{Bmatrix} = [N]\{\delta\}_e \tag{13.50}$$

These are to be determined from $\{\delta\}_e$ and position s; slope and displacement continuity are maintained throughout the element. The matrix $[N]$ is a function of position yet to be developed. On evaluating v and w at the nodes i and j, we can relate them to Equations 13.49 through use of a transformation matrix. At node i, for example,

$$\begin{Bmatrix} v_i \\ w_i \\ (dw/ds)_i \end{Bmatrix} = \begin{bmatrix} \cos\phi & \sin\phi & 0 \\ -\sin\phi & \cos\phi & 0 \\ 0 & 0 & 1 \end{bmatrix}\begin{Bmatrix} \bar{v}_i \\ \bar{w}_i \\ \beta_i \end{Bmatrix} = [\lambda]\{\delta_i\} \tag{13.51}$$

The following general expressions, employed for $\{f\}$, contain six constants:

$$\begin{aligned} v &= \alpha_1 + \alpha_2 s \\ w &= \alpha_3 + \alpha_4 s + \alpha_5 s^2 + \alpha_6 s^3 \end{aligned} \tag{13.52}$$

To determine the values of the αs, the coordinate s of the nodal points is substituted in the displacement functions (Equations 13.52). This will generate six equations in which the only unknowns are the coefficients. By so doing, we can solve for α_1 to α_6 in terms of the nodal displacements $v_i, \ldots, w_i$ and finally obtain [8]

$$\begin{Bmatrix} v \\ w \end{Bmatrix} = \begin{bmatrix} 1-s_1 & 0 & 0 & s_1 & 0 & 0 \\ 0 & 1-3s_1^2+2s_1^3 & s_1\left(1-2s_1+s_1^2\right)h & 0 & s_1^2(3-2s_1) & s_1^2(-1+s_1)h \end{bmatrix} \times \begin{Bmatrix} v_i \\ w_i \\ (dw/ds)_i \\ v_j \\ w_j \\ (dw/ds)_j \end{Bmatrix} \tag{13.53}$$

where

$$s_1 = \frac{s}{h} \quad (0 \le s_1 \le 1) \tag{a}$$

Denoting in Equations 13.53 the two-by-six matrix by $[\bar{P}]$, we have

$$\begin{Bmatrix} v \\ w \end{Bmatrix} = [\bar{P}] \begin{bmatrix} [\lambda] & 0 \\ 0 & [\lambda] \end{bmatrix} \{\delta\}_e = \left[[\bar{P}_i][\lambda], [\bar{P}_j][\lambda] \right] \{\delta\}_e = [P]\{\delta\}_e \tag{13.54}$$

Equations 13.47 then lead to

$$\{\varepsilon\} = [B]\{\varepsilon\}_e = [[B_i][\lambda], [B_j][\lambda]]\{\delta\}_e \tag{13.55}$$

in which

$$[B_i] = \begin{bmatrix} -1/b & 0 & 0 \\ (1-s_1)\sin\phi/r & (1-3s_1^2+2s_1^3)\cos\phi/r & bs_1(1-2s_1+s_1^2)\cos\phi/r \\ 0 & 6(1-2s_1)/b^2 & 2(2-3s_1)/b \\ 0 & 6s_1(1-s_1)\sin\phi/rb & (-1+4s_1-3s_1^2)\sin\phi/r \end{bmatrix} \tag{13.56a}$$

$$[B_j] = \begin{bmatrix} 1/b & 0 & 0 \\ s_1\sin\phi/r & s_1^2(3-2s_1)\cos\phi/r & bs_1^2(-1+s_1)\cos\phi/r \\ 0 & 6(-1+2s_1)/b^2 & 2(1-3s_1)/b \\ 0 & 6s_1(-1+s_1)\sin\phi/rb & s_1(2-3s_1)\sin\phi/r \end{bmatrix} \tag{13.56b}$$

The stiffness matrix for the element is given by Equation 7.24 in the form

$$[k]_e = \int_A [B]^T[D][B]dA \tag{b}$$

In the preceding, the area of the element equals

$$dA = 2\pi r\,ds = 2\pi rh\,ds_1 \tag{c}$$

Equation b then appears as

$$[k]_e = 2\pi h \int_0^1 [B]^T [D][B] r\,ds_1 \tag{13.57}$$

It is obvious that radius r must be expressed as a function of s prior to integration of the above.

Steps 1 through 3 of the *general procedure* described in Section 7.9 may now be applied to obtain the solution of the shell nodal displacements. We then determine the strains from Equations 13.55, the stress resultants from Equations 13.48, and the stresses from Equations 13.5. In the axisymmetrically loaded shells of revolution, "concentrated" or "nodal" forces are actually loads axisymmetrically distributed around the shells.

Clearly, if only the membrane theory solution is required, the quantities χ_s, χ_θ, β, M_s, M_θ are ignored, and the expressions developed in this section are considerably reduced in complexity.

Problems

Sections 13.1 through 13.7

13.1 A long gray, ASTM A-48 cast-iron pipe (Table B.3), 200 mm in diameter and 5 mm thick, carries a uniform line load $P = 150$ kN/m distributed over the circumference of the circular cross section at midlength (Figure 13.4). Calculate the values of maximum deflection and maximum stresses in the pipe. Let $\nu = 0.3$.

13.2 Determine the numerical value of the diameter for the pipe considered in Problem 13.1 for the case in which the maximum deflection is not to exceed 0.1 mm.

13.3 A long steel cylinder 120 mm in diameter and 2 mm thick is subjected to a uniform line load P distributed over the circumference of the circular cross section at midlength. Employing the energy of distortion theory, predict the value of the maximum load that can be applied to the cylinder without causing the elastic limit to be exceeded. Use $\nu = 0.3$ and $\sigma_{yp} = 210$ MPa.

13.4 Resolve Problem 13.3 using the maximum shear stress theory of failure.

13.5 If point O is taken to the right of the loaded portion of the cylinder shown in Figure 13.5, what is the deflection at this point?

13.6 A long circular pipe of diameter $d = 0.5$ m and wall thickness $t = 5$ mm is bent by a load P uniformly distributed along a circular section (Figure 13.4). The pipe is fabricated of a material of 200 MPa tensile yield strength and $\nu = 0.3$. Determine the value of P required according to the maximum shear stress failure criterion. Let $n = 1$.

13.7 A long cast-iron pipe of diameter $d = 0.8$ m and wall thickness $t = 10$ mm carries a load P uniformly distributed along a circular section (Figure 13.4). Knowing that $\sigma_u = 340$ MPa, $\sigma_{uc} = 620$ MPa, and $\nu = 1/3$, determine the allowable value of P using the following criteria: (a) the maximum principal stress; (b) Coulomb–Mohr.

13.8 Consider the cylinder loaded as shown in Figure 13.5. Determine the maximum values or (a) axial stress σ_s, (b) tangential stress σ_θ, and (c) shear stress $\tau_{x\theta}$.

13.9 A long steel pipe of diameter $d = 600$ mm and wall thickness $t = 8$ mm is subjected to a load as shown in Figure 13.5. Compute the value of axial stress, using $E = 200$ GPa, $\nu = 0.3$, $p = 10$ MPa, and $b = 61$ mm.

13.10 A long steel pipe of 0.75 m in diameter and 10 mm thickness is subjected to loads P uniformly distributed along two circular sections 0.05 m apart (Figure 13.4). For the midlength between the loads, obtain, by taking $\nu = 0.3$, (a) the radial contraction and (b) axial and hoop stresses at the outer surface.

Sections 13.8 through 13.14

13.11 Reduce the differential equations (Equations 13.25) to *two* expressions by eliminating shear force Q_ϕ.

13.12 Represent the equations of equilibrium of a conical shell (Equations 13.26) by two expressions containing N_s, N_θ, M_s, and M_θ.

13.13 Determine the forces N_θ and N_ϕ in a spherical dome carrying its own weight p using Equation b of Section 13.11 and Equations 13.27. Compare the results with those given by Equations 12.13. Take $\nu = 0.3$.

13.14 A spherical dome with a built-in edge and the half-angle $\alpha = 35°$ is subjected to uniform pressure of $p = 10$ kPa (Figure 13.8). Let $a = 2.25$ m, $t = 75$ mm, and $\nu = 1/6$. Calculate the meridional bending moment M_ϕ for $\phi = 35°, 30°, 25°, 20°, 15°, 10°$, and $5°$. Sketch the results.

13.15 Calculate the maximum meridional and tangential stresses, σ_ϕ and σ_θ in the spherical shell described in Problem 13.14.

References

1. W. Flügge, *Stresses in Shells*, 2nd ed., Springer, Berlin, 1990.
2. S. Timoshenko and S. Woinowsky-Krieger, *Theory of Plates and Shells*, 2nd ed., Chapters 15 and 16, McGraw-Hill, New York, 1959.
3. E. H. Baker, L. Kovalevsky, and F. L. Rish, *Structural Analysis of Shells*, Krieger, Melbourne, FL, 1981.
4. W. Flügge (ed.), *Handbook of Engineering Mechanics*, Sections 40.5 and 40.6, McGraw-Hill, New York, 1984.
5. J. E. Gibson, *Thin Shells: Computing and Theory*, Pergamon, New York, 1980.
6. H. Kraus, *Thin Elastic Shells*, Wiley, New York, 2003.
7. J. R. Vinson, *Structural Mechanics: The Behavior of Plates and Shells*, Chapter 7, Wiley, New York, 1974.
8. O. C. Zienkiewitcz and R. I. Taylor, *The Finite Element Method*, 6th ed., Elsevier, Oxford, UK, 2005.

9. T. Y. Yang, *Finite Element Structural Analysis*, Chapter 11, Prentice-Hall, Upper Saddle River, NJ, 1986.
10. J. N. Reddy, *Theory and Analysis of Elastic Plates and Shells*, 2nd ed., CRC Press, Boca Raton, FL, 2007.
11. J. H. Faupel and F. E. Fisher, *Engineering Design*, 2nd ed., Wiley, Hoboken, NJ, 1981.
12. W. C. Young, R. C. Budynas, and A. M. Sadegh, *Roark's Formulas for Stress and Strain*, 8th ed., McGraw-Hill, New York, 2011.
13. A. C. Ugural and S. K. Fenster, *Advanced Mechanics of Materials and Applied Elasticity*, 5th ed., Chapter 9, Prentice-Hall, Upper Saddle River, NJ, 2012.

14

Applications to Pipes, Tanks, and Pressure Vessels

14.1 Introduction

There are many examples of thin shells employed as structural components. In Chapter 12 we observe cases in which, when a cylinder, liquid tank, dome, or tower is subjected to a particular loading, consideration is given to the membrane stresses taking place over the entire wall thickness. We now discuss the general analysis of stress and deformation in several typical members, applying equations derived for plates and shells as well as the principle of superposition, as needed [1–3].

A degree of caution is necessary when employing formulas for which there is uncertainty as to applicability and restrictions of use. The relatively simple form of many expressions presented in this text result from severely limiting assumptions used in their derivation. Particular cognizance should be taken of the fact that high loadings, extreme temperature, and rigorous performance requirements present difficult design challenges [4–7].

The discussions of Sections 14.2 and 14.3 apply to uniform and reinforced cylinders. A comparison of the bending and membrane stresses for tanks with a vertical axis is made in Section 14.4. Sections 14.5 and 14.6 deal with uniform and compound heated shells. All the expressions derived for thin cylindrical vessels (Section 14.7 through 14.10) under uniform pressure apply to *internal pressure*. They pertain equally to cases of *external* pressure if the sign of p is changed.

However, the stresses thus determined are valid only if the pressure is not significant relative to that which would cause failure by elastic instability. That is, when a vessel such as a vacuum tank is to be constructed to withstand external pressure, the compressive stresses produced by this pressure in the shell must be lower than the critical stresses at which buckling of the walls might occur. Introduced in Section 14.11 of this chapter are some relationships employed in the design of plates and shell-like structures.

14.2 Pipes Subjected to Edge Forces and Moments

The circular *cylinder* or *pipe*, of special significance in engineering, is usually divided into long and short classifications. A *long cylinder* loaded at the middle (Figure 13.4) is defined as of length $L > (2\pi/\beta)$. It is clear from the discussion in Section 13.7 that the ends of a long cylinder are not affected appreciably by central loading; therefore, each end can be treated independently. For *short cylinders*, however, the influence at the ends of a central force is substantial. Discussed below is the bending of circular cylinders subject to edge forces and moments.

14.2.1 Long Pipes

Consider a long pipe (Figure 14.1a) loaded by uniformly distributed forces Q_1 and moments M_1 along edges $x = 0$, of length $L > (\pi/\beta)$. Constants C_1 and C_2 of Equation 13.18 can be determined by applying the following conditions at the left end of the shell:

$$\begin{aligned} M_x &= -D\frac{d^2w}{dx^2} = M_1 \qquad (x=0) \\ Q_x &= \frac{dM_x}{dx} = -D\frac{d^3w}{dx^3} = Q_1 \qquad (x=0) \end{aligned} \tag{a}$$

The results are

$$C_1 = -\frac{1}{2\beta^3 D}(Q_1 + \beta M_1) \qquad C_2 = \frac{M_1}{2\beta^2 D} \tag{b}$$

The deflection is now found by substituting C_1 and C_2 into Equation 13.18:

$$w = \frac{e^{-\beta x}}{2\beta^3 D}[\beta M_1(\sin\beta x - \cos\beta x) - Q_1\cos\beta x] \tag{14.1}$$

The maximum displacement occurs outward at the loaded section and is given by

$$w_{\max} = -\frac{1}{2\beta^3 D}(\beta M_1 + Q_1) \qquad (x=0) \tag{14.2}$$

The deflection is positive radially inward, hence the minus sign. The accompanying slope under the loadings is

$$\frac{dw}{dx} = \frac{1}{2\beta^2 D}(2\beta M_1 + Q_1) \qquad (x=0) \tag{14.3}$$

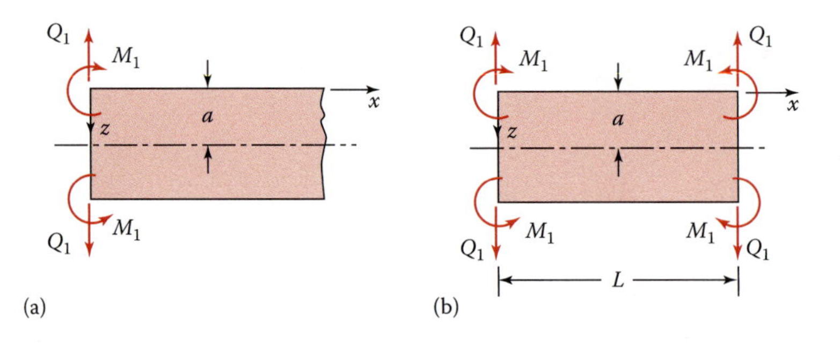

FIGURE 14.1
Edge loads: (a) On semi-infinite cylinder; (b) on short cylinder.

Finally, successive differentiations of Equation 14.1 yield expressions for the derivatives. Altogether, using the notation of Equations 13.22, we have

$$\begin{aligned} w &= -\frac{1}{2\beta^3 D}[\beta M_1 f_3(\beta x) + Q_1 f_4(\beta x)] \\ \frac{dw}{dx} &= \frac{1}{2\beta^2 D}[2\beta M_1 f_4(\beta x) + Q_1 f_1(\beta x)] \\ \frac{d^2 w}{dx^2} &= -\frac{1}{\beta D}[\beta M_1 f_1(\beta x) + Q_1 f_2(\beta x)] \\ \frac{d^3 w}{dx^3} &= \frac{1}{D}[2\beta M_1 f_2(\beta x) - Q_1 f_3(\beta x)] \end{aligned} \tag{14.4}$$

On substituting the above into Equations 13.20, we obtain expressions for N_θ, M_x, M_θ, and Q_x. The stresses are then found through the use of the definitions given in Section 13.4. As already observed in Section 13.7, functions $f_1(\beta x)$ through $f_4(\beta x)$, and thus the results employing them, become negligibly small for $x > (\pi\beta)$.

EXAMPLE 14.1: Variation of Deflection and Moment over Pipe Length

A long pipe supports a uniformly distributed end moment M_0 as depicted in Figure 14.2a. Develop and sketch the values of w and M_0 along the βx axis.

Solution

Through the use of Equations 14.4 and 13.22 with $M_0 = -M_1$ and $Q_1 = 0$, we obtain

$$w = -\frac{M_0}{2D\beta^2} e^{-\beta x}(\sin\beta x - \cos\beta x); \qquad M_1 = -M_0 e^{-\beta x}(\cos\beta x + \sin\beta x)$$

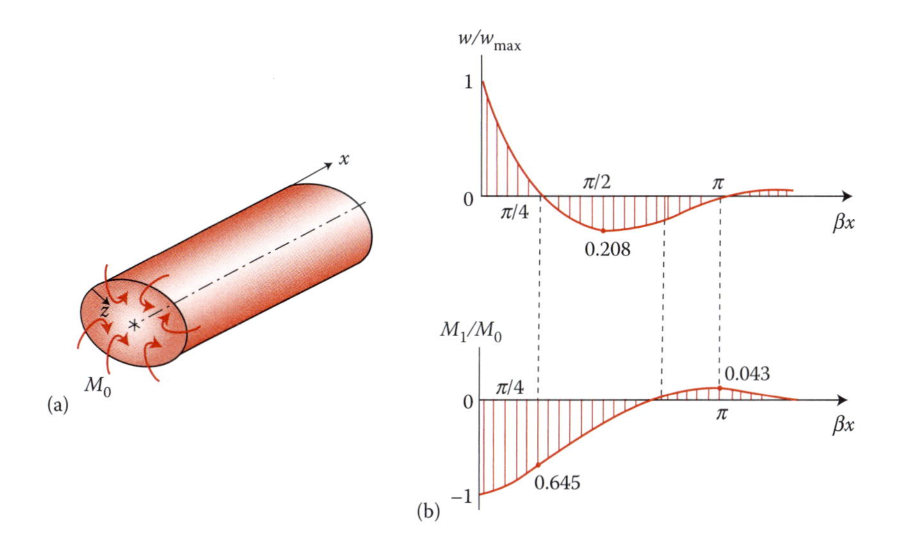

FIGURE 14.2
(a) Long cylinder carrying edge moment M_0; (b) variation of deflection and moment over its length.

On the loaded edge of the pipe:

$$w = w_{\max} = \frac{M_0}{2D\beta^2} \qquad (x = 0)$$

$$M_1 = M_{1,\max} = M_0 e^{-\beta x} = -M_0 \qquad (x = 0)$$

Figure 14.2b illustrates a dimensionless form of deflections and moments over the shell length.

Comment: Observe that all these values practically diminish at a distance $x > (\pi/\beta)$ over the edge of the pipe, as already noted in Section 13.7.

14.2.2 Short Pipes

The bending of this type of shell (Figure 14.1b) loaded along both edges, of length $L < (2\pi/\beta)$, may also be dealt with by applying the general solution (Equation 13.17). Now opposite end conditions interact. Equation 13.17 is rewritten in the following alternate form in terms of hyperbolic functions:

$$\begin{aligned} w &= C_1 \sin\beta x \sinh\beta x + C_2 \sin\beta x \cosh\beta x \\ &\quad + C_3 \cos\beta x \sinh\beta x + C_4 \cos\beta x \cosh\beta x \end{aligned} \tag{14.5}$$

Four constants of integration require evaluation. To accomplish this, two boundary conditions at each end may be applied, leading to four equations with four unknown constants C_1 through C_4. After routine but somewhat lengthy algebraic manipulations, the following expressions for end deflections and end slopes are obtained:

$$\begin{aligned} w &= -\frac{2\beta a^2}{Et}[\beta M_1 h_2(\beta L) + Q_1 h_1(\beta L)] \\ \frac{dw}{dx} &= \pm\frac{2\beta^2 a^2}{Et}[2\beta M_1 h_3(\beta L) + Q_1 h_2(\beta L)] \end{aligned} \tag{14.6}$$

where

$$\begin{aligned} h_1(\beta L) &= \frac{\cosh\beta L + \cos\beta L}{\sinh\beta L + \sin\beta L} \\ h_2(\beta L) &= \frac{\sinh\beta L - \sin\beta L}{\sinh\beta L + \sin\beta L} \\ h_3(\beta L) &= \frac{\cosh\beta L - \cos\beta L}{\sin\beta L + \sin\beta L} \end{aligned} \tag{14.7}$$

Table 14.1 lists values of $h_1(\beta L)$, $h_2(\beta L)$, and $h_3(\beta L)$ as a function of βL. Interestingly, in the case of long shells $h_1(\beta L) \approx h_2(\beta L) \approx h_3(\beta L) \approx 1$, Equations 14.6, as expected, reduce to Equations 14.2 and 14.3.

EXAMPLE 14.2: Analysis of Cylindrical Pressure Vessel

A long cylindrical shell is subjected to uniform internal pressure p (Figure 14.3a). Determine the stress and deformation of the vessel for two sets of conditions:

TABLE 14.1

Selected Values of the Functions Defined by Equations 14.7

βL	0.2	0.4	0.6	1.0	1.4	2.0	3.0	4.0	5.0
$h_1(\beta L)$	5.000	2.502	1.674	1.033	0.803	0.738	0.893	0.005	1.0170
$h_2(\beta L)$	0.0068	0.0268	0.0601	0.1670	0.3170	0.6000	0.9770	1.0580	1.0300
$h_3(\beta L)$	0.100	0.200	0.300	0.500	0.689	0.925	0.090	1.050	1.008

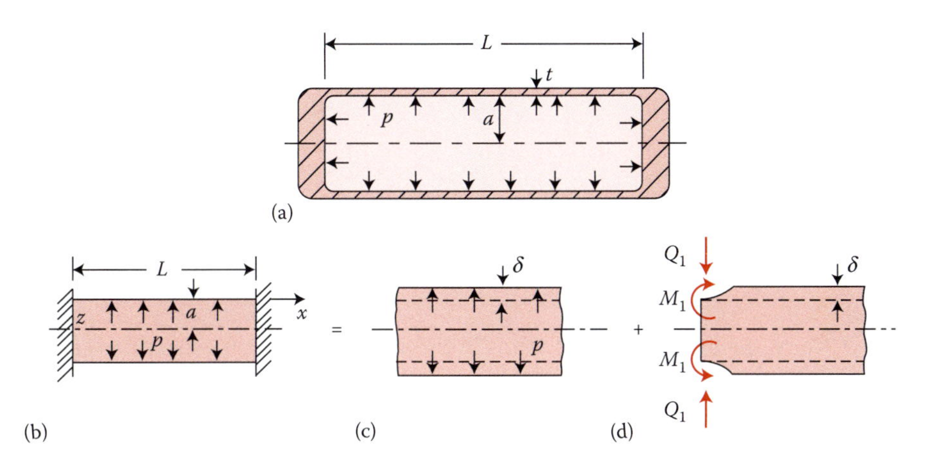

FIGURE 14.3
(a) Closed-end vessel; (b) schematic, compressed cylinder with rigid ends; (c) radial expansion owing to p; (d) edge loads.

(a) the ends are built in (Figure 14.3b) and the axial force $N_x = 0$, or the ends are welded to *thick* head plates as shown in Figure 14.3a; (b) the ends are simply supported.

Solution

a. The problem may be treated as the sum of the cases shown in Figures 14.3c and 14.3d. The radial expansion due to p (Figure 14.2c) is, from Equation 12.11,

$$\delta = \frac{pa^2}{Et} \tag{c}$$

The boundary conditions for the edge-loaded pipe (Figure 14.3d) are represented as follows, using Equations 14.2 and 14.3 together with Equation c:

$$-\frac{1}{2\beta^3 D}(\beta M_1 + Q_1) = \frac{pa^2}{Et}$$

$$\frac{1}{2\beta^2 D}(2\beta M_1 + Q_1) = 0$$

from which

$$M_1 = \frac{p}{2\beta^2} \qquad Q_1 = -\frac{p}{\beta} \tag{d}$$

In Equations d, the minus sign indicates a negative shear force, acting as shown in Figure 14.3d.

The expression for deflection, referring to Figure 14.3d and Equations 14.4 and c, is then

$$w = -\frac{p}{4\beta^4 D}[f_3(\beta x) - 2f_4(\beta x)] - \frac{pa^2}{Et} \tag{14.8}$$

The axial moment, from the first of Equations 13.20, is

$$M_x = -\frac{p}{2\beta^2}[f_1(\beta x) - 2f_2(\beta x)] \tag{e}$$

and $\sigma_x = 6M_x/t^2$. The circumferential stress

$$\sigma_\theta = -\frac{Ew}{a} - \nu\sigma_x \tag{f}$$

Clearly, the first and second terms in the above represent the membrane and the bending solutions, respectively.

b. Since in this case $M_x = 0$ along the edges, we have $M_1 = 0$. Equation 14.2 then yields

$$Q_1 = -2\beta^3 D\delta \tag{g}$$

where δ is given by Equation c. On introduction of Q_1 given by Equation g and $M_1 = 0$ into Equation 14.4, an expression may be derived for the displacement w.

14.3 Reinforced Cylinders

Cylindrical shells are often stiffened by *rings* or so-called *collars*. Examples are found in pipelines, submarines, and airplane fuselages (see Figure 8.1). We shall discuss the bending of a long pipe reinforced by *equidistant* rings and subjected to a uniform internal pressure of intensity p (Figure 14.4), applying Equations 14.6. The stiffness, size, and collar spacing are important in the analysis. It is presumed that each ring has *small cross-sectional dimensions* compared with shell radius a and does not resist shell rotation.

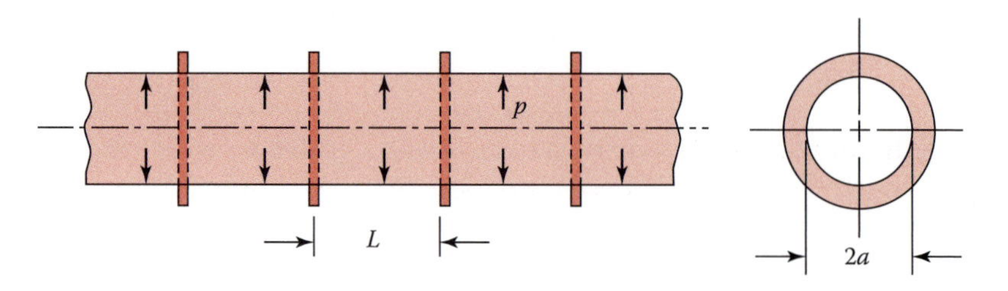

FIGURE 14.4
Long (infinite) pipe with many equidistant rings.

14.3.1 Cylinders with Collars That Prohibit Deflection

For this case, we assume the stiffening rings to be very rigid in comparison with the shell. That is, cylinder *deflection is zero* under the collars. If there were no ring, the pressure would produce a constant hoop stress pa/t, given by Equation 12.10. Referring to Equation 12.11, the radial displacement would then be

$$\delta_1 = \frac{pa^2}{Et}$$

Inasmuch as the rings are present, reaction forces P per unit circumferential length of pipe are produced between each ring and the shell.

To solve for P, we equate the pipe deflection under the ring, produced by the forces P, to the expansion δ due to p. The portion of the shell between two adjacent rings may be represented by the cylinder shown in Figure 14.1b. In the problem under consideration, $Q_1 = -P/2$ and M_1 is found by setting the slope $dw/dx = 0$ at an end point. From the second of Equations 14.6, we obtain

$$\frac{\beta^2 a^2}{Et}[4\beta M_1 h_3(\beta L) - Ph_2(\beta L)] = 0$$

or

$$M_1 = \frac{Ph_2(\beta L)}{4\beta h_3(\beta L)} \tag{a}$$

Then, applying the first of Equations 14.6,

$$\frac{P\beta a^2}{Et}\left[h_1(\beta L) - \frac{h_2^2(\beta L)}{2h_3(\beta L)}\right] = \delta_1 = \frac{pa^2}{Et}$$

from which

$$P\beta\left[h_1(\beta L) - \frac{1}{2}\frac{h_2^2(\beta L)}{h_3(\beta L)}\right] - \frac{\delta_1 Et}{a^2} = 0. \tag{14.9}$$

Note that when βL is large, the functions, $h_1(\beta L)$, $h_2(\beta L)$, and $h_3(\beta L)$ approach unity, and the above reduces to

$$\frac{P\beta a^2}{2Et} = \delta_1$$

This agrees with Equation 13.24, which was differently derived.

On introducing the value of P from Equation 14.9 into Equation a and simplifying, we obtain

$$M_1 = \frac{p}{2\beta^2} h_2(\beta L) \tag{14.10}$$

14.3.2 Cylinders with Collars That Resist Deflection

A second type of reinforced shell is encountered when the rings are relatively *flexible* compared with the flexibility of the shell. In this situation, the *deflection* of the shell *is not zero* under the collar. The interface forces P increase the inner radius of the ring by

$$\Delta\delta = \left(\frac{Pa}{AE}\right)a = \frac{Pa^2}{AE}$$

in which A and Pa are the area of the ring cross section and the tensile, respectively. Now δ_1 in Equation 14.9 is replaced by $\delta_2 = \delta_1 - \Delta\delta$ to obtain an equation used to evaluate the force P under a ring:

$$P\beta\left[h_1(\beta L) - \frac{1}{2}\frac{h_2^2(\beta L)}{h_3(\beta L)}\right] = p - \frac{Pt}{A} \tag{14.11}$$

Inserting $p = p - (Pt/A)$ into Equation 14.10, we also have

$$M_1 = \frac{h_2(\beta L)}{2\beta^2}\left(p - \frac{Pt}{A}\right) \tag{14.12}$$

for the moment under a collar.

14.3.3 Cylinders with Closed Ends

In the two cases discussed above, the ends of the shell are assumed to be open. A variation of the deflection and the stress in the pipe occurs when the ends are closed. The internal pressure acting at the ends creates an axial extension $pa/2Et$ of the pipe. The radial elongation of the shell is thus

$$\delta_2 = \frac{pa^2}{Et}\left(1 - \frac{1}{2}\nu\right) \tag{b}$$

For this case, p is replaced by $p(1 - \nu/2)$ in Equations 14.10 and 14.11.

14.4 Cylindrical Tanks

Many cylindrical tanks may be treated by a procedure similar to that described in the foregoing sections. Included among these are tanks with nonrigid or rigid bottom plates resting on the ground, tanks with an elastic roof and bottom, and tanks constructed of plates of several different constant thicknesses (Figure P14.14).

In the case of tanks of variable wall thickness, solution of the problem requires integration of Equation 13.13a. Flexural rigidity D and thickness t must be regarded no longer as constant but as functions of axial distance x. Thus, we must deal with a linear differential equation of fourth order and variable coefficient, requiring lengthy manipulation.

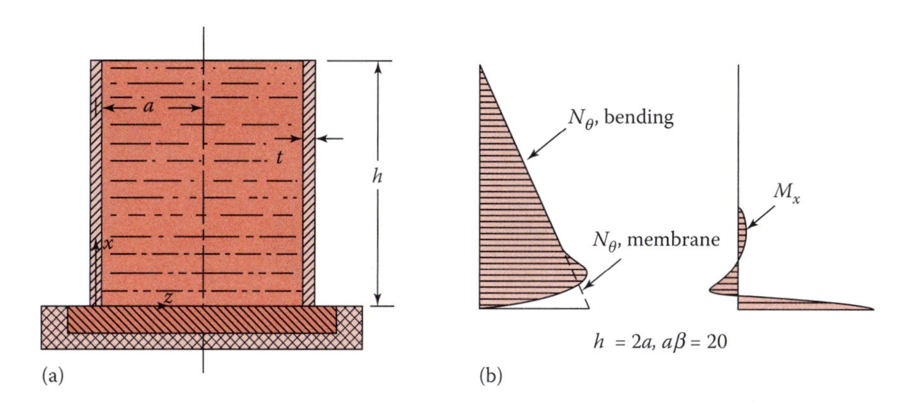

FIGURE 14.5
(a) Cylindrical water tank with clamped base; (b) stress resultants.

However, in concrete water tanks it is usually common to choose a linear variation of thickness increasing from top to bottom. The solution is thus rendered more manageable.

We here consider a *cylindrical tank* of *uniform thickness* entirely filled with liquid of specific weight γ (Figure 14.5a). The tank bottom is assumed to be built in, and the top is open. The physical conditions indicate that the upper edge is free to deform and that no force exists there ($N_x = 0$). At the lower edge, locating the coordinates in the figure, we have

$$w = 0 \qquad \frac{dw}{dx} = 0 \qquad (x = 0) \tag{a}$$

The differential equation is, from Equation 13.15b,

$$\frac{d^4 w}{dx^4} + 4\beta^4 w = -\frac{\gamma(h - x)}{D} \tag{b}$$

where the outward pressure acting at any point on the wall, P_r, is replaced by $-\gamma(h - x)$. The particular solution of Equation b is found to be $-\gamma(h - x)a^2/Et$. The general solution is thus

$$\begin{aligned} w &= e^{-\beta x}(C_1 \cos \beta x + C_2 \sin \beta x) \\ &\quad + e^{\beta x}(C_3 \cos \beta x + C_4 \sin \beta x) - \frac{\gamma(h - x)a^2}{Et} \end{aligned} \tag{c}$$

If t is small relative to a and h, as is usually the case, a longitudinal slice of unit width of the cylinder may be considered infinitely long. It is already observed in Section 13.7 that, because w must be finite for all x, $C_3 = C_4 = 0$ in the above expression.

We now satisfy the remaining conditions. On substitution of Equations a into Equation c,

$$w = C_1 - \frac{\gamma a^2 h}{Et} = 0$$

$$\frac{dw}{dx} = \beta(C_2 - C_1) + \frac{\gamma a^2}{Et} = 0$$

from which

$$C_1 = \frac{\gamma a^2 h}{Et} \qquad C_2 = \frac{\gamma a^2}{Et}\left(h - \frac{1}{\beta}\right)$$

The *radial deflection* of the tank is then

$$w = -\frac{\gamma a^2 h}{Et}\left\{1 - \frac{x}{h} - e^{-\beta x}\left[\cos\beta x + \left(1 - \frac{1}{\beta h}\right)\sin\beta x\right]\right\}$$

or

$$w = -\frac{\gamma a^2 h}{Et}\left[1 - \frac{x}{h} - f_4(\beta x) - \left(1 - \frac{1}{\beta h}\right) f_2(\beta x)\right] \tag{14.13}$$

where $f_2(\beta x)$ and $f_4(\beta x)$ are defined by Equation 13.22. Using Table 13.1, the deflection at any point is easily determined from the above equation. Equation 13.15a, after integration, provides the axial displacement

$$u = \int_0^h \nu \frac{w}{a}\,dx + u_0$$

in which the constant, from $u(0) = 0$, is $u_0 = 0$.

We next evaluate the *stress resultants* from Equations 13.11 and 13.12 together with 13.10a and 14.13:

$$\begin{aligned} N_\theta &= -\frac{Etw}{a} = \gamma ah\left[1 - \frac{x}{h} - f_4(\beta x) - \left(1 - \frac{1}{\beta h}\right) f_2(\beta x)\right] \\ M_x &= -D\frac{d^2 w}{dx^2} = \frac{2\beta^2 \gamma a^2 Dh}{Et}\left[-f_2(\beta x) + \left(1 - \frac{1}{\beta h}\right) f_4(\beta x)\right] \\ M_\theta &= \nu M_x \end{aligned} \tag{14.14}$$

The maximum bending moment occurs at the bottom of the tank, at $x = 0$:

$$M_{x,\max} = \left(1 - \frac{1}{\beta h}\right)\frac{2\beta^2 \gamma a^2 Dh}{Et} \tag{14.15}$$

The stress resultant N_θ, according to membrane theory, is found from Equation 12.30a by substituting $p_r = -\gamma(h - x)$:

$$N_\theta = \gamma(h - x)a \tag{d}$$

At $x = 0$, N_θ assumes its maximum value $N_{\theta,\max} = \gamma ha$. From membrane theory, the maximum stress is therefore

$$\sigma_{\theta,\max} = \frac{1}{t}\gamma ha \tag{e}$$

We observe that the force resultant corresponding to the particular solution, $w_p = -\gamma(h - x)a^2/Et$, from Equation 14.14, is

$$N_\theta = -\frac{Et\,w_p}{a} = \gamma(h - x)a$$

The above agrees with Equation d, which was derived by applying membrane theory. This conclusion is also valid for other cases [7].

According to bending theory, because $N_\theta = 0$ and M_x is a maximum at $x = 0$,

$$\sigma_{x,\max} = \frac{6M_{x,\max}}{t^2} = \frac{12\beta^2\gamma a^2 Dh}{Et^3}\left(1 - \frac{1}{\beta h}\right)$$

Introducing $D = Et^3/12(1 - \nu^2)$ and $\beta^2 = \sqrt{3(1-\nu^2)}/at$ into the above and treating the terms $1/\beta h$ and ν^2 as small compared with unity, we have

$$\sigma_{x,\max} = \frac{\sqrt{3}}{t}\gamma ha \tag{f}$$

Comment: Comparing Equations e and f, it is observed that the *true maximum stress* is in fact $\sqrt{3}$ times larger than that predicated by membrane theory.

A plot of Equations 14.14 and d is presented in Figure 14.5b, where the *dashed and solid lines* denote the stress-resultant distribution according to the *membrane and bending theories*, respectively. The figure shows that membrane theory is valid at sections away from the fixed end of a long, thin cylinder. For a comparatively thick-walled and short cylinder, however, the differences in the result given by the theories are pronounced in the lower half of the shell. In any case, membrane theory applies reasonably well to the upper portion of the tank.

14.5 Thermal Stresses in Cylinders

Thermal stresses are developed whenever the expansion or contraction that would normally be produced by the heating or cooling of a member is restricted. *Thermal stress* and deformation in *thin cylindrical shells* is a major factor in the design of structures such as boilers, heat exchangers, pressure vessels, and nuclear piping. In this section, consideration is given to two representative examples of temperature fields under which thermal stresses occur in cylindrical shells.

14.5.1 Uniform Temperature Distribution

A uniform temperature change in a cylinder with clamped or simply supported ends causes local bending stresses at the edges. These situations may be analyzed through application of Equations 14.2 and 14.3, following a procedure similar to that described in Example 14.1.

Examine, for example, a *long cylinder with fixed ends*, under a uniform temperature change ΔT (Figure 14.3b). For this case, free radial expansion of the pipe (Figure 14.3c) is described by

$$\delta = a\alpha(\Delta T) \tag{a}$$

Referring to Figure 14.3d, we thus have

$$-\frac{1}{2\beta^3 D}(\beta M_1 + Q_1) = a\alpha(\Delta T)$$
$$\frac{1}{2\beta^2 D}(2\beta M_1 + Q_1) = 0$$

or

$$M_1 = 2a\alpha(\Delta T)\beta^2 D \qquad Q_1 = -4a\alpha(\Delta T)\beta^3 D \tag{b}$$

Expressions for the *deflection* and the *moment* are then derived as illustrated in Example 14.2:

$$\begin{aligned} w &= -a\alpha(\Delta T)[f_3(\beta x) - 2f_4(\beta x)] - a\alpha(\Delta T) \\ M_x &= 2a\alpha\beta^2 D(\Delta T)[f_1(\beta x) - f_2(\beta x)] \end{aligned} \tag{14.16}$$

By applying Equations 14.16, the stresses at any point can readily be determined as in isothermal bending.

14.5.2 Radial Temperature Gradient

Consider a long *cylinder of arbitrary cross section* subjected to the uniform temperature T_2 at the outer surface and the uniform temperature T_1 at the inner surface (Figure 14.6).

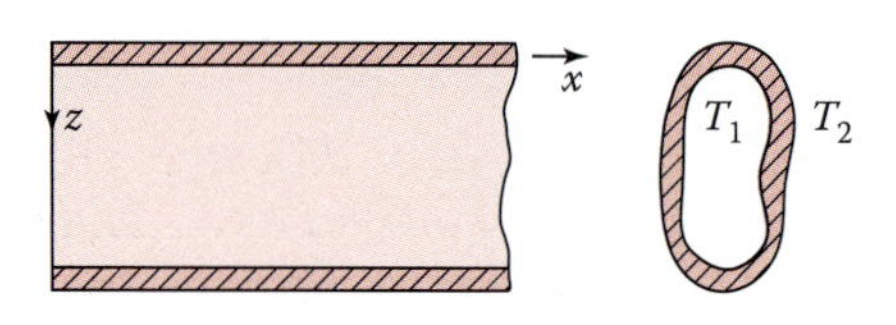

FIGURE 14.6
Semi-infinite cylinder subject to a linear temperature change through the thickness.

The *temperature gradient* through the thickness is assumed to be linear. Unless otherwise specified, we shall also take $\Delta T = T_1 - T_2$ as positive, that is, $T_1 > T_2$. In the shell described, *at points remote from the ends*, the thermal stress is the same as in a clamped-edge circular plate (Section 11.8). The *hoop and axial stresses*, from Equation 11.32, are thus

$$\sigma_\theta = \sigma_x = \sigma = \pm \frac{E\alpha(\Delta T)}{2(1-\nu)} \tag{14.17}$$

Note that the outer surface will be in tension and the inner surface in compression.

Next, we discuss the stress distribution *at the edges* of a *circular cylinder of uniform thickness* with *free ends*. It is observed from Figure 14.3d that the stresses given by Equation 14.17 must be balanced near the edge by distributed moments M_1. That is,

$$\sigma_x = -\frac{6M_1}{t^2}$$

from which

$$M_1 = -\frac{E\alpha(\Delta T)t^2}{12(1-\nu)} \tag{c}$$

Clearly, in order for an edge to be *free*, a moment equal to but opposite in sign to that defined by Equation c must be applied there. The deflection from such a moment ($+M_1$) can be determined by applying Equation 14.4. By doing so, substituting the resulting expression for w into Equations 13.19 and 13.20, we obtain the *deflection and stress resultants* applicable to the free edge ($x = 0$):

$$\begin{aligned} w &= \frac{M_1}{2\beta^2 D} \\ M_x &= \frac{E\alpha(\Delta T)t^2}{12(1-\nu)} \\ M_\theta &= \frac{\nu E\alpha(\Delta T)t^2}{12(1-\nu)} \\ N_\theta &= \frac{Et\alpha(\Delta T)}{2\sqrt{3}(1-\nu)}\sqrt{1-\nu^2} \end{aligned} \tag{14.18a–d}$$

Therefore, Equations 14.17, 14.18c, and 14.18d provide the resultant *thermal hoop stress* occurring at the free end. Observe that this stress is larger than the axial stress. Thus,

$$\sigma_{\theta,\max} = \frac{E\alpha(\Delta T)}{2\sqrt{3}(1-\nu)}\left[\sqrt{3}(1-\nu)+\sqrt{1-\nu^2}\right] \tag{14.19}$$

The stress defined by Equation 14.19, for $\nu = 0.3$, is 25% larger than the stress at points in the pipe away from the ends. This explains why *thermal cracks initiate at the free edge of cylinders.*

Comments: Expressions for stress and deformation in cylindrical shells with clamped or simply supported ends, as well as for shells under axial or circumferential temperature gradients, can also be obtained by modifying the cylindrical shell equations appropriate to isothermal bending [2,4]. Generally used in design computations is a system of nondimensional curves [6] based on these solutions or the finite element method.

14.6 Thermal Stresses in Compound Cylinders

The foregoing section and Section 11.2 provide the basis for design of composite or compound multishell cylinders, constructed of a number of concentric, thin-walled shells. In the development that follows, each component shell is assumed homogeneous and isotropic. Each may have a different thickness and different material properties and be subjected to different uniform or variable temperature differentials.

If a free-edged multishell cylinder undergoes a *uniform temperature change* ΔT, the free motion of any shell having different material properties is restricted by components adjacent to it. Only a hoop stress σ_θ and circumferential strain ε_θ are then produced in the cylinder walls. On applying Equations 11.2, we thus have

$$\sigma_\theta = E[\varepsilon_\theta - \alpha(\Delta T)] \tag{14.20}$$

In the case of a cylinder subjected to a *temperature gradient*, both axial and hoop stresses take place, and $\sigma_x = \sigma_\theta = \sigma$ (Section 14.5). Equations 11.3 give

$$\sigma = \frac{E}{1-\nu}[\varepsilon_\theta - \alpha(\Delta T)] \tag{14.21}$$

Consider a compound structure comprising three components, each under different temperature gradients (the ΔTs) (Figure 14.7). Equation 14.21 permits stress determination as follows:

$$\begin{aligned}
\begin{Bmatrix} \sigma_{a1} \\ \sigma_{a2} \end{Bmatrix} &= \frac{E_a}{1-\nu_a}\begin{Bmatrix} [\varepsilon_\theta - \alpha_a(\Delta T_1)] \\ [\varepsilon_\theta - \alpha_a(\Delta T_2)] \end{Bmatrix} \\
\begin{Bmatrix} \sigma_{b2} \\ \sigma_{b3} \end{Bmatrix} &= \frac{E_b}{1-\nu_b}\begin{Bmatrix} [\varepsilon_\theta - \alpha_b(\Delta T_2)] \\ [\varepsilon_\theta - \alpha_b(\Delta T_3)] \end{Bmatrix} \\
\begin{Bmatrix} \sigma_{c3} \\ \sigma_{c4} \end{Bmatrix} &= \frac{E_c}{1-\nu_c}\begin{Bmatrix} [\varepsilon_\theta - \alpha_c(\Delta T_3)] \\ [\varepsilon_\theta - \alpha_c(\Delta T_4)] \end{Bmatrix}
\end{aligned} \tag{14.22a–f}$$

where the subscripts a, b, and c refer to the individual components.

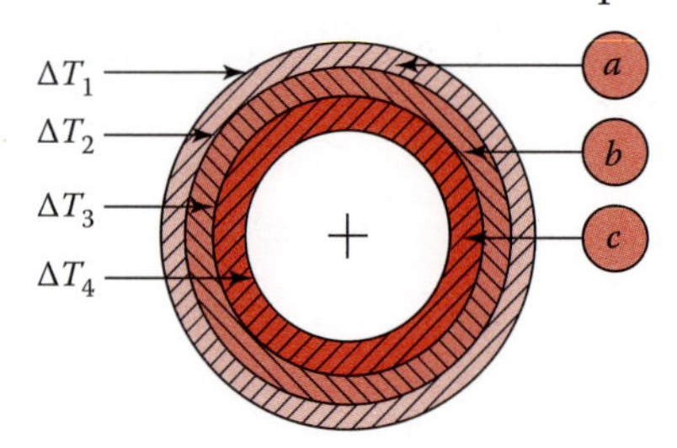

FIGURE 14.7
Compound cylinder subject to a uniform temperature change.

The *axial forces* corresponding to each layer are next ascertained from the stresses given by Equations 14.22:

$$N_a = \frac{1}{2}(\sigma_{a1} + \sigma_{a2})A_a$$
$$N_b = \frac{1}{2}(\sigma_{b2} + \sigma_{b3})A_b \quad \text{(a)}$$
$$N_c = \frac{1}{2}(\sigma_{c3} + \sigma_{c4})A_c$$

Here the *A*'s represent the cross-sectional area of each shell. The condition that the sum of the axial forces be equal to zero is satisfied if

$$N_a + N_b + N_c = 0 \quad \text{(b)}$$

From Equations 14.22, a, and b, assuming that $\nu_a = \nu_b = \nu_c$, the following expression for the *strain* is obtained

$$\varepsilon_\theta = \frac{A_aE_a\alpha_a(\Delta T)_a + A_bE_b\alpha_b(\Delta T)_b + A_cE_c\alpha_c(\Delta T)_c}{A_aE_a + A_bE_b + A_cE_c} \quad (14.23)$$

In the preceding, $(\Delta T)_a$, $(\Delta T)_b$, and $(\Delta T)_c$ are the *average* of the temperature differentials at the boundaries of elements *a*, *b*, and *c*, respectively, for example, $(\Delta T)_a = (1/2)(\Delta T_1 + \Delta T_2)$. The stresses at the inner and outer surfaces of each shell may be calculated readily on inserting Equation 14.23 into Equations 14.22, as illustrated in Examples 14.2 and 14.3.

Comments: Near the ends there will usually be some bending of the composite cylinder, and the total thermal stresses will be obtained by superimposing on Equations 14.22 such stresses as may be necessary to satisfy the boundary conditions, as shown in Section 14.5.

EXAMPLE 14.3: Stresses in Nonuniformly Heated Two-Layer Cylinder

The cylindrical portion of a jet nozzle is made by just slipping a steel shell over a brass shell (Figure 14.8). The radii of the tubes are $r_1 = 400$ mm, $r_2 = 398$ mm, and $r_3 = 392$ mm. If the uniform temperature differentials at the boundaries

$$\Delta T_1 = 200°\text{C} \qquad \Delta T_2 = 150°\text{C} \qquad \Delta T_3 = 40°\text{C}$$

must be maintained, what stresses will develop in the two materials? The properties of brass and steel are as follows:

$$E_b = 103\,\text{GPa} \qquad \alpha_b = 18.9 \times 10^{-6}\ \text{per°C} \qquad \nu_b = 0.3$$
$$E_s = 200\,\text{GPa} \qquad \alpha_s = 11.7 \times 10^{-6}\ \text{per°C} \qquad \nu_s = 0.3$$

Solution

We have $A_b = 2\pi r_2 t_b$ and $A_s = 2\pi r_2 t_s$, where t_b and t_s are the wall thicknesses of the brass and steel tubes, respectively. On applying Equation 14.23,

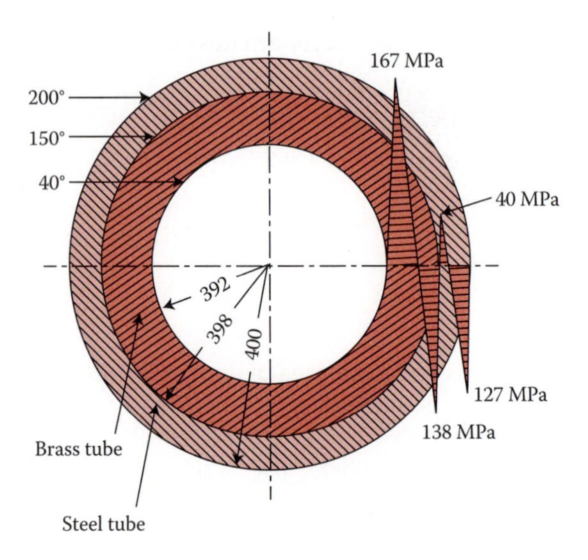

FIGURE 14.8
Two layer composite cylinder.

$$\varepsilon_\theta = \frac{t_s E_s \alpha_s \left[\frac{1}{2}(\Delta T_1 + \Delta T_2)\right] + t_b E_b \alpha_b \left[\frac{1}{2}(\Delta T_2 + \Delta T_3)\right]}{t_s E_s + t_b E_b} \tag{14.24}$$

and thus

$$\varepsilon_\theta = \frac{[0.002 \times 200 \times 10^9 \times 11.7 \times 10^{-6}(175) + 0.006 \times 103 \times 10^9 \times 18.9 \times 10^{-6}(95)]}{0.002 \times 200 \times 10^9 + 0.006 \times 103 \times 10^9} = 1.894 \times 10^{-3}$$

The expansion of the interfacial radius is then $\Delta r_2 = 1.894 \times 10^{-3}$ (0.398) = 0.754(10^{-3}) m or 0.754 mm. The radial growth at other locations may be found in a similar manner.

Through the use of Equations 14.22 and 14.24, the following values are obtained for the stresses at the *outer* and *inner surfaces* of each shell, respectively:

$$\begin{Bmatrix} \sigma_{s1} \\ \sigma_{s2} \end{Bmatrix} = \frac{200 \times 10^3}{1 - 0.3} \begin{Bmatrix} 1894 - 11.7(200) \\ 1894 - 11.7(150) \end{Bmatrix} = \begin{Bmatrix} -127 \\ 40 \end{Bmatrix} \text{MPa}$$

and

$$\begin{Bmatrix} \sigma_{b2} \\ \sigma_{b3} \end{Bmatrix} = \frac{103 \times 10^3}{1 - 0.3} \begin{Bmatrix} 1894 - 18.9(150) \\ 1894 - 18.9(40) \end{Bmatrix} = \begin{Bmatrix} -138 \\ 167 \end{Bmatrix} \text{MPa}$$

It is observed that

$$N_s + N_b = \frac{1}{2}(-127 + 40)(0.005) + \frac{1}{2}(-138 + 167)(0.015) = 0$$

as required according to Equation b. A sketch of the stresses throughout the wall thickness is given in Figure 14.8.

EXAMPLE 14.4: Stresses in Uniformly Heated Two-Layer Cylinder

Reconsider the preceding sample problem for the case in which a brass tube just slips over a steel tube with each shell *uniformly* heated and the temperature raised by ΔT°C. Determine: (a) the hoop stress that develops in each component on *cooling*; (b) the values of stresses in each components, using the material properties given in Example 14.3 and with $\Delta T = 110$°C.

Solution

For the case under consideration, the composite cylinder is *contracting*. Equation 14.20 then appears as

$$\sigma_\theta = -E[\varepsilon_\theta + \alpha(\Delta T)] \tag{14.25}$$

from which

$$\varepsilon_\theta = -\frac{1}{E}\sigma_\theta - \alpha(\Delta T) \tag{14.26}$$

The *hoop strain*, referring to Equations 14.24 and 14.26, is now expressed:

$$\varepsilon_\theta = -\frac{(\Delta T)[t_s E_s \alpha_s + t_b E_b \alpha_b]}{t_s E_s + t_b E_b} \tag{14.27}$$

a. Hence the *hoop stresses*, from Equations 14.25 and 14.27, after simplification are found to be

$$(\sigma_\theta)_b = -E_b[\varepsilon_\theta + \alpha_b(\Delta T)] = -\frac{E_b(\Delta T)(\alpha_b - \alpha_s)}{1 + (t_b E_b / t_s E_s)} \tag{14.28}$$

$$(\sigma_\theta)_s = -E_s[\varepsilon_\theta + \alpha_s(\Delta T)] = -\frac{E_s(\Delta T)(\alpha_s - \alpha_b)}{1 + (t_b E_b / t_s E_s)} \tag{14.29}$$

Comment: Thus, the stresses in the brass and steel tubes are tensile and compressive, respectively, as $\alpha_b > \alpha_s$ and ΔT is a negative quantity because of cooling.

b. Substituting the given numerical values into Equation 14.28, we obtain the stress in brass tube, we have

$$(\sigma_\theta)_b = \frac{103 \times 10^9(120)(18.9 - 11.7)(10^{-6})}{1 + (0.006 \times 103 / 0.002 \times 200)} = -\frac{88.992(10^6)}{2.545} = -35 \text{ MPa}$$

Similarly, Equation 14.29 gives

$$(\sigma_\theta)_s = -\frac{200 \times 10^9(110)(11.7 - 18.9)(10^{-6})}{2.545} = 62.2 \text{ MPa}$$

as the hoop stress in the steel tube.

14.7 Discontinuity Stresses in Pressure Vessels

Thin shell equations under uniform pressure apply to internal pressure p. They also pertain to cases of external pressure if the sign of p is changed. But, stresses so obtained are valid only if the pressure is not significant relative to that which causes failure by elastic instability. A degree of caution is necessary when using the formulas for which there is uncertainty as to applicability and restriction of use. Particular emphasis should be given to the fact that high loading, extreme temperature, and rigorous performance requirements present difficult design challenges.

The ever-broadening use of *vessels* for storage, industrial processing, and power generation under unique conditions of temperature, pressure, and environment has given special emphasis to analytical, numerical, and experimental techniques for obtaining appropriate operating stresses. The finite element method (Section 7.7) has gained considerable favor in the design of vessels relative to other methods.

A discontinuity of the membrane action in a vessel occurs at all points of external restraint or at the junction of the cylindrical shell and its *head* or *end* possessing different stiffness characteristics. Any incompatibility of deformation at the joint produces bending moments and shearing forces. The stresses due to this bending and shear are termed *discontinuity stresses.*

It is observed in Section 13.7 that the bending is of a local character. Hence, the discontinuity stresses become negligibly small within a short distance. The narrow region at the edge of *spherical-*, *elliptical-*, and *conical*-type vessel heads can be assumed as *nearly cylindrical* in shape. Cylindrical shell (see Figure 14.9) equations of Section 11.6 can therefore be employed to obtain an approximate solution applicable at the juncture of vessels having spherical, elliptical, or conical ends. In the case of *flat-end* vessels, expressions for circular plates and cylindrical shells are utilized.

Sections 14.8 through 14.10 can provide only a few basic relationships with an active area of contemporary pressure vessel analysis. Our treatment is applicable to thin-walled containers, pipes, and power cylinders when the fluid pressure on all walls may be considered uniform. Additional details and a list of references relevant to stresses in pressure vessels are to be found in a variety of publications [8,9]. The ASME unfired pressure-vessel code furnishes information of practical value for end design (Section 14.12).

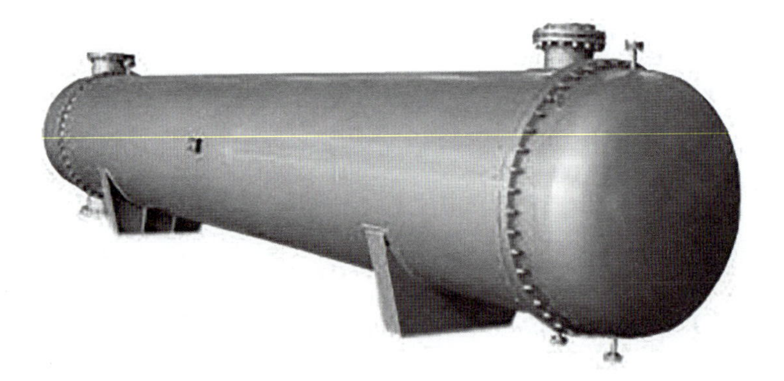

FIGURE 14.9
A cylindrical pressure vessel.

14.8 Cylindrical Vessel with Hemispherical Heads

Consider a *cylindrical vessel* of radius a, having *hemispherical ends*, under uniform internal pressure of intensity p (Figure 14.10a). Because of the action of p, the tube and its ends tend to expand by different amounts, as shown in exaggeration by the dashed lines in the figure. Because at some distance away from the joint membrane theory yields results of sufficient accuracy, for the cylindrical part of the vessel, we have, from Equations 12.10,

$$N_x = \frac{pa}{2} \qquad N_\theta = pa \tag{a}$$

For the spherical ends, Equation 12.4 yields

$$N = \frac{Pa}{2} \tag{b}$$

The growth in cylinder radius (Figure 14.10a) produced by membrane forces N_x and N_θ is expressed by Equation 12.11:

$$\delta_c = \frac{pa^2}{2Et}(2-\nu) \tag{14.30}$$

Similarly, extension of the radius of the spherical heads owing to N (Figure 14.10a), through application of Equation 12.5, is

$$\delta_s = \frac{pa^2}{2Et}(1-\nu) \tag{14.31}$$

It follows that, if the tube and its ends are disjointed (Figure 14.10b), the differential radial extension due to the membrane stresses would be $\delta = \delta_c - \delta_s$:

$$\delta = \frac{pa^2}{2Et} \tag{14.32}$$

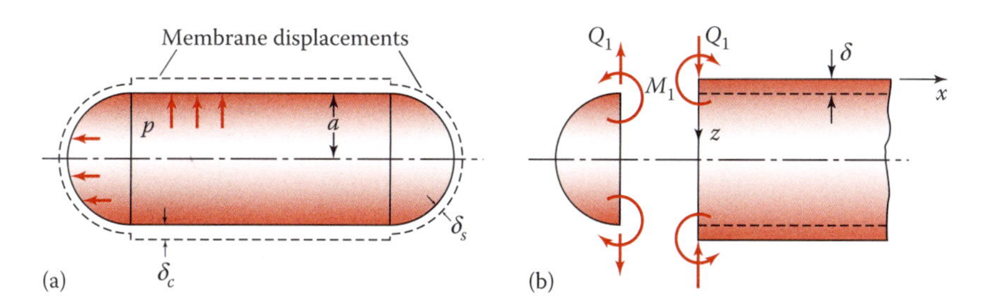

FIGURE 14.10
(a) Vessel with hemispherical ends; (b) cylinder and hemisphere cut apart to show redundant stress resultants.

In an actual vessel, this discontinuity of displacement is prevented by shearing forces Q_1 and bending moments M_1, both uniformly distributed along the circumference. As already mentioned, these discontinuity effects create bending stresses in the vicinity of the joint.

14.8.1 Cylinder with Semispherical and Heads of Equal Thickness

Often in practice, the *thicknesses* of the *cylinder* and its *ends* are usually the *same*. In this case, forces Q_1 produce equal deformations at the edges of each component at the joint. Hence, the conditions of continuity at the junction are satisfied if $M_1 = 0$ and Q_1 is of such magnitude as to result in deflections at the edges of the cylinder and its head whose sum is equal to δ. Applying Equations 14.2 and 14.32, we thus have

$$\frac{pa^2}{2Et} = \left(\frac{Q_1}{2\beta^3 D}\right)_c + \left(\frac{Q_1}{2\beta^3 D}\right)_s = \frac{Q_1}{\beta^3 D}$$

This, together with Equations 3.10 and 13.14, yields

$$Q_1 = \frac{pa^2\beta^3 D}{2Et} = \frac{p}{8\beta} \tag{c}$$

The displacement, from Equation 14.4, is then

$$w = \frac{Q_1}{2\beta^3 D} f_4(\beta x) = \frac{p}{16\beta^4 D} f_4(\beta x) \tag{14.33a}$$

and the bending moment is therefore

$$M_x = -D\frac{d^2 w}{dx^2} = -\frac{Q_1}{\beta} f_2(\beta x) = -\frac{atp}{8\sqrt{3(1-\nu^2)}} f_2(\beta x) \tag{14.33b}$$

The maximum moment takes place at $x = \pi/4\beta$, as $dM_x/dx = 0$ at this point.

The *highest axial stress* occurs on the outer face at $x = \pi/4\beta$ of the tube. By means of Equations a, 14.33b, and 13.5, for $\nu = 0.3$, it is found that

$$\sigma_{x,\max} = \frac{ap}{2t} + \frac{3}{4t}\frac{ap}{\sqrt{3(1-\nu^2)}} f_2\left(\frac{\pi}{4}\right) = 1.293\frac{ap}{2t} \tag{14.34a}$$

Hoop stress also occurs at the outer surface of the cylinder. This stress is composed of the membrane solution component pa/t, a circumferential stress produced by deflection w (see Section 13.6), and a bending stress resulting from the moment $M_\theta = -\nu M_x$:

$$\sigma_\theta = \frac{ap}{t} - \frac{Ew}{a} - \frac{6\nu}{t^2} M_x$$

Substituting Equations 14.33 into the above, we have

$$\sigma_\theta = \frac{ap}{t}\left[1 - \frac{1}{4} f_4(\beta x) + \frac{3v}{4\sqrt{3(1-\nu^2)}} f_2(\beta x)\right] \tag{d}$$

For $\nu = 0.3$, the *maximum tangential stress*, referring to Table 13.1, is found to be

$$\sigma_{\theta,\max} = 1.032\frac{ap}{t} \qquad (\beta x = 1.85) \tag{14.34b}$$

Comments: Inasmuch as the membrane stress is lower in the heads than in the tube, the maximum stress in the spherical ends is always smaller than the value given above. We are led to conclude therefore that the *hoop stress*, Equation 14.36, is *most critical in the design* of the vessel.

EXAMPLE 14.5: Computing Stresses and Deflections in the Vessel

Figure 14.10a shows a steel pressure vessel consisting of a cylinder joined to a semispherical head of the same thickness. What are the values of the maximum deflection and stresses in the vessel? *Given:* $a = 0.4$ m, $t = 8$ mm, $E = 200$ GPa, $\nu = 0.3$.

Solution

The geometric properties, from Equations 3.10 and 13.14, are

$$D = \frac{Et^3}{12(1-\nu^2)} = \frac{200(8^3)}{12(1-0.3^2)} = 9.38 \text{ kN}\cdot\text{m}$$

$$\beta^4 = \frac{3(1-\nu^2)}{a^2t^2} = \frac{3(1-0.3^2)}{(0.4^2)(0.008)^2} = 266.7(10^3)\text{ m}^{-1}$$

The largest expansion occurs at the junction ($x = 0$) of cylinder and head. Thus, Equation 14.33a and Table 13.1, leads to

$$w_{\max} = \frac{p}{16\beta^4 D^2}(0.049) = \frac{3(0.049)}{16(266.7)(9.38)} = 0.004 \text{ mm}$$

Through the use of Equation 14.34a, the largest axial stress equals

$$\sigma_{\theta,\max} = 1.293\frac{ap}{2t} = 1.293\frac{400(3)}{2(8)} = 97 \text{ MPa}$$

In a like manner, by Equation 14.34b, we obtain

$$\sigma_{\theta,\max} = 1.032\frac{ap}{t} = 1.032\frac{400(3)}{8} = 154.8 \text{ MPa}$$

as the maximum tangential stress in the cylinder.

14.8.2 Junction of a Cylinder and Sphere of Different Thickness

Consideration is now given to a cylindrical pressure vessel with semispherical heads of a different thickness. Figure 14.11 depicts the geometry and loads on a portion of such a tank. Since at the joint each element has different thickness, forces Q_1 moments M_1 cause unequal deformations.

This condition leads to a lengthy solution. Upon following a procedure similar to that described in Section 14.8.1, it can be verified that [4], the stress resultants are then expressed as follows:

$$Q_1 = \left\{ \frac{\left(\dfrac{1-\nu}{\sin\theta} - (2-\nu)\rho\right)\left(1-\rho^2\sqrt{\rho\sin\theta}\right)\dfrac{1}{\lambda_h} - \left(1+\rho^2+2\rho^2\sqrt{\rho\sin\theta}\right)\cos\theta}{(1+\rho^2)^2 + \left(2\rho^2/\sqrt{\rho\sin\theta}\right)(1+\rho\sin\theta)} \right\} \frac{pa'}{2} \tag{14.35a}$$

$$M_1 = \left\{ \frac{\left[\dfrac{1-\nu}{\sin\theta} - (2-\nu)\rho\right]\left(\dfrac{1-\rho^2}{2\lambda_h^2}\right) + \left[\left(1+\dfrac{1}{\sqrt{\rho\sin\theta}}\right)\dfrac{\rho^2}{\lambda_h}\right]\cos\theta}{(1+\rho^2)^2 + \left(2\rho^2/\sqrt{\rho\sin\theta}\right)(1+\rho\sin\theta)} \right\} \frac{paa'}{2} \tag{14.35b}$$

where

$$\rho = \frac{t_h}{t_c} \qquad \lambda_h = \sqrt[4]{\frac{3(1-\nu^2)(a')^2}{t_h^2}} \tag{14.36}$$

Comment: Note that, for the cylindrical portion of the vessel, the discontinuity stresses may be computed as described in Section 14.2.1.

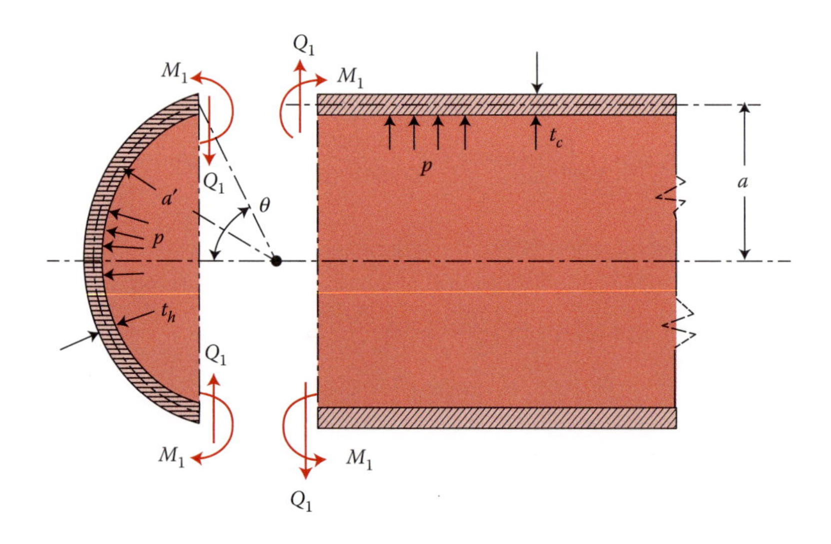

FIGURE 14.11
Cylinder and sphere disjoined to portray redundant forces and moments.

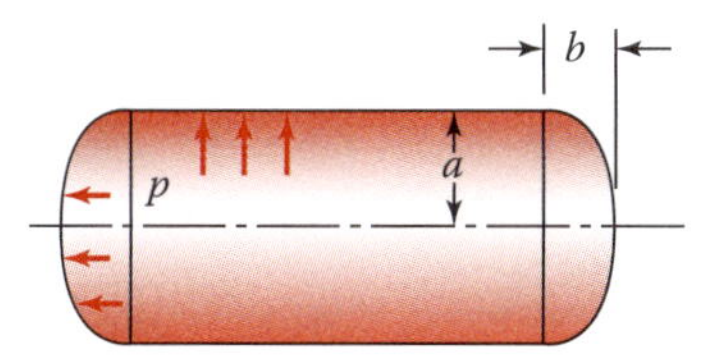

FIGURE 14.12
Cylinder having ellipsoidal ends.

14.9 Cylindrical Vessels with Ellipsoidal Heads

The general procedure for the determination of the stress in a *vessel with ellipsoidal ends* (Figure 14.12) follows the same pattern as employed in the foregoing section. The radial increment δ_c in the cylindrical portion is given by Equation 14.30. Referring to Equations 12.20, the increment at the edge of the ellipsoidal member is found to be

$$\delta_e = \frac{a}{E}(\sigma_\theta - \nu\sigma_\phi) = \frac{pa^2}{Et}\left(1 - \frac{a^2}{2b^2} - \frac{\nu}{2}\right) \tag{a}$$

The radial extension $\delta = \delta_c - \delta_e$ of the two components is thus

$$\delta = \frac{pa^2}{2Et}\left(\frac{a^2}{b^2}\right) \tag{14.37}$$

A comparison of Equations 14.37 and 14.32 reveals that the extension at the joint of an ellipsoidal-ended vessel is greater by the ratio a^2/b^2 than that for a vessel with hemispherical heads. Hence, the shearing force and the discontinuity stresses also increase in the identical proportions.

In the particular case of a vessel for which $a = 2b$, Equations 14.34 and d of Section 14.8 yield the following expressions for *maximum axial and the hoop stresses* in the vessel:

$$\begin{aligned}\sigma_{x,\max} &= \frac{ap}{2t} + \frac{3ap}{t\sqrt{3(1-\nu^2)}} f_2\left(\frac{\pi}{4}\right) = 2.172\frac{ap}{2t} \\ \sigma_{\theta,\max} &= 1.128\frac{ap}{t}\end{aligned} \tag{14.38}$$

Note that the last stress, as before, has the largest magnitude.

14.10 Cylindrical Vessel with Flat Heads

Cylindrical vessels are often constructed with *flat ends* as shown in Figure 14.13. In this case, each head is a thin circular plate. A solution may be obtained by assuming that the

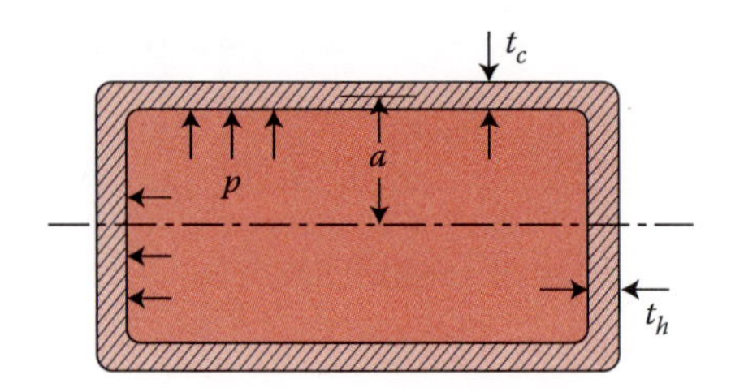

FIGURE 14.13
Cylinder with flat ends.

heads bend into spherical surfaces subsequent to the application of the uniformly distributed loading p. The internal pressure will produce extensions in the tube and its heads as observed in the cases of the vessels previously discussed. Therefore, at the joint there must act shearing forces Q_1 and bending moments M_1 to make the displacements compatible.

For a circular plate head, the *edge slopes* θ due to p and the edge moment M_1 are from Table 4.2,

$$\theta_h' = \frac{pa^3}{8D(1+\nu)} \qquad \theta_h'' = \frac{aM_1}{D(1+\nu)} \tag{a}$$

Here D represents the flexural rigidity of the end plate, defined by Equation 3.10. Similarly, the slope θ_c due to Q_1 and M_1 of the cylindrical portion at the joint is, referring to Equation 14.3,

$$\theta_c = \frac{1}{2\beta^2 D}(2\beta M_1 + Q_1) \tag{b}$$

The condition that the *edges* of the two parts *rotate the same amount* is satisfied by

$$\theta_h' - \theta_h'' = \theta_c \tag{14.39}$$

We next consider the *radial displacement*. The extension of the cylinder radius due to p is, according to Equation 14.30,

$$\delta_c' = \frac{pa^2}{2Et_c}(2-\nu) \tag{c}$$

It can be verified that the radial displacement of the circular plate head due to Q_1 is expressed by

$$\delta_h = -\frac{aQ_1}{Et_h}(1-\nu) \tag{d}$$

Equation 14.2 provides the increment in the cylinder radius at the joint:

$$\delta_c'' = -\frac{1}{2\beta^3 D}(\beta M_1 + Q_1) \tag{e}$$

The *compatibility of the radial displacements* of the edges of the two parts requires that

$$\delta_c' = \delta_h + \delta'' \tag{14.40}$$

Finally, the solution is completed by determining Q_1 and M_1 by using Equations 14.39 and 14.40. The results are presented in the following form:

$$Q_1 = -\frac{\rho^3\lambda_c^3 + 2(2-\nu)\rho^3\lambda_c + 2(2-\nu)(1+\nu)}{2\rho^3\lambda_c^2 + [(1-\nu)\rho^4 + (1+\nu)]\lambda_c + (1-\nu^2)\rho}\frac{pa}{4} \tag{14.41a}$$

$$M_1 = \frac{2\rho^3\lambda_c^3 + (1-\nu)\rho^4\lambda_c^2 + 2(2-\nu)(1+\nu)}{2\rho^3\lambda_c^2 + [(1-\nu)\rho^4 + (1+\nu)]\lambda_c + (1-\nu^2)\rho}\frac{pa^2}{8\lambda_c} \tag{14.41b}$$

where

$$\rho = \frac{t_c}{t_h} \qquad \lambda_c = a\beta = \sqrt[4]{\frac{3(1-\nu^2)a^2}{t_c^2}} \tag{14.42}$$

The displacement and the discontinuity stresses may now be obtained by following a similar procedure to that discussed in Section 14.8.

14.11 *Design Formulas for Conventional Pressure Vessels

It is observed in Sections 14.8 and 14.9 that the discontinuity stresses in the vicinity of the joint of a cylindrical shell and its head are quite a bit larger than the membrane stress in either portion. When a vessel is properly designed and constructed, however, these stresses are reduced greatly, and it becomes *unnecessary* to consider them. The ASME *Boiler and Pressure Vessel Code* [10–12] furnishes formulas for calculating the required minimum thickness of the shells and the ends. All the formulas, except those associated with the case of flat heads, are based solely on the membrane stresses, but the shape and proportions of the end and the manner of attachment to the shell are specified so as to avoid high discontinuity stresses.

The following *factors* and a host of others contributing to an ideal *vessel design* are described by the code: approved techniques for joining the end to the shell, formulas for computing the thickness of shell and end, materials in combination, temperatures ranges, maximum allowable stress values, corrosion, and types of closures. For some of the shells and the end types discussed in Chapters 12 through 14, Table 14.2 lists the *minimum required thickness*. The following symbols are employed in the table:

p = internal pressure (psi)

r = inside radius of shell or hemispherical head (in.)

σ = maximum allowable stress (psi)

e = lowest joint efficiency

TABLE 14.2

The ASME Design Code for Pressure Vessels

Geometry	Required Thickness (in.)	Geometry	Required Thickness (in.)
Cylindrical shell	$t = \dfrac{pr}{\sigma e - 0.6p}$	Flat head	$t = d\sqrt{\dfrac{cp}{\sigma}}$
Hemispherical head	$t = \dfrac{pr}{2\sigma e - 0.2p}$	Ellipsoidal head	$t = \dfrac{pD}{2\sigma e - 0.2p}$
Conical head	$t = \dfrac{pD}{2\cos\alpha(\sigma e - 0.6p)}$	Wholly spherical shell	$t = \dfrac{pr}{2\sigma e - 0.2p}$

D = inside diameter of conical head or inside length of the major axis of an ellipsoidal head (in.)

α = half the apex angle of the conical head

d = diameter of flat head (in.)

c = a numerical coefficient depending on the method of attachment of head; for example, $c = 0.5$ for circular plates welded to the end of the shell

It is mentioned that the required wall thicknesses for *tubes* and *pipes* under internal pressure are determined in accordance with the rules for shell in the code. In its present status, the code is applicable when the pressure *does not exceed* 3 ksi (21 MPa). Pressures in excess of this amount may require special attention in the design and construction of the vessels, closures, and branch connections of piping systems [13].

Note that the preceding is only a *partial description* of the code specifications, which, if complied with, provide assurance that discontinuity stresses may be neglected. For the complete requirements, reference should be made to the current edition of the code. The ASME publishes relevant books, conference papers, and a quarterly, the *Journal of Pressure Vessel Technology*.

Comments: Other vessel discontinuities include the openings, termed *nozzles*, required for the inward and outward flow of fluid; manholes for inspection and cleaning; attachments for tube sheets; supporting flanges; and changes in wall thicknesses. Openings can be reinforced by adding plates that are welded to the vessel shell. Nozzles often have tapering thickness providing reinforcement to the shell. Manholes and other openings, when their function allows, are placed in the lower-stressed center area of the heads. Most of the foregoing and some additional situations in pressure vessels are also covered in codes.

CASE STUDY 14.1 Filament-Wound Pressure Vessels

A unique group of composites, made by wrapping of high-strength filaments over a mandrel, followed by impregnation of the windings with a plastic binder and removal of the mandrel in pieces, is termed *filament-wound cylinders*. A typical system is the glass filament/epoxy resin combination. Filament materials of this type have an exceptional strength-to-weight ratio and reliability. Basic filament vessels consist of longitudinal, circumferential, or helical windings. A combination of these windings is employed if necessary.

The main application of filament-wound pressure vessels is in the transformation and storage of petroleum and chemical fluids. Glass filament or high-strength steel wires presently account for the most of these situations. They are also widely used as lightweight vessels and thrust chambers in spacecraft, rockets, and airborne vehicles.

Consider a *filament-wound vessel* with flat heads subjected to internal pressure p (Figure 14.14). The tangential and axial stresses due to p are, by Equation 12.10,

$$\sigma_\theta = \frac{pa}{t}, \qquad \sigma_x = \frac{pa}{2t}$$

Consequently,

$$\frac{\sigma_\theta}{\sigma_x} = 2 \tag{a}$$

In the foregoing, the quantities a and t denote the average radius and wall thickness, respectively, of the vessel composed entirely of filament of tensile strength σ_u and the binder.

The maximum tensile force carried by the filament is equal to

$$F = \sigma_u wt$$

in which w is the filament width wound at an helix angle ϕ. The associated circumferential force is

$$F_\theta = F \sin\phi$$

The filament cross-sectional area $A = wt/\sin\phi$. Then, tangential stress filament can carry equals

$$\sigma_\theta = \frac{F_\theta}{A} = \sigma_u \sin^2\phi \tag{14.43}$$

Similarly, the axial stress may readily be written in the form

$$\sigma_x = \sigma_u \cos^2\phi \tag{14.44}$$

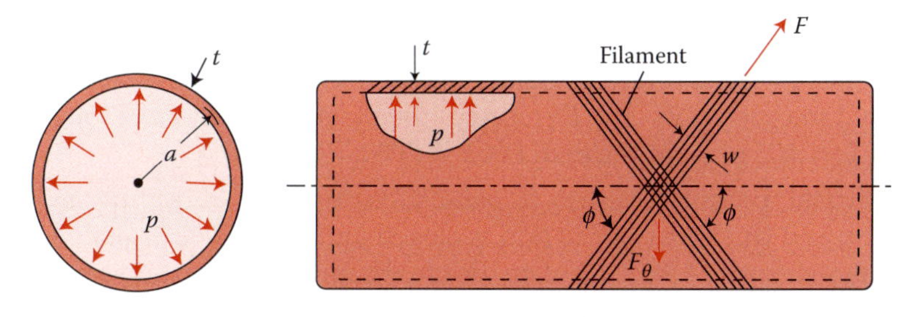

FIGURE 14.14
Filament-wrapped cylindrical pressure vessel with closed ends. *Note:* tension force in the filament, F = winding helix angle; w = filament width; F_θ = circumferential component of filament force; p = internal pressure.

It follows that

$$\frac{\sigma_\theta}{\sigma_x} = \tan^2\phi \tag{14.45}$$

The *optimum winding helix angle of filament*, from Equations a and 14.45, is thus

$$\frac{\sigma_\theta}{\sigma_x} = 2 = \tan^2\phi \tag{14.46}$$

Solving $\phi = 54.7°$, which represents the condition of helical wrapping to support an internal pressure.

Comments: We note that by additional use of circumferential filaments, the helix angle may be decreased for convenience in wrapping. Obviously, the preceding analysis *applies* only to the *cylindrical portion* of the vessel, away from the ends. Filament winding is also made by laying down a pattern over a base material and forming a so-called filament-overlay composite [4], for instance, thin-walled (polyethylene) pipe overlaid with (nylon) cord or a wire of the same material of the shell.

Problems

Sections 14.1 through 14.4

14.1 A long steel pipe is subjected to a circumferential load Q_1 at its end (Figure 14.1a). The radius and the thickness of the pipe are $a = 80$ mm and $t = 3$ mm. Determine: (a) the least distance from the end at which no deflection occurs; (b) the allowable value of Q_1 if diametrical expansion at the end is limited to 15×10^{-5} m. Assume $E = 200$ GPa and $\nu = 0.3$.

14.2 A steel pipe of 0.5-m radius and 10-mm wall thickness is subected to an internal pressure of 3.6 MPa. The pipe is joined to a pressure vessel. The juncture of the two parts is assumed to be rigid (i.e., $w = dw/dx = 0$ at the end of the pipe). Find, for practical considerations: (a) the minimum distance from the joint at which the pipe diameter attains its completely expanded value; (b) the maximum bending stress. Use $\nu = 0.3$.

14.3 A long steel boiler tube of 50-mm radius and 3-mm wall thickness is filled with hot water under 4-MPa uniform pressure. The tube is attached to a "header" in a simply supported manner at its end. Determine the deflection and the hoop stress at a distance of $L = 0.05$ m from the end. Let $E = 200$ GPA and $\nu = 0.3$.

14.4 A cylindrical pressure vessel ($E = 200$ GPa, $\nu = 0.3$) of radius $a = 1$ m, length $L = 1.5$ m, and thickness $t = 20$ mm is closed and welded at its ends to immovable, rigid plates (Figure 14.3a). Calculate the maximum deflection and stress in the vessel if it is subjected to an internal pressure of $p = 4$ MPa.

14.5 Redo Problem 14.4 for a cylinder that is constructed of gray, ASTM-48 cast iron (Table B.3) and subjected to a pressure $p = 5$ MPa. Let $\nu = 0.3$.

14.6 A short cylindrical shell subjected to uniform internal pressure is simply supported at both ends (Figure P14.6). Derive the following expression, setting $\alpha = \beta L/2$:

$$w_{\max} = -\frac{pL^4}{64D\alpha^4}\left(1 - \frac{2\cos\alpha\,\cosh\alpha}{\cosh\alpha + \cosh\alpha}\right) \quad \text{(P14.6)}$$

The foregoing represents the maximum displacement, occurring at the midlength of the shell. Employ Equation 14.5 as the homogeneous solution.

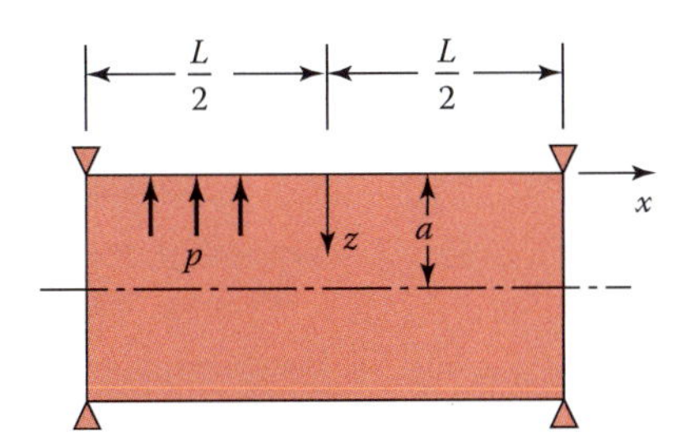

FIGURE P14.6

14.7 When the cylinder described in Problem 14.6 is long, show that Equation P14.6 reduces to Equation c of Section 14.2. What conclusion can be drawn from this result?

14.8 A circular steel pipe is reinforced by collars spaced L apart (Figure 14.4). The cylinder is under an internal pressure p. The cross-sectional area of the collar is 0.025 m². What is the maximum bending stress in the pipe if the collars are constructed of a very rigid material? Use $L = 1.3$ m, $a = 0.6$ m, $t = 10$ mm, $p = 1.4$ MPa, and $\nu = 0.3$.

14.9 A long thin-walled bronze pipe of radius a and wall thickness t has a bronze collar shrunk over it at its midlength. Verify that if the cross-sectional area of the collar is $A = 1.56\sqrt{at^3}$, the decrease of shell diameter equals the expansion of collar diameter at the time of the shrinkage. Let $\nu = 0.3$.

14.10 A narrow ring is given a shrink fit onto a long pipe of radius a and thickness t. The cross-sectional area of the ring is A. If the outer radius of the pipe is greater by δ than the inner radius of the ring, what is the shrink-fit bending stress in the pipe?

14.11 Redo Problem 14.8 for a ring made of a relatively flexible material as compared with the flexibility of the shell.

14.12 In a large *plate*, a hole of radius a is drilled and its edges rein forced by a *short cylinder*, as shown in Figure P14.12. Verify that the ratio of the plate thickness t_p to the cylinder thickness t is given by

$$\frac{t_p}{t} = \frac{1-\nu}{\sqrt[4]{3(1-\nu^2)}}\sqrt{\frac{t}{a}\,\frac{\sin \text{h}\beta L + \sin\beta L}{\cos \text{h}\beta L + \cos\beta L}} \quad \text{(P14.12)}$$

14.13 Compute the maximum bending and membrane stresses in a cylindrical tank wall with clamped base (Figure 14.5a). The tank is filled to the top with water (specific weight 9.81 kN/m³). The dimensions are $a = 2.7$ m, $h = 3.7$ m, and $t = 20$ mm. Use $\nu = 0.3$.

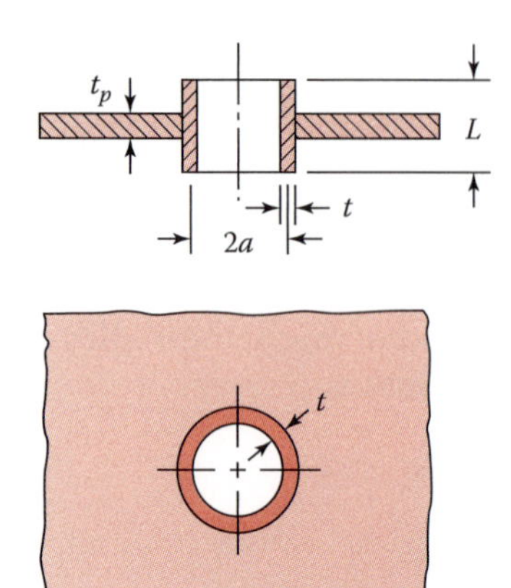

FIGURE P14.12

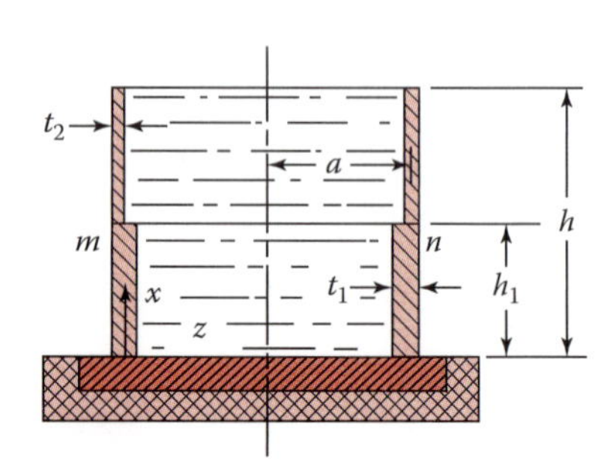

FIGURE P14.14

14.14 Figure P14.14 shows a water tower of radius a and height h filled to capacity with a liquid of specific weight γ. The upper half portion of the tank is constructed of sheet steel of thickness t_2 and the lower half of sheet of thickness t_1. Determine the values of the discontinuity moments and shearing forces along the joint at mn.

Sections 14.5 through 14.11

14.15 A long copper tube of mean diameter 0.60 m and thickness $t = 5$ mm is heated in oil to 120°C above room temperature 20°C. Assume that the ends of the tube are simply supported. Given: $E_c = 120$ GPa, $\alpha_c = 17 \times 10^{-6}$/°C, $\nu_c = 0.3$. What are the maximum thermal stresses?

14.16 Calculate the thermal deflection and stresses produced at a distance of $x = 40$ mm from an end in a long steel pipe with built-in edges, for an increase in the uniform temperature of the pipe 50°C. Use $a = 0.6$ m, $t = 12$ mm, $E = 200$ GPa, $\nu = 0.3$, and $\alpha = 11.7 \times 10^{-6}$/°C.

14.17 A long brass cylinder of mean radius $a = 0.5$ m and thickness $t = 10$ mm has the uniform temperatures $T_1 = 100$°C and $T_2 = 300$°C at the inner and the outer surfaces, respectively. Calculate the maximum stress if the ends of the shell are assumed to be free. Given: $E_b = 100$ GPa, $\alpha_b = 19 \times 10^{-6}$/°C, and $\nu_b = 0.3$.

14.18 Compute the stresses in the composite cylinder described in Example 14.3 (Figure 14.8) for uniform temperature differentials equal to $\Delta T_1 = \Delta T_2 = \Delta T_3 = 134$°C.

14.19 Determine the maximum discontinuity and the membrane stresses in the vessel shown in Figure 14.10 for $p = 1$ MPa, $a = 0.6$ m, $t = 10$ mm, and $\nu = 0.3$.

14.20 Find the maximum discontinuity and the membrane stresses in the vessel of Figure 14.12 for $p = 1$ MPa, $a = 0.6$ m, $b = 0.3$ m, $t = 10$ mm, and $\nu = 0.3$.

14.21 Figure P14.21 depicts a portion of cylindrical pressure vessel consisting of a cylinder joined to a semispherical ends of different thickness (see also, Figure 14.11). Compute the force Q_1 per unit length and moment M_1 per unit length for the shell. *Given*: $a = a' = 0.5$ m, $t_c = 10$ mm, $t_h = 7$ mm, $\nu = 0.3$, $p = 6$ MPa.

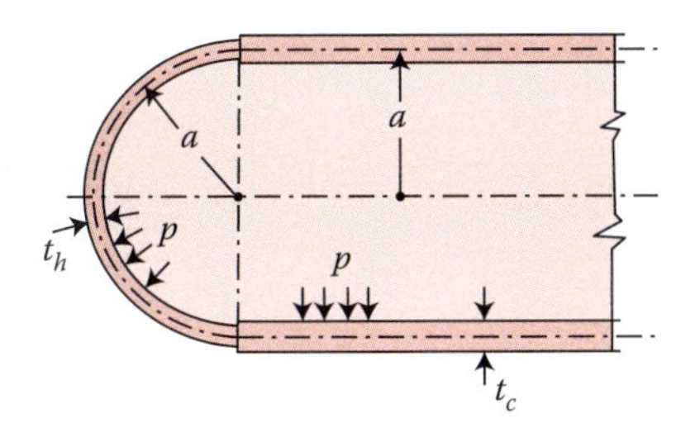

FIGURE P14.21

14.22 Find the maximum discontinuity and membrane stresses in the flat-head vessel of Figure 14.13 for $p = 1$ MPa, $a = 0.6$ m, $t_c = t_h = 10$ mm, and $\nu = 0.3$.

14.23 A spherical steel vessel is rated to operate at an internal pressure of up to 2 MPa. The radius and strength of the shell are $r = 0.5$ m, $e = 0.8$, $\sigma = 100$ MPa. Determine the minimum required thickness of vessel according to: (a) the membrane theory; (b) the ASME pressure-vessel code formula.

References

1. W. Flügge, *Stresses in Shells*, 2nd ed., Springer, Berlin, 1990.
2. S. Timoshenko and S. Woinowsky-Krieger, *Theory of Plates and Shells*, 2nd ed., Chapter 15, McGraw-Hill, New York, 1959.
3. J. E. Gibson, *Thin Shells: Computing and Theory*, Chapter 3, Pergamon, New York, 1980.
4. J. H. Faupel and F. E. Fisher, *Engineering Design*, 2nd ed., Wiley, New York, 1981.
5. A. C. Ugural, *Mechanics of Materials*, Wiley, Hoboken, NJ, 2008.
6. H. D. Tabakman and Y. J. Lin, Quick way to calculate thermal stresses in cylindrical shells, *Machine Design*, September 21, 1978.
7. A. H. Burr and J. B. Cheatham, *Mechanical Analysis and Design*, Chapter 8, Prentice-Hall, Upper Saddle River, NJ, 1995.
8. J. F. Harvey, *Theory and Design of Pressure Vessels*, 2nd ed., Van Nostrand Reinhold, New York, 1991.
9. W. C. Young, R. C. Budynas, and A. M. Sadegh, *Roark's Formulas for Stress and Strain*, 8th ed., Chapter 12, McGraw-Hill, New York, 2011.
10. *The ASME Boiler and Pressure Vessel Code, Sections I through XI*, ASME, New York, 2015.
11. *International Design Criteria of Boilers and Pressure Vessels*, ASME, New York, 2013.
12. *Fiberglass-Reinforced Plastic Vessels, Section X, ASME Boiler and Pressure Vessels Code*, ASME, New York, 1992.
13. J. P. Ellenberger, *Pressure Vessels, The ASME Code Simplified*, 8th ed., McGraw-Hill, New York, 2004.

15

Cylindrical Shells under General Loads

15.1 Introduction

This chapter is devoted to methods for determining the deflection and stress in *closed and open circular cylindrical shells* subjected to a *variety of loads*. In order to develop the relationships involving the applied-loading, cross-sectional, and material properties of a shell, internal stress resultants, and deformations, the approach applied earlier in Chapter 13 is again employed. This requires, first, that the equations governing the stress variation within the shell be developed (Section 15.2); second, that the deformation causing strain be related to the stress resultants through the appropriate stress–strain relationships as well as to the loading (Sections 15.3 through 15.5); and, finally, that the governing equations for deflection be solved by satisfying the edge conditions (Sections 15.6 through 15.9). Special cases of the above-described *general procedure* form the basis of the theories of plates and beams.

The *inextensional-shell theory* and the *shallow-shell theory* are *simplified* shell theories. The former is valid when no midsurface straining of the plate occurs because of a particular type of loading (Section 15.10). The latter applies to thin shells of very large radius or small depth. Topics on shells made from orthotropic and composite materials are included in Sections 15.11 and 15.12.

To complete the analysis of shell deformation and stress, the critical loads must also be considered. *Buckling* action underscores the difference in physical behavior of a thin shell under compression and tension. The critical load of a cylinder subjected to axial compression will be developed in Sections 15.15 and 15.16. The buckling is also of practical interest under torsion, bending, pressure, or combined loading. The literature treating bending and stability of cylindrical circular shells is extensive [1–12].

15.2 Differential Equations of Equilibrium

To establish the *differential equations of equilibrium* for a *circular cylindrical shell,* we proceed as in Section 13.6, this time taking into account *all* stress resultant and surface-loading components. An element separated from a cylinder of radius a will now have acting on it the internal and surface-force resultants (Figure 15.1a) and moment resultants (Figure 15.1b). Clearly, the force and moment intensities vary across the elements. To simplify the diagrams, the notation N_x^+ is employed to denote $N_x + (\partial N_x/\partial x)dx$ and so on.

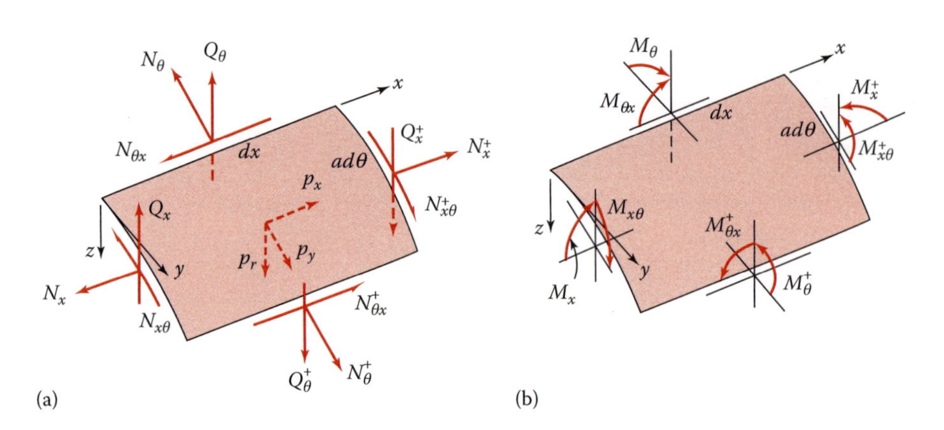

FIGURE 15.1
Cylindrical shell element: (a) With internal force resultants and surface loads; (b) with internal moment resultants.

Equilibrium of forces in the x, y, and z directions, after dividing by $dx\,d\theta$, now requires that

$$
\begin{aligned}
&a\frac{\partial N_x}{\partial x}+\frac{\partial N_{\theta x}}{\partial \theta}+p_x a=0\\
&\frac{\partial N_\theta}{\partial \theta}+a\frac{\partial N_{x\theta}}{\partial x}-Q_\theta+p_y a=0\\
&\frac{\partial Q_\theta}{\partial \theta}+a\frac{\partial Q_x}{\partial x}+N_\theta+p_r a=0
\end{aligned}
\tag{15.1a–c}
$$

In a similar manner, summation of moments, relative to the x and y coordinate directions, respectively, yields the expressions

$$
\begin{aligned}
&\frac{\partial M_\theta}{\partial \theta}+a\frac{\partial M_{x\theta}}{\partial x}-aQ_\theta=0\\
&a\frac{\partial M_x}{\partial x}+\frac{\partial M_{\theta x}}{\partial \theta}-aQ_x=0
\end{aligned}
\tag{15.1d, e}
$$

The sixth equilibrium equation, $\Sigma M_z=0$, leads to an identity and provides no new information.

The transverse shears Q_z and Q_θ may be eliminated from the above expressions. As already noted in Section 13.2, for thin shells, $N_{x\theta}=N_{\theta x}$ and $M_{x\theta}=M_{\theta x}$. Collection of the *equilibrium equations* and rearrangement then yield

$$
\begin{aligned}
&a\frac{\partial N_x}{\partial x}+\frac{\partial N_{x\theta}}{\partial \theta}+p_x a=0\\
&\frac{\partial N_\theta}{\partial \theta}+a\frac{\partial N_{x\theta}}{\partial x}-\frac{1}{a}\frac{\partial M_\theta}{\partial \theta}-\frac{\partial M_{x\theta}}{\partial x}+p_y a=0\\
&\frac{1}{a}\frac{\partial^2 M_\theta}{\partial \theta^2}+2\frac{\partial^2 M_{x\theta}}{\partial x\partial \theta}+a\frac{\partial^2 M_x}{\partial x^2}+N_\theta+p_r a=0
\end{aligned}
\tag{15.2a–c}
$$

These three expressions contain six unknown stress resultants. It is therefore necessary to identify three additional relationships, based on the deformation of the shell as discussed in Section 15.3.

We note that in the derivation of Equations 15.2, *stretching* of the midsurface and change in curvature of the shell element were neglected. Thus, the results obtained hold *only* if the forces N_x, N_y, and N_{xy} are *small* in comparison with their *critical* values, at which buckling of the shell may take place.

15.3 Kinematic Relationships

It is seen in Section 13.3 that the elongation at any distance normal to the midsurface of a thin shell is due to the midsurface stretch and the change in curvature associated with deformation. Hence, the components of the *strain at any point* through the thickness of the shell, from Equations 13.2, may be written in the form

$$\begin{Bmatrix} \varepsilon_x \\ \varepsilon_\theta \\ \gamma_{x\theta} \end{Bmatrix} = \begin{Bmatrix} \varepsilon_{x0} \\ \varepsilon_{y0} \\ \gamma_{xy0} \end{Bmatrix} - z \begin{Bmatrix} \chi_x \\ \chi_\theta \\ 2\chi_{x\theta} \end{Bmatrix} \tag{15.3}$$

For the cylindrical shells under consideration, the definition of each term in the preceding expressions is given in the discussions to follow.

In studying the *geometry* of strains and curvatures of a nonsymmetrically deformed cylindrical shell, we may begin as in the analysis of plates and shells of revolution. Kinematic expressions relating the *midsurface* strains to the displacements are thus

$$\varepsilon_{x0} = \frac{\partial u}{\partial x} \qquad \varepsilon_{\theta 0} = \frac{1}{a}\frac{\partial v}{\partial \theta} - \frac{w}{a} \qquad \gamma_{x\theta 0} = \frac{1}{a}\frac{\partial u}{\partial \theta} + \frac{\partial v}{\partial x} \tag{15.4a–c}$$

The components ε_{x0} and $\gamma_{x\theta 0}$ are the same as in the case of the plates except that $a\,d\theta$ is replaced by dy. Circumferential strain $\varepsilon_{\theta 0}$ is obtained from Equation 12.25a by setting $r_1 = a$.

Changes in curvatures, χ_x and χ_θ, and twist $\chi_{x\theta}$ are expressed by

$$\chi_x = \frac{\partial^2 w}{\partial x^2} \qquad \chi_\theta = \frac{1}{a^2}\left(\frac{\partial v}{\partial \theta} + \frac{\partial^2 w}{\partial \theta^2}\right) \qquad \chi_{x\theta} = \frac{1}{a}\left(\frac{\partial v}{\partial x} + \frac{\partial^2 w}{\partial x \partial \theta}\right) \tag{15.4d–f}$$

The terms of χ_θ are determined by considering the deformation of a circumferential element (Figure 15.2a). Solid line AB of length ds represents the side view of the midsurface prior to deformation, having a curvature in the circumferential direction: $\partial\theta/\partial s = \partial\theta/a\partial\theta = 1\,/a$. It is then displaced to the position $A'B'$, which is now ds' long. Its curvature then becomes

$$\frac{\partial \theta'}{\partial s'} \approx \frac{d\theta + (\partial^2 w/\partial s^2)ds}{(a-w)d\theta}$$

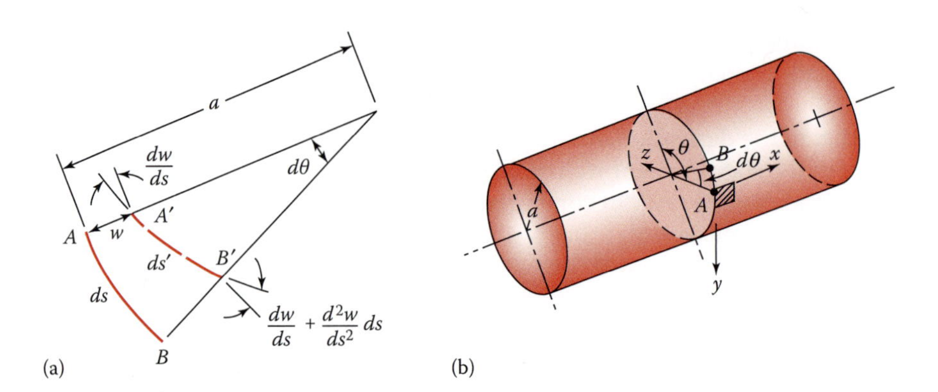

FIGURE 15.2
(a) Deformation of a circumferential shell element; (b) twist of a shell element.

Hence,

$$\chi_\theta = \frac{\partial\theta'}{\partial s'} - \frac{1}{a} \approx \frac{1}{a^2}\left(w + \frac{\partial^2 w}{\partial\theta^2}\right)$$

Note that an increase in circumferential length dv due to the radial displacement w is only $w\,d\theta$. Neglecting the effect of circumferential midsurface straining, we can therefore replace w in the above expression by $\partial v/\partial\theta$ to obtain Equation 15.4e.

With respect to the terms of the $\chi_{x\theta}$ in Equations 15.4f, representing the *twist* of the midsurface of each *shell element*, as indicated at point A in Figure 15.2b, additional *explanation* is required. During deformation, this element rotates through an angle $-\partial w/\partial x$ about the y axis and through an angle $\partial v/\partial x$ about the z axis. Here the sign convention used is based on the right-hand-screw rule. Now consider a similar element at point B. Because of displacement w, the second element experiences the following angular displacement about the y axis:

$$-\frac{\partial w}{\partial x} - \frac{\partial^2 w}{\partial\theta\partial x}d\theta \qquad \text{(a)}$$

The same element also experiences the following angular displacement in the plane tangent to the shell:

$$\frac{\partial v}{\partial x} + \frac{\partial(\partial v/\partial x)}{\partial\theta}d\theta$$

Neglecting a higher-order infinitesimal term, it is found that the latter rotation, due to angle $d\theta$, has a component with respect to the y axis:

$$-\frac{\partial v}{\partial x}d\theta \qquad \text{(b)}$$

Comment: We observe from Equations a and b that the total angle of twist between the element at A and B is equal to

$$-\chi_{x\theta} a\, d\theta = -\left(\frac{\partial^2 w}{\partial\theta\partial x} + \frac{\partial v}{\partial x}\right) d\theta$$

This expression yields the result given by Equation 15.4f.

15.4 The Governing Equations for Deflections

The expressions governing the deformation of cylindrical shells subjected to direct and bending forces can now be developed. This is accomplished by first introducing Equations 15.4 into the constitutive equations (Equations 13.4). In so doing, we have the following *stress–resultant displacement* relations:

$$\begin{aligned}
N_x &= \frac{Et}{1-\nu^2}\left[\frac{\partial u}{\partial x} + \nu\left(\frac{1}{a}\frac{\partial v}{\partial \theta} - \frac{w}{a}\right)\right] \\
N_\theta &= \frac{Et}{1-\nu^2}\left[\frac{1}{a}\frac{\partial v}{\partial \theta} - \frac{w}{a} + \nu\frac{\partial u}{\partial x}\right] \\
N_{x\theta} &= \frac{Et}{2(1+\nu)}\left(\frac{\partial v}{\partial x} + \frac{1}{a}\frac{\partial u}{\partial \theta}\right) \\
M_x &= -D\left[\frac{\partial^2 w}{\partial x^2} + \frac{\nu}{a^2}\left(\frac{\partial v}{\partial \theta} + \frac{\partial^2 w}{\partial \theta^2}\right)\right] \\
M_\theta &= -D\left[\frac{1}{a^2}\left(\frac{\partial v}{\partial \theta} + \frac{\partial^2 w}{\partial \theta^2}\right) + \nu\frac{\partial^2 w}{\partial x^2}\right] \\
M_{x\theta} &= -D(1-\nu)\frac{1}{a}\left(\frac{\partial v}{\partial x} + \frac{\partial^2 w}{\partial x\partial \theta}\right)
\end{aligned} \tag{15.5a–f}$$

Then, on substituting Equations 15.5 into Equations 15.2, there is derived a set of three expressions in three displacements u, v, and w:

$$\frac{\partial^2 u}{\partial x^2} + \frac{1-\nu}{2a^2}\frac{\partial^2 u}{\partial \theta^2} + \frac{1+\nu}{2a}\frac{\partial^2 v}{\partial x\partial \theta} - \frac{\nu}{a}\frac{\partial w}{\partial x} = -\frac{p_x(1-\nu^2)}{Et} \tag{15.6a}$$

$$\begin{aligned}
&\frac{1+\nu}{2a}\frac{\partial^2 u}{\partial x\partial \theta} + \frac{1-\nu}{2}\frac{\partial^2 v}{\partial x^2} + \frac{1}{a^2}\frac{\partial^2 v}{\partial \theta^2} - \frac{1}{a^2}\frac{\partial w}{\partial \theta} + \frac{t^2}{12a^2}\left(\frac{\partial^3 w}{\partial x^2\partial \theta} + \frac{1}{a^2}\frac{\partial^3 w}{\partial \theta^3}\right) \\
&+ \frac{t^2}{12a^2}\left[(1-\nu)\frac{\partial^2 v}{\partial x^2} + \frac{1}{a^2}\frac{\partial^2 v}{\partial \theta^2}\right] = -\frac{p_y(1-\nu^2)}{Et}
\end{aligned} \tag{15.6b}$$

$$\begin{aligned}
&\frac{\nu}{a}\frac{\partial u}{\partial x} + \frac{1}{a^2}\frac{\partial v}{\partial \theta} - \frac{w}{a^2} - \frac{t^2}{12}\left(\frac{\partial^4 w}{\partial x^4} + \frac{2}{a^2}\frac{\partial^4 w}{\partial x^2\partial \theta^2} + \frac{1}{a^4}\frac{\partial^4 w}{\partial \theta^4}\right) \\
&- \frac{t^2}{12a^2}\left[(2-\nu)\frac{\partial^3 v}{\partial x^2\partial \theta} + \frac{1}{a^2}\frac{\partial^3 v}{\partial \theta^3}\right] = -\frac{p_r(1-\nu^2)}{Et}
\end{aligned} \tag{15.6c}$$

These are the *governing equations for the displacements* in thin-walled circular cylindrical shells under *general* loading.

15.5 *Approximate Relations

Equations 15.6 contain a number of terms that in many applications yield numerical results of negligible practical importance [1,2]. The last two terms on the left-hand side of Equation 15.6b and the last term on the left-hand side of Equation 15.6c are of this type. Such quantities are of the same order as those that have already been neglected in Sections 13.3 and 15.2 for the thin shells ($t \ll a$). By dropping these terms of small magnitude, much time and effort is saved in the solving for the displacements. By doing so, the following *simplified* set of *governing equations* for displacements of circular cylindrical shells are obtained:

$$\begin{aligned}
\frac{\partial^2 u}{\partial x^2}+\frac{1-\nu}{2a^2}\frac{\partial^2 u}{\partial \theta^2}+\frac{1+\nu}{2a}\frac{\partial^2 v}{\partial x \partial \theta}-\frac{\nu}{a}\frac{\partial w}{\partial x} &= -\frac{p_x(1-\nu^2)}{Et}\\
\frac{1+\nu}{2a}\frac{\partial^2 u}{\partial x \partial \theta}+\frac{1-\nu}{2}\frac{\partial^2 v}{\partial x^2}+\frac{1}{a^2}\frac{\partial^2 v}{\partial \theta^2}-\frac{1}{a^2}\frac{\partial w}{\partial \theta} &= -\frac{p_y(1-\nu^2)}{Et}\\
\frac{\nu}{a}\frac{\partial u}{\partial x}+\frac{1}{a^2}\frac{\partial v}{\partial \theta}-\frac{w}{a^2}-\frac{t^2}{12}\left(\frac{\partial^4 w}{\partial x^4}+\frac{2}{a^2}\frac{\partial^4 w}{\partial x^2 \partial \theta^2}+\frac{1}{a^4}\frac{\partial^4 w}{\partial \theta^4}\right) &= -\frac{p_r(1-\nu^2)}{Et}
\end{aligned} \tag{15.7}$$

When a circular cylindrical shell is loaded symmetrically with respect to its axis, as expected the solution (Equations 15.6 or 15.7) can readily be reduced to the form given by Equations 13.15. Several examples of such a case are given in Section 13.7 and Chapter 14.

We observe that the terms corresponding to the moments M_θ and $M_{x\theta}$ in Equation 15.2b disappeared in the second of the approximate set. These moments result from the transverse shear force Q_θ in Equation 15.1b. It seems plausible therefore that we can take $Q_\theta = 0$ on Equation 15.1b. Then the *equations of equilibrium* (Equations 15.1) can be rewritten in the *simplified* form:

$$\begin{aligned}
a\frac{\partial N_x}{\partial x}+\frac{\partial N_{x\theta}}{\partial \theta}+p_x a &= 0\\
\frac{\partial N_\theta}{\partial \theta}+a\frac{N_{x\theta}}{\partial x}+p_y a &= 0\\
\frac{\partial Q_\theta}{\partial \theta}+a\frac{\partial Q_x}{\partial x}+N_\theta+p_r a &= 0\\
\frac{\partial M_\theta}{\partial \theta}+a\frac{\partial M_{x\theta}}{\partial x}-aQ_\theta &= 0\\
a\frac{\partial M_x}{\partial x}+\frac{\partial M_{x\theta}}{\partial \theta}-aQ_x &= 0
\end{aligned} \tag{15.8}$$

Hence, Equation 15.2 becomes

$$
\begin{aligned}
a\frac{\partial N_x}{\partial x}+\frac{\partial N_{x\theta}}{\partial \theta}+p_x a=0 \\
\frac{\partial N_\theta}{\partial \theta}+a\frac{\partial N_{x\theta}}{\partial x}+p_y a=0 \\
\frac{1}{a}\frac{\partial^2 M_\theta}{\partial \theta^2}+2\frac{\partial^2 M_{x\theta}}{\partial x \partial \theta}+a\frac{\partial^2 M_x}{\partial x}+N_\theta+p_r a=0
\end{aligned}
\tag{15.9}
$$

Based on the simplifications that lead to Equations 15.7, the effect of displacement u and v on the bending and twisting moments must be regarded as negligible. From Equations 15.5, the *simplified elastic law* is then

$$
\begin{aligned}
N_x &= \frac{Et}{1-\nu^2}\left[\frac{\partial u}{\partial x}+\nu\left(\frac{1}{a}\frac{\partial v}{\partial \theta}-\frac{w}{a}\right)\right] \\
N_\theta &= \frac{Et}{1-\nu^2}\left(\frac{1}{a}\frac{\partial v}{\partial \theta}-\frac{w}{a}+\nu\frac{\partial u}{\partial x}\right) \\
N_{x\theta} &= \frac{Et}{2(1+\nu)}\left(\frac{\partial v}{\partial x}+\frac{1}{a}\frac{\partial u}{\partial \theta}\right) \\
M_x &= -D\left(\frac{\partial^2 w}{\partial x^2}+\frac{\nu}{a^2}\frac{\partial^2 w}{\partial \theta^2}\right) \\
M_\theta &= -D\left(\frac{1}{a^2}\frac{\partial^2 w}{\partial \theta^2}+\nu\frac{\partial^2 w}{\partial x^2}\right) \\
M_{x\theta} &= -D(1-\nu)\frac{1}{a}\frac{\partial^2 w}{\partial x \partial \theta}
\end{aligned}
\tag{15.10}
$$

Comments: It is seen that the problem of circular cylindrical shells reduces, in each particular case, to the solution of a set of three differential equations (Equations 15.6 or 15.7). The latter set is often preferred for practical applications. A particular example of such a solution is illustrated in the next section.

15.6 A Typical Case of Asymmetrical Loading

A pipe supported on saddles at intervals or at the edges and filled partially with liquid is a typical case of axially *unsymmetrically loaded* cylindrical shells.

Consider a circular *cylindrical tank filled* to a given level *with a liquid* of specific weight γ (Figure 15.3a). Both ends of the tank may be taken as supported by end plates so that, at $x=0$ and $x=L$,

$$
v=0 \qquad w=0 \qquad N_x=0 \qquad M_x=0 \tag{a}
$$

It is observed that Equations a and the conditions of symmetry of deformation are satisfied by assuming the following displacements:

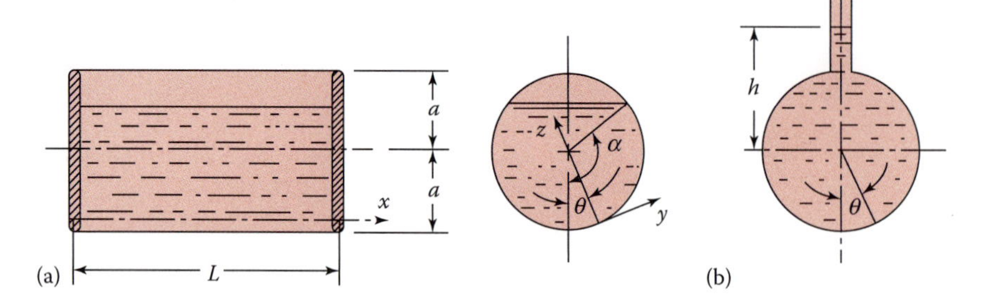

FIGURE 15.3
A cylinder supported at edges: (a) Filled partially; (b) filled to capacity.

$$
\begin{aligned}
u &= \sum_{m=1}^{\infty}\sum_{n=0}^{\infty} a_{mn}\cos n\theta \cos\frac{m\pi x}{L} \\
v &= \sum_{m=1}^{\infty}\sum_{n=0}^{\infty} b_{mn}\sin n\theta \sin\frac{m\pi x}{L} \\
w &= \sum_{m=1}^{\infty}\sum_{n=0}^{\infty} c_{mn}\cos n\theta \sin\frac{m\pi x}{L}
\end{aligned}
\tag{15.11}
$$

in which the angle θ is measured as shown in Figure 15.3a. The upper part ($\alpha \leq \theta \leq \pi$) of the tank carries no load. For the lower part ($0 \leq \theta \leq \alpha$),

$$
p_r = p = -\gamma a(\cos\theta - \cos\alpha) \qquad p_x = p_y = 0 \tag{15.12}
$$

Here the angle α defines the liquid level. It is useful to expand the foregoing load into a Fourier series:

$$
p = -\sum_{m}^{\infty}\sum_{n}^{\infty} p_{mn}\cos n\theta \sin\frac{m\pi x}{L} \tag{15.13}
$$

The coefficients p_{mn}, determined in the usual manner (Section A.3), are of the form

$$
p_{mn} = \frac{8\gamma a(\cos\alpha \sin n\alpha - n\cos n\alpha \sin\alpha)}{mn\pi^2(n^2-1)} \qquad \begin{pmatrix} m = 1,3,\ldots \\ n = 2,3,\ldots \end{pmatrix} \tag{15.14a}
$$

$$
p_{m0} = \frac{4\gamma a}{m\pi^2}(\sin\alpha - \alpha\cos\alpha) \qquad p_{m1} = \frac{2\gamma a}{m\pi}(2\alpha - \sin\alpha) \tag{15.14b}
$$

If the cylinder is *filled to capacity* (Figure 15.3b), the loading is expressed as

$$
p = -y(h + a\cos\theta)
$$

and the coefficients are

$$p_{mn} = 0 \qquad p_{m0} = \frac{4\gamma h}{m\pi} \qquad p_{m1} = \frac{4\gamma a}{m\pi} \qquad (m = 1,3,\ldots) \tag{b}$$

We now introduce the *notation*

$$\lambda = \frac{L}{a} \qquad \eta = \frac{t}{2L}$$

The *governing equations* (Equations 15.7), on substitution of Equations 15.11 and 15.13 and use of the above notation, are therefore

$$\begin{aligned} a_{mn}[2m^2\pi^2 + (1-\nu)\lambda^2 n^2] - b_{mn}(1+\nu)\lambda mn\pi + c_{mn}2\nu\lambda m\pi &= 0 \\ a_{mn}(1+\nu)\lambda mn\pi - b_{mn}[(1-\nu)m^2\pi^2 + 2\lambda^2 n^2] + c_{mn}2\lambda^2 n &= 0 \\ a_{mn}3\nu\lambda m\pi - b_{mn}3\lambda^2 n + c_{mn}[3\lambda^2 + \eta^2(m^2\pi^2 + \lambda^2 n^2)^2] &= -\frac{p_{mn}L^2 t^2}{4D} \end{aligned} \tag{15.15}$$

Inasmuch as p_{mn} is given by Equations 15.14 or b, the parameters a_{mn}, b_{mn}, c_{mn} can be determined in each particular case on application of Equations 15.15 for any m and n. The stress resultant and displacement at any cylinder location may then be ascertained using Equations 15.10 and 15.11.

EXAMPLE 15.1: Analysis of Liquid Filled Cylindrical Tank

Calculate the maximum values of lateral deflection and stress in the tank filled with liquid ($\alpha = \pi$), shown in Figure 15.3b. Use $a = h = 50$ cm, $L = 25$ cm, $t = 7$ cm, and $\nu = 0.3$.

Solution

As $a = h$, by setting

$$Z = \frac{2\gamma a t^2 t}{\pi^2 D}$$

for $n = 0$, $m = 1, 3, \ldots$ Equations 15.15 and b yield

$$c_{m0} = -\frac{m\pi}{\lambda\nu}a_{m0} \qquad b_{m0} = 0 \qquad a_{m0} = \frac{\nu Z\lambda t}{2m^2[3\lambda^2(1-\nu^2) + \eta^2 m^4\pi 4]} \tag{c}$$

For $n = 1$, the expressions for the coefficients become more complicated. The numerical values of the coefficients are furnished in Table 15.1 for $m = 1, 3, 5$ and $n = 0, 1$.

The largest values of the lateral deflection, force, and moment are found at $x = L/2$, $\theta = 0$. These will be computed by taking $m = 1, 3, 5$ and $n = 0.1$. On applying the third of Equations 15.11, together with the given data, we have

$$w_{\max} = c_{10} + c_{11} - c_{30} - c_{31} + c_{50} + c_{51} = -11{,}768.7\frac{\gamma}{E}\,\text{cm}$$

In a like manner, from Equations 15.11 and 15.10,

TABLE 15.1

Values of Coefficients of Equation c

m	$a_{m0}\dfrac{2(10^3)}{Zt}$	$c_{m0}\dfrac{2(10^3)}{Zt}$	$a_{m1}\dfrac{2(10^3)}{Zt}$	$b_{m1}\dfrac{2(10^3)}{Zt}$	$c_{m1}\dfrac{2(10^3)}{Zt}$
1	57.88	−1.212	49.18	−66.26	−1.183
2	0.1073	−6.742	0.1051	−0.0432	−6.704
5	0.00503	−0.526	0.00499	−0.00122	−0.525

$$N_{x,\max} = 19.68\gamma \text{ N/cm}$$
$$N_{\theta,\max} = 1607.7\gamma \text{ N/cm}$$
$$M_{x,\max} = -5657.5\gamma \text{ N}$$
$$M_{\theta,\max} = -1763.7\gamma \text{ N}$$

Through the use of Equation 13.5, we then obtain

$$\sigma_{x,\max} = \frac{19.68\gamma}{7} - \frac{6(5657.5\gamma)}{49} = -689.9\gamma \text{ N/cm}^2$$
$$\sigma_{\theta,\max} = \frac{1607.7\gamma}{7} - \frac{6(1763.7\gamma)}{49} = 13.7\gamma \text{ N/cm}^2$$

as the maximum axial and tangential stresses in the tank.

Comments: It is observed in the foregoing example that the coefficients a_{mn}, b_{mn}, and c_{mn} diminish very rapidly. We conclude therefore that only a few terms in the series (Equations 15.11) surface to yield fairly accurate results. However, in the case of longer and thinner shells a_{mn}, b_{mn}, and c_{mn} diminish rather slowly, and it is necessary to calculate a somewhat large number of coefficients in order to obtain the deformation and stress resultants with sufficient accuracy.

15.7 Curved Circular Panels

Consider now the bending of a *portion of a cylindrical shell* that is simply supported along the edges and under a distributed load $p(x, \theta)$. The midsurface of such a *curved circular panel* is defined by the angle α, the length L, and the radius a (Figure 15.4). In treating this problem, we can also apply the same method employed in Section 15.6.

The boundary conditions are represented as follows:

$$\begin{aligned} &w = 0 \qquad N_\theta = 0 \qquad M_\theta = 0 \qquad (\theta = 0, \theta = \alpha) \\ &w = 0 \qquad N_x = 0 \qquad M_x = 0 \qquad (\theta = 0, \theta = L) \end{aligned} \tag{a}$$

These requirements are satisfied by taking the displacements in the form

$$\begin{aligned} u &= \sum_{m=1}^{\infty}\sum_{n=1}^{\infty} a_{mn} \sin\frac{n\pi\theta}{\alpha}\cos\frac{m\pi x}{L} \\ v &= \sum_{m=1}^{\infty}\sum_{n=1}^{\infty} b_{mn} \cos\frac{n\pi\theta}{\alpha}\sin\frac{m\pi x}{L} \\ w &= \sum_{m=1}^{\infty}\sum_{n=1}^{\infty} c_{mn} \sin\frac{n\pi\theta}{\alpha}\sin\frac{m\pi x}{L} \end{aligned} \tag{15.16}$$

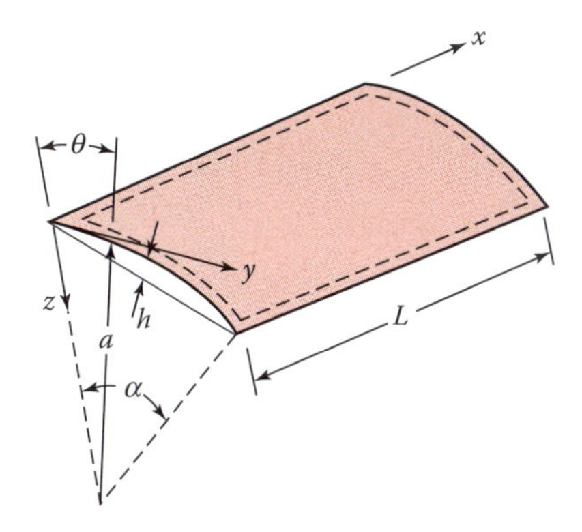

FIGURE 15.4
Simply supported curved panel.

The distributed load per surface area may be described by the Fourier series,

$$p_r = \sum_{m=1}^{\infty}\sum_{n=1}^{\infty} p_{mn} \sin\frac{n\pi\theta}{\alpha} \sin\frac{mn\pi}{L} \qquad p_x = p_y = 0 \tag{15.17}$$

where the coefficients are obtained in the usual way. Carrying Equations 15.16 and 15.17 into the *governing expressions* (Equations 15.7) gives

$$\begin{aligned}
&a_{mn}\pi\left[\left(\frac{am}{L}\right)^2 + \frac{(1-\nu)n^2}{2\alpha^2}\right] + b_{mn}\pi\frac{(1+\nu)amn}{2\alpha L} + c_{mn}\frac{\nu am}{L} = 0 \\
&a_{mn}\pi\frac{(1+\nu)amn}{2\alpha L} + b_{mn}\pi\left[\frac{(1-\nu)a^2m^2}{2L^2} + \frac{n^2}{\alpha^2}\right] + c_{mn}\frac{n}{\alpha} = 0 \\
&a_{mn}\nu\pi\frac{am}{L} + b_{mn}\frac{n\pi}{\alpha} + c_{mn}\left[1 + \frac{\pi^4 t^2}{12a^2}\left(\frac{a^2m^2}{L^2} + \frac{n^2}{L^2}\right)^2\right] = p_{mn}\frac{a^2(1-\nu^2)}{Et}
\end{aligned} \tag{15.18}$$

We see that the foregoing procedure and that used for the simply supported rectangular plates in Section 5.2 are alike. If the straight edges at $\theta = 0$ and $\theta = \alpha$ of the shell are simply supported and the curved edges are clamped or free, an approach similar to that of Section 5.4 can be employed. An analysis of this kind demonstrates that the solution is very involved.

Application of Equations 15.18 is illustrated in the solution of the following sample problem.

EXAMPLE 15.2: Deflection of Curved Panel

A curved panel of an aircraft is approximated as a portion of a thin cylinder having a small angle α and a small sag $h = a\,[1 - \cos(\alpha/2)]$ (Figure 15.4). The shell is simply supported along the edges and subject to a uniform radial pressure p_0. Outline the derivation of the expressions for the deflections.

Solution

In this situation, the coefficients p_{mn}, from Section 5.3, are

$$p_{mn} = \frac{16p_0}{\pi^2 mn} \qquad (m,n = 1,3,\ldots) \tag{b}$$

On introducing this into Equations 15.18, we can find the expressions for the constants a_{mn}, b_{mn}, and c_{mn}.

For any given numerical values a, L, α, t, E, and v, the components of the displacements are then obtained from the series (Equations 15.16). Calculations show that, for small values of the ratio h/t, the first few terms of this series lead to results of acceptable accuracy [2]. In the case of larger values of h/t, however, the series representing the moments does *not* converge.

15.8 *A Simple Theory of Bending of Curved Circular Panels

It is realized that the analysis of *bending of open shells*, in even the simplest cases, is very complicated (Section 15.7). Example 12.12 also clearly exhibits the failure of the membrane theory to accurately predict correct stress conditions in such shells. Thus, the solutions suitable for practical applications can be obtained only by introducing simplifying assumptions in the bending theory.

In this section, we discuss the *most simple method* developed by H. Schorer in 1935. This approach is applied to the *bending of open cylindrical shells* having the ends $x = \pm L/2$ (Figure 15.5) supported in such a manner that the displacements v and w at the ends vanish. Shells of this type exemplify common *roof structures*. *Schorer's solution* is based on the following *assumptions* [4]:

1. The bending in the axial planes is negligible. This implies that $M_x = 0$ and $Q_x = 0$.
2. The twisting moment $M_{x\theta}$ can be omitted.
3. The tangential strain ε_θ and the shearing strain $\gamma_{x\theta}$ are small compared with the axial strain and may be ignored.
4. Poisson's ratio may be neglected.

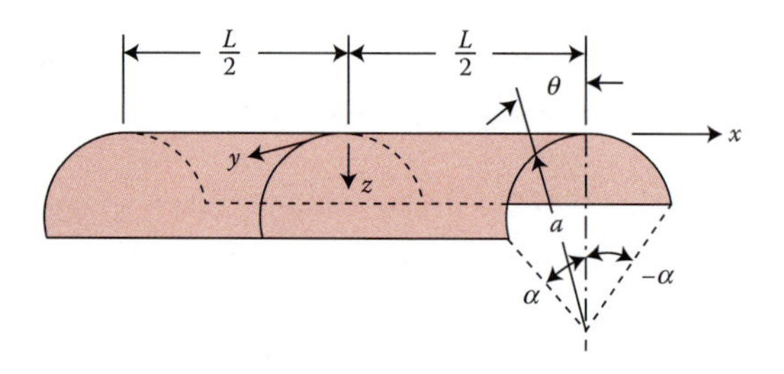

FIGURE 15.5
Curved circular panel with $v = w = 0$ at ends ($x = \pm L/2$); a typical roof shell.

Adequate experimental and theoretical justification may be found for the simplifications stated with respect to the state of stress resultants and deformations. With the foregoing presuppositions, the basic *shell equations* can be considerably *reduced*, as shown in the following paragraphs.

Substituting $M_x = 0$, $M_{x\theta} = 0$, and $\nu = 0$ into Equations 15.10, we obtain the *elastic law* in the form

$$
\begin{aligned}
N_x &= Et\frac{\partial u}{\partial x} \\
N_\theta &= \frac{Et}{a}\left(\frac{\partial v}{\partial \theta} - w\right) \\
N_{x\theta} &= \frac{Et}{a}\left(\frac{\partial v}{\partial \theta} + \frac{1}{a}\frac{\partial u}{\partial \theta}\right) \\
M_\theta &= -\frac{Et}{a^2}\frac{\partial^2 w}{\partial \theta^2}
\end{aligned}
\tag{15.19}
$$

where $I = t^3/12$. Insertion of $M_x = 0$ and $M_{x\theta} = 0$ into the *equations of equilibrium* (Equations 15.8) yields

$$
\begin{aligned}
&a\frac{\partial N_x}{\partial x} + \frac{\partial N_{x\theta}}{\partial \theta} + p_x a = 0 \\
&\frac{\partial N_\theta}{\partial \theta} + a\frac{\partial N_{x\theta}}{\partial x} + p_y a = 0 \\
&\frac{\partial Q_\theta}{\partial \theta} + N_\theta + p_r a = 0 \\
&\frac{\partial M_\theta}{\partial \theta} - aQ_\theta = 0
\end{aligned}
\tag{15.20a–d}
$$

The transverse shear force Q_θ can readily be eliminated from the last two of these equations

$$\frac{1}{a}\frac{\partial^2 M_\theta}{\partial \theta^2} + N_\theta + p_r a = 0 \tag{a}$$

Similarly, elimination of $N_{x\theta}$ between Equations 15.20a and 15.20b leads to

$$\frac{\partial^2 N_x}{\partial x^2} - \frac{1}{a^2}\frac{\partial^2 N_\theta}{\partial \theta^2} + \frac{\partial p_x}{\partial x} - \frac{1}{a}\frac{\partial p_y}{2\theta} = 0 \tag{b}$$

Then, eliminating N_θ from Equations a and b, we have

$$\frac{1}{a^3}\frac{\partial^4 M_\theta}{\partial \theta^4} + \frac{\partial^2 N_x}{\partial x^2} + \frac{\partial p_x}{\partial x} - \frac{1}{a}\frac{\partial p_y}{\partial \theta} + \frac{1}{a}\frac{\partial^2 p_r}{\partial \theta^2} = 0 \tag{15.21}$$

The preceding expresses the open shell equilibrium in terms of two unknowns M_θ and N_x. Reduction of unknowns to one is made next by using the stress–resultant displacement relationships.

According to assumption 3, Equations 15.4 become

$$\frac{\partial v}{\partial \theta} = w \qquad \frac{1}{a}\frac{\partial u}{\partial \theta} = -\frac{\partial v}{\partial x} \tag{15.22a, b}$$

Elimination of v from Equations 15.22 gives

$$\frac{1}{a}\frac{\partial^2 u}{\partial \theta^2} = -\frac{\partial w}{\partial x} \tag{c}$$

Differentiating Equation 15.21 twice with respect to θ, we have

$$\frac{1}{a^3}\frac{\partial^6 M_\theta}{\partial \theta^6} + \frac{\partial^4 N_x}{\partial x^2 \partial \theta^2} + \frac{\partial^3 p_x}{\partial x \partial \theta^2} - \frac{1}{a}\frac{\partial^3 p_y}{\partial \theta^3} + \frac{1}{a}\frac{\partial^4 p_r}{\partial \theta^4} = 0$$

Finally, substitution of M_θ and N_x using Equation 15.19 together with Equation c into the foregoing equation results in

$$\frac{\partial^8 w}{\partial \theta^8} + \frac{ta^4}{I}\frac{\partial^4 w}{\partial x^4} = \frac{a^4}{EI}\left(a\frac{\partial^3 p_x}{\partial x \partial \theta^2} - \frac{\partial^3 p_y}{\partial \theta^3} + \frac{\partial^4 p_r}{\partial \theta^4}\right) \tag{15.23}$$

This is the *governing equation for deflection* of *open cylindrical shells* with supported ends. When there is no load acting on the shell, we have

$$\frac{\partial^8 w}{\partial \theta^8} + \frac{ta^6}{I}\frac{\partial^4 w}{\partial x^4} = 0 \tag{15.24}$$

The formulas for the *stress resultants* and the *axial and the tangential displacements* may readily be obtained by applying Equations 15.19, 15.20, and 15.22b. These are given in the form

$$\begin{aligned}
M_\theta &= -\frac{EI}{a^2}\frac{\partial^2 w}{\partial \theta^2} \\
Q_\theta &= \frac{1}{a}\frac{\partial M_\theta}{\partial \theta} \\
N_\theta &= -\frac{\partial \theta_\theta}{\partial \theta} - p_r a \\
N_{x\theta} &= \int \left(\frac{\partial N_\theta}{\partial \theta} - p_y\right) dx \\
N_x &= -\int \left(\frac{1}{a}\frac{\partial N_{x\theta}}{\partial x} - p_x\right) dx
\end{aligned} \tag{15.25}$$

and

$$u = \int \frac{N_x}{Et} dx$$
$$v = -\frac{1}{aEI} \int \frac{\partial N_x}{\partial \theta} dx \tag{15.26}$$

Here, for simplicity, the constants of integration are taken as zero.

Comments: We observe from Equations 15.25 and 15.26 that the *key to determining* the stresses and the displacements u and v is the solution of Equation 15.23 for w. This is illustrated by considering a commonly referred-to case in Section 15.9. It is noted that Equation 15.23 is also used in the approximate stress analysis of cantilever *multiclosed* cylindrical shells [4]. Examples include the design of chimneys and cores of tall buildings under lateral loading conditions such as wind.

15.9 *Curved Circular Panels with Variously Supported Edges

Consider an *open cylindrical shell* with the ends at $x = \pm L/2$ simply supported and the remaining edges at $\theta = \pm\alpha$ free (Figure 15.5). It is a common practice in *design* of these shells to express the loading by

$$p = p_0 \cos\frac{\pi x}{L} \tag{a}$$

In the preceding, the constant p_0 represents the load intensity along the line passing through $x = 0$ and parallel to the y axis.

Using Equations 12.12 and a, we express the components of the load by

$$p_x = 0$$
$$p_y = p\sin\theta = p_0 \sin\theta\cos\frac{\pi x}{L} \tag{b}$$
$$p_z = p\cos\theta = p_0 \cos\theta\cos\frac{\pi x}{L}$$

Substitution of the preceding expressions into Equation 15.23 gives

$$\frac{\partial^8 w}{\partial \theta^8} + \frac{ta^6}{I}\frac{\partial^4 w}{\partial x^4} = \frac{2a^4}{I} p_0 \cos\theta\cos\frac{\pi x}{L} \tag{15.27}$$

Since the ends of the shell are supported, we have

$$w = 0 \qquad (x = \pm L/2) \tag{c}$$

At the unsupported edge α, the following resultants vanish:

$$M_\theta = 0 \qquad Q_\theta = 0 \qquad N_\theta = 0 \qquad N_{x\theta} = 0 \qquad (\theta = \alpha) \tag{d}$$

Conditions identical to that of the foregoing occur at the other free edge at $\theta = -\alpha$.

It is observed that the boundary conditions (Equation c) are satisfied if the *particular solution* is expressed by

$$w_p = kp_0 \cos\theta \cos\frac{\pi x}{L} \tag{15.28}$$

in which k is a constant. Carrying Equation 15.28 into Equation 15.27, we have

$$k = \frac{2a^4}{EI(1 + a^6\pi^4 t/IL^4)} \tag{15.29}$$

Substituting this into Equation 15.28, we obtain the particular solution for w.

We take the *homogeneous solution* in the form

$$w_h = W(\theta)\cos\frac{\pi x}{L} \tag{15.30}$$

that fulfills the edge restraints expressed by Equation c. Here the $W(\theta)$ is a function of θ only that must be obtained so as to satisfy the conditions (d) and Equation 15.24.

Introducing Equation 15.30 into Equation 15.24, we have the ordinary differential equation:

$$\frac{d^8W}{d\theta^8} + \lambda^8 W = 0 \tag{15.31}$$

in which

$$\lambda^8 = \frac{a^6\pi^4 t}{IL^4} \tag{e}$$

The solution of Equation 15.31 may be written in the form (Problem 15.3)

$$\begin{aligned} W = 2(&A\cos\beta\theta\cosh\gamma\theta - B\sin\beta\theta\sinh\gamma\theta + C\cos\gamma\theta\cosh\beta\theta \\ &- D\sin\gamma\theta\sinh\beta\theta + E\cos\beta\theta\sinh\gamma\theta - F\sin\beta\theta\cosh\gamma\theta \\ &+ G\cos\gamma\theta\sinh\beta\theta - H\sin\gamma\theta\cosh\beta\theta) \end{aligned} \tag{f}$$

Here

$$\begin{aligned} \gamma &= 0.923979\lambda \\ \beta &= 0.382683\lambda \end{aligned} \tag{15.32}$$

and $A, B, \ldots, H$ are arbitrary constants.

The condition that the shell deflection must be symmetrical with respect to the x axis (i.e., it must have the same values for $+\theta$ and $-\theta$ [Figure 15.5]) is fulfilled by Equation f if we let $E = F = G = H = 0$. The expression for W becomes then

$$W = 2(A\cos\beta\theta\cosh\gamma\theta - B\sin\beta\theta\sinh\gamma\theta + C\cos\gamma\theta\cosh\beta\theta - D\sin\gamma\theta\sinh\beta\theta) \tag{15.33}$$

Carrying the preceding into Equation 15.30, we arrive at the complementary solution for w.

Summing Equations 15.28 and 15.30 together with Equation 15.33 yields the *complete solution for the radial deflection*. In so doing, we have

$$w = 2[A\cos\beta\theta\cosh\gamma\theta - B\sin\beta\theta\sinh\gamma\theta + C\cos\gamma\theta\cosh\beta\theta - D\sin\gamma\theta\sinh\beta\theta + \frac{a^4 p_0\cos\theta}{EI(1 + a^6\pi^4 t / IL^4)}]\cos\frac{\pi x}{L} \tag{15.34}$$

This expression satisfied Equation 15.27 and the simple support restraints at $x = \pm L/2$. The coefficients A, B, C, and D in this equation will be ascertained *after finding* the moment, shear force, tangential force, and axial force.

The *complete expressions* for the stress resultants and the axial and the tangential displacements consist of the particular and the homogeneous solutions. The *procedure* used in determining the preceding is as follows:

1. Apply Equations 15.25 together with Equations 15.28 and b to obtain a *particular solution* (denoted by a single prime):

$$\begin{aligned}
M'_\theta &= \frac{EI}{a^2}\frac{\partial^2 w_p}{\partial\theta^2} = \frac{EIk}{a^2} p_0\cos\theta\cos\frac{\pi x}{L} \\
Q'_\theta &= -\frac{EIk}{a^3} p_0\cos\theta\cos\frac{\pi x}{L} \\
N'_\theta &= \left(\frac{EIk}{a^3} - a\right) p_0\cos\theta\cos\frac{\pi x}{L} \\
N'_{x\theta} &= \frac{cL}{\pi} p_0\sin\theta\sin\frac{\pi x}{L} \\
N'_x &= \frac{cL^2}{a\pi^2}\cos\theta\cos\frac{\pi x}{L} \\
u' &= -\frac{cL^3}{aE\pi^3 t} p_0\cos\theta\sin\frac{\pi x}{L} \\
v' &= -\frac{cL^4}{a^4 E\pi^4 t} p_0\sin\theta\cos\frac{\pi x}{L}
\end{aligned} \tag{15.35}$$

where

$$c = -\frac{EIk}{a^4} - 2 \tag{15.36}$$

2. Repeat step 1 with $p_x = p_y = p_r = 0$ and Equation 15.30 to obtain a *homogeneous solution* (denoted by a double prime):

$$
\begin{aligned}
M''_\theta &= -\frac{EI}{a^2}\frac{\partial^2 W_h}{\partial\theta^2} = -\frac{EI}{a^2}\frac{d^2W}{d\theta^2}\cos\frac{\pi x}{L}\\
Q''_\theta &= -\frac{EI}{a^3}\frac{d^3W}{d\theta^3}\cos\frac{\pi x}{L}\\
N''_\theta &= \frac{EI}{a^3}\frac{d^4W}{d\theta^4}\cos\frac{\pi x}{L}\\
N''_{x\theta} &= \frac{EIL}{a^4\pi}\frac{d^5W}{d\theta^5}\sin\frac{\pi x}{L}\\
N''_x &= -\frac{EIL^2}{a^5\pi^2}\frac{d^6W}{d\theta^6}\cos\frac{\pi x}{L}\\
u'' &= -\frac{EIL^3}{a^5\pi^3 t}\frac{d^6W}{d\theta^6}\sin\frac{\pi x}{L}\\
v'' &= -\frac{EIL^4}{a^6\pi^4 t}\frac{d^7W}{d\theta^7}\cos\frac{\pi x}{L}
\end{aligned}
\tag{15.37}
$$

3. Superimpose the expressions found in steps 1 and 2 to obtain the *complete solution*:

$$
\begin{aligned}
M_\theta &= \frac{EI}{a^2}\left(kp_0\cos\theta - \frac{d^2W}{d\theta^2}\right)\cos\frac{\pi x}{L}\\
Q_\theta &= -\frac{EI}{a^3}\left(kp_0\sin\theta + \frac{d^3W}{d\theta^3}\right)\cos\frac{\pi x}{L}\\
N_\theta &= \frac{EI}{L^3}\left[\left(k - \frac{a^4}{EI}\right)p_0\cos\theta + \frac{d^4W}{d\theta^4}\right]\cos\frac{\pi x}{L}\\
N_{x\theta} &= \frac{L}{\pi}\left(cp_0\sin\theta\sin\frac{\pi x}{L} - \frac{EI}{a^4}\frac{d^5W}{d\theta^5}\right)\sin\frac{\pi x}{L}\\
N_x &= \frac{L^2}{a\pi^2}\left(cp_0\cos\theta - \frac{EI}{a^4}\frac{d^6W}{d\theta^6}\right)\cos\frac{\pi x}{L}
\end{aligned}
\tag{15.38}
$$

and

$$
\begin{aligned}
u &= \frac{L^3}{a\pi^3 t}\left(\frac{c}{E}p_0\cos\theta - \frac{1}{a^4}\frac{d^6W}{d\theta^6}\right)\sin\frac{\pi x}{L}\\
v &= -\frac{L^4}{a^2\pi^4 t}\left(\frac{c}{E}p_0\sin\theta + \frac{1}{a^4}\frac{d^7W}{d\theta^7}\right)\cos\frac{\pi x}{L}
\end{aligned}
\tag{15.39}
$$

Here k, c, and W are defined by Equations 15.29, 15.36, and 15.33, respectively.

The stress resultants, Equations 15.38, must satisfy the straight edge conditions (Equations d). Substitution gives four equations for the constants A, B, C, and D. Solving these equations for the constants and introducing the results into Equations 15.34, 15.38, and 15.39 leads to the *final expressions* for w, u, v, M_θ, Q_θ, N_θ, $N_{x\theta}$, and N_x. Then, for any desired point (x, θ) of the shell, the displacements and the stress resultants can be determined in terms of the given shell dimensions (L, a, t, α), modulus of elasticity E, and the applied load intensity p_0. The described *computations are lengthy* and usually performed by a digital computer [4].

15.10 Inextensional Deformations

Associated with the *inextensional deformations* of shells, a *simplified* shell theory is useful under certain conditions. The *inextensional shell theory* is applicable to a variety of shell forms. However, we shall deal only with the inextensional deformation of circular cylindrical shells.

The inextensional theory is often preferred when shell structures resist loading principally through bending action. Such cases include a cylinder subjected to loads without axial symmetry and confined to a small circumferential portion (Figure 15.6a) and a cylinder with free ends under variable pressure $p(x, \theta)$ where $\int_0^{2\pi} p(x,\theta)d\theta \approx 0$ (Figure 15.6b). In both situations, shortening of the vertical diameter along which P or $p\ (x, \theta)$ act and lengthening of the horizontal diameter occur. Hence, there is considerable bending caused by the changes in curvature but *no stretching of midsurface length*. Deformations of these types are thus described as inextensional.

In inextensional-shell theory, the midsurface in-plane strain components given by Equations 15.4a, 15.4b, and 15.4c are taken to be zero. The *conditions of inextensibility* are therefore

$$\frac{\partial u}{\partial x} = 0 \qquad \frac{1}{a}\frac{\partial v}{\partial \theta} - \frac{w}{a} = 0 \qquad \frac{1}{a}\frac{\partial u}{\partial \theta} - \frac{\partial v}{\partial x} = 0 \tag{15.40a–c}$$

On introduction of the preceding in Equations 15.5a–c, we obtain

$$N_x = N_\theta = N_{x\theta} = 0 \tag{a}$$

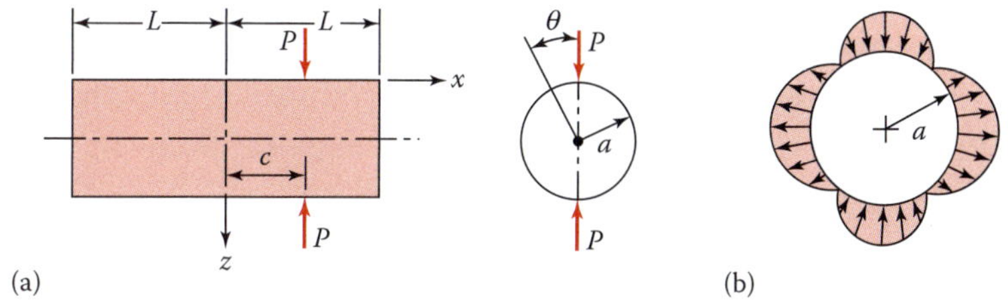

FIGURE 15.6
Asymmetrically loaded cylindrical shell: (a) Loading confined to a small circumferential portion; (b) variable pressure.

We also observe from Equation 15.40a that u depends on θ only, and Equation 15.40b leads to

$$w = \frac{\partial v}{\partial \theta} \tag{b}$$

The conditions (Equations 15.40) are thus fulfilled by assuming the displacements as follows [2]:

$$u = u_1 + u_2 \qquad v = v_1 + v_2 \qquad w = w_1 + w_2 \tag{15.41}$$

with

$$\begin{aligned} u_1 &= 0 \\ v_1 &= a\sum_{n=1}^{\infty}(a_n \cos n\theta - \bar{a}_n \sin n\theta) \\ w_1 &= -a\sum_{n=1}^{\infty} n(a_n \sin n\theta + \bar{a}_n \cos n\theta) \end{aligned} \tag{15.42a–c}$$

and

$$\begin{aligned} u_2 &= -a\sum_{n=1}^{\infty}\frac{1}{n}(b_n \sin n\theta + \bar{b}_n \cos n\theta) \\ v_2 &= x\sum_{n=1}^{\infty}(b_n \cos n\theta - \bar{b}_n \sin n\theta) \\ w_2 &= -x\sum_{n=1}^{\infty} n(b_n \sin n\theta + \bar{b}_n \cos n\theta) \end{aligned} \tag{15.43a–c}$$

In the above, a represents the radius of cylinder. The constants a_n, $\bar{a}_n$, b_n, and $\bar{b}_n$ are determined for each loading situation. According to Equations 15.42, all cross sections of the cylinder deform alike. These displacements varying along the length of the shell are given by Equations 15.43b and 15.43c.

A shell experiences inextensional deformations under a prescribed load in such a manner that a minimum strain energy is produced. Since Equations 15.41 yield $\partial^2 w/\partial x^2 = \chi_x = 0$, the required expressions for the bending-strain energy of the shell is given by (Section 13.5)

$$U = -\frac{D}{2}\int\int\left[\chi_\theta^2 + 2(1-\nu)\chi_{x\theta}^2\right] a\,d\theta\,dx \tag{c}$$

The *strain energy* of a *cylindrical shell undergoing inextensional deformation*, on substituting Equations 15.4 into Equation c, is then

$$U = \frac{D}{2a^4} \int\int \left[\left(\frac{\partial v}{\partial \theta} + \frac{\partial^2 w}{\partial \theta^2} \right) + 2(1-\nu)a^2 \left(\frac{\partial^2 w}{\partial x \partial \theta} + \frac{\partial v}{\partial x} \right)^2 \right] a\, d\theta\, dx \tag{15.44}$$

Inserting the displacements w and v from Equations 15.41 for a cylindrical shell of length $2L$ (Figure 15.6a), Equation 15.44 takes the form

$$U = \pi DL \sum_{n=2}^{\infty} \frac{\left(n^2-1\right)^2}{a^3} \left\{ \begin{array}{l} a^2 n^2 \left[a_n^2 + \left(\bar{a}_n\right)^2 \right] + \frac{1}{3} L^2 n^2 \left[b_n^2 + \left(\bar{b}_n\right)^2 \right] \\ +2(1-\nu)a^2 \left[b_n^2 + \left(\bar{b}_n\right) 2 \right] \end{array} \right\} \tag{15.45}$$

Omission of the term $n = 1$ in the preceding expression is explained as follows. For $n = 1$, we have

$$v_1 = a(a_1 \cos\theta - \bar{a}_1 \sin\theta)$$
$$w_1 = -a(a_1 \sin\theta + \bar{a}_1 \cos\theta)$$

The foregoing describes the rigid-body displacement of a circle in its plane and hence does not contribute to the strain energy of the shell. The same conclusion holds for displacements u_2, v_2, and w_2.

Equation 15.45 is useful in determining the components of shell displacement as is next observed.

EXAMPLE 15.3: Displacement of Cylinder Subjected to Concentrated Loads

A circular cylinder of length $2L$ is loaded by two equal and opposite concentrated forces a distance c from the midspan (Figure 15.6a). The ends of the shell are free. Derive the equations describing the elastic surface.

Solution

The radial displacements at $x = c$, in the direction of the loads, are $w(0)$ and $w(\pi)$. It is clear that the terms involving a_n and b_n in Equations 15.42c and 15.43c become zero at these points. The remaining terms with the coefficients $\bar{a}_n$ and $\bar{b}_n$ in w contribute *only* to the work produced by the forces. The potential energy $\Pi = U - W$ is therefore

$$\Pi = \pi DL \sum_{n=2}^{\infty} \frac{(n^2-1)^2}{a^3} \left\{ n^2 \left[a^2 (\bar{a}_n)^2 + \frac{1}{3} L^2 (\bar{b}_n)^2 \right] + 2(1-v)a^2 (\bar{b}_n)^2 \right\}$$
$$+ \left\{ a \sum_{n=1}^{\infty} n\bar{a}_n (1+\cos n\pi) + c \sum_{n=1}^{\infty} n\bar{b}_n (1+\cos n\pi) \right\} P$$

Applying Equations 3.53, from conditions $\partial\Pi / \partial\bar{a}_n = 0$ and $\partial\Pi / \partial\bar{b}_n = 0$, coefficients $\bar{a}_n$ and $\bar{b}_n$ are readily evaluated:

$$\bar{a}_n = -\frac{a^2(1+\cos n\pi)P}{2n\pi DL(n^2-1)^2}$$

$$\bar{b}_n = -\frac{nca^3(1+\cos n\pi)P}{2\pi DL(n^2-1)^2 \left[\frac{1}{3} n^2 L^2 + 2(1-\nu)a^2 \right]}$$

or

$$\bar{a}_n = -\frac{a^2 P}{n(n^2-1)^2 \pi DL} \qquad (n = 2, 4, \ldots)$$

$$\bar{b}_n = -\frac{nca^3 P}{(n^2-1)^2 \pi DL\left[\frac{1}{3} n^2 L^2 + 2(1-\nu)a^2\right]} \qquad (n = 2, 4, \ldots)$$

and $\bar{a}_n = \bar{b}_n = 0$ for $n = 1, 3, \ldots$. Equations 15.41 become, in this case,

$$\begin{aligned} u &= \frac{Pa^3}{\pi DL} \sum_n^{\infty} \frac{ac\cos n\theta}{n(n^2-1)^2[(1/3)n^2L^2 + 2(1-\nu)a^2]} \\ v &= \frac{Pa^3}{\pi DL} \sum_n^{\infty} \left\{ \frac{1}{n(n^2-1)^2} + \frac{ncx}{(n^2-1)^2[(1/3)n^2L^2 + 2(1-\nu)a^2]} \right\} \sin n\theta \\ w &= \frac{Pa^3}{\pi DL} \sum_n^{\infty} \left\{ \frac{1}{n(n^2-1)^2} + \frac{ncx}{(n^2-1)^2[(1/3)n^2L^2 + 2(1-\nu)a^2]} \right\} \cos n\theta \end{aligned} \tag{15.46}$$

where $n = 2, 4, \ldots$. The expressions for the moments can be obtained from the general solution (Equations 15.46) by applying Equations 15.5. The stresses are then determined from Equations 13.5.

When the loads are applied at midlength ($c = 0$), the *vertical reduction* of the cross section at the center, from $w(0) + w(\pi)$, is

$$w_v = \frac{2Pa^3}{\pi DL} \sum_n^{\infty} \frac{1}{(n^2-1)^2} = 0.149 \frac{Pa^3}{2DL} \tag{d}$$

At the center cross section the *horizontal increase*, as ascertained from $w(\pi/2) + w(3\pi/2)$, equals

$$w_h = \frac{2Pa^3}{\pi DL} \sum_n^{\infty} \frac{(-1)^{(n/2)+1}}{(n^2-1)^2} = 0.137 \frac{Pa^3}{2DL} \tag{e}$$

The displacements at any section of the cylinder may be found similarly.

Comments: It is observed in the foregoing solution that all coefficients a_n, b_n, $\bar{a}_n$, $\bar{b}_n$ are uniquely determined by the applied load. Hence, in inextensional-shell theory, *no* boundary conditions can be fulfilled at the free edges of the shell. However, the edge bending due to displacements (Equations 15.46) is of local character and does not have a pronounced effect on the accuracy of the results given by Equations d and e.

15.11 A Typical Layered Orthotropic Cylindrical Shell

The preceding sections have been concerned with isotropic shells. Now treated is the bending of a commonly used cylindrical shell consisting of three *cross-ply orthotropic* materials. The inner layer and the outer layers are considered to be oriented along the axial

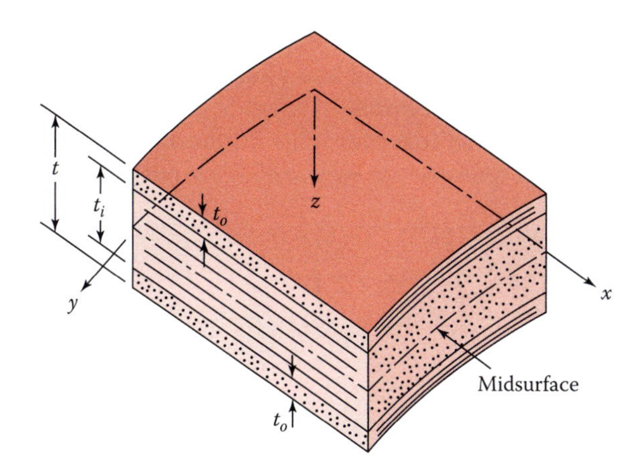

FIGURE 15.7
Element of a cylindrical three-cross-ply compound shell.

(x) and the tangential (y) directions, respectively (Figure 15.7). We also take all laminae to be one of the same type of material (e.g., as in plywood).

The modified Hooke's law for an orthotropic material is given in Section 8.2. Using Equations 8.1, the stresses in the *inner* and the *outer* layer are, respectively, as follows:

$$\begin{Bmatrix} \sigma_x \\ \sigma_\theta \\ \tau_{x\theta} \end{Bmatrix}_i = \begin{bmatrix} E_x & E_{x\theta} & 0 \\ E_{x\theta} & E_\theta & 0 \\ 0 & 0 & G \end{bmatrix}_i \begin{Bmatrix} \varepsilon_x \\ \varepsilon_\theta \\ \gamma_{x\theta} \end{Bmatrix} \tag{15.47a}$$

and

$$\begin{Bmatrix} \sigma_x \\ \sigma_\theta \\ \tau_{x\theta} \end{Bmatrix}_o = \begin{bmatrix} E_\theta & E_{x\theta} & 0 \\ E_{x\theta} & E_x & 0 \\ 0 & 0 & G \end{bmatrix}_o \begin{Bmatrix} \varepsilon_x \\ \varepsilon_\theta \\ \gamma_{x\theta} \end{Bmatrix} \tag{15.47b}$$

where E_x, E_θ, and $E_{x\theta}$ are defined by Equations 8.3. Carrying Equations 15.3 into the foregoing, we have

$$\begin{Bmatrix} \sigma_x \\ \sigma_\theta \\ \tau_{x\theta} \end{Bmatrix}_i = \begin{bmatrix} E_x & E_{x\theta} & 0 \\ E_{x\theta} & E_\theta & 0 \\ 0 & 0 & G \end{bmatrix}_i \begin{Bmatrix} \varepsilon_{xy0} \\ \varepsilon_{\theta 0} \\ \gamma_{x\theta 0} \end{Bmatrix} - z \begin{bmatrix} E_x & E_{x\theta} & 0 \\ E_{x\theta} & E_\theta & 0 \\ 0 & 0 & G \end{bmatrix}_i \begin{Bmatrix} \chi_x \\ \chi_\theta \\ 2\chi_{x\theta} \end{Bmatrix} \tag{15.48a}$$

and

$$\begin{Bmatrix} \sigma_x \\ \sigma_\theta \\ \tau_{x\theta} \end{Bmatrix}_o = \begin{bmatrix} E_\theta & E_{x\theta} & 0 \\ E_{x\theta} & E_x & 0 \\ 0 & 0 & G \end{bmatrix}_o \begin{Bmatrix} \varepsilon_{xy0} \\ \varepsilon_{\theta 0} \\ \gamma_{x\theta 0} \end{Bmatrix} - z \begin{bmatrix} E_\theta & E_{x\theta} & 0 \\ E_{x\theta} & E_x & 0 \\ 0 & 0 & G \end{bmatrix}_o \begin{Bmatrix} \chi_x \\ \chi_\theta \\ 2\chi_{x\theta} \end{Bmatrix} \tag{15.48b}$$

The stress resultant–displacement relations (Equations 15.5) and equilibrium conditions (Equations 15.2) may be employed without any change (Section 8.2). The same is true for the definitions (13.1) of the stress resultants.

On substituting Equations 15.48 together with Equations 15.4 into Equations 13.1 and neglecting z/r_x and z/r_y, with reference to Figure 15.7, we obtain

$$N = \int_{-t_i/2}^{t_i/2} \sigma_x^{(i)} dz + 2\int_{t_i/2}^{t/2} \sigma_x^{(o)} dz$$
$$= (E_x t_i + 2E_\theta t_0)\frac{\partial u}{\partial x} + \frac{E_{x\theta} t}{a}\left(\frac{\partial v}{\partial \theta} - w\right)$$

Similarly, expressions for the other forces and moments are derived as follows:

$$\begin{Bmatrix} N_x \\ N_\theta \\ N_{x\theta} \end{Bmatrix} = \begin{bmatrix} A_{11} & A_{22} & 0 \\ A_{12} & A_{22} & 0 \\ 0 & 0 & A_{33} \end{bmatrix} \begin{Bmatrix} \dfrac{\partial u}{\partial x} \\ \dfrac{1}{a}\left(\dfrac{\partial v}{\partial \theta} - w\right) \\ \dfrac{1}{a}\dfrac{\partial u}{\partial \theta} + \dfrac{\partial v}{\partial x} \end{Bmatrix} \tag{15.49a}$$

and

$$\begin{Bmatrix} M_x \\ M_\theta \\ M_{x\theta} \end{Bmatrix} = \begin{bmatrix} D_{11} & D_{12} & 0 \\ D_{12} & D_{22} & 0 \\ 0 & 0 & D_{33} \end{bmatrix} \begin{Bmatrix} \dfrac{\partial^2 w}{\partial x^2} \\ \dfrac{1}{a^2}\left(\dfrac{\partial v}{\partial \theta} + \dfrac{\partial^2 w}{\partial \theta^2}\right) \\ \dfrac{2}{a}\left(\dfrac{\partial v}{\partial x} + \dfrac{\partial^2 w}{\partial x \partial \theta}\right) \end{Bmatrix} \tag{15.49b}$$

In the preceding formulas, the shell properties are as follows:
Extensional rigidities

$$\begin{aligned} A_{11} &= E_x t_i + 2E_\theta t_o \\ A_{22} &= E_\theta t_i + 2E_x t_o \\ A_{12} &= E_{x\theta} t \end{aligned} \tag{15.50a}$$

Shearing rigidity

$$A_{33} = Gt \tag{15.50b}$$

Flexural rigidities

$$D_{11} = \frac{1}{12}\left[E_\theta\left(t^3 - t_i^3\right) + E_x t_i^3\right]$$
$$D_{22} = \frac{1}{12}\left[E_x\left(t^3 - t_i^3\right) + E_\theta t_i^3\right] \tag{15.50c}$$
$$D_{12} = \frac{1}{12}E_{x\theta}t^3$$

Torsional rigidity

$$D_{33} = \frac{1}{12}Gt^3 \tag{15.50d}$$

Now the elastic law (Equations 15.49) is substituted into equations of equilibrium (Equations 15.2). This, after proper rearrangement of the terms, results in

$$\begin{aligned}
&A_{11}\frac{\partial^2 u}{\partial x^2} + \frac{A_{33}}{a}\frac{\partial^2 u}{\partial \theta} + \frac{1}{a}(A_{12} + A_{33})\frac{\partial^2 v}{\partial x \partial \theta} - \frac{A_{12}}{a}\frac{\partial w}{\partial x} + p_x = 0 \\
&\frac{1}{a}(A_{12} + A_{33})\frac{\partial^2 u}{\partial x \partial \theta} + \left(A_{33} + \frac{2D_{33}}{a^2}\right)\frac{\partial^2 v}{\partial x^2} + \frac{1}{a^2}\left(A_{22} + \frac{D_{22}}{a^2}\right)\frac{\partial^2 v}{\partial \theta^2} \\
&- \frac{A_{22}}{a}\frac{\partial w}{\partial \theta} + \frac{1}{a^2}(D_{12} + 2D_{33})\frac{\partial^2 w}{\partial x^2 \partial \theta} + \frac{D_{22}}{a^4}\frac{\partial^3 w}{\partial \theta^3} + p_\theta = 0 \\
&\frac{A_{12}}{a}\frac{\partial u}{\partial x} + \frac{A_{22}}{a^2}\left(\frac{\partial v}{\partial \theta} - w\right) - D_{11}\frac{\partial^4 w}{\partial x^4} - \frac{2}{a^2}(D_{12} + 2D_{33})\frac{\partial^2 w}{\partial x^2 \partial \theta^2} \\
&- \frac{D_{11}}{a^4}\frac{\partial^4 w}{\partial \theta^4} - \frac{1}{a^2}(D_{12} + 4D_{33})\frac{\partial^3 w}{\partial x^2 \partial \theta} - \frac{D_{11}}{a^4}\frac{\partial^3 v}{\partial \theta^3} + p_r = 0
\end{aligned} \tag{15.51}$$

Equations 15.51 are the *governing equations* for the displacements in the circular *anisotropic shell*, which is made of three orthotropic layers.

In the case of *isotropy*, we have (Section 3.4)

$$E_x = E_\theta = \frac{E}{1-\nu^2} \qquad E_{x\theta} = \frac{E\nu}{1-\nu^2} \qquad G = \frac{E}{2(1+\nu)}$$

Consequently,

$$A_{11} = A_{22} = \frac{E}{1-\nu^2} \qquad A_{12} = \frac{E\nu t}{1-\nu^2} \qquad A_{33} = \frac{Et}{2(1+\nu)}$$
$$D_{11} = D_{22} = D \qquad D_{12} = \nu D \qquad D_{33} = \frac{Et^3}{24(1+\nu)}$$

and Equations 15.49 and 15.51 reduce to Equations 15.5 and 15.6. Another particular case is obtained by letting $t_o = 0$ and $t_i = t$ in Equations 15.50. Then, we have the rigidities of a solid orthotropic shell.

Comments: We observe that the orthotropic shell theory is *not* more involved than that of isotropic shells (see Section 8.9). Generally, added complication lies in the fact that the rigidities are no longer as simple as before.

15.12 Laminated Composite Cylindrical Shells

We shall now deal with *laminated cylindrical shells* made of orthotropic layers bonded together at *various orientations*. Each lamina may be of a single or a *composite* material. The material concerned here is thus a general kind as that discussed in Section 15.11.

Consider a laminate consisting of n orthotropic arbitrarily oriented layers (Figure 8.10). We follow a procedure similar to that employed in Section 8.10. To begin with, Equation 15.3 is substituted into Equation 8.42. This leads to the stress–strain–curvature relationships, for the kth layer, as follows:

$$\begin{Bmatrix} \sigma_x \\ \sigma_\theta \\ \tau_{xy} \end{Bmatrix} = [D^*]_k \begin{Bmatrix} \varepsilon_{x0} \\ \varepsilon_{\theta 0} \\ \gamma_{x\theta 0} \end{Bmatrix} - z[D^*]_k \begin{Bmatrix} \chi_x \\ \chi_\theta \\ 2\chi_{x\theta} \end{Bmatrix} \tag{15.52}$$

where the matrix D^* is given by Equation 8.52. Then, introducing the foregoing into the definitions (Equations 13.1) and neglecting z/r_x and z/r_y, we have

$$\begin{Bmatrix} N_x \\ N_\theta \\ N_{x\theta} \end{Bmatrix} = \sum_{k=1}^{n} \int_{t_{k-1}}^{t_k} [D^*]_k \begin{Bmatrix} \varepsilon_{x0} \\ \varepsilon_{\theta 0} \\ \gamma_{x\theta 0} \end{Bmatrix} dz \tag{15.53a}$$

and

$$\begin{Bmatrix} M_x \\ M_\theta \\ M_{x\theta} \end{Bmatrix} = -\sum_{k=1}^{n} \int_{t_{k-1}}^{t_k} [D^*]_k \begin{Bmatrix} \chi_x \\ \chi_\theta \\ 2\chi_{x\theta} \end{Bmatrix} z^2\, dz \tag{15.53b}$$

The *resultant forces and moments*, after integration of Equations 15.53 and substitution into Equations 15.5, can be written in the following convenient form:

$$\begin{Bmatrix} N_x \\ N_\theta \\ N_{x\theta} \end{Bmatrix} = \begin{bmatrix} A_{11} & A_{12} & A_{13} \\ A_{12} & A_{22} & A_{23} \\ A_{13} & A_{23} & A_{33} \end{bmatrix} \begin{Bmatrix} \dfrac{\partial u}{\partial x} \\ \dfrac{1}{a}\left(\dfrac{\partial u}{\partial \theta} - w\right) \\ \dfrac{1}{a}\dfrac{\partial u}{\partial \theta} + \dfrac{\partial v}{\partial x} \end{Bmatrix} \tag{15.54a}$$

and

$$\begin{Bmatrix} M_x \\ M_\theta \\ M_{x\theta} \end{Bmatrix} = \begin{bmatrix} D_{11} & D_{12} & D_{13} \\ D_{12} & D_{22} & D_{23} \\ D_{13} & D_{23} & D_{33} \end{bmatrix} \begin{Bmatrix} \dfrac{\partial^2 w}{\partial x^2} \\ \dfrac{1}{a^2}\left(\dfrac{\partial v}{\partial \theta} - \dfrac{\partial^2 w}{\partial \theta^2}\right) \\ \dfrac{2}{a}\left(\dfrac{\partial v}{\partial x} + \dfrac{\partial^2 w}{\partial x \partial \theta}\right) \end{Bmatrix} \tag{15.54b}$$

Here

$$\begin{aligned} A_{ij} &= \sum_{k=1}^{n} \left(D_{ij}^*\right)_k (t_k - t_{k-1}) \\ D_{ij} &= \frac{1}{3}\sum_{k=1}^{n} \left(D_{ij}^*\right)_k \left(t_k^3 - t_{k-1}^3\right) \end{aligned} \tag{15.55}$$

in which A and D are the *rigidity* matrices of the shell. For the case of the shell shown in Figure 15.7, Equations 15.55 reduce to Equations 15.50; see Problem 15.4.

Finally, the elastic law (Equations 15.54) is introduced into equations of equilibrium (Equations 15.2) to determine the governing expressions for the displacements in the multilayered anisotropic cylindrical shells. In so doing, we obtain the general form of Equations 15.51.

EXAMPLE 15.4: Calculating Rigidity Matrix of Three-Layer Composite Shell

Determine the rigidity matrix A for the three-ply laminate shown in Figure 8.12.

Solution

From Example 8.8, for the laminae 1, 2, and 3, we have

$$[D^*]_1 = [D^*]_3 = \begin{bmatrix} 10.437 & 0.437 & 2.812 \\ 0.437 & 10.437 & 2.812 \\ 2.812 & 2.812 & 3.937 \end{bmatrix} 10^9 \tag{a}$$

$$[D^*]_2 = \begin{bmatrix} 15 & 1.5 & 0 \\ 1.5 & 3.75 & 0 \\ 0 & 0 & 5 \end{bmatrix} 10^9 \tag{b}$$

Referring to Figure 8.12, the formula (Equations 15.55), with $(A_{ij})_1 = (A_{ij})_3$, results in

$$\begin{aligned} A_{ij} &= \sum_{k=1}^{3} \left(D_{ij}^*\right)_k (t_k - t_{k-1}) \\ &= \left\{\left(D_{ij}^*\right)_1 [-3-(-7)] + \left(D_{ij}^*\right)_2 [3-(-3)] + \left(D_{ij}^*\right)_3 [7-3]\right\} 10^6 \\ &= \left[8\left(D_{ij}^*\right)_1 + 6\left(D_{ij}^*\right)_2\right] 10^6 \end{aligned}$$

Carrying the terms in Equations a and b into the preceding, we obtain (in N/m)

$$[A] = \begin{bmatrix} 173.496 & 12.496 & 22.496 \\ 12.496 & 105.496 & 22.496 \\ 22.496 & 22.496 & 61.496 \end{bmatrix} 10^6$$

The rigidity matrix D is already found in Example 8.8.

15.13 *Symmetrical Buckling under Uniform Axial Pressure

In numerous examples in the preceding chapters, it is observed that thin shell structures are often subjected to compressive stresses at particular areas. To check the stability of the elastic equilibrium of compressed shells, the methods described in Sections 3.11 and 3.13 may be applied. In this section, we describe the fundamental concepts of the *energy approach* by considering the *symmetrical buckling* of a thin-walled *circular cylinder under uniform axial compression*. We thus provide only an introduction to shell instability [1,7], a critically important area of engineering design. In the section following, a more general case of buckling is discussed that employs the equilibrium approach.

When a circular cylindrical shell is under uniform axial compression, axisymmetrical buckling of the shell may take place at a particular value of compressive load (Figure 15.8). We shall determine the critical value of compressive force N per unit circumferential length. At the start of buckling, the shell-strain energy is increased by the following: midsurface straining in the circumferential direction, bending, and axial compression. At the critical value of load, this increase in energy must be equal to the work done by the compressive load because of axial straining and bending as the cylinder deflects because of buckling action. Thus, the critical load is that for which

$$\delta U = \delta W \tag{a}$$

This is the principle of virtual work (Section 3.12). It is therefore assumed that the axial compressive force *does not change during the buckling*.

If the cylinder is *simply supported* at both ends, the radial deflection satisfying the boundary conditions is

$$w = -c_0 \sin \frac{m\pi x}{L} \tag{b}$$

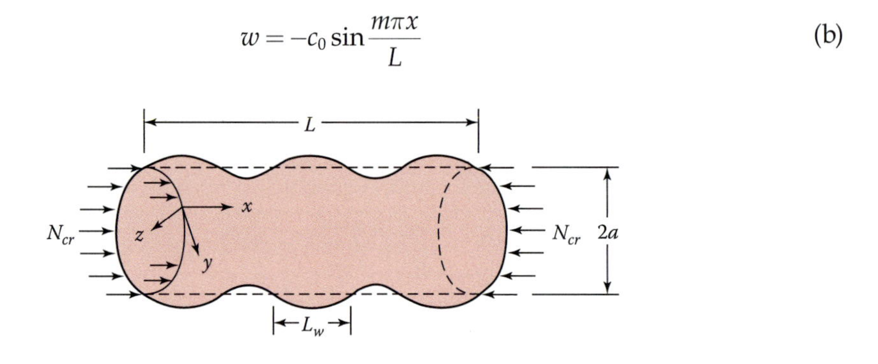

FIGURE 15.8
Buckled form of a cylindrical shell in axial compression, where the original shape is shown by the dashed lines.

In the preceding, c_0 is a constant, L the length of the shell, and m is the number of half-sine waves in the axial direction. As the cylinder deflects, it becomes shorter in length. The axial midsurface strain *before* buckling is given by

$$\varepsilon_{x1} = -\frac{N}{Et}$$

where t is the thickness of the shell. From Hooke's law,

$$\varepsilon_{x1} E = \frac{E}{1-v^2}(\varepsilon_{x2} + v\varepsilon_{\theta 2}) \tag{c}$$

or

$$\varepsilon_{x2} + v\varepsilon_{\theta 2} = (1-v^2)\varepsilon_{x1} \tag{d}$$

in which ε_{x2} and $\varepsilon_{\theta 2}$ are the axial and circumferential midsurface strains *after* buckling. Referring to Section 13.6, the latter strain is expressed as

$$\varepsilon_{\theta 2} = -v\varepsilon_{x1} = -\frac{w}{a} = -v\varepsilon_{x1} + \frac{c_0}{a}\sin\frac{m\pi x}{L} \tag{e}$$

Inserting the above into Equation d,

$$\varepsilon_{x2} = \varepsilon_{x1} - v\frac{c_0}{a}\sin\frac{m\pi x}{L} \tag{f}$$

Because of the axisymmetrical deformation,

$$\gamma_{xy} = \chi_y = \chi_{xy} = 0 \tag{g}$$

and the change in curvature in the axial plane is

$$\chi_x = c_0\frac{m^2\pi^2}{L^2}\sin\frac{m\pi x}{L} \tag{h}$$

As N remains constant during buckling, the work done is

$$\delta W = 2\pi N\left[v\int_0^L c_0 \sin\frac{m\pi x}{L}dx + \frac{a}{2}\int_0^L\left(c_0\frac{m\pi}{L}\sin\frac{m\pi x}{L}\right)^2 dx\right] \tag{i}$$

The components of this work are associated with the change of the axial strain $(\varepsilon_{x2} - \varepsilon_{x1})$ and the bending of the generators defined by Equation b. Next, the increase in strain energy during buckling must be determined. For this purpose, Equations e through h are introduced into Equations 13.7 and 13.8. After addition of δU_b and δU_m we have

$$\delta U = -2\pi t E \nu \varepsilon_{x1} \int_0^L c_0 \sin\frac{m\pi x}{L} dx + \frac{\pi c_0^2 EtL}{2a} + c_0^2 \frac{\pi^4 m^4}{2L^4} \pi a L D \tag{j}$$

The *criterion for buckling*, $\delta W = \delta U$, together with Equations i and j, is applied to find the critical value of N. It follows that

$$\left(\frac{\pi\, EtL}{2a} + \frac{\pi^5 a m^4 D}{2L^3} - \frac{\pi^3 a m^2 N}{2L}\right) c_0^2 = 0$$

For the preceding to be valid for any c_0, it is required that

$$N_{cr} = D\left(\frac{m^2\pi^2}{L^2} + \frac{EtL^2}{Da^2 m^2 \pi^2}\right) \tag{15.56}$$

As for columns and plates, for each value of m there is a unique buckling-mode shape and a unique buckling load. The lowest critical load is of greatest interest and is found by setting the derivative of N_{cr} with respect to L equal to zero for $m = 1$. We thus determine, for $\nu = 0.3$, the length at which the minimum buckling load has effect:

$$L_w = \pi \sqrt{\frac{a^2 t^2}{12(1-\nu^2)}} \approx 1.72\sqrt{at} \tag{15.57}$$

The above is the *length of half-sine waves* into which the shell buckles (Figure 15.8). The corresponding *minimum buckling load* is

$$N_{cr} = \frac{Et^2}{a\sqrt{3(1-\nu^2)}} = 0.605\frac{Et^2}{a} \tag{15.58}$$

We therefore have the equation

$$\sigma_{cr} = \frac{Et}{a\sqrt{3(1-\nu^2)}} = 0.605\frac{Et}{a} \tag{15.59}$$

for the *critical stress* for axially symmetric buckling of a simply supported circular cylindrical shell under uniform axial loading.

Comments: When the compression load N exceeds N_{cr}, given by Equation 15.58, the system is *unstable*, as the work done by N is greater than the increase in strain energy. We are led to conclude that the straight cylinder is unstable when $N > N_{cr}$. If $N = N_{cr}$, the cylinder is in *neutral equilibrium* for small displacements. If $N < N_{cr}$, the straight cylinder is in *stable equilibrium*. The solution $c_0 = 0$ therefore represents an unstable or a stable configuration according to whether N is greater or small than N_{cr}.

15.14 Nonsymmetrical Buckling under Uniform Compression

The preceding section concerned simple buckling of an axially loaded cylinder. But, cylindrical shells often buckle into *axially nonsymmetrical forms*. This case of instability is considered here through application of the *equilibrium method* [1,7,10].

We shall first modify the governing equations (Equations 15.1) for shell displacements to include axial load effects, as is done for plates and columns. This approach is based on the assumption that all surface forces, except N_x, are very small. Thus, the product formed of these forces and the derivatives of displacement are neglected. Following a procedure similar to that given in Section 9.2, it is found that the force N_x yields a y component $N_x(\partial^2 v/\partial^2 x)dx\, a\, d\theta$ in the expression $\sum F_y = 0$ and a z component $N_x(\partial^2 w/\partial^2 x)dx\, a\, d\theta$ in the expression $\sum F_y = 0$. These components, after division by $dx\, d\theta$, are added to Equations 15.1. The equations of equilibrium, assuming $p_x = p_y = p_z = 0$, are then

$$\begin{aligned} a\frac{\partial N_x}{\partial x} + \frac{\partial N_{x\theta}}{\partial \theta} &= 0 \\ \frac{\partial N_\theta}{\partial \theta} + a\frac{\partial N_{x\theta}}{\partial x} + aN_x\frac{\partial^2 v}{\partial x^2} - Q_\theta &= 0 \\ a\frac{\partial Q_x}{\partial x} + \frac{\partial Q_\theta}{\partial \theta} + aN_x\frac{\partial^2 w}{\partial x^2} + N_\theta &= 0 \\ \frac{\partial M_\theta}{\partial \theta} + a\frac{\partial M_{x\theta}}{\partial \theta} - aQ_\theta &= 0 \\ a\frac{\partial M_x}{\partial x} + \frac{\partial M_{x\theta}}{\partial \theta} - aQ_x &= 0 \end{aligned} \tag{15.60}$$

The shear forces Q_x and Q_θ may be eliminated from the above equations. In so doing, we have

$$\begin{aligned} a\frac{\partial N_x}{\partial x} + \frac{\partial N_{x\theta}}{\partial \theta} &= 0 \\ \frac{\partial N_\theta}{\partial \theta} + a\frac{\partial N_{x\theta}}{\partial x} + aN_x\frac{\partial^2 v}{\partial x^2} - \frac{\partial M_{x\theta}}{\partial x} - \frac{1}{a}\frac{\partial M_\theta}{\partial \theta} &= 0 \\ aN_x\frac{\partial^2 w}{\partial x^2} + N_\theta + a\frac{\partial^2 M_x}{\partial x^2} + 2\frac{\partial^2 M_{x\theta}}{\partial x \partial \theta} + \frac{1}{a}\frac{\partial^2 M_\theta}{\partial \theta^2} &= 0 \end{aligned} \tag{15.61}$$

We now employ the elastic law (Equations 15.5) to express all stress resultants in terms of u, v, w, and their derivatives. Taking the compressive force $N_x = N$ to be positive and introducing the dimensionless parameters

$$\alpha = \frac{t^2}{12a^2} \qquad q = \frac{N(1-\nu^2)}{Et} \tag{15.62}$$

we express the *differential equations of the buckling problem* as follows:

$$\frac{\partial^2 u}{\partial x^2}+\frac{1+\nu}{2a}\frac{\partial^2 v}{\partial x \partial \theta}-\frac{\nu}{a}\frac{\partial w}{\partial x}+\frac{1-\nu}{2}\frac{1}{a^2}\frac{\partial^2 u}{\partial \theta^2}=0$$

$$\frac{1+\nu}{2a}\frac{\partial^2 u}{\partial x \partial \theta}+\frac{1-\nu}{2}\frac{\partial^2 v}{\partial x^2}+\frac{1}{a^2}\frac{\partial^2 v}{\partial \theta^2}-\frac{1}{a^2}\frac{\partial w}{\partial \theta}$$
$$+\alpha\left[\frac{1}{a^2}\frac{\partial^2 v}{\partial \theta^2}+\frac{1}{a^2}\frac{\partial^3 w}{\partial \theta^3}+\frac{\partial^3 w}{\partial x^2 \partial \theta}+(1-\nu)\frac{\partial^2 v}{\partial x^2}\right]-q\frac{\partial^2 v}{\partial x^2}=0 \tag{15.63}$$

$$-aq\frac{\partial^2 w}{\partial x^2}+\nu\frac{\partial u}{\partial x}+\frac{1}{a}\frac{\partial v}{\partial \theta}-\frac{w}{a}$$
$$-\alpha\left[\frac{1}{a}\frac{\partial^3 v}{\partial \theta^3}+a\alpha(2-\nu)\frac{\partial^3 v}{\partial x^2 \partial \theta}+a^3\frac{\partial^4 w}{\partial x^4}+\frac{1}{a}\frac{\partial^4 w}{\partial \theta^4}+2a\frac{\partial^4 w}{\partial x^2 \partial \theta^2}\right]=0$$

It is noted that Equations 15.63 are satisfied for the particular case in which the solution is expressed in terms of constants c_1 and c_2,

$$u=\frac{c_1}{\nu a}x+c_2 \qquad v=0 \qquad w=c_1$$

This represents the cylindrical form of equilibrium wherein the compressed cylinder uniformly expands laterally. Also, when we assume $v=0$ and u and w to be functions of x only, a solution is obtained for the axisymmetrical buckling treated in Section 15.13.

The *general solution* of Equations 15.63, if the origin of coordinates is placed at one end of the shell, can be expressed by the series

$$u=\frac{c_1}{\nu a}x+c_2+\sum_m^\infty\sum_n^\infty a_{mn}\sin n\theta\cos\frac{m\pi x}{L}$$
$$v=\sum_m^\infty\sum_n^\infty b_{mn}\cos n\theta\sin\frac{m\pi x}{L} \tag{15.64}$$
$$w=c_1+\sum_m^\infty\sum_n^\infty c_{mn}\sin n\theta\sin\frac{m\pi x}{L}$$

For *long cylinders*, the results obtained from Equations 15.64 can be used *regardless of the type of edge supports*. This is attributable to the fact that the edge conditions have only a minor influence on the magnitude of the critical load provided that the shell length is not small (say $L > 2a$).

When we introduce the notation

$$\lambda=\frac{m\pi a}{L} \tag{15.65}$$

and substitute the solution (Equations 15.64) into Equations 15.63, the trigonometric functions drop out entirely, and we find that

$$
\begin{aligned}
&a_{mn}\left(\lambda^2+\frac{1-\nu}{2}n^2\right)+b_{mn}\frac{n(1+\nu)\lambda}{2}+c_{mn}v\lambda=0\\
&a_{mn}\frac{n(1+\nu)\lambda}{2}+b_{mn}\left[(1-\nu)\left(\frac{1}{2}+\alpha\right)\lambda^2+(1+\alpha)n^2-q\lambda^2\right]\\
&\quad+c_{mn}[n+\alpha n(n^2+\lambda^2)]=0\\
&a_{mn}\nu\lambda+b_{mn}n\{1+\alpha[n^2+(2-\nu)\lambda^2]\}+c_{mn}[1-q\lambda^2+\alpha(\lambda^2+n^2)^2]=0
\end{aligned}
\tag{15.66}
$$

The foregoing represent three linear equations with *buckling amplitudes* a_{mn}, b_{mn}, and c_{mn} as unknowns. The nontrivial solution, the buckling condition of the shell, is determined by setting the determinant of coefficients equal to zero. Usually α and q defined by Equations 15.62 are much smaller than unity, and we shall therefore neglect the terms containing the square of these quantities.

Observing that the minimum value of q takes place when λ^2 and n^2 are large numbers, the expanded determinantal equation then results in

$$
q=\frac{N(1-\nu^2)}{Et}=\alpha\frac{(n^2+\lambda^2)^2}{\lambda^2}+\frac{(1-\nu^2)\lambda^2}{(n^2+\lambda^2)^2}
\tag{15.67}
$$

We see that for $n = 0$, this expression reduces to Equation 15.56, the result obtained for symmetrical buckling.

To determine the minimum value of q, we let $\eta = (n^2 + \lambda^2)^2/\lambda^2$. Equation 15.67 becomes $q = \alpha\eta + (1 - \nu^2)/\eta$. The minimizing condition $dq/d\eta = 0$, leads to $\eta=\sqrt{(1-\nu^2)/\alpha}$. Hence,

$$
q_{\min}=2\sqrt{\alpha(1-\nu^2)}
$$

The corresponding *critical stress* is therefore

$$
\sigma_{cr}=\frac{N_{cr}}{t}=\frac{Et}{a\sqrt{3(1-\nu^2)}}
\tag{15.68}
$$

This coincides with Equation 15.59, which was derived in a different way for axisymmetrical buckling. Note that the critical stress *depends* on the material properties, thickness, and radius, while it is *independent* of cylinder length.

The value of critical stress defined by Equation 15.68, which is based on the small-displacement theory, often does *not* agree with experimental data. This discrepancy is explained by applying the large-deflection theory of buckling, which takes into account the squares of the derivatives of the deflection w, initial imperfections, and a host of additional factors. The foregoing could give a deeper insight into what really is happening to a shell subjected to axial compression. A further investigation of the reasons for the discrepancy between theory and experiment was made by studying the *postbuckling behavior* of an ideal compressed shell [7].

To relate the theoretical value obtained in this section to *actual test data*, it is necessary to incorporate an *empirical factor* in Equation 15.68. For example, based on the coefficients [9]

$$K = 1 - 0.901(1 - e^{-\psi}) \qquad \psi = \frac{1}{16}\sqrt{\frac{a}{t}}$$

Equation 15.68, for $\nu = 0.3$, becomes

$$\sigma_{cr} = 0.605K\frac{Et}{a} \tag{15.69}$$

Comment: The result, Equation 15.69, is in satisfactory agreement with the tests for cylinders having $L/a < 5$.

15.15 Buckling of Conical Shells

Consider a thin-walled *conical shell* is subjected to uniformly distributed load, as depicted in Figure 15.9. The situation described represents a complicated problem. However, on the basis of experiments the following formula has been suggested for the geometry shown in the figure [12]:

$$p_{cr} = \left(\frac{\pi^2}{14.45}\right)\frac{L^{1.1}Et^{1.45}}{(1-\nu^2)^{0.725}r_2^{1.55}} \tag{15.70}$$

In the foregoing, we have r_2 = mean radius of curvature, L = length, t = thickness, E = modulus of elasticity, and ν = Poisson's ratio.

15.16 Buckling of Cylindrical and Spherical Pressure Vessels

The determination of critical shell stresses is a very laborious process. The results are listed in Table 15.2 for four practical applications [7,8]. With them comes the off-repeated

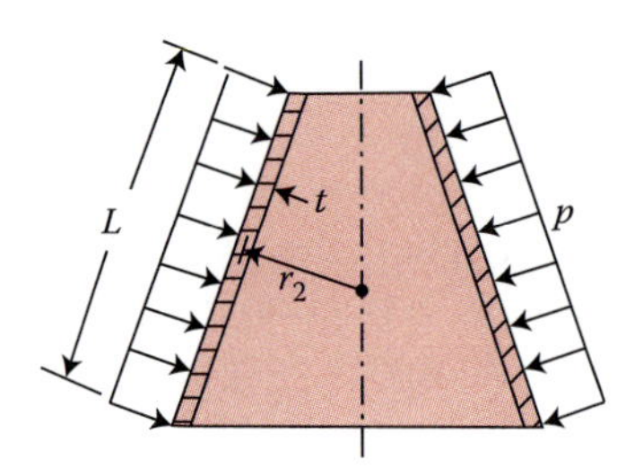

FIGURE 15.9
Conical shell under external pressure.

TABLE 15.2

Critical Pressure p_{cr} Formulas for Buckling of Thin-Walled Vessels

a. Long cylindrical shells under external pressure

$$p_{cr} = \frac{Et^3}{4a^3(1-\nu^2)} \quad (15.71)$$

b. Long cylindrical shells under internal pressure from ram action

$$p_{cr} = \pi^2 \frac{Eta}{L^2} \quad (15.72)$$

c. Spherical shells under external pressure

$$p_{cr} = \frac{2Et^3}{a^2\sqrt{3(1-\nu^2)}} \quad (15.73)$$

d. Hemispherical ends (see Figure 14.10a) under external pressure

$$p_{cr} = \frac{2Et^3}{a\sqrt{3(1-\nu^2)}} \quad (15.74)$$

Notes: t, wall thickness; a, mean radius; L, length; E, modulus of elasticity; ν, Poisson's ratio.

warning that they are valid only for pressure vessels of $a/t > 10$. There are many other kinds of shell-buckling problems, many of which are difficult to analyze. This is especially true, as in the case of plates, for shell structures with end restraints other than those of simple support.

Note that in some situations vessels can fail by buckling under internal pressure, such as hydraulic tubing loaded by rams. Various types of pressure vessel heads may also collapse when subjected to internal pressure due to the existence of large tangential compressive stresses in the knuckle region. The problems involving buckling by external pressure have been of considerable interest, particularly for vacuum systems and structures, like submarines. Interestingly, *spheres*, in addition to the desirable features noted in Case Study 12.1 of Section 12.6, have the *greatest* buckling resistance. In single or multisphere tandem configuration, they form the pressure-carrying structure and living space in most deep-submergence oceanographic vehicles.

Problems

Sections 15.1 through 15.12

15.1 Determine the maximum deflection w and stress in the tank described in Example 15.1. Use $t = 5$ mm. Assume that all the other data are unchanged.

15.2 Resolve Problem 15.1 for a tank of thickness $t = 10$ mm.

15.3 A cold-rolled, yellow brass cylinder (Table B.3) is loaded as described in Example 15.3. Compute vertical and horizontal deflections at the center of the cylinder. Use $a = 200$ mm, $L = 0.4$ m, $t = 5$ mm, $P = 4$ kN, and $\nu = 1/3$.

15.4 Redo Problem 15.3 for the case in which the cylinder is constructed of structural, ASTM-A36 steel (Table B.3). Let $\nu = 0.3$.

15.5 Verify the result for W given by Equation f of Section 15.9. (*Hint*:

$$\begin{aligned} W = {} & A_1 e^{(\gamma+i\beta)\theta} + A_2 e^{(\gamma-i\beta)\theta} + A_3 e^{-(\gamma+i\beta)\theta} + A_4 e^{-(\gamma-i\beta)\theta} \\ & + A_5 e^{(\beta+i\gamma)\theta} + A_6 e^{(\beta-i\gamma)\theta} + A_7 e^{-(\beta+i\gamma)\theta} + A_8 e^{-(\beta-i\gamma)\theta} \end{aligned} \tag{P15.5}$$

where $A_1, \ldots, A_8$ are arbitrary complex quantities. Use the relationships that exist between the exponential function and the trigonometric and hyperbolic functions to reduce the preceding to Equation f.)

15.6 Referring to Section 8.9, verify that, for the shell shown in Figure 15.7, Equations 15.55 reduce to the rigidities given by Equations 15.50.

15.7 Determine the rigidity matrix A for the two-ply laminate shown in Figure 8.12.

Sections 15.13 and 15.14

15.8 During a stage of firing, a long, thin cylindrical missile casing of 1-m radius and 25-mm thickness is loaded in axial compression. If $E = 200$ GPa, $\nu = 0.3$, and $\sigma_{yp} = 300$ MPa, what is the critical stress?

15.9 A 2.5-m-long steel pipe of 1.2-m diameter and 12-mm thickness is used in a structure as a column. What axial load can be tolerated without causing the shell to buckle? Use $E = 200$ GPa.

15.10 A water tower support, constructed of steel pipe ($E = 210$ GPa, $\nu = 0.3$, $\sigma_{yp} = 400$ MPa) of 0.8-m diameter, is to carry an axial compression load of 450 kN. Calculate the minimum thickness of the pipe: (a) for $L/a < 5$; (b) for $L/a > 5$.

References

1. W. Flügge, *Stresses in Shells*, 2nd ed., Springer, Berlin, 1990.
2. S. Timoshenko and S. Woinowskys-Krieger, *Theory of Plates and Shells*, 2nd ed., McGraw-Hill, New York, 1959.
3. L. H. Donnell, *Stability of Thin Walled Tubes under Torsion*, NACA Report 479, 1933.
4. J. E. Gibson, *Thin Shells: Computing and Theory*, Chapters 3, 4, and 9, Pergamon, New York, 1980.
5. E. H. Baker, L. Kovalevsky, and F. L. Rish, *Structural Analysis of Shells*, Krieger, Melbourne, FL, 1981.
6. H. Kraus, *Thin Shells*, Wiley, Hoboken, NJ, 2003.
7. S. Timoshenko and J. M. Gere, *Theory of Elastic Stability*, 2nd ed., McGraw-Hill, New York, 1961.
8. W. C. Young, R. C. Budynas, and A. M. Sadegh, *Roark's Formulas for Stress and Strain*, 8th ed., McGraw-Hill, New York, 2011.
9. J. R. Vinson, *Structural Mechanics: The Behavior of Plates and Shells*, Chapters 7 and 8, McGraw-Hill, New York, 1975.
10. D. O. Brush and B. O. Almroth, *Buckling of Bars, Plates, and Shells*, McGraw-Hill, New York, 1975.
11. W. Flügge (ed.), *Handbook of Engineering Mechanics*, Sections 40.5 and 44.7, McGraw-Hill, New York, 1984.
12. J. H. Fauppel and F. E. Fisher, *Engineering Design*, 2nd ed., Wiley, Hoboken, NJ, 1981.

Appendix A: Fourier Series Expansions

A.1 Single Fourier Series

The Fourier series are indispensable aids in the analytical treatment of many problems in the field of applied mechanics. The representation of periodic functions using trigonometric series is commonly called the *Fourier series expansion*. A function $f(x)$ defined in the interval $(-L, L)$ and determined outside of this interval by $f(x + 2L) = f(x)$ is said to be *periodic*, of period $2L$. Here L is a nonzero constant; for example, for $\sin x$ there are periodic $2\pi, -2\pi, 4\pi, \ldots$. The Fourier expansion corresponding to $f(x)$ is of the form [1]

$$f(x)=\frac{a_0}{2}+\sum_{n=1}^{\infty}\left(a_n\cos\frac{n\pi x}{L}+b_n\sin\frac{n\pi x}{L}\right) \tag{A.1}$$

in which a_0, a_n, and b_n are the *Fourier coefficients*.

To determine the coefficients a_n, we multiply both sides of Equation A.1 by $\cos(m\pi x/L)$, integrate over the interval of length $2L$, and make use of the *orthogonality relations*:

$$\begin{aligned}
&\int_{-L}^{L}\cos\frac{m\pi x}{L}\cos\frac{n\pi x}{L}dx=\begin{cases}0 & m\neq n\\ L & m=n\end{cases}\\
&\int_{-L}^{L}\cos\frac{m\pi x}{L}\sin\frac{n\pi x}{L}dx=0 \qquad \text{for all } m,n\\
&\int_{-L}^{L}\sin\frac{m\pi x}{L}\sin\frac{n\pi x}{L}dx=\begin{cases}0 & m\neq n\\ L & m=n\end{cases}
\end{aligned} \tag{A.2}$$

In so doing, we obtain

$$a_n=\frac{1}{L}\int_{-L}^{L}f(x)\cos\frac{n\pi x}{L}dx \qquad (n=0,1,\ldots) \tag{A.3a}$$

Similarly, multiplication of both sides of Equation A.1 by $\sin(m\pi x/L)$ and integration over the interval $(-L, L)$ results in

$$b_n=\frac{1}{L}\int_{-L}^{L}f(x)\sin\frac{n\pi x}{L}dx \qquad (n=1,2,\ldots) \tag{A.3b}$$

EXAMPLE A.1: Fourier Expansion of a Function

Develop the Fourier expansion of the periodic function defined by

$$f(x)=0 \quad \text{if} -\pi<x<0, \qquad f(x)=\sin x \quad \text{if } 0<x<\pi$$

Solution

The function is shown by a *solid* line in Figure A.1. The period $2L = 2p$ and $L = p$. On application of Equation A.3,

$$\begin{aligned} a_n &= \frac{1}{\pi}\int_{-\pi}^{0}(0)\cos nx\,dx + \frac{1}{\pi}\int_{0}^{\pi}\sin x\cos nx\,dx \\ &= -\frac{1}{2\pi}\left(\frac{-\cos n\pi}{1-n} + \frac{-\cos n\pi}{1+n} - \frac{2}{1-n^2}\right) \\ &= \frac{1+\cos n\pi}{\pi(1-n^2)} \qquad (n\neq 1) \\ a_1 &= \frac{1}{\pi}\int_{0}^{\pi}\sin x\cos x\,dx = 0 \end{aligned}$$

and

$$\begin{aligned} b_n &= \frac{1}{\pi}\int_{-x}^{0}(0)\sin\pi x\,dx + \frac{1}{\pi}\int_{0}^{n}\sin x\sin nx\,dx = 0 \qquad (n\neq 1) \\ b_1 &= \frac{1}{\pi}\int_{0}^{\pi}\sin^2 x\,dx = \frac{1}{2} \end{aligned}$$

The required Fourier series, for $n = 0, 1, 2, \ldots$ is thus

$$f(x) = \frac{1}{\pi} + \frac{\sin x}{2} - \frac{2}{\pi}\left(\frac{\cos 2x}{3} + \frac{\cos 4x}{15} + \frac{\cos 6x}{35} + \frac{\cos 8x}{63} + \cdots\right)$$

The first and the first three terms

$$y_1 = \frac{1}{\pi} \qquad y_3 = \frac{1}{\pi} + \frac{\sin x}{2} - \frac{2\cos 2x}{3\pi}$$

are sketched by *dotted* and *dashed* lines, respectively, in Figure A.1. We observe that as the number of terms increases, the approximating curves approach the graph of *f*(*x*).

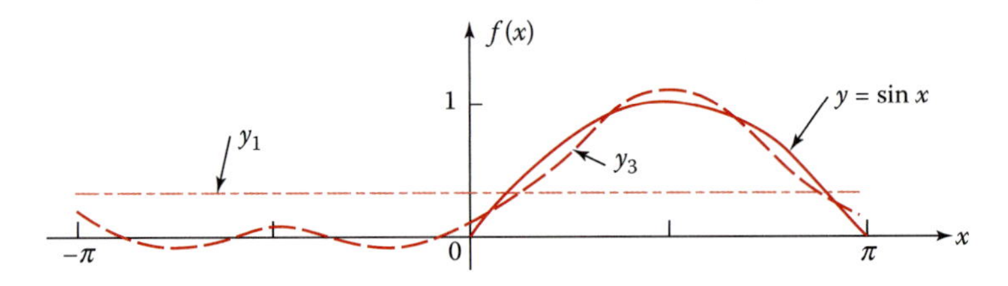

FIGURE A.1
Expansion of a periodic function.

A.2 Half-Range Expansions

In numerous applications, the evaluation of the *Fourier coefficients* can be simplified by employing a *half-range* series over the interval (0, L).

Assume that first $f(x)$ is an even function, that is,

$$f(-x) = f(x)$$

for all x. Viewed geometrically, a graph of an even function is symmetric with respect to the vertical, y axis. For example, x^2 and $\cos x$ are even functions. We can rewrite Equation A.3a as

$$a_0 = \frac{1}{L}\int_{-L}^{0} f(x)\cos\frac{n\pi x}{L}dx + \frac{1}{L}\int_{0}^{L} f(x)\cos\frac{n\pi x}{L}dx \tag{a}$$

In the first integral above, we substitute $x = -u$ and hence $dx = -du$. Inasmuch as $x = -L$ implies $u = L$ and $x = 0$ implies $u = 0$, this integral is therefore

$$\frac{1}{L}\int_{L}^{0} f(-u)\cos\left(\frac{-n\pi u}{L}\right)(-du) \tag{b}$$

Because

$$f(-u) = f(u) \qquad \cos\left(\frac{-n\pi u}{L}\right) = \cos\frac{n\pi u}{L}$$

expression (b) becomes

$$\frac{1}{L}\int_{0}^{L} f(u)\cos\frac{n\pi u}{L}du \tag{c}$$

Here, the dummy variable of integration u can be replaced by any other symbol, and in particular x. Substitution of Equation c into Equation a then yields, for an even, periodic function $f(x)$, the coefficients a_n:

$$a_n = \frac{2}{L}\int_{0}^{L} f(x)\cos\frac{n\pi x}{L}dx \qquad (n = 0,1,2,\ldots) \tag{A.4}$$

By using Equation A.3b, it is readily verified that $b_n = 0$.

In the case of an *odd* function,

$$f(-x)=-f(x)$$

for all values of x. A procedure identical with that described above yields

$$b_n=\frac{2}{L}\int_0^L f(x)\sin\frac{n\pi x}{L}dx \qquad (n=1,2,\ldots) \tag{A.5}$$

and $a_n=0$. The graph of an odd function (e.g., x and $\sin x$) is *skew-symmetric*.
The half-range single series expansion of various common loadings is listed in Table 5.2.

EXAMPLE A.2: The Half-Range Expansion of a Function

Determine the half-range expansion of the function

$$f(x)=x \qquad (0<x<2)$$

in (a) sine series and (b) cosine series.

Solution

The function is indicated by the *solid* line in Figure A.2.

a. We shall treat $f(x)$ as an odd function of period $2L=4$. The *extended function* is shown in Figure A.2a by the *dashed* lines. For the sine series $a_n=0$, and Equation A.5 gives

$$b_n=\frac{2}{2}\int_0^2 x\sin\frac{n\pi x}{2}dx=-\frac{4}{n\pi}\cos n\pi$$

Hence,

$$\begin{aligned} f(x)&=\sum_{n=1}^{\infty}\frac{-4}{n\pi}\cos n\pi\sin\frac{n\pi x}{2} \\ &=\frac{4}{\pi}\left(\sin\frac{\pi x}{2}-\frac{1}{2}\sin\frac{2\pi x}{2}+\frac{1}{3}\sin\frac{3\pi x}{2}-\cdots\right) \end{aligned} \tag{d}$$

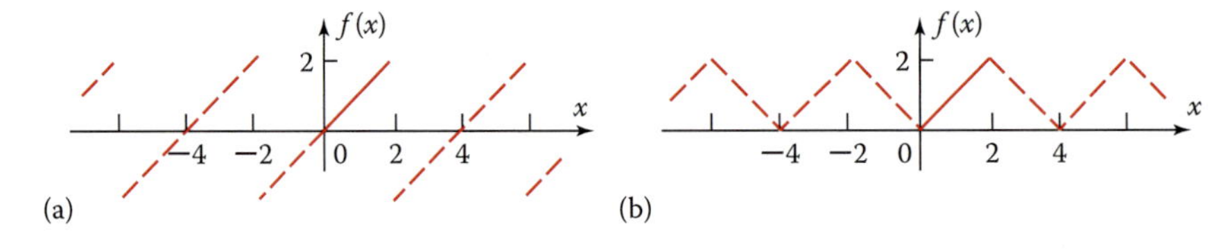

FIGURE A.2
Half-range expansion of: (a) An odd function; (b) an even function.

Observe that the odd function is discontinuous $x = 2$ and $x = -2$, and the sine series (d) converges to zero, the *mean* value of the function, at the points of discontinuity.

b. Now the definition of $f(x)$ is *extended* to that of an even function of period $2L = 4$ (Figure A.2b). As $f(x)$ is even, $b_n = 0$ and

$$a_n = \frac{2}{2}\int_0^2 x\cos\frac{n\pi x}{2}dx = -\frac{4}{n^2\pi^2}(\cos n\pi - 1) \qquad (n \neq 0)$$

When $n = 0$

$$a_0 = \int_0^2 x dx = 2$$

It follows that

$$\begin{aligned} f(x) &= 1 + \sum_{n=1}^{\infty} \frac{4}{n^2\pi^2}(\cos n\pi - 1)\cos\frac{n\pi x}{2} \\ &= 1 - \frac{8}{\pi^2}\left(\cos\frac{\pi x}{2} + \frac{1}{3^2}\cos\frac{3\pi x}{2} + \frac{1}{5^2}\cos\frac{5\pi x}{2} + \cdots\right) \end{aligned} \qquad \text{(e)}$$

Comments: The series d and e represent the extended function shown in Figure A.2a and b, respectively; however, our concern is only with the interval $0 < x < 2$. Note that the latter representation converges more rapidly than the former.

A.3 Double Fourier Series

The idea of a Fourier series expansion for a function of a single variable can be extended to the case of function of two or more variables. For instance, we can expand $p(x, y)$ into a *double Fourier sine series*, Equation 5.1a:

$$p(x,y) = \sum_{m=1}^{\infty}\sum_{n=1}^{\infty} p_{mn} \sin\frac{m\pi x}{a}\sin\frac{n\pi y}{b}$$

The above represents a half-range sine series in x, multiplied by a half-range sine series in y, using for the period of expansion $2a$ and $2b$, respectively. That is,

$$p(x,y) = \sum_{m=1}^{\infty} p_m(y)\sin\frac{m\pi x}{a} \qquad \text{(A.6)}$$

with

$$p_m(y) = \sum_{n=1}^{\infty} p_{mn} \sin \frac{n\pi y}{b} \tag{a}$$

Treating Equation A.6 as a Fourier series wherein y is kept constant, Equation A.5 is applied to yield

$$p_m(y) = \frac{2}{a} \int_0^a p(x,y) \sin \frac{m\pi x}{a} dx \tag{A.7}$$

Likewise, for Equation a:

$$p_{mn} = \frac{2}{b} \int_0^b p_m(y) \sin \frac{n\pi y}{b} dy \tag{b}$$

Equation b, together with Equation A.7, then leads to

$$p_{mn} = \frac{4}{ab} \int_0^b \int_0^a p(x,y) \sin \frac{m\pi x}{a} \sin \frac{n\pi y}{b} dx\,dy$$

This agrees with Equation 5.3 of Section 5.2, which was derived differently.

Similarly, the results can be obtained for cosine series or for series having both sines and cosines. The double-sine series expansion of some typical loadings is illustrated in Section 5.3.

Reference

1. E. Kreyszig, *Advanced Engineering Mathematics*, 10th ed., Wiley, Hoboken, NJ, 2011.

Appendix B: Tables

TABLE B.1

Conversion Factors between SI Units and U.S. Customary System (USCS) Units

Quantity	U.S. Customary to SI	SI to U.S. Customary
Acceleration	1 in./s^2 = 0.0254 m/s^2[a]	1 m/s^2 = 39.37 in./s^2
	1 ft/s^2 = 0.3048 m/s^2[a]	1 m/s^2 = 3.281 ft/s^2
Area	1 in.^{2} = 645.2 mm^2	1 mm^2 = 1.55(10^{-3}) in.^{2}
	1 ft^2 = 0.0929 m^2	1 m^2 = 10.76 ft^2
Distributed load	1 lb/ft = 14.59 N/m	1 kN/m = 68.52 lb/ft
Force	1 lb = 4.448 N	1 N = 0.2248 lb
Length	1 in. = 25.4 mm[a]	1 m = 39.37 in.
	1 ft. = 0.3048 m[a]	1 m = 3.281 ft
	1 miles = 1.609 km	1 km = 0.6214 miles
Mass	1 slug = 14.59 kg	1 kg = 0.06852 slug
	1 lb = 0.4987 kg	1 kg = 2.2051 lb
Moment of force; torque	1 lb · ft = 1.356 N · m	1 N · m = 0.7376 lb · ft
	1 lb · in. = 0.1130 N · m	1 N · m = 8.851 lb · in.
Moment of inertia of an area	1 in.^{4} = 0.4162(10^{-6}) m^4	1 m^4 = 2.402 (10^6) in.^{4}
Moment of inertia of a mass	1 ft · s^2 = 1.356 kg · m^2	1 kg · m^2 = 0.7376 ft · s^2
Power	1 lb · ft/s = 1.356 W	1 W = 0.7376 lb · ft/s
	1 hp = 745.7 W	1 kW = 1.341 hp
Pressure; stress	1 psi = 6.895 kPa	1 kPa = 0.145 psi
	1 ksi = 6.895 MPa	1 MPa = 145 psi
Specific weight	1 lb/in.^{3} = 271.444 kN/m^3	1 kN/m^3 = 3.684 (10^{-3}) lb/in.^{3}
Spring rate	1 lb/in. = 175.12 N/m	1 kN/m = 5.71 kip/in.
Velocity	1 in./s = 0.0254 m/s[a]	1 m/s = 39.17 in./s
	1 ft/s = 0.3048 m/s[a]	1 m/s = 3.281 ft/s
	1 miles/s = 1.6 km/h	1 km/h = 0.6241 miles/h
Volume	1 in.^{3} = 16.39(10^3) mm^3	1 mm^3 = 61.02(10^{-6}) in^3
	1 ft^3 = 0.02832 m^3	1 m^3 = 35.31 ft^3
Work; energy	1 lb · ft = 1.356 J	1 J = 0.7376 lb · ft
Temperature	$t_c = \frac{5}{9}(t_f - 32)$	$32t_f = 0\ t_c = 273.15t_k$

Note: See Table 1.1 for basic, derived, and supplementary SI units.

[a] Exact value.

TABLE B.2

SI Unit Prefixes

Prefix	Symbol	Factor
tera	T	10^{12} = 1,000,000,000,000
giga	G	10^{9} = 1,000,000,000
mega	M	10^{6} = 1,000,000
kilo	K	10^{3} = 1,000
hecto	H	10^{2} = 100
deka	da	10^{1} = 10
deci	d	10^{-1} = 0.1
centi	c	10^{-2} = 0.01
milli	m	10^{-3} = 0.001
micro	μ	10^{-6} = 0.000 001
nano	n	10^{-9} = 0.000 000 001
pico	p	10^{-12} = 0.000 000 000 001

Note: The use of the prefixes hecto, deka, and centi is not recommended. However, they are sometimes encountered in practice.

TABLE B.3

Typical Properties for Some Common Materials[a]

Material	Density (Mg/m^3)	Ultimate Strength (MPa)			Yield Strength[c] (MPa)		Modulus of Elasticity (GPa)	Modulus of Rigidity (GPa)	Coefficient of Thermal Expansion ($10^{-6}/°C$)	Elongation in 50 mm (%)
		Tension	Compression[b]	Shear	Tension	Shear				
Steel										
Structural, ASTM-A36	7.86	400	–	–	250	145	200	79	11.7	30
High strength, ASTM-A242	7.86	480	–	–	345	210	200	79	11.7	21
Stainless (302), cold-rolled	7.92	860	–	–	520	–	190	73	17.3	12
Cast Iron										
Gray, ASTM A-48	7.2	170	650	240	–	–	70	28	12.1	0.5
Malleable, ASTM-A47	7.3	340	620	330	230	–	165	64	12.1	10
Wrought Iron	7.7	350	–	240	210	130	190	70	12.1	35
Aluminum										
Alloy 2014-T6	2.8	480	–	290	410	220	72	28	23	13
Alloy 6061-T6	2.71	300	–	185	260	140	70	26	23.6	17
Brass, yellow										
Cold-rolled	8.47	540	–	300	435	250	105	39	20	8
Annealed	8.47	330	–	220	105	65	105	39	20	60
Bronze, cold-rolled (510)	8.86	560	–	–	520	275	110	41	17.8	10
Copper, hard-drawn	8.86	380	–	–	260	160	120	40	16.8	4
Magnesium alloys	1.8	140–340	–	165	80–280	–	45	17	27	2–20
Nickel	8.8	310–760	–	–	140–620	–	210	80	13	2–50
Concrete										
Medium strength	2.32	–	28	–	–	–	24	–	10	–

(*Continued*)

TABLE B.3 (*Continued*)

Typical Properties for Some Common Materials[a]

Material	Density (Mg/m^3)	Ultimate Strength (MPa) Tension	Ultimate Strength (MPa) Compression	Ultimate Strength (MPa) Shear	Yield Strength[c] (MPa) Tension	Yield Strength[c] (MPa) Shear	Modulus of Elasticity (GPa)	Modulus of Rigidity (GPa)	Coefficient of Thermal Expansion ($10^{-6}/°C$)	Elongation in 50 mm (%)
High strength	2.32	–	40	–	–	–	30	–	10	–
Timber[d] *(air dry)*										
Douglas fir	0.54	–	55	7.6	–	–	12	–	4	–
Southern pine	0.58	–	60	10	–	–	11	–	4	–
Glass, 98% silica	2.19	–	50	–	–	–	65	28	80	–
Nylon, molded	1.1	55	–	–	–	–	2	–	81	50
Polystyrene	1.05	48	90	55	–	–	3	–	72	4
Graphite	0.77	20	240	35	–	–	70	–	7	–
Rubber	0.91	14	–	–	–	–	–	–	162	600

[a] Properties may vary widely with changes in composition, heat treatment, and method of manufacture.

[b] For ductile metals the compression strength is assumed to be the same as that in tension.

[c] Offset of 0.2%.

[d] Loaded parallel to the gain.

TABLE B.4

Selected Plastics

Chemical Classification	Trade Name
Thermoplastic Materials	
Acetal	Delrin, Celcon
Acrylic	Lucite, Plexiglas
Cellulose acetate	Fibestos, Plastacele
Cellulose nitrate	Celluloid, Nitron
Ethyl cellulose	Gering, Ethocel
Polyamide	Nylon, Zytel
Polycarbonate	Lexan, Merlon
Polyethylene	Polythene, Alathon
Polystyrene	Cerex, Lustrex
Polytetrafluoroethylene	Teflon
Polyvinyl acetate	Gelva, Elvacet
Polyvinyl alcohol	Elvanol, Resistoflex
Polyvinyl chloride	PVC, Boltaron
Polyvinylidene chloride	Saran
Thermosetting Materials	
Epoxy	Araldite, Oxiron
Phenol-formaldehyde	Bakelite, Catalin
Phenol-furfural	Durite
Polyester	Beckosol, Glyptal
Urea-formaldehyde	Beetle, Plaskon

TABLE B.5

Properties of Common Areas

1. Rectangle

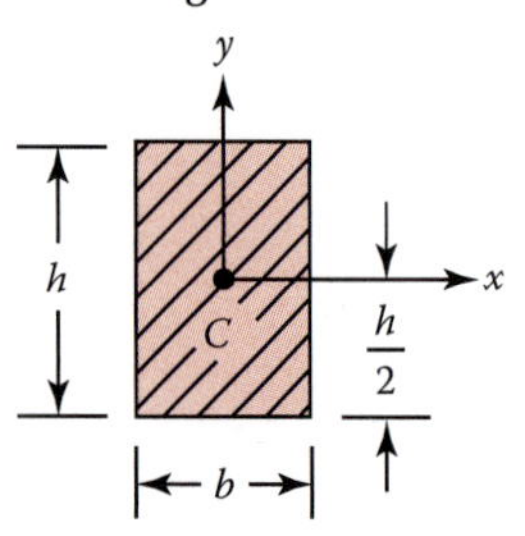

$$A = bh$$
$$I_x = \frac{bh^3}{12}$$
$$J_c = \frac{bh\left(b^2 + h^2\right)}{12}$$

2. Circle

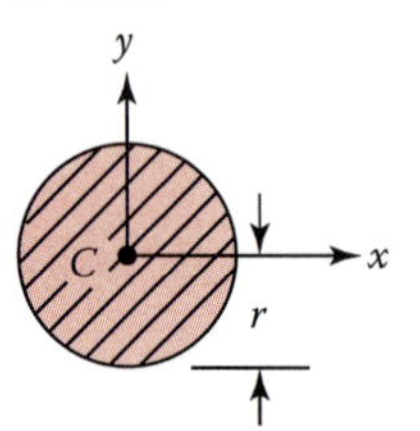

$$A = \pi r^2$$
$$I_x = \frac{\pi r^4}{4}$$
$$J_c = \frac{\pi r^4}{2}$$

3. Right triangle

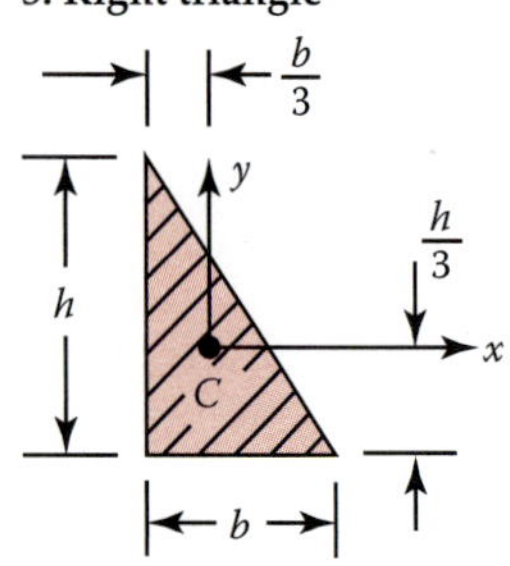

$$A = \frac{bh}{2}$$
$$I_x = \frac{bh^3}{36} \quad I_{xy} = -\frac{b^2h^2}{72}$$
$$J_c = \frac{bh\left(b^2 + h^2\right)}{36}$$

4. Semicircle

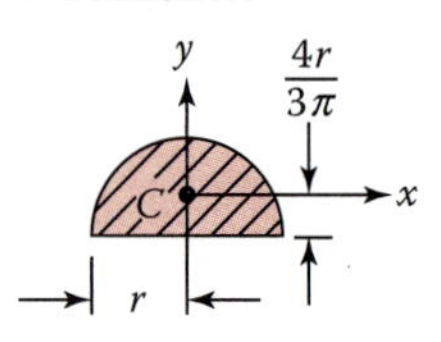

$$A = \frac{\pi r^2}{2}$$
$$I_x = 0.110r^2$$
$$I_y = \frac{\pi r^4}{8}$$

5. Ellipse

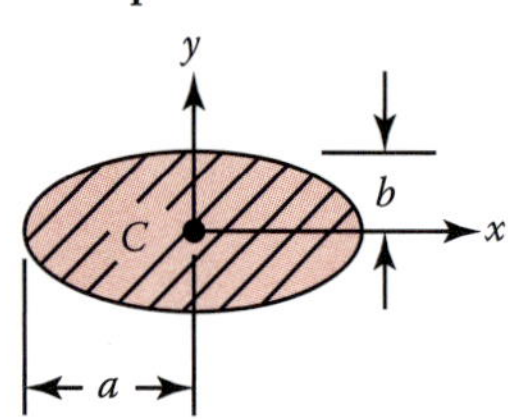

$$A = \pi ab$$
$$I_x = \frac{\pi ab^3}{4}$$
$$J_c = \frac{\pi ab\left(a^2 + b^2\right)}{4}$$

6. Thin tube

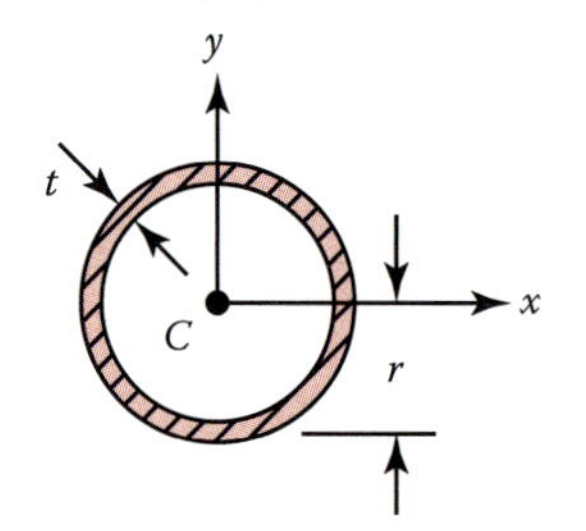

$$A = 2\pi rt$$
$$I_x = \pi r^3 t$$
$$J_c = 2\pi r^3 t$$

7. Isosceles triangles

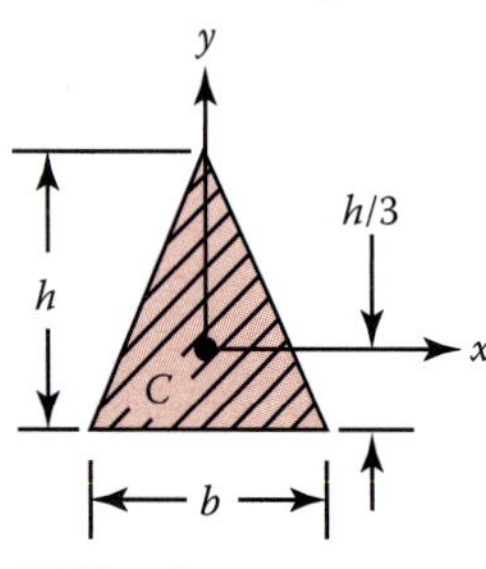

$$A = \frac{bh}{2}$$
$$I_x = \frac{bh^3}{36} \quad I_y = \frac{hb^3}{48}$$
$$J_c = \frac{bc}{144}\left(4h^2 + 3b^2\right)$$

8. Half of thin tube

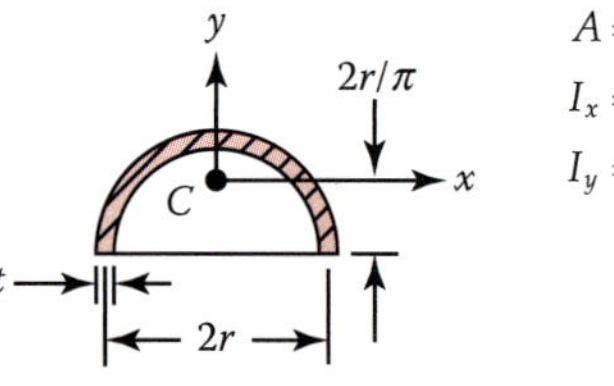
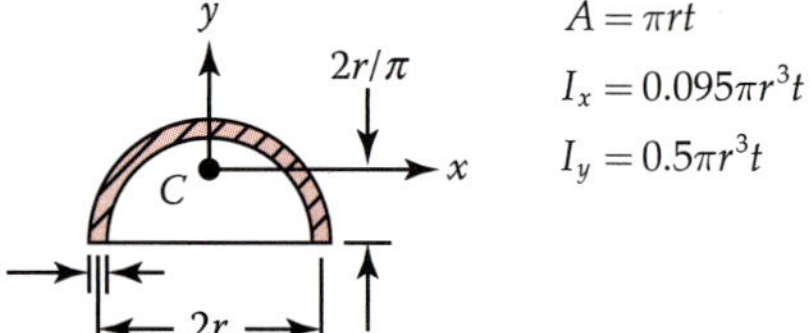

$$A = \pi rt$$
$$I_x = 0.095\pi r^3 t$$
$$I_y = 0.5\pi r^3 t$$

9. Triangle

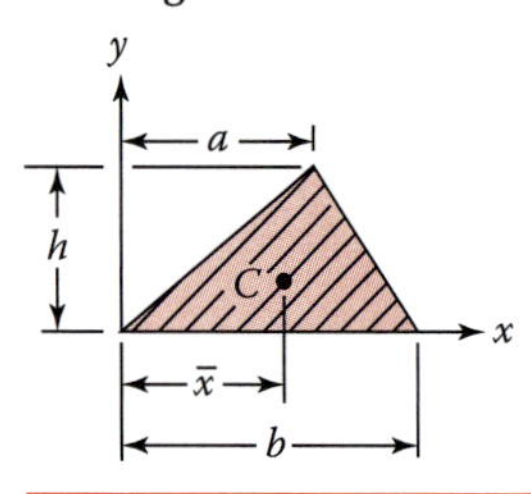

$$A = \frac{bh}{2}$$
$$\bar{x} = \frac{(a + b)}{3}$$

10. Parabola $(y = x^2)$

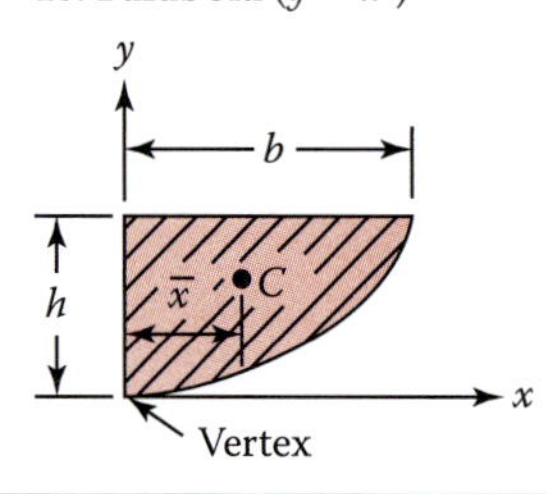

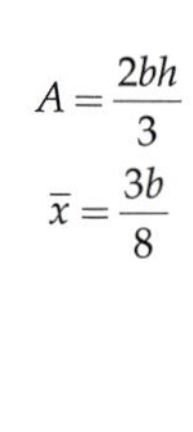

$$A = \frac{2bh}{3}$$
$$\bar{x} = \frac{3b}{8}$$

TABLE B.6

Maximum Shearing Stress for Some Typical Beam Cross-Sectional Forms

Cross Section	Maximum Shearing Stress	Location
1. Rectangle	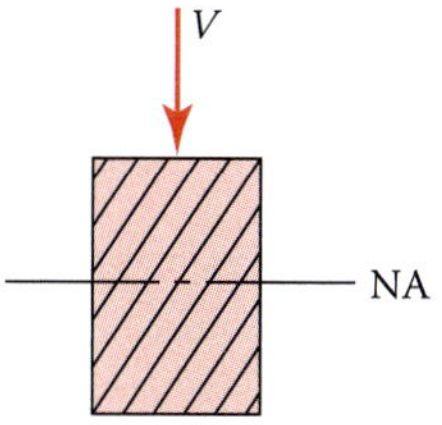 $\tau_{max} = \frac{3}{2}\frac{V}{A}$	NA
2. Circle	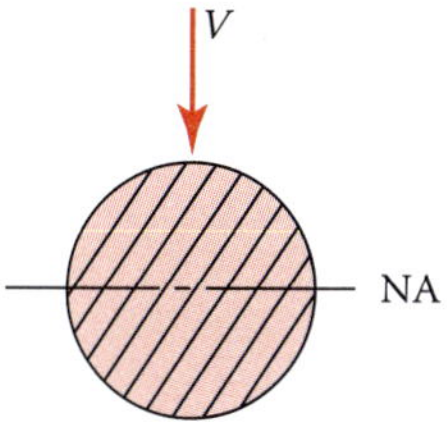 $\tau_{max} = \frac{4}{3}\frac{V}{A}$	NA
3. Hollow circle	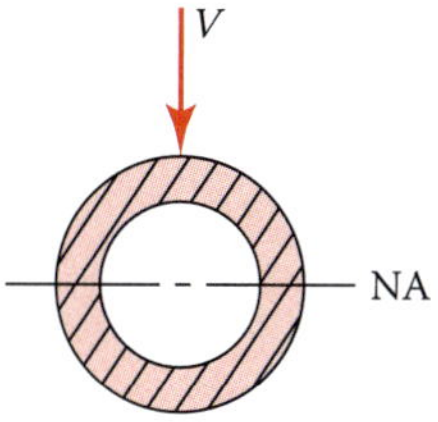 $\tau_{max} = 2\frac{V}{A}$	NA
4. Triangle	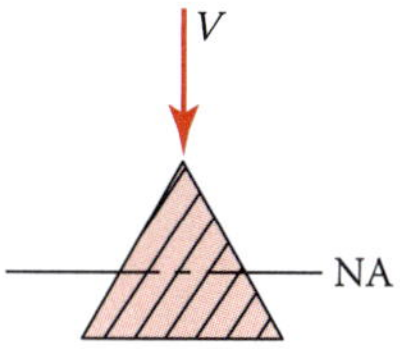 $\tau_{max} = \frac{3}{2}\frac{V}{A}$	Halfway between top and bottom
5. Diamond	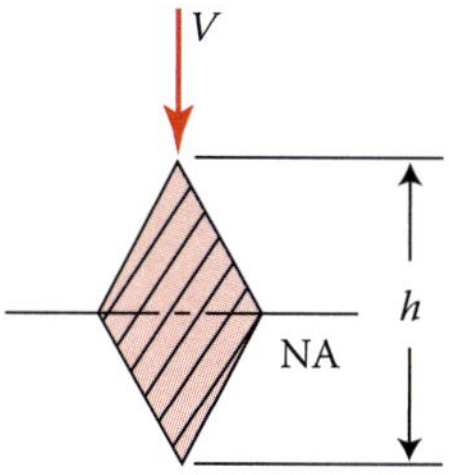 $\tau_{max} = \frac{9}{8}\frac{V}{A}$	At $h/8$ above and below the NA

Note: A, cross-sectional area; V, transverse shear force; NA, the neutral axis.

TABLE B.7

Beam Deflections and Slopes

Load and Support	Deflection	Slope (+ Counterclockwise)
1. 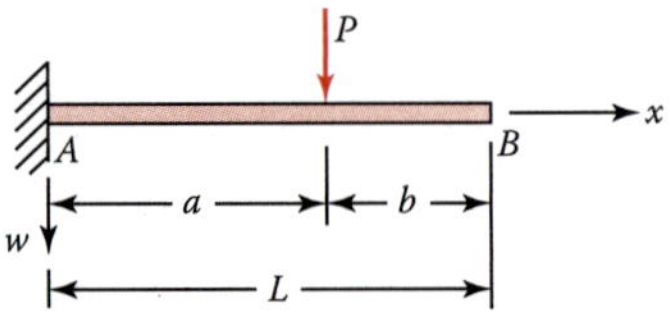	$w = \frac{Px^2}{6EI}(3a - x) \quad 0 \le x \le a$ $w = \frac{Pa^2}{6EI}(3x - a) \quad a \le x \le L$	$\theta_B = -\frac{Pa^2}{2EI}$
2. 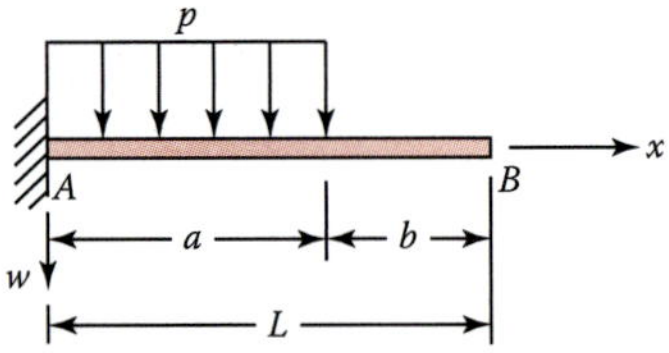	$w = \frac{Px^2}{24EI}(6a^2 - 4ax + x^2) \quad 0 \le x \le a$ $w = \frac{Pa^3}{24EI}(4x - a) \quad a \le x \le L$	$\theta_B = -\frac{Pa^3}{6EI}$
3. 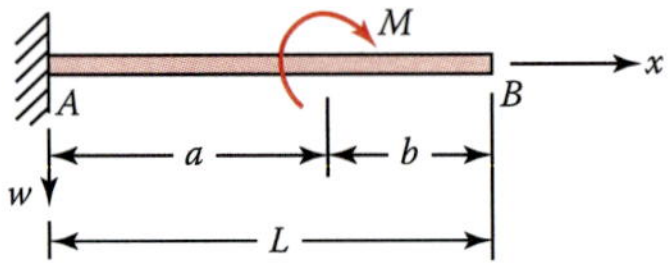	$w = \frac{Mx^2}{2EI} \quad 0 \le x \le a$ $w = \frac{Ma^2}{2EI}(2x - a) \quad a \le x \le L$	$\theta_B = -\frac{Ma}{EI}$
4. 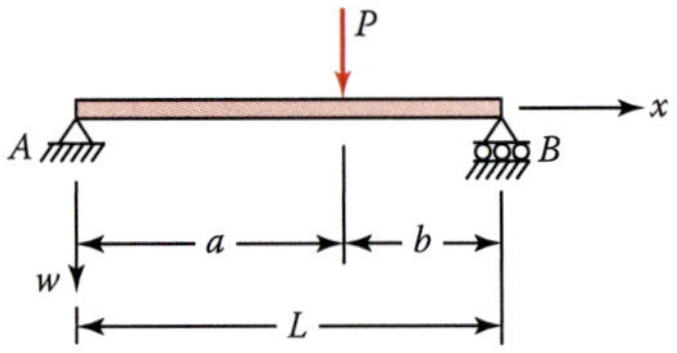	$w = \frac{Pbx}{6EIL}(L^2 - b^2 - x^2) \quad 0 \le x \le a$ For $a \ge b$: $w_{\max} = \frac{Pb(L^2 - b^2)^{3/2}}{9\sqrt{3}EIL}$ at $x = \sqrt{\frac{L^2 - b^2}{3}}$	$\theta_A = -\frac{Pab(L+b)}{6EIL}$ $\theta_B = \frac{Pab(L+a)}{6EIL}$
5. 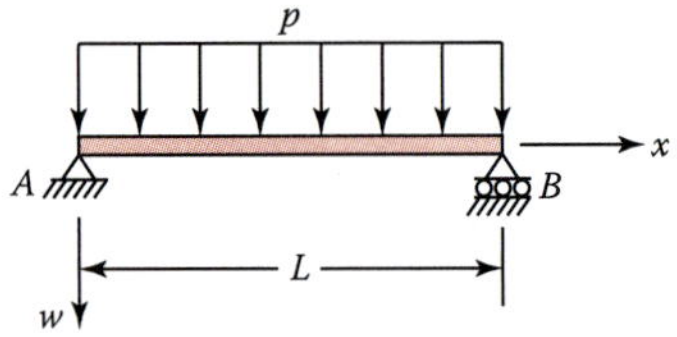	$w = \frac{px}{24EI}(L^3 - 2Lx^2 + x^3)$ $w_{\max} = \frac{5pL^4}{384EI}$ at $x = \frac{L}{2}$	$\theta_A = -\frac{pL^3}{24EI}$ $\theta_B = \frac{pL^3}{24EI}$
6. 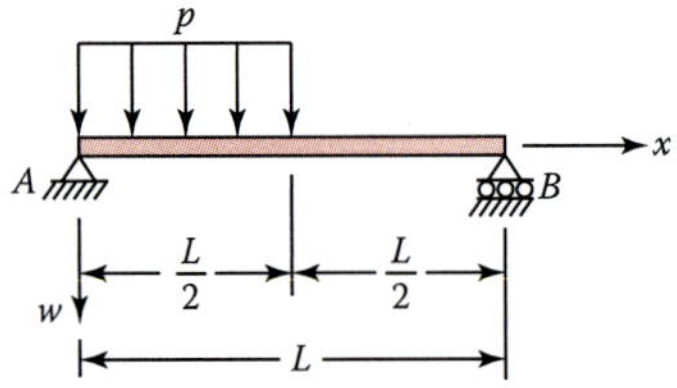	$w = \frac{px}{384EI}(9L^3 - 24Lx^2 + 16x^3) \quad 0 \le x \le \frac{L}{2}$ $w = \frac{5pL^4}{768EI}$ at $x = \frac{L}{2}$	$\theta_A = -\frac{3pL^3}{128EI}$ $\theta_B = \frac{7pL^3}{384EI}$
7. 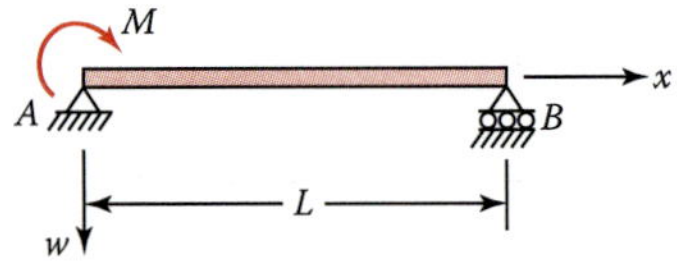	$w = \frac{Mx}{6EIL}(2L^2 - 3Lx + x^2)$ $w_{\max} = \frac{ML^2}{9\sqrt{3}EI}$ at $x = L\left(1 - \frac{\sqrt{3}}{3}\right)$	$\theta_A = -\frac{ML}{3EI}$ $\theta_B = \frac{ML}{6EI}$

TABLE B.8

Restrained Beam Reactions and Deflections

Load and Support	Reactions[a]	Deflections
1. 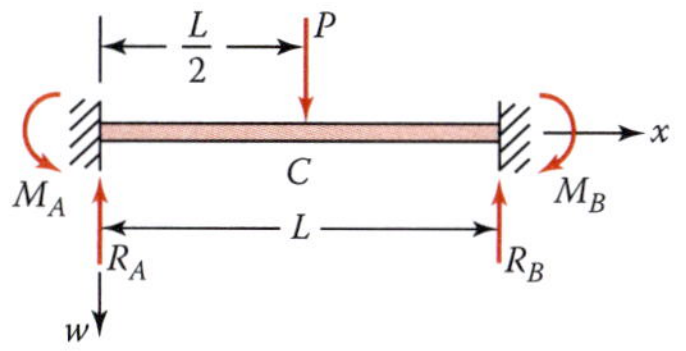	$R_A = R_B = \frac{P}{2}$ $M_A = M_B = \frac{PL}{8}$	$w_{max} = w_C = \frac{PL^3}{192EI}$
2. 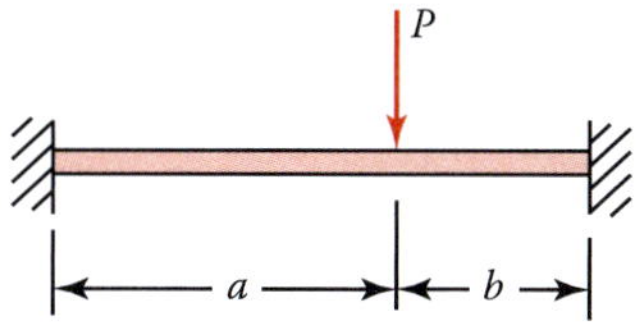	$R_A - \frac{Pb^2}{L^3}(3a+b) \quad R_B = \frac{Pa^2}{L^3}(a+3b)$ $M_A = \frac{Pab^2}{L^2} \quad M_B = \frac{Pa^2b}{L^2}$	For $a > b$: $w_C = \frac{Pb^2}{48EI}(3L-4b)$
3. 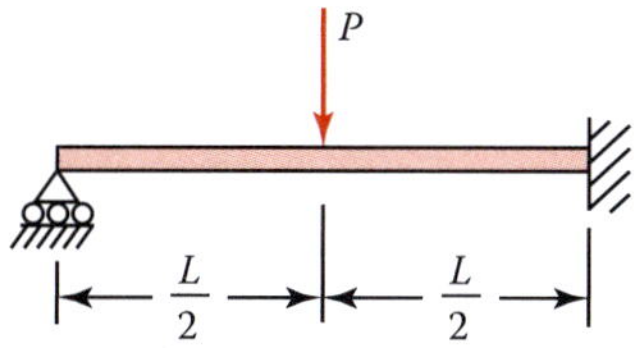	$R_A = \frac{5}{16}PR_B = \frac{11}{16}P$ $M_B = \frac{3}{16}PL$	$w_C = \frac{7PL^3}{768EI}$
4. 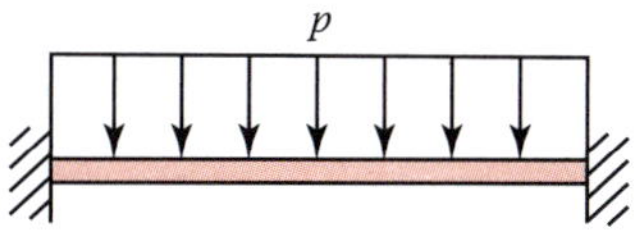	$R_A = R_B = \frac{pL}{2}$ $M_A = M_B = \frac{pL^2}{12}$	$w_{max} = w_C = \frac{pL^4}{384EI}$
5. 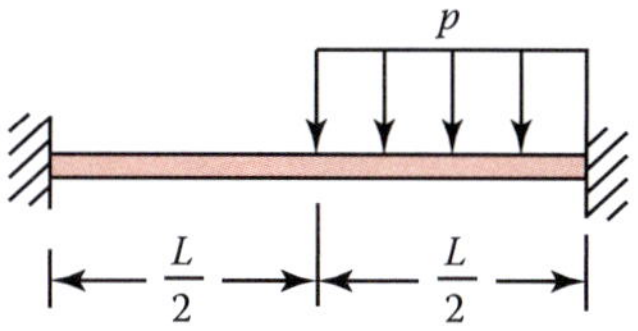	$R_A = \frac{3}{32}pL \quad R_B = \frac{13}{32}pL$ $M_A = \frac{5}{192}pL^2 \quad M_B = \frac{11}{192}pL^2$	$w_C = \frac{pL^4}{768EI}$
6. 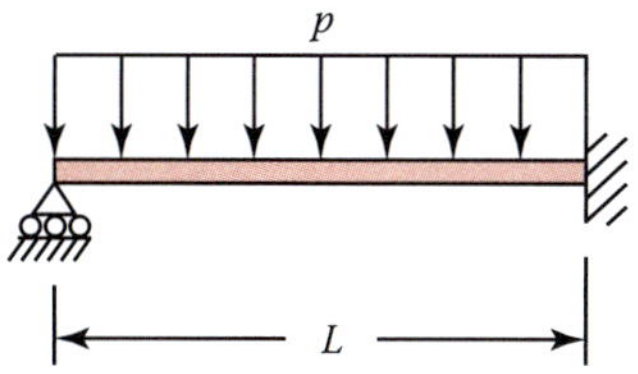	$R_A = \frac{3}{8}pL \quad R_B = \frac{5}{8}pL$ $M_B = \frac{1}{8}pL^2$	$w_C = \frac{pL^4}{192EI}$
7. 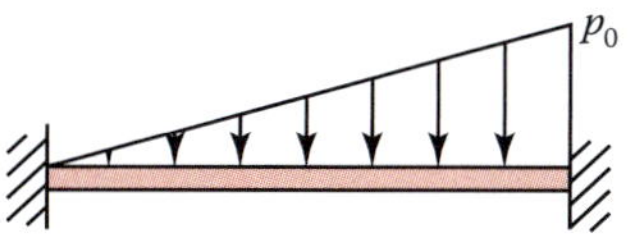	$R_A = \frac{3}{20}p_0L \quad R_B = \frac{7}{20}p_0L$ $M_A = \frac{p_0L^2}{30} \quad M_B = \frac{p_0L^2}{20}$	$w_C = \frac{p_0L^4}{768EI}$

[a] For all the cases tabulated, the senses of the reactions and the notations are the same as those shown in case 1.

Appendix C: Introduction to Finite Element Analysis

C.1 Bar Element

An *axial element*, also termed a *truss bar* or simply *bar element*, is the simplest form of structural finite element. An element of this kind having length L, modulus of elasticity E, and cross-sectional area A is designated by e (Figure C.1). The two ends or joints or nodes are numbered 1 and 2, respectively. We will develop a set of two equations in matrix form to relate the *joint forces* ($\bar{F}_1$ and $\bar{F}_2$) to the *joint displacements* ($\bar{u}_1$ and $\bar{u}_2$).

The following is a simple derivation by the *direct equilibrium method.* But, this method is practically applied only to truss and frame elements. Referring to Figure C.1, the equilibrium of the x-directed forces gives $\bar{F}_1 = -\bar{F}_2$. With the quantity AE/L representing the spring rate of the element, we have

$$\bar{F}_1 = \frac{AE}{L}(\bar{u}_1 - \bar{u}_2), \qquad \bar{F}_2 = \frac{AE}{L}(\bar{u}_2 - \bar{u}_1)$$

This may be written as in matrix form as

$$\begin{Bmatrix} \bar{F}_1 \\ \bar{F}_2 \end{Bmatrix}_e = \frac{AE}{L}\begin{bmatrix} 1 & -1 \\ -1 & 1 \end{bmatrix}\begin{Bmatrix} \bar{u}_1 \\ \bar{u}_2 \end{Bmatrix}_e \tag{C.1a}$$

or concisely

$$\{\bar{F}\}_e = [\bar{k}]_e\{\bar{u}\}_e \tag{C.1b}$$

Here $[\bar{k}]_e$ represents the *stiffness matrix* of the element, which relates the joint displacement to the joint forces on the element.

It should be noted that, alternatively the *energy method* is more general, easier to apply, and powerful than the direct method. The energy approach particularly applies conveniently for complex types of finite elements. To use this method it is necessary first to define a displacement function for the element (Section 7.11.1).

C.1.1 Global Stiffness Matrix

Let us consider an element oriented arbitrarily in a 2D plane. The *local coordinates* are chosen to conveniently represent each element, while the *global* or reference *coordinates* are selected to be convenient for the entire structure. We designate the local and global coordinate systems for an axial element by $\bar{x}$, $\bar{y}$ and x, y, respectively (Figure C.2).

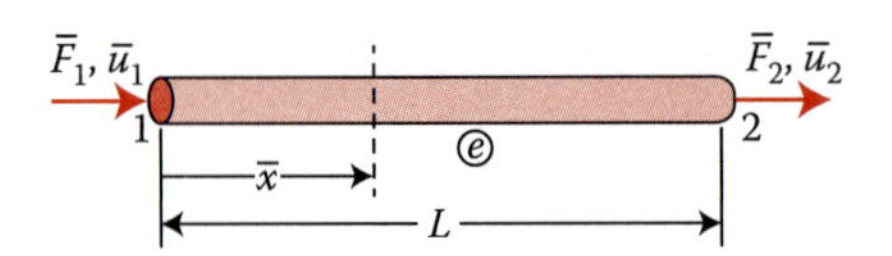

FIGURE C.1
One-dimensional axial element (e).

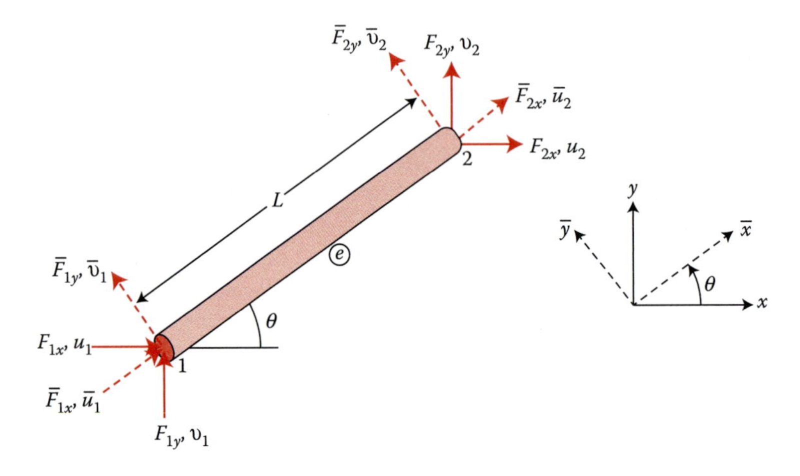

FIGURE C.2
Global coordinates (x, y) and local coordinates ($\bar{x}$, $\bar{y}$) for the plane bar element 1-2.

Observe from the figure that a typical axial element e lying along the $\bar{x}$ axis, which is oriented at an angle θ, is measured *counterclockwise*, from the reference axis x. In the local coordinate system, each joint has an axial force $\bar{F}_x$, a transverse force $\bar{F}_y$, an axial displacement $\bar{u}$, and a transverse displacement $\bar{v}$. Therefore, Equation C.1a is expanded as follows

$$\begin{Bmatrix} \bar{F}_{1x} \\ \bar{F}_{1y} \\ \bar{F}_{2x} \\ \bar{F}_{2y} \end{Bmatrix}_e = \frac{AE}{L} \begin{bmatrix} 1 & 0 & -1 & 0 \\ 0 & 0 & 0 & 0 \\ -1 & 0 & 1 & 0 \\ 0 & 0 & 0 & 0 \end{bmatrix} \begin{Bmatrix} \bar{u}_1 \\ \bar{v}_1 \\ \bar{u}_2 \\ \bar{v}_2 \end{Bmatrix}_e \tag{C.2a}$$

or symbolically

$$\{\bar{F}\}_e = [\bar{k}]_e \{\bar{\delta}\}_e \tag{C.2b}$$

where $[\bar{k}]_e$ and $\{\bar{\delta}\}_e$ designate the stiffness and nodal displacements in the local coordinate system.

C.1.2 Force and Displacement Transformations

It is seen from Figure C.2 that the two local and global forces at joint 1 may be related by the following expressions:

$$\bar{F}_{1x} = F_{1x}\cos\theta + F_{1y}\sin\theta$$
$$\bar{F}_{1y} = -F_{1x}\sin\theta + F_{1y}\cos\theta$$

Identical expressions apply at joint 2. For brevity, we denote

$$c = \cos\theta \quad \text{and} \quad s = \sin\theta$$

So, the local and global forces are related by

$$\begin{Bmatrix} \bar{F}_{1x} \\ \bar{F}_{1y} \\ \bar{F}_{2x} \\ \bar{F}_{2y} \end{Bmatrix}_e = \begin{bmatrix} c & s & 0 & 0 \\ -s & c & 0 & 0 \\ 0 & 0 & c & s \\ 0 & 0 & -s & c \end{bmatrix} \begin{Bmatrix} F_{1x} \\ F_{1y} \\ F_{2x} \\ F_{2y} \end{Bmatrix}_e \tag{C.3a}$$

or

$$\{\bar{F}\}_e = [T]\{F\}_e \tag{C.3b}$$

In the preceding, $[T]$ is the *coordinate transformation matrix:*

$$[T] = \begin{bmatrix} c & s & 0 & 0 \\ -s & c & 0 & 0 \\ 0 & 0 & c & s \\ 0 & 0 & -s & c \end{bmatrix} \tag{C.4}$$

and $\{F\}_e$ designates the *global nodal force matrix:*

$$\{F\}_e = \begin{Bmatrix} F_{1x} \\ F_{1y} \\ F_{2x} \\ F_{2y} \end{Bmatrix}_e \tag{C.5}$$

The displacement transforms in the same manner as forces. It follows that

$$\begin{Bmatrix} \bar{u}_1 \\ \bar{v}_1 \\ \bar{u}_2 \\ \bar{v}_2 \end{Bmatrix}_e = [T] \begin{Bmatrix} u_1 \\ v_1 \\ u_2 \\ v_2 \end{Bmatrix}_e \tag{C.6a}$$

or concisely,

$$\{\bar{\delta}\}_e = [T]\{\delta\}_e \tag{C.6b}$$

The quantity $\{\delta\}_e$ represents the global nodal displacements. Inserting Equations C.6b and C.3b into C.2b leads to

$$[T]\{F\}_e = [\bar{k}]_e[T]\{\delta\}_e$$

from which

$$\{F\}_e = [T]^{-1}[\bar{k}]_e[T]\{\delta\}_e$$

We note that the transformation matrix $[T]$ is an *orthogonal matrix*; that is, its inverse is the same as its transpose: $[T]^{-1} = [T]^T$, where the superscript T designates the transpose. The *transpose of a matrix* is found by interchanging the rows and columns.

C.1.3 Global Force–Displacement Relations

Governing Equations for a bar element e are given by

$$\{F\}_e = [k]_e\{\delta\}_e \tag{C.7}$$

in which

$$[k]_e = [T]^T[\bar{k}]_e[T] \tag{C.8}$$

Substitution of Equation C.4 and $[k]_e$ from Equation C.2a into Equation C.8, results in the *global stiffness matrix* for the element

$$[k]_e = \frac{AE}{L}\begin{bmatrix} c^2 & cs & -c^2 & -cs \\ cs & s^2 & -cs & -s^2 \\ -c^2 & -cs & c^2 & cs \\ -cs & -s^2 & cs & s^2 \end{bmatrix} = \frac{AE}{L}\begin{bmatrix} c^2 & cs & -c^2 & -cs \\ & s^2 & -cs & -s^2 \\ & & c^2 & cs \\ \text{Symmetric} & & & s^2 \end{bmatrix} \tag{C.9}$$

Clearly, the element stiffness matrix depends on its dimensions, orientation, and material property.

C.1.4 Axial Force in an Element

Let us refer to the general case of an axial element oriented arbitrarily in a 2D plane (Figure C.2). It can readily be verified that [Reference 1, Chapter 1], *equation for the axial force* is expressed as

$$F_{12} = \frac{AE}{L}[c \quad s]\begin{Bmatrix} u_2 - u_1 \\ v_2 - v_1 \end{Bmatrix} \tag{C.10}$$

This may be written for an element with nodes *ij* in the form:

$$F_{ij} = \left(\frac{AE}{L}\right)_{ij} [c \quad s]_{ij} \begin{Bmatrix} u_j - u_i \\ v_j - v_i \end{Bmatrix} \tag{C.11}$$

A positive (negative) value found for F_{ij} shows that the element is in tension (compression). The axial stress in the element is defined by $\sigma_{ij} = F_{ij}/A$.

C.2 Governing Equations for a Truss

A *truss* is an assembly of axial elements that are differently oriented. Figure C.3 depicts an example of a plane truss, for which the bars framing between the joints are considered as finite elements. Derivation of the governing equations appropriate to a truss illustrated the formulation of the FEM. To develop truss equations, the global element relations given by Equation C.7 must be assembled. This leads to force–displacement relations for the entire truss, the *system equations*:

$$\{F\} = [K]\{\delta\} \tag{C.12}$$

The *global nodal matrix* $\{F\}$ and the *global stiffness matrix* $[K]$ are given by

$$\{F\} = \sum_{1}^{n} \{F\}_e \tag{C.13a}$$

$$[K] = \sum_{1}^{n} [k]_e \tag{C.13b}$$

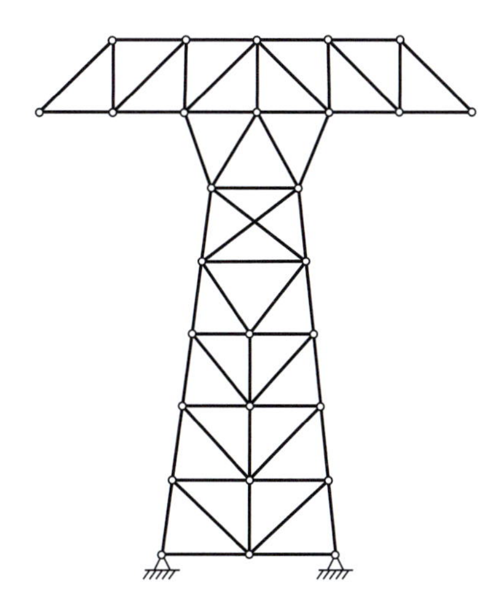

FIGURE C.3
Transmission line truss.

The symbol e denotes an element and n represents the number of elements making up the truss. It is seen that $[K]$ relates the global nodal force $\{F\}$ to the global displacement $\{\delta\}$ for the entire truss.

C.2.1 Assembling Element Stiffness Matrices

The element stiffness matrices in Equation C.13b must be properly added together or *superimposed*. To carry out proper summation, a simple way is to expand the $[k]_e$ for each element to the order of the truss stiffness matrix by adding rows and columns of zeros (see Example C.1). However, for the problem involving a large number of elements, it becomes tedious to apply this approach.

An alternative convenient method is to label the columns and rows of each element stiffness matrix according to the displacement components associated with it. In so doing, the truss stiffness matrix $[K]$ is found by adding terms from the individual element stiffness matrix into their corresponding locations in $[K]$. This approach of assemblage of the element stiffness matrix is given in Case study C.1.

EXAMPLE C.1: Axially Loaded Stepped Bar

Figure C.4a depicts an aluminum bar of two prismatic parts, fixed at the left end and subjected to a load P at the right end.

Find: The nodal displacements and nodal forces. *Assumptions*: The effect of stress concentrations at section 2 can be neglected. The deflections are elastic.

Solution

The *global* coordinates (x, y) *coincide* with *local* $(\bar{x}, \bar{y})$ coordinates. The bar discretized into elements with nodes 1, 2, and 3, as shown in Figure C.4b. The axial rigidities of the elements 1 and 2 are $2AE$ and AE, respectively. Consequently, $(AE/L)_1 = 2AE/L$ and $(AE/L)_2 = AE/L$.

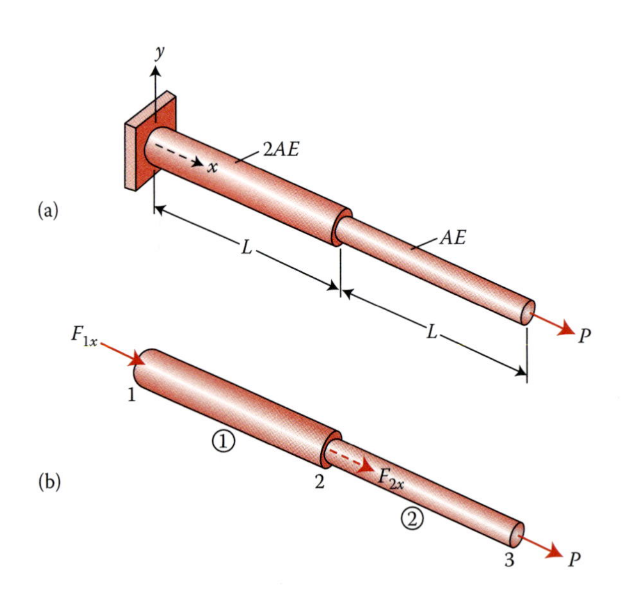

FIGURE C.4
(a) A stepped aluminum bar carries an axial load; (b) two-element model.

Element stiffness matrices. Referring to Equation C.1, the stiffness matrices for elements 1 and 2 are expressed as

$$[k]_1 = \frac{AE}{L}\begin{array}{c} \begin{array}{cc} u_1 & u_2 \end{array} \\ \begin{bmatrix} 2 & -2 \\ -2 & 2 \end{bmatrix} \end{array}\begin{array}{c} u_1 \\ u_2 \end{array} \qquad [k]_2 = \frac{AE}{L}\begin{array}{c} \begin{array}{cc} u_2 & u_3 \end{array} \\ \begin{bmatrix} 1 & -1 \\ -1 & 1 \end{bmatrix} \end{array}\begin{array}{c} u_2 \\ u_3 \end{array}$$

Note that the column and row of each stiffness matrix are labeled according to the nodal displacements associated with them. There are three displacement components (u_1, u_2, u_3) and hence the order of the system matrix must be 3×3. In terms of the bar displacements, the stiffnesses equal

$$[k]_1 = \frac{AE}{L}\begin{array}{c} \begin{array}{ccc} u_1 & u_2 & u_3 \end{array} \\ \begin{bmatrix} 2 & -2 & 0 \\ -2 & 2 & 0 \\ 0 & 0 & 0 \end{bmatrix} \end{array}\begin{array}{c} u_1 \\ u_2 \\ u_3 \end{array}$$

$$[k]_2 = \frac{AE}{L}\begin{array}{c} \begin{array}{ccc} u_1 & u_2 & u_3 \end{array} \\ \begin{bmatrix} 0 & 0 & 0 \\ 0 & 1 & -1 \\ 0 & 1 & 1 \end{bmatrix} \end{array}\begin{array}{c} u_1 \\ u_2 \\ u_3 \end{array}$$

Observe that in the foregoing matrices, the last row and the first column of zeros are added, respectively.

System stiffness matrix. The superposition of the terms of each stiffness matrix results in

$$[K] = \frac{AE}{L}\begin{array}{c} \begin{array}{ccc} u_1 & u_2 & u_3 \end{array} \\ \begin{bmatrix} 2 & -2 & 0 \\ -2 & 3 & -1 \\ 0 & -1 & 1 \end{bmatrix} \end{array}\begin{array}{c} u_1 \\ u_2 \\ u_3 \end{array}$$

Force–displacement relations. Equation C.7 are therefore

$$\begin{Bmatrix} F_{1x} \\ F_{2x} \\ F_{3x} \end{Bmatrix} = \frac{AE}{L}\begin{bmatrix} 2 & -2 & 0 \\ -2 & 3 & -1 \\ 0 & -1 & 1 \end{bmatrix}\begin{Bmatrix} u_1 \\ u_2 \\ u_3 \end{Bmatrix} \tag{a}$$

The boundary condition is $u_1 = 0$. We see that $F_{2x} = 0$ and $F_{3x} = P$. Then, with reference to Figure C.4b,

$$\begin{Bmatrix} F_{1x} \\ 0 \\ P \end{Bmatrix} = \frac{AE}{L}\begin{bmatrix} 2 & -2 & 0 \\ -2 & 3 & -1 \\ 0 & -1 & 1 \end{bmatrix}\begin{Bmatrix} 0 \\ u_2 \\ u_3 \end{Bmatrix}$$

Displacements. To determine u_2 and u_3, only portion of this equation is considered, as

$$\begin{Bmatrix} 0 \\ P \end{Bmatrix} = \frac{AE}{L}\begin{bmatrix} 3 & -1 \\ -1 & 1 \end{bmatrix}\begin{Bmatrix} u_2 \\ u_3 \end{Bmatrix}$$

Solving,

$$\begin{Bmatrix} u_2 \\ u_3 \end{Bmatrix} = \begin{Bmatrix} PL/2AE \\ 3PL/2AE \end{Bmatrix}$$

Comment: With the displacements available, axial forces and stresses in element 1 (or 2) can easily be found as illustrated in Case study C.1.

Nodal forces. Substituting the displacement obtained into Equation a, gives

$$\begin{Bmatrix} F_{1x} \\ F_{x2} \\ F_{3x} \end{Bmatrix} = \frac{AE}{L} \begin{bmatrix} 2 & -2 & 0 \\ -2 & 3 & 1 \\ 0 & -1 & 1 \end{bmatrix} \begin{Bmatrix} 0 \\ PL/2AE \\ 3PL/2AE \end{Bmatrix} = \begin{Bmatrix} -P \\ 0 \\ P \end{Bmatrix}$$

Comments: The results indicate that the reaction $F_{1x} = -P$ is equal in magnitude but opposite in direction to the applied force at node 3, $F_{3x} = P$. In addition, $F_{2x} = 0$ shows that no force is applied at node 2. Equilibrium of the bar assembly is thus satisfied.

CASE STUDY C.1 Analysis and Design of a Truss

A structural steel three bar truss 123 (Figure C.5a) supports a horizontal force P acting at joint 2. *Find:* Displacements, forces, and the required cross-sectional area of each member, based on a safety factor on yielding of n. *Assumptions:* All bars will have the identical yield strength σ_{yp}, length L, and axial rigidity AE. *Given:* $\sigma_{yp} = 250$ MPa (Table B.3), $P = 300$ kN, $n = 2$.

SOLUTION

The reactions are marked with dashed lines in Figure C.5a. The node numbering is arbitrary for each bar element.

Input data. Figure C.5b shows that at each node, there are two displacements and two nodal force components. The angle θ is measured counterclockwise from the positive x axis to each element (Table C.1). Since the terms in $[k]_e$ involve c^2, s^2, and cs, a change in angle from θ to $\theta + \pi$, causing both c and s to change sign, does not affect the signs of the terms in the stiffness matrix. For instance, in the case of member 3, $\theta = 60°$ if

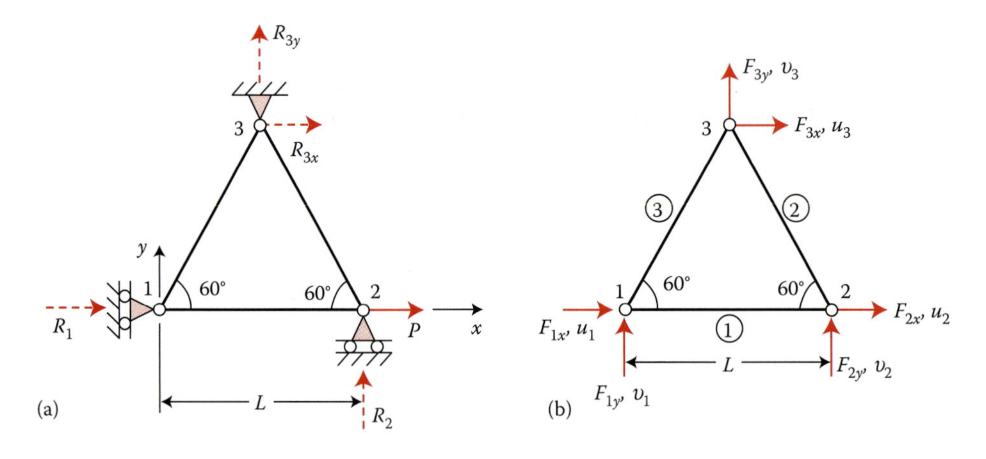

FIGURE C.5
(a) Plane truss; (b) finite element model.

TABLE C.1

Data for Truss of Figure C.5

Element	θ	C	s	c^2	cs	s^2
1	0°	1	0	1	0	0
2	120°	$-1/2$	$\sqrt{3}/2$	$1/4$	$-\sqrt{3}/4$	$3/4$
3	60°	$1/2$	$\sqrt{3}/2$	$1/4$	$\sqrt{3}/4$	$3/4$

measured counterclockwise at node 1 or 240° if measured counterclockwise at node 3. But, by inserting into Equation C.9, $[k]_e$ remains the same.

Element stiffness matrix. By Equation C.9 and Table C.1, we have for the bars 1, 2, and 3, respectively,

$$[k]_1 = \frac{AE}{L}\begin{bmatrix} 1 & 0 & -1 & 0 \\ 0 & 0 & 0 & 0 \\ -1 & 0 & 1 & 0 \\ 0 & 0 & 0 & 0 \end{bmatrix} \begin{matrix} u_1 \\ v_1 \\ u_2 \\ v_2 \end{matrix} \quad (\text{columns: } u_1\ v_1\ u_2\ v_2)$$

$$[k]_2 = \frac{AE}{4L}\begin{bmatrix} 1 & -\sqrt{3} & -1 & \sqrt{3} \\ -\sqrt{3} & 3 & \sqrt{3} & -3 \\ -1 & \sqrt{3} & 1 & -\sqrt{3} \\ \sqrt{3} & -3 & -\sqrt{3} & 3 \end{bmatrix} \begin{matrix} u_2 \\ v_2 \\ u_3 \\ v_3 \end{matrix} \quad (\text{columns: } u_2\ v_2\ u_3\ v_3)$$

$$[k]_3 = \frac{AE}{4L}\begin{bmatrix} 1 & \sqrt{3} & -1 & \sqrt{3} \\ \sqrt{3} & 3 & -\sqrt{3} & -3 \\ -1 & -\sqrt{3} & 1 & \sqrt{3} \\ -\sqrt{3} & -3 & \sqrt{3} & 3 \end{bmatrix} \begin{matrix} u_1 \\ v_1 \\ u_3 \\ v_3 \end{matrix} \quad (\text{columns: } u_1\ v_1\ u_3\ v_3)$$

We see that the column and row of each stiffness matrix are labeled in accordance with the nodal displacements related to them.

System stiffness matrix. There are a total of six components of displacement for the truss and thus the *order* of the trussstiffness matrix must be 6 × 6. By adding the terms from each element stiffness matrices into their corresponding locations in $[K]$, we have the global stiffness matrix for the truss as

$$[k] = \frac{AE}{4L}\begin{bmatrix} 4+1 & 0+\sqrt{3} & -4 & 0 & -1 & -\sqrt{3} \\ 0+\sqrt{3} & 0+3 & 0 & 0 & -\sqrt{3} & -3 \\ -4 & 0 & 4+1 & 0-\sqrt{3} & -1 & \sqrt{3} \\ 0 & 0 & 0-\sqrt{3} & 0+3 & \sqrt{3} & -3 \\ -1 & -\sqrt{3} & -1 & \sqrt{3} & 1+1 & \sqrt{3}-\sqrt{3} \\ -\sqrt{3} & -3 & \sqrt{3} & -3 & \sqrt{3}-\sqrt{3} & 3+3 \end{bmatrix} \begin{matrix} u_1 \\ v_1 \\ u_2 \\ v_2 \\ u_3 \\ v_3 \end{matrix} \quad (\text{columns: } u_1\ v_1\ u_2\ v_2\ u_3\ v_3) \tag{b}$$

System force and displacement matrices. Considering the applied load and support constraints (Figure C.5), the truss nodal force matrix equals

$$\{F\} = \begin{Bmatrix} F_{1x} \\ F_{1y} \\ F_{2x} \\ F_{2y} \\ F_{3x} \\ F_{3y} \end{Bmatrix} = \begin{Bmatrix} R_1 \\ 0 \\ P \\ R_2 \\ R_{3x} \\ R_{3y} \end{Bmatrix} \tag{c}$$

Similarly, accounting for the support conditions, the truss nodal displacement matrix is

$$\{\delta\} = \begin{Bmatrix} u_1 \\ v_1 \\ u_2 \\ v_2 \\ u_3 \\ v_3 \end{Bmatrix} = \begin{Bmatrix} 0 \\ v_1 \\ u_2 \\ 0 \\ 0 \\ 0 \end{Bmatrix} \tag{d}$$

Displacements. Introducing Equations b, c, and d into Equation C.13, results in the truss force–displacement relations:

$$\begin{Bmatrix} R_1 \\ 0 \\ P \\ R_2 \\ R_{3x} \\ R_{3y} \end{Bmatrix} = \frac{AE}{4L} \begin{bmatrix} 5 & \sqrt{3} & -4 & 0 & -1 & -\sqrt{3} \\ \sqrt{3} & 3 & 0 & 0 & -\sqrt{3} & -3 \\ -4 & 0 & 5 & -\sqrt{3} & -1 & \sqrt{3} \\ 0 & 0 & -\sqrt{3} & 3 & \sqrt{3} & -3 \\ -1 & -\sqrt{3} & -1 & \sqrt{3} & 2 & 0 \\ -\sqrt{3} & -3 & \sqrt{3} & -3 & 0 & 6 \end{bmatrix} \begin{Bmatrix} 0 \\ v_1 \\ u_2 \\ 0 \\ 0 \\ 0 \end{Bmatrix} \tag{e}$$

To determine v_1 and u_2, only the part of Equation e relating to the displacements v_1 and u_2 needs to be considered. Thus,

$$\begin{Bmatrix} 0 \\ P \end{Bmatrix} = \frac{AE}{4L} \begin{bmatrix} 3 & 0 \\ 0 & 5 \end{bmatrix} \begin{Bmatrix} v_1 \\ u_2 \end{Bmatrix}$$

Solving, the nodal displacements are

$$\begin{Bmatrix} v_1 \\ u_2 \end{Bmatrix} = \frac{4L}{15AE} \begin{bmatrix} 5 & 0 \\ 0 & 3 \end{bmatrix} \begin{Bmatrix} 0 \\ P \end{Bmatrix} = \frac{4PL}{5AE} \begin{Bmatrix} 0 \\ 1 \end{Bmatrix} \tag{f}$$

Reactions. The values of v_1 and u_2 are used to find reaction forces by Equation f as

$$\begin{Bmatrix} R_1 \\ R_2 \\ R_{3x} \\ R_{3y} \end{Bmatrix} = \frac{AE}{4L}\begin{bmatrix} \sqrt{3} & -4 \\ 0 & -\sqrt{3} \\ -\sqrt{3} & -1 \\ -3 & \sqrt{3} \end{bmatrix}\begin{Bmatrix} v_1 \\ u_2 \end{Bmatrix} = \frac{P}{5}\begin{Bmatrix} -4 \\ -\sqrt{3} \\ -1 \\ \sqrt{3} \end{Bmatrix} \tag{g}$$

These may be verified by applying the equilibrium equations to the free body diagrams (FBD) of the entire truss (Figure C.5a).

Axial forces in bars. From Equations C.11 and e and Table C.1, we have

$$\begin{aligned} F_{12} &= \frac{AE}{L}[1 \quad 0]\begin{Bmatrix} \dfrac{4PL}{5AE} \\ 0 \end{Bmatrix} = \frac{4}{5}P \\ F_{23} &= \frac{AE}{L}\left(-\frac{1}{2} \quad \frac{\sqrt{3}}{2}\right)\begin{Bmatrix} \dfrac{-4PL}{5AE} \\ 0 \end{Bmatrix} = \frac{2}{5}P \\ F_{13} &= \frac{AE}{L}\begin{bmatrix} \dfrac{1}{2} & \dfrac{\sqrt{3}}{2} \end{bmatrix}\begin{Bmatrix} 0 \\ 0 \end{Bmatrix} = 0 \end{aligned} \tag{h}$$

Stresses in elements. Dividing the preceding bar forces by the cross-sectional area, we obtain $\sigma_{12} = 4P/5A_1$, $\sigma_{23} = 2P/5A_2$, and $\sigma_{13} = 0$.

Required cross-sectional areas of bars. The allowable stress is $\sigma_{all} = 250/2 = 125$ MPa. We then have $A_1 = 0.8(300 \times 10^3)/125 = 1920\ \text{mm}^2$, $A_2 = 960\ \text{mm}^2$, and $A_3 =$ any area.

Problems

C.1 The truss element 4-1 of length L and the cross-sectional area A is oriented at an angle α clockwise from the x axis, as depicted Figure PC.1. *Determine:* (a) the global stiffness matrix of the bar; (b) the axial force in the bar; (c) the local displacements at the ends of the bar. *Given:* $A = 1000\ \text{mm}^2$, $L = 2$ m, $\alpha = 30°$, $E = 210$ GPa, $u_4 = -1.3$ mm, $v_4 = -1.4$ mm, $u_1 = 2.4$ mm, $v_1 = 1.8$ mm.

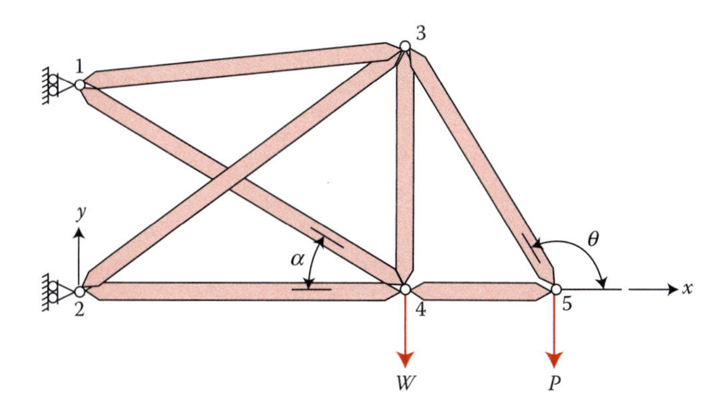

FIGURE PC.1

C.2 The bar element 5-3 of length L and diameter d is oriented at an angle counterclockwise from the x axis, as shown in Figure PC.1. *Calculate*: (a) the global stiffness matrix of the element; (b) the axial strain in the element. *Given:* $d = 30$ mm, $L = 1$ m, $\theta = 120°$, $E = 70$ GPa, $u_5 = -0.3$ mm, $v_5 = -0.15$ mm, $u_3 = 0.6$ mm, $v_3 = 0.1$ mm.

C.3 The stepped composite bar is held between two rigid supports and subjected to a concentrated load P at node 2 (Figure PC.3). The steel bar 1-3 has cross-sectional area A and modulus of elasticity E. The brass bar 3-4 is with cross-sectional area $2A$ and elastic modulus $E/2$. *Find*: (a) the system stiffness matrix; (b) the displacements of nodes 2 and 3; (c) the nodal forces and reactions at the supports.

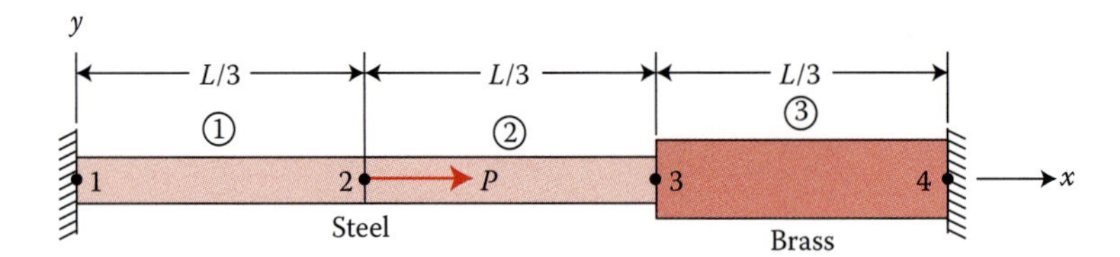

FIGURE PC.3

C.4 Resolve Problem C.3, for the case in which the force P is directed to the left and acts at node 3.

C.5 A composite stepped bar 1-4 is held between rigid supports and subjected to a concentrated load P at node 3 as illustrated in Figure PC.5. *Determine*: (a) the system stiffness matrix; (b) the displacements of nodes 2 and 3; (c) the nodal forces and reactions at supports.

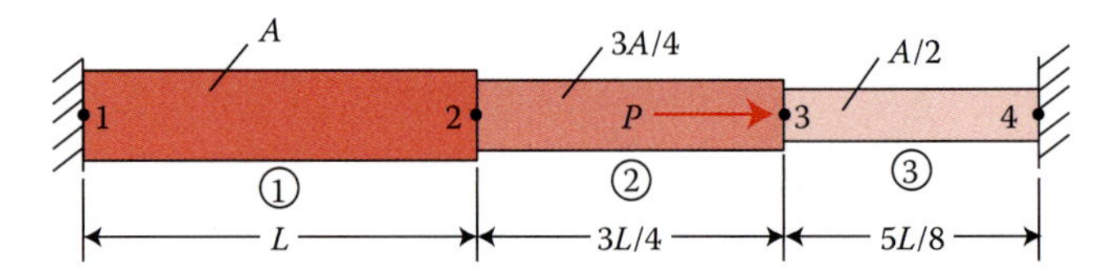

FIGURE PC.5

C.6 Redo Problem C.5, when the force P is directed to the left and applied at node 2.

C.7 A plane truss consists of 5 bars with axial rigidity AE is supported at joints 1 and 4 as portrayed in Figure PC.7. Calculate the global stiffness matrix for each element.

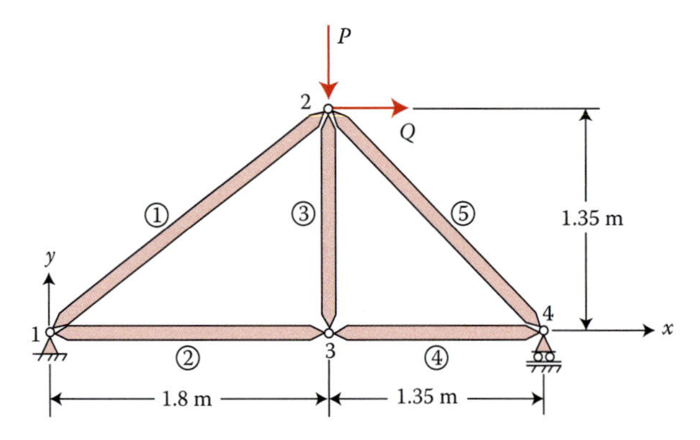

FIGURE PC.7

C.8 A vertical force $P = 20$ kN acts at joint 2 of the two-bar plane truss of Figure PC.8. Each member has axial rigidity $AE = 30$ MN. *Calculate*: (a) The global stiffness matrix of each bar; (b) the system stiffness matrix; (c) the nodal displacements; (d) the support reactions; (e) the axial forces in each bar.

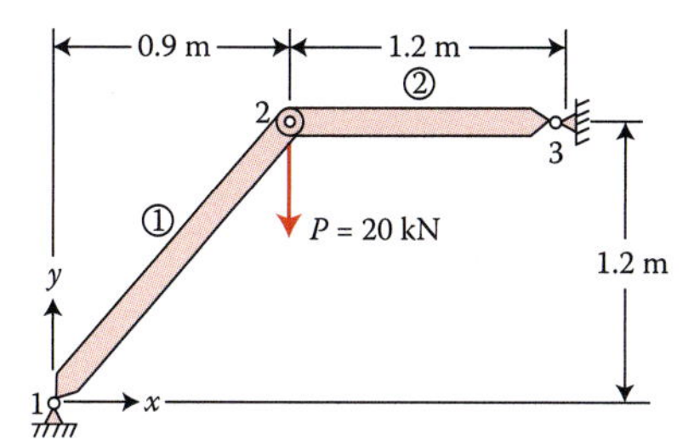

FIGURE PC.8

C.9 Redo Problem C.8 for the planar truss supported and loaded as portrayed in Figure PC.9. *Given:* $AE = 8$ MN for each bar element.

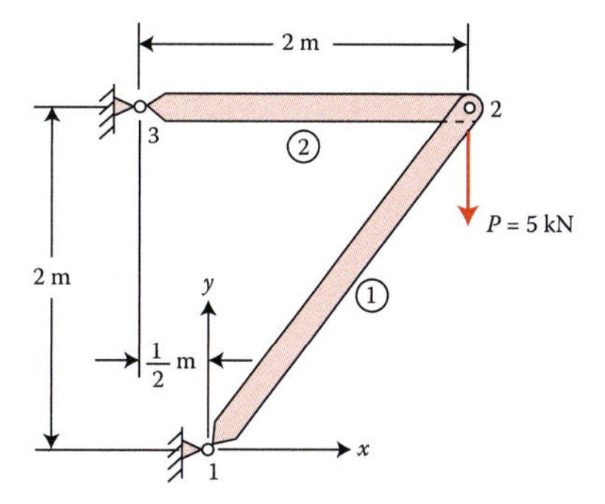

FIGURE PC.9

C.10 Figure PC.10 shows a two-bar plane truss has its support at joint 1 that settles vertically by an amount of Δ downward when loaded by a horizontal force *P*. *Find*: (a) the global stiffness matrix of each element; (b) the system stiffness matrix; (c) the nodal displacements; (d) the support reactions. *Given:* $E = 70$ GPa, $A = 35 \times 10^{-4}$ m^2, $P = 15$ kN, $\Delta = 12$ mm.

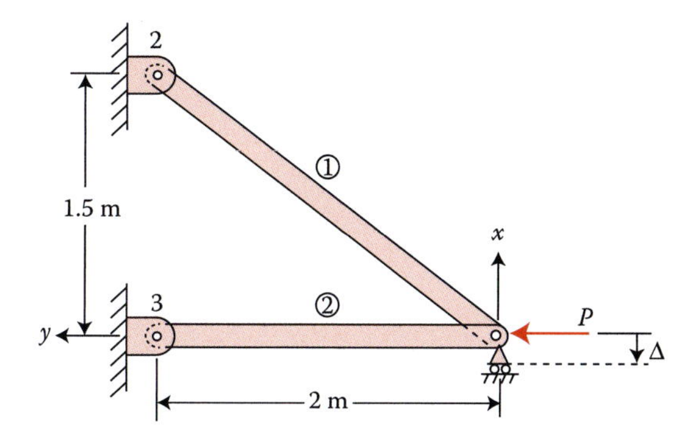

FIGURE PC.10

Appendix D: Introduction to MATLAB®

This appendix represents a brief introduction to MATLAB® (*matrix laboratory*). MATLAB is a programming medium for plotting of functions, data analysis, matrix manipulation and visualization, and numerical computation. Using MATLAB, technical computing problems can be solved more quickly than with traditional programming tools. MATLAB involves a wide variety of applications such as image processing, communications, design, test and measurement, and financial modeling and analysis. For more than a million engineers and scientists in industry and academia, MATLAB is the language of computing. Detailed information on the subject is available from MathWorks at www.mathworks.com/products/matlab/. There are also many printed publications on MATLAB for engineers.

Application of the MATLAB is built around the MATLAB language: *M-code*. The simplest way to execute *M*-code is to type it in at the prompt (≫) in the Command Window or in the Edit Window (recommended), which allows you to type all commends without executing them. The MATLAB teaching codes consists of 37 short text files containing MATLAB commands for performing basic linear algebra computations. The common *M-codes* include *determ.m* (matrix determinant), *cofactor.m* (matrix of cofactors), *cramer.m* (solve the system of equations $Ax = C$), *inverse.m* (matrix inverse by Gauss–Jordon elimination), *eigen2.m* (characteristic polynomial, eigenvalues, eigen vectors), and *plot2d.m* (2D plot). In MathWorks, new users can choose *Help* on the toolbar at the top of the MATLAB command window, then select *MATLAB Help* and *Getting Started*.

MATLAB (*professional or student version*) is now used in most universities. Among other features, it has the ability to deal with structural analysis and engineering design problems, allowing the iteration technique to proceed in an easy and rapid way. Students are able to do homework problems without employing a calculator. To learn the basics as well as to gain proficiency in MATLAB, many demonstrations are available with the *Help* feature. The functions often used in MATLAB can also be found at *Help* > *Function Browser* > *Mathematics* > *Elementary Math*. When editing a MATLAB program (as shown next), each step must be checked because any wrong result along the way will lead to an incorrect final solution.

A sample MATLAB solution of Example 1.1: Force analysis of an L-shaped pipe.

Code

```
% EXAMPLE 1.1-Force analysis of an L-shaped pipe.

% Determine: the axial, shear forces, and moments acting at point C.
% Given:
      m=1.1; L_AD=1.0; L_CD=0.8; g=9.81; F_v=60; T_AD=100; L_AD_half=L_
      AD/2.; L_BC=0.4;
% SOLUTION:
% The weight of segments AD and DC of the pipe are
      W_ad=m*L_AD*g
      W_cd=m*L_DC*g
% Considering the free-body diagram of ADC (Figure 1.5b)
% ΣF_x=0:, ΣF_y=0:, ΣF_z=0:, ΣM_z=0:
```

```
        P=0
        V_y=0
        V_z=-(W_ad+W_cd+F_v)
% and ΣM_x=0:, ΣM_y=0:, ΣM_z=0:
        T=-(W_ad*L_AD_half+F_v*L_AD)
        M_y=-(T_AD-F_v*L_CD-W_ad*L_CD-W_cd*L_BC)
        M_z=0

>>Solution EX_1_1
W_ad =10.7910
W_cd =8.6328
P =0
V_y =0
V_z =-79.4238
T =-65.3955
M_y =-39.9141
M_z =0
>>
```

For further illustration of this tool, various text examples and a case study (listed in Table D.1) are resolved using MATLAB on the CRC Website: http://www.crcpress.com/product/isbn/9781439887806. Observe the differences in numerical precision between

TABLE D.1

MATLAB Solution Contents

textbook and MATLAB answers. It is recalled from Section 1.17 that significant digits used in textbook calculations are on the basis of a usual engineering rule. MATLAB uses double-precision floating-point numbers in its calculations that give 16 decimal places. Altering the display format will not change the accuracy of the solution. Values with decimal fractions are printed in the default short format that indicates four digits after the decimal point. To quickly view MATLAB calculations for a problem solution, type in only the data and equations seen in bold, without the comment lines (these do not affect results).

Answers to Selected Problems

Chapter 1

1.2 (a) $M_{\max} = \frac{2}{9} pa^2$

(b) $M = \frac{1}{8} pa^2, V = \frac{5}{6} pa$

1.3 (a) $P = 17.5$ kN

(b) $F_{AB} = 18.2$ kN

1.8 (a) $\sigma_1 = 125$ MPa, $\sigma_2 = 0$

(b) $\tau_{\max} = 62.5$ MPa

$\theta_p'' = 26.6°$

1.10 (a) $\tau = 67.1$ MPa

(b) $\tau_{\max} = 90$ MPa

$\theta_s' = 20.9°$

1.14 (a) $\sigma_x = -20$ MPa

$\sigma_y = 60$ MPa

$\tau_{xy} = -30$ MPa

(b) $\sigma_1 = 70$ MPa

$\sigma_2 = -30$ MPa

$\theta_p'' = 18.43°$

1.18 (b) $\sigma_1 = 65.19$ MPa

$\sigma_2 = 24.81$ MPa

1.22 (a) $a' = 99.96$ mm

$b' = 49.98$ mm

$t' = 9.996$ mm

(b) $A'C'' = 111.7587$ mm

1.24 $w_x = 474.6$ kN/m

$w_y = 601.3$ kN/m

Chapter 2

2.2 (a) $T = 1.05$ kN · m

(b) $\sigma_{AB} = 20.88$ MPa

2.5 $T = 58.65$ kN · m

2.7 $d = 28.68$ mm

2.11 (a) $\tau_{\max} = 8.09$ MPa

(b) $\tau_{\min} = 6.22$ MPa

2.14 $w_B = \frac{19}{360} \frac{p_0 L^4}{EI}$

$\theta_B = \frac{1}{15} \frac{p_0 L^3}{EI}$

2.19 $(\tau_{\max})_A = 1.012$ MPa

$(\tau_{\max})_B = 599$ kPa

2.26 (a) $R = 1.45$ kN

(b) $R = 1.404$ kN

2.32 $R_A = -R_B = \frac{9}{16} \frac{M}{a}, M_B = -\frac{1}{8} M$

Chapter 3

3.2 $\varepsilon_{\max} = 1000\ \mu$

$\sigma_{\max} = 230.8$ MPa

3.8 (b) $w_{\max} = \frac{p_0 b^4}{\pi^4 D}\left(1 - \frac{\pi}{4}\right)$

$\varepsilon_{y,\max} = \frac{p_0 t b^2}{\pi^3 D}$

(c) $r_y = -151.2$ m

3.9 (b) $\sigma_{y,\max} = -0.75 p_0 \left(\frac{a}{t}\right)^2$

(c) $\varepsilon_{y,\max} = \frac{45}{128} \frac{p_0 b^2}{Et^2}$

3.11 (a) $w = \frac{p_0}{120bD}(y^5 - 2b^2 y^3 + b^4 y)$

(b) $\theta = \frac{p_0 b^3}{120D}$

3.17 (b) $\sigma_{x,\max} = \frac{4P}{\pi t^2}$

Chapter 4

4.3 (a) $t = 22.8$ mm

(b) $w_{\max} = 0.281$ mm

4.4 $n = 3.56$

4.10 $P = 7.87$ kN

4.14 (a) $t = 14.34$ mm

(b) $w_{\max} = -9.45$ mm

4.21 $w_{\max} = 0.0036 \dfrac{M_1 a^2}{Et^3}$

4.24 $w_{\max} = 0.670 \dfrac{p_0 a^4}{Et^3}$

Chapter 5

5.1 (a) $p_0 = 127.2$ kPa

(b) $w_{\max} = 11.66$ mm

5.4 (b) $p_0 = 3.633$ kPa

5.7 (a) $\sigma_{x,\max} = \sigma_{y,\max} = 0.16 p_0 \left(\dfrac{a}{t}\right)^2$

5.9 $R_c = 0.119P$

5.27 $w_{\max} = 0.00217 \dfrac{p_0 a^4}{D}$

$M_{\max} = 0.0694 p_0 a^2$

5.34 (a) $t = 70$ mm

(b) $t = 67.6$ mm

Chapter 6

6.1 (a) $w = 6.28(10^{-4}) \dfrac{p_0 a^4}{D}$

(b) $M_x = 0.02\, p_0 a^2$

6.5 (a) $t = 21.1$ mm

(b) $w = 0.099$ mm

6.9 (a) $t = 47.2$ mm

(b) $w_{\max} = 0.116$ mm

6.11 $w = 0.00588 \dfrac{p_0 a^4}{D}$

6.13 $(w_{\max})_e/(w_{\max})_c = 2.1695$

Chapter 7

7.2 $w_3 = 0.01852 \dfrac{PL^3}{EI}$

7.5 $w_c = 0.00684 \dfrac{p_0 L^4}{EI}, 5.1\%$

7.9 $\sigma_{\max} = 0.214 p_0 \left(\dfrac{a}{t}\right)^2$

7.10 $\varepsilon_1 = 211\mu$

7.11 $w_6 = 0.0062 \dfrac{p_0 a^4}{D}$

7.17 $w_2 = 0.03335 \dfrac{p_0 a^4}{D}$

7.21 (a) $\theta_1 = \dfrac{PL^2}{2EI}, v_1 = \dfrac{-PL^3}{3EI}$

7.23 $v_2 = \dfrac{-5PL^3}{48EI}, \theta_2 = \dfrac{-PL^2}{8EI}$

7.25 (c) $u_2 = -17.949$ mm

$\theta_2 = -0.003846$ rad

$\theta_3 = -0.015385$ rad

Chapter 8

8.1 $H = D_x = 146.52$ kN·m

$D_y = 1346.52$ kN·m

8.6 $p_0 = 1.72$ MPa

8.7 $w_{\max} = 1.012\ (10^{-7})\ p_0 a^4$

8.8 $w_{\max} = 4.83\ (10^{-7})\ p_0$

8.9 $w_{\max} = 6.706\ (10^{-6})\ p_0$

Chapter 9

9.5 $\sigma_{cr} = 2\pi^2 \dfrac{D}{a^2 t}$

9.12 $N_{cr} = 3.648\pi^2 \dfrac{D}{a^2}$

9.14 $N_{cr} = 6.08\pi^2 \dfrac{D}{a^2}$

9.15 $w_{\max} = \dfrac{a_0}{1+\alpha}$, where $\alpha = \dfrac{Na^2}{2\pi^2 D}$

Chapter 10

10.1 (a) $w_{\max} = 1.33$ mm

$\sigma_{\max} = -93.75$ MPa

(b) $w_{\max} = 3.85$ mm

$\sigma_{\max} = -104$ MPa

10.3 (a) $p_0 = 560$ kPa

(b) $w_{\max} = 21.5$ mm

Chapter 11

11.1 $N_x = N_y = 542.9\ \text{kN/m}$
$M_x = M_y = -208.8\ \text{N}$

11.4 $w_{\max} = 11.557(10^{-3})\alpha A a^2 t^2$
$M_{x,\max} = 0.693(10^{-3})\alpha E t^5 A$

11.7 $\sigma_{\max} = 1.5E\alpha T_1$
$$M_x = M_y = -\frac{1}{4}E\alpha T_1 t^2$$

Chapter 12

12.3 $p = 896\ \text{kPa}$

12.5 (a) $y = 2.815\ \text{m}$
(b) $y = 1.407\ \text{m}$

12.8 $\sigma_{\max} = 1.903\ \text{MPa}$

12.16 $\sigma_\phi = 25\ \text{MPa},\ \sigma_\theta = 50\ \text{MPa}$

12.21 $t = 7.46\ \text{mm}$

Chapter 13

13.3 $P = 57.3\ \text{kN/m}$

13.6 $P = 96.89\ \text{kN/m}$

13.7 (a) $P = 457.9\ \text{kN/m}$
(b) $P = 410.4\ \text{kN/m}$

13.10 (a) $2w = 218(10^{-7})\dfrac{P}{D}$

Chapter 14

14.1 (a) $L_{\min} = 18.93\ \text{mm}$
(b) $Q_1 = 42.37\ \text{kN/m}$

14.8 $\sigma_{\theta,\max} = -129.75\ \text{MPa}$

14.15 $\sigma_{x,\max} = 370.6\ \text{MPa}$
$\sigma_{\theta,\max} = -111.2\ \text{MPa}$

14.20 $\sigma_{x,\max} = 65.16\ \text{MPa}$
$\sigma_{\theta,\max} = 67.68\ \text{MPa}$
$\sigma_\theta = -\sigma_\phi = 60\ \text{MPa}$

14.23 (a) $t = 5\ \text{mm}$
(b) $t = 6.27\ \text{mm}$

Chapter 15

15.1 $w_{\max} = -164{,}761.2\dfrac{\gamma}{E}\ \text{cm}$
$\sigma_{x,\max} = -653.4\gamma\ \text{N/cm}^2$

15.8 $\sigma_{cr} = 3{,}025\ \text{MPa}$

15.9 $P_{cr} = 74{,}246\ \text{kN}$

15.10 (a) $t = 1.21\ \text{mm}$
(b) $t = 0.751\ \text{mm}$

Appendix C

C.2 $\varepsilon_1 = 666.5\ \mu$

C.3 (b) $u_2 = \dfrac{PL}{9AE}, u_3 = \dfrac{PL}{18AE}$
(c) $R_1 = \dfrac{2P}{3}, R_4 = \dfrac{P}{3}$

C.5 (b) $u_2 = \dfrac{PL}{2.6AE}, u_3 = 2u_2$
(c) $R_1 = \dfrac{P}{2.6}, R_4 = 1.6R_1$

C.9 (c) $u_2 = 0.938\ \text{mm}$
$v_2 = -3.145\ \text{mm}$
(d) $R_{1x} = 3.752\ \text{kN}, R_{1y} = 5.0\ \text{kN}$
$R_{3x} = -3.752\ \text{kN}$
(e) $F_{12} = -15.626\ \text{kN(C)}$
$F_{23} = 7.504\ \text{kN(T)}$

Index

C

R

T

U

V

W

Y

TABLE B.5

Properties of Common Areas

1. Rectangle

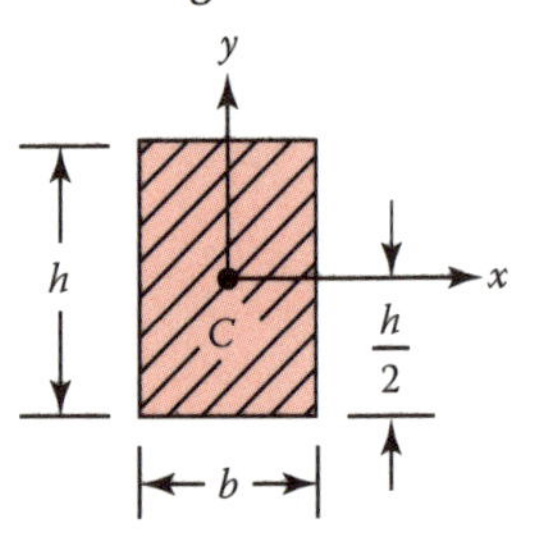

$$A = bh$$

$$I_x = \frac{bh^3}{12}$$

$$J_c = \frac{bh\left(b^2 + h^2\right)}{12}$$

2. Circle

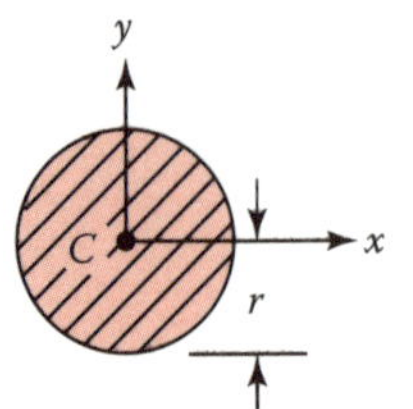

$$A = \pi r^2$$

$$I_x = \frac{\pi r^4}{4}$$

$$J_c = \frac{\pi r^4}{2}$$

3. Right triangle

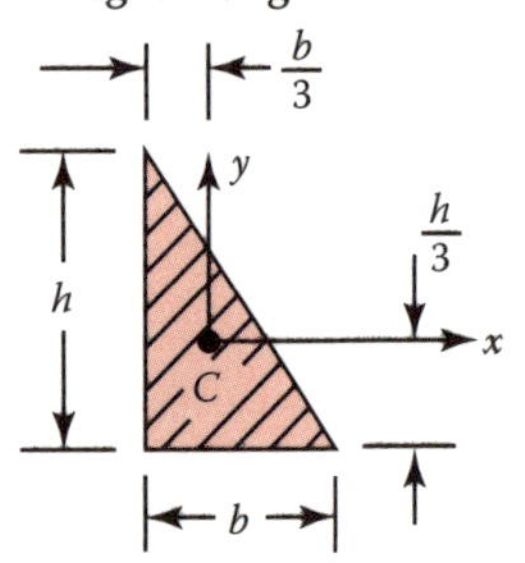

$$A = \frac{bh}{2}$$

$$I_x = \frac{bh^3}{36} \quad I_{xy} = -\frac{b^2h^2}{72}$$

$$J_c = \frac{bh\left(b^2 + h^2\right)}{36}$$

4. Semicircle

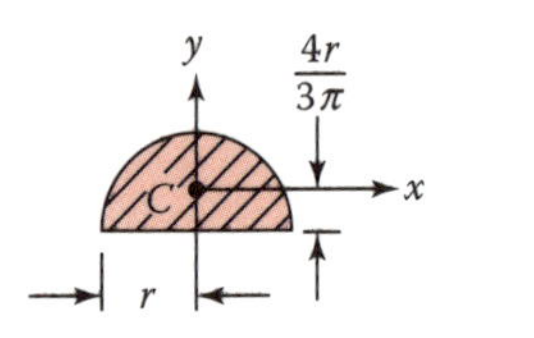

$$A = \frac{\pi r^2}{2}$$

$$I_x = 0.110r^2$$

$$I_y = \frac{\pi r^4}{8}$$

5. Ellipse

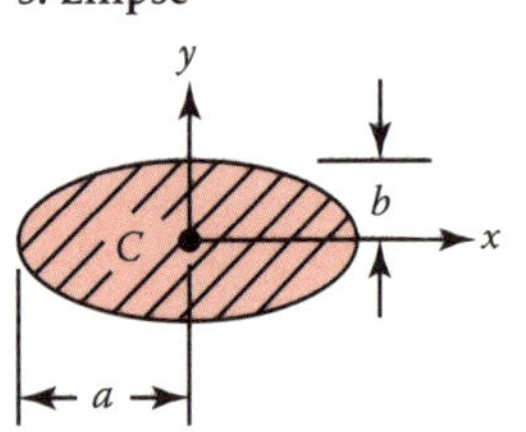

$$A = \pi ab$$

$$I_x = \frac{\pi ab^3}{4}$$

$$J_c = \frac{\pi ab\left(a^2 + b^2\right)}{4}$$

6. Thin tube

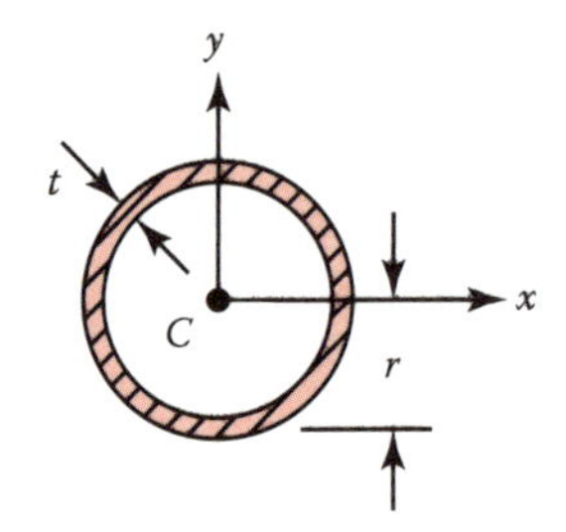

$$A = 2\pi rt$$

$$I_x = \pi r^3 t$$

$$J_c = 2\pi r^3 t$$

7. Isosceles triangles

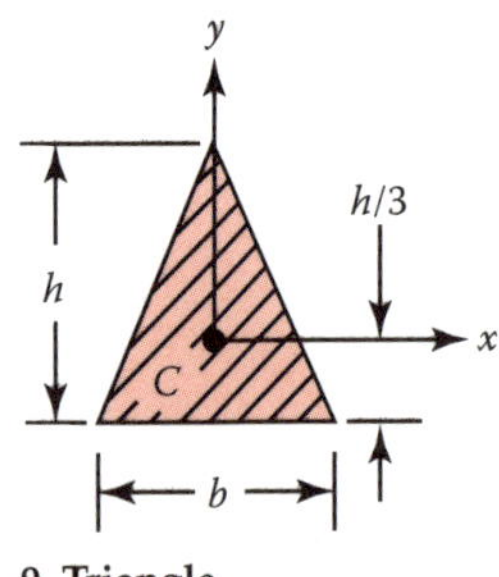

$$A = \frac{bh}{2}$$

$$I_x = \frac{bh^3}{36} \quad I_y = \frac{hb^3}{48}$$

$$J_c = \frac{bc}{144}\left(4h^2 + 3b^2\right)$$

8. Half of thin tube

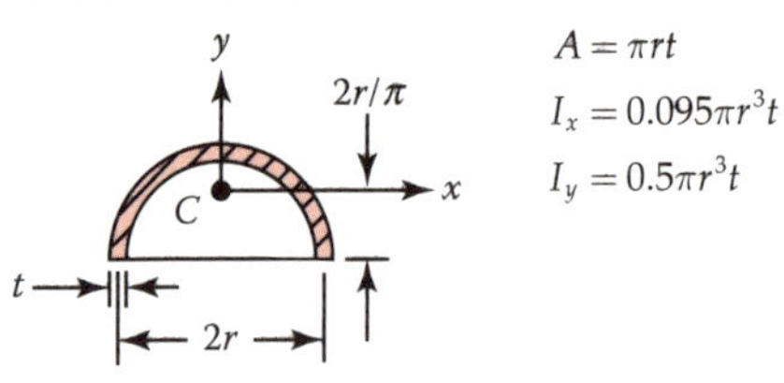

$$A = \pi rt$$

$$I_x = 0.095\pi r^3 t$$

$$I_y = 0.5\pi r^3 t$$

9. Triangle

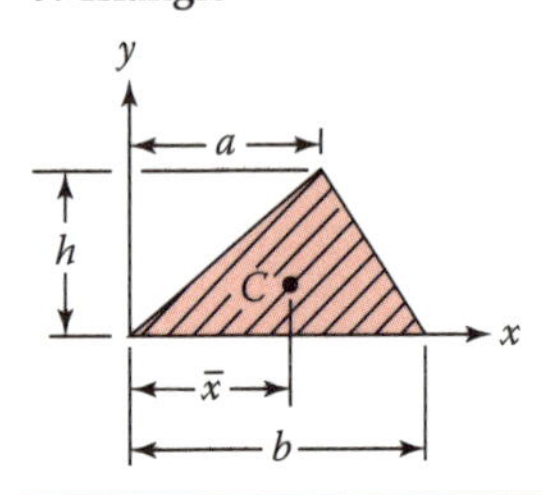

$$A = \frac{bh}{2}$$

$$\bar{x} = \frac{(a + b)}{3}$$

10. Parabola ($y = x^2$)

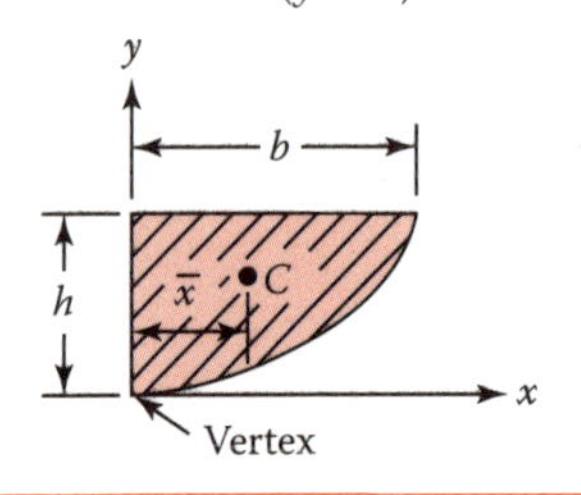

$$A = \frac{2bh}{3}$$

$$\bar{x} = \frac{3b}{8}$$